An Introduction

ASTROPHYSICS

SECOND EDITION

Baidyanath Basu

Formerly Professor and Head
Department of Applied Mathematics
University of Calcutta

Tanuka Chattopadhyay

Department of Applied Mathematics
University of Calcutta

Sudhindra Nath Biswas

National Awardee Teacher

PHI Learning Private Limited

New Delhi–110 001
2012

₹ 350.00

AN INTRODUCTION TO ASTROPHYSICS, Second Edition
Baidyanath Basu, Tanuka Chattopadhyay and Sudhindra Nath Biswas

ISBN-978-81-203-4071-8

The export rights of this book are vested solely with the publisher.

Ninth Printing (Second Edition) ... ... **May, 2012**

Published by Asoke K. Ghosh, PHI Learning Private Limited, M-97, Connaught Circus, New Delhi-110001 and Printed by Rajkamal Electric Press, Plot No. 2, Phase IV, HSIDC, Kundli-131028, Sonepat, Haryana.

To
the loving memory of my parents
Bisweswar Basu
Charushila Basu

— Baidyanath Basu

Contents

Foreword

The most important element in science teaching is the teacher himself, who has his own characteristic style of unfolding the storehouse of knowledge gathered through centuries of systematic, patient studies to inquisitive, uninitiated young minds. It requires great efforts to adapt existing textbooks written by other learned teachers to any individual Professor's characteristic style of delivering class lectures. This is the reason why in spite of multiplicity of introductory textbooks on Astrophysics another new one always fills up a gap. The present book represents the class which fills up the void in the context of contemporary endeavours in the Indian Universities, where the students come across this vast new subject almost for the first time in their postgraduate classes.

Professor B. Basu has been teaching Astrophysics in the Indian Universities for more than two decades; his rich experience is evident in the planning of the chapters where efforts to reach the hidden corners of the students' minds can be clearly seen. Being mostly unfamiliar with the range and diversity of the physical parameters in the Universe, the common student harbours many erroneous ideas about the processes which mould the cosmic structure and composition. It is absolutely essential that such doubts need be cleared at the beginning, and this appears to be the principal aim of this book.

In an introductory book of this type, equal emphasis cannot be placed on all aspects of the subject, and author's own preference comes to the fore. Professor Basu had chosen to concentrate on the stellar phenomena, the basic physical process of the visible universe. But he has also touched upon all essential components of the cosmos: the far away galaxies and clusters of galaxies, the exotic objects found by using modern instruments, the all pervading interstellar dust and gas, and even some aspects of the origin of everything—cosmology, the creation of the Universe. A very special feature has been the inclusion of Bioastronomy. All questions which may come to inquisitive young minds have been covered. The very basic question, how do we know so much about them, has been opened up in the very first chapter; details of techniques of observational astronomy have been given to inspire confidence in the facts and figures dealt in researches in Astrophysics. The chapterwise Exercise will train the students to go deeper into the subjects, and the suggested reading lists will cater to the more serious students to expose themselves to a wider horizon.

In essence, the topics covered in this book are spread over the entire length and breadth of the vast subject. I am confident that teachers in Indian Universities and institutions, and students embarking on a career in science will find this book very useful.

J.C. Bhattacharyya
Emeritus Professor
Indian Institute of Astrophysics
Bangalore

Preface

Since the first appearance of the book *An Introduction to Astrophysics* in 1997, significant progress has been made in different branches of astronomy and astrophysics. Through HST survey several observational missions have been taken over through which several data archives like VIZIER, ALADIN, SDSS, VOSTAT, VO etc., have been developed creating a solid platform for modern research in observational as well as in theoretical astrophysics. For the second edition, Tanuka Chattopadhyay has incorporated Sudhindra Nath Biswas and discussed several thrust areas such as the latest development of H-R diagram including L and T dwarfs, standard solar model, solar neutrino puzzle, cosmic microwave background radiation, Drake equation, dwarf galaxies, ultra compact dwarft galaxies, compact groups and clusters of galaxies.

Sri S.N. Biswas had assisted Dr. Baidyanath Basu in writing the first edition of the book under the UGC fellowship scheme for the preparation of university level books by Indian authors. His contribution has been recognized by including him as the third author of the book. Suggested readings at the end of each chapter have been complemented. We hope that the revised edition will help students and readers to get acquainted with the latest aspects of astrophysics which have come out so far in the transition period.

Baidyanath Basu
Tanuka Chattopadhyay
Sudhindra Nath Biswas

Preface to the First Edition

Astronomy, usually considered as the oldest observational science, can also claim to be the youngest of the modern scientific disciplines when combined with the spectacular development of its sister discipline of theoretical astrophysics. This apparently contradictory situation can be understood when one considers the fact that man's eternal quest to comprehend the splendour of the heavens had no beginning, and it will never cease. Modern astronomy and astrophysics therefore stand as a symbol for man's ceaseless endeavour to know the unknown, to see the unseen and to understand the origin and evolution of the physical universe. Gone are the days of Aristotle who emphatically declared—"Man can never unfold the mysteries hidden within the stars". Now, it has been made possible to explore the mysteries of the Universe by combining modern technological skills with the application of the laws of mathematics and modern physics.

Prior to the second decade of the twentieth century, the study and investigations in astrophysics were mostly limited to the affluent nations. But with Dr. Megnad Saha's works in 1920's and later, astronomical education began to spread in India slowly but surely for some decades and gained momentum over the last three decades. Simultaneously, the need for standard textbooks, particularly by Indian authors and catering to the needs of the students in Indian Universities was being badly felt. Although numerous books of foreign publication were available in the market, only a few could be used as textbooks by Indian students. These books were generally of two widely different levels. While the majority of these were of non-technical and descriptive nature, suitable for the first two years of undergraduate classes in American Universities, the others were of highly sophisticated and advanced level suitable for researchers and specialists. The need for books bridging this wide gap was seriously felt, and the present book is planned accordingly.

An Introduction to Astrophysics is the outcome of my lectures delivered to the M.Sc. students in the Department of Mathematics, Jadavpur University, and to the M.Sc. and M.Phil students in the Department of Applied Mathematics, Calcutta University. In this volume, I have discussed the basic astrophysics of stars, galaxies, clusters of galaxies and other heavenly objects of interest. Special attention has been given to the structure and content of the Milky Way Galaxy, and the various components of the Interstellar Matter and their physical behaviour. The mysteries that are Radio Galaxies and Quasars have been briefly introduced. To make the matter complete, I have also briefly introduced the basic concepts of Cosmology, and concluded with a brief discussion of Bioastronomy. Bioastronomy is an emerging discipline of modern science where efforts of astronomers, physicists, chemists, biologists, radio astronomers and engineers are converging. I have deliberately left untouched the Solar System Astrophysics.

This subject has developed to such a level over the past few decades that, in my opinion, it requires a separate volume for reasonable presentation.

The present text is suitable for use for a one year introductory course in these topics in undergraduate and/or postgraduate classes in Indian Universities. But the book will also be instructive to general scientific community who would like to know about splendours of the heaven. Problems have been added at the end of each chapter to help students develop their own thinking about the subject matter. Bibliography at the end of each chapter may be helpful for students and instructors alike.

I shall consider my humble effort successful if the book becomes useful to the students and the teachers and if it be well received by them. Any suggestion for improvement will be gratefully appreciated.

I am indebted to many persons and several organizations for helping me in preparing the manuscript. Firstly, I express my sincerest thanks to the University Grants Commission for awarding a book-writing project with funds under which the original version of the manuscript was prepared nearly two decades ago. Subsequently, the manuscript was revised and updated several times to bring it to the present form. My thanks are also due to the Inter-University Centre for Astronomy and Astrophysics (IUCAA) at Pune, where my visits as a Senior Associate have enabled me to improve and update the material. In this connection, I must register my thanks particularly to Prof. J.V. Narlikar, the Director of IUCAA, for awarding me a Senior Associateship.

It is a great pleasure to record my sincerest thanks to Prof. J.C. Bhattacharyya, who not only has written the Foreword for this book, but also has given many helpful suggestions for improvement, and supplied some photographs which I have used in the book with acknowledgement. I record my sincerest thanks to Prof. S.M. Alladin for his critical analysis and suggestions for improvement of the material which have been invaluable, and also to Prof. S.C. Dutta Roy of the IIT, Delhi, for his continuous encouragement and help at the publication stage. Many of my students have helped me in course of preparation of this manuscript. I register my thanks to them. But I shall be failing in my duty unless I mention in this connection the names of Dr. Tara Bhattacharyya and Dr. Tanuka Chatterjee. I have also the pleasure to thank Mr. Tusar Chakraborty who has helped me in preparation of the Index.

I must specially record my indebtedness to Mr. S.N. Biswas who, as the UGC Book-writing fellow, had worked with me for two years. Even after that, I received his continuous help during the subsequent years when the material was repeatedly modified and updated. But for his active help it would have hardly been possible to bring out this volume.

Lastly, it is with great pleasure that I record my heart-felt thanks to my wife *Bharati*, and my sons *Vedabrata* and *Vivekbrata* who have suffered a lot during the long years of writing this book and then revising it repeatedly.

Baidyanath Basu

Astronomical Instruments

1.1 LIGHT AND ITS PROPERTIES

The most important source of information for an astronomer is the light coming from a heavenly body. The light coming from a heavenly source is analyzed through instruments and conclusions are drawn regarding the physical nature of the source. Light has been recognized as the electromagnetic wave which is characterized by the moving transverse electric and magnetic fields, the direction of propagation being at right angles to both these fields. It can propagate through empty space and travels in the form of a sine-curve with respect to a line drawn in the direction of propagation, as shown in Fig. 1.1.

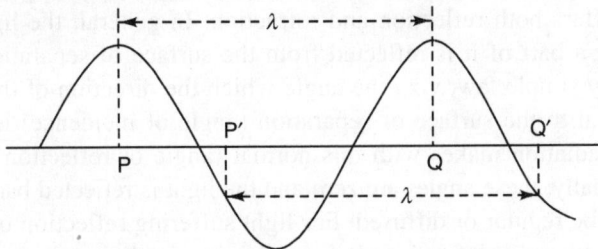

FIGURE 1.1 The wave motion of light.

The wavelength, λ of the sine-curve is measured by the length of one such wave, i.e., by the distance from crest to crest, or from through to through.

The electric field E and the magnetic field H, are in free space always numerically equal when measured in appropriate units. In free space, the energy per unit volume in an electric field is $E^2/8\pi$, and the energy per unit volume in a magnetic field is $H^2/8\pi$. So the energy W per unit volume in the electromagnetic wave field is

$$W = E^2/8\pi + H^2/8\pi = E^2/4\pi = H^2/4\pi \tag{1.1}$$

The intensity of electromagnetic radiation at any point in space is defined as the amount of energy that flows per unit time per unit area perpendicular to the direction of propagation. Since the energy flows with the speed c of light (a constant), if I be the intensity, then

$$I = \frac{cE^2}{4\pi} = \frac{cH^2}{4\pi} \qquad (1.2)$$

The number of oscillations in unit time of E and H at a given point is called the *frequency, v* of the electromagnetic radiation, so that

$$\lambda v = c \qquad (1.3)$$

The numerical value of v is very high. It is 6×10^{14} for green light.

The *wave number, n* is the number of waves per centimetre and is sometimes used in place of the wavelength and frequency. Thus,

$$n = \frac{1}{\lambda} \quad \text{or} \quad \frac{10^8}{\lambda} \qquad (1.4)$$

according as the wavelength λ is measured in centimetres or in Angstroms. [An angstrom is a very small unit of length used generally to measure wavelengths of visible light and those of higher frequency radiation (1 cm = 10^8 Å)].

The radiation having a single wavelength λ and vibrating with a single frequency v is called *monochromatic* (singly-coloured). The sine curve in Fig. 1.1 represents such a radiation. Strictly speaking, there is no finite amount of radiation which may be treated as monochromatic. So a non-monochromatic wave may be regarded as the union of a narrow band of frequencies for which appropriate sine-curve with suitable wavelengths may be drawn.

When an electromagnetic radiation, in the form of light travelling in the medium faces a new medium, it suffers both reflection and refraction. In general, the light is refracted into the new medium, but a part of it is reflected from the surface of separation between the two media, obeying a very simple law, viz., the angle which the direction of the incident radiation makes with the normal to the surface of separation (angle of incidence) is equal to the angle which the reflected radiation makes with this normal (angle or reflection). For light incident upon the surface normally, these angles are zero and the light is reflected back along the normal.

Reflection may be regular or diffused. For light suffering reflection on a smooth surface, the reflected light is sharply defined and the reflection is then called *regular*. The *rough surfaces* cause light to reflect irregularly in many directions and such reflection is termed *diffused*. Thus the reflection from the surfaces of planets and satellites is practically diffused.

The reflected light carries a part of the energy associated with the incident light. Also, a part of the energy is carried by the refracted rays and some may be lost at the surface. The *ratio* of the energies associated with the reflected radiation to the incident radiation is called the 'reflectivity' of the surface. If the surface is that of an astronomical body like planets and their satellites, the reflectivity is called *albedo*. The albedo is, in general, less than half for these bodies.

When light enters from one medium to another it suffers, in general, a change in the direction as well as in the velocity. The degree to which these changes may occur depends on the frequency of the light ray and on the media through which it travels.

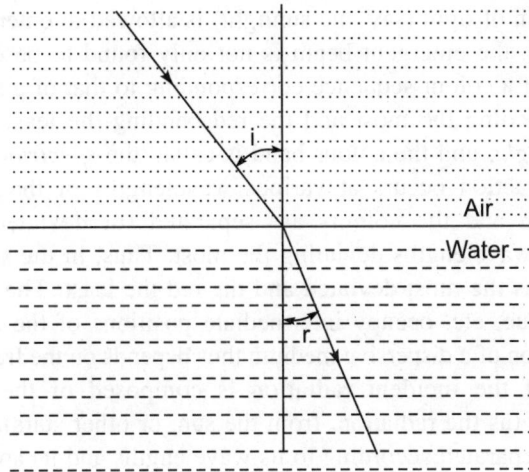

FIGURE 1.2 Refraction of light.

Let us suppose that a ray of light travelling from a rarer to a denser medium, e.g. from air to water, falls obliquely on the surface of water (Fig. 1.2). It is found that the light rays travel more slowly in water (denser) than in air (rarer) and its direction in water bends nearer to the normal drawn to the surface of separation at the point where the ray meets this surface.

If the ray of light falls perpendicular to the surface of separation, no bending occurs and the ray enters the second medium undeviated with its velocity reduced. If i be the angle which the direction of light travelling through a rarer to a denser medium makes with the normal to the common surface of these media, and r the angle which the direction of light makes with the normal on entering the new medium, then the ratio

$$\mu = \frac{\sin i}{\sin r} \qquad (1.5)$$

is called the *index of refraction* of the second medium with respect to the first. This is called *Snell's law*. The number μ is a constant for different values of i and r for the same pair of media. Physically, μ represents the ratio of the velocities of light before and after refraction in the two media. When light travels from a denser medium (of higher density) to a rarer medium (of lower density), the ray of light bends away from the normal after refraction, and the velocity of light is increased after refraction.

The value of μ is nearly 1.33 when light travels from air into water and is about 1.52 when it travels from air to glass (the latter value varying according to the type of glass). For light travelling in the reverse order, the values of μ will just be reciprocals of the above values. The phenomenon of refraction of light is important in positional astronomy. The light from a heavenly object reaches the earth after passing through layers of its atmospheres having steadily increasing densities. The light rays, therefore, in passing through these layers, continuously bend *towards* the normal to the surface of the earth. As a result, the observer will see object shifted towards his zenith, the amount of shift being proportional to the tangent of the *apparent zenith distance* of the object.

When a narrow beam of light (say, the sunlight) is allowed to enter obliquely the surface of a prism made of glass, the emergent beam is not only found to be deviated by refraction but also to form a band of a colour sequence corresponding to that of a rainbow (VIBGYOR), with the violet colour bending the most and the red bending the least. This phenomenon is known as *dispersion of light,* and the colour band is called the *spectrum* of the incident light. This is so since the refraction occurs at various wavelengths of the visible region of the electromagnetic spectrum and the colours are separated (or deviated) according to their wavelengths, the shorter wavelengths deviating the most. Thus, in the spectrum consisting of seven colours, the violet is the most deviated and the red the least. The other colours such as orange, yellow, green, blue, etc. occupy intermediate positions of the spectral pattern.

The index of refraction of a dispersive medium thus depends on the frequency or wavelength of the incident wave. If the incident radiation is composed of the entire range of the electromagnetic spectrum (as the radiation, from the sun, or other stars), then after dispersion each component will be separated according to its wavelengths and its angles of refraction will also depend on its frequencies or wavelengths. The dispersive medium does not abstract energy from the incident radiation but simply causes the change in the velocity of propagation. Mathematically, the variation $d\mu/d\lambda$ of the index μ with respect to the wavelength λ is termed *dispersion of the medium.*

When two or more beams of light of the same or nearly same wavelengths or frequencies are superposed, they combine to form a new wave pattern whose amplitude of vibration is different from that of the individual waves thus combined. This phenomenon is known as the *interference of light.* The interfering waves amplify one another at certain points and counteract one another at certain other points, thus forming a series of light or dark lines or bands called *fringes.* The amplitudes of the interfering waves are added when the crest of one wave coincides with that of another and the interference is called *constructive.* But when the resulting amplitude is less than the original amplitudes, the interference is called *destructive.* This occurs when the crest of one wave coincides with the trough of another. When the interfering amplitudes are the same and the phases are opposite, the waves annihilate completely and the resulting amplitude is zero. This might occur when two light beams of equal amplitudes interfere to give areas of darkness. For beams of white light interfering one another, the resulting beam might appear coloured. The phenomenon of interference affects the light coming from *double stars.* This light can be separated and studied by an instrument called the interferometer.

When the light is interrupted by an obstacle, the light waves bend and spread out slightly into the region of geometrical shadow of the obstacle. This phenomenon is called *diffraction of light.* The bending occurs around the edges of the obstacle and this causes a change in the amplitude or phase of the light waves. This results in the formation of alternate light and dark bands associated with the edge of the actual shadow of the obstacle. These bands are called *diffraction bands* or *diffraction rings.* The obstacle which causes the diffraction may be of different types. It may be a slit, a wire, a hole or simply a straight edge. Diffraction occurs when light passes through an objective or a refracting telescope. The amount of diffraction is proportional to the diameter of the objective.

The phenomenon of the diffraction is exhibited by all types of electromagnetic waves, such as radio waves, microwaves, infrared, visible and ultraviolet light, and X-rays. This effect is important in connection with the resolving power of the optical instruments, such as telescopes, spectroscopes and spectrographs. The devices used in the spectrographs for resolving light into its different constituent wavelengths are known as *diffraction gratings*.

Since light waves are transverse waves, they vibrate along straight lines perpendicular to the direction of propagation of light. A beam of ordinary light consists of innumerable such waves, each having its own plane of vibration. There are waves vibrating in all planes with equal probability so that there should be as many waves vibrating in one plane as there should be in any other. This is the case of perfect symmetry for the beam considered. If for any beam

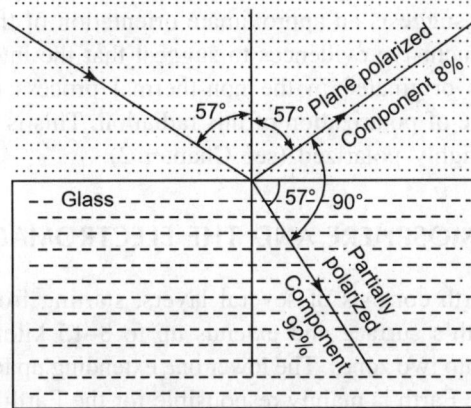

FIGURE 1.3 The polarization of light by reflection.

this symmetry of distribution of waves vibrating in different planes is somehow broken, the beam is said to be *polarized*. If by some means or other, all waves in a beam can be made to vibrate in planes parallel to each other, the beam is said to be *plane polarized*.

If a beam of ordinary light is incident on a glass surface at an angle of 57° with the normal to the surface, 8% of the light is reflected at the same angle to the normal while 92% is refracted through the glass. The reflected component is wholly plane-polarized while the refracted component is only partially plane-polarized. If the angle of incidence is other than 57°, the reflected component will also be only partially plane-polarized. The angle of incidence for which the reflected component of light is wholly plane-polarized is called the *polarizing angle*. Obviously, it varies from one medium to another, i.e., on the refractive index of the medium. Sir David Brewster, a Scottish physicist, showed that when the incident angle is the polarizing angle, the separation between reflected and refracted components is exactly 90°. This is known as *Brewster's law*. In this case, therefore, Snell's law can be written as (since sin r = cos i, here)

$$\frac{\sin i}{\cos i} = \mu \quad \text{or} \quad \tan i = \mu \tag{1.6}$$

This formula can be used to obtain the polarizing angle for any medium for which μ is known. For water, $\mu = 1.33$, so $i = 53°$; for glass $\mu = 1.52$, so $i = 57°$, and so on.

There are also other circumstances (e.g. double refraction) in which light is elliptically or even circularly polarized. The phenomenon of polarization of light and the electromagnetic radiation, in general, has found very useful applications in the field of astrophysics. For example, in the late 1940s, astronomers (Hiltner, Hall and others) observed that light from distant stars was polarized and the amount of polarization was found to vary with galactic longitude. The observation was interpreted as due to the scattering of starlight by interstellar grains oriented in some definite fashion by the influence of the magnetic field in the Galaxy. Thus, one of the most important discoveries in the field of astronomy, that the existence of a large scale magnetic field in our Galaxy, was the result. The varying amount of polarization in galactic longitudes also suggests an approximate orientation of the galactic magnetic field. Again, one of the most conclusive evidences to suggest that the intense radio energy coming from strong radio sources originated in the non-thermal process (synchrotron radiation) is supplied by the high degree of polarization of this radiation. This is because we know that the non-thermal radiation is highly polarized (see Chapter 2).

1.2   THE EARTH'S ATMOSPHERE AND THE ELECTROMAGNETIC RADIATION

The atmosphere of the Earth consists of several layers, starting from the *troposphere* which is in contact with the Earth's surface and extends up to 8–15 kilometres above the surface. This layer may be divided into two zones. The lower one extending up to a height of approximately 3 to 5 kilometres above the Earth is mainly responsible for the Earth's weather. The upper part of this layer is much less disturbed by atmospheric perturbations. The next 25 to 35 kilometres of the Earth's atmosphere belongs to the region known as stratosphere. It is a homogeneous layer and almost constant in temperature. There are two thin layers of gas molecules in the stratosphere. The lower one contains some sulphates which are thought to play some role in rainfall. The upper layer forms the ozone layer which absorbs most of the ultraviolet rays from the Sun and thus protects the life on the Earth.

Next comes the *mesosphere,* the layer where the meteors from the outer space come and burn themselves. The region extends roughly between 50 to 80 kilometres above the Earth's surface.

The *ionosphere* begins at a height of about 80 kilometres and extends up to a few hundred kilometres. In this layer there are different zones at different heights where the atmosphere is ionized by solar radiation. The lowest layer of the ionosphere is the D-layer which reflects the longer radio waves back to the Earth's surface, thus making the medium and long wave reception possible. The medium and shorter radio waves penetrate through this layer.

The E layer of the ionosphere is situated at a height of nearly 120 kilometres above the Earth, and plays the principal role in the reflection of the medium radio waves to the Earth. Meteors while approaching towards the Earth burn in this layer.

There are other layers like F_1 at a height of about 200 kilometres and F_2 at about 300 kilometres. Short radio waves are reflected back to the Earth by the F_2 layer of the ionosphere. Still shorter radio waves (e.g. metre and centimetre waves) as well as microwaves can pass unhindered through all the layers of the Earth's atmosphere.

The last layer is the *exosphere* where the atmosphere is extremely rarefied, even more rarefied than the vacuum we can achieve in the laboratory. This layer consists of a number of radiation belts which are believed to extend up to about 60,000 kilometres above the Earth's surface.

The various layers of the Earth's atmosphere block out most of the short wave radiation from the Sun and stars. Only two wavelength ranges of the electromagnetic spectrum reach the Earth unhindered. The first one is slightly broader in wavelength range than the spectrum of the visible light between 4000 Å to 7200 Å and is known as the optical window through the atmosphere. The second range lies between 3 mm and 50 metres. This is the range of microwaves and shorter radio waves and is known as the radio window. A third range, an extremely short one, lies between 8 and 14 μm in the infrared region of the spectrum. At the short end of the spectrum below 3000 Å, the radiation is completely blocked. So, before the advent of baloons, rockets and orbiting astronomical observatories, the astronomers could only guess about the nature and intensity of electromagnetic spectrum below 3000 Å which includes the ultraviolet rays (between 3000 Å and 300 Å), the X-rays (shorter than 300 Å but longer than 0.1 Å) and the γ-rays (shorter than 0.1 Å).

To study the nature of the ultraviolet and the infrared radiation, one should exactly know the heights at which the atoms and molecules absorb them. The infrared is absorbed chiefly by water vapour and CO_2, most of which is below 30 kilometres. This height can be exceeded by stratospheric baloons. The ultraviolet radiation is absorbed at much higher altitudes, which can be exceeded only by rockets. The ozone layer in the stratosphere formed by the action of ultraviolet radiation from the Sun on oxygen molecules absorbs the wavelength regions between 2900 Å and 2200 Å. The wavelengths between 2200 Å and 900 Å are absorbed by the molecular oxygen, most of which lies below 100 kilometres. Molecular nitrogen, atomic oxygen and atomic nitrogen in still higher layers absorb ultraviolet radiation of still shorter wavelengths, the maximum absorption occurring at about 500 Å. At his wavelength, very little radiation penetrates below 240 kilometres.

As the ground-based instruments could receive radiation only in the visible range and a part of the radio region of the spectrum, these two branches of astronomy have been by far the more developed as compared to the other branches. Since the beginning of Space Astronomy, other branches like ultraviolet astronomy, infrared astronomy, X-ray astronomy, and γ-ray astronomy have started developing. The Orbitting Astronomical Observatories have collected huge amount of information regarding the nature and intensity of the solar ultraviolet radiation. With the instruments of increasing refinement and complexities mounted in sounding rockets and in artificial satellites, shorter wavelength regions are being explored to study the physical nature of various celestial objects. Different branches of the observational astronomy have been created which deal with different regions of the electromagnetic spectrum. They have different observing techniques and instrumentation.

1.3 OPTICAL TELESCOPES

For astronomical observations, the telescope is the most important instrument in use today. Telescopes which are used for observations in the optical wavelengths of the electromagnetic spectrum are known as *optical telescopes*. Those which are used to detect electromagnetic radiation emitted in the radio wavelengths are known as the *radio telescopes*. In this section, we shall briefly discuss the principle and use of the optical telescopes. We shall deal with radio telescopes in Section 1.4.

Optical telescopes are of two kinds, viz. *refracting* and *reflecting,* having their objectives as a lens or a mirror respectively. The former is of more general use and smaller in size as a rule; however, the largest telescopes in the world are of the latter type. The fundamental principle is the same in both: a *real image* of a luminous object is formed at the focus of the large lens (refractor) or the mirror (reflector) of the instrument. This image is then magnified by an eyepiece which serves as a *magnifying glass.*

Magnifying Power. The main function of the eyepiece is to bring the image very close to the observer's eye with the help of a lens of very small focal length (usually 0″.5 to 1″), and thus magnify the image several tens to hundreds of times. The magnifying power Q of a telescope is the ratio of the focal length F of the objective to the focal length f of the eyepiece. Thus,

$$Q = F/f \qquad (1.7)$$

For example, the magnification Q in a telescope of focal length 600″ (e.g., the 120-inch reflector at Lick) and using an eyepiece of focal length 1″ is 600, and so on.

Brightness of Image. The brightness of the image of an object formed at the focus of a telescope depends on the aperture of the objective. The light-gathering power of a lens or a mirror is proportional to the square of its aperture. Thus more and more faint objects are accessible with telescopes of larger and larger apertures. The 200-inch reflector at Mount Palomar can penetrate through objects four times as faint as can be penetrated with the 100-inch reflector and 25 times as faint as can be done with the 40-inch reflector at Kavalur. In order to observe fainter and fainter objects, one has thus to increase the aperture of the telescope objective. There is no other known optical method to achieve the higher brightness of the image at the focus. The Keck telescopes at Mauna Kea, Hawaii are at present the largest optical telescopes in the world.

For extended objects like the moon, planets and nebulae, the brightness of the image depends on both the aperture as well as the focal length of the telescope. The area of the image of an extended object, formed at the focus is proportional to the square of the focal length. If two telescopes have focal lengths in the ratio of 1:2, the diameter of the image of, say, jupiter or moon at the focus will also bear the same ratio. The surface area of the image will therefore be four times. Now, if these two telescopes have the same aperture, the total light gathered from the object by both is the same. Thus, the same amount of light is spread in the image of one with larger focal length over an area four times that of the other with smaller focal length. The brightness of the image in the former will, therefore, be four times dimmer than in the latter, since the amount of light per unit area is four times less in the former than in the latter. This is the reason why moon's surface is visible in considerable detail with a telescope of small aperture than with the naked eye. The details are washed out to the naked eye by the excessive flood of light, but when a telescope is used, the light is dimmed by spreading over larger areas showing the details there. In the case of stars, on the other hand, since they produce point images, the maximum advantage is gained by using the largest apertures.

Resolving Power. Due to the wave properties of light, it is subjected to *diffraction* and *interference.* Thus the image of a star produced at the focus of a telescope is not just a point image, but a *diffracted image,* that is, a tiny spot of finite size surrounded by concentric diffraction rings as shown in Fig. 1.4 (in a much exaggerated scale). The net effect of this is that the stellar image is produced on a much larger surface area than it would be if it was just a point image and there were no effects of diffraction. As a result, the chance of overlapping of the images of two stars which are close together is much more increased if the diffraction pattern is present than if this pattern is absent. This increased chance of overlapping *limits the resolving power of a telescope.* If the two stars are so close that their diffraction discs overlap, we say that the stars are *not resolved.* The diffraction discs are rendered smaller by using a larger aperture and the stars may then be separated. Thus the resolving power of a telescope depends on its aperture. There is no other optical means which can be applied to increase the resolving power. No amount of magnification is of any use. The limit of resolution is given by the formula

$$\text{Limit of resolution} = \frac{4.5}{a} \text{ seconds of arc,}$$

where a is the aperture of the telescope in inches. This is known as *Dawes' rule.* According to this formula, the limit of resolution of the Kavalur 40-inch reflector is $0''.1125$, while that of the 200-inch reflector at Mount Palomar is $0''.0225$, i.e. two stars separated by $0''.0225$ will be resolved by the 200-inch telescope.

FIGURE 1.4   The diffracted image of a star.

The f/a Ratio.   The ratio of the focal length to the aperture of a telescope, called *the focal ratio* or *f-ratio,* is an important quantity to be designed in the construction of a telescope to serve some specific objectives. As we have already seen, the light gathering power of a telescope depends on a only, whereas the size of the image depends on f only. So the focal ratio will, in general, influence the brightness of the image. Large focal ratio indicates the spreading of the image over a larger area. Large reflectors generally have paraboloid mirrors with their focal ratios lying between 3 and 5, or somewhat larger. The focal ratio of the 200-inch Hale reflector is 3.3, which implies that its paraboloid mirror has a focal length of 660-inches. It is expressed as *f/*3.3. The 120-inch Lick telescope has a focal ratio of 5.0. In fact, *f/*5 is a traditional value for earlier large telescopes. The Magdonaid 82-inch telescope is *f/*4.

Of the various advantages that may be derived from different *f*-ratios, those obtained from small *f*-ratios are more important. To mention two of them, one is the smaller expenditure for the complete construction, and the other is the availability of more accurately defined optical system. There are many other points in the construction technicalities as well as good functioning of a telescope which are related to the *f*-ratio. Also, since different foci are designed in large telescopes, the *f*-ratio changes from one focus to the other.

Types of Reflecting Telescopes

Several different arrangements of optical parts are usually required for different types of work with reflecting telescopes. We shall illustrate this with the help of Fig. 1.5. The figure represents the focal arrangements of the Lick 120-inch reflector which can be considered as a typical sample of a large reflector. The incoming parallel rays of light from a source enter the telescopic tube to strike the *parabolic mirror* (the primary mirror) which has a hole at the centre. The rays are then reflected straight to the *prime focus* of the mirror when the convex reflector and the optical flat are removed. The observer sits in the observer's cage when the prime focus is used. This focus is used invariably for doing photography, the plateholder being attached then at the focus.

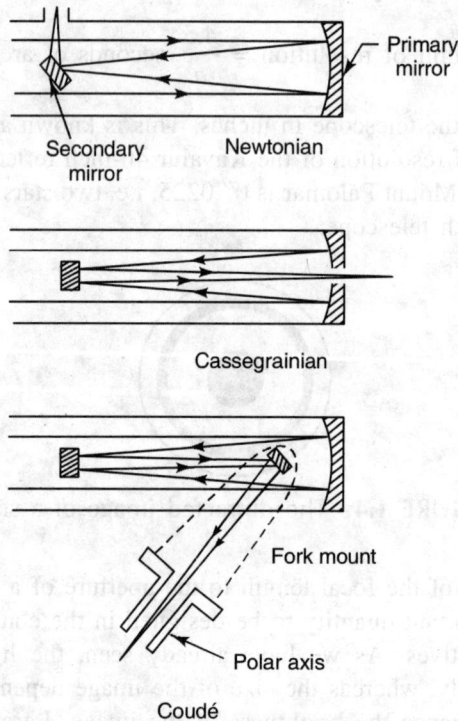

FIGURE 1.5 Focal arrangements in a reflecting telescope. (Newtonian, Cassegrainian and Coudé foci are shown.)

In the optical arrangement of a Cassegrain reflector, the light reflected by the primary mirror before converging to the prime focus is intervened by a *hyperbolic secondary* which is convex towards the primary. In the absence of the optical flat, the light reflected by the convex mirror may pass through the hole at the centre of the primary and converge to the *Cassegrain focus*. This focus may be conveniently used for spectroscopic and photoelectric works. Its principal advantage lies in the fact that the observations can be made at the lower end of the telescope where the auxiliary instruments can be easily attached.

Again, light reflected by the convex reflector, instead of being passed through the central hole, may be reflected far out by a flat mirror placed at some angle to the direction of rays.

The reflected rays in this case pass through a door in the tube and converge at a focus which is known as the *Coudé focus*.

This focus is used primarily for high dispersion spectroscopic work. The fourth kind of optical arrangements, which is of general use in large reflectors, is the Newtonian focus arrangement (Fig. 1.5). In this arrangement, the incoming parallel rays of light after being reflected by the primary mirror are intervened by a plane mirror placed at some angle to the direction of parallel rays and reflected by it to a point outside the telescopic tube. This converging point is called the *Newtonian focus*. This focus is particularly suited for photoelectric photometry.

Large reflectors are, in general, constructed in such a manner that they can be used in more than one of the above four forms. The 200-inch Hale telescope and the 120-inch Lick telescope can each be used at the prime, Cassegrain and Coudé foci. The 100-inch reflector at Mount Wilson can be used both at Newtonian and Cassegrain foci. Arrangements are made so that the auxiliary instruments such as *the filar micrometer, spectroscope* and *spectrograph* (see the following section), the *photometers*, the *photomultipliers, thermocouples, photoconductive* cells, etc. can be attached with the telescopes for various types of astronomical measurements.

The Refracting Telescopes

We shall conclude this section with a brief description of a *refracting* telescope. The essential parts of a simple refracting telescope are shown in Fig. 1.6.

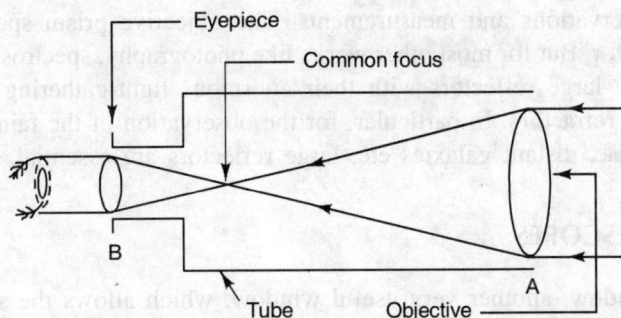

FIGURE 1.6 The simple refracting telescope.

A simple refracting telescope consists essentially of two convex lenses, one of which is the objective *A* with a large focal length and the other is the eyepiece *B* of short focal length. The two lenses are so adjusted that their distance is nearly equal to the sum of the focal lengths of the lenses. In refractors, the *f/a* ratio is generally much higher than that in reflectors. While 3.0 to 5.0 are the usual values of the ratio in reflectors, the values for the refractors may be as high as 15.0 to 20.0. Refractors are conveniently used for visual observations and for photography. They are also used sometimes for spectroscopy.

Although refractors are of more common use and smaller refractors are more easily constructed than the reflectors, many inherent problems are involved in the construction of large refractors. The principal among these are the difficulties faced in making good homogeneous large lenses and the requirement of very complicated and massive mounting structure. The weight of the large objective and of the long tube itself may cause the tube to sag. Because of these types of inherent difficulties, the largest telescopes that have been built are all reflectors. While one of the largest reflectors now in operation is of 236-inch diameter in the Soviet Union, the largest refractor is at the Yerkes Observatory, which has a 40-inch aperture. Among other large refractors, the 36-inch aperture refractor at the Lick Observatory may be mentioned.

Relative Advantages of Reflectors and Refractors

In the case of a refractor, the light passes through a lens while in a reflector, light is just reflected by a mirror. So the glass of the lens of a refractor must be *optically homogeneous,* while that of the mirror of a reflector needs only to be *mechanically homogeneous.* The latter means that the glass must not bend irregularly and this condition is easier to achieve. The chromatic aberration of lenses sometimes seriously affects the quality of photographic and spectroscopic works with refractors. But the mirrors being perfectly achromatic, high quality is always achieved by a reflector. On the other hand, the small differences in temperatures of the different parts of a large mirror often decrease the defining power of the mirror to a value much less than the theoretical value. The objective lenses of a refractor being much smaller in size, this difficulty never becomes serious, so that the refractors have wide field of good definition.

For the same size of the aperture, the refractors are more costly than the reflectors.

As regards the use of refractors and reflectors is concerned, refractors are almost exclusively used for visual observations and measurements. The objective prism spectrographs can be used only in a refractor. But for most other works like photography, spectroscopy, photoelectric photometry etc., the large reflectors with their enormous light-gathering power have great advantages over the refractors. In particular, for the observation of the faint objects like faint stars, gaseous nebulae, distant galaxies etc, large reflectors are essential.

1.4   RADIO TELESCOPES

Like the optical window, another very useful window, which allows the study of the nature of the electromagnetic radiation from the heavenly objects, is the *radio window* through which the radio waves from these objects can reach the ground-based instruments. The 1930's ushered in the study of the radio waves from space with their first detection by Karl Jansky in 1932. During the Second World War, in 1942, Sir Stanley Hey of England first discovered that the Sun was a radio emitter. The study received great impetus in the early post-war years and new radio telescopes of large sizes were built all over the world. In 1951, the work of great astrophysical significance by H.I. Ewen and E.M. Purcell of Harvard in their detection of the 21-cm hydrogen line in space whose possibility had been predicted earlier by H.C. Van de Hulst came into light. The result has been the development of a parallel subject called *radio astronomy.*

Being electromagnetic waves, radio waves are reflected according to the same laws as light. Shorter radio waves of millimetre range suffer absorption in the ionized hydrogen region, and in the Earth's atmosphere while the waves longer than about 50 metres are almost completely reflected back by the ionosphere of the Earth. So the most suitable wevelengths from space to study are those ranging from a few centimetres to a few metres which pass unhindered through space to reach the ground-based telescopes. This is precisely why most of the radio astronomical work has been done in this wavelength range and maximum work has been done in one single wavelength, viz., the 21-cm line of neutral hydrogen.

Radio telescopes of different sizes and forms have been developed and constructed according to their usefulness for various types of work. The basic principle is the same as in the optical telescopes, i.e., radio waves from the cosmic sources are reflected from, say, a *paraboloidal* dish to the focus of the dish where a dipole antenna is fixed. The energy is then transmitted to a receiver and is amplified by an amplifier. The amplified signal is then recorded by a pen on a moving sheet of tracing paper. A schematic diagram of such a radio telescope is shown in Fig. 1.7. The dish is made of conducting material such as metals. But since it reflects large waves, it need not be highly polished. For a radio telescope the material should not be *continuous* either. Wire meshes like those used in fences serve well as reflector, provided the separation between the consecutive wires is much less than the length of the waves. In this case also, the radiation is focussed on a receiver and amplifier at the principal focus from which the amplified signal goes to the recorder.

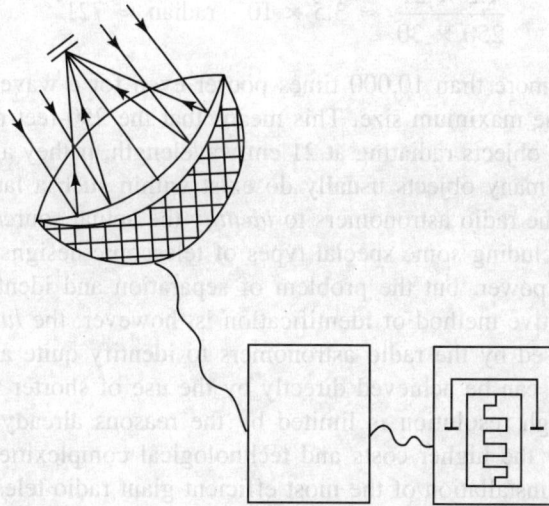

FIGURE 1.7 The schematic diagram of a paraboloidal dish radio telescope.

One fundamental difference between the optical and the radio telescopes is that while the former records simultaneously the entire range of optical frequencies, the latter can record only one frequency at a time. If work is done simultaneously with a number of frequencies, many telescopes have to be used at the same time. This puts a severe limitation on the efficiency of radio telescopes for working at several frequencies. A partial remedy for this problem is provided by specially constructed *tunable, receivers* which enable to *sweep over* a range of frequencies in a short time interval. This mechanism is particularly important when the radio spectrum over a frequency range of a particular source is desired. Such a study is very helpful to analyze the radio waves coming from the Sun and radio galaxies.

The resolving power of a radio telescope is very low compared to that of the optical telescopes. This is because the wavelengths of light are many orders of magnitude lower than those of the radio waves. Also, while the ratio of the longest to the shortest wavelengths of light never exceeds 2, the corresponding ratio for the radio waves may be as large as 10^3, because these waves are 10^4 to 10^7 times larger than the light waves. Since the limit of resolution of a telescope is directly proportional to the wavelength and inversely, proportional to the aperture, and the aperture is severely restricted by cost and technological complexities, long wavelengths bring in poor resolution. If the wavelength λ and the aperture a are measured in the same units, then the resolution of a telescope is given by 1.2 λ/a in radian measure.

Now suppose observations are made with the Lick 120-inch reflector at wavelength 6000 Å. The limit of resolution obtained will be

$$\frac{1.2 \times 6000}{120 \times 2.5 \times 10^8} = 2.4 \times 10^{-7} \text{ radian} = 0''.048$$

The same law applies to the radio waves. Suppose 21-cm waves are observed with the 250-feet parabolic reflector at Jodrell Bank, the resolution achieved in this case will be given by

$$\frac{1.2 \times 21}{250 \times 30} \approx 3.5 \times 10^{-3} \text{ radian} \approx 721''$$

The resolution is thus more than 10,000 times poorer even for a wavelength on shorter side and a dish of almost the maximum size. This means that the 250-feet parabolic dish will not be able to separate two objects radiating at 21 cm wavelength, if they are closer to each other than 12′ of arc. Since many objects usually do exist within such a large field, it remains a common problem for the radio astronomers to *identify* the actual source of radiation. Several different techniques including some special types of telescopic designs have been devised to increase the resolving power, but the problem of separation and identification is still to be solved. One very effective method of identification is, however, the *lunar occultation* which has been effectively used by the radio astronomers to identify quite a few radio sources.

Higher resolution can be achieved directly by the use of shorter wavelength and larger dish. But while the high resolution is limited by the reasons already mentioned, the latter condition is limited by the higher costs and technological complexities. Various difficulties are encountered in the installation of the most efficient giant radio telescopes. These must be steerable in order to achieve the maximum efficiency. But the construction of large steerable telescopes presents many difficult mechanical problems to the designer. As the telescope is tilted for use, the gravitational forces on the structure change, which put a severe test on the structure to remain unaltered. While the deflection of the structural members under their own weight roughly increases as the square of the linear dimension of the telescope, the cost goes roughly as the cube of the dimensions. So while on the one hand very large steerable dish-type telescopes become prohibitively expensive, on the other hand, their installation poses unmanageable technological complexities. One of the largest fully steerable parabolic dish-type telescopes is the 250-feet reflector at Jodrell Bank in England. It is a solid metal reflector with such a short focal length that its focus lies within the rim of the dish, thus shielding it from the radio interference from sources on the ground. The 210-feet parabolic dish at Parkes in Sydney, Australia, is another very efficient fully steerable radio telescope. While the former has been used extensively for observation and detection of extra-galactic radio sources, extremely valuable work has been done with 210-feet parabolic dish (at Parkes in Sydney) in the observation of the 21-cm radiation.

The largest spherical-reflector telescope at Puerto Rico has a diameter of 1000-feet which can be steered through a considerable range of zenith angles. Such a telescope, however, is inefficient since it has no true focal point to which the rays may converge.

It is clear, therefore, that to achieve high resolution with one parabolic dish reflector does not seem probable in foreseeable future. But necessity has prompted other alternative devices to be evolved for the purpose. One such device is the variable-spacing, two-antenna interferometer which yields higher resolution in *one direction*. In this arrangement, two antennae are spaced hundreds of feet apart and the interference pattern they produce in the radiation is studied to derive the resolution. The same result can be achieved by using long arrays of dipoles or of helical antennae. Higher resolution in *two directions* is yielded by two such perpendicular arrays. Such an arrangement is called the *Mills Cross*. Another cross-type antenna, called the *Benelux Cross,* has been designed, which consists of a large number of two-antennae interferometers, rather than as a horizontal cross. This antenna consists of 120 large circular paraboloids, each having a diameter of 100 feet which are spread over a length of 1.5 km. Several other designs have also been constructed while some others still are in the process of construction. The discussion of all these varieties of radio telescopes is beyond the scope of our discussion here. The inherent difficulties of various types involved is the process step by step give way to human endeavour. The design and construction of radio telescopes are still in the process of evolution.

1.5 THE HUBBLE SPACE TELESCOPE (HST)

However large in size and good in quality a ground-based optical telescope may be, it cannot avoid the degradation of images caused by atmospheric turbulence. An orbiting telescope being free from the effects of atmosphere can produce sharper images than its ground-based counterparts. But to make significant gains in angular resolution over the existing ground-based telescopes, the space telescope must be large with high quality optical components made and aligned with great precision. All these requirements make the launching of a large space telescope a high cost adventure. But NASA formally decided to undertake the programme in 1976, after years of deliberations on the feasibility of the programme. In 1983, NASA decided to name the satellite as Edwin P. Hubble Space Telescope and finally adopted the shortened name Hubble Space Telescope or HST.

The planned launch of the HST in late 1986 had to be postponed due to the disaster that took place in the loss of the Space Shuttle Challenger. The payload was finally launched in April, 1990. The HST carries on board the *f*/24 telescope with a 2.4 m primary and a 0.3 m secondary mirror. There are five scientific instruments on board the HST, each located behind the primary mirror in an instrument module. In what follows, these five scientific instruments are described in a nutshell.

The Wide Field Planetary Camera (WFPC) can survey selected areas for very faint objects and can also image smaller areas at increased angular resolution. The Camera works with four sets of 800×800 CCDs, is capable of detecting objects over a magnitude range 9–28 and has wavelength sensitivity in the 115 nm to 1 μm range. It is equipped with a variety of filters, gratings.and polarizers, which make it suitable for a wide range of observations.

The Faint Object Spectrograph (FOS)will perform low resolution spectroscopy, spectropolarimetry, and time resolved spectroscopy. The limiting magnitude of FOS varies with wavelength and spectral resolution. It can attain 21st magnitude at moderate resolution and 25th magnitude at low resolution with a 3000 second exposure (see Chapter 3).

The High Resolution Spectrograph (HRS) is meant to perform spectral investigations with a resolution of very high order at wavelengths between 110 and 320 nm. The limiting magnitude range is from 11 to 17 for a 2000 second integration time, depending on wavelength, resolution mode and exposure time.

In order to study the brightness fluctuations over a wide spectral range, a High Speed Photometer (HSP) has been placed in the system. The instrument is so powerful that it can resolve events only a few microseconds apart. Such resolution is unthinkable by any system on the ground because of the atmospheric turbulence. The detectors are essentially high efficiency photomultipliers covering a spectral range of 120 nm to 800 nm.

The fifth and the last instrument is the Faint Object Camera (FOC) which is a contribution of the European Space Agency (ESA). This instrument is specially designed to study very faint objects at high angular resolution. The two separate camera systems attached to the FOC operate at *f*/96 and *f*/48, producing angular resolutions of 0.022 and 0.044 arcseconds respectively. The instrument is provided with a variety of filters, polarizers and other accessories, together with an occulting coronagraph and a long slit spectrograph.

Since its launching in 1990, HST has been peering through different regions of the Universe, both far and near and acquiring thousands of superb quality images and sending them to the Earth. When these images are properly analyzed and interpreted, astronomers are likely to acquire the picture of a new Universe, a Universe more mysterious and sophisticated than was previously conceived of. The HST will continue to work through this century and beyond, in the 21st century, keeping the zeal and hope of the present and future generation of astronomers. The efficient functioning of the mission will be kept going by sending repeated servicing missions as was done in December 1993 to obviate the effects of spherical aberration. More servicing missions are in progress. In 1997, an advanced spectrograph and a near infrared camera will be installed, and in 1999, a third servicing mission will restore the spacecraft orbit. With repeated servicing and installation of newer and more improved instruments, HST will continue to make unique and important astronomical discoveries in the early decades of the 21st century. No ground-based telescope, however large and well equipped with allied instruments, can match the HST in terms of wavelength coverage and image quality.

1.6 ASTRONOMICAL SPECTROGRAPHS

The spectroscope or spectrograph is an instrument that can separate a non-monochromatic beam of light into its constituent wavelengths and can record the intensities of the radiation according to the wavelengths. The instrument is so called according to the way it is used. If the observations are made with the naked eye with an eyepiece, the instrument is called a spectroscope. The same instrument is called a spectrograph if the record is photographed at the focus. For astronomical observations, it is used to analyze a starlight from which important information like the chemical composition, the physical conditions of the stellar atmospheres as also the evolutionary stage of the star may be known. From the study of the Doppler shifts of the spectral lines, we can calculate the radial velocities of stars and other objects in the sky.

As the stars differ greatly in brightness, different types of spectrographs have been developed for their studies, starting from the faintest visible stars to those of about billion times brighter than them. With the modern spectrographs, spectra have been recorded for the faintest stars which are about 10^8 times less bright than the Sirius, the brightest star in the sky. Sirius, in turn, is about 10^{10} times fainter than the Sun.

There are several other factors which have to be considered while choosing the suitable spectrograph for a particular purpose. The range of wavelengths over which the spectroscope would serve, its power of resolving two close lines in the spectrum, the nature of the dispersing material and its dispersive power, the variation of the dispersion with wavelengths of the incident radiation, the brightness and the speed of recording of the spectra of faint celestial sources, are some of the factors which should be taken as primary considerations. Other factors like the absorption and scattering in the instrument and the size and shape of the instrumental profiles as well as those of the actual line profiles should also be observed with great care.

The principal part of a spectrograph is the dispersing medium, which disperses light to different extents, depending on the wavelengths of the incident radiation and of the dispersing material used. All that we need is a good spectrum even for a very faint object, and the other parts of the spectrograph are designed accordingly. Sometimes scales are attached with the spectrographs to measure the wavelengths directly from the spectrum. These scales may be calibrated in microns, millimicrons or in angstroms to study the fine details.

The dispersing medium in a spectrograph may be mainly of two types: (a) the prism made of glass or quartz, and (b) the diffraction grating. These are now discussed.

The Prism

The prism is generally made of glass or quartz and is used in the form of an equilateral triangle (see Fig. 1.8). When white light is made to fall upon the face AB of the prism, it gets refracted and dispersed into its component colours. The amount of dispersion is proportional

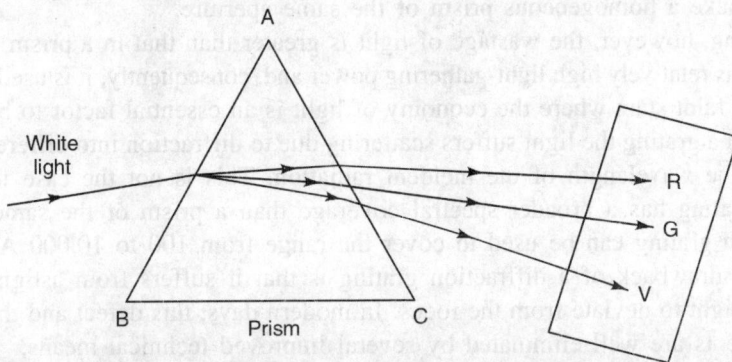

FIGURE 1.8 Dispersion of light of different wavelengths through a prism.

to the component wavelengths and depends on the material (or refractive index) of the prism. It is found that the dispersion occurs as a result of the change of speed of the constituent wavelengths on entering the face AB and this change is greater for the shorter wavelengths. On emerging from the face AC of the prism, the rays are further refracted and dispersed to form a spectrum on a screen or on a photographic plate. In this spectrum, the colours are found to be placed in the order of their wavelengths and the violet ray (of shortest wavelength) is found to be the most deviated from its original direction and the red ray (of greatest wavelength) the least deviated.

The Diffraction Grating

A diffraction grating is a surface consisting of a large number of finely placed close equidistant lines called the *diffraction lines*. There are mainly two types of gratings which are used in practice for astronomical spectroscopy. They are: (a) the transmission grating, and (b) the reflection grating.

The transmission grating consists of a transparent surface (e.g. glass) ruled with a large number of fine parallel equidistant lines or grooves in a small surface (thousands of lines per inch, may be 10,000 lines or more per inch), so that light can only pass through and disperse by the surface between the lines (or scratches). This type of grating is generally used in small grating spectrometers and spectroscopes.

In a *reflection grating*, light encounters a polished surface (e.g. a mirror) which has similar fine, equidistant, close parallel lines or grooves, so that light is reflected by the polished surface between the lines or scratches. Reflection grating may be either plane or concave. The Rowland's concave grating named after H.A. Rowland is in much use as a dispersing medium of a spectrograph. Reflection gratings containing 5000 to as many as 30,000 lines or grooves per inch have been made possible. They are used mostly in large spectrographs.

The resolving power of a grating spectrograph is indicated by the number of lines ruled within a fixed width on the surface of grating. The greater the number, the more is the dispersing power of the grating.

A grating has, in general, greater dispersive power and resolving power than a prism of the same aperture and cost. It is also a much easier task to rule a grating surface a few inches wide than to make a homogeneous prism of the same aperture.

In a grating, however, the wastage of light is greater than that in a prism. On the other hand, a prism has relatively high light-gathering power and, consequently, it is used in measuring the intensity of faint stars where the economy of light is an essential factor to be considered.

Further, in a grating the light suffers scattering due to diffraction into different directions, depending on the wavelength of the incident radiation. This is not the case for the prism. Besides, the grating has a broader spectral coverage than a prism of the same aperture. A single reflection grating can be used to cover the range from 100 to 10,000 Å.

The main drawback of a diffraction grating is that it suffers from astigmatism which causes rays of light to deviate from the focus. In modern days, this defect and the light losses by the instruments are well eliminated by several improved technical means.

A Slit Spectrograph

In an astronomical spectrograph, the photographic record of the stellar spectra is done mainly by two methods. The first method employs a slit spectrograph which essentially consists of the following arrangements. The light from a star is focussed on a narrow slit which is in the focal plane of the telescope (e.g. a Cassegrain reflector) objective. Most of the light is then passed through a collimator lens, which renders the rays parallel. The light rays are then allowed to pass through a dispersing medium which may be either a prism or a grating. The emergent spectral lines, which are monochromatic images of the slit, are then focussed by a camera lens and a curved photographic plate. The slit spectrograph with all its components is illustrated in Fig. 1.9.

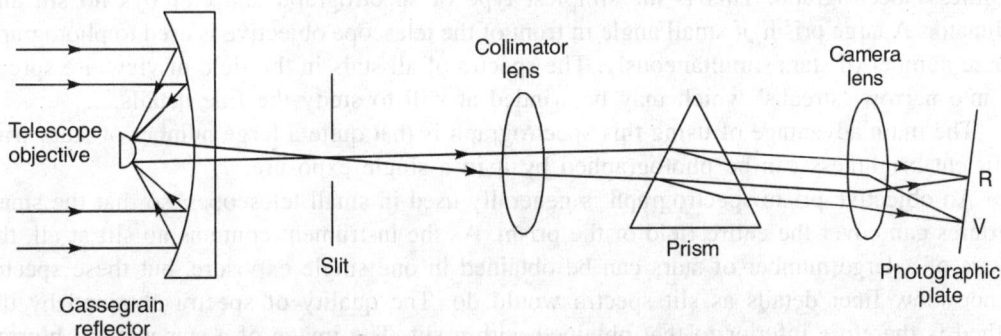

FIGURE 1.9 The optical parts of a slit spectrograph.

Some light from the telescope objective is intercepted and reflected back from the polished surfaces of the slit jaws. A reflecting prism placed near the slit jaw can send this light through a guiding eyepiece, by means of which we can observe the star-image.

For extended objects like Sun and the planets, the spectrograph is used in the same position with the slit crossing the image of the object.

The main advantages of the slit spectrograph are that the slit can be narrowed at will, so that it can eliminate the night sky light, and the spectrum yielded can be studied in great detail. Its main drawback is that it can record only one star-image at a time. So it becomes more time consuming. Even the 200-inch reflector at the Mount Palomar can record a low-dispersion spectrum of star not fainter than the 18th magnitude with an exposure of several hours. Light is also wasted by the reflection of the star light by the slit jaws.

One main objective of using the slit spectrograph is to measure the wavelengths of the spectral lines with high precision. This is done by exposing a comparison spectrum on either side of the stellar spectrum. The comparison spectra may be produced by either a spark or an arc between two metal terminals. Generally, iron-arc is used in the ultraviolet and blue regions of the spectrum. Silicon and titanium are also used. The comparison spectra of a neon discharge tube are used in the yellow, red and infrared regions. For low dispersion spectra of celestial sources, the comparison spectra of discharge tubes containing helium or mercury are generally used.

The comparison spectrum from a spark or an arc, after passing through suitable diffusers to reduce intensity, is sent through the slit after reflection by two quartz prisms placed near the slit jaws. This is done during the exposure of the starlight from the telescope objective. The comparison spectrum appears adjacent to the stellar spectrum and by knowing the wavelengths of lines given by the particular element used, the wavelengths of the lines of the stellar spectrum can be ascertained by interpolation. By a comparative study of the spectral lines from the two sources, the stellar lines can be identified and the Doppler shifts of the lines in the stellar spectrum can be calculated. This technique is used for measuring, among several things, the radial velocities of stars.

The Objective Prism: Slitless Spectrograph

The second type of astronomical spectrograph which is used in practice is the objective prism or slitless spectrograph. This is the simplest type of spectrograph and employs no slit and collimator. A large prism of small angle in front of the telescope objective is used to photograph a large number of stars simultaneously. The spectra of all stars in the field of view are spread out into narrow 'streaks' which may be winded at will to study the fine details.

The main advantage of using this spectrograph is that quite a large number of stars with sufficient brightness can be photographed by it in a single exposure.

An objective prism spectrograph is generally used in small telescopes, so that the small apertures can cover the entire field of the prism. As the instrument contains no slit at all, the spectra of a large number of stars can be obtained in one single exposure, but these spectra do not show finer details as slit spectra would do. The quality of spectra obtained by the method is therefore inferior to that obtained with a slit. The image of a star may be blurred due to bad seeing or due to improper guiding of the telescope, which can be avoided in a slit spectrograph by setting it to long exposures. Also, no comparison spectra can be exposed along with the stellar spectra for studying the Doppler shift of lines. It happens, therefore, that when a large number of stars require to be studied spectroscopically in a short time, but with lesser details, the objective prism spectra become more useful than the slit spectra. The former are therefore very useful for a quick survey type study.

Objective prisms of large size are quite difficult to make. Objective prisms as large as 24-inches diameter have been made possible for use over the telescope objective of 24-inches aperture. Now-a-days, objective prisms are being used with large Schmidt cameras.

Resolving Power of Spectrographs

In a spectrum if two close lines of equal intensity are seen at wavelengths λ and $\lambda + d\lambda$, then the resolving power of the spectrograph which resolves the lines at wavelength λ is given by $\lambda/d\lambda$, a pure number. The resolving power of different spectrographs varies from 5000 to 200,000. The resolving power of a spectrograph actually depends on the material, shape and quality of the dispersing medium of the spectrograph. The highest possible resolving power is required to study the spectral lines with high precision so that the actual physical condition of a celestial object may be known in great detail.

Although a high resolution spectrum is required for its analysis in great detail which only can yield a better knowledge of the physical state of the object concerned, such a good spectrum can be obtained only for bighter sources. For faint sources, high resolution will render the spectral lines so diffused that they can hardly be distinguished and separated from the background continuous spectrum. Thus, two very close lines can be well separated only in spectra of bright sources for which even quite high resolution will yield spectra with clear conspicuous lines. But for the lines to be clearly recognized in the spectra of faint sources, sufficiently low dispersion has to be used. For example, two lines whose separation is only 5.5×10^{-3} Å can be resolved in the yellow region around 5500 Å of the solar spectrum. But for two lines to be resolved in a spectrum of a faint source, their separation should be nearly 5 or 6 Å.

Nebular Spectrograph

For studying the spectra of faint and diffuse nebulae occupying a large area of the sky, a special type of spectrograph has been developed by O. Struve at the Yerkes and Macdonald Observatories. This instrument essentially consists of a wide slit (a few inches), a prism and a small camera. The slit is placed at a large distance (150-feet for the 82-inch reflector at the Macdonald Observatory) from the prism, so that light falling on the latter surface becomes almost parallel. Consequently, no collimator is needed for the combination. A screen may be placed between the slit and the prism to reduce the stray sky light.

Such a simple type of spectrograph is most conveniently used for studying the spectra of faint nebulae whose diameters exceed 1/4°. It can photograph areas of about 2° × 1/4° in the sky without the help of any telescope. Large telescopes are needed only when one wants to photograph areas much less than 1/4° in diameter.

1.7 PHOTOGRAPHIC PHOTOMETRY

Photometry is the means for recording the brightness of stars and other heavenly objects with the help of suitable instruments and accessories placed at the focus of a telescope. Different accessories are used for different types of photometry, serving different astronomical objectives. Photographic photometry is chiefly used for its high sensitivity and the panoramic property of the photographic emulsions. The principal uses of this method lie in the fact that it integrates the light over the given exposure and that a good number of stars can be simultaneously photographed. This gives a permanent record of a large number of stars in a given region of the sky and the technique does not require high precision. The saving in telescopic time is also an advantage of this method. It is particularly useful in the field of astrophysics such as stellar statistics, variable star surveys or investigations of rich star clusters, galaxies and clusters of galaxies.

The fundamental principle of photographic photometry is that equal intensities of two sources should produce equal photographic effects on the emulsions under identical conditions. These conditions involve factors like the spectral energy distribution of the exposing light, the exposure time, the processing of the photographic plate, the pre- and past-exposure treatment of the plate, the size and structure of the optical image, the temperature and luminosity of the exposed source, etc. The main disadvantage of using this method is that it is difficult to control the above factors carefully and account for the changes involved in some of these conditions. Also, the extreme variation of sensitivity of the photographic emulsions with wavelengths, the delay in photographic processing etc. are some of the factors which degrade the photographic photometry to some extent. In spite of these disadvantages, photographic photometry is widely used in astronomy. The intensity of a starlight is measured by the amount of blackening of the star image on the photographic negative. This blackening can be measured by passing a collimated beam of light through the star image on the negative. The attenuation of the beam is then measured with a photocell. The brighter the star, the darker is its image and more is the attenuation of the beam, while for a faint star, the attenuation is small and the image is thus less dark.

Photographic photometry requires a standard magnitude scale against which the measured intensity of the stellar image or the image of any other source can be accurately matched. So the main object of using this method is to establish a suitable magnitude scale (discussed in Chapter 3) which could be used in any region of the sky. This magnitude determination is of extreme importance to various branches of stellar and galactic astrophysics. The more accurate method for determining the stellar magnitudes is however the photoelectric method, as will be discussed in Section 1.8.

1.8 PHOTOELECTRIC PHOTOMETRY

Today, the stellar magnitudes are determined much more accurately by modern photoelectric methods. Photoelectric photometry provides such measurements where higher sensitivity and greatest possible accuracy is demanded. This employs a photoelectric photometer by which the intensity of a star light can be accurately measured. The principal parts of an astronomical photoelectric, photometer are:

1. Finding-guiding eyepiece
2. A focal plane diaphragm containing a hole isolating the object to be viewed
3. A field lens
4. Suitable filters
5. A suitable detector with amplifier and a recording system.

The last one includes a photomultiplier. The photomultiplier is a photoelectric cell placed at the focus of the telescope. The star light is focussed by telescope on the photoemissive cathode of the photocell through its transparent envelope. When light from a source hits the photocell, electrons are emitted from the inner surface of the photocathode which is usually coated with potassium or antimony-cesium alloys or with cesium oxide. The electrons emitted are accelerated by the voltage existing between the terminals of the photoelectric cell and are then made to strike a second photosensitive surface called a dynode. When the photoelectrons from the cathode encounter this dynode, they dislodge more electrons from the surface than their original number. This emission of the secondary electrons is proportional to the number of primary electrons striking the dynode which again are proportional to the intensity of the celestial source.

The secondary electrons, in turn, encounter a second dynode which gives off more electrons than the secondary ones, and this process is repeated through successive dynodes until a considerably strong current is obtained. This current is amplified by an external amplifier and then recorded by a pen on a continuously rolling sheet of paper. It has been experimentally found that if we use a dozen dynodes, we might record a current whose strength would be a million times that produced by the primary electrons from the photocathode.

Let δ be the ratio of the average number of primary electrons emitted to that emitted from the first dynode. Now, if the photomultiplier contains n dynodes for each of which δ is the multiplication factor, then we may have the total multiplication for the electron tube, $M = \delta^n$, Further, if we take I_0 to be the current produced by the primary electrons from the photocathode, then the total current output from the tube is

$$I = I_0 \delta^n \qquad (1.8)$$

This enhanced current is recorded by the pen on a rolling sheet of paper. Consequently, high accuracy in the measurement of even quite faint sources is achievable by the use of photoelectric photometry. The method has therefore ensured a highly efficient technique in astronomical measurements.

The photoelectric photometry has been most successfully and widely used for the study of variable and eclipsing binary stars and also for finding the colour magnitude diagrams of clusters of stars. These points will be further discussed in the respective chapters.

Several filters are usually used successively to record the intensity at several different narrow bands of wavelengths. The most popular are the U, B, V bands that give rise to the three-colour photometry. This has been very extensively used and is being used still to study the colour, magnitude or other aspects of stars. In particular, U, B, V photometry has excelled in the study of the light and colour curves of the variable stars. The V magnitude in the three-colour system nearly corresponds to the usual visual magnitude of a star (see Chapter 3).

The three-colour photometry has been extended to include the broader wavelength regions. This gives rise to six-colour photometry which ranges from the ultraviolet to infrared regions of the spectrum and covers the range of wavelengths from 3300 to 12,500 Å.

The six-colour system is chiefly used for the determination of colours of stars (specially the bright ones having the magnitude value up to +7), from which their energy distributions are known. The disadvantage of this method is the excessive loss of telescopic time.

The photometric observations yield chiefly the brightness of celestial objects as a function of certain parameters, such as wavelength, and time. For each set of observations, these parameters have to be determined in order to get the brightness of objects. The wavelength is almost an essential parameter. The direct measures of brightness which are obtained in any photometric study are functions not only of the above mentioned parameters but also of various other parameters related to the equipment which receives the light and to the atmosphere through which the light passes. Also, the particles in the interstellar-space dim or extinct the light during its journey through space and this dimming is proportional to the distance of the source of light. For any useful measure of a set of observational material, corrections must be made for all these various factors.

1.9 SPECTROPHOTOMETRY

In photographic and photoelectric photometry, the total intensity of radiation is measured from a star and accordingly the magnitude of the star is assigned. These methods do not always require the actual spectral energy distribution of the source of light, whereas, *spectrophotometry* is used solely for measuring the spectral energy distribution of the light source. This involves the relative measurement of the intensities of the spectral lines with those due to a standard laboratory source in the same spectral region. By a comparative study with the known lines of the standard source, the energy distribution of the desired object at each wavelength may be known.

Stellar spectrophotometry involves both the relative and absolute measurements of the spectral energy distribution of the stars in all observable regions. Such measurements should be made accurate for studying the physical conditions of the atmosphere of a star. The most frequent problem in stellar spectrophotometry is to measure the continuous energy distributions in stellar spectra and to determine the intensities of the spectral lines relative to that of the continuum.

The intensity of the spectral energy distribution of a star is measured either by photographic or by photoelectric methods. The photographic or photoelectric detector is placed at the focus of a spectrograph attached to a telescope. In the photoelectric method, the detector is a photometer which scans the stellar spectrum according to the wavelengths and thus provides an accurate measurement of stellar energy distribution.

Photographic spectrophotometry, although less accurate, has some advantages over the photoelectric method, the principal one among these being that it covers a wide range of wavelengths in a single exposure and integrates the light received over a given exposure. This method is also useful for the sensitivity of the photographic plate over a large range of wavelengths (starting from UV up to 12,000 Å) and for the permanency of the photographic negatives, which can be used at any time. There are, however, many disadvantages of photographic spectrophotometry, decided chiefly by the limitations of the photographic emulsions. The variation of the spectral sensitivity curve of a photographic plate with wavelengths and with the type of emulsion used is one of the principal disadvantages of this method.

Photoelectric spectrophotometry, on the other hand, is unrivalled for its high accuracy which is demanded by the advanced astrophysics of today. Precise measurements of the stellar energy distribution and the line profiles are done by this method. The chief advantages of this method are the linearity of response of the photoemissive surfaces (i.e. the linear relation existing between the number of incident photons and the electrical output), their high sensitivity and speed of response (quantum efficiency). The main disadvantage is, however, its inability to record more than one picture at a time in a single exposure. Thus, a lot of valuable telescopic time has to be spent on each single source.

1.10   DETECTORS AND IMAGE PROCESSING

The light gathered by telescopes from heavenly objects, either single or many at a time within the field of view of the telescope, is recorded by astronomers with a detector, and the recorded images are analyzed and studied conveniently under desired conditions, the classical and most extensively used detector over many decades is the photographic plate, usually glass plate coated with light-sensitive emulsion. Photography still serves as the old standby when recording of a large amount of storable information is desired in a short time. The plates are often specially treated to increase their sensitivity to light, but still their *quantum efficiency* (i.e., percentage of photons striking the detector which activate it, relative to the total incoming photons striking the detector) remains very low, only to the level of a few per cent.

Phototubes which work on the principle of the *photoelectric effect* (Chapter 2) have been used as a detector by astronomers for several decades. Phototubes do not produce images. The incident light knocks out electrons from certain materials inside the tube which later multiply to many times ($\approx 10^5$ times) by repeating the process inside the tube. Such tubes are therefore called *photomultipliers.* The displaced electrons flow in the form of a current, the measured strength of which tells about the intensity of the incident light. In such a device, wavelength bands can be isolated by using filters in the light path in front of the photomultiplier (Chapter 3). The quantum efficiency of photomultipliers is much higher than that of the photographic plates. Its response is smoothly linear—twice the flux produces twice the current. The accuracy of measurement of the current or counting the rate of production of the individual photons may be very high.

At present, however, the dream detector of astronomers is the Charge-Coupled Device (CCD) which has been made possible by the great advancement in solid-state microelectronics. A CCD is a thin silicon wafer a few to 10 mm on each side. The chip consists of a large number of small regions, each of which makes up a picture element called a *pixel,* A typical CCD chip may contain 1000 by 1000 pixels (10^6 total) arranged in rows and columns that can be controlled electronically. Each pixel converts photons to electrons (thus behaves like a small phototube) and builds, up charge over a very long time. The integrated charges are then moved in a regimented way to preserve the spatial pattern of light intensity falling on the chip in computers, which then process the image for astronomers to study at ease. The quantum efficiency of CCDs is very high, nearly 100 per cent, so that even small telescopes equipped with CCDs can imitate the efficiency of large telescopes. Also, CCDs are smoothly linear; hence light intensity can be measured accurately. Like photographic plates, CCDs also gather a large amount of information in one exposure. But the advantage over photographic plates is that CCDs collect information in digital format for easy manipulation by computers.

Whatever detectors are used by astronomers for recording the images of heavenly objects, these images can be later converted to a digital form for manipulation by computers. Any specific aspect of the original data can thus be reproduced in an enhanced form for careful study and analysis. This process is called *image processing* which plays a fundamental role in present-day astronomical investigations.

EXERCISES

1. The indices of refraction (μ) for some substances are given in the following Table. Calculate their polarization angles.

Substance	μ
Air, at 0°C	1.00030
Water, at 0°C	1.333
Quartz	1.544
Crown glass	1.51714

If the polarization angle of a medium is 60°, find the index of refraction of the medium.

2. Draw a sketch of the Earth's atmospheric layers, indicating the major constituents and their interaction with electromagnetic radiation of different wavelengths.

3. What are the advantages that are derived by astronomers by using telescopes? Compare these advantages in using radio and optical telescopes.

4. Calculate the theoretical resolution of a reflector 1 m in diameter and of human eye pupil of 5 mm diameter (use Dowes' rule).

5. The limit of resolution of a telescope can be taken as 1.2 λ/a, where a is the aperture and λ is the wavelength of radiation in the same units. Use this to calculate the resolutions of a 1 m reflector in H_α and a 100 m paraboloid in 21 cm radio line.

6. Compare the brightness of images of the Moon produced by two telescopes—one with f = 200 cm, a = 40 cm, and the other with f = 600 cm and a = 100 cm.

7. Two sources emitting 21 cm radiation are situated at an angular distance of 5° from each other. Obtain the minimum diameter of a parabolic dish that will just separate the two sources.

8. Compare the light gathering powers of the pupil of the human eye and a 40 cm diameter reflector. If the eye can just see a 6 mag. star, what will be the faintest mag. that can be observed with the above reflector?

9. Compare the advantages and disadvantages of reflecting and refracting telescopes.

10. What important considerations are needed to choose a good observatory site?

11. Discuss the advantages of a space telescope over a ground based one, for optical and infrared observations?

12. Discuss the points of similarity and difference between optical and ultraviolet observations.

13. Why short waves are used for radio transmission at far-off places?

14. Comment on the importance of high resolving power of a spectrograph. Why high resolution cannot be used for faint objects?

SUGGESTED READING

1. Abell, CO., *Exploration of the Universe,* Holt, Rinehart and Winston, Inc., New York, 1969.

2. Huffer, Charles M., Trinklin, F.E., and Bunge, M., *An Introduction to Astronomy,* Holt, Rinehart and Winston, Inc., New York, 1972.

3. Kutner, Marc L., *Astronomy,* Harper & Row, New York, 1987.

4. Mc Laughlin, Dean, B., *Introduction to Astronomy,* Houghton Mifflin Company, Boston, 1961.

5. Zeilik, Michael and Gaustad, John, *Astronomy,* Harper & Row, New York, 1983.

Basic Physics

2.1 PLANCK'S THEORY OF BLACKBODY RADIATION

It is a common experience that a hot body radiates electromagnetic radiation in the form of heat or light or both and this radiation takes place at *all wavelengths*. Such a body also absorbs a fraction of radiation that may fall on it. In 1859, G.R. Kirchhoff verified that the ratio of the emissive power to the absorbtivity at a given temperature T is *the same for all bodies,* irrespective of the material of which the body is made. So Kirchhoff's laws demand that this ratio should be a universal function depending only on the temperature T of the body and the frequency v of radiation. Now, a *blackbody is one that absorbs all the radiation falling on it,* implying that absorbtivity of a blackbody is unity. Therefore, the emissive power of a blackbody must be a universal function of v and T. It is required therefore, to find a function depending on v and T that will produce the experimentally observed energy distribution in frequency by a blackbody at any given temperature T.

With this end in view, W. Wien proposed in 1895 a law for blackbody radiation in the form

$$U(v)dv = Av^3 e^{-hv/kT} dv \quad \text{(Wien's third law)} \tag{2.1}$$

where $U(v)$ is the energy density of the blackbody radiation in the frequency range from v to $v + dv$, A is a constant, k is the Boltzmann constant whose value is 1.38×10^{-16} erg deg^{-1} and h is the Planck constant having the value 6.625×10^{-27} erg sec. It is found that Wien's law agrees quite well with the blackbody radiation spectrum in the high frequency (ultraviolet-violet) region (in fact, for large values of v/T) but in the low frequency region the law gives hopelessly inadequate agreement. Another law proposed by Wien in connection with the blackbody radiation is known as *Wien's displacement law* which is given by

$$\lambda_{max} \, T = \text{Constant} \tag{2.2}$$

This states that the wavelength of the maximum intensity of the blackbody radiation is inversely proportional to the temperature of the blackbody. For higher temperature of a body the *peak of the distribution curve* will shift towards shorter wavelength, and vice versa. The constant on the right-hand of Eq. (2.2) is known as the *displacement constant* with its value nearly 0.2897 cm deg.

In 1900, Lord Rayleigh proposed, a new blackbody radiation law which was later modified by J. Jeans in 1905 and is now known as the Rayleigh-Jeans formula. This is given by

$$U(v)\, dv = \frac{8\pi v^2 kT\, dv}{c^3} \tag{2.3}$$

This law, unlike Wien's law, was found to agree well with the experimental results of the blackbody radiation curve only at the low-frequency region (in fact, for small values of the ratio v/T) corresponding to the micro waves and radio waves. But the law is found to be miserably at variance with the experimental curve in the high frequency region. In fact, the energy density asymptotically goes to infinity according to this law as v becomes very large.

At this stage, the exact law was hit upon empirically by Max Planck in 1901, which gave a perfect agreement with the experimental blackbody curve for all values of v/T. This law, known as *Planck's law of blackbody radiation,* is given by

$$U(v)\, dv = \frac{8\pi h v^3}{c^3}\, \frac{1}{e^{hv/kT} - 1}\, dv, \tag{2.4}$$

Planck however derived this law empirically. He could neither give the mathematical derivation of the law nor any physical explanation as to why the law was of that particular form. The logical derivation of the law came much later, in 1924, from the great Indian scientist S.N. Bose, which finally gave birth to a new discipline of modern physics, viz., the *statistical physics.* But in the process, Planck introduced in physics a revolutionary concept that the energy exchanges between the oscillators of frequency v and the radiation field always took place in integral multiples of hv, that is

$$E = nhv \tag{2.5}$$

where n = 1, 2, 3, ..., and so on, and h is Planck's constant introduced by himself. The electromagnetic radiation of frequency v can thus be regarded as a bundle of energy of unit hv. This unit was later called the *energy quantum* or *light quantum* by Albert Einstein and is now known as *Photon.* The introduction of the *light quantum hypothesis* by Einstein thus implied the, *particle nature of the electromagnetic radiation* which was verified by the *photoelectric effect* by Einstein himself. During the next quarter of the century, Planck's original postulate regarding the quantum nature of the electromagnetic radiation embodied in Eq. (2.5) developed into a new branch of modern physics, *the Quantum Physics.* The energy distribution curves obtained by using the laws given by Wien, Rayleigh Jeans and Planck are illustrated in Fig. 2.1, of which Planck's curve only agrees fully with the experimental curve.

Planck's formula as given in Eq. (2.4) also indicates that the intensity of a blackbody radiation increases in all wavelengths as the temperature of the body increases. And, according to Wien's displacement law, the peak of the intensity curve shifts towards the shorter wavelengths as the temperature of the body increases. These two properties of the blackbody radiation are well illustrated in Fig. 2.2. These properties can be exploited in calculating the unknown temperature of a blackbody by fitting its energy curve with the theoretical blackbody radiation curves. This is done, in fact, to find the temperatures of stars, assuming that stars radiate, like

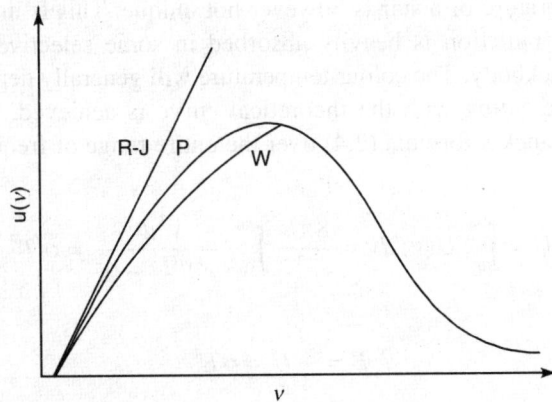

FIGURE 2.1 The energy distribution curves for the three (1. R-J—Rayleigh-Jean's formula, 2. P—Planck's formula, 3. W—Wien's formula) different laws of blackbody radiation.

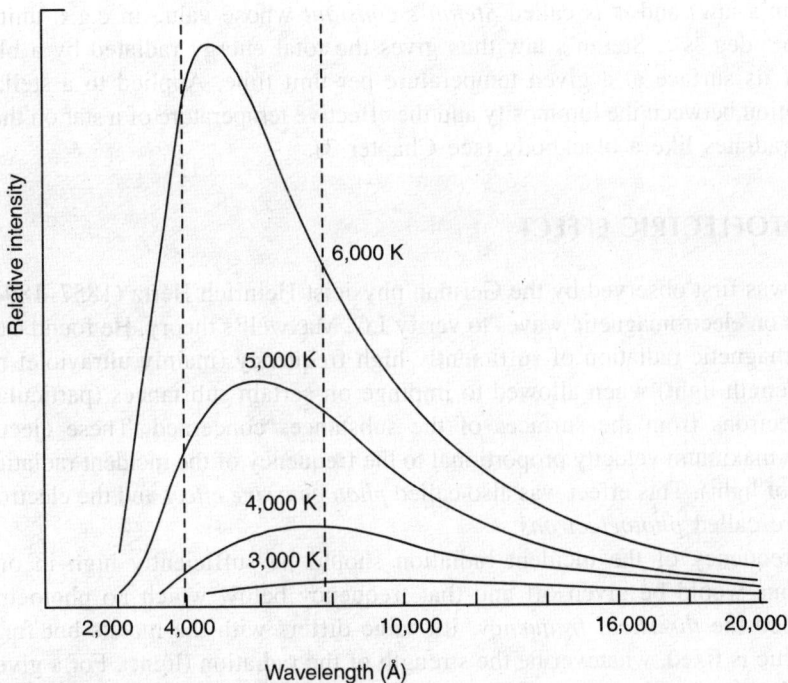

FIGURE 2.2 The total intensity of radiation is represented by the area under the energy curves and varies as Stefan's law. The visible region of the spectrum is approximately shown by the two vertical dotted lines.

blackbodies. The energy distribution curve of a star is fitted approximately with that of a blackbody of known temperature and the temperature of the star is thus deduced. This derived stellar temperature is known as the *colour temperature*.

The colour temperature of a star is however not unique. This is mainly because the stars in whose atmospheres radiation is heavily absorbed in some selective wavelengths, deviate substantially from a blackbody. The colour temperature will generally depend on the wavelength regions over which the fitting with the theoretical curve is achieved.

If we integrate Planck's formula (2.4) over the entire range of frequencies, we get energy density

$$U = \int_0^\infty U(v)\, dv = \frac{8\pi h}{c^3} \int_0^\infty \frac{v^3 dv}{e^{hv/kT} - 1} = \sigma' T^4 \tag{2.6}$$

so that

$$E = \frac{c}{4} U = \sigma T^4 \tag{2.7}$$

and

$$\sigma' = 8\pi^5 k^4/(15h^3 c^3) = 7.569 \times 10^{-15} \text{ erg cm}^{-3} \text{ deg}^{-4} \tag{2.8}$$

is the constant of radiation density. Formula (2.7) is known as the *Stefan-Boltzmann law* (often called Stefan's law) and σ is called *Stefan's constant* whose value in c.g.s. unit is 5.6687×10^{-5} erg cm^{-2} deg^{-4}s^{-1}. Stefan's law thus gives the total energy radiated by a blackbody per unit area of its surface at a given temperature per unit time. Applied to a stellar body, this yields a relation between the luminosity and the effective temperature of a star on the assumption that a star radiates like a blackbody (see Chapter 3).

2.2 PHOTOELECTRIC EFFECT

This effect was first observed by the German physicist Heinrich Hertz (1857–1894) during his experiments on electromagnetic waves to verify J.C. Maxwell's theory. He found experimentally that electromagnetic radiation of sufficiently high frequency (mainly ultraviolet radiation and short wavelength light) when allowed to impinge on certain substances (particularly, metals), dislodge electrons from the surfaces of the substances concerned. These ejected electrons move with a maximum velocity proportional to the frequency of the incident radiation (generally in the form of light). This effect was also called *photoemissive effect* and the electrons produced thereby were called *photoelectrons*.

The frequency of the incident radiation should be sufficiently high in order that the photoelectrons would be given off and that frequency below which no photoemission takes place is called the *threshold frequency*. Its value differs with the metals but for a particular metal its value is fixed, whatever be the strength of the radiation (light). For a given frequency above the threshold, the photoemission takes place within a short time interval of 10^{-8} to 10^{-9} seconds, irrespective of the value of the frequency. C. Huygen's wave theory of light could not explain the photoelectric effect, as it contradicted experimental facts. According to the wave theory, the energy necessary for ejection of the photoelectrons is supplied at the expense of the energy of light uniformly distributed over wavefronts. This supply of energy might be possible with enough intensity and long exposure of light. There was no need of any threshold frequency whatsoever.

But experimentally, it was found that the frequency of the radiation should be taken into account, as it plays the vital role than does the intensity in the production of photoelectrons. This led Einstein to think about the quantum nature of the radiation energy (which was originally suggested by Planck in his empirical deduction of the blackbody radiation law). According to this theory, the radiant energy is emitted only as integral multiples of quanta, each of energy $h\nu$ and that the energy of each quantum increases with increasing frequency. Light radiation was thus supposed to take place in discontinuous units of energy propagating in some fixed direction without spreading uniformly as a wave does. Such a quantum will also have a linear momentum. The quantum has, therefore, all the properties of a particle with a linear momentum and a fixed direction of motion. This light particle was named by Einstein as *photon.* By this theory, Einstein actually revived Isaac Newton's corpuscular theory of light.

By Maxwell's theory the momentum associated with a plane electromagnetic wave moving with a velocity c is given by

$$p = \frac{E}{c} = \frac{h\nu}{c} = \frac{h}{\lambda}, \quad E = h\nu \qquad (2.9)$$

where E is the energy and ν the frequency of the radiation of wavelength λ. The relations in Eq. (2.9) are called *Einstein's* relations for a photon.

According to Einstein's photon theory the energy of the photoelectrons is abstracted from the photons striking the photoemissive surfaces (metals) and there is a conservation of energy and momentum. This is given by *Einstein's equation of photoelectric effect,*

$$h\nu = \frac{1}{2} m_e v^2 + w \qquad (2.10)$$

where $\frac{1}{2} m_e v^2$ is the K.E. of the photoelectron and w is the work function, which is the energy needed to remove a photoelectron from the surface of the metal, and is given by

$$h\nu_0 = w \qquad (2.11)$$

ν_0 being the threshold frequency. Hence,

$$h\nu - h\nu_0 = \frac{1}{2} m_e v^2 \qquad (2.12)$$

Thus the difference between the incident energy and the energy lost during the process is equal to the kinetic energy of the photoelectron.

The photoelectric effect has been one of the most useful principles of physics applied to the astronomical observations. The photoelectric photometry developed on this principle has achieved a high degree of accuracy over the photographic photometry in various types of observations of stars and other astrophysical bodies.

2.3 PRESSURE OF RADIATION

The photoelectric effect is a clear demonstration of the *particle nature* of the radiation energy which consists of photons, that is, the discrete quanta. The photons have with them associated a momentum p and energy E which are given by Eq. (2.9). It is natural to assume therefore that when a photon impinges on a particle, the particle will experience a pressure on it, the strength of the pressure being proportional to the frequency associated with the photon. This pressure is very important in the hydrostatic balance of stars. In the case of a *nonluminous* gaseous sphere, the balance at any point is reached between the inward *gravitational attraction* and the *outward pressure*. This outward pressure is the gas pressure given by

$$p_g = \frac{\rho k T}{\mu} \qquad (2.13)$$

where ρ is the gas density, T its temperature, and μ the mean molecular weight, depending on the composition of the gas, k being the Boltzmann constant. The analysis however changes in the case of *luminous* gaseous objects like stars. Here the emergent radiation of all wavelengths flow outward, imparting on the superincombent gas layers a pressure of radiation. This radiation pressure can play an important role in the hydrostatic balance of the stellar body in the sense that a part of the inward attraction is now balanced by the outward pressure due to radiation. The hydrostatic balance is now reached by the *joint influence* of the gas pressure and radiation pressure. The outward pressure is now given by

$$p_g + p_r = \frac{\rho k T}{\mu} + p_r \qquad (2.14)$$

Statistical mechanical calculations show that the radiation pressure at any point in a radiating body is given by

$$p_r = \frac{U}{3} = \frac{1}{3} \sigma' T^4 \qquad (2.15)$$

by using Eq. (2.6), where U is the total energy density at the point, given by

$$U = \sum_{i=0}^{\infty} n_i h v_i \qquad (2.16)$$

Eq. (2.14) thus reduces to

$$p = \frac{\rho k T}{\mu} + \frac{1}{3} \sigma' T^4 \qquad (2.17)$$

The pressure given by Eq. (2.17) must be used in the consideration of the hydrostatic equilibrium of a star. It is found that the second term on the right-hand side of Eq. (2.17) is small compared to the first in the surface layers of cooler stars. But in the hot stars, the second term must be taken into account. It is further observed that since in hot stars, the radiation pressure plays a significant role, the role played by the gas pressure diminishes accordingly. The gas in the hot stars is therefore less compressed than in cool stars. So cooler stars have in general higher density than the hotter ones. We shall see in Chapter 10 that the *radiation pressure drives out streams of particles from surfaces of stars of sufficiently high temperatures.*

2.4 CONTINUOUS, ABSORPTION AND EMISSION SPECTRA— KIRCHHOFF'S LAWS

An element in the gaseous state can absorb or emit light only of certain wavelengths characteristic of its own. So, the nature and composition of a gas may be known by studying the pattern of lines of the spectrum produced by the gas. Generally, line spectra are produced by the elements in the glowing or vapour state or they may be produced by artificial means such as in electric bulbs, fluorescent tubes, mercury vapour lamps, etc. The spectra which are obtained from different luminous sources are mainly of three types: (a) *the continuous spectrum,* (b) *the bright-line or emission-line spectrum,* and (c) *the dark-line* or *absorption-line spectrum.* The formation of these line spectra can be well explained by three laws given by the German physicist Gustav Kirchhoff (1824–1887) in 1859. These are:

1. When the source is an incandescent solid, liquid, or a compressed gas, the radiation given out is a continuous emission in all wavelengths, thus producing a continuous spectrum.
2. When the source is a glowing gas under low pressure, the radiation emitted is selective. The spectrum produced thereby consists of discrete bright lines or emission lines superposed upon a faint continuous spectrum.
3. If the actual source of emission be a hotter source emitting radiation continuously and if, it be viewed through a comparatively cooler gas, then the latter may absorb some wavelengths of the incoming radiation of the actual source. The spectrum thus produced consists of dark lines or absorption lines superposed upon a continuous background. This is called the dark line or absorption line spectrum.

The three different types of spectra are shown in Fig. 2.3.

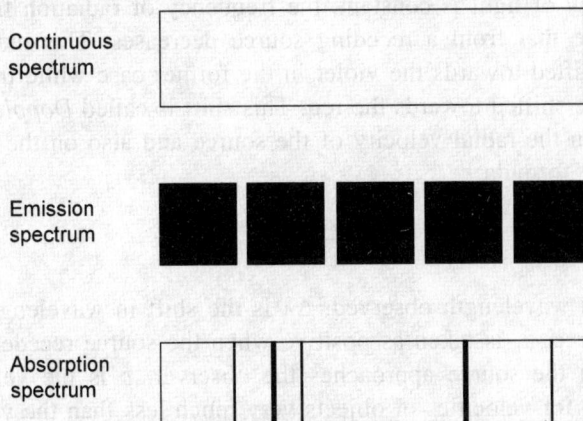

FIGURE 2.3 The continuous, emission and absoption line spectra.

A continuous spectrum, therefore, is an array of *all wavelengths* of white light or can be obtained from a source emitting all wavelengths of the visible light. The bright line or emission line spectrum is produced by a glowing gas which emits selectively only in certain definite wavelengths characteristic of the chemical composition of the gas.

The dark line or absorption spectrum is a series of dark lines or Fraunhofer lines named after the German physicist, Joseph Fraunhofer (1787–1826), superposed upon a continuous spectrum. These lines were first observed by Fraunhofer in the solar spectrum and the name has since been used to describe the absorption line spectrum of any source. The gas absorbs from the radiation source the wavelengths, that they would emit if they had been in the glowing condition. As a result, the dark lines are formed in characteristic patterns depending on the elements present in the gas.

In recent years, many thousands of Fraunhofer lines have been observed in the solar spectrum which have been used to determine the composition of the Sun. Most of the stars also produce the absorption line spectra while a minority of these show emission lines superimposed on the continuous spectrum. These stars have, in general, extended low density atmospheres.

The study of stellar spectra helps to know the actual picture of the universe. The temperature, pressure, the physical condition of the atmosphere, the radial velocity, existence of magnetic and electric fields and their strengths and similar other important information may be obtained by studying the different types of spectra of stars (see Chapter 4).

2.5   DOPPLER EFFECT

In 1842, Christian J. Doppler (1803–1853), an Austrian physicist, pointed out that when a light source approaches an observer, the waves are crowded together and the lengths of the waves are decreased. Similarly, when the source recedes from the observer, the waves spread out and wavelengths are increased. This phenomenon is known as *Doppler Effect*. The same effect is observed when the source is stationary and the observer is moving. In actual practice the relative motion along the line of sight, that is, the radial component of motion between the source and the observer is considered.

Since the velocity of light is constant, the frequency of radiation from an approaching source increases while that from a receding source decreases. The entire spectrum of the source will thus be shifted towards the violet in the former case while in the latter case, the entire spectrum will be shifted towards the red. This shift is called *Doppler shift*. The amount of shift will depend on the radial velocity of the source and also on the frequency observed which is given by the formula

$$\frac{\Delta\lambda}{\lambda} = \frac{v_r}{c}$$

(2.18)

where λ is the natural wavelength observed, $\Delta\lambda$ is the shift in wavelength due to the radial velocity v_r of the source. v_r is taken as positive when the source recedes from the observer and as negative when the source approaches the observer, c is the velocity of light. The formula (2.18) applies for velocities of objects very much less than the velocity of light. For objects with velocities comparable to that of light, the modified relativistic formula is

$$\sqrt{\frac{1 + v_r/c}{1 - v_r/c}} = \frac{\lambda + \Delta\lambda}{\lambda} = 1 + Z$$

(2.19)

where $Z = \dfrac{\Delta\lambda}{\lambda}$.

Although the radial motion of the source causes a Doppler shift of the entire spectrum, this shift is difficult to measure in continuous spectrum. But the shift in the position of spectral lines can be measured by matching with the unshifted lines superimposed on the spectrum which are produced in the laboratory by applying spark to some element or elements. After the shift of any known line has been measured in the spectrum of the source, Eq. (2.18) can be used to calculate the radial velocity of the source. Applying this process the radial component of motions of the heavenly bodies has been measured. In particular, this has been done for thousands of stars and hundreds of distant galaxies. The same principle is used to calculate the rotational velocities of stars and the thermal states of the stellar atmospheres and other radiating gaseous astronomical bodies, as we shall discuss later.

2.6 ZEEMAN EFFECT

In 1896, Pieter Zeeman (1865–1943), a Dutch physicist at the Physics Institute of Leiden University, exposed a Bunsen sodium flame to the poles of a strong electromagnet. He analyzed the radiation emitted by the flame by means of a Rowland's concave grating and found that the spectral lines obtained got broadened as the electromagnet was turned on; the broadening disappeared when the current was put off. The presence of the strong magnetic field splits the spectral lines into different component lines, the amount of splitting depending on the frequency or wavelength of the radiation and on the strength of the magnetic field. The component lines are found to be polarized in different ways although the original radiation remains unpolarized. This phenomenon is known as *normal Zeeman effect*.

The splitting of lines may be observed in two ways. First, when the light is viewed perpendicular to the direction of the magnetic field, three component lines are observed. But when observations are made parallel to the magnetic field, only two component lines are found. These cases are illustrated in Fig. 2.4.

In the first case when three component lines are observed, one is formed at the same wavelength as that of the original line and is called the central line. The other two components are formed at equal distances on either side of the central line. When observed through a Nicol prism, the central line is found to be polarized in a direction parallel to the magnetic field. The other two components are found to be polarized perpendicular to the magnetic field.

In the second case when two component lines are observed, the central line is absent. These two components are observed in the same shifted positions as in the first case. These are found to be circularly polarized in opposite directions (see Fig. 2.4).

Zeeman effects are observed in the absorption lines of the sunspot spectrum. Measurements of the amount of splitting of the component lines prove the presence of a strong magnetic field of the order of 10^3 to 10^4 gauss in the sunspots.

The mathematical formula which relates the amount of splitting of lines to the strength of the magnetic field to which the source of lines is exposed is given by

$$\Delta\lambda = \pm\, 4.67 \times 10^{-5} \, g\lambda^2 H \text{ cm} \qquad (2.20)$$

where '*g*' is called *Landé factor* (for the solar lines its value is 3), H is the magnetic field in gauss, and λ is the wavelength measured in centimetres. In terms of frequency, the formula for normal Zeeman splitting is given by

$$\nu = \nu_0 \pm eH/(4\pi m_e) \qquad (2.21)$$

Original line

For observation
⊥ to the magnetic
field H

H →

v_2 v_0 v_1

(a)

For observation
‖ to the magnetic
field H

v_2 v_1

H

(b)

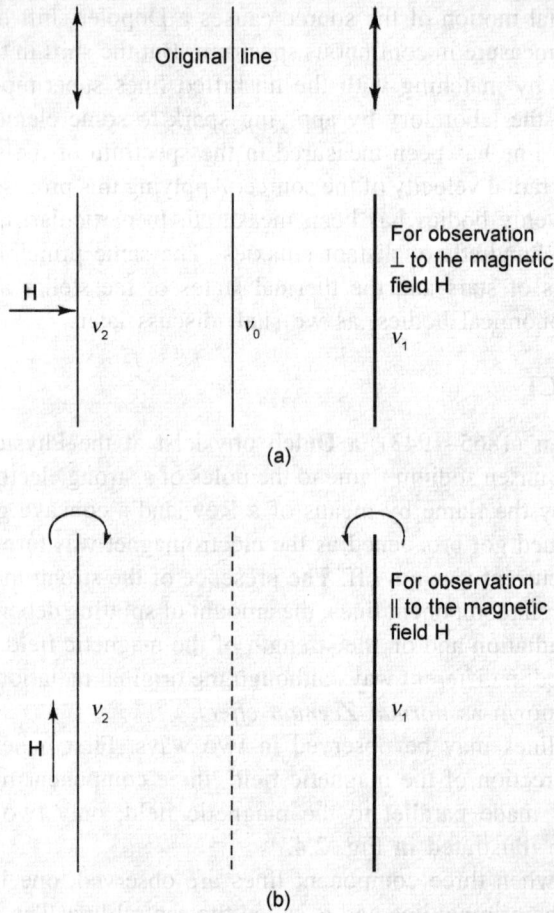

FIGURE 2.4 The normal Zeeman splitting of spectral line: (a) For observations perpendicular to the magnetic field H; (b) for observation parallel to the magnetic field.

The quantity $eH/(4\pi m_e)$ is called the normal Zeeman separation in a magnetic field of strength H, e being the electronic charge, and measured in electromagnetic unit its value is 1.6×10^{-20} and m_e is the mass of an electron equal to 9.1085×10^{-28} gm.

Besides, the absorption lines of the sunspot, large Zeeman splitting of the absorption lines in some strong lined stars indicate the presence of very strong magnetic field of ~10^3 to 10^4 gauss at the surfaces of these stars. The magnetic fields are also found to vary in time. These stars are known as the *magnetic variable stars*. From Zeeman splitting of the 21-cm hydrogen line in the direction of Tauras A, R.D. Davies and his co-workers (1962) have calculated the strength of the galactic magnetic field to be about 2×10^{-5} gauss. Analysis of Zeeman splitting thus gives valuable information about the strength and structure of the magnetic fields in astrophysical bodies.

2.7 BOHR'S CONCEPT OF THE H-ATOM

In the later part of the nineteenth century and in the beginning of the present one, the understanding of the formation of line spectra by an element was an intriguing problem to the scientists. The line spectra are produced when an element is excited by physical means so as to absorb or emit *light* consisting of different-wavelengths characteristic of the element. Line spectra can therefore be explained only on the assumption that the energy exchange involved in producing a line is always fixed and this requires a stable structure of the atom.

In those days the structure of the atom was not known. The classical electromagnetic theory of Maxwell could not explain the stable structure of atom. According to this theory, the electrons orbiting about the nucleus would continually radiate electromagnetic energy and approach the nucleus steadily in spiral path, emitting radiation of increasing frequency, thereby losing energy continuously according to the law of conservation of energy. The electrons on losing energy will continue to travel in orbits of smaller dimensions and ultimately spiral into the nucleus. But this situation cannot occur inside the stable structure of the atom. As a consequence the formation of line spectra could not be explained with the help of this theory.

However, in 1885 J.J. Balmer, a Swiss physicist demonstrated that the wave length of any line, then known in the spectrum of hydrogen can be determined by a simple empirical formula

$$\lambda = B \frac{n^2}{n^2 - 4} \tag{2.22}$$

where B is a constant whose value is 3645.6 and n is an integer taking the values 3, 4, 5, 6 and so on. With the help of this formula, physicists were able to calculate the wavelengths of other lines in the hydrogen spectrum which were not known previously. The series of lines that are obtained from the above formula were later known as the *Balmer series* of hydrogen lines.

In 1890, J.R. Rydberg, a Swedish physicist modified the Balmer formula in the form:

$$\bar{v} = \frac{1}{\lambda} = R_H \left[\frac{1}{2^2} - \frac{1}{n^2} \right] \tag{2.23}$$

where $n = 3, 4, 5$ and so on, and $\bar{v}$ is the wave number. He showed that the wavelengths of the Balmer lines could be well determined by this equation. Rydberg also determined several unknown wavelengths in the series of lines of the hydrogen spectrum and for a number of elements with hydrogen-like spectra. The constant R_H is called the *Rydberg constant* for hydrogen and is equal to 109,677.58 cm^{-1}.

Although the above formula located many lines of the hydrogen spectrum and measured the wavelengths thereof, the actual structure of the atom was still eluding the scientists. They were also puzzled by the complexity of the line spectra formed by different elements.

In 1901, Planck introduced the quantum hypothesis of electromagnetic radiation which actually assumed that the energy is radiated in bundles rather than as continuous chain. This assumption had definitely some bearing on the fixed energy requirement for the production of line spectra.

In 1911, E. Rutherford from his famous experiment on the scattering of α-particles by a thin gold foil concluded that the nucleus of an atom contains the positive charge of the atom and this charge deflects the positively charged α-particles to different extent. The α-particles penetrating the atom near the nucleus are more deviated than those passing a little distance apart. The α-particle which strikes the nucleus right through was found to turn back straight. But this occurs only for a small fraction of the total number of incident α-particles, that is about 1 in 20,000, This observation led Rutherford to come to the conclusion that most part inside the atom was void. The electrons may be found revolving about the positive nucleus in orbits whose distances from the nucleus are much greater compared to the dimensions of the electron and the proton. The present theory of atomic structure owes greatly to this planetary system of atomic model given by Rutherford.

Ultimately, in 1913 Neils Bohr, a Danish physicist, established his classical theory of hydrogen atom combining the essential elements of the quantum hypothesis of Planck and atomic model of Rutherford. He made use of certain postulates to establish his idea.

In one of his postulates, he assumed that the nucleus of the hydrogen atom contains one proton and the electron revolves about the nucleus as much like a planet revolving about the Sun. The centripetal force that holds the electron in an orbit about the nucleus is supplied by the Coulomb force of attraction between the nucleus and the electron. The electron does not radiate electromagnetic energy according to the Maxwell equation even if it is accelerating.

In his second postulate, Bohr suggested that the electron may be found at different times only in certain possible stationary circular orbits about the nucleus. These orbits are defined by the condition that the angular momentum of the electron in one of such orbits is quantized and can have only certain discrete set of values, Mathematically, the formula is

$$m_e vr = n\left(\frac{h}{2\pi}\right) \tag{2.24}$$

where h is the Planck's constant, n takes the integral values 1, 2, 3 etc., and is called *the principal quantum number, r* is the radius of the orbit described by the electron and m_e is the mass of the electron.

Since the angular momentum of an electron can have only discrete set of values, the energy associated with the electron in a permissible orbit is found only to have certain discrete set of values. Moreover, the electron has potential energy depending on the radius of the orbit together with the kinetic energy due to the mass and the velocity about the nucleus. So the total energy associated with the atom is found with its electron residing on certain permissible discrete orbits.

The atom does not absorb or emit radiation as long as the electron remains in one of such well-defined orbits. The absorption or emission occurs in the form of discrete amount of *light quanta* or *photons,* when the electron makes a transition from one orbit to another. The atom absorbs the photon from the radiation which strikes it and this absorption takes place only when it can absorb the right quantity of energy to raise its electron from an orbit of lower energy (smaller radius) to one of higher energy (larger radius). The atom emits the photon when the electron falls back to lower orbits.

Bohr postulated that whenever transition of electron takes place it occurs according to the relation:

$$E_i - E_f = h\nu \quad \text{(for emission)} \quad \text{or} \quad E_f - E_i = h\nu \quad \text{(for absorption)} \tag{2.25}$$

where E_i is the initial value of the energy of the atom and E_f is the final value of this energy, ν is the frequency of radiation emitted or absorbed by the atom and h is the Planck's constant.

The atom absorbs or emits the energy, whenever there is difference between the energy values of the two orbits and the electron jumps as a result of transition from one orbit to another. When $E_i > E_f$ the energy is radiated by the atom and if $E_i < E_f$ the energy is absorbed by it. The frequency ν of the radiation emitted as the electron jumps from an initial stationary orbit n_i (of radius r_i) to another final orbit n_f (of radius r_f) is given by

$$\nu = \frac{E_i - E_f}{h} = \frac{2\pi^2 m_e e^4}{h^3} \left[\frac{1}{n_f^2} - \frac{1}{n_i^2} \right] \tag{2.26}$$

We can also write, in terms of wave number, since $\dfrac{1}{\lambda} = \dfrac{\nu}{c}$,

$$\bar{\nu} = \frac{1}{\lambda} = \frac{2\pi^2 m_e e^4}{ch^3} \left[\frac{1}{n_f^2} - \frac{1}{n_i^2} \right] \tag{2.27}$$

Substitution of the values of m_e, e, c and h yields

$$\frac{2\pi^2 m e^4}{ch^3} = 109{,}740 \text{ cm}^{-1} \tag{2.28}$$

Bohr compared this value with that of R_H and found a close agreement in the two values. He later applied Eq. (2.27) to find the other series of lines of the hydrogen spectrum and succeeded quite remarkably.

If we put $n_f = 1$ and $n_i = 2, 3, 4$ etc. in Eq. (2.27) the Lyman series of spectral lines (in emission) are obtained. If we take $n_i = 1$ and $n_f = 2, 3, 4$ etc., the formula (2.27) modifies to

$$\bar{\nu} = \frac{1}{\lambda} = \frac{2\pi^2 m_e e^4}{ch^3} \left[\frac{1}{n_i^2} - \frac{1}{n_f^2} \right] \tag{2.29}$$

which yields the Lyman series of absorption lines of hydrogen. These lines were discovered by Theodore Lyman (1874–1954), an American physicist, in the deep ultraviolet region of the hydrogen spectrum. These lines correspond to the transition of the electron from level $n = 1$ to any higher level or vice versa. The lines in succession are called Lyman $\alpha(L_\alpha)$, Lyman $\beta(L_\beta)$, Lyman $\gamma(L_\gamma)$ etc.

For $n_f = 2$ and $n_i = 3, 4$ etc., the spectral lines of the Balmer series (in emission) are obtained. These lines are produced as a result of transition from level $n = 2$ to any higher level (absorption) or vice versa (emission) and lie in the visible region of the hydrogen spectrum, starting in the red (H_α) and merging in the near ultra-violet region of the spectrum. Other lines in succession, viz. H_β, H_γ etc. lie in the intermediate region. The Lyman and Balmer series of hydrogen lines are very important in astrophysical context.

Other series of lines in the hydrogen spectrum for $n_f = 3$ and $n_i = 4$, 5, etc., are known as Paschen series formed due to transition of the electron to level $n = 3$ from any higher level or vice versa. The German physicist, F. Paschen, in 1908, first predicted that these lines lie in the near infrared region of the spectrum.

The series corresponding to the transition from level $n = 4$ to any higher level and vice versa called Brackett series was observed in 1922 by Brackett and the series produced by transition from $n = 5$, to any higher level or vice versa called the Pfund series, was discovered by Pfund in 1924. Both of these series of hydrogen lines lie in the far infrared region of the spectrum.

All the different series of lines of the hydrogen spectrum corresponding to the quantum numbers $n = 1, 2, 3, 4$ etc. are shown in Fig. 2.5 which is known as the energy level diagram of the hydrogen atom. The energy required by the atom for transition from the ground level ($n = 1$) or normal state to any higher level or excited state is called the *excitation potential* of that level, and the atom that has absorbed this energy is said to have been *excited*. Thus, the excitation potential of the second level of the hydrogen atom is 10.2 eV, meaning that the energy required in transition from $n = 1$ to $n - 1$ level is 10.2 eV. The energy difference in successive levels upwards (or outwards) gradually decreases, so that the levels converge at the last quantized orbit, i.e. $n = \infty$. The energy difference between the last quantized orbit of the hydrogen atom and its ground level is 13.6 eV, This energy is the ionization potential energy for the hydrogen atom. The absorption of any greater amount of energy by the atom will result in the ejection of the electron from the Coulomb field of the nucleus. The electron will then become free. It will no longer be bounded by the quantum condition of the orbits and the atom

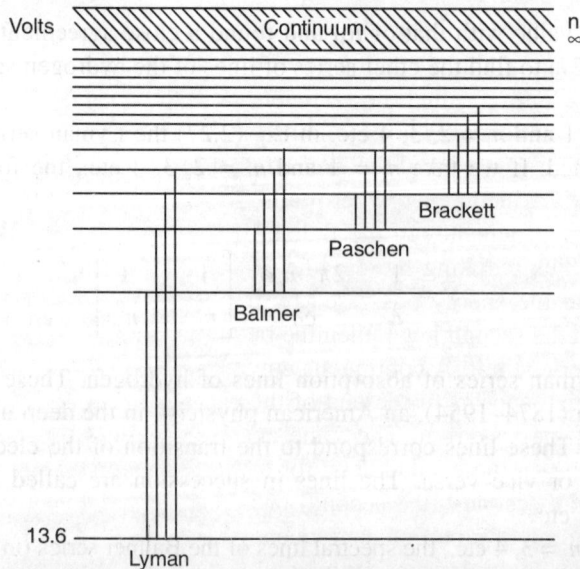

FIGURE 2.5 The energy level diagram of hydrogen spectrum.

is then said to be *ionized*. If a gas is exposed to strong radiation field, the atoms may be stripped of two, three or more electrons, and these are then said to be doubly, triply or multiply ionized. This holds true for atoms of higher atomic number.

Since the higher levels merge at $n = \infty$ corresponding to an energy value of 13.6 eV (for hydrogen), the higher lines of each series must converge at the frequency (or wavelength) corresponding to this energy. These frequencies are different for different series. The lines of the series crowd together at these frequencies and approach certain limits known as the *Series Limits*. Thus the Lyman series limit is at the wavelength that corresponds to an energy of (13.6 − 0) = 13.6 eV (since the energy of the inner lowermost orbit, $n = 1$ is taken as zero). The entire Lyman series lies between the wavelengths 912 Å to 1215 Å, the first corresponding to the Lyman series limit while the last to the L_α line (that is, the first line of the Lyman series). The Balmer series limit corresponding to an energy difference of $n = 2$ and $n = \infty$ levels is at 3650 Å, the first Balmer line (H_α) being at 6563 Å.

The energy values of the different orbits defined by the quantum numbers $n = 1, 2, 3$ 4 etc. may be used to calculate the wavelength of any line of any series of the hydrogen spectrum.

In the energy level diagram (Fig. 2.5) the shaded region, called the continuum, represents continuous energy values and extends beyond the last quantized orbit $n = \infty$. The energy corresponding to the last quantized orbit is called the threshold energy. When the atoms are ionized, the electron is free from the nuclear binding force and moves as a free particle in space. The kinetic energy of this electron is derived from the excess energy absorbed by the atom during ionization and is given by $1/2\ m_e v^2 = h\nu - h\nu_0$, where ν is the frequency of the absorbed photon and ν_0 is the threshold frequency. This excess energy therefore, appears as the energy of motion (kinetic energy) of the free electron. This motion is not periodic as it is not bounded by the quantum conditions of the Bohr orbits. The kinetic energy values of the aperiodic motion of the free electron depends on the energy of the incident photons absorbed by the atom.

A continuum of wavelengths in emission is produced when a free electron moving close to an ion falls into one of the permitted energy levels of the neutral atom. This process is opposite of ionization and is called the *recombination* of the electron by the ion. During this process, the atom radiates energy corresponding to the *kinetic energy* possessed by the electron during the process of ionization and the *excitation energy* of the level to which it falls. When the electron with initial kinetic energy zero falls into the second energy level (i.e., $n = 2$), the atom will radiate at the frequency of the Balmer series limit. The electron may cascade down to the orbit $n = 1$, thus radiating the L_α line. If the initial kinetic energy of the electron is greater than zero and the energy possessed by the electron is in excess of that necessary to ionize the atom from the second level, then the radiation takes place at a frequency higher than that of the Balmer series limit. A continuous emission spectrum, called the *Balmer continuum* is then produced which appears at the end of the Balmer series limit and extends far into the ultraviolet region of the spectrum. The same analysis is true for every other series.

An atom remains in the excited state for an extremely short interval of time, about one hundred millionth of a second (10^{-8} second) or so and comes down to the ground state with the emission of a photon of a particular frequency (or wavelength). This process of transition may occur in single or multiple steps with simultaneous emission of the corresponding single or multiple lines of definite wavelengths. The energy emitted during transition corresponds to the difference of energies of the initial and final orbits occupied by the electron. For instance, the electron in $n = 3$ level can drop back down to the $n = 1$ level by a single transition radiating the L_β line or by cascading down successively to $n = 2$ and then to $n = 1$ level, thus emitting the H_α and L_α lines respectively.

The Bohr model of hydrogen atom is however an extremely simplified one. Bohr assumed that the electron revolves in circular orbits around the nucleus but in fact, the orbits are elliptical similar to the planetary orbits around the Sun. Furthermore, the nucleus of the atom is fixed in his model which is also not exactly so. The correction was introduced by A. Sommerfeld to derive the exact model of the hydrogen atom. Due to this correction minute changes were introduced in the energy values and wavelengths of the hydrogen lines. The detailed discussion however, is beyond the scope of this book.

2.8 MOLECULAR SPECTRA

In what follows, we shall be concerned with a brief discussion of the spectra of diatomic molecules which only are important in astrophysics. In the discussion of the simple model of the hydrogen atom and its spectrum analysis, we have found that the prominent features of the atomic spectra consist of a regular line pattern superimposed on a continuous spectrum. The lines form a series and the difference between two successive members decreases gradually. These lines ultimately converge at some wavelength called the *series limit*. Although different atoms have different characteristic line patterns, they agree in general features mentioned above. But an entirely different type of spectra is obtained when they are produced by molecular gases. Instead of single sharp lines as produced by atoms, *bands* of absorption or emission are observed over broad wavelength ranges. These bands are the most prominent features of the molecular spectra. Accordingly, they are called the *band spectra*.

A detailed discussion of different characters of the bands produced by different molecules is beyond the scope of the book. We shall discuss here only the important features relevant to the text. Each band in molecular spectra covers a sufficiently large wavelength range. On one side of this range the intensity falls suddenly to zero at certain wavelength which is called the *band head* or *band edge*. Starting from the band head, towards the other side of the wavelength range, the intensity of the band falls off gradually and appears to merge into the continuum. If the intensity of the band gradually falls off towards the shorter wavelengths of the spectrum, the band is said to be *shaded* or *degraded to the violet*; if the falling off is towards the longer wavelengths of the spectrum, it is said to be *shaded to the red*.

In spectra of some molecules such as H_2, instead of well-developed bands, numerous closely-spaced lines are observed which have the appearance of a many-lined spectrum of an atom of complicated structure. Besides the lines and bands, an extended continuous region is also an important characteristic of a molecular spectrum.

Most of the molecular spectra in general consist of a series of bands with slowly changing separation between them. The bands in any particular series are called a *progression*. The wave numbers of the bands with slowly decreasing separation in any particular progression can be conveniently represented by a formula,

$$\tilde{v} = \tilde{v}_m - (an - bn^2) \qquad (2.30)$$

where a and b are positive constants with $b \ll a$, and n is zero or a *positive integer*. With $b = 0$, the formula gives a progression of equidistant bands with separation a. If $b \neq 0$ but $\ll a$, bands with slowly decreasing separations are obtained, $n = 0$ corresponds to the band of highest wave number which is $\tilde{v}_m$. Sometimes this band is not observed at all.

The bands in the molecular spectra as discussed above are so observed, only if low dispersion is used. In a low dispersion spectrum, two very close lines cannot be resolved and appear to be one. But when spectrographs of higher resolving power are used, the close lines that appear overlapping in a low dispersion spectrum are resolved into a large number of individual lines which are arranged in a regular pattern. This is the fine structure of molecular bands. But the regularity of the arrangement of lines of a band is of quite different nature from that of a series of lines produced by an atom. The separation between two successive lines of a band changes *nearly linearly* and the wave numbers of these lines can be approximately represented by a formula of the type,

$$\tilde{v} = a + bn + cn^2 \qquad (2.31)$$

where a, b and c are constants, and n is an integer which can take both positive and negative values as also the value zero, by suitable choice of the constants a, b and c. For $n = 0$, $\tilde{v}_0 = a$, which can be considered to correspond to the missing lines in the series. This missing line, often called the *zero line* or *null line* is a general characteristic of the band spectra of molecules. On one side of the null line lies the lines of the band corresponding to the positive values of n, and on the other side lie those corresponding to the negative values of n. The former are called the *positive branch* or *R branch* and the latter are called the *negative branch* or *P branch*. These two branches together constitute one simple series of lines forming a band. A more detailed discussion of molecular spectra is beyond the scope of this book.

2.9 BREMSSTRAHLUNG

It is a well known fact that accelerated charged particles give rise to electromagnetic radiation. This property of charged particles has very great significance in astrophysics where both charged particles and accelerating mechanisms for them are known to be present. When non-relativistic free electrons are accelerated moving in the Coulomb field of an ion, the electron emits radiation. This radiation is called *bremsstrahlung*. It is also called *free-free* emission, because the radiation is emitted by free electrons which are accelerated by the field produced by free ions. The total power radiated by an electron having an acceleration $\ddot{\mathbf{x}}$ is given by

$$s_e = \frac{2e^2}{3c^3} |\ddot{\mathbf{x}}|^2 \qquad (2.32)$$

where e is the charge of an electron. This is Larmor result for a single non-relativistic, accelerated electron. If an electron is accelerated in the field of an ion of charge Ze at a distance $\mathbf{r}$ from it, then

$$\ddot{\mathbf{x}} = - \frac{Ze^2}{r^3 m_e} \mathbf{r} \qquad (2.33)$$

If an electron moves with velocity v having its impact parameter b, then it experiences the strong Coulomb force for a time approximately equal to b/v. Therefore, the total power radiated by an electron of velocity v and impact parameter b is obtained by integrating the power over the appropriate time. If the electron is assumed to move with a constant acceleration during this period the total energy emitted by such an electron is approximately given by

$$\mathcal{E}_e = \frac{2e^2}{3c^3} \int_{-1/2\,b/v}^{1/2\,b/v} |\ddot{x}|^2 \, dt = \frac{2Z^2 e^6}{3m_e^2 c^3 b^3 v} \text{ erg} \tag{2.34}$$

which has been obtained by using Eqs. (2.32) and (2.33). The frequency v of the photon emitted by the electron with impact parameter b and moving with velocity v, is given approximately by

$$2\pi b v = v \tag{2.35}$$

If the velocity distribution function of the electrons with respect to the ions be Maxwellian, then the probability that an electron will be found in the element $dv_x \, dv_y \, dv_z$ of velocity space is given by

$$f(v) dv_x dv_y dv_z = \left(\frac{m_e}{2\pi kT} \right)^{3/2} \exp \left[\frac{-m_e v^2}{2kT} \right] dv_x dv_y dv_z \tag{2.36}$$

where

$$\iiint f(v) \, dv_x dv_y dv_z = 1 \tag{2.37}$$

the integration being taken over the entire range of possible velocities. Combining Eqs. (2.34), (2.35) and (2.36) and integrating over the whole velocity space, it is found that the emission coefficient for free-free transition is given by

$$j_v = \left(\frac{2\pi}{3m_e c^3} \right) \left(\frac{2\pi}{kTm_e} \right)^{1/2} Z^2 e^6 n_e n_i e^{-hv/kT} \text{ erg cm}^{-3} \text{ s}^{-1} \text{ str}^{-1} \text{ Hz}^{-1} \tag{2.38}$$

where n_e, n_i are respectively the number densities of electrons and ions. This approximate result is obtained by purely classical considerations. Quantum mechanical calculations give more exact result but that does not significantly differ from Eq. (2.38) in numerical value. The expression obtained for j_v from quantum mechanical considerations is

$$j_v = \left(\frac{128\pi m_e}{27kT} \right)^{1/2} \frac{Z^2 e^6}{m_e^2 c^3} g_{ff} n_e n_i e^{-hv/kT} \text{ erg cm}^{-3} \text{ s}^{-1} \text{ str}^{-1} \text{ Hz}^{-1} \tag{2.39}$$

where g_{ff} is called the Gaunt factor for free-free transitions whose numerical value does not differ much from unity. Since hydrogen is overwhelmingly abundant in nature, we can consider the ions to be protons so that $Z = 1$ and n_p replaces n_i. Equation (2.39) then becomes

$$j_v = \left(\frac{128\pi m_e}{27kT} \right)^{1/2} \frac{e^6}{m_e^2 c^3} g_{ff} n_e n_p e^{-hv/kT} \text{ erg cm}^{-3} \text{ s}^{-1} \text{ str}^{-1} \text{ Hz}^{-1} \tag{2.40}$$

Multiplying Eq. (2.40) by $4\pi d\nu$ and integrating over the entire frequency range we can calculate what is called the volume emissivity in free-free transitions, given by

$$\varepsilon_{ff} = 4\pi \int_0^\infty j_\nu d\nu = \left(\frac{2\pi kT}{27 m_e}\right)^{1/2} \frac{2^5 \pi e^6}{hm_e c^3} g_{ff} n_e n_p$$

$$= 1.435 \times 10^{-27} g_{ff} n_e n_p \sqrt{T} \text{ erg cm}^{-3} \text{ s}^{-1} \quad (2.41)$$

Equation (2.41) shows that the volume emissivity in free-free transitions is strongly dependent on the particle density in the plasma and depends but quite weakly on the plasma temperature. In astrophysics, bremsstrahlung is important in H II regions, atmospheres of hot blue stars, the upper chromosphere and corona of the Sun, etc. The radiation is thermal, since it is dependent on the temperature of the gas.

Larmer's formula (Eq. (2.32)) for the radiation by a non-relativistic, accelerated electron can be generalized also to the case of an accelerated electron moving with relativistic velocities. This is possible because the electromagnetic radiation energy behaves under Lorentz transformation as the fourth component of a 4-vector. The total power radiated by an accelerated electron in this case is

$$(s_e)_{rel} = \frac{2e^2}{3 m_e^2 c^3} \gamma^2 \left[\left(\frac{d\mathbf{P}}{dt}\right)^2 - \frac{1}{c^2}\left(\frac{d\varepsilon}{dt}\right)^2\right] \quad (2.42)$$

where $\varepsilon = \gamma m_e c^2$, the total energy, and $\mathbf{p} = \gamma m_e \mathbf{v}$, the momentum of the particle and γ has the usual definition

$$\frac{1}{\gamma} = (1 - \beta^2)^{1/2}, \beta = \frac{\upsilon}{c} \quad (2.43)$$

Equation (2.42) shows that the total power radiated by the accelerated charge is proportional directly to the square of its charge and inversely to the square of its mass. Since the charge is the same in both protons and electrons, but the mass of the proton exceeds that of the electron by about a factor of 1840, the power radiated by electrons is very much important compared to that by protons.

2.10 SYNCHROTRON RADIATION

The radiation emitted by charged particles spiraling around magnetic lines of force in an applied magnetic field is called *synchrotron radiation*. The name is derived from the way the motion in a synchrotron takes place. In this case, as the particle revolves in circular orbits, its momentum changes much more, rapidly compared to its energy change per revolution.

Thus, $\left|\dfrac{d\mathbf{p}}{dt}\right| \gg \dfrac{1}{c^2}\dfrac{d\varepsilon}{dt}$ in this case, so that the total radiated power by an electron may be approximated by (using Eq. (2.42)).

$$(S_e)_{syn} = \frac{2 e^2}{3 m_e^2 c^3} \gamma^2 \left|\frac{d\mathbf{p}}{dt}\right|^2 \quad (2.44)$$

or since $\left|\dfrac{d\mathbf{p}}{dt}\right| = \omega|\mathbf{p}|$, where ω is the instantaneous angular velocity and since $\omega\rho_c = c\beta$,

where ρ_c is the instantaneous radius of curvature of the orbit, Eq. (2.44) can be written as

$$(S_e)_{\text{syn}} = \frac{2e^2c}{3\rho_c^2}\,\gamma^4\beta^4 \qquad \text{(using } |\mathbf{p}| = \gamma m|\mathbf{v}|) \tag{2.45}$$

For highly relativistic electrons $\beta \sim 1$; so γ may be very large and Eq. (2.45) suggests that the radiated power may be quite significant. Since the total power radiated as given in Eq. (2.45) is for a number of revolutions performed in one second which is actually $v/2\pi\rho_c$, the energy radiated per revolution is

$$\delta_\varepsilon = \frac{2\pi\rho_c}{v}\,(S_e)_{\text{syn}} = \frac{4\pi e^2}{3\rho_c}\,\gamma^4\beta^3 \text{ eV rev}^{-1} \tag{2.46}$$

where the relation $v = c\beta$ has been used. Since for a given energy and charge, γ is about 1840 times larger for electrons and positrons than for protons, it is only for the former that synchrotron mechanism is important.

Two of the basic properties of synchrotron radiation by electrons are:

1. The energy is radiated in a very narrow cone, the cone angle depends on the velocity of the electron in its circular orbit.
2. Very high frequency radiation may be emitted by electrons in this mechanism. The cone angle is related to the electron velocity by

$$\langle \sin^2\theta \rangle = 1 - \beta^2 = 1/\gamma^2 \tag{2.47}$$

or equivalently, since $\sin\theta = \theta$ when θ is small,

$$\langle \theta^2 \rangle^{1/2} = \frac{1}{\gamma}, \tag{2.48}$$

which in appropriate units can be written as

$$\theta = 1.76/\varepsilon \text{ (BeV) min of arc} \tag{2.49}$$

This implies that the radiated power by synchrotron mechanism reaches the observer as quick pulses of short duration. This duration is actually $\delta t = \rho_c/\gamma c$ in the frame of reference of the radiating particle, but for the observer the duration will be modified due to Doppler effect. This modified duration is given by $\delta t' = \rho_c/\gamma^3 c$. Since the *gyrofrequency* (number of revolutions per second) of the particle is $\omega \sim c/\rho_c$, so $\delta t' \sim \dfrac{1}{\omega\gamma^3}$, and using the Fourier theory relating the

duration of radiation and frequency range, viz., $\delta\omega\,\delta t' = 1$ it turns out that the spectral frequency range of synchrotron radiation spreads upto $\omega_c \sim \omega\gamma^3$, ω_c being considered as the critical frequency. The radiation is, however, negligible at frequencies higher than ω_c. A relativistic electron gyrating in circular orbits thus emits radiation over a very large frequency range.

In astrophysical cases, the accelerating agent for the particles are the local or galactic magnetic fields. So introducing the magnetic field through the Lorentz force by $\frac{d\mathbf{p}}{dt} = \frac{q}{c}\mathbf{v} \times \mathbf{B}$ the Eq. (2.44) for the total power radiated transforms, in the case of an electron, to

$$(S_e)_{\text{syn}} = \frac{2e^2}{3m_e^2c^3}\,\gamma^2\left(\frac{e^2}{c^2}\,v^2B^2\,\sin^2\theta'\right) \tag{2.50}$$

where θ' is the angle between the directions of $\mathbf{v}$ and $\mathbf{B}$. For highly relativistic particles gyrating around magnetic lines of force one can take $v \simeq c$ and $B\sin\theta' = B_\perp$, the component of magnetic field perpendicular to the direction of velocity, so that Eq. (2.50) transforms to

$$(S_e)_{\text{syn}} = \frac{2e^2}{3m_e^2c^3}\,\gamma^2B_\perp^2 = 1.58 \times 10^{-15}\,\gamma^2B_\perp^2\,\text{erg s}^{-1} \tag{2.51}$$

expressed in cgs units. In terms of the energy of particle Eq. (2.51) reduces to

$$(S_e)_{\text{syn}} = 3.79 \times 10^{-6}\,\varepsilon^2\,B_\perp^2\,\text{s}^{-1} \tag{2.52}$$

where ε is now expressed in BeV and $B_\perp$ in gauss. Since the power radiated is just the rate of losing energy by particle, the left-hand side of Eq. (2.52) is nothing but $-d\varepsilon/dt$, and on integration, it leads to the result that half of the energy of the particle is radiated away in a time-scale of about

$$t_{1/2} = 2.63 \times 10^5\,\frac{1}{B_\perp^2\,\varepsilon(\text{BeV})}\,\text{s} \tag{2.53}$$

It is to be noted that the energy of the particle is rapidly lost in a high magnetic field, and also if the particle possesses high energy. Thus extreme relativistic particles gyrating in a strong magnetic field are likely to lose most of their energy within a very short time-scale. Calculations also show that the spectral frequency at which the maximum energy is radiated by the particle is given by

$$\nu_{\text{max}} = \frac{3e}{4\pi m_e c}\,\gamma^2B_\perp = 1.61 \times 10^{13}\,B_\perp\varepsilon^2\,(\text{BeV})\,\text{c/s} \tag{2.54}$$

So far we have considered the case of power radiated by a single particle only. When an assemblage of radiating particles is considered, the calculations become a little more involved. The radiating particles in this case will have an energy distribution of the form

$$n(\varepsilon) = K\varepsilon^{-\beta'}\,d\varepsilon, \tag{2.55}$$

where $n(\varepsilon)$ is the number of particles per unit volume in the energy range ε to $\varepsilon + d\varepsilon$. K is a constant to be evaluated by observation of flux of particles (for example, flux near the earth). β' is the spectral index of particle energies measured generally in BeV. The total power radiated by such a distribution of particles having their energy in the range ε_1 to ε_2 is calculated to be

$$P = (3.79 \times 10^{-6})\,N\int_{\varepsilon_1}^{\varepsilon_2} B_\perp^2\varepsilon^{2-\beta'}\,d\varepsilon \tag{2.56}$$

where N is the total number of radiating particles (electrons), and the total energy of the electrons is given by

$$\varepsilon_e = \int_{\varepsilon_1}^{\varepsilon_2} n(\varepsilon)\varepsilon \, d\varepsilon = N \int_{\varepsilon_1}^{\varepsilon_2} \varepsilon^{1-\beta'} \, d\varepsilon \qquad (2.57)$$

These formulae may be used to determine any of the unknown parameters involved when others are known, either from observations or from theory. In astrophysical cases, β' and P are generally predetermined from observation. So one can calculate the total number of particles, N, as a function of $B_\perp$. Or, if either of the two can be determined by some other means the second is obtained.

Synchrotron radiation (alternatively called magnetobremsstrahlung) plays a key role in the understanding of many of the current astrophysical phenomena. The radiation in a wide frequency range by supernova remnants such as the Crab Nebula, the radio energy flux from radio galaxies and quasars, the galactic radio background radiation etc., are now satisfactorily explained in terms of synchrotron radiation. The very high degree of polarization of synchrotron radiation enables it to be easily discriminated from other types of radiation. Since the mechanism of this radiation does not involve any kind of *thermal transitions,* it is often termed as *non-thermal radiation,* as opposed to the thermal radiation which always involves some type of electronic transition.

2.11   SCATTERING OF RADIATION

In this section we shall consider *scattering by free electrons* and *scattering by a harmonic oscillator.* The scattering by free electrons is important in atmospheres of hot stars where hydrogen is heavily ionized. This is known as *Thomson scattering* and we shall consider it first.

A plane polarized monochromatic light wave of frequency v falling upon a charged particle imparts acceleration to it, causing the charged particle to radiate. Suppose the light wave falls on an electron of mass m_e and charge e vibrating about its mean position which is stationary, and suppose the electric field **E** associated with the wave is given by

$$\mathbf{E} = \mathbf{E}_o \, e^{-ivt} \qquad (2.58)$$

The *self-force* experienced by the electron as it is accelerated is $\dfrac{2e^2}{3c^3}\,\dddot{\mathbf{x}}$. The equation of motion of the particle can therefore be written as

$$\ddot{\mathbf{x}} = \frac{e}{m_e} \mathbf{E}_o \, e^{-ivt} + \frac{2e^2}{3c^3 m_e} \dddot{\mathbf{x}} \qquad (2.59)$$

This equation has the solution

$$\mathbf{x} = -\frac{e\mathbf{E}_o}{m_e v^2} \frac{1}{1 + i\xi} e^{-ivt} \qquad (2.60)$$

where

$$\xi = \frac{2e^2}{3 m_e c^3} v \qquad (2.61)$$

Equation (2.60) shows that the electron after being perturbed by the incident wave of frequency v oscillates with the same frequency. This oscillating electron, in turn, gives rise to radiation of the *same frequency* v. Light is thus just scattered by electrons, the secondary waves being generated at the expense of the primary (incident) waves. The quantity ξ arises out of the damping effect and calculation shows that compared to unity, ξ is negligibly small even for very large values of v which includes even the very high-frequency γ-rays. So we shall neglect ξ. Calculation then shows that the power radiated by the electron in a solid angle $d\Omega$. averaged over time can be written as

$$ds = \frac{c}{8\pi} |\mathbf{E}_0| r_e^2 \sin^2\theta \, d\Omega \tag{2.62}$$

where θ is the angle between the directions of acceleration and observation respectively and r_e is the classical electronic radius, given by

$$r_e = \frac{e^2}{m_e c^2} = 2.818 \times 10^{-13} \text{ cm} \tag{2.63}$$

The quantity

$$d\sigma_e = r_e^2 \sin^2\theta \, d\Omega \tag{2.64}$$

is called the *differential scattering cross-section* for the electron. Using

$$\langle \sin^2\theta \rangle = 1/2(1 + \cos^2\theta') \tag{2.65}$$

where θ' now represents the angle of scattering, combination of Eqs. (2.64) and (2.65) yields

$$d\sigma_e = \frac{1}{2} r_e^2 (1 + \cos\theta') \, d\Omega \tag{2.66}$$

as the differential scattering cross-section. Integrating Eq. (2.66) over all solid angles and over all angles of scattering θ', the total *scattering cross-section* is obtained to be

$$\sigma_e = \frac{8\pi}{3} \left(\frac{e^2}{m_e c^2} \right)^2 = 6.655 \times 10^{-25} \text{ cm}^2 \tag{2.67}$$

This is called *Thomson scattering cross-section* for scattering of radiation by an electron. The scattering by proton is less than this value by a factor of $\left(\dfrac{m_p}{m_e} \right)^2$ which is $\sim 10^6$. So scattering by electron only is important (when free charges are considered). The relation (2.67) shows that scattering by free electrons is an absolute constant, not depending on any variable quantity whatsoever.

We now consider the scattering by a bound charge harmonically oscillating with a natural frequency v_0. If incident waves of frequency v fall on this oscillator, the electric field of this wave will tend to cause the oscillator to vibrate with its own frequency v. The case is therefore that of a forced oscillation having the equation of motion as

$$\ddot{\mathbf{x}} + v_0^2 \mathbf{x} = \frac{e}{m_e} \mathbf{E}_0 e^{-ivt} = \frac{e\mathbf{E}}{m_e}, \tag{2.68}$$

where the damping force has been neglected. We see that $\mathbf{E} = 0$ gives the oscillation of the
bound charge with its natural frequency v_0. The solution of Eq. (2.68) yields

$$\mathbf{x} = \frac{e\mathbf{E}}{m_e} \frac{1}{(v_0^2 - v^2)} \tag{2.69}$$

The power radiated by the oscillator in a solid angle $d\Omega$ and averaged over time is in this case
given by

$$s_{\text{osc}} = \frac{c}{8\pi} |\mathbf{E}_0|^2 \, r_e^2 \, \frac{v^4 \sin^2\theta}{(v_0^2 - v^2)^2} \tag{2.70}$$

where $\dfrac{c}{8\pi} |\mathbf{E}_0|^2$ is the time average of the intensity of primary radiation and

$$d\sigma_s = r_e^2 \sin^2\theta \, \frac{v^4 \, d\Omega}{(v_0^2 - v^2)^2} = r_e^2 \sin^2\theta \, \frac{d\Omega}{[(v_0/v)^2 - 1]^2}$$

is the differential cross-section for scattering by bound oscillator. The cross-section σ_s, for
total radiation scattered is obtained by integration as before to be

$$\sigma_s = \frac{8\pi}{3} \left(\frac{e^2}{m_e c^2} \right) \frac{1}{[(v_0/v)^2 - 1]^2} \tag{2.71}$$

From where

$$\sigma_s = \sigma_e \frac{1}{[(v_0/v)^2 - 1]^2} \tag{2.72}$$

Two limiting cases are important. When $v_0 \ll v$, $\sigma_s \to \sigma_e$; that is, the electron in this
case behaves as a free electron and Thomson scattering results. When on the other hand,
$v_0 \gg v$ we get, using wavelengths

$$\sigma_s = \sigma_e \, v^4/v_0^4 = \frac{8\pi e^4 \lambda_0^4}{3 m_e^2 c^4} \frac{1}{\lambda^4} \tag{2.73}$$

as the scattering coefficient per electron. This is *Rayleigh-Jeans scattering* by bound charges
which is strongly colour-dependent ($\propto 1/\lambda^4$). Scattering of sunlight at day time by air molecules
obeys this law. According to this law, since the wavelengths of blue light are about one-half
of those of red light, blue light should be scattered 16 times more than red light. This is
actually displayed by blue colour of the sky at daytime. Most of the blue sunlight being
scattered at random the sky appears blue. Red light, on the other hand, can travel in its course
almost unhindered. Physically, this scattering condition is realized because the strongly bound
electrons to the molecules have oscillation frequencies much higher than those of visible light.

When $v_0 = v$, the value of σ_s becomes very large which gives the case of *resonance fluorescence*. In this case we can write

$$\sigma_s = \sigma_e \frac{v^4}{(v_0^2 - v^2)^2} = \frac{\sigma_e}{4} \frac{v^2}{(v_o - v)^2} \tag{2.74}$$

since

$$v_0 + v \simeq 2v$$

If the damping force is taken into account then the expressions corresponding to Eqs. (2.72) and (2.74) will respectively be

$$\sigma_s = \sigma_e \frac{v^4}{(v_0^2 - v^2)^2 + v^2 \Gamma^2} \tag{2.75}$$

and

$$\sigma_s = \frac{\sigma_e}{4} \frac{v^2}{(v_o - v)^2 + \Gamma^2/4} \tag{2.76}$$

where Γ is the damping constant.

In astrophysics, scattering by intersteller grains is important (see Chapter 13). The grain sizes are not definitely known. But the scattering cross-section for grains has been found to approximately follow the λ^{-4} law. This suggests that the radii of grains are much smaller than the wavelengths λ in visible region.

2.12 COMPTON EFFECT AND INVERSE COMPTON EFFECT

We have thus seen how Thomson scattering arises when electromagnetic waves fall upon free electrons. This scattering process is, however, characteristic of relatively low energy incident electromagnetic radiation. But when a high energy photon, such as a γ-ray or an X-ray, impinges on a free electron (a charged particle), the former imparts momentum to the latter at an angle θ with the original direction of its propagation. The result is that the incident photon scatters off an electron as if it is a two-particle collision, but the frequency of the scattered photon in this case is different from that of the incident photon, unlike in Thomson scattering. This scattering of high energy photons by charged particles is known as *Compton effect* (after the name of A.H. Compton who first observed it). For the mathematical formulation of the problem, viz., the transfer of energy and momentum to the electron from the photon, we have to consider the conservation laws of mass-energy and momentum.

Suppose a photon of initial energy $h\nu$ impinges on an electron at rest. The collision will impart to the electron a velocity $\mathbf{v}$ at the expense of the photon energy, and the electron will move at an angle ϕ, say, with the direction of motion of the incident photon. The incident photon on its part, will lose energy and will be scattered in a different direction. Let $h\nu'$ be the final energy of the incident photon and it travels at an angle θ with the original direction of propagation of the primary photon as shown in Fig. 2.6.

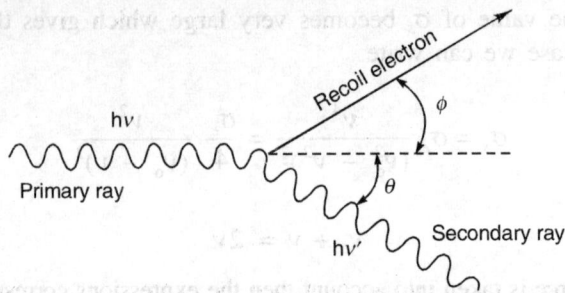

FIGURE 2.6 The Compton effect.

The momentum of a photon of energy $h\nu$ is $h\nu/c$ and its mass is $h\nu/c^2$. The rest mass energy of electron is $m_e c^2$ and its energy when moving relativistically with a velocity $\mathbf{v}$ is $m_e c^2 \gamma$ where

$$\gamma = (1 - v^2/c^2)^{-1/2} \qquad (2.77)$$

So energy conservation law yields

$$h\nu = h\nu' + m_e c^2 (\gamma - 1) \qquad (2.78)$$

where we must note that $\nu > \nu'$, and so $\lambda' > \lambda$. Considering the conservation of linear momentum in the direction of propagation of the incident photon (longitudinal) and in the transverse direction, we get respectively:

$$\frac{h\nu}{c} = m_e \gamma v \cos \phi + \frac{h\nu'}{c} \cos \theta \qquad (2.79)$$

and

$$0 = m_e \gamma v \sin \phi - \frac{h\nu'}{c} \sin \theta \qquad (2.80)$$

Equations (2.79) and (2.80) can each be squared and written as

$$\left(\frac{h\nu}{c} - \frac{h\nu'}{c} \cos \theta \right)^2 = (m_e \gamma v)^2 \cos^2 \phi \qquad (2.81)$$

and

$$\left(\frac{h\nu'}{c} \right)^2 \sin^2 \theta = (m_e \gamma v)^2 \sin^2 \phi. \qquad (2.82)$$

Adding Eqs. (2.81) and (2.82) we get

$$(h\nu)^2 + (h\nu')^2 - 2(h\nu)(h\nu') \cos \theta$$
$$= (m_e \gamma v c)^2 = (m_e \beta \gamma c^2)^2 = m_e^2 c^4 (\gamma^2 - 1) \qquad (2.83)$$

since $\beta^2 \gamma^2 = \gamma^2 - 1$. Equation (2.78) can be written as

$$(h\nu - h\nu')^2 + 2(h\nu - h\nu')\, m_e c^2$$

$$= -m_e^2 c^4 + m_e^2 \gamma^2 c^4 = m_e^2 c^4\, (\gamma^2 - 1). \tag{2.84}$$

Equations (2.83) and (2.84) therefore taken together, yield (since their right hand sides are equal)

$$m_e c^2\, (\nu - \nu') = h\nu\nu'\, (1 - \cos\,\theta)$$

whence

$$\frac{c(\nu - \nu')}{\nu\nu'} = \frac{h}{m_e c}\, (1 - \cos\,\theta) \tag{2.85}$$

since $\nu = c/\lambda$ and $\nu' = c/\lambda'$, Eq. (2.85) can be written in terms of wavelengths as

$$\lambda' - \lambda = \frac{h}{m_e c}\, (1 - \cos\,\theta) \tag{2.86}$$

Equation (2.86) gives the change in wavelengths of the incident and scattered photons in terms of the scattering angle. The quantity $\lambda_c = h/m_e c$ is called the *Compton wavelength* of the electron. Its numerical value is 2.426×10^{-2} Å or 2.426×10^{-10} cm. Equation (2,86) shows that the maximum change in wavelengths that can be produced by Compton effect is $2\lambda_c$, or about 0.05 Å and this is obtained for back-scattered radiation. So for visible light of wavelength ~ 5000 Å the effect is negligible. This is why the Compton effect can be safely neglected for low-energy incident photons. Thomson scattering is relevant in this case. But the situation drastically changes when the incident photons are X-rays or γ-rays. For example, for hard X-ray of wavelength 0.5 Å the shift is 10 per cent of the incident wavelength and for γ-rays the effect becomes enormously large. So Compton effect is very important in high energy astrophysics.

Physically, an inverse process in which a highly energetic particle impinging upon a relatively low energy photon transfers momentum and energy to the latter, is also possible. This process is known as the *inverse Compton effect*. The photon can by this process acquire large energy and momentum at the expense of those of the impinging particle. The processes of Compton effect and inverse Compton effect are exactly alike. Whether an observer will notice a Compton effect or an inverse Compton effect in a particular case will depend upon his choice of the reference frame with which he is attached. The inverse Compton effect will appear to him as Compton effect if he moves with the highly relativistic impinging electron. On the other hand, if the observer moves with the high energy photon producing the ordinary Compton effect, the phenomenon will appear to him as the inverse Compton effect in which a high energy particle will appear to impinge on a low energy photon to transfer momentum and energy to the latter.

The inverse Compton effect, like synchrotron radiation, appears to play a vital role in high-energy astrophysics. The two processes have similar characteristics in that while the total radiated power in synchrotron process is proportional to the magnetic energy density in space, that in the inverse Compton scattering is proportional to the square of the energy density of the electromagnetic radiation in space, the constant of proportionality in both the cases being the same. The close similarity between these two processes are borne by the facts that both are caused by highly energetic particles and both are capable of producing highly-energetic photons. This last property, in fact, accounts for the very great usefulness of the inverse Compton effect in explaining the high energy radiation from many astrophysical objects.

EXERCISES

1. (a) Using Wein's displacement law, find the temperature of an object whose blackbody spectrum peaks at the wavelength of (i) 4000 Å, and (ii) 6563 Å.
   (b) If the Earth be considered as a blackbody at a temperature of 300 K, at what wavelength will its radiation peak?

2. Using Planck's blackbody radiation law, obtain Wien's radiation law for high frequencies and Rayleigh-Jeans law for low frequencies.

3. Find the expression for $u(\lambda)d\lambda$ using Planck's law for $u(v)dv$.

4. Calculate the frequency of the highest intensity radiation of a blackbody whose temperature is (i) 6000 K, and (ii) 25000 K.

5. Calculate the wavelengths of radiation corresponding to photon energies of $E = 13.6$ eV, 10.2 eV, 3.4 eV and 1.8 eV. Comment on these wavelengths in relation to the spectrum of the hydrogen atom. Deduce the frequencies of these photons.

6. A photon of 20 eV energy strikes a hydrogen atom in its ground level and knocks out the electron. What will be the velocity of the free electron?

7. A photon of 25 eV energy strikes an atom in its ground level and removes the electron with a velocity of 5000 km s^{-1}. What is the ionization potential of the atom?

8. The spectrum of a star shows a Doppler shift of 10^{-2} Å of a line whose natural wavelength is 5000 Å. Calculate the velocity of the star along the line of sight.

9. An object moves towards the observer with a velocity of 1000 km s^{-1}. What will be the shift of the H_α line in its spectrum.

10. In the spectrum of a distant galaxy, the lines observed are Doppler-shifted by 20 per cent towards the longer wavelengths. Calculate its radial velocity.

11. Calculate the Zeeman splitting of the Ca II, K line in the spectrum of a sunspot where the strength of the magnetic field is 1000 gauss.

12. If the strength of the galactic magnetic field was 10^{-2} gauss, what would be the splitting of the 21 cm line of neutral hydrogen?

13. Using formula (2.22), calculate the wavelengths of the first five absorption lines of the Balmer series of hydrogen spectrum.

14. Using formula (2.29), calculate the wavelengths of the first five emission lines of the Lyman series of hydrogen spectrum.

15. Calculate the Compton effect on a photon of energy *hv* scattered at right angles to its original path.

16. Obtain the percentage shift of a back-scattered photon of 10 keV energy.

SUGGESTED READING

1. Born, Max, *Atomic Physics,* G.E. Stechart & Company, New York, 1951.

2. Herwit, M., *Astrophysical Concepts,* John Wiley & Sons, New York, 1973.

3. Jackson, J.D., *Classical Electrodynamics,* John Wiley & Sons, New York, 1962.

4. Lehnert, B., *Dynamics of Charged Particles,* North Holland, Amsterdam, 1964.

5. Semat, Henry, *Introduction to Atomic and Molecular Physics,* Holt, Rinehart and Winston, Inc., New York, 1962.

6. Zeilik, Michael and Elske, V.R. Smith, *Introductory Astronomy and Astrophysics,* Saunders College Publishing, New York, 1987.

3 Magnitudes, Motions and Distances of Stars

3.1 STELLAR MAGNITUDE SEQUENCE

Stars, in general, greatly differ in brightness from one another. The magnitude scale had been developed on the basis of measurement of the brightness of stars. We can observe some stars with naked eye. The brightness of the star thus appears to us is called *the apparent brightness* of the star. This brightness depends on the actual (or intrinsic) brightness as well as on the distance of the star from us. The magnitude based on the apparent brightness of the star is called its *apparent magnitude* and is usually denoted by *m*.

During the 2nd century B.C., Hipparchus first classified the stars according to their apparent brightness or magnitudes and catalogued about 1000 stars into 6 groups. The first magnitude group contained the brightest stars, the second magnitude the next bright ones and so on. In this system the faintest visible stars were contained in the sixth magnitude group. A similar magnitude scale was also used in the C. Ptolemy's catalogue in the 2nd century A.D.

In about 1830, William Herschel, by his experiment on stellar photometry determined that a star of first magnitude was about 100 times brighter than a star of sixth magnitude. Later in 1856, N.R. Pogson constructed the quantitative scale on a precise physical basis. He defined the magnitude scale by taking the ratio of brightness of a first magnitude star to that of sixth magnitude one exactly as 100 : 1.

Pogson assumed that equal ratios of brightness would give equal differences in the magnitudes. By this assumption, the ratio of brightness of two stars whose magnitudes differ by unity is the fifth root of 100, i.e. 2.512.

Pogson also calibrated the zero-point of the magnitude scale with reference to the faintest detectable stars, whose apparent brightness could be measured accurately. This roughly corresponds to the brightness of sixth magnitude stars proposed by Hipparchus and Ptolemy. The zero-point of the modern magnitude scale has been fixed by the international polar sequence, which is a group of stars in the vicinity of the north pole whose brightnesses have been measured accurately.

For a mathematical formulation of the magnitude sequence, let us suppose that B_1 and B_6 denote the brightnesses of two stars which are of the first and sixth magnitude respectively. Then, we have $B_1/B_6 = 100$. We also have,

$$\frac{B_1}{B_2} = \frac{B_2}{B_3} = \frac{B_3}{B_4} = \frac{B_4}{B_5} = \frac{B_5}{B_6} = x \text{ (say)},$$

yielding

$$\frac{B_1}{B_6} = x^5 = 100. \text{ Thus, } x = \sqrt[5]{100} = 2.512$$

Therefore, the brightness of two stars whose apparent magnitudes differ by unity will differ by a factor of 2.512.

If B_m and B_n $(n > m)$ be the brightnesses of two stars having magnitudes m and n respectively, then

$$\frac{B_m}{B_n} = (2.512)^{n-m}$$

Therefore,

$$\log \frac{B_m}{B_n} = (n - m) \log (2.512) = 0.4 \, (n - m)$$

yielding

$$\frac{B_m}{B_n} = 10^{0.4(n-m)} \tag{3.1}$$

Equation (3.1) shows that the higher the magnitude of a star, the less is its brightness and vice versa. The apparent magnitude of the brightest star Sirius is –1.4. Among the planets, Venus is the brightest with a magnitude of – 4.4 at its maximum. The apparent magnitude of the Sun is –26.8. This means that the Sun appears to be more than 10 billion times brighter than the brightest star Sirius. The faintest limiting magnitude star which can be photographed with the 200-inch reflector at Mount Palomar observatory is about +23.

3.2 ABSOLUTE MAGNITUDE AND THE DISTANCE MODULUS

The apparent magnitudes of stars depend on their intrinsic brightnesses or luminosities (the total amount of radiant energy emitted per second from the surfaces of stars) and also on their distances from us. Since the distances vary considerably, their apparent magnitudes do not provide their actual brightnesses or luminosities. If, therefore, we know the actual distances of stars, we can compare their luminosities by referring them to any chosen distance from us from where they are assumed to emit their light. For such a comparison of luminosities of stars, astronomers have chosen a standard distance of 10 parsecs (i.e. 32.6 Light years). The magnitude of the star is then called the *absolute magnitude. Thus, the absolute magnitude M of a star has been defined as its apparent magnitude would be if it were placed at a standard distance of 10 parsecs.* The absolute magnitude of a star, therefore, depends on its absolute or intrinsic brightness and not on its actual distance from us.

For two stars, let m be the apparent magnitude of one whose brightness is B_m and distance d. For the second star let M, B_M and D be the corresponding quantities. Then we have

$$\frac{B_m}{B_M} = 10^{0.4(M-m)} \tag{3.2}$$

But the brightness of a star is inversely proportional to the square of its distance. Therefore, we get

$$\frac{B_m}{B_M} = \frac{D^2}{d^2} = 10^{0.4(M-m)} \tag{3.3}$$

which yields

$$m - M = 5\,[\log d - \log D] \tag{3.4}$$

If the distance is measured in parsecs and we take $D = 10$, then by definition, M is the absolute magnitude and we have the relation

$$m - M = 5 \log d - 5 \tag{3.5}$$

This is the fundamental relation between the apparent magnitude m, the absolute magnitude M and the distance d of a star measured in parsecs. We shall see later (in Section 13.9) that another term should be added to the right-hand side of Eq. (3.5) to account for the extinction of starlight by the interstellar grains. The quantity $m - M$ depends only on the distance d and is called the *distance modulus*. This being known, the actual distance of a star can be easily derived from Eq. (3.5).

The absolute magnitude may be either visual (M_v) or photographic (M_{pg}). When not otherwise stated, it is taken to be a visual magnitude. The absolute magnitudes of the normal stars lie between –10 to +20. This range corresponds to a brightness ratio of about 10^{12}. The absolute magnitude of the apparently brightest star Sirius is +1.5; for the Sun it is about +4.8. Among the entire population of stars in our Galaxy, therefore, the Sun figures just as a mediocre member in intrinsic brightness. Stellar, magnitudes are based on the colour or spectral energy distribution of the stars and on the wavelength sensitivity of the detector, which may be either an eye or a photographic plate (or film) or a photoelectric cell.

The retina of our eye is particularly sensitive to the green and yellow light. It has lower sensitivity to other wavelengths, and is absolutely insensitive to ultraviolet and infrared light. On the other hand, basic photographic emulsion is particularly sensitive to the blue and violet light as also to the ultraviolet light. It is less sensitive to other wavelengths. When stars are observed visually, the magnitudes (apparent or absolute) assigned to them according to their brightness are termed *visual magnitudes* (or absolute visual magnitudes) and are usually symbolized by m_v or absolute visual magnitude M_v. A star can be photographed, by using a yellow filter and a green and yellow sensitive photographic emulsion to imitate the spectral sensitivity of the human eye. The magnitudes assigned to stars on the basis of such measurements are termed *photovisual magnitudes m_{pv}* or absolute photovisual magnitude M_{pv}. These magnitudes can also be determined accurately by using a suitable filter and a photocell.

During 1904 to 1908, K. Schwarzschild first used ordinary blue sensitive plates in his experiments on photographic photometry. Since then, experiments have been made with various types of photographic emulsions. When the basic types of blue and violet sensitive emulsions are used to photograph the stellar images, the magnitudes so determined are called *photographic magnitudes* m_{pg} or absolute photographic magnitudes M_{pg}. In the photographic method the magnitude of a star is generally ascertained by observing the size of the stellar image on the photographic plate and comparing it with that of a star of known magnitude. Also, it can be measured directly by observing it with an eyepiece. In both the cases, it is found that greater brightness of stars corresponds to larger star images.

3.3 THE BOLOMETRIC MAGNITUDE

All the stellar magnitudes defined so far cover only some limited regions of the stellar spectrum. The stellar magnitudes based on the radiations measured over the entire range of the electromagnetic spectrum, are called the bolometric magnitudes (m_{bol}), and the absolute bolometric magnitudes (M_{bol}) are accordingly defined. As some of the wavelengths of the electromagnetic radiation from the stars are blocked either partially or completely by the earth's atmosphere, modern observations provide measurements of star's spectrum from orbiting satellites. But, since no detector is sensitive to all wavelengths of the stellar spectrum, to obtain the bolometric magnitudes from any other magnitudes, some corrections need to be applied. In particular, the difference between the photovisual and bolometric magnitudes of a star is called the bolometric correction, BC, that is,

$$BC = m_{bol} - m_{pv} = M_{bol} - M_{pv} \tag{3.6}$$

The BC for the Sun is nearly -0.11 as referred to the bolometric magnitude scale. Since the bolometric magnitudes are always brighter than the photovisual magnitudes, bolometric corrections are always negative and these values gradually increase along both the hotter and cooler sides of the spectral classes of stars. Mathematically, BC is calculated from

$$BC = 2.5 \log \frac{\int P_\lambda F_\lambda d_\lambda}{\int F_\lambda d_\lambda} + \text{constant} \tag{3.7}$$

where P_λ is the sensitivity function of the eye, πF_λ is the flux at the wavelength λ and the constant is determined from the condition that for Sun, $BC = -0.11$. Instead of using P_λ, the sensitivity function of the eye, in modern work the sensitivity function P_λ for the V magnitude system is used. Again, the flux being a function of the effective temperature (T_{eff}) of a star, Eq. (3.7) ensures that BC is a function of T_{eff}. Thus, bolometric magnitudes are deduced theoretically from the observed data. They represent the total luminosity of stars.

3.4 DIFFERENT MAGNITUDE STANDARDS: THE UBV SYSTEM AND SIX-COLOUR PHOTOMETRY

From early days, the apparent magnitudes of stars were being measured on the basis of brightness ratios of stars. But gradually the astronomers realized, through their experience, the necessity of a standard magnitude scale with reference to which stars over all parts of the sky could be compared conveniently. They decided to set up a standard of magnitude sequence by selecting some limited regions of the sky in the vicinity of the north celestial pole. These regions contained groups of stars which were visible from all observatories on the Northern Hemisphere and were known as the Mount Wilson North Polar Sequence of F.H. Seares. For a long time, this group served as the primary standard for the measurement of the photovisual and photographic magnitudes.

The magnitudes of stars in the small regions of the sky were determined accurately and their brightnesses were assigned in small steps in order to avoid the atmospheric absorption or reddening. These magnitudes were determined with reference to the earlier convention of assigning them by means of brightness ratios.

With the advent of the photoelectric photometry, newer sequences of magnitude standards (the Mount Wilson and Palomar sequences) have been set up on the basis of precise measurements of the stellar magnitudes by modern photoelectric methods. Among the various photometric techniques which have been developed during the last few decades, the U, B, V magnitude system has been the most advantageous. and useful. This standard of magnitude sequence was established photoelectrically by H.L. Johnson and W.W. Morgan in 1953. This system refers to the measurement of starlight in the ultraviolet (U), blue or photographic (B), and green-yellow or visual (V) regions of the spectrum. The UBV magnitude scale is so fixed that for a star of spectral class A0 V, U = B = V. With the development of the photoelectric photometry, the U, B, V system has become very popular. This system has later been extended to include various other spectral regions for studying different types of celestial objects like faint stars, nebulae, star clusters etc. The three-colour magnitude system has the unique advantage for the study of the interstellar extinction and reddening, the knowledge of which is a prerequisite for studying the problems of galactic structure.

The stellar magnitude measurement in different colours actually depicts its energy distribution at various wavelength ranges through wide band-pass filters. The centres of the bands for the U, B, V magnitudes are respectively at about λ 3500 Å, λ 4300 Å and λ 5500 Å. The method therefore has one great advantage that it avoids the region of the Balmer jump around the Balmer series limit at λ 3650 Å. The centre of the U band pass is much shortward while that of the B band-pass is much longward of the region of the Balmer jump. The three-colour photometry thus eliminates the great uncertainty produced in the energy distribution curve near the Balmer series limit, particularly in cases of hot stars in which the attenuation of energy near the series limit is very great.

Another very useful photometric magnitude system is the six-colour photometry introduced by J. Stebbins and A.E. Whitford. This system covers the wavelength range 3300 Å to 12,500 Å. The effective wavelengths of the filters used in this system are λ 3530, λ 4220, λ 4880, λ 5700, λ 7190 and λ 10,300, all measured in angstrom unit. Many advantages are derived from such a broad-based wavelength range. This has been particularly useful to study the nature of the grains that are sprinkled throughout the interstellar space. The principal disadvantage of the system is however that it consumes large amount of telescope time.

3.5   RADIOMETEIC MAGNITUDES

As the photoelectric cells and photographic plates are insensitive to infra-red radiation mainly from the cool stars, thermocouples are used for the measurement of such radiation. The corresponding magnitudes are called radiometric magnitudes. Photoconductive cells and bolometers can also be used in this region.

The difference between the photovisual and radiometric magnitudes of a star is called the heat index, that is,

$$\text{Heat index} = m_{pv} - m_{\text{rad}}$$

$$= M_{pv} - M_{\text{rad}}$$

In the radiometric magnitude scale the heat index is zero for AO stars. Its value is large and positive for cool stars.

3.6 THE COLOUR-INDEX OF A STAR

The intensity of stellar radiation is different in different wavelength regions and higher intensity shifts towards the shorter wavelengths for higher temperatures, in accordance with Planck's formula of the blackbody radiation. Therefore, stars of higher temperature will emit much more light of shorter wavelengths (blue-violet) than those of lower temperatures which will emit more light in longer wavelengths (red). Since the colour of light is dependent on its wavelength, stars of different temperatures will appear to be of different colours. Again, since the photographic plates are more sensitive to blue light than yellow or green, the photographic and photovisual magnitudes of the same star are not the same. The difference between the photographic and photovisual magnitudes of a star is known as the Colour Index, C.I. of the star, or,

$$m_{pg} - m_{pv} = \text{C.I.} \tag{3.8}$$

Since the light of all wavelengths or colour vary inversely as the square of the distances from sources, the C.I. of a star does not vary if the distance of the star varies. Thus, we can have the relation

$$M_{pg} - M_{pv} = m_{pg} - m_{pv} = \text{C.I.} \tag{3.9}$$

where M_{pg} and M_{pv} represent the absolute photographic and absolute photovisual magnitudes of a star respectively.

The U, V, B system provides two colour indices of stars, viz. B–V and U–B. The B–V colour index roughly approximates the usual C.I. defined in Eq. (3.8) and its value varies with the colours of stars observed. A hot star radiates more in blue and violet light than in the yellow or red. Therefore, the blue magnitude (B) of a hot star will be brighter than its yellow or visual magnitude (V). The colour index B–V is therefore negative for a blue or a blue-white star that corresponds to a high temperature. On the other hand, a cooler star radiates more energy in yellow or red light than in the blue or violet and will have brighter magnitude in yellow (V) than in the blue (B). For a cooler star therefore, the quantity B–V will be positive. The same arguments are applicable to the colour index U–B. A hot star will have negative value of U–B while this value will be positive for a cooler star. The scales are so defined that for a A0 V star, both U–B and B–V are zero, which follows from the definition already mentioned that for a A0 V star, U = B = V.

Colour indices provide a numerical measure of colour or the spectral-energy distribution of the star which depends on its temperature. Thus if the photographic and photovisual magnitudes of a star are known, one can determine its colour or temperature. Colour indices of stars may have values ranging between $- 0.4$ for the bluest stars to more than $+ 2.0$ for the reddest ones. In practice, values of $+ 5.0$ or more have been found for a few red stars. For Sun, the value of C.I. is $+ 0.53$.

In determining the colour indices of stars, the space between the star and the observer is assumed to be transparent. But in fact, the interstellar space is not void but contains some microscopic particles. These particles dim the light from the star on its way to the observer. This dimming is due to absorption of light by the particles. Also, since blue light is absorbed more than the red light, therefore, the stars appear redder than if they were observed through a completely transparent medium. The amount of dimming or reddening of a starlight is proportional to the distance of the star from the observer and can be determined numerically by knowing the spectral type of the star and comparing the measured value of the C.I. with that which would be for an unreddened star of that particular spectral class.

3.7 LUMINOSITIES OF STARS

The luminosity of a star is a measure of the total amount of energy that is radiated from its surface in one second. It is usually expressed in terms of the luminosity of the Sun and is measured by the absolute bolometric magnitude of the star. Thus, if M_{bol} and L are respectively the absolute bolometric magnitude and the luminosity of the star and $M_{bol\odot}$ and $L_{\odot}$ are corresponding quantities for the Sun, then the following relation holds true:

$$M_{bol} - M_{bol\odot} = 2.5 \log \frac{L_{\odot}}{L} \qquad (3.10)$$

The bolometric magnitude of the Sun is $M_{bol\odot;} = + 4.72$ and its luminosity is $L_{\odot} = 3.84 \times 10^{33}$ erg s^{-1}. Thus, L can be calculated when M_{bol} for the star is determined from its apparent photovisual magnitude.

From the value of $M_{bol\odot}$, it is observed that the Sun is a star whose absolute magnitude is a little brighter than the fifth. The other stars in our Galaxy have a range of luminosities from nearly 10^{-6} to 10^{+6} relative to that of the Sun. This means that the brightest blue stars are about 10^{12} times brighter than the faintest red stars.

The total luminosity L or the bolometric magnitude of a star can also be expressed in terms of the effective temperature T_{eff} of the star by the relation

$$L = 4\pi r^2 \sigma T_{eff}^4 \qquad (3.11)$$

where $4\pi r^2$ represents the surface area of the star of radius r, and σ is the Stefen-Boltzmann constant. The above formula gives the total amount of radiant energy passing perpendicular to the entire surface of the star in one second. Here the effective temperature T_{eff} represents the average value for the entire range of the electromagnetic spectrum.

3.8 STELLAR PARALLAX (TRIGONOMETRIC) AND THE UNITS OF STELLAR DISTANCES

Stellar parallax is a measure of the star's distance. The diameter of the Earth's orbit around the Sun is used as a basic line of such measurements and the corresponding parallax is called the trigonometric parallax. Thus, the trigonometric parallax of a star may be defined as the angle p (in seconds of arc) subtended at the star σ by the mean radius 'a' of the Earth's orbit round the Sun. This is also called heliocentric or annual parallax. A nearby star σ is photographed from the position E of the Earth against the background of the far off stars. Similar observations are also made after six months, from the position E' of the Earth on the other end of the base line. From these observations, the angle subtended by the diameter EE' (or base line) at the star σ is measured, comparing the change in the position of σ during the six months with respect to the fainter far off stars whose shift in position due to the motion of the Earth is negligible. One-half of this angle gives the measure of annual parallax p of the star. The sets of observations at E and E' are repeated when the Earth comes back to this position after a year and the average of a number of measurements is taken as the parallax of the star. As the stars are too far away, the parallax of a star is a very small angle and its value has been found to be always less than $1''$ of arc even for the nearest stars.

From Fig. 3.1, we have

$$p_{rad} = \frac{a}{d}, \text{ where } d = \text{distance of the star.}$$

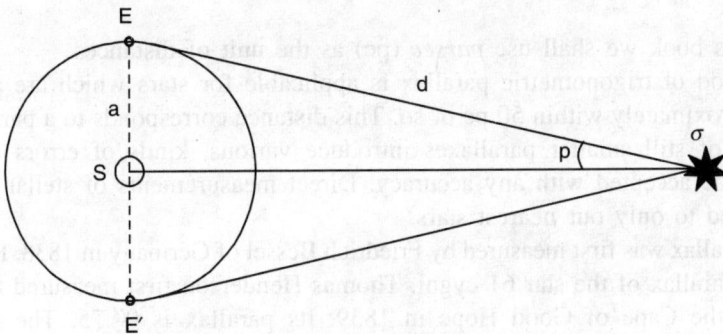

FIGURE 3.1 The annual parallax of a star.

But

$$1_{rad} = 206,265'' \text{ of arc.}$$

Therefore,

$$\frac{p''}{206,265} = \frac{a}{d},$$

since the parallax of a star is expressed in seconds of arc, or

$$d = \frac{206,265\, a}{p''} \tag{3.12}$$

The radius 'a' of the Earth's orbit is called the *astronomical unit* (a.u.) and its value is approximately equal to 150×10^6 km. This is the simplest unit of astronomical distances. If in Eq. (3.12) we take the numerator, i.e., 206,265 a as the unit of distance, then in this unit

$$d = \frac{1}{p''} \qquad (3.13)$$

This unit of distance is called a *parsec* (*pc*) and its value in cgs unit is about 3×10^{18} cm. If now in Eq. (3.13) we put $p = 1$, we get $d = 1$. *Thus, a parsec is a distance at which the radius of the Earth's orbit (1 a.u.) subtends an angle of one second of arc.* In other words, parsec is a distance at which the star must be situated in order to exhibit 1 second of parallax, using 1 a.u. as base line. The relation (Eq. 3.13) further shows that the stellar parallaxes are inversely proportional to their distances. Thus, we have

$$1 \text{ parsec} = 206,265 \text{ a.u.} \approx 3.26 \text{ light years} \approx 3 \times 10^{18} \text{ cm} \qquad (3.14)$$

(To have a proper idea of this distance, the reader may note that if he starts in a vehicle moving at the rate of 1 km s^{-1}, he will take one million years to travel one parsec.)

The relation between a parsec and a light year shows that if distances are measured in light years, Eq, (3.13) will change to

$$d = \frac{3.26}{p''} \qquad (3.15)$$

Throughout this book we shall use *parsec* (pc) as the unit of distances.

The method of trigonometric parallax is applicable for stars which are astronomically nearer, i.e. approximately within 50 pc or so. This distance corresponds to a parallax of 0''.02. Measurements of still smaller parallaxes introduce various, kinds of errors and thus such results cannot be accepted with any accuracy. Direct measurements of stellar parallaxes are therefore limited to only our nearest stars.

Stellar parallax was first measured by Friedrich Bessel of Germany in 1838. He successfully measured the parallax of the star 61 cygni. Thomas Henderson first measured the parallax of α Centauri at the Cape of Good Hope in 1839. Its parallax is 0''.75. The nearest star is Proxima Centauri and has got the largest known parallax of 0''.785 with a distance of about 1.3 pc (or 4.24 light year). The parallax of the brightest star, Sirius, is about 0''.379 and its distance is about 2.67 pc (or 8.7 light year).

3.9 STELLAR POSITIONS: THE CELESTIAL COORDINATES

An observer at any place on the Earth apparently sees all the bright heavenly objects in the sky to be fixed on the inner surface of a sphere whose centre is at the observer's eye. This apparent sphere is called the *celestial sphere,* the apparent dome on which the stars and other bright heavenly bodies *appear to lie,* just as the artificial stars lie on the dome of a planetarium.

If through an observer at any point of the Earth a straight line is drawn in the direction in which the gravity acts and extended at both ends, it will meet the celestial sphere at two point, one vertically above and the other vertically below the observer. The former is called the *Zenith* and the latter the *Nadir* (Z & Z′ in Fig. 3.2). The plane through the observer perpendicular to the line joining the Zenith and Nadir cuts the celestial sphere in a great circle. (Every plane section of a sphere is a circle. A plane section passing through the centre of the sphere is called a great circle, other sections are called small circles). This great circle is called the *celestial horizon* (SENW in Fig. 3.2).

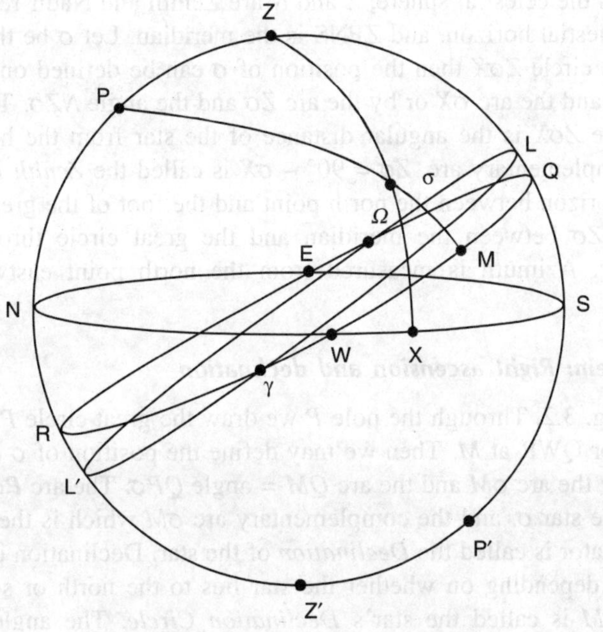

FIGURE 3.2 The celestial sphere.

If a line is drawn through the observer parallel to the Earth's rotational axis, it will meet the celestial sphere in two points, P and P′. The point P lying above the horizon is called the *North Celestial Pole* or the *North Pole* and the point P′ lying below the horizon is called the *South Celestial Pole* or the *South Pole*. The great circle having P and P′ as its poles is called the *Celestial Equator*. This is actually the great circle in which the plane of the Earth's equator meets the celestial sphere. In Fig. 3.2 QERW is the celestial equator. *Meridian* is the great circle (ZPZ′P′) passing through the Zeneith, Nadir and the Celestial Poles.

Due to the annual motion of the Earth around the Sun, the latter appears to move relative to the stars and returns to the same apparent position in the heavens relative to the stars in the course of a year, as the Earth completes its revolution. This relative path of the Sun on the celestial sphere, traces out the great circle LγL′Q which is inclined to the celestial equator at an angle of about 23°27′. The great circle LγL′Q, is called the *Ecliptic*. The ecliptic intersects the equator at two points called the *Equinoctial Points*. The point through which the Sun crosses the equator from south to north is called the *First Point of Aries* and is denoted by the symbol γ. The other point through which the Sun crosses the equator from north to south is called the *First Point of Libra* and is denoted by the symbol Ω.

Just as the position of a place on the Earth is defined by the two coordinates, latitude and longitude, referred to the *Prime Meridian* and *Earth's Equator,* similarly a position of a heavenly object can be defined by referring to its celestial coordinate systems. In the following, we shall briefly discuss three such coordinate systems.

The Horizontal System: Altitude (or Zenith Distance) and Azimuth

Figure 3.2 represents the celestial sphere, Z and N are Zenith and Nadir respectively, the great circle NXS is the celestial horizon, and ZPNS is the meridian. Let σ be the position of a star. If we draw the great circle ZσX then the position of σ can be defined on the celestial sphere either by the arc *NX* and the arc σX or by the arc $Z\sigma$ and the angle NZσ. The arc σX measured along the great circle ZσX is the angular distance of the star from the horizon and is called its *Altitude*. The complementary arc, $Z\sigma = 90° - \sigma X$ is called the *Zenith Distance* of the star. The arc *NX* of the horizon between the north point and the foot of the great circle through the star or the angle NZσ between the meridian and the great circle through cris called the Azimuth of the star. Azimuth is measured from the north point eastwards or westwards through 0° to 180°.

The Equatorial System: Right ascension and declination

We again refer to Fig. 3.2. Through the pole P we draw the great circle PσM through the star σ to meet the equator QWR at M. Then we may define the position of σ either by the arc Pσ and angle QPσ or by the arc σM and the arc QM = angle QPσ. The arc Pσ is called the *North Polar Distance* of the star σ, and the complementary arc σM which is the angular distance of the star from the equator is called the *Declination* of the star. Declination of a star is measured positive or negative depending on whether the star lies to the north or south of the equator. The great circle PσM is called the star's *Declination Circle.* The angle QPσ between the meridian and the star's declination circle is called the Hour Angle (H.A.) of the star. The hour angle of a star is measured from the meridian towards the west through 0° to 360°. The *Right Ascension* (R.A.) is the arc of the equator γM between the First Point of Aries and the foot of the declination circle. Since the First Point of Aries shares common diurnal motion as the stars, both right ascension and declination of a star remain unchanged during the diurnal motion.

The Ecliptic System: Celestial latitude and longitude

If a great circle is drawn through the pole of the ecliptic and the star σ (not shown in Fig. 3.2) then the angular distance of σ from the ecliptic measured along this great circle is called the *Celestial Latitude* of the star σ. The arc of the ecliptic intercepted between the First Point of Aries and the foot of the above great circle measures the Celestial Longitude of σ. Like right ascension and declination, the celestial latitude and longitude of a star also remain unchanged during its diurnal motion. While the latitude is measured positive or negative depending on whether the star is to the north or south of the ecliptic, the longitude is measured eastwards from 0° to 360°.

We shall discuss galactic coordinate in Section 16.2.

3.10 STELLAR MOTIONS

Apparently stars seem to be fixed in space. But this is not really the case as they move in space with respect to one another as well as with respect to the Sun. Since the distances of stars are much larger as compared to those of the planets, the stars seem to be stationary with respect to the motion of the planets. Actually the stars are in motion through space with almost equal velocities as those of the planets.

As the stars are at great distances, their relative change of position in the sky is quite small in an interval of one year. The change is noticeable only after a lapse of several years. For this reason stellar motions are observed at an interval of 20 to 50 years and this is measured with respect to the motion of the Sun. The motion of a star through space may be considered to be composed of two components, one along the line of sight of the observer and the other perpendicular to the line of sight. The first component, called the *radial motion, V_r,* determines the speed of recession or approach of the star with respect to the Sun. This is determined by measuring the 'Doppler shift' of the lines in the stellar spectra by the formula $\dfrac{\Delta\lambda}{\lambda} = \dfrac{V_r}{c}$, where $\Delta\lambda$ represents the amount of Doppler shift in the spectral line at wavelength λ and c the velocity of light. As the radial velocity of a star is the relative motion between the star and the Sun, necessary corrections for the Earth's motion must be applied. This consists of the corrections due to both the orbital as well as the rotational motion of the Earth. The radial velocity is taken as positive if the star recedes from the Sun, and it is negative when the star approaches it. The velocity is usually expressed in kilometers per second.

The component of the space motion of a star, which is perpendicular to the line of sight is called the *transverse velocity component* or the *tangential velocity component, V_t.* The space velocity, V of a star which is the resultant of the radial velocity (V_r) and the tangential velocity (V_t) is represented in Fig. 3.3.

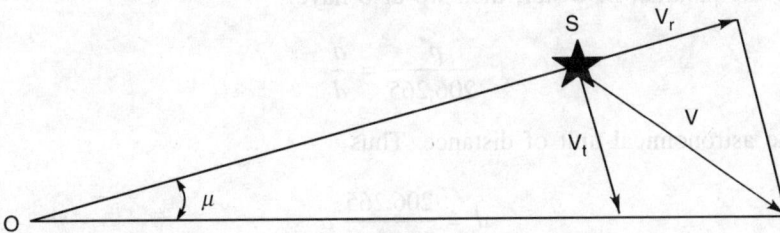

FIGURE 3.3 The diagram showing the different components of the motion of star. The observer is at O and the star is at S.

Thus, V represents the total velocity of a star in kilometers per second and is given by

$$V^2 = V_r^2 + V_t^2 \tag{3.16}$$

The transverse velocity V_t represents the star's movement across the line of sight, i.e. across the direction of V_r. As a result of such a motion, the component V_t describes an angle with the direction of total velocity V. If μ be the angle described at the observer O in one year, then μ is defined as the *proper motion* of the star and is measured in seconds of arc per year. This motion describes the angular rate of displacement in the position as well as in the direction of the star in the sky. Like the radial and the tangential velocities, the proper motion of a star is also considered as observed from the Sun and hence the effect due to Earth's motion must be taken into account. Proper motions of stars are in general very small and range from zero to a few seconds of arc. The largest known proper motion is that of Barnard's star (a star of tenth magnitude) amounting to $10''.25$ per year. Large proper motions indicate the high velocity and relative nearness of the star with respect to the Sun so that the parallax can be conveniently measured. Proper motions of the stars in general, are larger than their parallaxes. The mean for all stars which are visible to the naked-eye amounts to about $0''.1$.

To calculate the space velocity or total velocity V of a star, we must first know the radial velocity V_r and the transverse velocity V_t. V_r can be calculated from the Doppler formula by measuring the Doppler shift of lines. To calculate V_t, we require the proper motion μ and the parallax p. The former may be known from the value of $N\mu$ taken over a period of N years. The latter can be calculated by applying a suitable method for determining the parallax of a star.

If V_t be the transverse velocity in kms^{-1} of a star at a distance d km and n be the number of seconds in a year, then from Fig. 3.3, we get

$$\frac{\mu''}{206,265} = \frac{nV_t}{d}$$

Therefore,

$$V_t = \frac{d\mu''}{206,265} \cdot \frac{1}{n} \qquad (3.17)$$

If p'' be the parallax of a star, then we also have

$$\frac{p''}{206,265} = \frac{a}{d}$$

where a is the astronomical unit of distance. Thus

$$d = \frac{206,265}{p''} a$$

Therefore,

$$V_t = \frac{206,265}{p''} a \frac{\mu''}{206,265} \frac{1}{n} = \frac{\mu''}{p''} \frac{a}{n} \qquad (3.18)$$

Using $a \simeq 1.49 \times 10^8$ km, and $n \simeq 3.16 \times 10^7$ seconds, we get

$$V_t = 4.74 \frac{\mu''}{p''} \text{ km s}^{-1} \qquad (3.19)$$

From Eq. (3.19) the transverse velocity V_t can be calculated, if the parallax p'' and proper motion μ'' of a star are known. It may be noted that the radial velocity V_r depends only on the shift $\Delta\lambda$ of a line of natural wavelength λ and it is independent of the distance of the star.

Note: In order to determine the proper motion, a star is observed and photographed at an interval of about two decades or more. The photographs are taken against the background of more remote stars or galaxies which remain practically stationary over the observed time periods in the region of the location of the star. From a comparative study of the star images on the two photographic plates, the change in positions of the star during the observed period can be calculated. These observations can however be made only for those stars which are comparatively nearer.

3.11   THE SOLAR MOTION AND THE PECULIAR VELOCITIES OF STARS

In our previous section, we have defined the radial velocity, the transverse velocity and the proper motion of stars, all with respect to the Sun. But the Sun itself is a star and moves like other stars through space in out Galaxy. We should, therefore, consider the Sun's motion as a whole and define a suitable reference system with respect to which we could define the stellar motions. From these considerations astronomers have defined the *Local Standard of Rest* (LSR). This is a hypothetical origin of a reference system with respect to which the motions of all stars in some neighbourhood of the Sun, say, within 50 or 100 pc average out to zero. The local standard of rest with respect to a group of stars may therefore be considered as the centroid of motion of this group of stars. This point in the local region of the galaxy is not infact at rest, but is also moving with the nearby stars about the galactic centre. This motion need not be considered in defining the Sun's motion, as such motions average out to be the same for all the members of the system. The motion of an individual star with respect to the LSR is called its *peculiar motion* and that of the Sun with respect to the LSR is called the *local solar motion* or the Sun's *peculiar motion*.

Let the components of the peculiar motion of the star be u, v, w and those of the Sun be $u_\odot$, $v_\odot$, $w_\odot$ respectively. Then the observed motion of the star which is a combination of its own peculiar motion and that of the Sun, is given by the components $\dot{x}, \dot{y}, \dot{z}$ where

$$u = \dot{x} + u_\odot, \quad v = \dot{y} + v_\odot, \quad w = \dot{z} + w_\odot. \tag{3.20}$$

From the definition of the LSR, we have

$$\frac{\Sigma u}{n} = \frac{\Sigma v}{n} = \frac{\Sigma w}{n} = 0$$

where n is the number of stars considered.

Therefore,

$$\Sigma u = \Sigma v = \Sigma w = 0 \tag{3.21}$$

Hence from Eq. (3.20), summing up for n stars, we obtain

$$u_\odot = \frac{\Sigma \dot{x}}{n} = -\dot{\bar{x}} \quad v_\odot = \frac{\Sigma \dot{y}}{n} = -\dot{\bar{y}} \quad w_\odot = \frac{\Sigma \dot{z}}{n} = -\dot{\bar{z}} \tag{3.22}$$

Thus, we find that the components of the local solar motion with respect to a group of stars are equal in magnitude but opposite in sign to the average velocity components ($\bar{\dot{x}}, \bar{\dot{y}}, \bar{\dot{z}}$) of the group of stars measured relative to the Sun. Since the stars of different spectral classes move with different velocities in space, the value of the local solar motion differs when we consider the average observed motions of the stars of different spectral classes.

To define clearly the Sun's peculiar motion, we must consider the motions of all stars in the solar neighbourhood. If we analyze the observed radial velocities of stars with respect to the Sun in different directions over the celestial sphere, we find that the radial velocities of stars in the constellation Hercules average to -20 km s^{-1} in the direction of solar apex which is the direction of Sun's motion in the sky. The Sun is thus moving towards Hercules at the rate of 20 km s^{-1}. Similarly, the stars in the opposite direction in the constellation Columba seem to be receding from the Sun at a speed of $+20$ km s^{-1} in the direction of the solar antapex which is the direction opposite to that of the solar apex. The average radial velocities of stars in the direction perpendicular to the Sun's motion turn out to be zero. This situation has been illustrated by means of a schematic diagram (Fig. 3.4) which contains the cross-section of the celestial sphere.

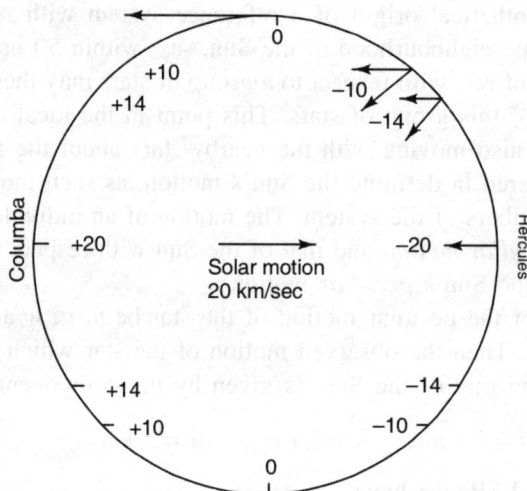

FIGURE 3.4 The radial velocity diagram of the solar motion.

Again, the analysis of the proper motions of these groups of stars in the solar neighbourhood also suggest a relative motion of the Sun towards Hercules. The average proper motions of the group of stars as seen from the Sun seem to be moving outward from Hercules and circling round the sky to meet at Columba.

The arrows in Fig. 3.5 represent the average proper motions of the group of stars all over the celestial sphere. The maximum measures of proper motions are observed in a direction perpendicular to Sun's motion and the minimum along the solar apex and the antapex. Figures 3.4 and 3.5 demonstrate that the Sun is moving with respect to all stars in its neighbourhood towards the direction of the constellation Hercules with a velocity of about 20 km s^{-1}. The

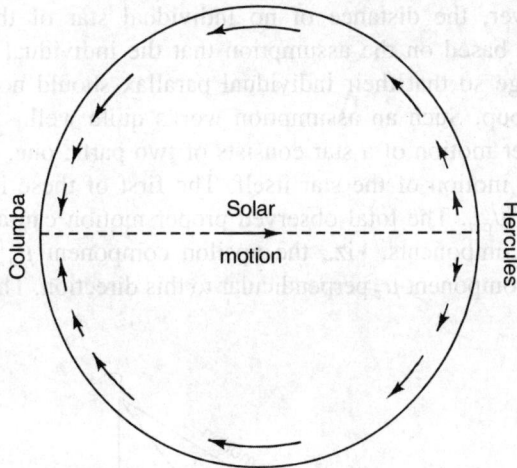

FIGURE 3.5 Average proper motion of stars due to the local solar motion towards Hercules.

analysis of solar motion is very important from the point of view of the galactic dynamics. In fact, the solar motion with respect to any group of stars can be rigourously derived by using their radial velocities or proper motions or even the space motions.

3.12 THE VELOCITY DISPERSION

When observation is carried out on the motion of a group of stars, it is found that the velocity components of the group of stars relative to the LSR are randomly distributed about the velocity components of the LSR itself. Each velocity component of the stellar group follows gaussian distribution and the parameters σ_1, σ_2 and σ_3 of the distribution in three mutually perpendicular directions provide a measure of the dispersion of velocities along those directions. It can be easily shown that the root-mean-square random velocities in the assumed directions are given by

$$\langle u^2 \rangle^{\frac{1}{2}} = \frac{\sigma_1}{\sqrt{2}}, \qquad \langle v^2 \rangle^{\frac{1}{2}} = \frac{\sigma_2}{\sqrt{2}}, \qquad \langle w^2 \rangle^{\frac{1}{2}} = \frac{\sigma_3}{\sqrt{2}} \tag{3.23}$$

These root-mean-square velocities which actually measure the deviations from the motion of the LSR of any group of stars are known as the *velocity dispersions* in the respective directions of the group of stars considered. The velocity dispersion components are different for groups of stars of different spectral types.

3.13 STATISTICAL PARALLAX

Proper motions of a group of stars of the same class and of a narrow magnitude range, occupying a small region of the sky can yield the mean distance of this group, as a whole.

By this analysis, however, the distance of no individual star of the group considered is obtained. The method is based on the assumption that the individual members all belong to a narrow magnitude range so that their individual parallax should not differ much from the mean parallax of the group. Such an assumption works quite well.

The observed proper motion of a star consists of two parts: one, due to the solar motion and the other due to the motion of the star itself. The first of these is called the *parallactic motion*. We denote it by μ_{par}. The total observed proper motion can also be broken into two mutually perpendicular components, viz., the upsilon component μ_v in the direction of the solar apex and the "tau" component μ_τ perpendicular to this direction. These various components are shown in Fig. 3.6,

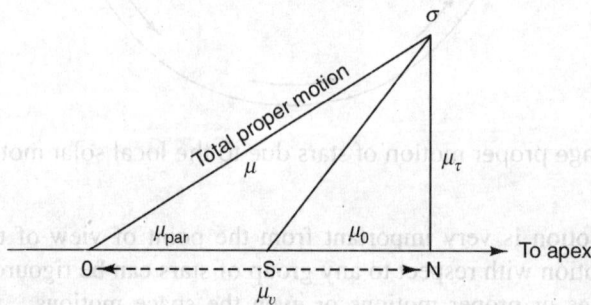

FIGURE 3.6   The diagram showing various parts of the total proper motion.

where

   $O\sigma = \mu$ = total proper motion,

   $ON = \mu_v$ = upsilon component,

   $OS = \mu_{\text{par}}$ = parallactic motion,

   $SN = \mu_0$ = part of the upsilon component due to the actual motion of the star.

Thus,

$$\mu_v = \mu_{\text{par}} + \mu_0 \qquad (3.24)$$

for any star. If a sufficiently large number N of stars are observed, then summing up for N stars,

$$\Sigma\mu_v = \Sigma\mu_{\text{par}} + \Sigma\mu_0 = \Sigma\mu_{\text{par}} \qquad (3.25)$$

since $\Sigma\mu_0$ should be zero for the group.

If $S_\odot$ km s^{-1} be the solar motion towards apex and p'' be the parallax of a star at an angular distance X from the solar apex, then we have, as in Fig. 3.7,

$$(\mu_{\text{par}})_{\text{rad}} = \frac{nS_\odot}{d} \sin \lambda \qquad (3.26)$$

where n is the number of seconds per year and d is the distance of the star in parsec. But

$$p_{\text{rad}} = \frac{a}{d}, \text{ therefore } \frac{1}{d} = \frac{p_{\text{rad}}}{a}$$

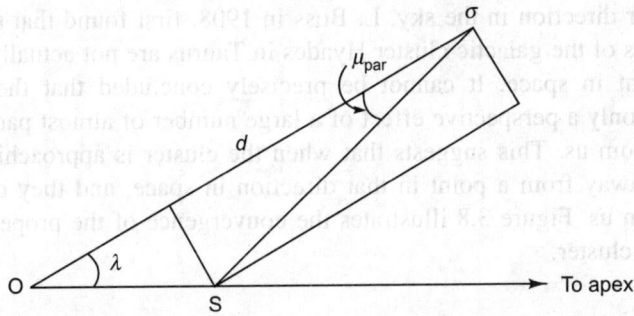

FIGURE 3.7 The diagram showing relation between the solar motion, the parallactic motion of a star and its distance.

By Eq. (3.26)

$$(\mu_{par})_{rad} = n S_{\odot} \sin \lambda \; \frac{p_{rad}}{a}$$

and

$$(\mu_{par})'' = S_{\odot} \sin \lambda \; \frac{n}{a} \, p'' = S_{\odot} \; \frac{\sin \lambda}{4.74} \, p'' \tag{3.27}$$

Summing for the N stars, we get

$$\Sigma \mu_v'' = \Sigma \mu_{par}'' = \frac{S_{\odot} \sin \lambda}{4.74} \Sigma p'' \approx \frac{S_{\odot} \sin \lambda}{4.74} N \overline{p}''$$

where $\overline{p}''$ is the average parallax of the group of N stars measured in seconds of arc. Thus,

$$\overline{p}'' = \frac{4.74}{S_{\odot} \sin \lambda} \cdot \frac{\Sigma \mu_v''}{N} \tag{3.28}$$

$\Sigma \mu_v''$ can be measured for the group of stars considered, and the other quantities on the right-hand side of Eq. (3.28) are known. Thus, $\overline{p}''$, the mean or *statistical parallax* for the group of stars, can be calculated. Statistical parallax is very useful in supplementing trigonometric parallaxes, particularly for the stars at distances greater than 50 pc. It has been used to determine the distances of far off bright stars. In particular, the distances of the cepheid variables and the long-period variables both of which are supergiants and usually at great distances, have been calculated by using the method of statistical parallax. The method has also been used to determine the distances of compact groups or clusters of stars.

3.14 MOVING CLUSTER PARALLAX

Generally, the stars in a cluster all have a common motion in space and this motion defines the motion of the star cluster as a whole. The radial velocities of the member stars of a moving cluster is found to be almost similar, and most of their proper motions appear to converge towards a particular direction in the sky. L. Boss in 1908, first found that the proper motions of the group of stars of the galactic cluster Hyades in Taurus are not actually parallel, but they converge to a point in space. It cannot be precisely concluded that the motions actually converge as this is only a perspective effect of a large number of almost parallel motion either towards or away from us. This suggests that when the cluster is approaching us the motions appear to diverge away from a point in that direction in space, and they converge when the cluster recedes from us. Figure 3.8 illustrates the convergence of the proper motions of some members of a star cluster.

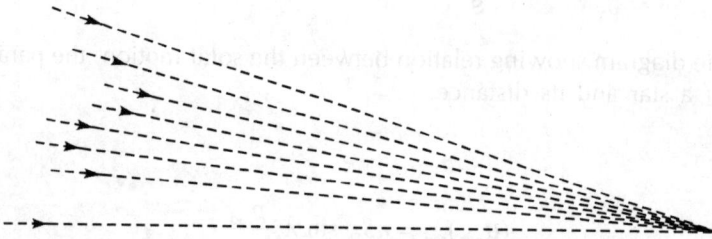

FIGURE 3.8 Convergence of proper motions of a group of stars.

The point towards which the motions appear to converge is called the *convergent point* of the particular group of stars. When the stars of the cluster *are sufficiently spread over the sky,* their proper motions can be conveniently analyzed and so the convergent point can be defined accurately. For a compact cluster it becomes difficult to ascertain the convergent point accurately. Once the convergent point is known, the distance of the cluster can be calculated.

We proceed to calculate the parallax of a cluster by observing the space motions of its member stars relative to the Sun.

Let us consider a star S in the cluster moving in the direction SA relative to the Sun and parallel to the direction of the convergent point, as viewed from the Sun. Let θ be angular distance of S from the convergent point (Fig. 3.9). SB is the radial velocity of S in the line of sight from the Sun, and this can be measured. Also θ can be observed. Then space velocity SA = SB sec θ, and the tangential velocity SC = BA = SB tan θ can be calculated. But

$$BA = SC = 4.74 \frac{\mu''}{p''} \text{ km s}^{-1} \qquad (3.29)$$

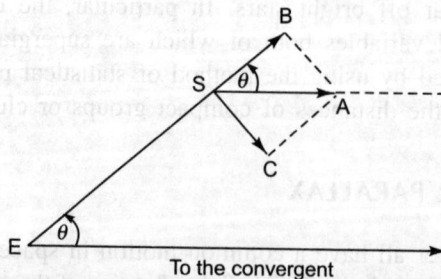

FIGURE 3.9 The relation between the radial velocity, proper motion and the convergent of a moving cluster.

Now μ'' can be calculated from the observed proper motions, so the parallax p'' of S, i.e. of the cluster may be known from the formula

$$p'' = \frac{4.74\,\mu''}{SC} = \frac{4.74\,\mu''}{SB\,\tan\theta} \qquad (3.30)$$

since all quantities on the right-hand side are now known.

The parallax of Hyades in Taurus has been calculated to be $0''.025$ and hence, the distance of this nearest cluster is about 40 pc. At this distance or at still greater distances, the moving cluster method of parallax determination is much more accurate than the trigonometric method.

This independent evaluation of the distance of the Hyades cluster has great astrophysical importance. It has practically served as the key to the calibration of the distance scale, thereby enabling us to standardise the absolute magnitude scale of the stars of different spectral classes. This standardisation essentially forms the basis for the calibration of the Hertzsprung-Russel diagram, as we shall find in Chapter 4.

EXERCISES

1. Define parsec; how is it related to astronomical unit (a.u.).

2. The parallax of our nearest star Proxima Centauri is $0''.785$. Find its distance in parsecs, light years, astronomical units, miles and kilometres.

3. Calculate the time taken by light to travel to the Earth from the following stars:
   (a) α Centauri with $p = 0''.75$;
   (b) Proxima Centauri with $p = 0''.785$;
   (c) 61 Cygni with $p = 0''.3$;
   (d) Vega with $p = 0''.123$.

4. A visual binary star has a parallax of $0''.025$, and the angular distance between the component stars is $2''.5$. Calculate the linear separation between the two members of the binary.

5. Define *apparent* and *absolute* magnitudes of a star and establish a relationship between them.

6. Define *luminosity* of a star. Derive the relationship between the luminosity and the absolute magnitude of a star.

7. The apparent magnitude of a star is observed to be $+3.3$ and its parallax is $0''.025$. Find the absolute magnitude of the star. Compare the luminosity of this star with that of the Sun ($M_{v\odot} = +5.0$).

8. The luminosity of a star of apparent magnitude $+3.0$ is 100 times that of the Sun ($M_{v\odot} = +5.0$), and its proper motion is $0''.05$. Find the tangential velocity of the star. If the star's distance doubles in one million years, find the space velocity of the star.

9. Define bolometric magnitude and bolometric correction. Why is bolometric correction negative for every star?

10. The luminosity of a star is 1.005×10^{40} erg s^{-1}, its distance is 1 kpc, and its bolometric correction is $-3^m.2$. Find m_{pv} and m_{pg} for the star, if its colour index is $-0^m.6$.

11. If the absolute magnitude of the Sun is +5.0, find its apparent magnitude.

12. Prove that the altitude of a heavenly body is greatest when on the meridian.

13. Show that the altitude of the pole is equal to the latitude of the place of the observer.

14. What is the latitude of a place for which the celestial equator coincides with the horizon?

15. Represent in one diagram, the horizon, the celestial equator, the ecliptic, and the latitude, longitude, declination, right ascension, hour angle, zenith distance and azimuth of a star.

16. Explain: A massive star has a much shorter lifespan than the Sun.

SUGGESTED READING

1. De, K.K., *A Text Book of Astronomy,* Book Syndicate, 1964.

2. Green, Robin, *Spherical Astronomy,* Cambridge University Press, Cambridge, 1985.

3. Mc Laughlin, Dean, B., *Introduction to Astronomy,* Houghton Mifflin, Boston, 1961.

4. Shu, Frank, H., *The Physical Universe,* University Science Books, California, 1982.

5. Wyatt, Stanley, P., *Principles of Astronomy,* Allyn and Bacon, Inc., Boston, 1977.

6. Zeilik, Michael and Gaustad, John, *Astronomy,* Harper & Row, New York, 1983.

Spectral Classification
of Stars

4.1 INTRODUCTION

Stellar spectral studies actually began with the discovery of many dark absorption lines of different elements in the spectrum of the Sun by the German physicist Joseph Fraunhofer, in 1814. These lines are superposed against the bright continuous spectrum and have been named in honour of the discoverer as the *Fraunhofer lines*. The spectrum of any other star manifests the general characteristics of the Fraunhofer spectrum with many dark absorption lines, superposed on the brighter background of a continuous spectrum. Another important property common to all stellar spectra is that each one of these shows lines of some element or elements to be much stronger compared with the lines of other elements, and also the strength of the lines of the same element varies continuously in the spectra of different stars. Thus, while lines of He I and He II are the principal features of some spectra, these lines are either completely absent or are very weak in some others where instead, the Balmer series of hydrogen lines dominate over the entire spectrum. In some others, the helium lines are totally absent, the hydrogen lines are weak, while Lines of neutral and ionized metals constitute the predominant feature. Again in some others, the hydrogen lines are very faint or even absent, the ionized metal lines are weak, while the neutral metal lines and molecular bands appear to be the most important characteristics.

We, now know that the sequence described above is clearly a temperature sequence, running from the high to the low value of temperature. But at the beginning of the present century and previous to that at the end of the last century, when the atomic theory and the theory of radiation were still unknown, the observational facts as described above naturally tempted many astrophysicists, to interpret the sequence as the result of difference in the initial compositions of the stellar material. Thus, the stars showing prominent He lines were believed to be composed mostly of helium, those having very broad and prominent H lines were believed to be composed mostly of hydrogen, and so on.

There were other interpretations too, but generally two lines of arguments were pursued by the astrophysicists. While some of these astrophysicists favoured the idea of a real difference to exist in the chemical composition of stellar material to explain the wide variety of spectral types, others argued that the variety actually manifested the different stages of evolution of stellar life. Both groups of scientists however were unable to establish the merits of their theories with any conclusive proof.

Amidst this confusion, it was only the genius of Sir Norman Lockyer that rightly suggested the temperature dependence of the spectral sequence of the stars. His conclusions were based on the laboratory experiments in which conspicuous difference was discovered between the arc and the spark spectra of elements. This difference was recognized to be due to the difference in the temperatures involved in the processes (higher temperature is associated with the spark spectra). Lockyer's work therefore was a forerunner of successful interpretation of stellar spectra with the help of the Ionization Theory of Dr. M.N. Saha. But because the mystery of the atomic structure was completely unknown at that time, Lockyer could hardly establish his theory on any solid ground.

It was in 1920, about thirty years after Lockyer's work, when the scientists all over the world were struggling hard to solve the riddle of the stellar spectral variety, that Saha came forward with his famous Ionization Theory. By that time Niels Bohr (1913) established his classical theory of the hydrogen atom by combining the essential elements of the quantum hypothesis of Max Planck (1901) and the atomic model of Rutherford (1911). Saha's Ionization Theory demonstrated the most important application of the atomic theory of Bohr. He established a mathematical formula (Section 4.3) which for the first time demonstrated conclusively that the state of excitation and ionization in the atmospheres of the Sun and stars was infact dependent on the temperature and pressure prevailing in their atmospheres. His first two papers published in the *Philosophical Magazine* in 1920 explained the Fraunhofer spectrum and the chromospheric lines of the Sun. In these papers, he categorically stated that the Sun was composed of the same elements as the Earth, and contained all the 92 elements known to the chemists at that time. He pointed out that *"pressure has a very great influence on the degree of ionization, which does not seem to have been anticipated"*. For the same stimulus a higher ionization will be favoured at a low pressure, and vice versa. In a subsequent paper published in 1921, in the *Proceedings of the Royal Society,* Saha employed his theory to explain the stellar spectral sequence. Here he argued: "We are not justified in speaking of a star as a hydrogen, helium or carbon star, thereby suggesting that these elements form the chief ingredients in the chemical composition of the star. The proper conclusion would be that under the stimulus prevailing in the star, the particular element or elements are excited to radiation of their characteristic lines, while other elements are either ionized or the stimulus is too weak to excite the lines by which we can detect the element".

Saha's work therefore, conclusively established the theoretical basis of stellar spectral classification which was so earnestly sought for by the astrophysicists during the past decades. The intuitive arguments thus gave way to mathematical formulation and physical understanding.

4.2 BOLTZMANN'S FORMULA

In Astrophysics, we are concerned with different atomic and nuclear processes occurring in the atmosphere and in the interior of a variety of celestial objects having a wide range of temperatures. To study the intensities of different spectral lines due to different celestial sources, we must know the relative amount of ionization and excitation of atoms of different elements, present under different physical conditions existing in those objects. Different atomic processes that are going on inside and in different layers of the outer envelope of stars are considered as steady-state phenomena, which primarily account for the particular surface temperatures of the objects concerned. This steady-state consideration has led to the formulation of ionization and excitation equations, as approximations of the thermal equilibrium between the atoms and ions of different elements. The Boltzmann equation has been so formulated on the basis of statistical thermodynamics. The general Boltzmann formula is

$$\frac{N_{i,r}}{N_i} = \frac{g_{i,r}}{u_i(T)} \exp\left(-X_{i,r}/kT\right) \qquad (4.1)$$

which gives the fraction of the atoms in rth state (level) of excitation of the total number N_i in the ith stage of ionization; $g_{i,\,r}$ is called the statistical weight of the level r at the ith stage of ionization and is given by

$$g_{i,\,r} = 2J + 1 \qquad (4.2)$$

where J is the total angular momentum quantum number. The relation gives $2J + 1$ sample quantum states into which the level r is split up due to the presence of an external magnetic field (thus represents the number of Zeeman states into which the rth level of the jth stage ion is split by a magnetic field). $X_{i,\,r}$ is the excitation energy associated with the level r in the ith stage of ionization. The quantity

$$u_i(T) = \sum_{r=1}^{\infty} g_{i,\,r} \exp\left(-X_{i,\,r}/kT\right) \qquad (4.3)$$

is called the *partition function* for the absolute temperature T and k is the Boltzmann constant.

The general Boltzmann formula can also be expressed in a logarithmic form by

$$\log \frac{N_{i,\,r}}{N_i} = -\frac{5040}{T} X_{i,\,r} + \log \frac{g_{i,\,r}}{u_i(T)} \qquad (4.4)$$

where $X_{i,\,r}$ is expressed in electron volts (eV).

If again, we take $r = 1$ in Eq. (4.4) and subtract the resulting relation from the above formula (note that in the ground level $X_{i,\,r} = 0$), we have the relation

$$\log \frac{N_{i,\,r}}{N_{i,\,1}} = -\frac{5040}{T} X_{i,\,r} + \log \frac{g_{i,\,r}}{g_{i,\,1}} \qquad (4.5)$$

This relates the number of atoms in the rth level with that in the ground level ($r = 1$) at the ith stage of ionization.

4.3 SAHA'S EQUATION OF THERMAL IONIZATION

In a certain volume of a gas, the excitation and ionization of atoms increase with the temperature. At a particular temperature, in a steady state, the fraction of the ionized or excited atoms, the electrons and the neutral atoms are in thermal equilibrium with one another. The rate of ionization of an atom is counteracted by an equal rate of recombination of the electron and the ion. But it is also found that the above equilibrium is dependent on the density of the gas at a particular temperature. This suggests that the ionization processes also depend on the relative densities of the electrons and the ions, i.e. on their partial pressures. For instance, for a given temperature, the higher density of a gas implies that the electrons and the ions are relatively closely packed together so that there is greater chance of recombination of the two to form neutral atoms, although the chance of ionization remains unaltered so long as the stimulus is the same. In other words, the greater partial pressures of the gas components help in recombination processes and thereby relatively decrease the ionization. On the other hand, the electrons and ions move more freely in gas having low density. There is lesser chance of an electron coming close to an ion to reunite and form a neutral atom. Low gas density therefore favours relatively higher ionization at the same temperature.

At a given temperature and density (or partial pressure) the probability of ionization of an atom depends on its ionization potential (I.P.). The greater the I.P. of an atom, the more energy will be needed to remove an electron from its outermost orbit. Thus at a given temperature, Cs which has the least known I.P. of 3.89 eV will be ionized more than the Si atoms with I.P. = 8.11 eV.

In 1920, M.N. Saha formulated the theoretical ionization equation, considering the thermodynamical equilibrium between the atoms and ions of a perfect gas. This equation actually laid the foundation of the modern stellar spectral studies which is a very important branch of the modern astrophysics. The equation gives quantitative estimates of the fraction of the neutral and ionized atoms, or that of an atom in two consecutive ionization states in the atmospheres of stars. The basic parameters of the equation are the temperature T and the electron pressure P_e or the electron density N_e.

The Saha equation as is now used has the form

$$\frac{N_{q+1}N_e}{N_q} = \frac{(2\pi m_e kT)^{3/2}}{h^3} \cdot \frac{2u_{q+1}(T)}{u_q(T)} \exp\left(-X_q/kT\right) \tag{4.6}$$

where N_q is the number of atoms in the qth stage of ionization, N_{q+1}, the number in the $(q + 1)$th stage, X_q is the energy necessary to ionize the atom from the qth stage to $(q + 1)$th stage, and m_e is the mass of an electron. If we want the ratio of the number of singly ionized atoms to that of the neutral atoms and express the equation in terms of the partial pressure of the electrons, then substituting for N_e its value from the equation of state

$$P_e = N_e kT \tag{4.7}$$

in Eq. (4.6), we get (taking $q = 0$)

$$\frac{N_1}{N_0} P_e = \frac{(2\pi m_e)^{3/2}}{h^3} \cdot (kT)^{5/2} \cdot \frac{2u_1(T)}{u_0(T)} \cdot \exp\left(-X_0/kT\right) \tag{4.8}$$

where X_0 gives the first ionization potential of the atom, $u_1(T)$ and $u_0(T)$ are respectively the partition functions for the ionized and neutral atoms.

Equations (4.6) and (4.8) are more conveniently used in their logarithmic forms, in which case Eq. (4.8) becomes

$$\log \frac{N_1}{N_0} P_e = - I\theta + 2.5 \log T - 0.48 + \log \frac{2u_1(T)}{u_0(T)} \tag{4.9}$$

where $I = X_0$ is expressed in electron volt, $\theta = \dfrac{5040}{T}$, P_e is in dyne cm^{-2} and N_1 and N_0 are numbers respectively of singly ionized and neutral atoms per unit volume. The term involving $u_1(T)$ and $u_0(T)$ is small and was omitted in the original Saha equation.

The Boltzmann and the Saha equations can be combined into a suitable form (by dividing the Saha equation with partial pressure by the Boltzmann equation) as follows:

$$\frac{N_1}{N_{0,r}} P_e = \frac{(2\pi m_e)^{3/2}}{h^3} (kT)^{5/2} \frac{2u_1(T)}{g_{0,r}} \exp[-(I - X_r)/kT] \tag{4.10}$$

where $N_{0,\,r}$ is the number per unit volume of the neutral atoms in the rth level of excitation and N_1 that of the singly ionized atoms, $X_r = X_{0,\,r}$ is the excitation energy of the level r for the neutral atom and I is the first ionization potential for the atom in question. This equation is generally applicable to outer envelope of hot stars in which atoms are found in higher excited states close to ionization limit.

The logarithmic form of Eq. (4.10) is given by

$$\log \frac{N_1}{N_{0,r}} P_e = - \frac{5040}{T} (I - X_r) + 2.5 \log T - 0.48 + \log \frac{2u_1(T)}{g_{0,r}} \tag{4.11}$$

Figure 4.1 is a schematic diagram indicating the different energy states of the neutral atoms.

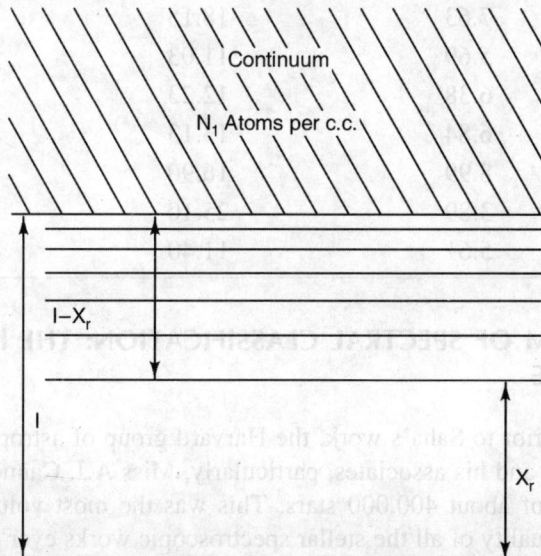

FIGURE 4.1 Schematic diagram indicating the different energy stages of the neutral atom.

Table 4.1 shows the Ionization Potentials (I.P.) of some common elements of great astrophysical importance.

TABLE 4.1 Ionization Potentials of Some Common Elements

Atom	First I.P.	Second I.P.	Third I.P.
H	13.60	—	—
He	24.58	54.40	—
Li	5.39	75.62	—
Be	9.32	18.21	—
C	11.26	24.38	47.87
N	14.53	29.59	47.43
O	13.61	35.11	54.89
Ne	21.56	41.07	63.50
Na	5.14	47.29	71.65
Mg	7.64	15.03	80.12
Al	5.98	18.82	28.44
Si	8.15	16.34	33.46
S	10.36	23.40	35.00
Ar	15.76	27.62	40.90
K	4.34	31.81	46.00
Ca	6.11	11.87	51.21
Ti	6.82	13.57	27.47
Fe	7.87	16.18	30.64
Ni	7.63	18.15	35.16
Sr	5.69	11.03	—
Y	6.38	12.23	—
Zr	6.84	13.13	—
Cd	8.99	16.90	—
Cs	3.89	25.10	—
La	5.61	11.40	—

4.4 HARVARD SYSTEM OF SPECTRAL CLASSIFICATION: THE HENRY-DRAPER (HD) CATALOGUE

During the three decades prior to Saha's work, the Harvard group of astrophysicists under the direction of E.C. Pickering and his associates, particularly, Miss A.J. Cannon and H. Shapley, had classified the spectra of about 400,000 stars. This was the most voluminous and at the same time outstanding in quality of all the stellar spectroscopic works ever done. The outcome of the long years of devoted work of many scientists was the *Henry Draper Catalogue,* after the name of the pioneer spectroscopist Henry Draper. The catalogue included *all* stars up to magnitude 8.25 and *some* stars upto magnitude 10.0 that were observable from the Harvard observatory. Even to-day every astrophysicist has to rely heavily on this work. From the long experience of the Harvard scientists, they could recognize that the huge multitude of stars could be classified into a small number of spectral classes and that this classification actually represented a temperature sequence. This idea was particularly stressed upon by H.N. Russell.

In the H.D. catalogue, the letters A, B etc. have been used to denote stars of different spectral classes. Each class has again been subdivided into tenths by using a number after it. Thus, the class B contains ten sub-classes B0, B1, B2, ..., B9 and then the next class A starts. The subclass B5 is thus intermediate between B0 and B9. After some modification of the original scheme, the sequence now stands as follows:

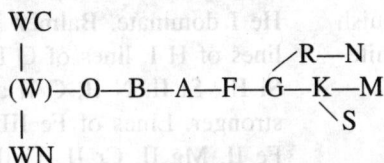

$$
\begin{array}{c}
\text{WC} \\
| \qquad\qquad\qquad\quad \nearrow \text{R—N} \\
\text{(W)—O—B—A—F—G—K—M} \\
| \qquad\qquad\qquad\qquad \searrow \text{S} \\
\text{WN}
\end{array}
$$

The class W, called the Wolf-Rayet stars, is a later discovery and was not included in the Harvard catalogue. This class again has two sub-classes—WC in which the emission lines due to carbon dominate the spectrum, and WN in which the emission lines due to nitrogen are the dominant feature. We shall later discuss more about the Wolf-Rayet stars (Chapter 10). The side branches R, N and S represent rare stars whose temperatures are similar to those of classes G to M. Their spectra contain some prominent bands due to molecules which are not found in the spectra of G to M classes. Types W to A are called *early* and those from F to M including the side branches are called *late*.

The Harvard classification recognized only the classes O, B, A, F, G, K and M and they recognized these classes to be a high to low temperature sequence. However, the Harvard astrophysicists developed the remarkable criterion through experience only, without any well-conceived physical foundation of their ideas. The physical explanation came from M.N. Saha who wrote in his Royal Society paper:

> The work thus corroborates Russell's view that the continuous variation of spectral types is due to the varying values of the temperature of the stellar atmosphere and the classification B, A, F, G, K, M which has been adopted by the Harvard astrophysicists, as the result of long years of study and observation, is therefore seen to acquire a new physical significance.

The Harvard scientists also recognized the conspicuous differences in the widths of the spectral lines in the stars of the same Draper classes. The stars with narrow sharp lines were arbitrarily called by A.C. Maury as those having the c-characteristic. They are actually the stars with low surface pressure and high luminosity, now known as the *giant* and *Supergiant* stars. This characteristic also finds explanation in Saha's formula in which the partial electron pressure is a parameter. The effect of luminosity difference on the spectra of stars which was recognized by Russel as resulting from variation in surface density thus gained a theoretical background by Saha's work. In the following Section, we shall discuss this point in more detail.

The main spectral features of the Harvard sequence may be given as in Table 4.2.

TABLE 4.2 The Principal Features of the Harvard Spectral Sequence

Spectral type	Temperature	Colour	Spectral features	Remarks
O5–O9	40,000 K to 25,000 K	Blue to bluish white	Emission lines of He II and O II. Absorption lines of He II, He I, O II, Si IV, N III, C III, dominate the spectrum. Lines of neutral H I are relatively weak.	No stars are found earlier than O5.
B0–B9	25,000 K to 11,000 K	Blue to Bluish white	Absorption lines due to He I dominate. Balmer lines of H I, lines of O II, Si III, Si II, N II, C II are stronger. Lines of Fe III, Fe II, Mg II, Cr II, Ti II appear.	Some stars in the constellation Orion belong to these classes.
A0–A9	11,000 K to 7,500 K	White	Balmer lines of HI very strong. He lines are absent. Absorption due to Mg II, Si II is strong. Lines of Fe II, Ti II, Ca II are relatively weak.	"Hydrogen stars." Vega and Sirius are of these types.
F0–F9	7,600 K to 6,000 K	White to yellowish white	Balmer lines of hydrogen are weaker than in class A. Lines of Ca II, Fe II, Ti II are strong. Those of Ca I, Fe I are less conspicuous.	Procyon is typical of the class.
G0–G9	6,000 K to 5,000 K	Yellowish white to yellow stars	Balmer lines of hydrogen are still weaker. Lines of Ca II are at maximum in these classes. Also, lines of neutral metals like Fe I, etc. are conspicuous. Lines due to CH radical are prominent.	Sun belongs to the class G2 and the notable star is Capella.
K0–K9	5,100 K to 3,500 K	Deep yellow to orange to red	Balmer lines of hydrogen are very weak. Lines of neutral metals are stronger than in class G. CH and other molecular bands strong.	Examples are Arcturus and Aldebaran.

(*Contd.*)

TABLE 4.2 The Principal Features of the Harvard Spectral Sequence

Spectral type	Temperature	Colour	Spectral features	Remarks
M0-M5	3,600 K to 3,000 K	Red	Hydrogen lines scarcely present. Neutral metal lines are at maximum strength (e.g. those of Ca I). Molecular bands of TiO dominate the spectrum.	Stars like Betelgeuse, Antares belong to these classes.
R(C)	3,000 K	Red	Similar to the classes G to K, and characterised by the presence of C_2 bands and also some CN bands. TiO bands absent.	Carbon stars
N(C)	3,000 K	Red	Similar to M stars with the exception of the presence of strong bands of C_2, CH, CN. TiO bands are absent.	Carbon stars
S	3,000 K	Red	Much like the M classes. Exception is due to the presence of strong bands of ZrO, YO, LaO instead of TiO bands.	

4.5 THE LUMINOSITY EFFECT ON STELLAR SPECTRA

Saha's ionization formula is conveniently written as

$$\log \frac{N_{r+1}}{N_r} = \log \frac{2u_{r+1}}{u_r} + 2.5 \log T - \frac{5040}{T} I_r - \log P_e - 0.48 \qquad (4.12)$$

where the symbols have their usual meanings. The formula shows that the degree of ionization of an element in the atmosphere of a star is dependent inversely as the electron pressure, which again directly depends on the surface density in the atmosphere. Other things being equal, an increase in P_e reduces the ratio N_{r+1}/N_r and vice versa. But this reduction or increase *is the same for all elements*. On the other hand, the temperature T in Eq. (4.12) is locked with the ionization potential I_r which is different for different elements. Thus, the ratio (N_{r+1}/N_r) is not related to T in the same way as with P_e. The change in N_{r+1}/N_r with respect to T is *not the same for all elements*. The change in the ratio for a given change in T will be greater for small I_r and smaller for large I_r. The sign of the term $5040\ I_r/T$ implies that for the same temperature the ionization will be stronger in atoms with small I_r than in those with large I_r. Since the surface gravity (or equivalently P_e) in giants and supergiants is several orders less than in dwarfs, the terms in I_r and P_e in Eq. (4.12) interplay in such a manner that atoms with small I_r will have a higher ionization in the atmospheres of giants than in those of dwarfs, and the atoms with large I_r will behave in the opposite way. This theoretical prediction finds support in the observational facts that the lines of Ca I (I.P. = 6.09 eV) and Sr I (I.P. = 5.67 eV) are stronger in dwarfs than in giants of the same spectral class, while the reverse is observed for lines of Ca II (I.P. = 11.82 eV) and Sr II (I.P. = 10.98 eV). Hydrogen lines serve as interesting examples. Hydrogen has a sufficiently high ionization potential (13.6 eV). These lines appear as much stronger in the dwarfs of classes B and A than in giants of these classes. The reverse is observed in later classes K and M, where relatively stronger lines of hydrogen appear in giants. Sr II at λ 4077 serves as an important line for distinguishing between giants and dwarfs. This line is stronger in giants than in dwarfs of classes F to M. The I.P. of Sr being low, it is more easily ionized than atoms like Fe (I.P. = 7.87 eV) which remains mostly neutral through these later classes. The ratio of Sr II (λ 4077)/Fe I (λ 4045) is sometimes used to differentiate the luminosity effect.

Again, we have seen that for the same temperature higher ionization is favoured in lower gas density. This is illustrated, in Fig. 4.2. So since the *atmospheres* of giants are much rarer than those of the dwarfs, and since the spectral classes are determined by the relative ionization of a particular element or elements, it follows that the giants will have lower effective temperatures than the dwarfs of the same spectral class.

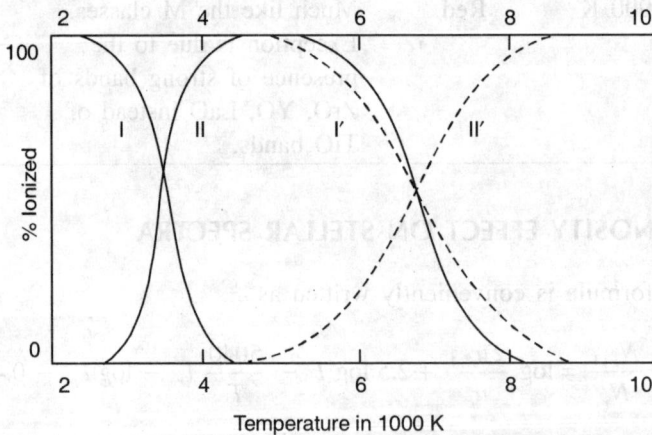

FIGURE 4.2   Ionization of Ca. The full curves show percentage Ca I and Ca II with $P_e = 10^{-6}$ atmosphere. The dashed curves show similar percentage with $P_e = 0.1$ atmosphere.

Theoretical ionization curves as in Fig. 4.2 can be utilized to determine the electron pressure in the atmosphere of a star if its temperature can be known. The observed intensities of the stellar lines are fitted with theoretical curves produced in the laboratory for various values of P_e at the temperature considered and the value of P_e is derived from the best fit.

The understanding of the current concept of the luminosity effect on stellar spectra evolved over the decades. As we have already mentioned, Miss Maury in her work with classification of spectra at Harvard noticed that some stars of the same spectral class differed in width and intensity of spectral lines. She observed some stars with sharp H-lines peculiar to the corresponding class and called them c-stars. Later in 1905, H.N. Russell recognized these stars to be of high luminosities compared to the other stars of that particular class. He compared their luminosities by studying their parallaxes and motions and ultimately named these as "Supergiants". This discovery of the supergiants led the astronomers to classify spectra on the basis of two parameters, namely, the temperature and the luminosity.

In 1914, W.S. Adams and A. Köhlschutter of the Mount Wilson observatory first used spectroscopic criteria to determine the absolute magnitudes of some individual stars of the early spectral types. This effort ultimately resulted in the publication of the Mount Wilson Catalogue (MW types) in 1935. In this classification the *Spectroscopic absolute magnitudes* and *spectroscopic parallaxes* of 4179 stars have been recorded on the basis of some spectrograms of greater resolving power with larger scales.

The two-dimensional spectral classification developed at the Yerkes observatory by W.W. Morgan, P.C. Keenan and E. Kellman in "An atlas of stellar spectra with an outline of spectral classification" was published in 1943. This classification is based on the HD system and on some empirical criteria which have been observed on slit spectrograms of sufficiently low dispersion (≈ 115 Å/mm) to extend the classification to faint celestial sources. According to Morgan and Keenan, the Harvard stars from classes O9 to M3 have been further subdivided (MK classification) into five main luminosity classes denoted by Roman numerals I, II, III, IV and V as follows:

Class	Symbol
Bright supergiants	Ia
Supergiants	Ib
Bright giants	II
Giants	III
Subgiants	IV
Dwarfs or Main sequence	V

Thus, a star is now designated as B5V or M2Ia meaning respectively a dwarf star of spectral class B5 or a bright supergiant of spectral class M2.

4.6 IMPORTANCE OF IONIZATION THEORY IN ASTROPHYSICS

The Ionization Theory is one of the few scientific works ever done that have laid broad foundations of new disciplines of science. It established *astrophysics* on a firm physical basis. The age of imagination and 'good guesses' merged into the age of experimental discipline of science.

In the first place, the theory definitely ascertained that the Harvard spectral sequence of stars from O through M was essentially a temperature sequence which could be quantitatively explained by the formula. According to the theory, the spectra with bands of molecules such as hydrocarbon (CH), cyanogen (CN), carbon molecules (C_2), titanium oxide (TiO), etc. should be produced by the coolest stars in whose atmospheres these molecules can exist without being dissociated. We know that as the theory predicts, these bands are actually observed in spectra of red and yellow stars with the lowest surface temperatures. In the atmospheres of hotter stars the molecules dissociate resulting in the absence of any band spectrum in these stars.

The study of the metallic lines also provides a good example. Numerous metallic lines are revealed in the spectra of coolest stars. The metals have low ionization and excitation potentials. So, even the weak stimulus in the atmospheres of cool stars is sufficient to excite metallic lines. With the result, strong neutral metal lines appear as the principal feature of these spectra. At higher temperatures, the metals become partially ionized. Thus, as we pass on from spectra of lower temperature to those of higher temperature stars, the neutral metal lines start weakening while the ionized metal lines start appearing in greater strength. The case of calcium may be very appropriately cited as an example. The resonance line, that is, a line excited from the ground level of the atom (the line excited from a higher level is called a subordinate line) with λ 4227 of Ca I appears in greatest strength in the spectra of the coolest M stars. In hotter classes K or G, calcium becomes ionized to some extent. So λ 4227 of Ca I is weak in these spectra and the resonance H and K lines at λ 3968 and λ 3933 of the ionized calcium (Ca II) appear in increased strengths from M through G, and in G stars these lines are the most dominant feature of the spectrum. At higher temperatures, calcium becomes doubly ionized so that H and K lines of Ca II start fading in the spectra of early F and A stars. The general trend of this variation of the neutral and ionized metal lines with changing temperature for any particular element is well explained by Fig. 4.3 in which intensities of lines have been plotted against the temperature. The same arguments apply to any other metal.

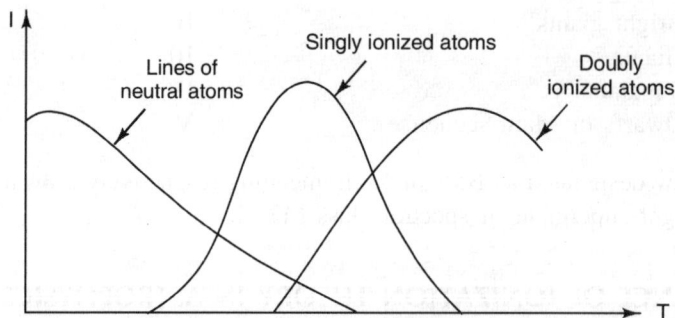

FIGURE 4.3   Intensities of the lines of an element at various ionization stages with changing temperature are shown.

The Balmer series of hydrogen lines provide an excellent example to this effect. These are subordinate lines arising in transitions from the second to higher levels. The intensities of the series lines will chiefly depend on the number of hydrogen atoms in the second excited level which is capable of absorbing the energy corresponding to these lines. Since the excitation potential of the second level is quite high, viz., 10.15 eV, sufficiently high temperature is needed to raise the atoms from the ground level to the second excited state. So by the stimulus present in the atmospheres of cool stars, only relatively small number of H-atoms are excited to the second level. Since the strength of a line depends on the number of atoms capable of absorbing (or emitting) it, and only a very small fraction of H atoms are excited to the second level in the atmospheres of cool stars, the Balmer series of lines are very weak in the spectra of these stars. With increasing temperatures, more and more atoms are excited to the second level resulting in the stronger and stronger Balmer series of lines in the spectra of these stars. In agreement with this prediction, the Balmer lines of H steadily increase in strength as we pass from M through K, G, F and A stars in whose spectra these lines are the strongest. At still higher temperatures, like those in the atmospheres of B and O stars, hydrogen becomes ionized and the number of atoms in the excited levels including the second decreases, thus resulting in weakening of the Balmer series of lines in the spectra of these stars. In the atmospheres of O stars only about a millionth fraction of hydrogen remains neutral, but due to the overwhelming abundance of this element, hydrogen lines are still visible in the spectra of O stars, although relatively much weaker. The relation between the temperature (or spectral class) and the strength of hydrogen lines is illustrated in Fig. 4.4.

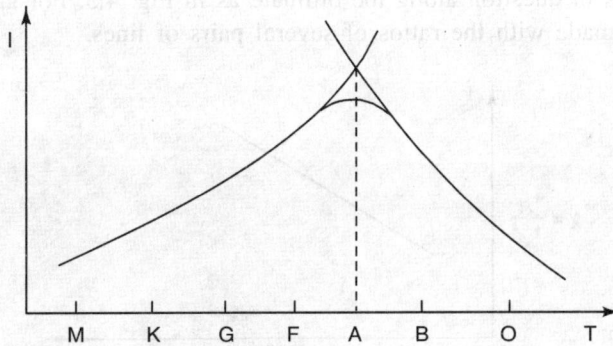

FIGURE 4.4   The intensity of Balmer lines shown against temperature, [Note that the intensity is maximum in AO stars.]

Finally, we consider the case of He lines. He has the highest ionization potential 24.54 eV. The excitation potential of He atoms are also so high that the subordinate lines in the optical wavelengths can be produced only in atmospheres of quite high temperatures. In agreement with this, He I lines appear first in the early B stars. In fact, He I and H lines are the most prominent features of the spectra of B stars as a class. The O stars are so hot that hydrogen is almost completely ionized. The hydrogen lines therefore, almost fade away in these stars. Lines of neutral and ionized helium and those of the light elements O, N, C and Ne at various stages of ionization appear as the most conspicuous features in the spectra of O stars. Thus, Saha's ionization theory clearly unfolded the mystery of the temperature effect on the stellar spectral sequence. We have already seen in the previous section how the theory also explains the absolute magnitude effect on the stellar spectra, thus forming the basis of the two-dimensional spectral classification by Morgan and Keenan.

4.7 SPECTROSCOPIC PARALLAX

We have seen that certain spectral lines are sensitive to the absolute magnitude differences of stars belonging to the same spectral class. This means that a progressive variational relation exists between the equivalent width ω_λ of the line and the surface gravity g of the star. This relation is changed for more convenient use to that between ω_λ and the absolute magnitude M_v or the Bolometric luminosity L of the star. Thus, intensities of certain lines or more appropriately, the ratios of the intensities of certain lines, may serve as indicators of absolute magnitudes of stars, which in turn, yield the distances of these stars. Since the method uses spectral line intensities as the basis for calculation of distances of stars, it has been known as the method of *spectroscopic parallax.*

Let us now discuss the working principle of the method. For a given spectral class one selects a pair of lines whose intensity ratio varies progressively with the change in absolute magnitude of the star. The ratio is first determined for a number of stars of known absolute magnitudes but belonging to the same spectral class. A standard calibrated curve is thus obtained for each spectral class, with the absolute magnitudes plotted against the abscissa and the ratio of the lines in question along the ordinate as in Fig. 4.5. For greater accuracy, the calibration may be made with the ratios of several pairs of lines.

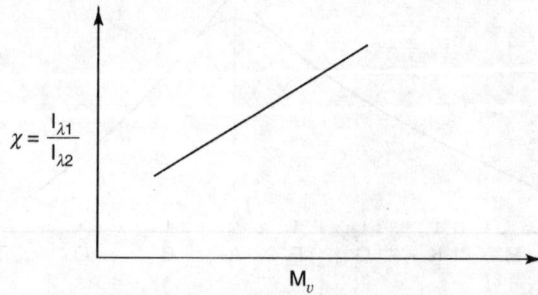

FIGURE 4.5 The Standard calibrated curve for measuring the absolute magnitude.

When this curve has been calibrated for any spectral class, one can compare the intensity ratio of the same lines of any individual star but belonging to the same spectral class, of unknown absolute magnitude and thus its absolute magnitude may be calculated. The distance of the star is then yielded by Eq. (3.5). Thus, a very important physical basis which depends on measurable quantities in stellar spectra was discovered to calculate the distances of stars. Since the intensity ratios of lines in the stellar spectra are used in the method as the basis for stellar distances, the prerequisite for applying this method is a good spectrum. Since with modern instruments a good spectrum can be obtained even of far off stars, spectroscopic parallaxes can be determined even for stars at very great distances at which all other methods become uncertain. Again, one common difficulty that is faced with the distant stars is to determine the absorption of its light and its consequent reddening by the intervening dust particles between the star and the observer (discussed in Chapter 13). But since a pair of nearby lines will be almost equally contaminated by the particles, the intensity ratio of the pair will remain mostly unaffected. Spectroscopic parallaxes are therefore hardly affected by the *interstellar absorption* which generally poses serious problems to various types of astrophysical observations, particularly those based on the colours of the stars.

4.8 THE HERTZSPRUNG-RUSSELL DIAGRAM

Early this century E. Hertzsprung of Denmark and H.N. Russell of the U.S.A., the two foremost astrophysicists of the time, discovered through their experience, independently of each other, that the spectral types of stars (or equivalently, their colours) were closely correlated with their intrinsic luminosity (or equivalently, their absolute magnitudes). Hertzsprung and Russell independently plotted the absolute magnitudes versus spectral types for stars whose distances were known at that time (in 1913). Surprisingly, a regular pattern was shown by the position of the stars in their diagrams now known as the *Hertzsprung-Russell diagram* (or H–R diagram), which relates two fundamental physical parameters of stars as shown in Fig. 4.6. With the discovery of the Ionization Theory, the spectral sequence gained a firm theoretical background and the Hertzsprung-Russell diagram acquired primary importance in astrophysics.

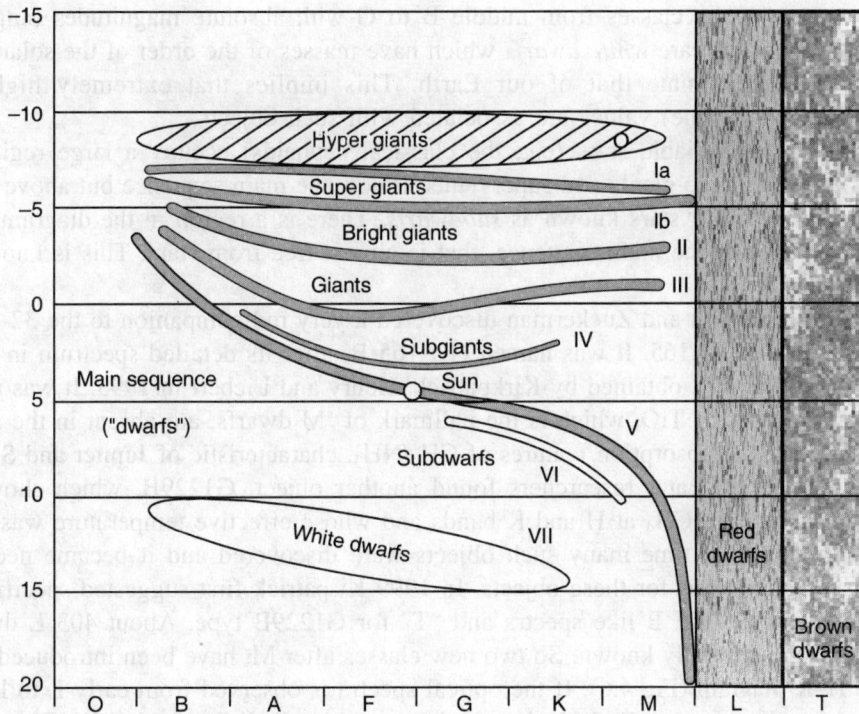

FIGURE 4.6 The Hertzsprung–Russell (H–R) diagram.

In the plot of absolute magnitude versus spectral type for all stars, it is found that almost 90% of the stars lie along a narrow and long band which runs diagonally from the top left corner consisting of the hot O and B stars to the lower right corner below, which contains mostly the cool and faint red M stars. This main band of stars is called the *Main sequence* or *dwarf sequence,* which extends from the absolute magnitude of about –7 or –8 to +15. Many famous stars like Sirius, the Sun, Procyon, 61 Cygni, Barnard's star etc. belong to this main sequence. From Fig. 4.6, it is evident that both the luminosities and the surface temperatures fall steadily along the main sequence with a temperature range from more than 30,000 K to about 3,000 K. The Sun figures as a mediocre member of this sequence which belongs to the class G2 with an absolute magnitude of about + 4.8 and surface temperature of about 6,000 K.

Besides the main sequence which contains the overwhelming majority of the total multitude of stars, there is a group of luminous stars which is found to occupy some definite areas of the diagram. This is a short branch which extends upwards to the right of the main sequence in the classes F to M. This is the group of *giant stars,* which is more luminous than the dwarfs of the same spectral classes and whose absolute magnitudes range between –1.0 to +1.0. Above the "giants" lies the group of *bright giants* with absolute magnitude ranging upto –3.0. Throughout the top of the diagram, in the absolute magnitude range –3.0 to –8.0, there is a sparse distribution of highly luminous *super giants.* These are found in almost all the spectral classes. Between the giants and dwarf sequence, a small group of stars of absolute magnitude between +1 to +5 form the *subgiant* branch of the class F to K.

At the lower left corner of the diagram, well below the main sequence, lies a group of very faint stars of the classes from middle B to G with absolute magnitudes ranging from +10.0 to +15.0. These are *white dwarfs* which have masses of the order of the solar mass but whose sizes approximate that of our Earth. This implies that extremely high density ($\simeq 10^6$ times solar value) values are associated with such objects.

A variety of variable stars (e.g. the classical cepheids) occupy a large region of the diagram that belongs to giants and supergiants. Below the main sequence but above the white dwarfs, lies a group of stars known as *subdwarfs.* There is a region in the diagram, between the giant branch and the main sequence, that is almost free from stars. This is known as the *Hertzsprung gap.*

In, 1988, Becklin and Zuckerman discovered a very red companion to the 32-pc distant DA4 white dwarf GD 165. It was named GD 165 B, after its detailed spectrum in the range 6400–9000 Å was first obtained by Kirkpatrick, Henry and Liebert in 1993. It was noted that the absorption lines of TiO, which is the hallmark of M dwarfs, are absent in the spectra of this object. Also the absorption features of CH_4/NH_3, characteristic of Jupiter and Saturn (i.e. planets), are absent. Later researchers found another object, G1229B, which showed clear absorption features of CH_4 at H and K bands and whose effective temperature was less than 1000 K. In course of time many such objects were discovered and it became necessary to provide separate classes for these objects. In 1999 Kirpatrick first suggested specifically that "L" be used for GD165 B like spectra and "T" for G1229B type. About 403 L dwarfs and 62 T dwarfs are currently known. So two new classes after M, have been introduced as L and T in the H–R diagram (Fig 4.6). If the optical spectra is observed from early-L to late-T, the following features emerge. Early-L dwarfs show neutral alkali lines (NaI, KI, RbI, CsI, LiI), oxide bands (TiO, VO), hydride bands (CrH, FeH, CaOH). Mid-L, spectra show prominent NaI and KI lines, hydrides but not TiO and VO. Late-L, early-T spectra have strong H_2O lines, neutral alkali lines but hydrides disappear. For Late-T, H_2O, NaI and KI have the highest strength (Fig. 4.7).

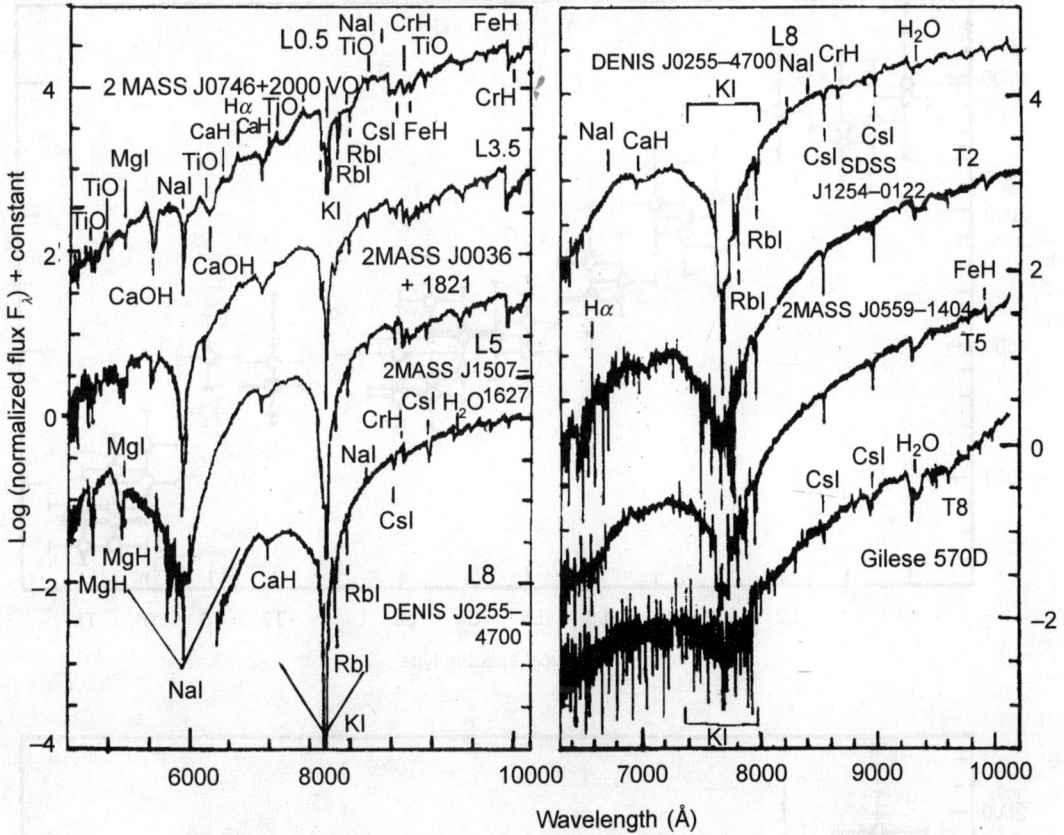

FIGURE 4.7 The optical spectra of early –L to late –T dwarfs.

Question arises what is the underlying physical phenomenon governing the spectral sequence for L to T dwarfs. Now, for MS (Main Sequence) stars the most important parameter leading to such spectral features is the temperature. Whether it is also true for L and T dwarfs? For this, T_{eff} is calculated for L and T dwarfs using Stefan-Boltzmann law, $L = 4\pi R^2 \sigma T_{eff}^4$, where L is the bolometric luminosity, R is the radius and σ is the Stefan-Boltzmann constant.

Now Monte Carlo simulations using Burrows evolutionary models for L and T dwarfs with 1 Myr < age < 1 Gyr and 1 M_{Jup} < mass < 100 M_{jup} show a total range of radius ~ 0.90 ± 0.15 R_{Jup}, assuming a constant birth rate and mass function of the form $dN \propto M^{-1} dM$ (Burgasser 2001). Thus to first order of approximation T_{eff} does not depend on R but on L. So the fourth root of luminosity gives the value of T_{eff}. For lumininosity measurement, the trigonometric parallaxes of field L and T dwarfs have been measured by Dahn et al. (2002), Tinney et al. (2003) and Vrba et al. (2004). In addition, distances to several L and T dwarfs have been estimated from common proper motion with an objective of higher luminosity, whose trigonometric parallaxes have also been measured. From these methods, luminosity (*L*), hence T_{eff} is determined. They are shown in Fig. 4.8. The upper diagram is for optical spectra and the lower one is for near infrared. The correlation is much tighter for the optical

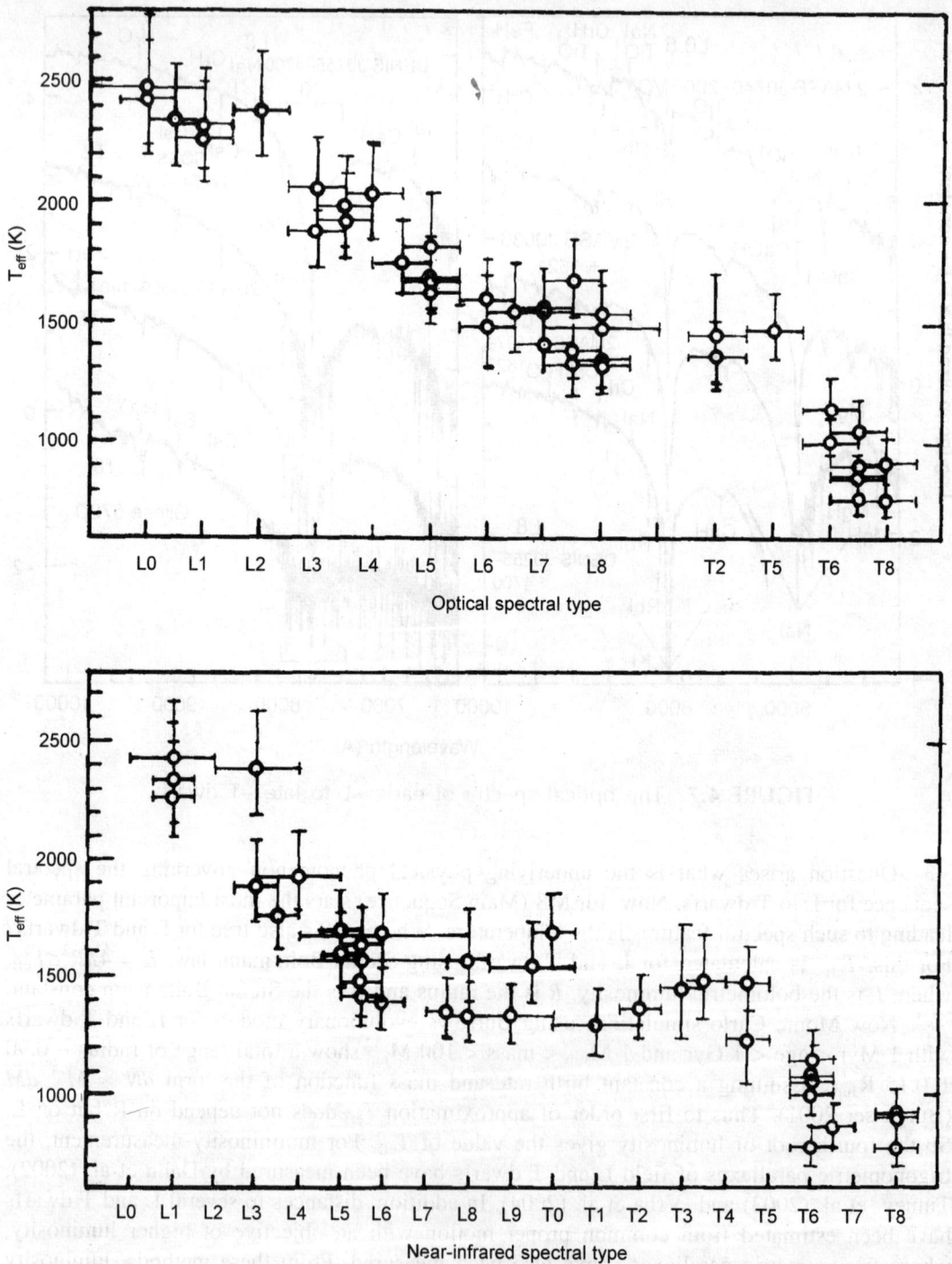

FIGURE 4.8 Luminosity vs T_{eff} of L and T dwarfs in optical (top) and near infrared (bottom).

one. The discrepancy is explained as due to the influence of clouds or gravity. Clouds are product of condensation and sedimentation and their pressure have the effect of both veiling features in spectra and reddening the near infrared colours. So the other par meter besides temperature, might be related to the inhomogeneity of clouds and/or disappearance of clouds below the photosphere. So in a nutshell, the spectral sequence accounts for the temperature sequence except from mid/late–L to mid–T where cloud physics becomes important as temperature in determining the form of spectral energy distribution.

Regarding the nature of objects under the L–T umbrella, evolutionary tracks for T_{eff} vs age have been constructed by various authors. It shows that (Fig. 4.9) early to mid-L dwarfs are a mixture of low mass stars (< 85 M_{Jup}) that are fairly older (> 300 Myr) and brown dwarfs, from high mass ($= 70$ M_{Jup}) to very low mass and all T dwarfs are sub stellar.

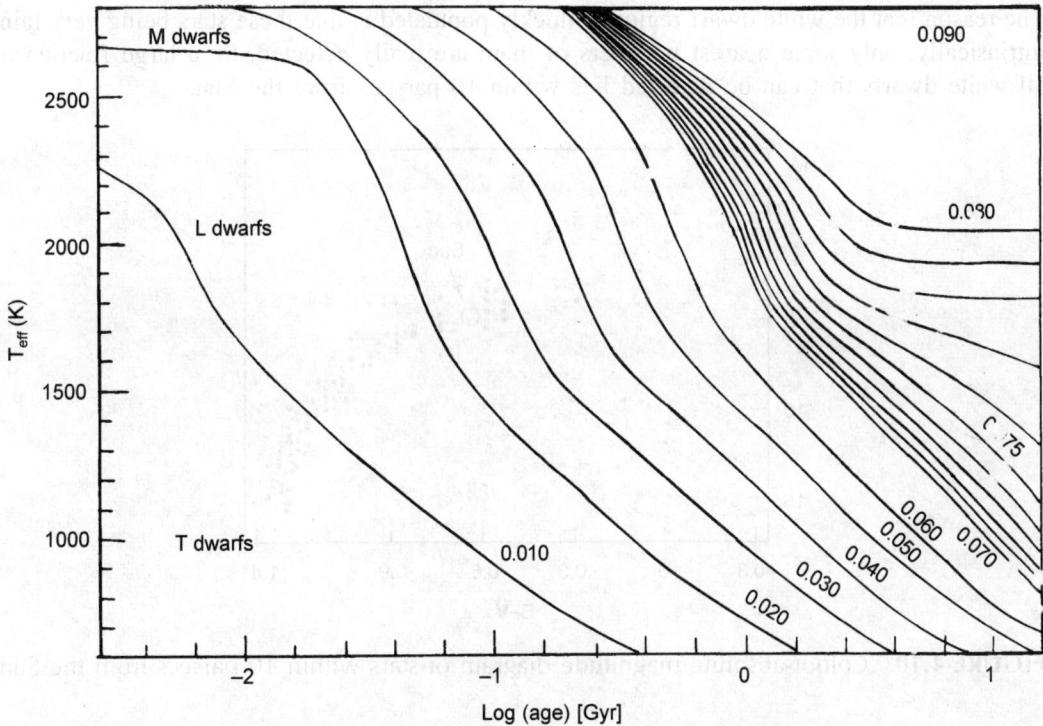

FIGURE 4.9 Evolutionary tracks of L and T dwarfs.

We differentiate stars of the same spectral class, assigning them the names (following E. Hertzsprung who first used the terms in 1905) "dwarfs" or "giants", which refer to their luminosity or intrinsic brightness. Giants are more luminous than the dwarfs of the same temperature and spectral class. Accordingly, it follows from Eq. (3.11) that the former (giants) have larger surface areas (i.e., larger radii) compared to the latter. Thus Capella, which is a giant with almost the same temperature as the Sun, has diameter sixteen times that of the Sun. Supergiants are much more luminous than the giants of the same temperature, their radii being still many times larger. Examples are α–Cygni, α–Scorpii, Antares and Betelgeuse. The last three stars are red supergiants of class M, where the difference in luminosities is found to be greatest. The classes late F to late K are also marked by the presence of various giants and supergiants and also the intermediate members. Equation (3.11) also indicates that the blue giants and supergiants of spectral classes O to A should be smaller in size than red counterparts.

If instead of plotting the spectral class of stars we plot the colour B–V along the abscissa, we get what is known as a *colour-magnitude diagram* (or C–M diagram). Figure 4.10 gives the colour magnitude diagram of the stars lying within 10 parsecs from the Sun. We find that the general appearance of the diagram is quite similar to the H–R diagram illustrated in Fig. 4.6. Here the main sequence is considerably populated as also the white dwarf region. The giant branch is hardly present with a few exceptions and there is no supergiant stars at all. This is because there are very few intrinsically luminous stars as close as 10 parsecs from the Sun. The reason that the white dwarf region is thickly populated is that these stars being very faint intrinsically, only some nearest members of them are really detected. So a large fraction of all white dwarfs that can be detected lies within 10 parsecs from the Sun.

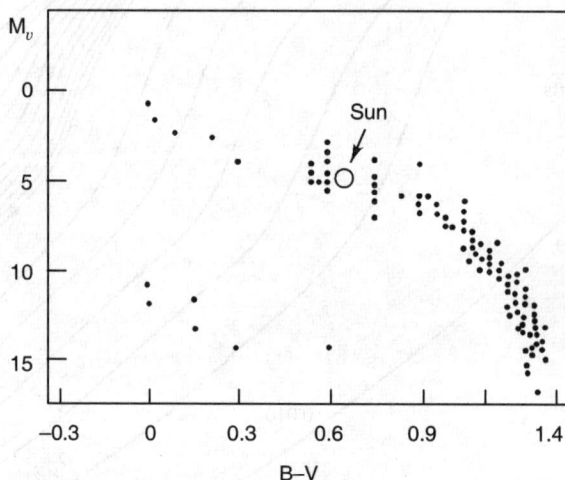

FIGURE 4.10 Colour-absolute magnitude diagram of stars within 10 parsecs from the Sun.

Since the spectral classes of stars are intrinsically related to their colours in a manner that the earlier spectral classes are associated with smaller values of the colour B–V and later classes with larger values, it is obvious that the C–M diagram will resemble the H–R diagram in general appearance. With the modern photoelectric techniques, it is easier and less time-consuming to find the colour of a star than to have a good spectrum of it and then classify it. This advantage becomes more and more pronounced as we go towards fainter stars. Thus the colour-magnitude diagram serves as a much more convenient alternative to the H–R diagram now a days. One difficulty is however inherent in the former that the colours are affected by interstellar absorption and reddening. This has to be corrected beforehand and the corrected values $(B–V)_0$ have to be used in the diagram.

EXERCISES

1. Discuss briefly how Saha's work of the Ionization Theory ushered in a breakthrough in the understanding of the puzzling variety of stellar spectrum. What were the earlier interpretations?

2. Give reasons for the following:
   (a) Lines of ionized helium are found only in O stars.
   (b) Balmer lines of hydrogen are weak both in the hottest and the coolest stars.
   (c) Ionized metal lines are the predominant feature of stars like the Sun.
   (d) Strong lines of neutral metals and bands of molecules are present in M stars.
   (e) Among stars of the same spectral class giants are cooler than the dwarfs.
   (f) Stars having narrow and sharp spectral lines have larger radii and lower temperatures.

3. What elements (neutral as well as ionized) will show strong lines in spectra of stars having the following temperatures:

   (i) 35,000 K; 25,000 K; 15,000 K; 11,000 K; (ii) 8,000 K; 6,000 K; (iii) 5,000 K and 3,000 K.

   What is the predominant colour of each of these stars?

4. Draw a neat sketch of the Hertzsprung-Russell (H–R) diagram showing the positions of:

   (a) the main sequence stars with the Sun;
   (b) the blue and red supergiant stars;
   (c) the red giants and the cepheid variables;
   (d) the subdwarfs;
   (e) the white dwarfs.

5. Show how the positions of stars as enumerated in Exercise 4 in the Hertzsprung-Russell diagram can be explained on the basis of the stellar evolution theory (refer to Chapter 14).

6. Plot the stars having the following spectral classes on the Hertzsprung-Russell diagram:

   O7V; B0V; B5III; B9Ia; A0V; F5V; G2V; K0Ib; K5Ia; K9V; M5V and M0Ia.

7. Using Eq. (4.5) compute the relative populations of the $n = 1$ and $n = 2$ levels for hydrogen for $T = 30,000$ K, 25,000 K, 20,000 K, 15,000 K, 11,000 K, 8,000 K, 6,000 K, and 4,500 K. The statistical weight of the level n of hydrogen is $2n^2$.

8. Using Eq. (4.9), compute the fraction of calcium atoms in the singly ionized state in the following atmospheres:

   (a) $T = 10,000$ K, $P_e = 300$ dynes cm^{-2};
   (b) $T = 6,000$ K, $P_e = 300$ dynes cm^{-2};
   (c) $T = 6,000$ K, $P_e = 30$ dynes cm^{-2};

   For calcium log $2u_1(T)/u_0(T) = 0.18$.

9. Using Eq. (4.9) compute the proportion of neutral aluminium in an atmosphere of $T = 6,000$ K, $P_e = 300$ dynes cm^{-2}. For aluminium, log $2u_1/u_0 = 0.34$.

10. Define *spectroscopic parallax*. Explain how spectroscopic parallax can be used to determine distances of far-off stars.

11. Explain the equivalence of C–M and H–R diagrams. Why C–M diagram is used more often than the H–R diagram in the stellar photometry?

SUGGESTED READING

1. Aller, L.H., *Astrophysics: The atmospheres of the sun and stars,* 2nd Ed., Ronald Press, New York, 1963.

2. Ambertsumyan, V.A., *Theoretical Astrophysics,* Pergamon Press, New York, 1958.

3. Bohm-Vitense, Erika, *Introduction to Stellar Astrophysics,* Vol. 1, Cambridge University Press, Cambridge, 1989.

4. Burgasser, A.J., PhD thesis, California Institute of Technology, 2001.

5. Dahn, C.C., et al., AJ, **124**, 1170, 2002.

6. Kaler, J.B., *Stars and Their Spectra,* Cambridge University Press, Cambridge, 1989.

7. Kirkpatrick, J.D., Henry, T.J. and Liebert, J., ApJ, **406**, 701, 1993.

8. Saha, M.N., Ionization in the solar chromosphere, *Phil Mag.,* **40**, VI, p. 472, 1920.

9. _____, Elements in the sun, *Phil. Mag.,* **40**, VI, p. 809, 1920.

10. _____, On a physical theory of stellar spectra, *Proc. Royal Society of London,* **A99**, p. 135, 1921.

11. Smith, E.V.P., and Jacobs K.C., *Introductory Astronomy and Astrophysics,* W.B. Saunders, Philadelphia, 1973.

12. Struve, O., and Zebergs, V, *Astronomy of the 20th Century,* Macmillan, New York, 1962.

13. Thackery, A.D., *Astronomical Spectroscopy,* Macmillan, New York, 1961.

14. Tinney, C.G., Burgasser, A.J. and Kirkpatrick, J.D., AJ, **126**, p. 975, 2003.

15 Vrba F.J. et al., AJ, **127**, p. 2948, 2004.

5 The Sun

5.1 SUN—A TYPICAL STAR

In the vast stellar system to which we belong, i.e. our Galaxy consisting of some 10^{11} stars, the Sun acts just as a mediocre member in every physical aspect. Nevertheless, it is the most important star from our point of view, the "effulgent being" around which moves with 'awe' and 'reverence' the Earth and other major and minor planets and comets, and which controls, directly or indirectly, the motion of every member of the solar system, big and small. Being the nearest star, it has offered scope for detailed physical studies. The Sun is the only star which shows a disc the different parts of which can be studied in isolation, unlike any other star which looks just like a point source. Being the source of almost all the energy which we use, the Sun maintains life on Earth and also in any other planets, if there happens to be any living being. The use of modern sophisticated instruments and efficient techniques of observation coupled with the physical laws, have enabled us to gather considerable insight into the structure and true physical characteristics of the Sun. We, now know fairly accurately the values of pressure, density and temperature all the way from the surface to the centre of the Sun. Its radiant energy is generated by thermonuclear transmutation of hydrogen into helium at a central temperature of about 16 million degrees Kelvin. Using the fairly accurate model of the interior and atmospheric structures of the Sun, we can construct models for distant average stars and can thus understand the physical nature of those remote objects for which even a good spectrum can hardly be obtained with the most precision instruments ever made.

It should be made clear, however, that although the Sun offers scope for detailed and most accurate observations by virtue of its nearness, these observations are restricted only to its atmosphere and extreme superficial layers. The main body of the solar interior cannot be directly observed. One has to extend knowledge to the interior by applying the laws of physics governing the equilibrium of a radiating gaseous sphere, in combination with the observed results of the upper layers. The problems of the overall equilibrium structure, thermonuclear energy generation at the core, and the mechanisms for transporting this energy through the interior to the surface layers etc. have been discussed in Chapter 14. The interaction of the outgoing radiation with the relatively tenuous atmosphere has been considered in Chapter 6. We shall therefore discuss in this Chapter only the observable aspects of the Sun and some of the physical theories which have been applied to find satisfactory interpretation of the observed results. The observed phenomena are many and varied. Sunspots and faculae in the photosphere, spicules and plages in the chromosphere, great prominences, flares and streamers in the corona, the solar wind and sporadic radio bursts from the outer atmosphere of the Sun, and the most mysterious 11-year cycle of solar activity—all these call for their explanation a host of physical theories. But our discussion here will be limited to only some of these varied solar phenomena in their simpler aspects. Figure 5.1 gives the schematic description of the interior and outer structures of the Sun.

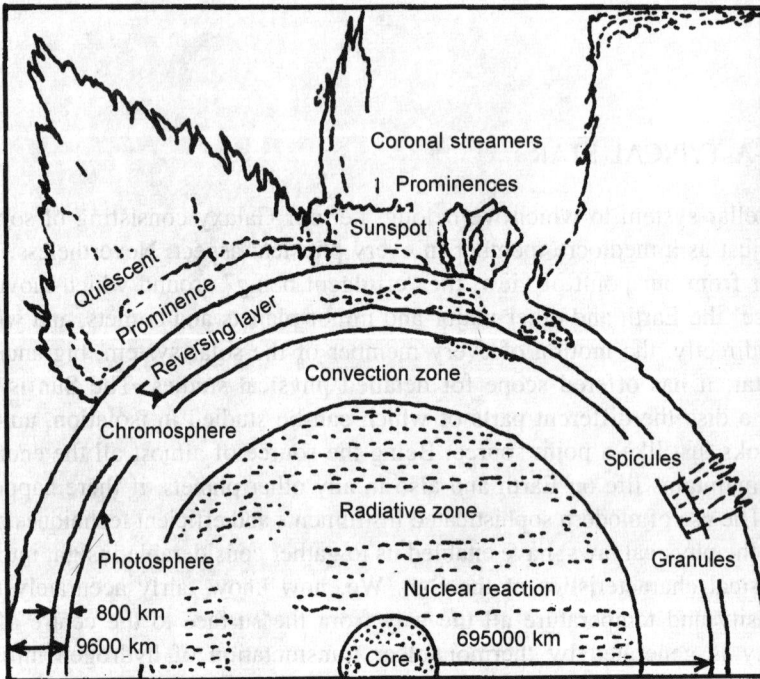

FIGURE 5.1   A schematic description of the interior and outer layers of the Sun, together with some atmospheric phenomena.

Although it is now generally believed that most of the solar phenomena and their physical explanations are very well understood, some recent discoveries about the Sun, if confirmed, indicate that many physical processes in the Sun still remain shrouded in mysteries. It has been found that the Sun is pulsating with a period of nearly 2^h40^m. The amplitude of pulsation is very small, only about 10 km, the average radial velocity ranging from 2 to 4.5 m s^{-1}. If this is confirmed by further experiments, it will have far-reaching consequences in our present understanding of the physical processes in the Sun. Analyzing the various consequences of this new discovery, A.B. Severny and his co-workers have suggested that the solar energy generation does not occur by the hitherto well known proton-proton reactions at the core. This conclusion seems to be supported, by another entirely different and independent experiment by Davis and his co-workers. If proton-proton reactions were operative in the core of the Sun then certain amount of neutrino flux should emerge and a reasonable amount should be capable of being detected by a favourable set of experiments. Such an experiment was carried out by Davis, Jr., and his co-workers for several years in Brookhaven; however, the observed neutrino flux was considerably lower than should have been if the proton-proton reactions were operative.

All these recent discoveries indicate that time may not be remote when our present understanding of the solar physics may have to be drastically revised. It is now too early, however, to draw any firm conclusion in this regard. In the following sections, we shall, therefore, pursue the discussion of the solar phenomena as are currently known to be established.

5.2 THE PHOTOSPHERE: LIMB-DARKENING

The atmosphere of the Sun is composed of layers of hot gases. These layers, however, are not sharply defined. In fact, each one of these is separated by transition zones of several kilometres thickness and they greatly differ from each other in physical characteristics. Starting from the visible solar surface upwards, these layers are known as the *photosphere,* the *chromosphere* and the *corona,* respectively, the last one merging into the interplanetary space (see Fig. 5.1). The temperature minimum is attained at the transition layer between the photosphere and the chromosphere. The temperature increases rapidly both ways from this layer which is often described as the *base of the chromosphere* (see Fig. 5.2). This minimum temperature is about 4200 K while the temperature at the base of the photosphere where our view is obstructed is around 5800 K.

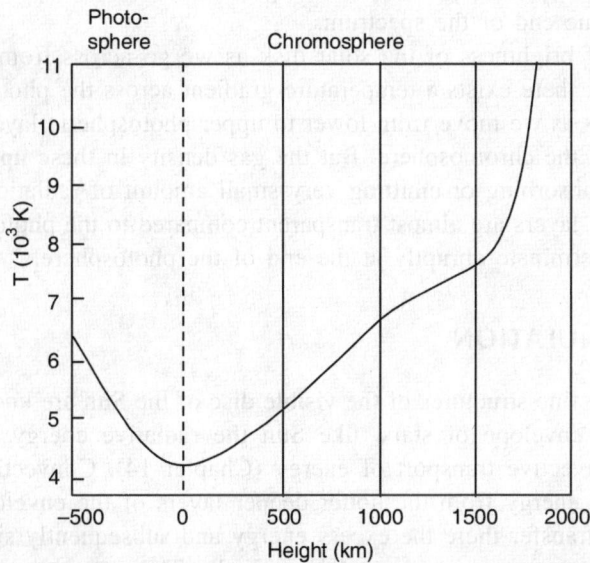

FIGURE 5.2 The temperature distribution near the photosphere-chromosphere boundary.

The visible disc of the Sun which we recognize as the solar surface may be considered as the base of the photosphere. This is the innermost layer of the solar atmosphere to which our eyes can penetrate through the superficial transparent layers. Our view is obstructed beyond this layer by the rapidly increasing opacity of the denser layers of the gas inwards. The thickness of the entire photospheric layer which runs from completely transparent to perfectly opaque layers of gas is quite small, only of the order of 200–300 km. The continuous absorption spectrum of the Sun produced by H⁻ ions originates entirely in the photosphere. It is now believed that most of the Fraunhofer lines also originate in this layer. Previously it was believed that these lines form in the lower part of the chromosphere which has therefore been given the name "reversing layer", meaning that most of the "reversed" lines (absorption lines) of the solar spectrum originate in this layer. The current observations however indicate that most of the solar absorption lines originate in the photosphere—the weaker lines in the lower photosphere and the stronger lines in the upper photosphere. Only, some of the strongest lines, such as the Balmer lines of hydrogen and H and K lines of Ca II have their origin primarily in the lower chromosphere.

The pressure and density in the photosphere are much lower in comparison to those in the terrestrial atmosphere at sea-level. These values are of the order of 10^{-2} and 10^{-4} of the respective terrestrial values at the sea-level (the density of molecules at standard pressure at sea level is $= 2.688 \times 10^{19}$ cm^{-3}). The depth of the photosphere through which our view can penetrate depends on the part of the disc we are observing and the opacity of the photospheric gas. Deeper and hotter photospheric layers are observed when we look to the Sun's disc near the centre. When we see the disc near the limb our view penetrates through cooler superficial layers of the photosphere. This causes the well-known *limb darkening effect* in solar observation. Light from near the edge of the disc comes from upper, cooler and more tenuous layers of the photosphere than that from near the centre (see Fig. 5.3). So we see the limb of the solar disk redder and dimmer as compared to the bright and hot central part. The effect is most pronounced at the blue end or the spectrum.

The decrease of brightness of the solar disk as we go across from centre to the edge, points to the fact that there exists a temperature gradient across the photospheric layers. The temperature decreases as we move from lower to upper photospheric layers. The temperature again rises rapidly in the chromosphere. But the gas density in these upper layers is so low that it is capable of absorbing or emitting very small amount of radiation. The opacity thus being very low, these layers are almost transparent compared to the photosphere. This is why the disk appears to terminate abruptly at the end of the photosphere.

5.3   SOLAR GRANULATION

The typical grain-like fine structures of the visible disc of the Sun are known as *photospheric granulations*. In the envelope of stars, like Sun the radiative energy transport has to be supplemented by convective transport of energy (Chapter 14). Convective cells of gaseous mass carrying higher energy from the hotter deeper layers of the envelope rise through the upper cooler layers, transfer there the excess energy and subsequently sink down again into the deeper layers. The process is repeated incessantly. These rising and falling convective

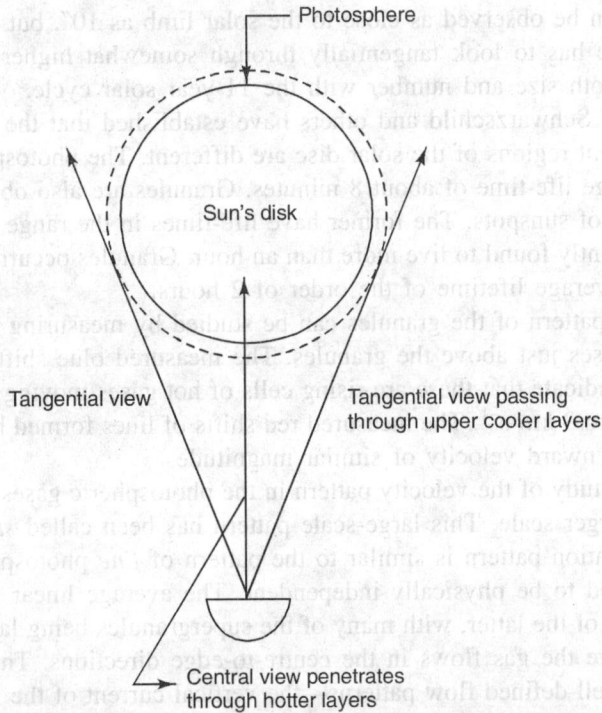

FIGURE 5.3 A schematic diagram explaining the cause of the limb-darkening in the Sun.

cells appear in the photographs of the solar disk as bright *granules* of various sizes bordered by darker regions.

The granulation of Sun was known to observers for many decades, but its detailed visual studies were difficult, because the fine structures were not revealed due to bad "seeing" due to Earth's atmospheric turbulence. In 1959, a part of the effect of bad "seeing" was overcome when D.E. Blackweli, D.W. Dewhirst and A. Dollfus obtained photographs of the Sun from a balloon sent to a height of 18,000 ft. above the Earth's surface. The complex structures of the granules were clearly revealed in these photographs. Far better photographs with finer and distinct structures were, however, obtained soon after in the same year by Martin Schwarzschild and J.B. Rogerson of Princeton, who sent instruments to photograph the Sun's disk to a height of 80,000 ft. above the Earth's surface in a stratospheric balloon. Analyses of these and subsequent materials revealed an enormous amount of information regarding the physical and dynamical properties of the granules.

The diameters of the granules vary from 300 km to 18,000 km. with a mean value of about 700 km. Their shapes resemble irregular polygons or cells which are separated from one another by narrower and darker filamentary regions. K. Schwarzschild suggested that the cellular structures arise from a state of non-stationary convection below the solar surface. The narrow dark areas, on the other hand, characterize the inflow of cool materials from the surrounding photospheric regions. Granules are found everywhere on the solar surface including the regions of sunspots. The total number of granules on the visible disk of the Sun is about

3.5×10^6. They can be observed as close to the solar limb as $10''$, but are not observed still closer as there one has to look tangentially through somewhat higher layers. Granules are found to vary in both size and number with the 11-year solar cycle.

The works of Schwarzschild and others have established that the lifetimes of granules occurring in different regions of the solar disc are different. The photospheric granules (quiet region) have average life-time of about 8 minutes. Granules are also observed in umbral and penumbral regions of sunspots. The former have life-times in the range 15–30 minutes while the latter are frequently found to live more than an hour. Granules occurring in facular regions have the highest average lifetime of the order of 2 hours.

The velocity pattern of the granules can be studied by measuring the Doppler shifts of lines formed by gases just above the granules. The measured blue shifts of lines formed by gases in granules indicate that these are rising cells of hot gases moving upwards with speeds ranging between 1 to 3 km s^{-1}. The measured red shifts of lines formed by gases in dark lines also indicate a downward velocity of similar magnitude.

The detailed study of the velocity pattern in the photospheric gases has revealed cellular pattern of much larger scale. This large-scale pattern has been called *supergranulation*. The *coarse* supergranulation pattern is similar to the pattern of *fine* photospheric granulation but the two are believed to be physically independent. The average linear size of the former is about 20 times that of the latter, with many of the supergranules being larger than 30,000 km. Within this structure the gas flows in the centre-to-edge directions. The photospheric gases thus possess two well-defined flow patterns—the vertical current of the fine granules and the horizontal current of the supergranules. Superimposed on these two patterns of motion there is also a third pattern of slow oscillatory motion up and down which performs a complete cycle in about 5 minutes.

One can study the nature of convection which gives rise to granulation by applying Reynolds or Rayleigh numbers. For granulation these numbers are high, indicating that turbulence beneath the photosphere is probably the basic cause of the entire manifestation. Let us consider the Rayleigh number defined by

$$R = \frac{g\alpha|\beta|\,l^4}{Kv_K} \tag{5.1}$$

where g is the acceleration due to solar gravity, α is the coefficient of volume expansion of the material, β is the temperature gradient in the level considered, l is the linear scale of granules, K is the coefficient of thermoelectric conductivity of the gas and v_K is its kinematic coefficient of viscosity. When R exceeds some critical value, say R_o, one obtains the onset of convection. R can be increased by slowly increasing β.

In subphotospheric layers the appropriate values of the parameters in cgs units are:

$$l \sim 3 \times 10^7, \ \alpha \sim 10^{-4} \ \text{(for perfect gas)}, \ \beta \sim 10^{-4}$$
$$g \sim 2.7 \times 10^4, \ v_K \sim 10^3 \ \text{(atomic viscosity), and}$$
$$K \sim 10^{12} \ \text{(determined by the radiation field)}$$

When these values are substituted in Eq. (5.1), one gets $R \sim 10^{11}$ indicating the existence of a random convection generated by turbulent motions. Thus, photospheric granulation is but the manifestation of a subphotospheric random convective motion originated by the high opacity in the hydrogen convection zone of the solar envelope.

5.4 FACULAE

The bright areas around sunspots occupying at one time a substantial fraction of the solar disc are known as faculae. These are phenomena belonging to the transition region between the photosphere and the chromosphere. As a result, *both photospheric faculae* and *chromospheric faculae* are recognized. The former are visible in white light near the limb only. Closer to the centre of the solar disk, they lose contrast with respect to the photospheric brightness and so cannot be recognized. The chromospheric faculae, on the other hand, are visible in bright chromospheric lines, particularly in H and K lines or Ca II. The brightness of faculae remain almost unchanged over the entire disc of the Sun.

Although photospheric faculae are definitely associated with sunspots, they are found to appear before the spots and outlive the latter by several months. In fact, whereas the lifetime of spots hardly exceeds a month, that of faculae generally lies in the range 200–300 days. There are also faculae which are not associated with sunspots, but these are invariably smaller, fainter and relatively shortlived. So, observation of very bright and compact masses covering large areas but unassociated with spots, invariably points to the conclusion that either it was associated with a spotgroup which has just recently disappeared, or that a spotgroup is soon going to appear. Photospheric faculae are thus closely correlated with the solar activity. Largest individual faculae generally occur at the time of maximum activity. Two to three years after maximum, the individual faculae become smaller in size and also fainter, although they may be more numerous. Photospheric faculae, like the photosphere itself, exhibit a granular structure. The average lifetime of these granules is about an hour which is much larger than the average lifetime of about 8 minutes for the photospheric granules.

As has been already mentioned, the chromospheric faculae have been mostly observed in H_α and Ca II lines. Of these, the latter have been more extensively studied because of their greater brightness. The faculae observed in these two lines also differ in general structure. Those in Ca II lines appear more compact and even-shaped, while faculae in H_α are more patchy. Like photospheric faculae, chromospheric faculae also exhibit close correlation with sunspot cycle. These two classes of faculae also bear close similarity in respect of shape, size, lifetime and frequency of occurrence.

Like many other solar phenomena, the real cause of formation and development of faculae is still only poorly known. They exhibit a peculiar temperature distribution, cooler below and hotter above than the surrounding gases, which can be maintained only by some kind of forced supply of energy. Whether this supply is maintained by convection currents or by magnetic field in the subphotospheric layer is not definitely known. Although faculae are associated with sunspot activity, they don't seem to bear cause-effect relationship. On the one hand, the patterns of their motion are not exactly the same. On the other, faculae usually precede sunspot groups and outlive their several cycles. These and other considerations indicate that both faculae and sunspot activity are manifestations of some third agent. The subphotospheric magnetic field is believed to be a likely candidate.

5.5   THE CHROMOSPHERE

Above the photosphere lies the second major layer of the solar atmosphere known as *chromosphere*. The nomenclature is derived from its display of reddish colour when the photosphere is eclipsed by the moon. The layer extends to nearly 20,000 km above the photosphere with gas density gradually thinning out but the temperature rising rapidly upwards. Since the hot chromospheric gas is viewed against the dark background of the sky when the solar disk is hidden by the Moon, we get at the instant the chromospheric emission spectrum in place of the photospheric absorption spectrum of the Sun. Most of the photospheric Fraunhofer absorption lines are observed for the few moments as the chromospheric emission lines, together with many additional lines of He and ionized metals which are absent in Fraunhofer spectrum. This emission spectrum is known as the *flash spectrum* as earlier when great advancement in instrumentation was not achieved, this spectrum could be obtained only during the few seconds of the totality of solar eclipse. Now, however, a solar observer can study the chromosphere at his convenience by artificially eclipsing the Sun with the use of a *Coronagraph*. The reddish hue of the chromosphere is mostly imparted by the Balmer H_α emission in the red (λ 6563 Å).

The lower layers of the chromosphere, about 500 km in thickness, is simply an extension of the photosphere where the strongest absorption lines, in particular, the Balmer series of hydrogen and H and K lines of Ca II are formed. The gas in the chromosphere is fairly inhomogeneous. The density of H$^-$ which is mainly responsible for producing the continuous absorption in the photosphere, being very low in the chromosphere, the medium is transparent to most of the radiation from the photospheric layer. The lower chromosphere extends to about 8000 km above the photosphere where temperature is not high enough to substantially ionize hydrogen and Helium. The temperature increases gradually to very high values in the upper layers of the chromosphere. At a height of about 20,000 km above the photosphere, where the chromosphere merges into the *Corona*, the temperature is $\sim 10^6$ K.

The increasing temperature along the height is clearly reflected in the spectra produced at different levels of the chromosphere. At the lowest level near the base of the chromosphere, the spectrum shows lines of various neutral metals, such as, Fe I, Na I etc. which are characteristic of low excitation. Higher above, lines of ionized metals such as, Fe III, Ca II, Sr II, Ti II constitute the chief spectral feature. Balmer lines of hydrogen and He I lines are very prominent in spectra of still higher layers, say, at a height of around 10,000 km. Still higher, layers of the chromosphere are characterized by strong He II line at λ 4686 Å. Because of decreasing density with increasing height all the chromospheric lines fade away gradually with height. But the effect of temperature rise with height is not uniform for all the elements. Lines of low excitation fade most rapidly with height, while the rate of fading is slow for lines of H and He. In fact, He II line at λ 4686 Å is observed to fade most slowly with height.

Spicules

Having thus discussed briefly some of the large-scale observed features of the chromosphere, we shall now study some of the fine structure features of chromospheric phenomena. During a total solar eclipse or with the use of a coronagraph "spiky" structures of hot tenuous gas are seen to rise from lower chromosphere and extend upwards to 10,000 km, or more, sometimes even to 20,000 km. The diameters of these spikes are, on the average, between 500–600 km with larger ones having diameters ~ 2000–5000 km. These projected chromospheric streamers are called *spicules*. They are best observed in H_α radiation. The velocity of their upward movement is ~ 30 km s^{-1}, and their average lifetime is ~10 minutes. Spicules are observed as dark mottles in H_α spectroheliograms to cover a good fraction of the solar disk. When viewed near the solar limb, such a large number of these is seen together in projection as to give the appearance of a "forest of spikes".

Spicules exhibit a network pattern which rise from the boundaries of the supergranule cells. They are therefore, believed to be the manifestation of the same common underlying cause, viz. the enhanced magnetic activity. They appear to be the moving disturbance through solar chromosphere, driven by enhanced magnetic field, and probably play a vital role in transferring mass from lower chromosphere to the corona.

Plages and Filaments

Large bright and dark clouds are revealed in the chromosphere when spectroheliograms in the monochromatic light of H_α and Ca II K are examined. These are respectively called *Plages* and *Filaments*. Plages are alternatively called Flocculi. In *spectroheliograms* they appear as *clouds* of hydrogen and calcium, but in fact they are just the regions of chromospheric gas which radiate more intensely at the wavelengths of H_α and Ca II K line. Faculae are similar regions which radiate intensely in many different wavelengths constituting white light in which they are observed.

Like faculae and spicules, plages are also associated with enhanced magnetic field of the active Sun. They are mostly seen in regions of sunspots, but sometimes they are also seen where there are no visible sunspots. They are likely to be associated with regions of higher magnetic field.

5.6 SOLAR CORONA

The corona is the natural and smooth extension of the chromosphere, but yet differ greatly from the latter in physical state. It is seen very clearly during a total solar eclipse as the effulgent halo embracing the solar limb by the inner end while the outer end stretches far beyond the solar disk, gradually merging and losing its identity into the interplanetary space.

Unlike chromosphere, the corona was known to people for many centuries, but in early days it was recognized generally as an optical illusion. Its real existence and particularly, its physical nature could hardly be guessed. Even an astronomer like Kepler who studied the corona in some detail could hardly recognize its actual nature. The spectrum of the corona was first studied in 1869 by the American astronomers, W. Harkness and C.A. Young, and its First successful photograph was obtained in 1930 by the French physicist B. Lyot, with his newly discovered instrument, the Coronagraph. With the advent of this instrument, it is now possible to observe and analyze the inner corona even without a solar eclipse. The outer corona, however, can be still observed at the time of total solar eclipse only. It may be remembered, of course, that the duration for which the corona can be observed at an eclipse ranges from only a fraction of a second to a few minutes, the actual time depending on the extent of the eclipse and on the position of the observer.

The corona extends up to several solar radii above the photosphere. The transition between the chromosphere and corona takes place at about a radial distance of $R = 1.03 \ R_\odot$, $R_\odot$ being the radius of the Sun. Although the corona emits half as much light as the full moon, because of its enormously large surface area, the surface brightness is quite low which is completely lost in the brilliance of the photospheric radiation. Three components of the corona can be recognized on the basis of the nature of radiation emitted by it.

The Inner Corona or K-Corona

The inner corona which is also sometimes called as the real corona extends between $1.03 \ R_\odot$ $< R < 2.5 \ R_\odot$. This part of the corona, known as the K-corona, imitates the continuous spectrum of the photosphere (the name K-corona is derived from the German word "Kontinuum", after W. Grotrian) but the Fraunhofer lines are absent. This can be understood if one considers that the free electrons in the highly ionized coronal gas scatter photospheric light to produce the observed continuum. It has been suggested that the Thomson scattering is operative in the process. If, however, the scattering electrons move extremely fast, Doppler broadening may be so high as to diffuse the Fraunhofer lines completely and render these unrecognizable. The lines are then said to be "washed out" by the fast moving particles. The temperature of the medium required to impart such high velocities to particles is higher than 10^6 K, leading to the conclusion that the coronal temperature is extremely high.

The Outer Corona or F-Corona

The F-corona which lies at $R > 2.5 \ R_\odot$ displays the solar spectrum with Fraunhofer lines superimposed on the continuum. This is sometimes called the "false" part of the corona and the prefix F stands for Fraunhofer. The spectrum of the F-corona is produced by the Sun's light scattered by tiny dust particles of the interplanetary space. The dust is concentrated in the place of the ecliptic and the outer part of the F-corona is observed to merge into the zodiacal light. These lines are assumed to be formed by forward scattering by particles with sizes somewhat larger than the wavelengths scattered. These particles lie far enough from the Sun and are not therefore heated to the degree of vaporization. Grotrian has suggested that the F-corona is essentially an inner extension of the zodiacal light. This has been confirmed by Blackweli's measurements of the brightness distribution and polarization from the F-corona.

The F-corona together with the K-corona constitute what is called the white corona which merges into interplanetary space with decreasing brightness.

The Emission Corona or E-Corona

In the optical range of the coronal spectrum, about two dozen emission lines are found to be superimposed on the continuous background. The total light of these emission lines formed by highly ionized atoms in the extremely hot inner part of the corona constitute what may be called the E-corona or emission corona. The total radiation in these lines is, however, quite small, that is less than even 1% of the total coronal radiation, but the study of the physical conditions under which this radiation is emitted is very interesting. Until 1942, these lines remained unidentified and were called *coronium lines*. Then W. Grotrian of Germany and

B. Edlen of Sweden identified these with *the forbidden lines* of highly ionized atoms like Ca, Fe, Ni and some other abundant elements. The strongest among these lines are those of Fe XIV at λ 5303 Å in the green and of Fe X at λ 6374 Å in the red. Some very strong lines as λ 7892 Å of Fe XI, and λ 10746.8 Å and λ 10797.9 Å of Fe XIII are also observed in the near-infrared. The formation of these forbidden lines of extremely high order ions of various elements yields two definite informations about the physical state of the coronal gas: first, the gas density is very low and second, the temperature of the corona is extremely high, as high as 2×10^6 K or even higher in some layers. The density of particles (mostly electrons) is only of the order of 10^6 to 10^8 cm^{-3} as against the values of 10^{10} to 10^{12} for chromosphere and of 10^{16} to 10^{17} for the photosphere.

Highly ionized atoms as are observed in the corona also possess metastable level transition which will give rise to ultraviolet lines. These lines could not, however, be observed previously as they are shielded by Earth's atmosphere. A major programme was undertaken however in late sixties in order to study the Sun in ultraviolet radiation by sending instruments in Orbiting Solar Observatory. The first few launching attempts failed, but the fourth attempt successfully placed the Observatory in the orbit, which has subsequently been called the Orbiting Solar Observatory IV, A huge amount of observational material has been sent from this Observatory which gave invaluable information about the ultraviolet radiation from the Sun. Among other things, many forbidden lines' as well as permitted lines of highly ionized elements have been detected as a result, their wavelengths ranging all the way from extreme near to extreme far ultraviolet. The electrons being strongly bound in these highly ionized atoms permitted transitions correspond to very high excitation potentials so that the emitted photons give rise to lines in the far ultraviolet. As a result, most of the permitted lines lie in the extremely far ultraviolet region of the spectrum.

We have thus seen that the upper layers of the solar atmosphere is tenuous but extremely hot. The temperature rises from ~ 4500 K at the base of the chromosphere to ~ 10^6 K at the upper layers of the chromosphere. Still higher values of temperature, in the range between $(1–3) \times 10^6$ K, are encountered in the various layers of the corona. The temperature becomes still higher at the time of maximum solar activity. Such high temperature in the higher tenuous layers of the solar atmosphere are now physically well understood. It has been suggested that hydromagnetic waves and acoustic waves generated in the convection zone beneath the photospheric layers are damped while propagating through the upper layers of gradually decreasing density, and thus dissipate much of the wave energy to heat the gas. Theoretical studies have indicated that this mechanism of heat transport is sufficient to explain the extremely high temperature of the corona.

It is found that like other physical parameters, the shape of the corona also undergoes sufficient change with the solar activity. The corona becomes sufficiently elongated and coronal streamers are driven at the time of minimum solar activity, while during the maximum activity of the Sun the corona looks almost spherical (see Fig. 5.4). The values of the ellipticity ε defined by

$$\varepsilon = \frac{\text{equatorial diameter} - \text{polar diameter}}{\text{equatorial diameter}}$$

are ~ 0.05 at maximum activity and ~ 0.25 during the minimum activity of the Sun. Such ellipticities can, of course, be measured only for the inner corona. Since the boundary of the

FIGURE 5.4   Solar Corona near maximum activity. (*Courtesy:* Indian Institute of Astrophysics, Bangalore).

outer corona cannot be defined because of its extremely low density it eludes any such measurements. The outer corona probably extends as far as the Earth and even beyond. This is indicated by the scintillation of radio sources lying in space 90° away from the Sun.

5.7   PROMINENCES

During a total solar eclipse or while photographing the Sun with the help of a coronagraph the solar disk exhibits the well known *red flames* protruding from the chromosphere through the corona (see Fig. 5.5). These flame like structures are called prominences which are sometimes seen to rise to a height of more than a million kilometres above the solar surface. Typical

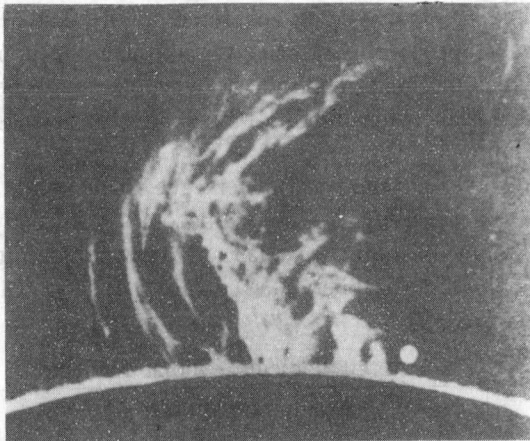

FIGURE 5.5   A huge spectacular solar prominence.

dimensions of a prominence can be considered'as being 30,000 km in height, 200,000 km in length and 5,000 km in thickness yielding a total volume of 3×10^{28} cm^3. Though the prominences are best seen at the limb of the Sun as bright structures against the comparatively darker corona, they are also observed for longer intervals of time as dark irregular filaments against the surrounding bright background of the chromosphere from which they are thought to originate. Observations are generally made in monochromatic lights (most often in H$_\alpha$ and Ca II lines) on spectroheliograms or on filtergrams.

The changing structures and mass motions in prominences are now-a-days studied with the help of motion picture records, a very useful method first devised by R.R. McMath. The method provides a continuous record of motions of a prominence from its birth to death. Such motion pictures show that the gaseous material in prominences moves downwards along curved paths called *arches* as if the coronal materials are being continually poured into the chromosphere. From their structures and pattern of motion of material, several different types of prominences can be recognized. In the following, we briefly discuss these various types pointing out the basic feature or features of each of them.

Quiescent Prominences

These are characteristic prominences in less active Sun and as such are marked by slower mass motions and greater longevity. They generally appear as dark but stable filaments against the bright disk of the Sun and survive for periods in the range from a few hours to several days or even months. Lifetimes equal to a few solar rotations are very common with the quiescent prominences. A typical filament when fully developed may have dimensions of 50,000 km in height, 200,000 km in length and about 10,000 km in thickness. The length increases, on the average, by about 100,000 km per rotation of the Sun.

After a "quiet" period of life a quiescent prominence may evolve into an active or eruptive prominence and merge into space with violent velocities (~ 700 km s^{-1}). This transformation is generally accompanied by the development of a *facular region* or a *centre of activity* in the vicinity of the prominence.

Even in the 'quiet' or 'stable' phase of a quiescent prominence there is, of course, always some turbulent motions present in it.

Two basic problems to understand about quiescent prominences are how can they form and how can they maintain their structures at a temperature of ~ 10^4 K surrounded by coronal gas at a temperature ~ 10^6 K. The first problem relates to condensation of coronal material by cooling which can be caused by expansion of a magnetic region. Magnetic force must also be the chief agent for supporting the filaments in horizontal as well as in vertical equilibrium. In the regions where filaments occur the magnetic field is invariably parallel to the solar surface. Such a situation has important bearing in maintaining equilibrium because the mass motion is prevented only along perpendicular to the field. Thus, Coronal materials are being continually condensed in the prominence region by expansion, and the equilibrium is maintained by a magnetic field parallel to the Sun's surface. The magnetic energy density in quiescent prominences is comparable to the thermal energy density.

Active Prominences

These prominences usually occur in sunspot zones but are different in nature from those of the *sunspot type* of prominences. Occasionally, they are found to develop from the quiescent types and, in later phases, they often become eruptive ones. Active prominences are characterized by huge mass motions which are sometimes joined by curved filaments with the photosphere. Long narrow filaments are seen to be ejected continuously from the main massive part of these prominences.

Eruptive Prominences

Eruptive type of prominences are the most violent of all solar prominences. Sometimes they develop from active prominences where material is thrown violently from the solar body with velocities as high as 1200 km s^{-1}. Since the escape velocity on the Sun's surface is 618 km s^{-1}, this means that material thrown in some eruptive prominences escape from the Sun and mingles into interplanetary space. The eruptive prominences of June 4, 1946 observed by Pettit and Hickox were found to rise to a height of ~ 1.7×10^6 km or ~ $2.44 \, R_\odot$ above the solar surface.

Sunspot Type Prominences

These prominences generally appear in regions above sunspot activity in the form of *curved arches* or *loops* having mean projected lengths of the order of 60,000 km. They consist of condensed coronal material moving downward along trajectories just above spot groups. The curved paths are thought to be guided by lines of force of a strong magnetic dipole which probably lies immediately below the surface of activity. In sunspot prominences, the magnetic energy density is much larger than the energy density of thermal or random motions.

Condensation or *Knot* type prominences (also called coronal prominences) form yet another variety of sunspot prominences. These are seen as bright structures which apparently condense high up in the corona as Knots or series of arches. The material condenses at heights between 50,000 to 100,000 km above Sun's surface and flows down along arches into the chromosphere. The Knots are found to be joined by long but narrow streamers to some regions of sunspot groups which suggests their classification among sunspot prominences. The condensed materials flow down through these delicate streamers into the chromosphere with velocities of the order of 100 to 200 km s^{-1}. It is believed that the structure of Knot is defined by the equilibrium between the rate of condensation of material in the corona and the outflow of condensed material from the Knot into the chromosphere (along the magnetic lines of force). This type of prominence is believed to be associated with solar flare events.

Surges belong to another special class of sunspot prominences as were first classified by McMath and Pettit. These develop from sudden eruption of material from spot groups which scatter as splinters in all directions, rising to heights of several hundred thousands of kilometres and ultimately mingling into space. Occasionally, the materials fall back to the chromosphere with the same velocity with which they were initially projected. The velocity of projection may be as high as 1300 km s^{-1}. The eruption of surges is seldom repeated from the same region of the spotgroup. Surges are also found to be sometimes associated with flare activity.

Tornado type prominences belong to most rare types of solar prominences. These are twisting (helical) columns of gas rotating with violent velocities. Their diameters range from 5,000 to 20,000 km, with heights attained between 25,000 to 100,000 km. When rotational motion is sufficiently high these prominences may lose their structure. Also, sometimes they may evolve into an eruptive type. The cause of the development of such cyclonic prominences is still not well understood.

Various other types of prominences have been proposed by many authors. Divisions of these types into further sub-types has also been suggested according to the details in the structure and other physical characteristics. New schemes of classification have also been developed with different basic parameters like the downward motions from the corona ("coronal rains"), early phases of evolution, and others. Any further discussion on this topic is beyond the scope of the book.

Prominence spectrum has been studied by various authors in some detail. The principal feature of the spectrum of quiescent prominences is the presence of emission lines of H_α, He I and H and K lines of Ca II. Some sunspot prominences show strong emission lines of Fe II, Mg II and He I. Temperature measured from the analyses of spectral lines yield $\sim 10^4$ K for the quiescent prominences and $\sim 3 \times 10^4$ K for transient features associated with flares. Gas kinetic temperatures as determined from Balmer and Paschen continua are of the order of 1.2×10^4 K for the quiescent prominences, whereas for the active loop prominences connected with flares the values are of the order of 1.5×10^4 K. Kinetic temperature as obtained from measurement of line widths appear to be quite high, of the order of 5×10^4 K, because the large scale mass motions in flares or in active prominences would cause sufficient broadening of spectral lines. Electron densities in the prominences are of the order of $(0.5-3) \times 10^{11}$ cm^{-3} whereas in the surrounding coronal regions the gas densities are $(0.7-5) \times 10^8$ cm^{-3}, where the temperature is $\sim 2 \times 10^6$ K. Thus the densities in prominences are about 2 to 3 orders of magnitude higher than in the corona, whereas the temperature in the former is about 2 orders of magnitude lower than in the latter.

5.8   THE 11-YEAR SOLAR CYCLE AND SUNSPOTS

Among the photospheric phenomena sunspots are the most striking. These are regions of strong magnetic fields and of lower temperature, the average continuum temperature in the spots being ~ 3800 K, about 2000 K cooler than the surrounding photospheric temperature. Hence sunspots look darker against the brighter background of the photosphere although the spot regions are about as hot as K0 stars. In fact, the spectrum of sunspots resembles that of average K0 stars. Like other transient phenomena such as granulation, spicules, prominences, plages etc., sunspots also are manifestations of the *activity* of the Sun.

Sunspots were first observed by G. Galileo in 1610 with his newly discovered telescope, from which he estimated the rotation period of Sun to be nearly equal to a lunar month. The early observers like S.A. Wilson, William Herschel etc. in the late 18th and early 19th centuries believed that the spots were "holes" through which the "cooler interior" of the Sun could be seen. Now we know that these are all wild thinking but in Herschel's time such wild ideas were given to easy belief.

A spot first appears as a small "pore" of diameter of the order of 1000 km. The size and shape change gradually as the spot grows. A developed sunspot consists of two well defined regions—the dark central region called *umbra* which is surrounded by the relatively lighter region called *penumbra* (Fig. 5.6). In small spots only the umbral part is recognizable, while both the features are clearly visible in larger spots. Both the umbral and penumbral parts are found to contain granulations having lifetimes much longer than those found elsewhere. The reason for such difference is not yet well understood. The penumbra shows a structure composed of bright long filaments radiating outward from the umbra into the brighter surroundings. For a typical sunspot the umbral and penumbral diameters are of the order of 20,000 km and 40,000 km respectively, but for the largest umbral diameters, as large as 30,000 km with twice as large penumbral diameters are observed.

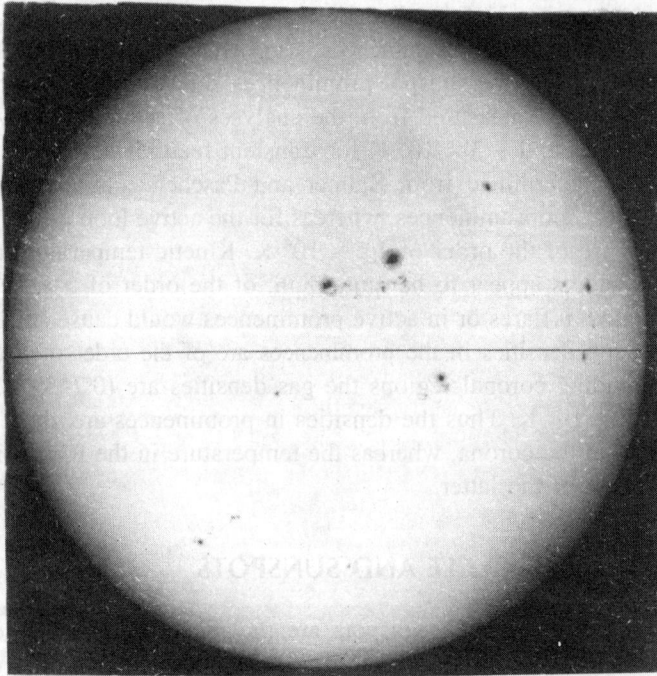

FIGURE 5.6 Sunspots showing umbra and penumbra. (*Courtesy:* Indian Institute of Astrophysics, Bangalore).

The spots first appear as pores in groups of many in the intergranular regions. The pores usually disappear within a period ranging from a few hours to a few days. Some of these last for about a week or more and may develop into large sunspot groups. These groups usually contain two principal members which can be easily recognized. The one that leads the group in its motion in the direction of Sun's rotation is called *leader* and the other which dogs the former is known as the follower in the group. In general, the leader spot is found to be the larger of the two. The follower spot, becomes fully developed within 3 to 4 days but the leader takes a week or more before it attains its maximum size. In the mean time, several other individual pores and larger spots begin to develop in areas surrounding and in between the main spot group. A well-grown group with all its members persists for a week or more after which it begins to decay gradually. At first, the follower spot breaks up into smaller fragment spots which disappear from the group within a few days or weeks. But the leader spot survives as an isolated, round-shaped spot for another few weeks or months and ultimately disappears from the Sun's disk.

From his long years of investigation the German amateur astronomer, Heinrich Schwabe, concluded in 1851 that the number of visible spots on the Sun's disk varied with time. He also discovered a periodicity associated with such variations. The period proposed by Schwabe was 10 years. Later observations, however, established that the period is in fact 11.2 years (but commonly described as 11-year cycle). The cycle of activity of the Sun is repeated nearly over this period, which is therefore known as the *solar cycle*. Within a solar cycle which is defined by the period between one minimum to the next, the number of sunspots and the intensity of other transient phenomena change appreciably. The maximum solar activity is characterized by the maximum number of spots which may be 100 or so or about 10 groups of spots. This is the time of sunspot maximum. The activity of the Sun and so the sunspot number decreases subsequently until it reaches a minimum. This is the time of sunspot minimum when scarcely any spot is visible on the solar disk. At any particular phase of the cycle, the total number of sunspots is fairly represented by *Zurich sunspot number R,* given by

$$R = K(10g + f) \tag{5.2}$$

a formula given by M. Wolf. Here g is the number of groups and f is the number of individual spots, K-being a normalizing factor equal to 1 for Zurich observations. The period between two successive maxima or two successive minima is not always the same. The average value between a maximum to the next minimum is about 6.7 years while that between a minimum and the next maximum is about 4.6 years. Two successive maxima may occur in an interval of as short as 8 years (1830 to 1838), and as long as 16 years (1888 to 1904). Thus quite a large variation to the commonly described 11-year cycle may occur in any particular cycle.

Like the sunspot numbers, variation is also found with the phase of the solar cycle in the latitude distribution of the spots. At the beginning of a cycle some spots or pores first appear in two zones at about ± 30° solar latitudes. These spots disappear within a few hours or a few days, and when new spot groups appear they do so in lower latitudes. These groups again give way to new groups of spots which originate at still lower latitudes. In this manner, the spot zones continue to shift towards lower heliocentric latitudes as the cycle advances till the centres or spot zones migrate to ± 15° solar latitudes at the time of sunspot maximum. During the next part of the cycle, the number of spots continue to decrease while the spot zones continue to migrate to lower latitudes. The last few spots at the minimum are visible at about ± 8°. Even prior to the minimum new spotgroups start appearing at latitudes ± 30° ushering in a new cycle of sunspot activity.

So the following rules have been found for polarity by G.E. Hale (1919) observing three 11-year cycles (Fig. 5.7). They are:

(i) The magnetic orientation of leader and follower spots and the uni-bipolar groups remain the same in each hemisphere over each 11-year cycle.

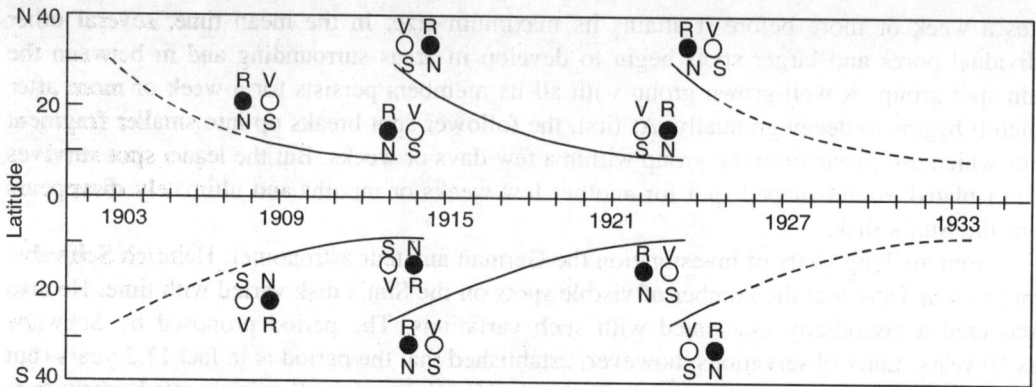

FIGURE 5.7   Theoretical Butterfly diagram showing latitude migration (Maunder 1922).

(ii) The bipolar groups in the two hemispheres have opposite magnetic orientation.

(iii) The magnetic orientation of bipolar groups reverses from one cycle to the next.

(iv) The latitude migration of the sunspot zones. The latitude migration is beautifully documented in the famous 'butterfly diagram' (Maunder 1922) and more recent example in Fig. 5.7. A theoretical butterfly diagram is shown in Fig. 5.8 using an oscillatory kinematical '$\alpha\Omega$' dynamo model (Steenbeck & Krause 1969).

Here, $\alpha \sim \pm\, l\Omega$, where Ω is the mean angular velocity of sun and l is the scale of variation of turbulent motion of Sun.

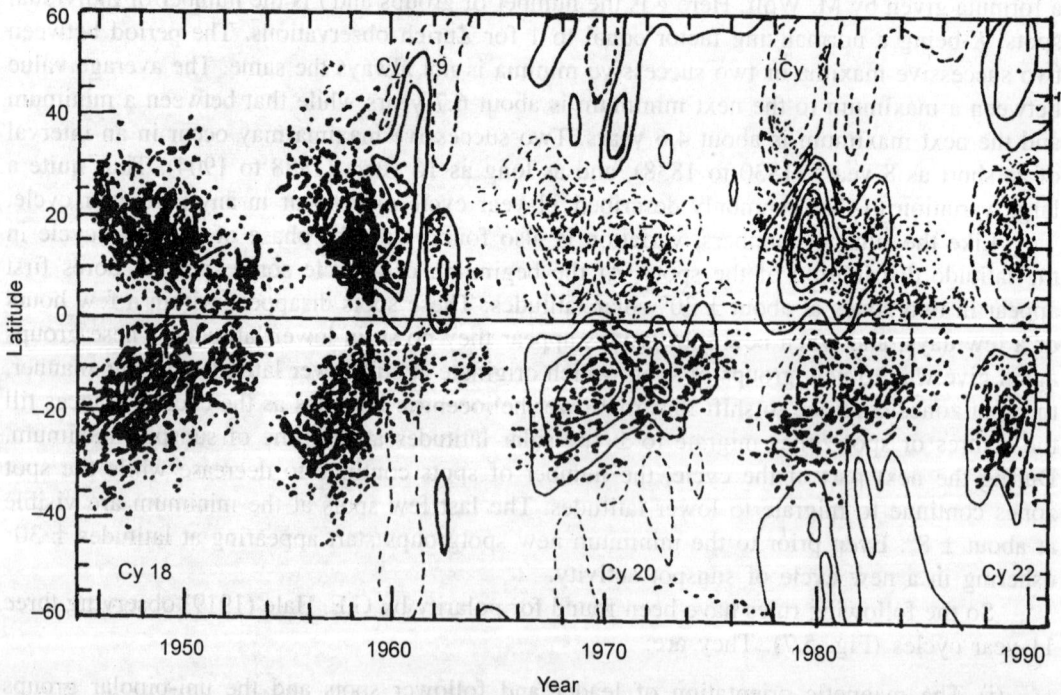

FIGURE 5.8   Recent theoretical Butterfly diagram using dynamo model (Steenback & Krause 1969).

The extreme higher and lower latitudes beyond which sunspots are never seen are respectively $\pm 45°$ and $\pm 5°$. Although the zones move simultaneously in both the hemispheres about the solar equator, the period of a cycle in either hemisphere may sometimes vary.

Motions of material within a sunspot can be derived from analysis of spectroheliograms of the spot. Within a spotgroup motions appear to be very complex. But within a single spot, simple model of gas motion has been derived. From an analysis of the H_α spectroheliograms of spots, Hale first suggested the motion to be *vertical* in which the gas was observed to move along filaments through penumbra *into* the umbra. Hale thus observed the *inflow* of gas into the spot. The more correct picture of the gas flow was proposed soon after by J. Evershed who found that close to the photospheric level, the gas flows outward from umbra *into* the penumbra, while at the upper levels of the solar atmosphere a reverse flow is observed. This has since been known as the *Evershed effect*. Measurements of intensity and Doppler shifts in the spot spectra reveal that Evershed motions vary with the distance from the centre of spots. The value increases from almost zero at the centre to about 3 km s^{-1} at the end of the penumbra (R_p) and then again decreases to nearly zero at about 1.5 R_p from the centre. It has also been noticed that the Evershed effect increases with the size as well as with the magnetic field strength in the spot.

5.9 THE SOLAR MAGNETIC FIELDS

The existence of large magnetic fields in sunspots probably provides the most important characteristic of the active Sun. It occurred to Hale that the observed *vertical* motions of gas in sunspots might be explained in terms of some magnetic phenomena. So he set himself for the measurement of magnetic fields in sunspots and in 1908, actually could measure the Zeeman splitting between the two circularly polarised components of a line. The resulting magnetic fields measured by Hale were in the range from about 100 gauss for small spots to about 3000 gauss or more for the larger ones. For the largest spots, the field strength may be as high as 4000 gauss. Even when the spots disappear high magnetic fields persist in spot regions.

The sunspot groups that are generally observed have been classified into three classes by G.E. Hale and S.B. Nicholson, according to the nature of magnetic polarity observed in them:

1. The *unipolar* groups consist of individual spots or groups of spots with similar magnetic polarity.
2. The *bipolar* groups generally have two main spots in each group having opposite polarity. One of these is called the *preceding* spot (p) or the *leader* spot, while the other *following* spot (f) or the *follower* according to the positions they occupy in course of the motion of the group across the *solar disc*.
3. The *complex* groups of spots contain spots with opposite polarity mixed together. Rough statistics indicate that about 90% of the spot groups are bipolar, about 10% are unipolar while complex spot groups are very rare.

The overwhelming preponderence of bipolar groups indicate that these are fundamental to the process in which the spots originate. The unipolar spots are generally identified as the last visible preceding spots in the bipolar groups when the following spots have already died out.

The most remarkable feature observed in a bipolar spot group is the reversal of magnetic polarities in either hemisphere with the beginning of a new cycle. The preceding spots of bipolar groups possess the same polarity in a particular hemisphere during a particular cycle. Whatever be the sign of this polarity, the preceding spots in the other hemisphere will always possess the opposite polarity during the same cycle. For example, if in a particular cycle the preceding spots in the Northern Hemisphere have the North polarity (positive) then the preceding spots in the Southern Hemisphere will have the South polarity (negative). The polarities of the preceding spots will just be reversed in the two hemispheres when new spots of the following cycle begin to appear in higher latitudes. Thus a reversal of polarities of the bipolar spot groups occurs in each hemisphere at each new cycle, that is, after an average period of 11.2 years. So spot groups of each hemisphere exhibit the same polarity at *every alternate cycle,* that is, after a lapse of 2×11.2 years. This may, therefore, be considered as the duration of a solar cycle (taken as 22-year cycle for convenience), judging from the point of view of the repetition of magnetic polarity. Several other features such as a higher number of sunspots accompanied by shorter period for attaining the peak of the spot activity, etc. are also repeated in a 22-year cycle.

From his long years of investigations of magnetic field directions, Nicholson of Mount Wilson found that the angles made by lines of force emerging from a spot with the normal to the surface could be quantitatively represented by the formula.

$$\theta = \frac{\pi}{2} \frac{r}{R_p} \tag{5.3}$$

where r is the distance from the centre of the spot and R_p represents the penumbral radius. The radial variation of the magnetic field intensity outward from the centre of the spot can approximately be represented by the simple formula

$$H(r) = H_0 \left(1 - \frac{r^2}{R_p^2} \right) \tag{5.4}$$

where H_0 is the maximum field strength at the centre of the spot. The maximum strength H_0 is found to be approximately related to the area of the spot by the relation,

$$H_0 = \frac{3700\, A_0}{66 + A_0} \text{ gauss} \tag{5.5}$$

where A_0 is the area measured in millionths of the visible hemisphere. H_0 remains almost constant during the full-grown stage of the spot. The high magnetic field inside the spot contributes its own pressure $H^2/8\pi$ in the maintenance of the mechanical equilibrium inside the spot. The gas pressure P_{spot} inside the spot must therefore be less than the gas pressure P_{phot} of the surrounding photosphere, these two pressures being related by the formula

$$P_{spot} + \frac{H^2}{8\pi} = P_{phot} \tag{5.6}$$

This last relation is due to H. Alfvén,

From what we have discussed in preceding discussions the following facts are apparent:

1. Sunspots are the seats of large magnetic Field and they exhibit many magnetic properties.
2. The magnetic field of the sunspots underlies the manifestations of many other transient solar phenomena, viz. plages, supergranulations, etc.
3. These manifestations of various phenomena are essentially caused by localized and transient magnetic activity but do not establish the existence of a general and large-scale magnetic field in the Sun.

Establishing the existence of a general magnetic field in the Sun was not an easy problem and it could not be done until 1950, although the localized magnetic affects were known since the beginning of this century. Although strong observational support in this direction came much earlier from the phenomena like the appearance of coronal streamers at sunspot minimum and the observation of chromospheric network in spectroheliograms around regions outside sunspots, it was not until 1953 when H.W, Babcock and H.D, Babcock of the Hale observatories first successfully measured a general solar field with their newly invented device, the *magnetograph*. The instrument uses the Zeeman splitting of the spectral lines which is automatically recorded photoelectrically for each point of the solar disk. The recorded splitting is proportional to the intensity of the magnetic field. A magnetic field as low as 0.3 gauss can be recorded by the instrument and a general solar poloidal field of the order of 1 or 2 gauss has been derived from measurements. R.B. Leighton of Mount Wilson devised in 1959 an alternative method of measuring solar magnetic field by using spectroheliograms, but this method was not suitable for measuring very weak fields.

Both the methods as devised by Babcocks and by Leighton, however, measure only the longitudinal component (line of sight component) of the solar magnetic field; these do not give an estimate of the transverse component of the field. For the measurement of the latter V.E. Stepanov and A.B. Severny of the Crimean Astrophysical Observatory developed a magnetograph in 1962. Alternative methods of measuring the transverse component of solar magnetic field were devised among others, by A. Dollfus and Le Roy.

The solar poloidal field (dipole) was discovered by the Babcocks from a routine observation of the polar fields. The mean value of the field was found to be 1–2 gauss. The general field was also observed to vary in intensity and in gross as well as in fine structures within a period of a few days. The most, interesting feature of the Sun's poloidal field as observed by the Babcocks was that for more than a year, from July 1957 to November 1958, the two magnetic poles of the Sun had the same polarity, viz., the north polarity. The polarity of this field was opposite to that of the Earth's field during the years 1953–57. Then during the period between March and July 1957, when the sunspot cycle attained its maximum, the south polarity gradually changed to north polarity and maintained its sign, till November, 1958. The north polar field, however, remained unchanged during this period so that the Sun possessed a dipole field having the same polarity. By the end of 1958, however, the original north pole also reversed its sign and the Sun's poloidal field became aligned parallel to the Earth's magnetic field. This reversal of magnetic polarity in Sun can be explained in terms of the sunspot maximum. This becomes apparent in view of the observation that during the cycle in question, the maximum of the sunspot activity was delayed in the Northern Hemisphere of the

Sun by about a year than the maximum in the Southern Hemisphere. Such observations therefore lead to the unavoidable conclusion that the general field (poloidal) and the sunspot field (toroidal) of the Sun are intimately correlated with each other. Such a conclusion is further supported by M. Waldmeier's observation that either pole of the general polar field reversed its sign at the maxima of the solar activity in the two hemispheres.

Babcock and Babcock first proposed in 1955 that the entire solar magnetic region could principally be divided into two main regions: (a) the Bipolar Magnetic Region (BMR) with two opposite polarities which constitutes the sunspot groups and the plage areas; and (b) the Unipolar Magnetic Region (UMR) with only one polarity where the measured magnetic field is of the order of 3 gauss.

Many theories have been proposed to explain the observed properties of the general solar magnetic field. One such theory suggests that the field is primordial and frozen-in with the highly conducting material of the Sun. The diffusion of such a field, if the material be at rest, is given by the equation

$$4\pi\sigma \frac{\partial \mathbf{B}}{\partial t} = \nabla^2 \mathbf{B} \tag{5.7}$$

where

$$\sigma = 2 \times 10^{-14} \frac{T^{3/2}}{z} \tag{5.8}$$

is the conductivity of the material expressed in terms of the temperature T in K taken constant in the solar interior and z is the ionic charge. In dimensional form, Eq. (5.7) can approximately be written as

$$\frac{4\pi\sigma B}{T_0} \approx \frac{B}{L^2} \quad \text{or} \quad T_0 \approx 4\pi\sigma L^2 \tag{5.9}$$

where L is the characteristic linear dimension of the field and T_0 is the approximate decay time of the field. For the general solar field $L \sim 2 \times 10^{10}$ cm ($\approx R_\odot/3$) and for the solar material $\sigma \sim 10^{-4}$ emu, leading to a decay time $T_0 \sim 10^{10}$ years, which is larger than the age of the Sun. This difficulty can, however, be overcome by H.W. Babcock's model of the solar poloidal field in which the field lines are considered submerged in the upper layers ($L \approx R_\odot/10$) as shown in Fig. 5.9.

In this layer $T \sim 2 \times 10^5$ K and $\sigma \sim 2 \times 10^{-6}$ emu, leading to a value of $T_0 \sim 10^7$ years for the decay time.

Many other theories have been proposed to explain the production and maintenance of the general solar magnetic field but none with complete success. The discussion of these theories is, however, beyond the scope of this book.

5.10 THEORY OF SUNSPOTS

Any satisfactory theory of sunspots must be able to explain all the fundamental aspects that are observed to be associated with them. For example, such aspects as the coolness of the spot

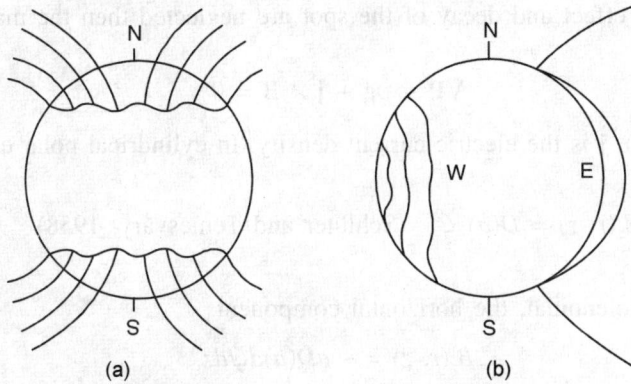

FIGURE 5.9 (a) babcock's model of the general solar magnetic field showing the polar field lines, (b) submerged field lines together with the surface of constant angular velocity *w*.

regions, their sharply defined boundaries, the associated strong magnetic fields, Evershed motions, their polarity law and reversal of polarity with the solar cycle, etc. and many others need to be explained by a comprehensive theory. Many theories have been proposed so far for the purpose but no such comprehensive one has yet been established. Each theory has its own standpoint and seeks to explain some of the observed properties of spots, but fails to account for some other properties. Again, the very basis of some of the older theories has been knocked down by information provided by later observations. Broadly speaking, however, all these theories may be classified into two groups. The first group of theories considers *coolness* of the spot as fundamental and the resulting magnetic fields as the effect of cooling in the spot. The second group of theories accepts the *magnetic field* as fundamental, all other observable properties being manifestations of that. The former group may be called the *convective and hydrodynamic theories* while the latter the *magnetic theories*.

Theories are many and even a brief discussion of all of them will be quite lengthly. Moreover, after it has been recognized by theoretical investigations that the lifetime of sunspots is negligibly small compared to the time of growth and decay of the magnetic field, the first group of theories has lost its physical significance and is now of historical importance only. We shall therefore, omit their discussions altogether. The second group of theories, the magnetic theories, stand on stronger physical basis.

These theories have been proposed by L. Biermann (1941), F. Hoyle (1949), H. Alfve'n (1943), Meyer et al. (1974), Meyer et al. (1979), Schmidt (1991) and Jahn and Schmidt (1994). According to Magneto hydrostatic model, it is assumed that a sunspot consists of a tube of magnetic flux embedded in the solar plasma and the flux tube is brought to the surface, perhaps in small portions, by the action of magnetic and convective forces and it has been swept together by the super granular flow. Also the magnetic field greatly inhibits convective forces. The inhibition of convective transport decreases the temperature of the spot. Thus the sunspot represents a 'dip' in the solar surface. This is known as Wilson effect.

If the dynamic effect and decay of the spot are neglected then the magnetohydrostatic equilibrium leads to

$$\nabla P + \rho \mathbf{g} + \mathbf{j} \times \mathbf{B} = 0 \qquad (5.10)$$

where $\mathbf{g} = (0, 0, -g)$, $\mathbf{j}$ is the electric current density. In cylindrical polar coordinate system (r, θ, z),

$$B_z(r, z) = D(a)\ \zeta^2 \quad \text{(Schlüter and Temesváry, 1958)} \qquad (5.11)$$

where, $a(r, z) = r\zeta(z)$.

If the field is solenoidal, the horizontal component

$$B_z(r, z) = -aD(a)d\zeta/dz \qquad (5.12)$$

Substitution into the horizontal component of (5.10) and integration over 'a' yields an ordinary differential equation for

$$y(z) \equiv (B_z(0, z))^{1/2} \quad \text{as} \quad fyy'' - y^4 + 2\mu\Delta P = 0 \qquad (5.13)$$

where f is a constant which depends on the spot's total magnetic flux and $\Delta P(z)$ is the pressure difference between the spot's exterior and its axis. If $\Delta P(z)$ is known, then Eq. (5.13) can be integrated. The above self-similar sunspot model demonstrates that the convective heat transport in sunspots should be substantially reduced but does not account for the discontinuous transitions between the umbra and the penumbra, and between the umbra and the surrounding free gas. In recent studies (Schmidt 1991; Jahn and Schmidt 1994), two interfaces, tangential to the magnetic field, separating three domains of different convective energy transport, the umbra, penumbra and the exterior zone, have been considered. Here thermodynamic properties of the three domains are determined such that the observed magnetic field configuration,

$$B(r) = B(0)/(1 + r^2/r_*^2) \qquad (5.14)$$

(where r is the distance to the sunspot's centre and r^* is the spot radius) and magneto-hydrostatic balance are satisfied.

Stability

A perturbation of a magnetohydrostatic system typically propagates with Alfve'n velocity. The time of traverse across the spot is 1 hour or smaller which is very small compared to the lifetime of the spot (~ several months). So a sunspot is stable against perturbation.

If the convective motion increases, it would tend to destroy the magnetic inhibition. As a result, the field would try to spread out and thereby get weakened. This will lead to a condition of instability between the gaseous material inside and outside a spot in presence of a vertical magnetic field. This condition of instability may be expressed by the inequality

$$\rho_0\alpha\beta\delta > \pi H_0^2/\lambda^2 \qquad (5.15)$$

where λ is the vertical wavelength of disturbances, $\alpha = 1/T$ is the reciprocal of temperature and β is the excess of the actual temperature gradient over the adiabatic gradient in layers immediately below the photosphere. Since the adiabatic gradient is a few degrees km^{-1}, we can choose $\beta = 2°$ km^{-1} as the typical value. The other typical values in the layer considered are T = 20,000 K, H_0 = 3,000 gauss and $\rho_0 \sim 10^{-5}$ gm cm^{-3} leading to the result $\lambda < 3000$ km as the condition of instability. This shows that the observed typical magnetic field in spots can prevent outflow of material from layers just below the base of the spot. We have already seen Eq. (5.6) that the magnetic field in sunspots greatly influences the mechanical equilibrium in the spot. The gas pressure within the spot region at any level is less than that at the corresponding level of the surrounding photosphere. The surface of the Sun should therefore have a depression in the spot region. Such *depressions* of the scale of 500 km have been proposed to be real by many observers.

5.11 SOLAR FLARES

Flares are the most energetic phenomena of highly transient nature occurring in the Sun. While observing the Sun on spectroheliograms or filtergrams, sometimes the chromospheric emission lines, particularly the H and K Ca II lines originating in the sunspot and surrounding plage regions are observed to brighten up intensely within a few seconds or minutes. Such occurrences are known *as flares*. The intensity of emission lines increases as much as four to five times of those originating in plages within a few minutes, and sometimes within only 10 sees, (for microflares). The strengths of lines then decay rather slowly within a time interval of 10 to 100 minutes. A very large flare can also be observed in the white light of the Sun when it appears as an intense bright spot against the continuous background of the spectrum of the surrounding solar disc.

Flares are believed to originate in chromosphere but rise far up into the lower corona. They have a tendency to occur near the complex groups of spots and usually the occurrence is repeated many times in the same place. Occasionally, they are seen to occur when a spot group is rapidly developing. Some of these have also been observed to occur from the umbrae of large full-grown spots with complicated magnetic field structures. These observations clearly associate flares with sunspot activity.

Flares are classified according to their brightness and the areas covered by them. The area seems to be more important criterion of classification because usually, the larger the flare the brighter it is and with longer lifetime. A vast range in size and brightness is observed among flares and their *importance* is assigned accordingly. The importance is labelled 1, 2, 3 and 3$^+$ according as the areas covered by the flare lie in the ranges 100–250, 250–600, 600–1200 and > 1200 respectively, which are measured in units of a millionth part of the solar disk. Besides the above classes, there are *subflares* labelled with importance 1$^-$ and the microflares. Although the flare statistics partly depend on the observer, rough analysis indicate that about 80 per cent of flares are of importance 1 and about 3 per cent are of importance 3, while the remaining flares are of importance 2, the flares of importance 3$^+$ are of extremely rare occurrence.

Flares, are events of frequent occurrence in the Sun and if we include subflares, there may be one flare every hour but flares of importance 3$^+$ hardly occur even once in a month. Flares are correlated with sunspot activity and the number of flares is observed to bear an approximate mathematical relation with the number of sunspots. It is found that the number of flares per day of importance higher than or equal to 1 is ~ *R*/25, where *R* is the sunspot number given by (5.2). Thus, if *R* = 100 on any day, four flares of importance ≥ 1 can be expected. Since subflares are several times more numerous, if these are taken into account, a flare every hour will be more natural. Since flares originate almost always within about 100,000 km of a spot group, the flare distribution on the solar disk bears close correspondence with the distribution of spot groups. Further, quantitatively, observations indicate that the rate of flare production associated with any particular spot group depends on the area *A* of the spot group and on the time rate of change of the area *dA/dt*.

A typical flare has a surface area ~ 5×10^{19} cm^2 (assuming radius = 4×10^9 cm and circular shape) and a height of ~ 2×10^9 cm so that the typical volume is ~ 10^{29} cm^3. The total energy output for such a flare is ~ 2×10^{32} erg, so that the average energy density in a flare is ~ 2×10^3 erg cm^{-3}. Most of this energy Is emitted in corpuscular radiation. Although flares radiate in most of the frequencies of the electromagnetic spectrum, from X-rays and γ-rays to long-wavelength radio waves, and also relativistic particles (cosmic rays), most of the energy is emitted in X-rays and ultraviolet radiation. The energy emitted in optical range is not too large, may be about 10 per cent and that in radio waves and cosmic rays is insignificant.

Like size and intensity, the spectrum of flares also differ greatly from one another. In large flares the continuum is strongly brightened, but in normal flares of moderate importance the brightening of the continuum is not very marked, except over a region surrounding H$_\alpha$. H$_\alpha$ itself is observed in strong emission with large symmetrical emission wings. The wings, however, fade away quickly after the onset of the flare, while the flare may still linger. The central intensity of H$_\alpha$ becomes three to four times that of the continuum and the width may be as large as 15 Å. Among other lines seen in emission are those of ionized metallic lines, the D$_3$ line of He I at λ 10,832 Å and some H lines of the Paschen series. D$_3$ line in small flares is seen in absorption. Lines of Fe II and those of low excitation Fe I are strongly brightened in flares. All these observations are consistent with higher excitation in flares. Electron densities in the range 10^{11} to 10^{13} cm^{-3} and electron temperatures in the range 10,000 to 17,000 K have been suggested by various authors to be the most favourable state under which the observed spectra are produced.

During his study of flare spectra with high dispersion instruments, Severny observed some fine structure granules which develop during the initial stage of the flares. These granules are short-lived with diameters ~ 400 km. Gas moves within granules with speeds as high as 1000 km s^{-1}, and it greatly influences the *wings* of lines of flare spectrum. Severny called these phenomena as "moustaches". *Moustaches* are believed to be different from flares. While flares are manifestations of release of large energies within a relatively short-time, moustaches are believed to be associated with small-explosion like events.

Flares eject material with speeds ranging from a few hundred to as high as 1000 km s^{-1}. The streaming material perturbs prominences and filaments. The latter are sometimes destroyed or disrupted by flares, but strangely enough, reappear again after hours or days. Spectrographic studies indicate that large gaseous blobs which are manifestation of flare surges, rise high up into the solar atmosphere, undergoing steady deceleration due to solar gravity and come down into the Sun again with approximately the same speed. Besides this Sun-bound mass motion, flares also eject streams of corpuscular radiation in a wide range of energies. These high-speed particles leave the influence of solar gravity and after a certain period depending on their speeds, impinge on the Earth's magnetosphere giving rise to geomagnetic storms and aurorae. This phenomenon, commonly known as *solar wind,* will be discussed in greater details in the last section of this Chapter.

As we have already mentioned, solar flares are always accompanied by enhanced X-ray and ultraviolet radiation. The first X-ray measurements during a flare were performed by H. Friedman in 1956 by sending rockets to heights in the range of 60 km. These data were extended by several other flights in subsequent years. Such measurements, in general, show two components of X-ray radiation: (a) a less energetic slow component lasting for a longer time; usually of order of an hour, and (b) a *sudden* short-lived high energetic component lasting for a few minutes. The first component originates from natural transitions in the coronal gas heated by flares to a temperature $\sim 5 \times 10^6$ K. The second component is associated with *bursts of nonthermal* emission. The Earth's ionosphere and terrestrial radio communications are disrupted by flare X-rays. The solar ultraviolet radiation is greatly enhanced during solar flares. Sometimes, a single flare may outshine the entire Sun in ultraviolet radiation. Ultraviolet lines are generally enhanced but the L_α line remains remarkably unchanged during flares. It is interpreted that this line probably is radiated from the entire chromosphere.

The Sun is the nearest source of cosmic rays. During large flares some of the solar particles (protons, electrons and heavier nuclei) are accelerated to very high energies—particles with energies even higher than 10^{10} eV have been observed. The highest flux of particles occurs, of course, in the MeV range. These high energy solar particles are the solar *cosmic rays* whose composition has been found not to differ significantly from the normal composition of the Sun itself. The most abundant particles are protons. Next comes the Alpha particles, and so on. The fraction of the flare energy contained in the flare cosmic rays is, of course, quite small, usually less than one per cent. The particles are believed to be accelerated in regions of high magnetic field gradient but no entirely convincing explanation of the origin of solar cosmic rays has yet been established.

Several theories have been proposed by various authors to explain the origin of solar flares. Flares have been speculated as manifestations of electrical discharge. It has also been conjectured that flare energy might be produced by nuclear reactions in the surface layer of the Sun. Since both these theories have later been shown to be highly improbable, we shall not discuss these any further.

Although none of the flare theories so far proposed appear completely satisfactory, we shall briefly discuss two of the theories which can explain many of the observed characteristics of flares. Both these theories are based on the high magnetic fields in sunspot regions. Observationally, it is found that flares occur mostly in regions of complex spot groups, where magnetic fields are high. Magnetic fields of a few hundred gauss can supply energy of the order of a few times 10^3 erg cm^{-3}, as flares are observed to possess (using $\varepsilon_m = B^2/8\pi$). But the theory must also explain the extreme suddenness of the release of this energy. Severny proposed that flares originate as a result of magnetic discharge near an X-type neutral point which may arise by interaction of fields of two neighbouring bimagnetic spots in a complex spot group (Fig. 5.10). The magnetic field gradient is very high near neutral points. The intermotion of spot groups may initiate a kind of perturbation, which may trigger the magnetic discharge, thereby suddenly dissipating large amount of magnetic energy and neutralizing the

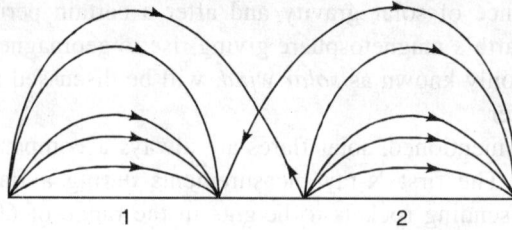

FIGURE 5.10   A schematic diagram showing the formation of an X-type neutral point.

high gradient of fields. In fact, substantial modifications in the magnetic field structure are actually observed in the flare region after the flare has died out. Moreover, large magnetic field gradient is capable of accelerating particles to high energies which can subsequently emit radiation in wavelengths ranging from X-rays to long radio waves, as are actually observed in flares. But there are other observations of magnetic fields which Severny's mechanism cannot explain, and so Severny's theory has not emerged as full proof.

A plausible theory of the origin of flares has also been proposed by Gold and Hoyle. Their theory relies for the energy source on coalescing and cancellation of two bundles of twisted magnetic lines of force. They imagine a filament consisting of a bundle of lines of force which emerges from a point in the photosphere and re-enters at another point thus forming an arch. This bunch of lines of force *may be twisted* due to highly convective motion in the subphotospheric layer. The energy density of this convective mass motion is $\sim 10^5$ erg cm^{-3} only a few per cent of which is associated with flare phenomena. The twisting of the field can greatly increase the intensity of the initial moderate field. Calculation shows that an amount of magnetic energy equal to $\sim 10^{30}$ erg may be associated with such a twisted filament. This is the right amount of energy that is involved with flare phenomena. In order that this energy may be released abruptly, a pair of such filaments with initial magnetic fields in opposite directions but the same sense of twisting is hypothesized. Two such filaments placed close to each other will attract each other, interpenetrate, and a part of the field will be annihilated in a time-scale of the order of 10^2 seconds. The theory of Gold and Hoyle thus predicts the release of the right amount of energy within the right amount of time as are observed in flares, through the mechanism of the magnetic field annihilation. The theory therefore implies the operation of a mechanism by which the twisted magnetic field which develops on the solar surface may become untwisted through flare discharge. This transformation from twisted to untwisted field may have important bearing on the equilibrium configuration of the Sun.

5.12   RADIO EMISSION FROM THE SUN

Soon after the production of radio frequency waves in the laboratory by Heinrich Hertz in 1888, some scientists started thinking that the Sun might be a source of radio waves. Valuable work was done in this line by Wilsing, Scheiner, Nordmann, Jansky and others during the first few decades of this century and the end of the last century. But the actual detection of Sun as the radio source was missed primarily due to poor technological advancement in those years. The discovery was finally made during the second world war independently by Stanley Hey of Britain, and by G.C. Southworth and G. Reber of the U.S.A. The subsequent decades were marked by intensive as well as extensive studies of the solar radio emission by scientists all over the world.

The study of the radio-spectrum of the Sun has revealed that the radio emission from the Sun can be classified into two well-defined components: (a) a slowly varying component with sufficiently long characteristic times, and (b) the sporadic *burst* component having much shorter characteristic times. The first component is of *thermal origin* and is recognized as the characteristic of the *quiet Sun,* while the second component is *nonthermal* which is characteristic of the *disturbed Sun.*

As a hot body the Sun must emit radiations in radio wavelengths, but since the temperature of the Sun is not uniquely defined (being different at different levels of the atmosphere) the nature of the solar thermal radio emission is altitude-dependent. The highly ionized corona is completely opaque to longer radio waves just as the case with the Earth's ionosphere. Thus, observing the Sun at different radio wavelengths means observing at different levels of its atmosphere. This also enables us to distinguish between different levels in chromosphere and corona from which various disturbances give rise to radio emission.

The shorter radio waves of wavelength ~ 1 cm come from the lower chromosphere which corresponds to a radiation temperature of 10^4 K. Meter and decameter wavelengths which are generated at the lower chromosphere are absorbed by ionized gas in the upper chromosphere and the corona, and therefore, cannot reach the Earth. The longer wavelengths from Sun that we receive on Earth originate in the corona where the characteristic temperature is high. The temperature relevant to the measured intensity of meter wavelengths is ~ 10^6 K. The intensity of this radiation undergoes slow variation, corresponding to the variations in the coronal structure. The measured radiation actually corresponds to a temperature of 10^6 K and a density nearly equal to three times the normal coronal density. The slowly varying component of the solar radio radiation thus fits with the concept that—it is thermal radiation from regions of coronal condensations. The mechanism which gives rise to this radiation is free-free transition of electrons in the hot coronal gas.

The burst component of solar radio emission is sporadic, intense and *nonthermal* in character. This component is associated with the activity of the Sun. Different types of bursts have been recognized and the physics behind them is quite complex. As the name implies, bursts are manifested as the enhancement of the radio-emission by orders of magnitude within periods as short as a few seconds. If the intensities of bursts are interpreted as of thermal origin, the corresponding equivalent blackbody temperatures would be of the order of 10^{12} K, which is absurd.

The bursts are of different nature and they have been broadly classified into types I, II, III, IV and V. Of these, all but the Type I bursts are correlated with flares. Type I bursts are not associated with flares or centres of activity. Type I bursts originate in noise storms that are produced high up in the corona usually above larger spots, their polarization showing no correlation with that of the spots. These bursts usually have life-times of 0.1 to 0.2 seconds and during this period the wavelength of radiation remains fairly constant. The radiation shows sharp peaks superimposed on a continuum. It is observed to undergo strong circular polarization assuring that its origin is nonthermal. It has been suggested that amplification (or negative absorption), of gyroradiation may generate all types of rapidly varying bursts in which Type I bursts are included.

All other types of bursts, viz., Type II to Type V are in some way or other correlated with solar flares. Careful analyses have shown that Types II and IV, and also Types III and V are closely associated with each other. Types II and III bursts originate in plasma oscillations in higher layers of the corona excited by rising streams of highly energetic ionized particles ejected by flares. It appears that such flares drive two components of ejected particles. The relativistic component consisting mainly of electrons and moving upwards with average speeds in the range from 0.1 to 0.3 c excites coronal plasma oscillations to give rise to Type III bursts. The shower component of flares ejecta consisting mainly of protons and electrons and moving with velocities of the order of 1000 km s^{-1} in a shock front, also similarly excites plasma oscillations in coronal layers to generate the type of bursts that is identified as Type II.

A Type III burst which lasts about a minute is composed of a bunch of isolated bursts, each of duration of about 10 seconds, with their frequencies changing continuously with time, indicating that these are the results of rising disturbances through the corona. Also, that the mechanism of plasma oscillations is definitely operative becomes evident from the occasional existence of two harmonics, almost nearly in the frequency ratio 2:1. Closely following the Type III bursts are observed the Type V events which are characterized by continuous emission over a wide band, particularly in metre wavelengths. They result from synchrotron radiation from the relativistic electrons which excite the Type III bursts via plasma oscillations, and last for several minutes after the associated Type III bursts have died out. The synchrotron origin of these bursts is confirmed by their high degree of polarization.

The slow component of flare ejecta generates Type II bursts in the corona. These are less frequent than Type III events and about one-half of these are preceded by the latter. Type II bursts are narrow in bandwidth and, like Type III events, they also sometimes show two harmonically related frequency bands. As the disturbance moves from lower to higher levels in corona, the bands drift from high to low frequencies, thg rate of drift being about 1 Me per second. The corresponding propagation velocities of disturbances range around 1000 km s^{-1}. It is observed that each drifting band sometimes splits into two or more nearly parallel components, which might be interpreted as due to a coupling between gyrofrequency and plasma frequency in a magnetic field of a few gauss. But the polarization measurements do not confirm such coupling effect. Type II bursts are sometimes followed by Type IV events. The latter originate in the regions behind the shocks which accompany Type II bursts. Type IV bursts are characterized by continuous emission rising to maximum within about half-hour and lasting for about four hours. The continuous emission in these events exhibits a low frequency cut-off. The observed radiation can therefore, be interpreted as that due to a combination of synchrotron radiation and plasma oscillations produced by accelerated particles behind the shock. Observations have shown that Type II events usually lead to geomagnetic storms after about 1.5 to 3 days provided that they are followed by Type IV events. But those Type II events which do not generate the associated Type IV events are uncorrelated with geomagnetic disturbances. Such observations may be interpreted physically by assuming that sufficient amount of accelerated material must be created behind the shock accompanying Type II events, in order that the Type IV continuum may develop.

The structure of a Type IV burst is, however, quite complex. Several sub-types have been suggested according to structural differences the discussion of which will be omitted here. A U-type burst is a fast burst resembling in many ways a Type III burst except that the former undergoes a frequency inversion. U bursts at first propagate with decreasing frequencies and then again with increasing frequencies, indicating that they are guided by magnetic fields to move first upwards in the solar atmosphere and then again downwards. The turning frequencies usually lie in the range 100 to 150 Me. U bursts have velocities around 1000 km s^{-1} and are believed to be associated with flares. In conclusion, it may be mentioned that the physical processes underlying the burst phenomena are quite complex which are still only poorly understood. Only *plausible* pictures of the phenomena in their simplest fashion have been described here.

5.13 SOLAR WIND

The Sun emits not only the electromagnetic radiation in all wavelengths from X-rays to decameter radio waves, but also a continuous stream of plasma with high velocities in all directions. This plasma stream composed mainly of protons and electrons ejected by the Sun, has been called by L. Biermann as *corpuscular radiation* and by E.N. Parker as *solar wind*. The last name has currently become more popular after its origin in 1958. At the high coronal temperature ($\sim 2 \times 10^6$ K), if the particles are assumed to have attained Maxwellian distribution, particles at the tail of the distribution will have velocities greater than the escape velocity from the solar gravitational field. Some particles will therefore continuously escape from the Sun and stream outwards to engulf all the nearby planets, and quite likely, the influence extends as far as Saturn. Our Earth is thus immersed in the stream of plasma radiated from the Sun.

A flow of particles from the Sun has been suggested by several early observers during the 19th century. Works of R.C. Carrington, Oliver Lodge, G.F. FitzGerald and other solar physicists definitely established a generic relationship between the flare activity in the Sun and the phenomena like geomagnetic storms, intense aurorae and enhanced airglow. Stream of charged particles ejected from a flare and moving with velocities of the order of 300–400 km s^{-1} was thought to be responsible for the observed phenomena. These theories, however, did not gain general acceptance at that time because many foremost scientists of those days (even Lord Kelvin) considered this relationship to be coincidental.

By 1960, however, the theory of the continuous solar corpuscular emission had been firmly established, mainly through the investigations of Birkeland, Stöermer, Chapman, Bertels, Hoffmeister, Biermann, Parker and others. Evidences multiplied in such observations as periodical change in the intensity of magnetic storms in 27 days (Sun's rotation period), intense auroral display at the peak of the solar activity, sudden decrease in the flux of galactic cosmic rays about a day after large flares ("Forbush decreases"), the definite fashion of orientation of the ionic comet tails with respect to the Sun, large accelerations produced in the Knots (clouds of CO^+ gas) in the ionic tails, etc. Careful analyses have shown that ionic comet tails always point to the direction of the *relative velocity* as would be obtained by combining the velocity of the comet in its orbit and that of the stream of plasma emitted radially away from the Sun. Streaming velocities in the range of 400–500 km s^{-1} generally fit the observations. Again, photographic measurements of the velocity and acceleration of knots along the ionic tails indicate large repulsive accelerations of ~ 10^2 times the local solar gravity. Such large accelerations cannot be produced by radiation pressure. Repulsive pressures produced by high speed streams of particles can only explain these observations. The pioneering work on the orientation of ionic comet tails was by Hoffmeister and that on the acceleration of Knots was by Biermann. On the basis of his work and of other evidences, Biermann proposed in 1957 that "Corpuscular radiation" flows from the Sun essentially in all directions at all times. Parker in 1958, arrived at essentially the same conclusion by a quite independent approach. He proposed a hydrodynamical model in which he showed that the usual consequence of a temperature of ~ 2×10^6 K of the solar corona will be to eject continuous stream of plasma from it. He favoured the *name solar wind* for this stream.

These theoretical advances were soon put to observational verification by sending instruments in rockets far away from the Earth's surface. For best results, observations should be made further away from the Earth's magnetosphere, so that the solar wind may be *undisturbed*. The early observations made from the Soviet rockets Lunik III and Venus I and the American rocket Explorer 10 yielded valuable data. These were superseded by those supplied by Mariner 2 in 1962 which probed far away into the space. The results confirmed that the solar wind was an essentially continuous stream of plasma, mainly protons and electrons but with an admixture of normal abundance of alpha particles and other heavy nuclei. The average speed of particles is ~ 500 km s^{-1}, the speed varying in the range'from 300 to 860 km s^{-1}, and the average proton density is ~ 5 cm^{-3}, which leads to an average proton flux of 2.5×10^8 cm^{-2} s^{-1}, a result agreeing excellently with the earlier Soviet observations. A correlation was observed between the particle speeds and geomagnetic activity, and high velocity plasma streams were found to occur periodically at twenty-seven day intervals.

From his hydrodynamical model of the solar wind, Parker predicted that because the highly conductive plasma of the wind would be frozen-in with the lines of force of the solar magnetic field, the latter will be drawn into an Archimedes spiral due to solar rotation. Such a configuration of the solar magnetic field was subsequently verified by Mariner 2 measurements.

5.14 THE SOLAR NEUTRINO PUZZLE

The Sun is mostly composed of hydrogen gas. According to the standard solar model the sun is supposed to be in hydrostatic equilibrium and the radiation pressure, generated from nuclear reactions at its centre balances the gravity. According to standard solar model, proposed by Eddington in 1920s, if the radius and mass are known, then the central temperature can be calculated, provided the assumption of hydrostatic equilibrium is considered. The temperature is of the order of 10 to 20 million degree kelvin. The most important reactions at this high temperature are those of proton proton chains, ppI, ppII and ppIII. The pp chains are (according to Bahcall and Ulrich 1988, and Caughlin and Fowler 1988):

	Reaction	Q'(Mev)	Q_ν(Mev)
ppI	$p + p \rightarrow {}^2D + e^+ + \nu_e$	1.177	0.265
	${}^2D + p \rightarrow {}^3H + \gamma$	5.494	
	${}^3H + {}^3H \rightarrow {}^4He + 2p$	12.860	
ppII	${}^3He + {}^4He \rightarrow {}^7Be + \gamma$	1.586	
	${}^7Be + e^- \rightarrow {}^7Li + \nu_e$	0.049	0.815
	${}^7Li + p \rightarrow {}^8B + \gamma$	17.346	
ppIII	${}^7Be + p \rightarrow {}^8B + \nu_e$	0.137	
	${}^8B \rightarrow {}^8Be + e^+ + \nu_e$	8.367	6.711
	${}^8Be \rightarrow 2{}^4He$	2.995	

From the usual understanding of the p-p reaction, about 1.8×10^{38} neutrinos are produced by Sun per second. So at Earth's distances, some 400 trillion neutrinos go through our bodies every second, but most of the neutrinos produced as a result of p-p reactions have energy which is too low for detection. However higher energy neutrinos come from the last neutrino reaction whose frequency of occurrence is 2 out of 10,000 completions of the p-p reaction, so these neutrinos are rare. To detect these higher energy electron neutrinos, a large vessel (400 cubic metres) filled with dry-cleaning solvent (perchloro ethylene) was placed 1.5 km underground in a gold mine in South Dakota—free from all other cosmic radiation. A few of the ${}^{37}Cl$ are expected to react with electron neutrinos to form ${}^{37}Ar$ and an electron which then reverts to ${}^{37}Cl$ and a neutrino.

The ${}^{37}Ar$ atoms are purged with helium gas and the decay is counted. According to standard model, there will be 8 SNU, but neutrino detector has counted only 2.2 SNU with a deviation of 0.3 SNU.

Solutions and Contradictions

In 2001, the results from Sudbury Neutrino Observatory (SNO) in Canada, confirmed that electron neutrinos produced by nuclear reactions inside sun, 'oscillate' or change flavour on their journey to earth, i.e. they have been transformed into muon and tau neutrinos and this is possible only if neutrinos have mass. Although Super Kamiokande experiment in Japan has seen strong discrepancy in the observed and predicted neutrinos, the SNO results when combined with solar neutrino data from Super Kamiokande, showed that the disappearance of one neutrino flavour is accompanied by the appearance of another. But there are contradictions too to SNU results. First of all there can be no confirmation of oscillation of neutrino flavour between sun and earth without simultaneous measurements being made near the sun. On the other hand, an electrical model for stars has been proposed by Ralph Jurgens. According to this model, the sun is not a sphere of neutral gas. Due to a large difference in mass between the electrons and the protons, the hydrogen atoms act like a dipole with positive aimed at the Sun's centre. Since the electric force outguns gravity by 10^{39}, its omission from standard model makes it unrealistic. As like charges repel, so the interior of sun, full of excess positive charges, resist the compression due to gravity, making it more homogeneous with a small core. So, it is not necessary for an internal nuclear furnace to bloat the sun to size we see. Also, if neutrinos do have mass it will tend to confirm electric model. If they have mass they are comprised of the same charged subparticles, that make up all matter. When a positron and electron annihilate, the orbital energy in both is radiated as a gamma ray and the subparticles assume a new stable orbital configuration of low energy or mass. The difference between the neutrino "flavours" is merely the different quantum states and therefore different masses. The electric sun model expects heavy element synthesis so that various neutrino flavours are all generated in sun and do not need to oscillate in their way to earth. Here fluctuations in neutrino counts are expected to be correlated with various solar activities, e.g. sunspot numbers, solar wind activity, etc. It has been observed that the standard solar model has no such correlation as there is a lag of million years between nuclear reactions in the core and its final expression at the surface of sun.

Standard Solar Model

In the standard solar model, the sun is considered as a spherically symmetric body of hydrogen gas whose most of its luminosity originates from the nuclear reactions. The different features like luminosity, radius, age and composition (hydrogen-to-heavy elements ratio) are calculated using helium abundance and mixing length parameter as free parameters and are matched with their observed values though, both the sun's neutrino flux and the 'p' mode oscillation spectrum, predicted by standard model do not match with the observed values (Super KamioKande experiment in Japan and the experiment held by Davis in 1978 in South Dakota) but the free parameters are adjusted so that the derived solar mass and radius match with the observed values. So the purpose of standard solar model is to provide an estimate of the solar model free parameters, as well as to put a bench mark to compare "improved" solar models which have mere complicated physics, like, rotation, magnetic field, diffusion, overshooting and metal rich cores.

The fundamental equations consisting of standard model are, conservation of mass, momentum, energy equations, energy transport equations and nuclear reaction network. It is assumed that the system is in hydrostatic equilibrium, i.e. the weight of any volume element is supported by the sum of all pressure forces acting on the element.

The energy balance of any shell in the sun is given by

$$\frac{dL}{dr} = 4\pi r^2 \rho \left(\varepsilon + \frac{T ds}{dt} \right)$$

where L is the luminosity, r is the outer radius of the shell, ε is the nuclear energy generation rate per unit mass, s is the entropy per unit mass.

The radiative energy transport equation is

$$L = \frac{16 \, \pi \sigma c r^2 T^3 \left(\dfrac{dT}{dr} \right)}{3 \kappa \rho}$$

where, σ is the Stefan's constant, c is the speed of light, and κ, the Rosseland mean opacity. When the temperature gradient exceeds the adiabatic temperature gradient, convection starts and energy is transported by convection. This is not true near the surface where the density is low enough, so radiative transport must be taken into account. These four equations (mass, momentum, energy and energy transport equations) and equation of state are used to find five variables P, T, r, $M(r)$ and L.

There are two major nuclear chain reactions, ppI, ppII, ppIII, and CNO cycle that occur in the core region. They are given below:

pp chains

	Reaction	Q'(Mev)	Q_v(Mev)
ppI	$p + p \rightarrow {}^2D + e^+ + \nu_e$	1.177	0.265
	${}^2D + p \rightarrow {}^3H + \gamma$	5.494	
	${}^3H + {}^3H \rightarrow {}^4He + 2p$	12.860	
ppII	${}^3He + {}^4He \rightarrow {}^7Be + \gamma$	1.586	
	${}^7Be + e^- \rightarrow {}^7Li + \nu_e$	0.049	0.815
	${}^7Li + p \rightarrow {}^8B + \gamma$	17.346	
ppIII	${}^7Be + p \rightarrow {}^8B + \nu_e$	0.137	
	${}^8B \rightarrow {}^8Be + e^+ + \nu_e$	8.367	6.711
	${}^8Be \rightarrow 2{}^4He$	2.995	

CNO Chains

$${}^{12}C + {}^1H \rightarrow {}^{13}N + \gamma$$
$${}^{13}N \rightarrow {}^{13}C + e^+ + \nu_e$$
$${}^{13}C + {}^1H \rightarrow {}^{14}N + \gamma$$
$${}^{14}N + {}^1H \rightarrow {}^{15}O + \gamma$$
$${}^{15}O \rightarrow {}^{15}N + e^+ + \nu_e$$
$${}^{15}N + {}^1H \rightarrow {}^{12}C + {}^4He$$

or

$${}^{15}N + {}^1H \rightarrow {}^{16}O + \gamma$$
$${}^{16}O + {}^1H \rightarrow {}^{17}F + \gamma$$
$${}^{17}F \rightarrow {}^{17}O + e^+ + \nu_e$$
$${}^{17}O + {}^1H \rightarrow {}^{14}N + {}^4He$$

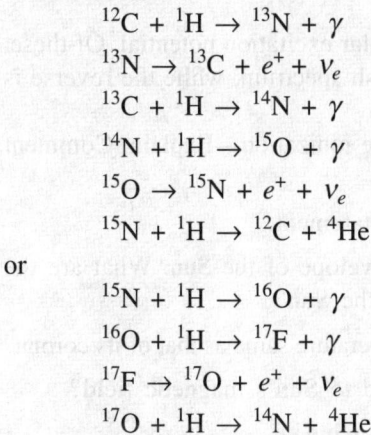

The nuclear reactions have two results: (i) they determine the energy output per unit time of a given shell which is used in the energy balance equation and (ii) they determine the abundances of elements involved in the nuclear reactions. The former is used in the energy balance equation and the latter is used to find the evolution of mean molecular weight.

The standard model is constructed through an iterative method. First an initial guess of mixing length parameter and helium abundance are considered and their evolution are found at the current age of the sun. Then these are compared to the observed values, and the discrepancy is adjusted by adjusting the mixing length and abundance.

EXERCISES

1. Compute the volume of the Sun and compare this with that of the Earth and hence compute the average density of the Sun.

2. Assuming the standard values of Sun's luminosity and distance from the Earth, compute the value of the solar energy that strikes the Earth's surface, and hence compute the approximate value of the solar constant.

3. Compute the solar constant if the temperature of the Sun were: (i) 10,000 K; (ii) 20,000 K; (iii) 4,000 K. What would have been the Sun's colours with these temperatures?

4. Compute the Solar Constant if the distance between the Sun and Earth were: (i) 0.4 A.U.; (ii) 0.7 A.U.; (iii) 1.6 A.U.; (iv) 5 A.U.

5. Draw a schematic diagram showing the variations of density and temperature in layers of the Sun's atmosphere starting from the photosphere.

6. What would have been Sun's luminosity if its photospheric temperature were: 3,000 K; 4,500 K; 6,000 K; 10,000 K; and 16,000 K?

7. The flash spectrum shows strong H and K lines of Ca II in high level chromosphere, but these lines are weak in low level chromosphere where instead, lines of Ca I are stronger—Explain.

8. The two lines Fe I λ 4144 and Fe II λ 4173 have similar excitation potential. Of these, λ 4144 is stronger in the photosphere than in the flash spectrum, while the reverse is true for the line λ 4173—Explain.

9. The coronal spectrum shows emission lines of intense ionization—Explain. Comment on the sources of the coronal heating.

10. The *radio Sun* is the Solar Corona—Establish the statement.

11. Explain why convection current originates in the envelope of the Sun. What are the effects of the current on the atmospheric layers of the Sun?

12. Why cannot we see the Sun as a blackbody with a temperature same as that of its corona?

13. How are the various forms of solar activities related to Sun's magnetic field?

14. Compute the rate of energy carried off by the solar wind.

SUGGESTED READING

1. Alfven, H., Arkiv f. Mat., *Astron. O. Fys.*, Vol. 29A(12), 1, 1943.

2. Aller, L.H., *Astrophysics: The atmospheres of the sun and stars,* 2nd ed., Ronald Press, New York, 1963.

3. Bahcall, J.N. and Ulrich, R.K., *Review Modern Physics*, 60, 297, 1988.

4. Biermann, L., Viertelzahrsschr. Astr. Gesellsch., 76, 194, 1941.

5. Brandt, J.C. and Hodge, P.W., *Solar System Astrophysics,* McGraw-Hill, New York, 1964.

6. Giovanelli, Roland, *Secrets of the Sun,* Cambridge University Press, Cambridge, 1984.

7. Hale, G.E., Ellerman, F., Nicholson, S.B., and Joy, A.H., *Astrophysical Journal,* **49**, 153, 1919.

8. Hoyle, F., Godwin, H., Manley, G., Brooks C.E.P. and Schove, D.J., *Quarterly Journal of the Royal Meteorological Society,* 75:324, 169, 1949.

9. Jahn, K. and Schmidt, H.U., *Astronomy & Astrophysics,* **290**, 295, 1994.

10. Kuiper, Gerard, P. (Ed.), *The Sun,* The University of Chicago Press, Chicago, 1953.

11. Kundu, Mukul, R., *Solar Radio Astronomy,* Interscience Publishers, New York, 1965.

12. Menzel, Donald, H., *Our Sun,* Revised edition, Harvard University Press, Cambridge (Mass.), 1959.

13. Menzel, Donald, H., Whipple, Fred L., and De Vaucauleurs, Gerard, *Survey of the Universe,* Prentice-Hall, Inc., Englewood Cliffs, New Jersey, 1970.

14. Pap, J.M., Fröhlich, C, Hudson, H.S., and Solanki, S.K. (Ed.), *The Sun as a Variable Star,* Cambridge University Press, Cambridge, 1994.

15. Phillips, Kenneth, J.H., *Guide to the Sun,* Cambridge University Press, Cambridge, 1992.

16. Meyer, T.W. and Rhodes, C.K., *Physical Review Letters*, 32, 637, 1974.

17. Meyer, et al., *Physical Review A.*, 19, 515, 1979.

18. Sagan, Carl, *The New Solar System,* Cambridge University Press, Cambridge, 1990.

19. Schüssler, Manfred, and Schmidt, Wolfgang, *Solar Magnetic Fields,* Cambridge University Press, Cambridge, 1994.

20. Schlüter, A., Temesváry, S., *In Lehnert;* 263, 1958.

21. Schmidt, H.U., *Geophys. Astrophys. Fluid Dynamics,* 62, 249, 1991.

22. Shklovskii, I.S., *Physics'of the Solar Corona,* Pergamon Press, New York, 1965.

23. Steenbeck, M. and Krause, F., *Astron. Nachr.,* 291, 271, 1969.

24. Zheleznyakov, V.V. (Massey, H.S.H., (Tr.), Hey, J.S., Ed.), *Radio Emission of the Sun and the Planets,* 1st ed., Pergamon Press, New York, 1970.

6. Atmosphere of Stars

6.1 INTRODUCTION

The layers of stars which produce the continuous spectrum, absorption line spectrum and the emission line spectrum are together said to constitute the stellar atmosphere. For Sun, these layers are respectively the photosphere, the reversing layer and the chromosphere. The physical conditions in these layers determine the nature of the continuous and the line spectra of stars. These physical conditions again are governed by the temperature distribution on the surface layers of stars and their chemical composition.

In the extremely high temperature and pressure conditions at the core of a star, energy is generated by nuclear fusion (see Chapter 14). This energy is transferred through the interior layers to the superficial layers of the star and emerges through its surface at some constant rate (for stable stars) which is called the luminosity of the star. In problems of stellar atmosphere we are interested to know how this outflowing energy interacts with stellar material in the surface layers. We observe certain characteristics and by applying other known physical laws on these we have to infer the actual physical processes that underlie the observed phenomena. Study of these underlying physical processes constitutes an important part of the theoretical astrophysics.

The temperature at the surface of the star is determined by the flux of radiation that reaches the surface layers through the interior. The flux F and the effective temperature T_{eff} are related by

$$\pi F = \sigma T_{eff}^4 \tag{6.1}$$

where σ is the Stefan-Boltzmann constant. We have already seen that T_{eff} for normal stable stars may range from about 50,000 K to less than 3000 K. For Sun, $T_{eff} \sim 5800$ K. The flux of energy through the surface layers thus determines the temperature distribution in the star. Again, the stable structure of the star is governed by the equation of hydrostatic equilibrium which relates the pressure gradient with density and surface gravity (see Eq. 14.10). Thus the surface gravity g_* of the star determines the pressure distribution in its atmospheric layers. For Sun, $g_\odot = 2.74 \times 10^4$ cm-s^{-2}. The observation of spectroscopic and eclipsing binaries enables us to determine the *mass-radius* relation for main sequence stars. It is found that the surface gravity of these stars does not, in general differ from that of the Sun by a large factor. If, together with the temperature and pressure, the chemical composition of the stellar material also is known, then application of some known physical theories enables us to compute as to how the various measurable quantities in stars are related to their temperature-gravity-chemical-composition parameters. This is the technique for construction of what has been called a *model stellar atmosphere*. Once this problem of determining the relationship between the various observable quantities and the assigned parameters of T_{eff}, g_* and the chemical composition has been solved, the problem then can be reversed. That is, the measurable quantities, viz. the energy distribution in the continuum (which is equivalent to the colour indices), the strengths of various lines of atoms and ions, the widths of spectral lines and shape of line profiles etc., can be used to compute the parameters T_{eff}, g_* and the chemical composition. This latter approach, in fact, underlies the practical aspect of the theory of stellar atmospheres.

We have already discussed the main observable features of stellar spectra and their physical interpretation. In this chapter, we shall discuss in a quantitative manner the physical theories which underlie the manifestation of the observed spectral characteristics. The basic physical process is the interaction of matter in the star's atmosphere with the emerging radiation flux through the interior layers. Three distinct processes are recognized by which thermal energy can be transferred through these layers. These are *conduction, convection* and *radiation*. All theoretical investigations have established the fact that the energy transfer by conduction in stellar atmospheres is negligible. Energy transfer by convection is important in layers beneath the photosphere of stars like the Sun. But this process is supplementary to the more important mechanism of energy transfer by radiation. All the investigations have led to one definite conclusion that radiative energy transport is by far the most important mechanism not only in photospheres but also in the interior of stars (see Chapter 14). In the following sections, therefore, we shall consider some basic properties of the transfer of energy by radiation through layers of stellar material, and show how these can be applied to interpret some of the measurable characteristics of stellar atmosphere. The transport of energy by conduction and convection is important only in deeper, layers of some stars and this will be briefly discussed when we consider the interior structure of stars in Chapter 14.

6.2 SOME IMPORTANT DEFINITIONS

Let $d\sigma$ be a surface element within the radiation field and **n** be the normal to the element $d\sigma$. Let dE_v be the radiant energy flowing through $d\sigma$ in time dt within a cone of solid angle $d\Omega$ whose axis is inclined at an angle θ to the direction of **n**, within the frequency interval v and $v + dv$. Then the intensity I_v of the radiation field is defined by the relation

$$dE_v = I_v \cos\theta \, d\Omega \, dt \, d\sigma \, dv \tag{6.2}$$

The solid angle $d\Omega$ is defined in terms of the polar angles (θ, ϕ) by

$$d\Omega = \sin\theta \, d\theta \, d\phi \tag{6.3}$$

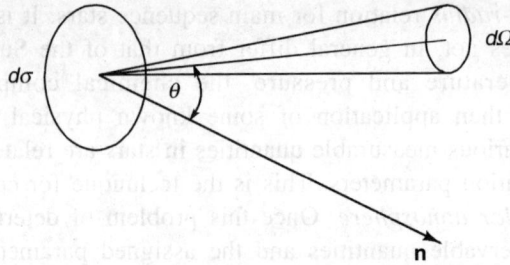

FIGURE 6.1 Radiation intensity defined.

The intensity $I_\nu(\theta, \phi)$ thus measures the energy flow about the direction (θ, ϕ) per unit time per unit frequency interval per unit solid angle across unit area, oriented perpendicularly to the direction (θ, ϕ).

Integrating over the entire frequency interval, we get the total intensity I of the radiation field, given by

$$I = \int_0^\infty I_\nu \, d\nu = \int_0^\infty I_\lambda d\lambda \qquad (6.4)$$

the last equality following from symmetry.

The *mean intensity* of radiation at frequency ν is defined by

$$\bar{I}_\nu = \frac{1}{4\pi} \int_0^{2\pi} \int_0^\pi I_\nu(\theta, \phi) \sin \theta \, d\theta \, d\phi \qquad (6.5)$$

The total *flux* of radiation in the direction **n** across unit area of the surface element in unit time at the frequency ν is defined by

$$\int_0^{2\pi} \int_0^\pi I_\nu(\theta, \phi) \cos \theta \sin \theta \, d\theta \, d\phi \qquad (6.6)$$

Integrating over the azimuthal angle ϕ we get the flux along the axis of symmetry, given by

$$\pi F_\nu = 2\pi \int_0^\pi I_\nu(\theta) \cos \theta \sin \theta \, d\theta \qquad (6.7)$$

If the radiation field is isotropic, then I_ν is independent of θ and ϕ, and so $F_\nu = 0$. But the stellar radiation field is never isotropic and there is always a net flux of energy outward. Integrating over all frequency interval we get the total radiation flux as

$$\pi F = \int_0^\infty \pi F_\nu \, d\nu = \int_0^\infty \pi F_\lambda \, d\lambda \qquad (6.8)$$

The radiation flux πF_ν and πF are in general functions of position and orientation of the area.

We, now introduce the concepts of *emission* and *absorption* of radiation by matter. Suppose that the thermal radiation is emitted in all directions by a mass element $dm = \rho \cos \theta \, d\sigma \, ds$, which is the mass contained in a cylindrical element of cross-section $d\sigma \cos \theta$ and height ds. The total radiation ε_v emitted by dm within the solid angle $d\Omega$ in time dt and in the frequency interval dv is proportional to $dm, dt, d\Omega, dv$ and is given by

$$\varepsilon_v = J_v \, dm \, dt \, d\Omega \, dv \tag{6.9}$$

The coefficient J_v which makes the proportionality equal is defined as the *coefficient of emission*. It is therefore the radiation emitted per unit mass, per unit time, per unit solid angle and unit frequency interval. The total energy emitted by mass dm in time dt in all solid angle and over the entire frequency interval is given by

$$\varepsilon_{em} = 4\pi \, dmdt \int_0^\infty J_v dv \tag{6.10}$$

An element of material, in general, does not only emit radiation, it also absorbs radiation, and thus reduces the intensity of a passing beam. Let a narrow pencil of radiation of intensity I_v traverse a layer of material of thickness ds, the density of the absorbing matter being ρ. It is found that the quantity dI_v by which the intensity of radiation decreases due to absorption by unit mass is proportional to the initial intensity of the beam. Quantitatively, the proportionality is expressed by

$$dI_v = -I_v K_v \rho \, ds \tag{6.11}$$

where K_v is defined as the *coefficient of absorption* at frequency v. It is, in general, dependent on the direction of the passing beam, the frequency of radiation, and the chemical composition and physical state of the absorbing matter, K_v thus defined is also called the *mass absorption coefficient*. One can also define the absorption coefficient K'_v per unit volume which will be related to K_v by

$$K'_v = \rho K_v \tag{6.12}$$

If N be the number of atoms per unit volume, then the atomic absorption coefficient K''_v is defined by

$$K'_v = NK''_v \tag{6.13}$$

If the total thickness of the absorbing layer be s, then the Eq. (6.11) can be written in the form

$$\frac{dI_v}{I_v} = -\rho K_v \, ds \tag{6.14}$$

and integrating over the entire thickness of the layer we get

$$I_v = I_{v,0} \exp\left[-\int_0^s \rho K_v \, ds\right] \tag{6.15}$$

Where $I_{v,0}$ is the intensity of the incident beam at $s = 0$. The attenuation of the incident beam passing through an absorbing layer thus takes place exponentially. If we use τ_v for the index of the exponential factor so that

$$\tau_v = \int_0^s \rho K_v \, ds \qquad (6.16)$$

then Eq. (6.15) can be written as

$$I_v = I_{v, \, 0} \, e^{-\tau_v} \qquad (6.17)$$

τ_v so defined is called the *optical depth* or the *optical thickness* of the layer. We can see from Eq. (6.17) that the intensity of the incident beam is reduced by the factors $1/e$ (= 0.368) at a point in the absorbing layer where $\tau_v = 1$. τ_v thus provides with a measure of extinction of radiation passing through an absorbing material. We also see from the same relation that τ_v will, in general, depend on the density and composition of the absorbing material as well as on the frequency of radiation. Equation (6.17) gives the extinction if the beam is incident normally to the absorbing slab. If the direction of the beam makes an angle z_0 with the normal to the absorbing surface, the *extinction law* becomes

$$I_v = I_{v, \, 0} \, \exp \left(- \tau_v \sec z_0 \right) \qquad (6.18)$$

at the frequency v. For example, assuming the Earth's atmosphere to be composed of plane-parallel layers, the intensity of radiation from a star having Zenith distance Z_0 is reduced according to the law (Eq. (6.18)), where τ_v now represents the optical thickness at frequency v of the Earth's atmosphere, measured vertically.

We would like to introduce now the concepts of *thermodynamic equilibrium* and of *local thermodynamic equilibrium* (LTE). The thermodynamic equilibrium in an arbitrary volume-element of gas is characterized by a temperature T and a frequency v such that the rates of emission and absorption of radiation by the volume-element is equal to each other. Thus, in *thermodynamic equilibrium* we must have:

Emission per second = Absorption per second

This relation can be quantitatively, represented as

$$J_v = K_v B_v(T) \qquad (6.19)$$

where $B_v(T) = 1/4\pi \, cu(v)$ is the intensity of radiation of a black body determined by Planck's law.

The relation given in Eq. (6.19) is called *Kirchhoff's law* which states that *the ratio of the emission coefficient to the absorption coefficient in a thermodynamic equilibrium is the Planck's black body intensity function, which is a universal function of v and T.* Thus, Kirchhoff's law holds in thermodynamic equilibrium.

The condition of strict thermodynamic equilibrium, however, does not hold inside the star because in any arbitrary element of volume of stellar material at any level, there is a net outward flux of radiation. As a simplifying measure, however, it has been customary to assume a restricted type of thermodynamic equilibrium, called the *local thermodynamic equilibrium* (LTE), to hold in the stellar photosphere. This is the assumption that in any small region of the stellar photosphere, Kirchhoff's law (Eq. (6.19)) holds, i.e., the assumption that the ratio of J_v and K_v is equal to $B_v(T)$ holds locally. The assumption of LTE is of very fundamental nature because most of the results relating to stellar atmospheric phenomena have usually been computed with this assumption.

Before concluding this section we briefly discuss the concept of what is known as the *radiative equilibrium*. We consider an element of mass *dm* in the photosphere of a star. For a stable star (to which our consideration here will be confined), the temperature of the photosphere does not change with time. Therefore, the total energy gained, $\bar{\varepsilon}_{abs}$, by the mass element *dm* per unit time must be equal to the total energy lost, $\bar{\varepsilon}_{em}$, by the element. For, if $\bar{\varepsilon}_{abs} > \bar{\varepsilon}_{em}$, there is a net gain of energy by the mass which will cause an increase in temperature of the element. On the other hand, $\bar{\varepsilon}_{em} > \bar{\varepsilon}_{abs}$ will imply a net loss of energy in the element causing a consequent fall of temperature of the element. Thus, for the temperature to remain constant in *dm* with time, the equality

$$\bar{\varepsilon}_{em} = \bar{\varepsilon}_{abs} \tag{6.20}$$

must hold. Equation (6.20) embodies the *condition of energy equilibrium* in the stellar photosphere. Since we are here considering that any energy gain or loss in any element of mass of the photosphere is accomplished by *radiative transfer* alone. Eq. (6.20) can also be considered as embodying the *condition of radiative equilibrium*.

6.3 THE EQUATION OF TRANSFER

We consider an elementary right circular cylinder of length *ds* and cross-section $d\sigma$ with its axis directed along **s** inside the photosphere of a star. Let a pencil beam of radiation of intensity I_v enter the face 1 of this cylinder, pass through the matter within the cylinder, and leave it through the face 2, the beam being directed along the axis of the cylinder so that $\theta = 0$ (see Fig. 6.2). While passing through the cylinder of matter, the intensity of the beam will suffer a change, because in course of interaction of the pencil with the matter inside, in general, all the processes of absorption, scattering and emission will take place. Let the changed intensity of the emerging pencil be $I_v + dI_v$. The amount of radiation entering into the cylinder through the face 1 in time *dt*, inside the solid angle $d\Omega$ and in the frequency interval *dv* at *v* is given by

$$\varepsilon_i = I_v \, d\Omega \, dt \, d\sigma \, dv \tag{6.21}$$

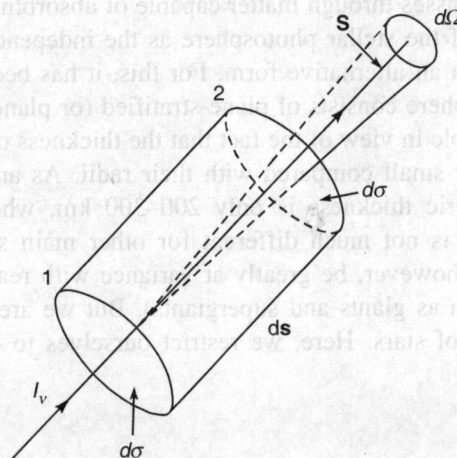

FIGURE 6.2 Change of intensity of a beam in passing through a cylinder of matter.

and the amount leaving the cylinder through the face 2 under the same specifications will be

$$\varepsilon_e = (I_\nu + dI_\nu)\, d\Omega\, dt\, d\sigma\, d\nu \tag{6.22}$$

It must be noted, however, that the quantities ε_i, and ε_e are not equal, because the radiation undergoes sufficient changes by the processes of emission, absorption and scattering inside the cylinder. The amount of change due to each of these processes must therefore have to be considered. The radiation *emitted* by the matter within the cylinder in time interval dt within the solid angle $d\Omega$ in the direction **s** in the frequency interval $d\nu$ at ν is ε_ν as given by Eq. (6.9) where dm is now equal to $\rho d\sigma\, ds$, since $\cos\theta = 1$. The energy absorbed (true absorption) within the cylinder is given by, following Eq. (6.11),

$$\Delta\varepsilon_{abs} = -I_\nu K_\nu \rho\, ds\, d\Omega\, dt\, d\sigma\, d\nu \tag{6.23}$$

and the energy scattered within the cylinder is similarly given by

$$\Delta\varepsilon_s = -I_\nu \sigma_\nu \rho\, ds\, d\Omega\, dt\, d\sigma\, d\nu \tag{6.24}$$

both the absorption and scattering being reckoned in time dt within the solid angle $d\Omega$ and in the frequency interval $d\nu$ at ν. Taking into consideration all the changes undergone within the cylinder as the beam passes through the matter, the equation that expresses the energy balance within the cylinder is given by

$$\varepsilon_e - \varepsilon_i = \varepsilon_\nu + \Delta\varepsilon_{abs} + \Delta\varepsilon_s \tag{6.25}$$

which yields after cancelling the common factor $dt\, d\sigma\, d\Omega\, d\nu$,

$$dI_\nu = -(K_\nu + \sigma_\nu)\, I_\nu \rho\, ds + j_\nu \rho\, ds \tag{6.26}$$

from which we deduce that

$$\frac{dI_\nu}{ds} = -(K_\nu + \sigma_\nu)\, \rho\, I_\nu + j_\nu \rho \tag{6.27}$$

Equation (6.27) is called the equation of transfer of radiation. It represents the change in intensity of radiation as it passes through matter capable of absorbing, scattering and emitting. By introducing the depth of the stellar photosphere as the independent variable instead of s, Eq. (6.27) can be written in an alternative form. For this, it has been customary to make the assumption that the photosphere consists of plane-stratified (or plane-parallel) layers. Such an assumption is highly plausible in view of the fact that the thickness of the photosphere of main sequence stars is extremely small compared with their radii. As an example, we may recall that in Sun, the photospheric thickness is only 200–300 km, whereas its radius is nearly 700,000 km. The situation is not much different for other main sequence stars. The plane stratified assumption may, however, be greatly at variance with reality in evolved stars with extended atmospheres (such as giants and supergiants). But we are not considering here the cases of these latter types of stars. Here, we restrict ourselves to only solar-type and other main sequence stars.

Turning then to the plane-parallel photosphere as shown in Fig. 6.3 (this is analogous to the assumption of plane-parallel atmosphere of the Earth in reckoning the refraction of starlight), we deduce the relation

$$dz = - ds \cos \theta \qquad (6.28)$$

where θ is the angle between the directions of the pencil and the normal to the plane photosphere, and the negative sign arises out of the fact that while the pencil moves outwards from some depth of the atmosphere, the depth z is measured inwards. Now, replacing ds by $- dz \sec \theta$, Eq. (6.27) becomes

$$\cos \theta \, \frac{dI_v(\theta)}{dz} = (K_v + \sigma_v) \rho I_v(\theta) - \rho j_v \qquad (6.29)$$

where the intensity I_v is now a function of θ.

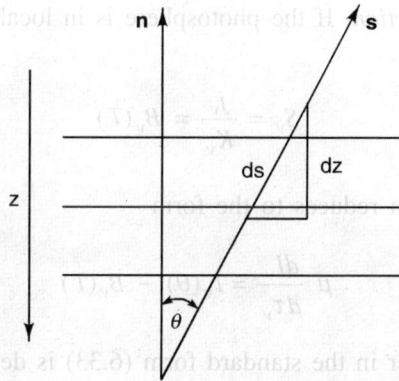

FIGURE 6.3 The relation between the depth z and length s of any direction of the pencil in a plane parallel photosphere.

If, we now assume that to each volume element of the atmospheric gas, Kirchhoff's law (Eq. (6.19)) holds (that is, the LTE holds) then $\sigma_v = 0$ and $j_v = K_v B_v(T)$. Thus, the equation of transfer (6.29) takes the form

$$\cos \theta \, \frac{dI_v(\theta)}{dz} = \rho K_v (I_v(\theta) - B_v(T)) \qquad (6.30)$$

It is to be remembered that both $I_v(\theta)$ and $B_v(T)$ are functions of the depth z in the photosphere and the frequency v of radiation, besides their dependence on radiation angle θ and temperature T respectively. It often becomes more convenient to use *optical depth* τ_v for v of radiation instead of the geometrical depth z, as a measure of the depth in the photosphere where the analysis is considered. In the general case of both absorption and scattering, the optical depth is given by

$$\tau_v = \int_{-\infty}^{Z} (K_v + \sigma_v) \, dz \qquad (6.31)$$

whence

$$d\tau_v = \rho(K_v + \sigma_v)\,dz \quad \text{and} \quad dz = \frac{d\tau_v}{\rho(K_v + \sigma_v)} \tag{6.32}$$

Thus dividing Eq. (6.29) by $\rho(K_v + \sigma_v)$, using Eq. (6.32) and writing $\cos\theta = \mu$, the equation of transfer takes the form

$$\mu\,\frac{dI_v(\theta)}{d\tau_v} = I_v(\theta) - S_v \tag{6.33}$$

This is the standard form of the equation of transfer where

$$S_v = \frac{j_v}{K_v + \sigma_v} \tag{6.34}$$

is defined as the *source function*. If the photosphere is in local thermodynamic equilibrium, then $\sigma_v = 0$, and

$$S_v = \frac{j_v}{K_v} = B_v(T) \tag{6.35}$$

The equation of transfer then reduces to the form

$$\mu\,\frac{dI_v}{d\tau_v} = I_v(\theta) - B_v(T) \tag{6.36}$$

The equation of transfer in the standard form (6.33) is derived under the conditions of (a) radiative equilibrium, and (b) plane stratified atmosphere. Under the additional condition of local thermodynamic equilibrium, Eq. (6.36) is derived. It can be proved that under the conditions stated in (a) and (b) above, the total integrated flux πF of radiation is independent of depth z of the photospheric layer. The transfer equation is a first order differential equation which has been solved by various authors under different assumptions. Some of these methods will be discussed in the next section.

6.4 THE SOLUTION OF THE EQUATION OF TRANSFER

The equation of transfer obtained in forms (6.33) or (6.36) relates the directional dependence of the intensity of radiation at any point in the stellar photosphere with the monochromatic optical depth at that point. In order to know this intensity distribution in the photosphere as a function of (μ, τ_v) one has therefore to solve the equation of transfer by some method. One can then express the total integrated flux of radiation at any depth of the photosphere in terms of the optical depth and the source function. Using then the Stefen-Boltzmann law which relates the integrated source function with the temperature in any level of the photosphere, one gets the (T, τ_v) relation which gives the temperature distribution in terms of the optical depth.

This last relation plays a vital role in analyzing the radiation characteristics and other physical properties of the steller photosphere.

The smooth deductive procedure enumerated above is, however, so easily achieved *only in principle. In reality,* however, due to the *frequency dependence* of the variables involved, the complete analytical solution of transfer equation poses great complexities. Only approximate solutions with simplifying assumptions have therefore been derived by various authors. The results have then been improved by iterative processes using the approximate solutions as initial values. Works of numerous authors over the last fifty years have now resulted in a fairly good understanding of the physical structure of stellar photospheres. What once was a formidable problem can now be handled in a simpler manner. The literature in the field has been enormous, and even a brief discussion of these is out of scope here. We shall just mention the results of a few simple cases and briefly discuss how these results fit with observations of certain types of stars. Let us consider the equation of transfer in the form of Eq. (6.36). This is the form for the photosphere satisfying both LTE and radiative equilibrium. It is a linear differential equation of the first order, and as such, can be solved easily, in principle. The general solution will yield the intensity $I(\mu, \tau_v)$ of emergent radiation coming from below at a point in the level whose optical depth is τ_v, in the direction θ. Treating τ_v as constant, being the optical depth at a fixed level, and t_v as the variable of integration, the solution of Eq. (6.36) can be written as

$$I_v(\mu, \tau_v) = \int_{\tau_v}^{\infty} B_v\left(T(\tau_v)\right) \exp\left\{- (t_v - \tau_v) \sec \theta\right\} \sec \theta \, dt_v \qquad (6.37)$$

Equation (6.37) expresses the intensity of radiation arriving at an angle θ to the normal direction at the level of optical depth τ_v from all other layers with $t_v > \tau_v$. On the boundary of the photosphere, $\tau_v = 0$. So the intensity of radiation emerging from the boundary of the photosphere at an angle θ is given by

$$I_v(\mu, 0) = \int_{0}^{\infty} B_v(T) \exp\left\{- (t_v \sec \theta\right\} \sec \theta \, dt_v \qquad (6.38)$$

The problem is sometimes simplified by removing the θ-dependence of the emergent radiation. For this the average value, $\cos \theta = 2/3$, is used. The solution (6.38) can then be expressed as

$$I_v = \frac{3}{2} \int_{0}^{\infty} B_v(T) \exp\left(- \frac{3}{2} t_v\right) dt_v \qquad (6.39)$$

Equation (6.39) gives a sort of average intensity of the emergent radiation at the boundary of the photosphere. The theoretical result in this case, shows large deviation from observed values near the limb of the Sun.

The basic problem in the study of the structure of stellar photospheres is, however, to find the (T, τ_v) relation; that is, the run of temperature inside the photosphere with optical depth. This problem becomes fairly simple in the case when the absorption coefficient K_v is assumed to be independent of the frequency v at every depth of the photosphere; that is $K_v = K =$ constant. This assumption often becomes far from reality. It is done mainly to achieve mathematical simplicity for handling the analytical procedure more easily, but at the cost of accuracy from physical point of view. Nevertheless, comparison between the theoretical and observed results has revealed that this approximation becomes fairly good for stars like the Sun. For this reason, great importance has been attached historically to this simple solution.

The material in which K_v is independent of v has been called *grey material*, and the approximation that is made by such an assumption is called the *grey-body approximation*. The solar photosphere closely represents grey material. Since, then

$$K_v = K = \text{Constant} \tag{6.40}$$

We can write

$$d\tau_v = d\tau = \rho K dz \tag{6.41}$$

by Eq. (6.31) so that the optical depth is also independent of v. Further, we have

$$\bar{I} = \int_0^\infty \bar{I}_v \, dv \tag{6.42}$$

and

$$B = \int_0^\infty B_v(T) \, dv \tag{6.43}$$

The condition of radiative equilibrium takes the form

$$\int_0^\infty B_v(T) K_v \, dv = \int_0^\infty \bar{I}_v \, K_v \, dv \tag{6.44}$$

in a photosphere at LTE. By virtue of Eqs. (6.40), (6.42) and (6.43), $\bar{I} = B$.

Calculations show that the mean intensity $\bar{I}$ over all the frequencies is related to the integrated flux πF by the relation

$$\frac{d\bar{I}}{d\tau} = \frac{3}{4} F \tag{6.45}$$

from which integration yields

$$\bar{I} = B = \frac{3}{4} F\tau + \text{Constant} \tag{6.46}$$

Using the appropriate boundary condition, it can be shown that the constant of integration in Eq. (6.46) is equal to 1/2 F, so that Eq. (6.46) can finally be written as

$$B = \frac{1}{2} F\left(\frac{3}{2} \tau + 1\right) \tag{6.47}$$

Thus, the intensity of radiation at a point of the photosphere where the optical depth is τ is related to the integrated flux πF by Eq. (6.47). Also, by Stefan-Boltzmann law, we have

$$\pi B = \pi \int_0^\infty B_v(T) = \frac{c}{4} U = \sigma T^4 \tag{6.48}$$

where σ is Stefan-Boltzmann constant and T is the temperature of the radiating surface. Equations (6.47) and (6.48) can be combined together to yield

$$T^4 = \frac{\pi}{2} \frac{F}{\sigma} \left(\frac{3}{2} \tau + 1 \right) \tag{6.49}$$

Again, the *effective temperature* T_e of a star is defined by the relation

$$\pi F = \sigma T_e^4 \tag{6.50}$$

so that the temperature T at any optical depth τ is expressed in terms of the effective temperature T_e by the relation

$$T^4 = \frac{1}{2} \left(\frac{3}{2} \tau + 1 \right) T_e^4 \tag{6.51}$$

Thus, the temperature distribution in the photosphere is expressed as some constant multiple of the effective temperature of the star. Let us consider *two* simple cases. In the first case, we see that at the stellar surface defined by $\tau = 0$, the temperature is given by

$$T_0^4 = \frac{1}{2} T_e^4 \tag{6.52}$$

from which

$$T_0 = 0.841 \, T_e \tag{6.53}$$

Secondly, the level at which the temperature equals the effective temperature of the star is given by

$$\frac{3}{2} \tau + 1 = 2 \tag{6.54}$$

by Eq. (6.51), whence, $\tau = 2/3$. Thus, in a *grey* photosphere in *radiative* and *local thermodynamic* equilibria, the temperature at the level $\tau = 2/3$ is equal to the effective temperature of the star. The above results are due to Eddington's approximation. Historically, this is the first solution of transfer equation of significant astrophysical importance and has been extensively used to study the nature of radiation flow in stellar photosphere. In this approximation, the limb-darkening effect in integrated light can be computed from Eq. (6.38) to yield

$$I(0, \mu) = \frac{1}{2} F \left(1 + \frac{3}{2} \cos \theta \right) \tag{6.55}$$

Equation (6.55) yields $I(0, 1) = \frac{1}{2} F \frac{5}{2}$, which at once gives the limb-darkening effect in white light for grey atmosphere as

$$\frac{I(0, \mu)}{I(0, 1)} = \frac{2}{5} + \frac{3}{5} \cos \theta = A_1 + A_2 \cos \theta \tag{6.56}$$

This theoretical value agrees fairly well with the observed value in the case of the Sun where A_2 observed is nearly 0.56.

More general and accurate solution of the transfer equation has been obtained by Chandrasekhar in the case of grey atmosphere. The equation of transfer can be written as

$$\mu \frac{dI}{d\tau} = I - \frac{1}{2} \int_{-1}^{+1} I d\mu \qquad (6.57)$$

if the radiation field is due to *pure, isotropic scattering*, because in that case $S_v = \overline{I}_v = \frac{1}{2} \int_{-1}^{+1} I\, d\mu$, and thus Eq. (6.57) is obtained from Eq. (6.33) if we write $\tau_v = \tau$ and note that $\overline{I}_v$ is in this case independent of direction (for isotropic scattering). Chandrasekhar replaces Eq. (6.57) by $2n$ ordinary linear differential equations,

$$\mu_i \frac{dI_i}{d\tau} = I_i - \frac{1}{2} \sum_{i=-n}^{n} a_i I_i \qquad (6.58)$$

Here $i = \pm1, \pm2, ..., \pm n$, $i = 0$ is omitted, and

$$\sum_{i=1}^{n} a_i = 1 \qquad (6.59)$$

Using a fairly long analytical procedure, Chandrasekhar ultimately arrived at a solution for the source function, given by

$$B(\tau) = \frac{3}{4} F[\tau + q(\tau)] \qquad (6.60)$$

where $q(\tau)$ is a monotonically increasing function of τ, such that,

$$q(\tau = 0) < q(\tau) < q(\tau \to \infty) \qquad (6.61)$$

$q(\tau)$ increases very slowly, such that

$$q(\tau = 0) = \frac{1}{\sqrt{3}}, \ q(\tau \to \infty) = 0.7104 \qquad (6.62)$$

Again, using Eqs. (6.48) and (6.50), Eq. (6.60) stands as

$$T^4(\tau) = \frac{3}{4} [\tau + q(\tau)] T_e^4 \qquad (6.63)$$

which gives the *local temperature* in terms of the effective temperature of the star. Equation (6.60) or (6.63) represents the exact solution for the *grey photosphere* in LTE and radiative equilibrium. We note that in this case, on the boundary $\tau = 0$ of the photosphere, $q(\tau) = 1/\sqrt{3}$, and

$$T_e^4 = \frac{\sqrt{3}}{4} T_e^4$$

from which

$$T_0 = 0.81 \, T_e \qquad (6.64)$$

Also, Eddington's solution (Eq. (6.51)) is related from Eq. (6.63) as a particular case, when $q(\tau) = 2/3$.

The limb-darkening is given in Chandrasekhar's first approximation $q(\tau) = 1/\sqrt{3}$ by

$$\frac{I(0, \mu)}{I(0, 1)} = \frac{1 + \sqrt{3}\mu}{1 + \sqrt{3}} \qquad (6.65)$$

This result differs somewhat from that in Eq. (6.56). It is to be noted further that Eddington's approximation becomes by and large accurate in deeper layers of the photosphere, whereas the exact solution of Chandrasekhar gives better fit with observations of the upper layers of the photosphere.

Many other authors have devised solutions of the transfer equation for the grey-body problem under different approximations. Besides the analytical methods summarized above, some authors have also used iterative methods which yielded improved results over the initial values chosen. Such iterative processes, in general, converge rapidly only for small values of the optical depth. For larger optical depths the convergence is poor and other methods, in general, work better. With these remarks, we shall restrict any further discussions of these methods. The principal point to be stressed here is that the grey-body approximation, particularly, the grey-body temperature distributions cannot fit with very great accuracy with any *real* stellar atmosphere. This is because atmospheres of all real stars, even that of the Sun and of other solar-type stars, differ significantly from a grey-body. In most cases, not only that the continuous absorption is highly frequency dependent, but also the line absorption at certain fixed frequencies causes significant deviation of the actual temperature distribution from that of the theoretical grey-body temperature distribution. So the subsequent emphasis was naturally placed on the solution of the transfer equation for non-grey atmospheres with which the actual stellar atmosphere resemble more closely. We shall therefore confine our discussion very briefly to some of the principal works on this problem.

The direct calculation of the source function in the case of highly frequency dependent absorption coefficient K_v is very difficult. So various approximations have been adopted by authors for solutions of non-grey problems. Instead of taking a constant absorption coefficient as in the grey-body problem, some sort of *mean absorption coefficient* $\bar{K}$ has generally been used by authors for approximate solutions of non-grey problems. For the purpose, various types of mean absorption coefficient have been devised and used by authors. Among these various means, we mention here only the Rosseland mean absorption coefficient $\bar{K}_R$. This is the most extensively used of all the mean absorption coefficients and is defined by

$$\frac{1}{\bar{K}_R} = \frac{\displaystyle\int_0^\infty \frac{1}{K_v} \frac{dB_v}{dT} \, dv}{\displaystyle\int_0^\infty \frac{dB_v}{dT} \, dv} \qquad (6.66)$$

Rosseland mean, in fact, is a measure of the mean *transparency* of radiation through the stellar material, as it yields the reciprocal of the mean absorption coefficient in terms of some weighted average. Strictly speaking, Rosseland mean is a quite good approximation in stellar interiors where the optical depth is large. In fact, the mean is defined precisely on the basis of satisfaction of this last condition (large optical depth). So it is used most extensively for the computation of opacity in the *stellar interiors*. Nevertheless, Rosseland mean absorption coefficient has been used by many authors for the computation of the structure of stellar atmosphere. It should be emphasized, however, in this connection, that in the superficial layers of stellar atmosphere, Rosseland mean absorption coefficient gives only gross results.

The main object of these solutions is to derive the temperature distribution in the atmosphere with respect to the optical depth $\tau(T - \tau$ relationship) which can then be compared with the empirical models, which are generally constructed on the basis of observed limb-darkening. It is found that the theoretical models agree well with the empirical models.

When the important $(T - \tau)$ relationship has been computed from the solution of the transfer equation by using some form of the mean absorption coefficient $\bar{K}$, other physical parameters in the photosphere can be obtained from additional equations and the structure of the photosphere can be determined completely. For example, using the hydrostatic equation

$$\frac{dP}{dz} = - \rho g \tag{6.67}$$

and the equation for the mean optical depth

$$\frac{d\bar{\tau}}{dz} = - \rho \bar{K} \tag{6.68}$$

we get

$$\frac{dP}{d\bar{\tau}} = g/\bar{K} \tag{6.69}$$

Equation (6.69) determines the pressure distribution with respect to $\tau(P - \tau$ relationship), because it can be integrated if the composition and ionization structure of the atmosphere is known, the last two information giving $\bar{K}$, in principle. Knowing the temperature and pressure distribution, it becomes easier to calculate the density distribution. It is possible also to compute the relationship between the optical and geometrical depths with Eq. (6.68). A detailed specification of a model atmosphere is thus possible. Many such models have been computed for the Sun which agree fairly well with the corresponding observed features in the solar atmosphere. The facts underlying one of the most difficult but interesting problems of vital importance in astrophysics have thus been known.

6.5 PROCESSES OF ABSORPTION IN STELLAR ATMOSPHERES

In the preceding section we have seen how the frequency dependence of the absorption coefficient in stellar photosphere complicates the solution of the equation of transfer and as a consequence, how we are restricted to only approximate knowledge of the photospheric structure. Interaction of matter in the photosphere with the emergent radiation of the star presents a whole set of complicated phenomena. One such very important phenomenon is the absorption of radiation by various processes by various types of atoms and molecules which constitute the atmosphere of a star. The complete solution of the problem requires a knowledge of the chemical composition of the material, the ionization and excitation condition in it, and the particular process of absorption that is mainly operative. The temperature distribution in the photosphere depends on these and various other factors. But all these informations can be obtained only through complex physical analyses. In this section, we shall discuss some of the important factors that contribute to absorption of radiation in the stellar photosphere. Since the absorption of radiation by various elements in the photosphere due to various processes operating cuts off a fair amount of the emergent radiation of the star, the process is commonly known as the opacity in stellar photosphere.

We shall consider here only *true absorption* of radiation which is accompanied by transition of atoms or ions from one energy state to another. The absorbed radiant energy in this case is converted into thermal energy of atoms and ions in the photospheric material. This thermal energy is again lost after sometime by re-emission as radiant energy, but now at some entirely different frequencies. Thus, true absorption is associated with radiative transfer mechanism, and it is through these co-existing processes that the stellar radiation from the deep interior layers reach the superficial layers and finally leave the stellar body. True absorption is thus different from scattering of radiation by electrons or molecules. In scattering, no conversion of radiant into thermal energy is involved and so no change of frequencies of photons takes place. Photons only change directions by scattering.

True absorption is again of two kinds, viz., true selective absorption and true continuous absorption. We shall discuss both these types of absorption with reference to hydrogen (or hydrogen-like) atoms which are overwhelmingly predominant in stellar photospheres. *True selective absorption* is responsible for the production of Fraunhofer lines in stellar spectrum. It is associated with bound-bound (b-b) transitions of atoms in which the electron jumps from one energy state i to another energy state k, thereby absorbing a photon of fixed energy

$$h\nu_k - h\nu_i = h\nu_{ik} \qquad (6.70)$$

and producing a spectral line corresponding to the photon energy $h\nu_{ik}$. Thus the transition from the ground level of hydrogen atoms to any higher level is associated with the absorption of Lyman photon which contributes to the formation of the corresponding Lyman line. Similarly, the transition from the second energy state to any higher state of hydrogen atom is accompanied by the absorption of a Balmer photon which contributes to the formation of the corresponding Balmer line, and so on. Every Fraunhofer line in the stellar spectrum is thus associated with the absorption of a large number of photons by a particular type of atoms responsible for the line and undergoing the specific transition which corresponds to that line. In general, any bound-bound transition will result in a spectral line because such transitions always involve fixed energy differences.

It must be mentioned, however, that although line absorption has very great astrophysical significance such as in spectral classification of stars, abundance determination etc., calculations show that it plays minor role in the overall energy balance in stellar photospheres. Also, the line absorption is significant only over a small fraction of the total width of the spectrum. But there occur, indeed, small regions of the spectrum where numerous absorption lines concentrate, such as at the long-wavelength side of the series limits of hydrogen atoms. In these regions the normal outflow of radiant energy is seriously obstructed and intensity of radiation is greatly diminished. This diminution of radiation intensity in regions of spectral lines is known as the *blanketing effect* in the atmosphere of stars. The continuous energy distribution in the far-ultraviolet region of the spectrum beyond the Lyman series limit also is seriously influenced by absorption lines in stars like Sun and also in those hotter than the Sun. For these stars, the region of spectrum mentioned possesses a large concentration of strong resonance lines of various elements. However, for Sun the overall radiation intensity in the far-ultraviolet region is quite small any way, and thus the influence of the far-ultraviolet region as a whole is not significant in the total energy balance of the solar photosphere. On the other hand, in stars like Sun and those cooler than it, numerous absorption lines of metals and molecular bands shortward of 4000 Å play a significant role in the overall energy balance in photospheres of these stars.

Quantitatively, the selective absorption coefficient per atom at frequency v, according to the classical theory, is given by

$$s_v = \frac{e^2}{m_e c} \frac{\Gamma_0}{4\pi} \frac{1}{(v - v_0)^2 + (\Gamma_0/4\pi)^2} \tag{6.71}$$

where v_0 is the *natural frequency* of the line and Γ_0 is the *damping constant,* given by

$$\Gamma_0 = \frac{8\pi^2 e^2 v_0^2}{3 m_e c^3} \tag{6.72}$$

Γ_0 measures the exponential loss of energy of a naturally radiating classical oscillator which follows the law:

$$\varepsilon = \varepsilon_0 \exp(-\Gamma_0 t) \tag{6.73}$$

The relation (6.71) shows that the absorption takes place not in an infinitely sharp monochromatic line in its natural frequency, but in a small frequency range equally spread on both sides of the natural frequency. Thus, radiation damping imparts *a finite width* to the absorption line. The value of s_v diminishes by half its value at v_0, at a distance of $1/2\,\Gamma_0$ from v_0 on either side. Considering both sides of the natural frequency v_0, the value of s_v becomes one-half within a spread of Γ_0 around v_0. For this reason Γ_0 has sometimes been called the total *half-width of the absorption coefficient.* It is also called the *natural line width.*

The damping effect is enhanced by collision in atmospheres of stars having fairly high pressure. Collisional damping is important in stars like Sun and in cooler stars. Collisions broaden the spectral line, thereby increasing the non-monochromatic nature of the absorption coefficient. With *oscillator strength f,* the absorption coefficient per atom at frequency v is given by

$$s_v = \frac{e^2}{m_e c} \frac{1}{4\pi} \frac{\Gamma f}{(v - v_0)^2 + (\Gamma/4\pi)^2} \tag{6.74}$$

where Γ is now the sum of the radiation and collisional damping constant, that is, $\Gamma = \Gamma_0 + \Gamma_{\text{coll}}$.

Although, as has been already mentioned, selective absorption plays some role over the spectral energy distribution of stars in some restricted regions, continuous absorption arising out of the bound-free (b-f) and free-free (f-f) transitions is overwhelmingly more important in the problem of stellar energy balance. Of these two processes again, the bound-free absorption resulting from photo-ionization of atoms is much more important in almost all types of stars. Only in the hottest stars in whose atmospheres hydrogen is almost completely ionized, free-free absorption process plays a significant role. We again consider the case of hydrogen atoms. Lines of each of the series of hydrogen atom crowd closer and closer together towards the limit of the series until they coalesce into a continuum at the series limit. Continuous absorption starts beyond this limit. We thus have Balmer continuum shortward of the Balmer series limit at $\lambda = 3650$ Å, Lyman continuum shortward of the Lyman series limit at $\lambda = 912$ Å, and so on. In fact, each of the hydrogen series is associated with a continuum. This is also true for atoms of every other species. Figure 6.4 illustrates the different kinds of absorption in the case of hydrogen atoms.

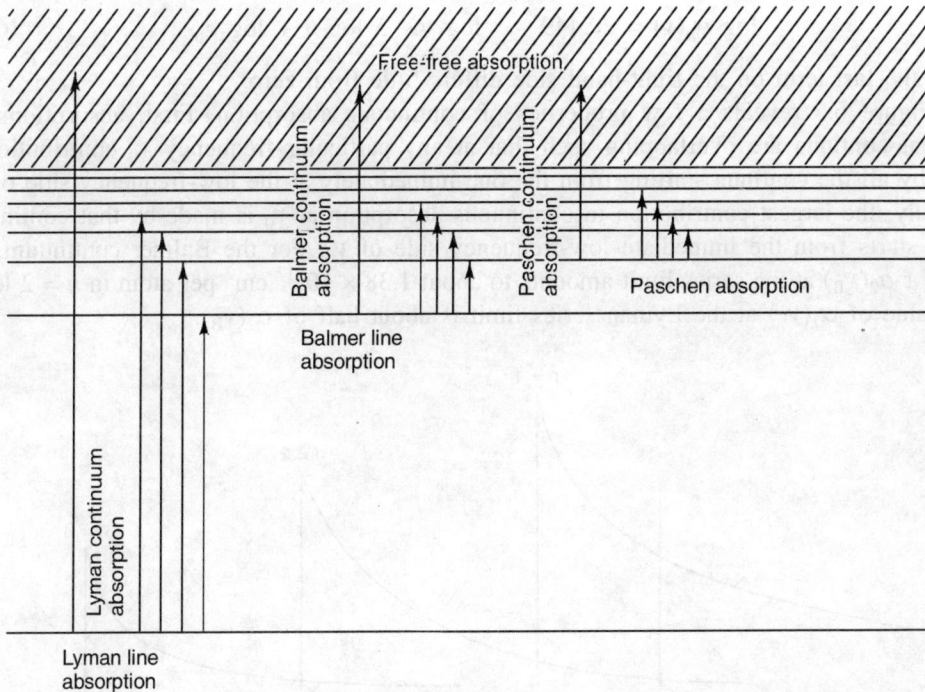

FIGURE 6.4 Continuous, selective and free-free absorption by hydrogen.

Quantum mechanical calculations reveal that the continuous (b-f) absorption coefficient per atom of neutral hydrogen (or any hydrogen-like atom) in the nth state of excitation is

$$\alpha_n(v) = \frac{32}{3\sqrt{3}} \; \frac{\pi^2 e^6}{ch^3} \cdot \frac{R}{n^5} \; \frac{Z^4}{v^3} \; g_{bf} \qquad (6.75)$$

where Z is the atomic number with nuclear charge Ze and g_{bf} is the Gaunt factor for bound-free transition whose value does not differ much from unity. R is the Rydberg constant given by

$$R = \frac{2\pi^2 e^4 m_e}{h^3} \tag{6.76}$$

For hydrogen, $R = R_H = 109{,}677.58$ cm^{-1}. The relation (6.75) shows strong dependence of the absorption coefficient on n and v for a particular species of atoms. Each value of n corresponds to a definite absorption band of the atom followed by a continuum. In each of these continua (associated with a fixed n) the absorption is strongest at the series limit (corresponding to the smallest value of v for the particular series) and falls off gradually to shorter wavelengths according to the law $1/v^3$. Figure 6.5 gives a schematic representation of absorption in different continua up to $n = 3$. The curves of variation are cubical parabolas in $[\alpha_n(\lambda), \lambda]$ coordinates. Writing the values of all the constants involved, Eq. (6.75) for hydrogen atom can be written as

$$\log \alpha_n(v) = 29.449 - 3 \log v - 5 \log n + \log g_{bf} \tag{6.77}$$

where the last term on the right-hand side differs little from zero.

Figure 6.5 reveals a few properties of continuous absorption. First, absorptions for different continua are overlapping such that at any particular frequency v_0 contribution is made by all the continua starting from the one immediately at the low-frequency side of v_0. Secondly, the largest contribution to continuous absorption at v_0 is made by that continuum which starts from the immediate low-frequency side of v_0. For the Balmer continuum, the value of $\alpha_2(v_B)$ at the series limit amounts to about 1.38×10^{-17} cm^2 per atom in $n = 2$ level. The value of $\alpha_1(v_L)$ at the Lyman series limit is about half of $\alpha_2(v_B)$.

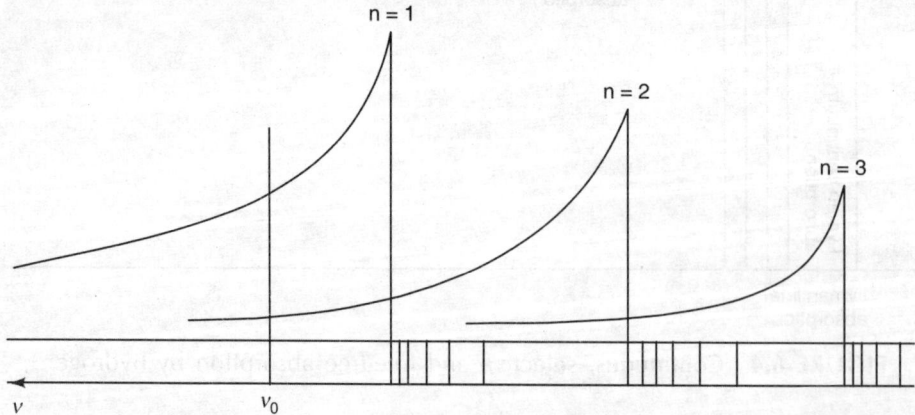

FIGURE 6.5 A schematic representation of absorption in different continua up to $n = 3$.

The above results hold for one atom of hydrogen which is assumed to occupy a particular energy level. The situation becomes more complicated, however, if we want to compute the total absorption coefficient for a gram of hydrogen, because in this case we have to have a pre-possessed knowledge of the distribution of atoms in different energy levels. The latter depends mainly on the temperature to which the atoms are exposed and also on some other physical factors. The continuous absorption coefficient per gram therefore involves the temperature, and we have to sum up the contributions of all levels which can take part in producing the absorption at the frequency concerned. For example, if we are interested in the continuous absorption at λ 5000 Å, we have to consider the contributions by all atoms occupying the levels $n \geq 3$, because the level $n = 2$ can contribute only shortward of λ 3650 Å and the level $n = 1$ can contribute only shortward of λ 912 Å. Calculations show that the continuous absorption coefficient per gram of neutral hydrogen at the ground level at the temperature T is given by

$$K(v)_{bf} = \frac{32}{3\sqrt{3}} \frac{\pi^2 e^6 R_H Z^4}{m_H ch^3} \frac{\exp(-X_1)}{v^3} \sum \frac{1}{n^3} \exp(X_n) g_{bf} \qquad (6.78)$$

where

$$X_n = \frac{hR_H Z^2}{n^2 kT} = \frac{158{,}000}{n^2 T} \qquad (6.79)$$

where k is the Boltzmann constant and m_H is the mass of a hydrogen atom so that the number of atoms per gram is $1/m_H$. The summation is extended to all levels which can contribute to the absorption at the frequency of interest.

Absorption by free-free transition is important, in general, only in the infrared and microwave regions of the spectrum except in stars of spectral classes O and B. In the latter stars free-free absorption may be significant in the visible region of the spectrum. Calculations show that the absorption coefficient for free-free transitions per gram of neutral hydrogen in the ground level is

$$K(v)_{ff} = \frac{32}{3\sqrt{3}} \frac{\pi^2 e^6 R_H Z^4}{m_H ch^3} \cdot \frac{\exp(-X_1)}{v^3} \frac{g_{ff}}{2X_1} \qquad (6.80)$$

where g_{ff} is the Gaunt factor for free-free transition and X_1 is defined by Eq. (6.79). In most cases g_{ff} is not far from unity. The total absorption coefficient per gram of neutral hydrogen at ground level at temperature T is therefore given by the sum of Eqs. (6.78) and (6.80), i.e.

$$K(v) = K(v)_{bf} + K(v)_{ff}$$

$$= \frac{32}{3\sqrt{3}} \frac{\pi^2 e^6 R_H Z^4}{m_H ch^3} \frac{\exp(-X_1)}{v^3} \left[\sum \frac{1}{n^3} \exp(X_n) g_{bf} + \frac{1}{2X_1} g_{ff} \right] \qquad (6.81)$$

The presence of the exponential factor on the right-hand side of Eq. (6.81) indicates stiff dependence of the continuous absorption of hydrogen atoms on temperature. At low temperatures, almost all hydrogen atoms occupy the ground level and the continuous absorption almost entirely lies shortward of the Lyman series limit at λ 912 Å. As the temperature rises, absorption shortward of other series limits begins to become important as more and more atoms begin to populate higher levels at these temperatures. In Sun, continuous absorption by atomic hydrogen is nowhere very important, as we shall see in the next section.

6.6 CONTINUOUS ABSORPTION BY THE NEGATIVE HYDROGEN IONS (H⁻) IN COOLER STARS

When the theoretical values of the continuous absorption by neutral hydrogen atoms are compared with the actually observed continuous absorption in stars, it is found that except for the hottest stars of types O and B the former turns out to be insignificantly small to explain the latter. That the neutral hydrogen might not be responsible for the observed continuous absorption in cooler stars was further supported by relatively smaller *Balmer Jump* as observed in these stars. Because, theoretically, as temperature decreases the ratio of the continuous absorption on the shortward and longward sides of the Balmer series limit should gradually increase to very large values. Such an increase to the extent as theoretically predicted is not actually observed.

Initially, an attempt to explain the discrepancy was made by assuming that perhaps it was due to continuous absorption by metals. It was soon found, however, that if the absorption by metals was held responsible for this and the observed Balmer Jump was computed by introducing the right amount of metal abundance, the required abundance would be so high as to make the metallic lines in cooler stars much more strong than are actually observed. So an alternative source of continuous absorption had to be found in order to explain the observation in cooler stars. The actual source was ultimately found to be the *negative hydrogen ions* by R. Wildt in 1939, who suggested that most of the continuous absorption in late-type stars was produced by these ions. Due to the incomplete shielding of the proton charge in a neutral hydrogen atom, the latter can capture an additional electron to form a negative hydrogen ion to which the symbol H⁻ has been assigned. The additional electron is loosely bound to the atom, the binding energy being only 0.747 eV.

The continuous absorption coefficients due to bound-free and free-free transitions of H⁻ have been known mainly through the works of S. Chandrasekhar and his co-workers. The binding energy of 0.747 eV corresponds to the ionization limit at nearly λ 16500 Å. The continuous absorption produced by H⁻ thus starts at wavelength about λ 16500 Å, rises gradually to a maximum at about λ 8500 Å and then falls again gradually towards the shorter wavelengths. The absorption is therefore very high in the visible region of the spectrum. The continuous absorption due to bound-free transition of one H⁻ ion is shown in Fig. 6.6. This is the only important absorption band due to H⁻. Besides the bound-free absorption, H⁻ is also

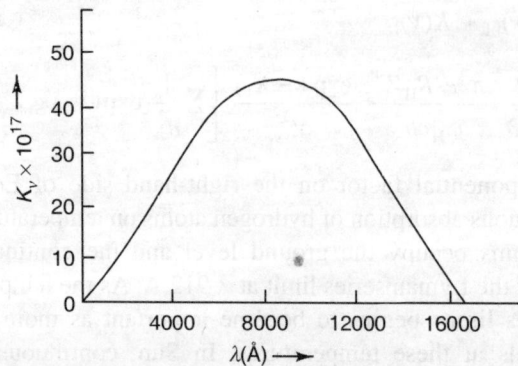

FIGURE 6.6 The bound-free absorption due to H⁻.

capable of producing free-free absorption. The latter occurs when an electron moves near a neutral hydrogen atom in a hyperbolic orbit. The contribution due to both these processes must be taken into account while computing the total continuous absorption produced by H⁻.

The works of Wildt and of Chandrasekhar and his co-workers have revealed that for stars like Sun the most dominant source of continuous absorption is the negative hydrogen ion. In Sun, the continuous absorption of radiation in the visible region of the spectrum is mostly produced by H⁻ the next important source being H. Continuous absorption by metals contribute only about 20 per cent of that jointly produced by H and H⁻ in the blue-violet region. The contribution due to metals increases in ultraviolet. In fact, continuous absorption by H⁻ begins to be significant starting from the spectral class A2 and becomes more and more important for stars of later spectral classes, until in solar-type stars it becomes overwhelmingly dominant.

Continuous absorption by any other atom or molecule is not at all comparable with that produced by H or H⁻ when the entire spectral region is considered. There may, however, be some special regions of the spectrum where some other source of absorption may be significant. He I and He II might be quite important sources of continuous absorption in hot stars. But on account of their lower abundance and higher ionization and excitation potentials, they can never attain comparable importance with hydrogen in the visible region. Metals are important sources of continuous absorption but only in the ultraviolet.

6.7 ANALYSIS OF SPECTRAL LINE BROADENING

If the energy levels of an atom were as sharp as a horizontal straight line, as we generally depict in an energy level diagram, then any transition between such two levels would have been accompanied by a precise amount of energy change thereby producing an extremely sharp spectral line. But partly due to the inherent structural property of the atom and partly due to the conditions present in the stellar atmospheres, the energy levels are not quite sharp, but each level is as if composed of a continuous set of sub-levels as illustrated in the Fig. 6.7. When the transition takes place to produce the line, it is most likely to take place between the sharp heavy lines *P* and *Q*, but some transitions also take place in which the sub-levels of both *P* and *Q* are involved. These transitions involve energies which are slightly more or slightly

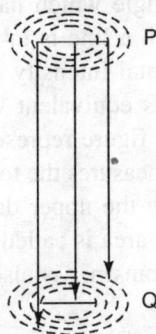

FIGURE 6.7 The widths of energy levels and transitions.

less than that corresponding to the natural transitions between the sharp thick lines P and Q. As a result, the spectral lines instead of being precisely sharp, are broadened on both sides of the natural wavelength. The line therefore has some finite width having a *deep dark core* at the central part that corresponds to the most likely transitions between the sharp thick lines P and Q together with less dark portions on both sides of this core, which correspond to the rarer transitions between the levels concerned. These portions form the wings of the line. The profile of an absorption line is shown in Fig. 6.8. The measurement of the width and a careful study of the line profiles is of extreme importance in that, it may provide us valuable information regarding the physical structure of the stellar envelopes, their motions, the chemical compositions and other related features.

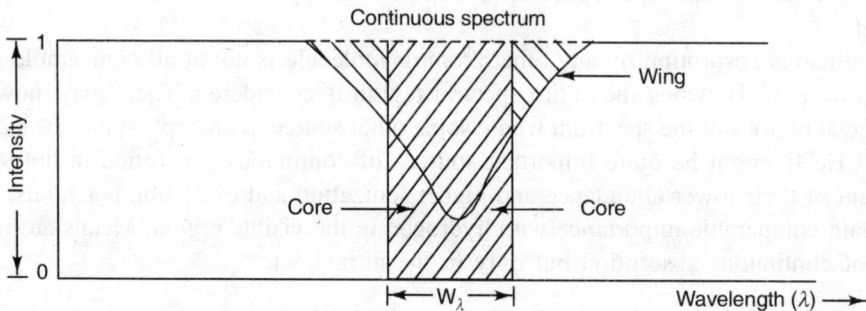

FIGURE 6.8 The profile of an absorption line and its equivalent width.

The shape of the profile of a spectral line differs with the intensity of the source at various wavelength and with the resolving power of the spectroscope used. It is necessary to have a spectroscope of sufficiently high resolving power so that the finite width of strong spectral lines (e.g. the wide Balmer line of hydrogen) can be measured directly with sufficient accuracy. As the resolution of a spectroscope is also finite, one has to correct for the *instrumental profile* in the observed spectrum. Thus, the actual profile could be accurately measured knowing the instrumental resolution. Generally, stellar spectra are observed with low-dispersion spectrographs so that their line-widths cannot be always measured directly from the observed spectrum. The width of a line in a spectrum is generally measured in terms of its "*equivalent width*". This is the width w_λ of a rectangle which has the same area as that under the line-profile in question. The intensity of such a line is taken to be zero along the widths, that is, the height of the rectangle covers the total intensity of the continuum. Figure 6.8 illustrates the profile of an absorption line with its equivalent width w_λ.

The shaded rectangular area in the figure represents the area equivalent to that bounded by the absorption line profile. This area measures the total intensity subtracted by the absorption line from the continuum (represented by the upper dashed line) at the particular wavelength considered. The width of the rectangular area is called the equivalent width, w_λ of the spectral line and is generally measured in angstroms. w_λ is also called the *total absorption* in the line. This is mathematically represented as

$$R_\lambda = 1 - I_\lambda, \qquad w_\lambda = \int R_\lambda \, d\lambda \qquad (6.82)$$

or

$$w_v = \int R_v \, dv \tag{6.83}$$

where R_λ is the *depth* of the profile for the particular wavelength and the integration being taken throughout the observed profile, w_λ varies with the intensity and strength of the absorption lines. It is large for strong lines. Theoretically, the equivalent width should not change with the resolutions used in different spectrograms. But practically it is found to vary appreciably with different observers working with different instruments. Two sources of error generally creep in the measurement of the equivalent widths. The first error arises due the error in ascertaining the continuum, and the second due to the blending of other fainter lines in the wings of the line in question.

We now discuss the various factors that may be responsible for the broadening of an observed spectral line.

Natural Broadening

As we have already seen at the beginning of this section, the principal energy states of the atom are composed of a number of sub-states that contain somewhat different energies than that of the principal energy-states. The transitions between such sub-states give rise to many weaker line of slightly different wavelengths (shorter or larger) in the neighbourhood of the strong and sharp principal line. This results in the formation of the "natural width" of the spectral line corresponding to a small range of wavelengths. In Fig. 6.8, the deep core of the line profile is due to the most probable transition between the principal energy states and the side "wings" are the result of fewer transitions between the substates having slightly different energies.

Generally, an atom spends very little time (generally a small fraction of a microsecond) in states of higher energies. After being excited to higher energy states it comes down immediately to the ground state releasing a definite amount of energy associated with the orbit just left. The time spent by an atom in an excited state is the factor which determines the amount of the natural broadening of the absorption line corresponding to that state in accordance with the Heisenberg's uncertainty principle ($\Delta E \Delta t \approx h/2\pi$). The relation demonstrates that shorter the time spent by the electron in a level, the longer is the broadening of that level, and the larger is the width of the line. Since the atom tends to spend most of the time in the ground level, this level is hardly broadened. Consequently, resonance lines (lines arising out of the transitions from the ground level) are much sharper compared to the subordinate lines.

The natural broadening of a spectral line depends on a constant Γ_0 known as *damping constant,* which determines the radiation damping of a classical oscillator according to the equation,

$$E = E_0 \exp(-\Gamma_0 t) \tag{6.84}$$

where Γ_0 is given by Eq. (6.72).

In wavelength scale it is independent of λ and equal to 1.18×10^{-4} Å. If τ_0 be the damping time of the oscillator, then it is the reciprocal of the damping constant Γ_0, that is

$$\tau_0 = \frac{1}{\Gamma_0} \qquad (6.85)$$

It follows from Eq. (6.84) that τ_0 is the time in which the energy of the oscillator is reduced by the factor e. For the sodium lines D_1 and D_2 with mean wavelength λ 5893 Å, the damping time is 1.58×10^{-8} sec.

Doppler Broadening

In the outer envelopes of stars, thermal motions of atoms cause appreciable broadening of spectral lines. Atoms in the hot photospheric layers move at random with high velocities, absorbing and emitting radiations continually from all possible directions. The Doppler effects of such motions result in the shift of absorption lines slightly on both sides of the natural wavelengths of these lines causing a finite width of them. This is known as the Doppler broadening of the spectral lines. Such a broadening effect is very important in case of the "weak" lines in the spectrum. Doppler broadening is much pronounced in the hot stars in whose atmospheres the thermal motions of the atoms are quite high. At any temperature, such effect is much larger for the lines of light elements such as H and He and is relatively small for heavy elements like Ca and Fe.

Effect due to turbulence.   Like the thermal motions, the turbulent motions of gaseous masses also lead to broadening of lines in the spectra of certain stars. For some supergiant and giant stars (specially of early type) it is found that if we consider the Doppler effect due to thermal motions of atoms to be the chief source of line-widths in the spectrum, we would have obtained an exceedingly high value of Kinetic temperature that would fit to such high thermal velocities. Thus, O. Struve had measured the temperature of α-Persei (T_{eff} = 7000 K) to be as large as 300,000 K. From a study of the deviations of the Doppler branch of the *curves of growth* constructed from theory and observations, it was found that Doppler broadening in some early type supergiants and giants was large enough to be attributed merely to the thermal motions of the atoms. The excess width is principally due to the turbulent motions of the gas particles in the stellar atmosphere.

The thermal velocity of the atoms of a gaseous mass is related to the gas temperature by

$$v_{th}^2 = \frac{2RT}{\mu} \qquad (6.86)$$

where R is the universal gas constant and μ the molecular weight of the gas. If the gas has also turbulent velocity given by v_{turb}; then due to the law of superposition of the dispersion velocities being valid, the total random velocity v_{tot} will be given by

$$v_{tot}^2 = v_{th}^2 + v_{turb}^2 = \frac{2RT}{\mu} + v_{turb}^2 \qquad (6.87)$$

If all the velocities mentioned above are assumed to represent the corresponding line-of-sight components only, then v_{tot} is associated with the total Doppler shift of the frequency (or the wavelength) concerned, due both to the thermal as well as the turbulent motions of the gas. If v_0 is the natural frequency (or λ_0, the natural wavelength) of a line and v the displaced frequency (or λ the displaced wavelength), and if Δv_D or $\Delta \lambda_D$ is the Doppler-shift of the line in frequency or wavelength, then we have

$$\frac{\Delta v_D}{v_0} = \frac{|v - v_0|}{v_0} = \frac{\Delta \lambda_D}{\lambda_D} = \frac{|\lambda - \lambda_0|}{\lambda_0} = \frac{v_{\text{tot}}}{c} \tag{6.88}$$

c being the velocity of light. The relative frequency distribution or the relative wavelength distribution inside the line, for the *Maxwellian velocity distribution* of the absorbing or emitting atoms, is given respectively by

$$\frac{dn}{n} = \frac{1}{\sqrt{\pi}} \exp\left[-\left(\Delta v / \Delta v_D\right)^2\right] \frac{dv}{\Delta v_D} \tag{6.89}$$

and

$$\frac{dn}{n} = \frac{1}{\sqrt{\pi}} \exp\left[-\left(\Delta \lambda / \Delta \lambda_D\right)^2\right] \frac{d\lambda}{\Delta \lambda_D} \tag{6.90}$$

where we have written $d(\Delta v) = dv$ and $d(\Delta \lambda) = d\lambda$.

In Eq. (6.89), the quantity Δv represents any frequency interval within the line and dn the fraction of atoms taking part in the formation of that part of the line. Similar explanation holds for Eq. (6.90) also. In the case $\Delta v - \Delta v_D$ the fraction reduces by the factor i/e. This fact leads to the definition of Δv_D or $\Delta \lambda_D$ as the *Doppler widths*.

It follows from Eq. (6.86) that for purely thermal motion, the Doppler width is proportional to $1/\sqrt{\mu}$ for the same temperature. This supports the statement we have already made that for the lighter elements (that is, for small μ) the Doppler width is larger while for the heavier elements, it is smaller. Thus in the Sun ($T_{\text{eff}} = 5710$ K), $\Delta \lambda_D \approx 0.212$ Å for H_α (λ 6563 Å) line and $\Delta \lambda_D = 0.04$ Å for D_1 (λ 5896 Å) line of sodium. At a temperature of 10,000 K (e.g. in Sirius) these values are respectively = 0.281 Å and 0.053 Å. Recalling that the approximate size of the natural width is of the order of 10^{-4} Å, it is clear that the Doppler effect plays a much more important role in broadening the spectral lines formed in the atmospheres of stars.

As long as the intensity of a line at any frequency within it is proportional to the number of atoms absorbing (or emitting) that frequency, the line is said to be *unsaturated*. It is said to be *saturated* at a frequency when the intensity no longer increases proportionally with the increase of atoms absorbing that frequency. As a rule, the core of the line generally becomes first saturated as the number of atoms absorbing the line gradually increases. In an unsaturated *line* the intensity distribution within the line is given by

$$I(v)\,dv = \frac{I_0}{\sqrt{\pi}} \exp\left\{-\frac{c^2}{\alpha^2}\left(\frac{v - v_0}{v_0}\right)^2\right\} \frac{c\,dv}{\alpha v} \tag{6.91}$$

where α represents the most probable speed and is given by

$$\alpha^2 = \frac{2kT}{m} \tag{6.92}$$

m being the mass of the atom and k the Boltzmann constant equal to 1.38×10^{-16} erg deg^{-1}. I_0 represents the total intensity integrated over the entire line, in wavelength units, Eq. (6.91) becomes

$$I(\lambda)\, d\lambda = \frac{I_0}{\sqrt{\pi}}\, \exp\left\{ -\frac{c^2(\lambda - \lambda_0)^2}{\alpha^2 \lambda_0^2} \right\} \frac{c\, d\lambda}{\alpha\lambda} \tag{6.93}$$

The intensity at the centre of the line, I_C, is related to the total intensity I_0 by the equation

$$I_C = \frac{cI_0}{\lambda\alpha\sqrt{\pi}} \tag{6.94}$$

which can also be written as

$$I_C = \frac{cI_0}{\lambda} \sqrt{\frac{m}{2\pi kT}} \tag{6.95}$$

Calculation of the line profile with Eq. (6.91) or (6.93) shows that the profile is more or less bell-shaped with round bottom and intensity away from the line centre rapidly falling off as shown in Fig. 6.9.

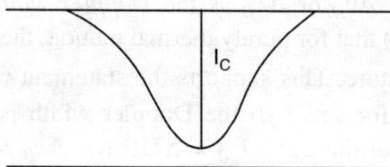

FIGURE 6.9   Bell-shaped profile of a Doppler broadened line.

The half-width of a line is defined as the width where the intensity I has fallen to one-half the value of the maximum intensity I_C. The intensity at half-width is given by

$$I = \frac{1}{2} I_C \tag{6.96}$$

Using Eq. (6.93), Eq. (6.96) yields

$$\frac{cI_0}{\lambda\alpha\sqrt{\pi}}\, \exp\left\{ -\frac{c^2(\lambda - \lambda_0)^2}{\alpha^2 \lambda_0^2} \right\} = \frac{1}{2}\, \frac{cI_0}{\lambda\alpha\sqrt{\pi}}$$

that is,

$$\exp\left\{-\frac{mc^2(\lambda - \lambda_0)^2}{2kT\lambda_0^2}\right\} = \frac{1}{2} \tag{6.97}$$

by using Eq. (6.92). The total half-width of the line is thus given by

$$2\delta\lambda = 2(\lambda - \lambda_0) = 2\sqrt{\frac{2kT\lambda_0^2}{mc^2}\ln 2} \tag{6.98}$$

Equation (6.98) can be used to compute the total half-width of any line of an element at any given temperature. It should however be borne in mind in this connection that the total half-width is an entirely different quantity from the Doppler width.

Collisional Broadening or Pressure Broadening

Collisions between radiating atoms and nearby ions or electrons sometimes cause broadening of spectral lines. When a neutral atom encounters a radiating atom or ion its energy states are generally perturbed to some extent. As a result, the atom might exhibit different energies and radiate at different frequencies from that it would do if it were in the normal energy states. The effect of such encounters may result in the smearing out of the absorption lines into some finite widths. Such phenomena are known as collisional broadening of lines and are generally observed in the strong lines of the spectrum. This is also called *"Pressure broadening"* because the broadening which is caused by collisions between atoms and ions is proportional to the gas pressure in the medium.

In the spectrum of cool dwarf stars like the Sun, the broadening of the strong lines of neutral hydrogen is an important phenomenon. This is explained by the fact that the neutral hydrogen atoms in the solar atmosphere interact with other radiating atoms by the van der Waal forces [i.e., forces of the order of $(distance)^{-6}$] and that many of such collisions lead to the broadening of the hydrogen lines.

The collision of the perturbing and radiating atoms are dependent on a constant known as the *collisional damping constant*, Γ_{coll}. This is defined as twice the number of effective collisions per second. In the van der Waal type of collisions in the solar atmosphere Γ_{coll}, is proportional to pressure of the gas, P_g. The mean time τ_C between two collisions may be expressed by the relation

$$\Gamma_{coll} = \frac{2}{\tau_C} \tag{6.99}$$

which is much like the relation between Γ_0 and τ_0 in the case of radiation damping. When the line-broadening results from collisions of the neutral hydrogen atoms with other radiating atoms, i.e., in the case of van der Waal type of interactions, the damping constant is expressed as

$$\Gamma_{coll} = 17.0\ C^{2/5}\ \bar{v}^{3/5} n \tag{6.100}$$

where C is a constant determined from quantum-mechanical considerations, $\bar{v}$ is the mean relative velocity of the radiating and perturbing particles determined from the kinetic theory of gases by the relation

$$\bar{v} = \sqrt{\left[\frac{8}{\pi} RT \left(\frac{1}{\mu_1} + \frac{1}{\mu_2} \right) \right]},$$ (6.101)

where R is the universal gas constant having the value of 8.314×10^7 erg deg^{-1} mole^{-1}, T is the kinetic temperature of the medium, μ_1 being the atomic weight of the radiating atom and $\mu_2 = 1.008$, that for the H-atom. n in Eq. (6.100) gives the number of neutral H atom/cc.

If the radiating and perturbing particles belong to the same elements, the perturbing force field varies as (distance)$^{-3}$, the resulting broadening of the spectral lines is called the self-pressure broadening and in this case we have

$$\Gamma_{\text{coll}} = 4\pi^3 \, Cn$$ (6.102)

where C is a constant and n is again the number of perturbing particles/cc.

In the case where the perturbing particles are electrons or ions of an element different from the radiating particles, the perturbing force field varies as (distance)$^{-4}$, and in this case

$$\Gamma_{\text{coll}} = 38.8 \, C^{2/3} \, \bar{v}^{1/3} n$$ (6.103)

where $\bar{v}$ is given by Eq. (6.101) and other symbols have their usual meaning.

In all the above three cases, the change in the frequency of radiation due to interaction of the perturbing particles at a distance can be represented by

$$\Delta v = C/r^j$$ (6.104)

where $j = 6, 3$ and 4 respectively.

The Stark Effect

The broadening of spectral lines by the interaction of the radiating atoms or ions with the electrostatic field created by the ions and electrons present in the stellar atmosphere is known as the *Stark effect,* after the name of J. Stark who first observed the effect in 1913. Stark found that when an electrostatic field is applied to hot hydrogen gas, the Balmer lines were split into a number of components. Due to the interaction with the Field, an atom radiating at its normal frequencies will have these frequencies shifted. This shift is dependent on the configuration of the radiating atom with respect to the surrounding ions and electrons to which the field is due. Different atoms having different orientation with respect to the field will have different frequency shifts. As there is a large number of atoms responsible for absorbing a spectral line, the individual Stark effect for each of these would combine together to result in an appreciable widening of the line. This broadening is markedly observed in the Balmer lines of hydrogen and in some lines of He in the spectra of B and A-type dwarfs. It is relatively small in the spectra of giants of the same classes. This can be explained by the low density of ions and electrons in the atmospheres of these stars and the consequent weakness of the field produced. But it is very difficult to use this criterion for distinguishing between giants and dwarfs of the same spectral class.

The *average* electrostatic field produced by ions and electrons in an atmosphere of temperature T is given by

$$E = 46.8 \left(- \frac{p_i}{T} \right)^{2/3} \tag{6.105}$$

where p_i is the pressure of the ions in dynes cm^{-2}. In most stellar conditions however, $p_i = p_e$.

The number of levels into which a hydrogen line corresponding to the principal quantum number n splits due to stark effect is $2n - 1$, and this splitting is symmetrical.

Rotational Broadening

The wide and diffuse absorption lines in the spectra of some stars of the classes B and A are found to arise from their rapid axial rotations. The axes of stars are generally oriented in different directions in space so that their rotational motions could affect the absorption lines in various manner. However, generally we observe the broadening effect, assuming the axis of rotation lying more or less perpendicular to the line of sight. From such an observation (Fig. 6.10), we notice that the different portions of the visible disc of the star approaches us or recedes from us with different velocities and the combined effect leads to widening of each absorption line in the spectrum according to Doppler's principle. When the Doppler velocity is measured, the observed line-contours in the spectrum are raised and broadened towards both the violet and red ends of the spectrum, but the equivalent-width of each line remains the same. Velocity measurements from the observed line-widths yield large equatorial velocity of the order of 200 km s^{-1} or more for some rapidly rotating stars. The Balmer lines in the spectrum of these hot stars with such a high rotational velocity is broadened by as much as 6 Å or more.

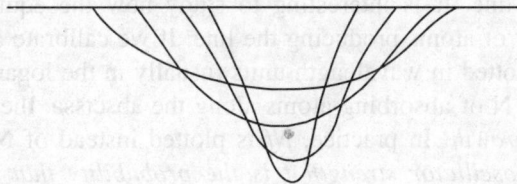

FIGURE 6.10   Line profiles caused by rotation of stars.

Magnetic Effect

The splitting of a special line in the presence of a magnetic field is called the *Zeeman effect* and has already been discussed. H.W. Babcock first observed this effect in the broadening of lines in the spectrum of some stars. If the field is so weak that the Zeeman components cannot be separated, the line will appear as a single broadened line. Magnetic field was first detected in the peculiar A-type star 78 Virginis in 1947. The peculiar spectral feature of the star is due to the presence of some strong and sharp lines of Sr, Cr, and Europium. Some other peculiar A-type stars also show the presence of magnetic field in them by their strong lines due to metals like Mn, Si, etc.

Magnetic fields as large as 5000 gauss have been detected in some stars by Babcock and others. In the spectra of some stars, the absorption lines due to some elements are found to vary in strengths as a result of the variations in magnetic fields existing in those stars. These are known as the magnetic variable stars.

Hyperfine Structure in Spectral Lines

When observed with high resolution instrument a spectral line sometimes shows peculiar structures different from those formed by the broadening mechanisms discussed above. These complicated multiple structures are what are known as Hyperfine Structures.

The Hyperfine Structures are explained to be due to the different orientations and masses of the nuclei of atoms. Firstly, if we consider a nucleus spinning about its own axis and the spin-energy that is being emitted (or absorbed) may counteract with the orbiting electrons, such interactions might result in the broadening of the line. Thus, an intrinsic property of an atom may cause appreciable line-widths.

Secondly, different isotopes of an element might give rise to line profiles that have slightly different structures. For two isotopes of the same element, the observed shift in wavelength is small and is determined by the ratio of the masses of nuclei and the electrons. The shift is markedly observed in the H_α line which is shifted by 1.79 Å by the presence of Deuterium, the isotope which is twice as heavy as ordinary H atom.

The famous radio line at 21 cm wavelength of which we will have occasions to discuss in later chapters arises out of a hyperfine transition in the ground level of the hydrogen atom.

6.8   THE CURVE OF GROWTH

The intensity of an absorption line varies according to the total number of atoms that participate in the formation of the line. It is interesting to study how the equivalent width of a line changes with the number of atoms producing the line. If we calibrate a criterion in which the equivalent width w_λ is plotted in wavelength units (usually in the logarithmic scale) along the ordinate and the number N of absorbing atoms along the abscissa, then the resulting curve is known as the *curve of growth*. In practice, Nf is plotted instead of N, where f is called the oscillator strength. *The oscillator strength f is the probability that an atom in a state of absorbing a line will absorb one particular line in preference to other lines in its vicinity.* It is expressed mathematically as

$$n = Nf \qquad (6.106)$$

where n is the number of oscillators and N the total number of atoms in a particular absorbing state.

In constructing theoretical curves of growth, different theories of model atmospheres have been proposed. The original curve of growth was constructed by M. Minnaert and C. Slob in 1931. D.H. Menzel and A. Unsold first used the fact that the absorption and scattering in the stellar atmospheres were produced by the atoms above the photospheric layers. This fact was later extended in the Schüster-Schwarzchild model of stellar atmospheres. It suggests that the photosphere produces a continuous radiation and the absorptions and scatterings arise in a reversing layer which lies above the photosphere. This is explained by the Schüster-Schwarzschild approximation formula:

$$r_v = \frac{1}{1 + N\alpha_v} \tag{6.107}$$

where r_v is called the residual intensity, and gives the ratio of the flux F_v in the line concerned and the flux F_v^0 in the nearby continuum, i.e., in other words,

$$r_v = \frac{F_v}{F_v^0} \tag{6.108}$$

α_v is the atomic coefficient for line absorption, which is expressed in the form (as proposed by Menzel, 1931),

$$\alpha_v = \frac{\pi e^2}{m_e c} f \left[\frac{1}{\sqrt{\pi}} \frac{c}{vv} \exp\left\{ -\frac{(v - v_0)^2 c^2)}{v_0^2 v_0^2} \right\} + \frac{\Gamma}{4\pi^2} \frac{1}{(v - v_0)^2} \right] \tag{6.109}$$

where

$$\Gamma = \Gamma_0 + \Gamma_{\text{coll}} \tag{6.110}$$

is the effective damping constant for simultaneous radiation and collision dampings; v_0 is the most probable velocity of the atoms and v_0 is the frequency at the centre of the observed line-profile, e and m_e being the charge and mass of the electron respectively.

The first term in the expression of α_v refers to the absorption near the centre of the weak lines where the Doppler broadening dominates. The second part represents the absorption from the wings which results mainly from the simultaneous radiation and collision dampings.

According to the above theory of the model atmosphere, the equivalent width is given by the formula:

$$w_v = \int_0^\infty [1 - r_v] \, dv = \int_0^\infty \frac{N\alpha_v}{1 + N\alpha_v} \, dv \tag{6.111}$$

In the above formula, w_v is expressed in frequency units. But observers generally measure w_λ in wavelength scale in angstroms and milliangstroms. So, remembering that $\lambda_0 = c/v_0$ and $\Delta\lambda_0 = c/v_0^2 \, \Delta v_0$, we can express in angstroms as

$$w_\lambda = \frac{c}{v_0^2} w_v \tag{6.112}$$

Since the values of w_λ and w_v are relatively small, we can have the relation

$$\frac{w_\lambda}{\lambda} = \frac{w_v}{v} \tag{6.113}$$

It also follows from the above theory of the curve of growth, that we should plot along the ordinates the values of w_λ/λ instead of merely w_λ so that the resulting curve becomes independent of wavelenghts.

Another well-known theory of the model stellar atmosphere which has been used for the construction of the theoretical curves of growth, is the so-called Milne-Eddington Model. To obtain the curve of growth according to this theory, we plot log (w_λ/λ) (c/v) along the ordinates instead of (w_λ/λ) and log η_0 along the abscissa, η_0 being the ratio of the absorption coefficient at the centre of the line to the continuous absorption coefficient $(\eta_0 = l_0/\bar{k})$. The basic assumption in this theory is that both continuous and the absorbing radiations are produced from the same layer of the star's atmosphere.

It has been found both theoretically and empirically that the equivalent width increases directly with N, the number of atoms in absorbing layers, when the latter is not very large. This is the case for faint absorption lines in which Doppler broadening plays the vital role. As the number of atoms increases initially, the core of the line profile becomes deeper, the line becomes wider and thus corresponds to the initial straight portion of the curve of growth, where $w_\lambda \propto Nf$ as shown in Fig. 6.11. Then, with further increase in the number of atoms, the intensity as well as the width of the line increases, but this time much less steadily as the number of atoms for absorption goes on increasing. The core of the line now becomes saturated and any further increase in the intensity of the core ceases and the atoms which could contribute to the further increase in the intensity of the core just become "hidden". After the absorption in the Doppler core is complete, the line develops its damping 'wings' on both sides of the core as a result of collisions among the increased number of atoms. The corresponding part of the curve is somewhat flat and is called the 'damping branch', whose position in the curve depends on the value of the damping constant. Now, with a further increase in the number of atoms, the equivalent width increases again but now the increase is almost proportional to $\sqrt{Nf}$ as the absorption now occurs mainly in the 'wings'. The corresponding part of the curve results from the simultaneous effects of radiation as well as collisional damping together with the Doppler broadening. This portion again is a straight line slightly inclined to the origional direction of increase of w_λ. For this part $w_\lambda \propto \sqrt{Nf}$. These are observed with most of the strong lines in the spectrum. A typical curve of growth is shown in Fig. 6.11.

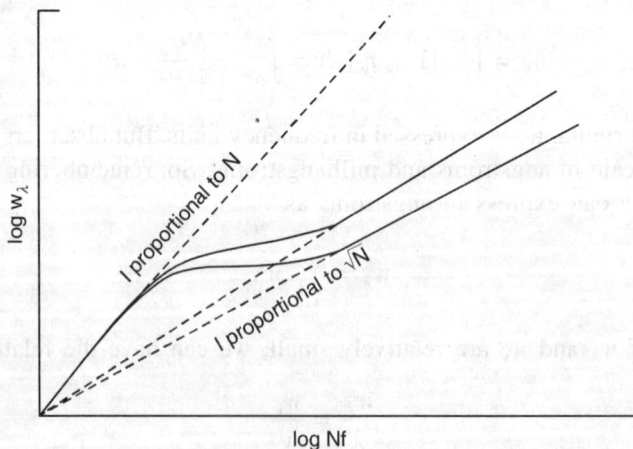

FIGURE 6.11   The curve of growth dependence of absorption line strength on the number of absorption atoms.

The height of the flat portion of the curve of growth determined by the Doppler broadening of the lines in the spectrum depends on the temperature as well as the element producing the lines. For higher temperature the random motion of the atom and consequently the Doppler width of the line increase appreciably. The effect of saturation will show up after a sufficient increase in the number of atoms. In this case, therefore, the initial straight portion will reach higher. Secondly, at any given temperature, the Doppler broadening in the lines of lighter elements like H, He, etc. is much larger than in those of heavier elements like Ca, Fe, etc. For example, the broadening of an Fe line in the solar atmosphere is only about 14% of that of the H_α line, the two lines being in close wavelength region. Again, the height of the right-hand branch of the curve of growth which is a product of the simultaneous effect of the Doppler broadening as well as the natural and collisional damping mainly depends on the damping constant. For different damping constants the curve will have branching at different heights, as shown in Fig. 6.11. Only the initial straight branch of the curve ($w_\lambda \propto Nf$) which results from the weak lines of the spectrum is free from the effect of both damping and Doppler broadening. For this reason, the analysis of weak spectral lines gives more useful and correct information regarding the chemical composition in stars.

6.9   STELLAR TEMPERATURES

The study of the various spectral properties of stars yields a wealth of information regarding their physical and chemical characteristics. One such very important information regarding their physical characteristics consists of the knowledge of the different scales of stellar temperatures which can be gathered from the spectral analysis of stars. We assign definite temperature to a star according to its radiation that we record with our spectrograms. As the radiation arises out of various depths in the stellar atmosphere, it is not reasonable to assign a definite temperature to all the layers of the atmosphere. Actually, we measure only one temperature at a time and the way in which we measure it corresponds to different depths in the atmosphere to which we assign the values. Thus, different types of stellar temperatures have been proposed.

Ionization Temperature

From the observed spectrum of a star we can determine the ratio of the number of ionized to neutral atoms of an element measuring the corresponding equivalent widths of the lines. If now, the electron pressure at that particular ionization level is known, then temperature can be determined from Saha's formula. The temperature of a star thus determined is called the *ionization temperature* (T_{ion}) of the corresponding atmospheric level.

Excitation Temperature

From a comparative study of different lines of an atom, the relative number of atoms in different excited levels can be calculated. Measurement of the equivalent width yields the number of absorbing atoms per gram of the atmospheric gases for each of these excited lines. The temperature in Boltzmann's formula that fits these relative number of atoms is defined to be the *excitation temperature* (T_{ex}) of the levels in the stellar atmosphere.

Colour Temperature

The observed energy distribution of a star is sometimes compared with the continuous distribution of a black-body for a wavelength interval (λ_1, λ_2) (say), given by Planck's formula. The temperature at which the energy curve due to the star well fits the theoretical curve of the black body, is defined as the *colour temperature,* T_c of the star over the wavelength region considered. T_c may be determined from the intensity ratios of the derived energy distribution at λ_1 and λ_2 expressed by the relation

$$\log \frac{F(\lambda_1)}{F(\lambda_2)} = \left(\frac{\lambda_2}{\lambda_1}\right)^5 \frac{\exp\left(\dfrac{hc}{\lambda_2 k T_c}\right) - 1}{\exp\left(\dfrac{hc}{\lambda_1 k T_c}\right) - 1} \tag{6.114}$$

Since a star does not radiate exactly as a black-body in all wavelength regions, the above ratio varies with different wavelength regions for which different values of T_c are obtained for a given star.

Effective Temperature

The effective temperature is that temperature of a star which is associated with a black-body of similar size having an equal amount of total energy output as that of the star. If L denotes the luminosity of a black-body of radius R, then the effective temperature T_{eff} of the star is given by Eq. (3.11).

When we mention the temperature of a star without specifying any particular scale, we generally mean the *effective temperature* of a star. It should be noted that while the colour-temperature of a star depends on the particular wavelength region over which the stellar radiation is fitted with the black-body radiation and thus differs at different wavelength regions, the effective temperature being scaled down to the total radiation of the star, is uniquely assigned.

6.10 THE CHEMICAL COMPOSITION OF STARS

The determination of the chemical composition of stars and other heavenly bodies (or abundance of elements in them) is extremely important for the understanding of the theories of cosmogony and the evolution of the stars and stellar systems. The problem has therefore drawn attention of many foremost scientists—chemists, physicists and astrophysicists during the last half-century. The work of these scientists has now yielded a fairly good knowledge of the chemical composition of stars and other heavenly bodies which was once though to be an unsolvable problem. We now know from the spectral analysis of stars and other objects that excepting a few special types of stars, most stars as well as the gaseous nebulae and the ocean of interstellar matter possess almost similar chemical composition which is close to that of the Sun. Gone are the days when the puzzling variety of stellar spectra misled people to believe that almost every star has different composition—some of them were helium stars, some hydrogen stars, and so on.

The first step in the development of the knowledge of the abundance was from the understanding of the fact that hydrogen was superabundant in the material of the Sun and other stars. This idea first came from the theoretical work of Eddington in 1924 while detennining the luminosity of stars. According to him, the luminosity was very sensitive to the mean molecular weight of the stellar material. The mean molecular weight of an atom which is almost fully ionized in the stellar interior conditions, can be defined as

$$\mu = \frac{\text{no. of protons in the atom} + \text{no. of neutrons}}{\text{no. of electrons in the atom} + 1 \ (\text{nucleus})}$$

for He^1,

$$\mu = \frac{1+0}{1+1} = 0.50, \text{ for } He^4, \ \mu = \frac{2+2}{2+1} = 1.33,$$

for O^{16},

$$\mu = \frac{8+8}{8+1} \approx 2.0$$

For any other heavier element it can be shown similarly that μ lies approximately between 2.0 to 2.1. It was found that the observed values of the luminosity of known stars could be realised from A.S. Eddington's formula if μ was close to 0.5. The natural conclusion was that among the chemical elements which the stars were made of, hydrogen must be overwhelmingly abundant. This conclusion was confirmed soon after, in 1929, by Russell from his study of the lines of the solar spectrum recorded in H.A. Rowland's Catalogue. Russell's work established not only the superabundance of hydrogen in the solar material, but also estimated the relative abundance of other elements whose lines are present in the solar spectrum. Russell's estimation was so accurate that even to-day with much improved knowledge of fainter lines and use of sophisticated instruments it could hardly be surpassed in quality. H.N. Russell's estimated values of the abundance of material composing the Sun has since been known as the *Russell's mixture.*

For the determination of the chemical composition of the stars the elementary theory of curves of growth is generally used. The observed curve of growth for any particular element is fitted with the theoretical curve of growth for that element which then yields the value of $N_i f_i$ for the i-th element in the experimental star. Knowing the f-value (oscillator strength) for the element from the laboratory experiments, the value of N is calculated. In this way, by constructing the curves of growth for various elements for which the f-values have been experimentally determined in the laboratory, and fitting with the corresponding theoretical curve of growth, the relative abundance of each element can be calculated. But several sources of uncertainty are inherent in the procedure. First, the procedure is based on the basic assumption that the composition of the reversing layer of the star, where the absorption lines are formed and that of all the interior layers is the same. This means that the material in different layers of a star from the centre to the surface is well mixed. This assumption is generally believed to be true; but it may not be so at all stages of evolution of a star. Moreover, we now know that as the evolution of the star proceeds, hydrogen is converted to helium or carbon in a large portion of the interior of the star. So it will hardly be justified to assume the composition of the surface layer as the representative of the whole star.

Secondly, the curves of growth are constructed on the assumption that the thickness of the reversing layer for all the lines (that is, for all the elements) is the same. This amounts to the assumption that the absorption coefficient of the stellar material is independent of the frequency v. This approximation, known as the *grey body approximation,* is only roughly true in solar type stars. But for other types of stars this approximation introduces large uncertainty. Thirdly, one must be very sure and careful while choosing the lines of an element for the construction of the curve of growth. Strong lines are generally saturated at the core, thereby making a percentage of the participating atoms "hidden". Medium strong lines are contaminated by damping effect which is quite difficult to calculate. Thus, for the best results, one must use *the faint lines* which are unaffected by any such effects as above. But we know that the calculation of the equivalent widths of faint lines is a difficult task. Particularly in the spectra of cool stars, the presence of too many strong lines almost entirely obliterates the very faint lines. Thus, inherent difficulties in properly using the spectral lines for achieving the best results from the curve of growth heavily interfere with the accuracy of the chemical composition calculation in stars.

Other methods, including some variations and modifications of the method of the elementary theory of curve of growth, have been used by various authors for the determination of the abundance of elements in the Sun and stars. Among these, the method of determining the chemical composition of a star by studying the contours of its spectral lines of different elements is important, since it generally yields improved results over those obtained by the application of the elementary theory of curves of growth.

All these methods have been extensively applied to determine the chemical composition of the Sun and other stars. The Sun being the most extensively observed star has been studied by different methods. Due to the availability of the better choice of lines in the spectra of F, B and O type main sequence stars, these stars have also been extensively studied for their chemical composition and the results obtained thereby are generally accepted as reliable. Although in several cases discordant results have been obtained, as is expected in the case of studies by various authors and with various assumptions, these discordances are not as prominent as to yield a conclusion that the chemical composition varies from one star to another. On the other hand, the results point to a general identity of chemical composition of the overwhelmingly large number of stars with that of the Sun. The same approximate identity has been also found to extend even to the planetary nebulae, the diffuse nebulae and interstellar gas in general. The discovery of this identity of the chemical composition in great majority of the heavenly objects is a triumph of the modern astrophysical investigations. It certainly has very great bearing on the better understanding of the cosmological theories and problems of the evolution of the galaxies as well as of the universe. The average results of the various investigations regarding the solar abundance of elements may be summarised in Table 6.1. For convenience, we shall call it the normal cosmic abundance and any variation from it will be called the deviation from the normal abundance.

It must be mentioned here that the determination of the helium abundance involves some inherent difficulties. This is due to the facts that in cool stars up to class A, no helium lines

TABLE 6.1 Cosmic Abundance of Astrophysically Important Elements [mainly based on the abundance in the solar system]

Element	Symbol	Atomic no.	Atomic weight	Log abundance	
				Number	Mass
Hydrogen	H	1	1.0080	12.00	12.00
Helium	He	2	4.0026	10.93	11.53
Lithium	Li	3	6.941	0.7	1.6
Beryllium	Be	4	9.0122	1.1	2.0
Boron	B	5	10.811	<3	<4
Carbon*	C	6	12.0111	8.52	9.60
Nitrogen	N	7	14.0067	7.96	9.11
Oxygen	0	8	15.9994	8.82	10.02
Neon	Ne	10	20.179	7.92	9.22
Sodium	Na	11	22.9898	6.25	7.61
Magnesium	Mg	12	24.305	7.42	8.81
Aluminium	Al	13	26.9815	6.39	7.78
Silicon	Si	14	28.086	7.52	8.97
Phosphorus	P	15	30.9738	5.52	7.01
Sulphur	S	16	32.06	7.20	8.71
Potassium	K	19	39.102	4.95	6.54
Calcium	Ca	20	40.08	6.30	7.90
Titanium	Ti	22	47.90	5.13	6.81
Vanadium	V	23	50.9414	4.40	6.11
Manganese	Mn	25	54.9380	5.40	7.14
Iron	Fe	26	55.847	7.60	9.35
Nickle	Ni	28	58.71	6.30	8.07
Copper	Cu	29	63.546	4.5	6.3
Zinc	Zn	30	65.37	4.2	6.0
Strontium	Sr	38	87.62	2.85	4.79
Yttrium	Y	39	88.9059	1.8	3.8
Zirconium	Zr	40	91.22	2.5	4.5

*Atomic Weight Standard is $C^{12} = 12.00$

Source: Adopted from C.W. Allen, *Astrophysical Quantities,* 1973, pp. 30–31.

are observed, and the excitation or the effective temperature of the B and O stars in whose spectra the helium lines appear, are difficult to determine correctly. Consequently, the controversy about the correct value of the helium abundance has not yet been settled;

Let us now discuss some cases of observed deviations from the normal abundance:

1. One striking example lies in the presence of the two parallel sequences among the *Wolf-Rayet* stars. It is found that the spectra of one group of these stars are dominated by strong and wide emission lines of carbon in its various stages of ionization. The

other group shows similar lines of nitrogen (Chapter 10). In fact, the spectral features of the two groups are completely different except that both possess wide emission bands characteristic of very high surface temperature as well as of high velocity ejection of matter. Accordingly, the classes WC and WN have been differentiated among these stars. It is difficult to believe that stars of similar chemical composition can produce spectra of such widely different characteristics. It is most likely that the two groups possess compositions largely dominated by carbon and nitrogen respectively.

2. The second group of stars that may be mentioned as deviation from normal abundance in composition are the *carbon stars*. There are a variety of these stars one of which, WC, we have already mentioned. The other varieties, classes N and R are cool carbon stars. These stars, otherwise belonging to the spectral class K to M in which the bands of TiO are prominent, show strong bands of C_2 (the swan band), CN (cyanogen) and CH. Such prominent difference in the spectra of stars having similar physical conditions in their atmospheres can be understood only if one assumes that carbon is more abundant than oxygen in R and N stars, while the opposite is the case in the normal K and M stars. Also, in some N stars of variable luminosity (WZ Cas, WX Cyg and T Arae, for example) very strong resonance lines of lithium in the red part of the spectrum have been detected, indicating that these stars possess a genuine excess of lithium.

3. Another group of giant variable stars of intermediate temperature range is called the R Coronae Borealis stars to which the star R. Cor. Bor. itself belongs. They show strong bands of C_2 in their spectra while the band of CH is weak meaning that hydrogen is underabundant in these stars. The R Coronae Borealis itself shows very strong lines of neutral C. These spectral features of the R Coronae Borealis group of stars strongly indicate that they have overabundance of carbon but underabundance of hydrogen.

4. The prominent bands of ZrO in the spectra of cool S stars and the unusually strong lines of Ba II (ionized barium) in some so called carbon stars, also point to the conclusion that the composition anomaly does exist in a variety of objects. The discovery of the lines of technetium, an unstable element, by P.W, Merill in the S stars had aroused much interest among the astronomers about these objects.

5. The metallic-line stars which is the name given to a large group of stars due to the presence of exceptionally strong lines of some metals in their spectra also serve as examples of composition anomalies in stars. These are known as peculiar A stars. Their spectra belong to the class A if H and Ca II lines are taken as criteria. But they fall to the class F when judged by the strengths of lines of metals like Na, Sr, Zn, etc., so that these metals must be overabundant in their atmospheres. On the other hand, it has been found that the metals like Mg, Ca, Ti, Zr and V are very much underabundant in these stars. Such anomaly in the metal content of these stars has not yet been well understood. Selective excitation and ionization of some elements to the suppression of others may be the cause of the anomalous manifestation.

6. Many stars have so far been observed which indeed appear to be *hydrogen-poor*. Hydrogen lines are either weak or altogether absent in the spectra of these stars.

These stars belong to classes O, B and A. The extreme examples of such stars are HD 124448, 160641 and 168476 which belong to classes O and B. Among these the first and the third show no hydrogen lines at all in their spectra. Similar examples are HD 30353 and V sagittarii which belong to class A. We have already mentioned the case of R Coronae Borealis stars which are H-deficient. Among other hydrogen-poor stars, the case of white dwarfs must be mentioned. These stars probably possess wide variations of abundances which are expected if they are indeed the end product of a stellar life.

7. We conclude this discussion by considering the case of the extreme Population II stars. The subdwarfs, the R-R Lyrae variables and the stars comprising the globular clusters belong to this group. The most striking spectral feature of these stars is that their metal lines are extremely weak indicating that these stars must be extremely *metal-poor*. Spectroscopic evidences suggest that the metal content of these stars is only one to ten per cent of that in stars like the Sun. The spectroscopic peculiarity of these stars is again accompanied by their dynamical peculiarity that they deviate largely from the galactic rotational motion. Their orbits with respect to the galactic centre are highly elongated. Some of these may even perform pendulum motions through the centre of the Galaxy. Both these spectroscopic as well as the dynamical peculiarities are most satisfactorily explained, if they are assumed to have been formed during the initial phases of the formation of the Galaxy. The Galaxy was then larger and of more spherical shape than it is now and the velocities of the gas in it were more or less isotropic. So stars formed at this stage will possess more non-circular velocities than those formed at later stages when the Galaxy contracted and its circular velocity increased according to the *principle of conservation of the angular momentum.* Also, these being the *first generation stars* when the creation of heavy elements in the interior of the stars just began, they had little scope of possessing much heavy elements in their initial composition. The composition anomaly of these stars compared to the normal composition therefore seems to be real.

The discussions in the previous paragraphs thus substantiate the facts that although most of the cosmic objects show near-identity of their chemical compositions, there are certain objects which illustrate real composition anomalies, these anomalies being generally manifested in the form of the overabundance or underabundance of certain element or elements in them with respect to the solar abundance. It must be mentioned also that for a variety of practical difficulties encountered in the way of determination of the chemical composition of stars, the study may be said to be still in its initial stage. The coming years hold promise of more sophisticated investigation in this line with the consequent improvement of the knowledge over that which we have at present.

EXERCISES

1. Define *coefficient of emission*, *coefficient of absorption* and *optical thickness* of a matter.
   Compute the fraction of energy transmitted through layers of matter whose optical depths are respectively e, 1 and $1/e$.

2. Interpret each term of the transfer equation

$$\frac{dI_v}{ds} = - (K_v + \sigma_v) \, \rho I_v + \rho J_v$$

   What further assumptions are made to reduce the above equation to the form

$$\mu \frac{dI_v}{d\tau_v} = I_v(\theta) - B_v(T)$$

3. Explain the concept of *grey atmosphere*. Deduce the relation between the *effective* (T_e) and *surface* (T_0) temperatures of a star having grey photosphere in radiative equilibrium and local thermodynamic equilibrium. Calculate the optical depth in the layer of such a photosphere where $T_e = T_0$.

4. Draw a graph of temperature versus effective temperature in a grey photosphere in radiative and local thermodynamic equilibria, using values of τ in ten steps in the interval (0, 1) and read the value of τ where $T = T_e$.

5. Explain *blanketting effect* in atmospheres of stars. In which wavelength bands of spectrum this effect is important in stars like the Sun and the Sirius?

6. Calculate the values of the damping constant Γ_0 for H_α, H_β lines and the K line of Ca II. Hence obtain the selective absorption coefficients s_v for these lines where $\lambda - \lambda_0 = 0.1$ Å and 0.5 Å.

7. Calculate the bound-free absorption coefficients per atom of neutral hydrogen at the $n = 1$ and $n = 2$ levels. Calculate the corresponding absorption coefficients for a singly ionized helium atom.

8. Compute the total absorption coefficient at λ 3000 Å for one gram of neutral hydrogen in the ground level at temperature 10,000 K, due to bound-free transitions from $n = 2$ and $n = 3$ levels.

9. Compute the total absorption coefficient at λ 5000 Å for one gram of neutral hydrogen in the ground level at temperature 25,000 K, due to free-free transition.

10. Discuss the role of continuous absorption by negative hydrogen ions in stars like the Sun.

11. Define *Doppler broadening* and *turbulent broadening* of spectral lines. In the absence of turbulence, calculate the *Doppler width* of H_α line in atmospheres of stars having temperatures 25,000 K, 20,000 K, 11,000 K and 6,000 K.

12. If turbulence contributes to 50 per cent of the total random velocity of atoms, calculate the Doppler width of the line λ 5000 Å in the atmosphere of Sirius ($T = 10,000$ K).

13. Define *equivalent width* and *total half-width* of a spectral line. Calculate the total half-widths of D_1 line in the Sun ($T - 6000$ K) and of H_α line in Sirius.

SUGGESTED READING

1. Aller, L.H., *Astrophysics: The atmospheres of the sun and stars,* 2nd ed., The Ronald Press Company, New York, 1963.

2. Ambertsumyan, V.A., *Theoretical Astrophysics,* Pergamon Press, New York, 1958.

3. Böhm-Vitense, Erika, *Introduction to Stellar Astrophysics:* Vol. 2, *Stellar Atmospheres,* Cambridge University Press, Cambridge, 1990.

4. Chandrasekhar, S., *Radiative Transfer,* Dover Publications, New York, 1960.

5. Gary, David F., *The Observation and Analysis of Stellar Photospheres,* 2nd ed., Cambridge University Press, Cambridge, 1992.

6. Sen, K.K. and Wilson, S.J., *Radiative Transfer in Curved Media,* World Scientific, Singapore, 1990.

7. Swihart, Thomas L., *Astrophysics and Stellar Astronomy,* John Wiley & Sons, New York, 1968.

8. Ünsold, A. and Baschek, B., *The New Cosmos,* Springer-Verlag, Berlin, 1991.

SUGGESTED READING

1. Aller, L.H. *Astrophysics: The atmosphere of the sun and stars*, Ronald Press Company, New York, 1963.

2. Ambartsumyan, V.A., *Theoretical Astrophysics*, Pergamon Press, ...

3. ..., Cambridge University Press, Cambridge 19...

4. Chandrasekhar, S., *Radiative Transfer*, Dover Publications, New ...

5. Gray, David F. *The Observation and Analysis of Stellar Ph...* Cambridge University Press, Cambridge 2005.

6. Sen, K.K. and Wilson, S.J. *Radiative Transfer in Curved M...* World Scientific, Singapore 1990.

7. Swihart, Thomas L. *Astrophysics and Stellar Astronomy*, John Wiley & Sons, New York, 1968.

8. Unsöld, A. and Baschek, B. *The New Cosmos*, Springer Verlag, Berlin, 1991.

7 Binary and Multiple Stars

7.1 INTRODUCTION

The study of binary stars has been of great importance in the progress of astronomy. The knowledge of their orbital motion combined with Kepler's third law enabled us to determine the masses of stars which could not be estimated previously. Thus the knowledge of binary stars led astronomers to progress through two steps—the first, in determining masses of stars, and the second, in finding a relation between the mass and luminosity of stars.

Two classes of binaries are generally recognized in the sky viz., the optical binaries and the physically associated pairs or multiple systems. Sometimes two stars actually at a great distance from each other, are lined up by chance so that they appear as a close system like double stars. But these stars are not gravitationally bound. They are known as optical double stars. But the real binary systems are physical pairs which describe orbits round their common centre of gravity under the influence of their mutual gravitational attraction. In what follows we shall call these only as binary stars. This latter category of stars is again divided into three separate classes on the basis of their closeness and mutual orientations as seen by the observer on Earth. These are:

1. Visual binary
2. Spectroscopic binary
3. Eclipsing binary

Visual binary is a pair of stars whose binary nature can actually be detected through a telescope. This is possible either when the distance of the pair from the Earth is not large or when their mutual separation is appreciably large; may be as large as several hundred or even thousands of a.u. In angular measure the separation must be greater than 0".50 at some point of the orbit in order that the components may be identified. The period of revolution of visual binaries ranges from about a year to thousands of years. More than 65,000 visual binaries have been discovered so far.

Some binary stars are too close together to be resolved visually into two separate components. That they belong to binary systems undergoing orbital motions round their common centre of mass is revealed by the periodic oscillation in spectral lines. If both components are bright, two different sets of lines are observed to oscillate in opposite phases in Doppler shifts. If however one component is faint then its spectrum is suppressed by that of the brighter one which alone is observed to oscillate in phase with Doppler shifts. These systems are known as *spectroscopic binaries* because their binary character is revealed through spectroscopic analysis only. About 1000 spectroscopic binaries are known. A small subclass of spectroscopic binaries is *spectrum binaries.* In these latter systems, the orbital motions are not revealed through Doppler shifts of lines but two distinct spectra are found to be superimposed. It is concluded therefore that there must be two stars very close together producing the composite spectrum.

When the orbital plane of a binary system is not perpendicular to the line of sight, each component may be eclipsed by its companion in the course of their orbital motion. Such a pair of stars is called an *eclipsing binary.* Eclipsing binaries may also include visual and spectroscopic binaries. The schematic illustration of classification of double stars is given in Fig. 7.1.

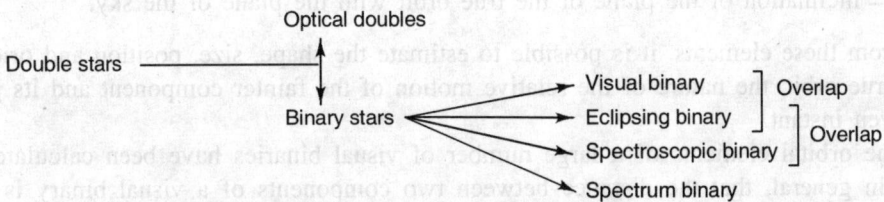

FIGURE 7.1 Classification of binary stars.

7.2 VISUAL BINARY

Visual binaries have been recognized by their wide separation in orbits. The two stars in a Visual binary system are revolving about their common centre of gravity. Generally, the motion of the fainter star about its brighter component is observed and the apparent relative orbit of the fainter star is traced out.

To determine the relative orbit of the visual binary we have to proceed step by step. We can determine the following data by observation:

1. Angular separation between the stars measured in seconds of arc;
2. Their *position angle,* i.e. the direction of the fainter star (called secondary) relative to the brighter one (called primary). This direction is measured by the angle subtended by the line joining the two stars with the north-south line, in the eastward sense.

These data can be determined by observation when the angular distance of the component stars is not very small. When these observed data are plotted graphically the apparent relative orbit is determined. This apparent orbit is found to be elliptical and the line joining the two components describes equal areas in equal intervals of time. So their attractive force is assumed to obey Newtonian Law. In the elliptical orbit thus traced out, the brighter star is not necessarily found to be at its focus, as may be thought to be obvious in case of a two-body system obeying Newtonian Law of gravitation. The deviation is explained by the fact that the orbit thus obtained is not the true orbit but the projection of the true orbit in the plane of the sky. The true orbit is an ellipse with the brighter star at its focus.

When the apparent orbit is determined and the inclination of the true orbital plane with the plane of the sky is known by applying geometrical technique, we can obtain the elements of the true relative orbit of a binary system. These elements are:

P = period of revolution of the secondary round the primary;

T = the time from the periastron;

e = eccentricity of the true orbit;

a = semi-major axis of the true orbit in seconds of arc;

Ω = position angle of the line of nodes (i.e. the line of intersection of the true orbital plane with the plane of the sky;

ω = the angle between the major axis of the true orbit and the line of nodes, measured in the true orbital plane; and

i = inclination of the plane of the true orbit with the plane of the sky.

From these elements, it is possible to estimate the shape, size, position and orientation of the true orbit, the nature of the relative motion of the fainter component and its position at a given instant.

The orbital elements of a large number of visual binaries have been calculated. It is found, in general, that the distance between two components of a visual binary is several hundred times that between the Earth and the Sun and the secondary star moves slowly around the primary. As examples of visual binaries we shall discuss the following stars in some details:

Sirius

It is a visual binary star discovered by F.W. Bessel in 1834. The primary star is so luminous (it is the brightest star in the sky) that the secondary star is almost invisible. R.G. Aitken calculated the elements of its orbit in the year 1918. Aitken's observations were later confirmed by Volet who obtained almost the same results. The period of revolution of the star was found to be 49.5 years. The maximum and the minimum angular separation between the primary and the secondary is 11″.2 and 0″.2 respectively.

The primary belongs to A0 spectral class with apparent photographic magnitude −1.6 and mass 3.4 $M_\odot$. The fainter star is a white dwarf having mass 0.94 $M_\odot$ and apparent magnitude +7 which belongs to the spectral class F. Both stars are revolving about their common centre of mass.

61 Cygni

This was recognized by Herschel as a binary star. In the year 1905, however, S.W. Burnham declared it to be an optical pair of double stars. But Osten Bergstrand and later F. Schlesinger and D. Alter showed that the two components were physically connected.

Different astronomers calculated their orbital elements at different times and found different results. According to R.M. Petrie

$$a = 29''.5,$$
$$p = 782.6 \text{ years},$$
$$e = 0.17.$$

Alan Fletcher found its semi-major axis to be $24''.5$, eccentricity 0.4 and period of revolution 696.63 years. More recently, the period of revolution has been calculated to be 720 years. Both stars of the system have nearly equal brightness, their apparent magnitudes being 5.6 and 5.3 respectively. Most probably they have similar masses but their actual masses are not determined. They probably belong to the solar mass group.

It has been suggested by several authors that widely separated visual binary pairs may be disrupted by encounters with field stars moving at random. The orbital velocity of the wider pair is very small, but in comparison, the random velocity of the passing star may be quite high. This star may impart enough energy to the binary to break it off. So it is thought that visual binaries are diffused by encounters. Calculation shows that the energy increase in the binary system due to encounter with field stars in time Δt is given by

$$\Delta E = 8\pi G^2 m^3 N \, \Delta t \, \frac{1}{u} \ln \frac{au^2}{2Gm} \qquad (7.1)$$

where m is the mass of each star, N the number of stars per unit volume, u is the velocity of the passing star, a is the separation between the components of the binary and G is the constant of gravitation.

The initial binding energy of the binary system is negative and is given by

$$E_i = -\frac{Gm^2}{2a} \qquad (7.2)$$

The binary system will break up when

$$\Delta E = -E_i = \frac{Gm^2}{2a} \qquad (7.3)$$

The corresponding time $\Delta t = T_d$ is the time for disruption. This time is therefore given by

$$T_d = \frac{Gm^2}{2a} \, \frac{u}{8\pi G^2 m^3 N \ln(au^2/2Gm)} = \frac{u}{16\pi Gm \, Na \ln \dfrac{a}{a_e}} \qquad (7.4)$$

where

$$a_e = \frac{2Gm}{u^2} = \text{Semi-major axis of the hyperbolic orbit of the passing star.}$$

If we take $a = 10^4$ a.u.; $a_e = 5$ a.u.; and $u = 20$ km s^{-1}, then $T_d \sim 3 \times 10^9$ years. Thus all visual binaries with $a \geq 10^3$ a.u. should break off in a time-scale of 10^{10} years which is believed to be the age of the Galaxy.

7.3   SPECTROSCOPIC BINARY

We have just discussed that the visual binary components are widely separated from one another and can be detected visually through telescope. Other types of binary components were unknown until the end of the nineteenth century. In the year 1889, Pickering found a peculiar phenomenon while studying the spectrum of the bright component of the visual binary star Mizar. Generally, the spectrum of a star reveals absorption lines shifted towards the shorter or the longer wavelength depending on whether the star is approaching towards or receding away from the observer; but the spectrum of the bright component of Mizar showed absorption lines shifted alternately towards shorter and longer wavelengths. This led astronomers to recognize new types of binary stars. It was concluded that the bright component of Mizar consists of two close stars of almost equal brightness. Due to their closeness, they cannot be separated by the aid of the telescope. Both stars are revolving about their common centre of gravity owing to the effect of the force of gravitation. In the course of their motion when one of them is approaching towards the earth and the other is receding away, the spectral lines of the approaching star are shifted towards violet and those of the receding one towards red; so two separate line patterns in opposite phases are observed. When both move perpendicular to the line of sight, there is no Doppler shift and the spectrum of two stars coincide giving single lines. The spectrum of a pair of stars will reveal alternately single and double lines provided the stars are sufficiently close to one another and have almost equal brightness. If one is much fainter than the other, the spectrum of the fainter star will be suppressed by that of the brighter one. The spectroscopic binary systems are separated by relatively small distance, generally less than 1 a.u. As a result, their orbits are small and orbital speeds are high leading to the consequent short periods of revolution, usually lying in the range from a few hours to a few months. Measuring the Doppler shifts of lines of the components of a binary system, one can get what is known as the velocity curves of these stars (Fig. 7.2).

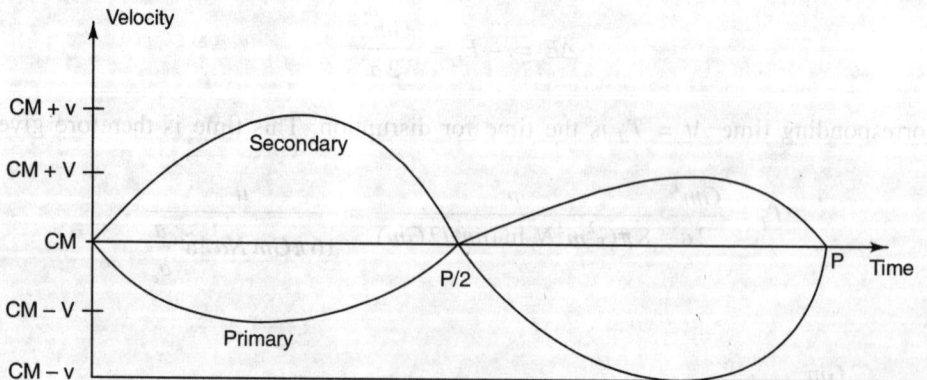

FIGURE 7.2   Schematic velocity curve of a spectroscopic binary. The zero line represents the velocity of centre of mass. The sinusoidal curves are the velocities of the component stars relative to the centre.

Figure 7.2 shows the schematic velocity curve of a spectroscopic binary. It can be shown that the amplitude of the velocity curve of a star is inversely proportional to its mass. This fact is utilised to determine the mass-ratio of the component stars of a spectroscopic binary system. In the simplest case the mass-ratio is determined as follows:

Let M_p be the mass and r_p be the radius of the orbit of the primary whose velocity in its orbit is v_p. Let m_s, r_s and v_s be the corresponding quantities for secondary ($r_s > r_p$, $v_s > v_p$ and $m_s < m_p$). Since both the primary and secondary complete the orbit (assumed circular and edge-on, that is, $i = 90°$) in the same period τ we have

$$\left.\begin{array}{c} 2\pi r_p = v_p\tau \\ 2\pi r_s = v_s\tau \end{array}\right\} \tag{7.5}$$

Also

$$m_p/m_s = r_s/r_p = v_s/v_p \tag{7.6}$$

Aitken gave comparative estimates for the orbits of the spectroscopic and visual binaries. The eccentricity of the orbit of a visual binary is larger than that of a spectroscopic binary. Visual binaries have much longer periods of revolution. The shortest period for a visual binary is $4\frac{1}{2}$ years whereas for spectroscopic binaries, periods vary from a few hours to 150 days.

7.4 ECLIPSING BINARY

An eclipsing binary is a special type of spectroscopic binary. When the orbit of a spectroscopic binary is actually lined up with the line of sight or makes some small angle with it, one component is eclipsed periodically by the other and variation of light takes place. Since two stars are revolving about their common centre of gravity, during one complete revolution each component is eclipsed once by the other. So two eclipses occur within the period of one complete revolution. If the two stars are of equal brightness, light is reduced by equal amount in both eclipses. If one of them is brighter, one eclipse is deeper than the other. The deeper one is called primary *eclipse* and the other is the *secondary eclipse.*

Let us discuss the conditions under which eclipses are possible. When the orbital plane of the binary is perpendicular to the line of sight no eclipse occurs. When the orbit is lined up with the line of sight the eclipse is either total or annular. In any other position partial eclipses are possible, which are more probable than total and annular eclipses.

In the last quarter of the eighteenth century, J. Goodricke first recognized Algol as an eclipsing binary star. This inspired many astronomers to probe further about this type of newly discovered mysterious stars. Subsequent observations led to the discovery of many other eclipsing binaries. In the beginning of the 20th century H.N. Russell and H. Shapley made thorough investigations of the nature of eclipsing binaries. P. Gaposchkin catalogued more than three hundred eclipsing binaries and gave an estimate for their mass, density, luminosity, spectral class, orbit and galactic concentration. Most of the brighter components are main sequence stars ranging from B8 to A5 in spectral class, though main sequence stars of other spectral types are also observed among these objects. Fainter stars are, in general, late-type giants. White dwarfs are almost absent in the eclipsing binary systems. The eccentricity of the orbit of eclipsing binaries increases with the increase of the period of revolution which varies from several hours to 40 years. It increases also with spectral class of the stars as they pass from A to M class. It has been estimated that the average mass density of an eclipsing binary star is one-third that of the Sun.

Light Curve

The light variation curve of an eclipsing binary depends on the size, luminosity and relative orientation of the component stars and also on the inclination of the orbital plane with the plane of the sky. McLaughlin considered different types of eclipsing binaries and described their light curve with utmost skill and efficiency. Let us follow McLaughlin to explain the same. The nature of light variation can be divided mainly into four classes:

(a) *Class I.*   Provided that the component stars are spherical in shape, the same amount of light is received in every position except at eclipses. When the brighter star is eclipsed there is a deep minimum; a shallow minimum is obtained when the fainter one is hidden behind the brighter. The light curve of such a system is shown in Fig. 7.3(a).

(b) *Class II.*   When the orbital plane is parallel to the line of sight, both total and annular eclipses occur alternately. The flat minimum in the Fig. 7.3(b) indicates the total eclipse whereas the annular eclipse is represented by the other minimum.

(c) *Class III.*   When two stars are unequal in brightness, the light of the brighter star is reflected upon the fainter, so sharp diminution of light does not occur even at eclipses. Hence the rise and fall of light curve is rather gradual as shown in Fig. 7.3(c).

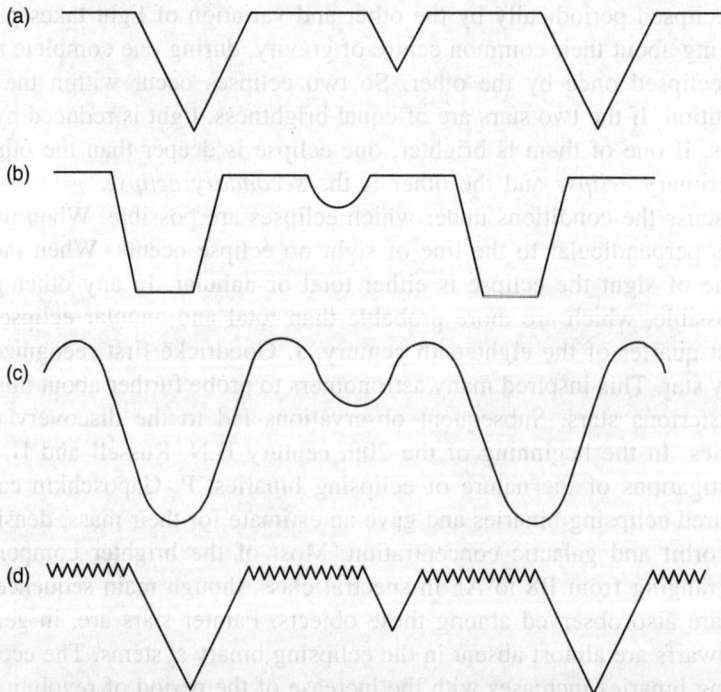

FIGURE 7.3   Schematic light curves of eclipsing binaries.

(d) *Class IV.* For massive close eclipsing binaries, the spherical configurations of the component stars are affected severely by their strong mutual gravitational pull and they become prolate spheroidal in shape. The light variation curve in this case becomes wavy as in Fig. 7.3(d).

Some Typical Eclipsing Binaries

Since the discovery of eclipsing binaries at the end of the eighteenth century, many astronomers have devoted their attention to the study of these stars. As a result numerous eclipsing binaries have been discovered. A few prominent members of these stars are discussed in the following.

Algol. Algol (β Persei) was the second known spectroscopic and the first known eclipsing binary star. In 1782, Goodricke studied the light variation curve of Algol. The curve attains two consecutive minima at an interval of 2 days 20 hours. Light remains almost constant for most of the period (nearly 2 days 4 hours), then rapidly falls to minimum within 8 or 9 hours and takes an equal interval of time to regain its full strength as shown in Fig. 7.4.

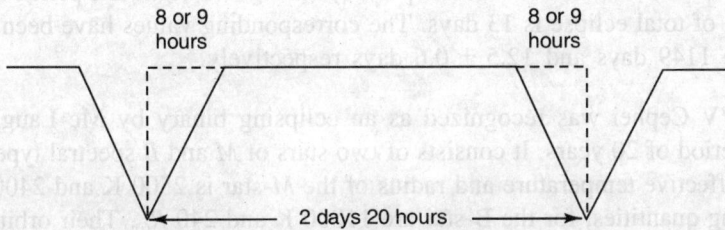

FIGURE 7.4 Schematic light curve of Algol.

Goodricke interpreted the light curve with a suggestion that Algol was a two-body system containing a dark and a luminous star both revolving about their common centre of mass, and the brighter one was being eclipsed partially by its dark companion. His view was proved to be correct after a long interval of time. H.C. Vogel, in 1889, proved Algol to be a spectroscopic binary possessing the same property as suggested by Goodricke. In 1906, A.A. Belopolsky found variation in radial velocity of the centre of mass in a period of 1.8 years which was due to the existence of a third body in the system as suggested by Schlesinger in 1912.

The inclination of the orbit of Algol is 81.8°. The radii of the brighter and the fainter bodies are respectively 3.12 $R_\odot$ and 3.68 $R_\odot$ and their corresponding masses are 4.72 $M_\odot$ and 0.95 $M_\odot$. The orbit of the system is almost circular.

V 453 Cyg. In 1939, Wachmann reported it to be an eclipsing binary star with a sharp minimum in its light variation curve indicating partial eclipses. According to J.A. Pearce, it is composed of two massive and very early type stars. Two hot stars have nearly the same effective temperature of about 28,000 K and their absolute magnitudes are –4.2 and –3.0 respectively. V 453 Cyg has great astronomical importance. Study of this binary enables us to understand several facts about early-type stars.

ζ Aurigae.   W.E. Harper and K.F. Bottlinger predicted this star to be an eclipsing binary and this prediction was proved to be correct by later observations. The mass ratio of the two components cannot be determined very accurately. This is because the spectral lines of the fainter component are partially suppressed by the Balmer lines of the brighter star. O.C. Wilson and W.H.M. Christie found the ratio to be 1.85 which they subsequently revised to 2.04. J. Hopmann found the mass-ratio to be 2.25. D.M. Popper calculated it to be 1.8. According to Wellmann, the spectral classes of the component stars are K4 II and B7 V having respective effective temperatures of 3480 K and 13,500 K.

31 Cygni.   It consists of two stars belonging to spectral classes K0 and B8 having absolute magnitudes –4.0 and –2.0, masses 18.0 $M_\odot$ and 9.0 $M_\odot$ and radii 174 $R_\odot$ and 4.7 $R_\odot$, respectively. Its period of revolution has been estimated by A. Mckellar and R.M. Petrie to be (3781 ± 8) days.

32 Cygni.   It has been recognized as an eclipsing binary in 1949, containing two stars of spectral classes K0 and A3. K.O. Wright as well as Wellmann calculated the period of revolution as well as the duration of total eclipse. According to Wright, the period is 1140.8 days and the duration of total eclipse is 13 days. The corresponding values have been estimated by Wellmann to be 1149 days and 12.5 ± 0.6 days respectively.

VV Cephei.   VV Cephei was recognized as an eclipsing binary by Mc Laughlin in 1936, having a long period of 20 years. It consists of two stars of *M* and *B* spectral types. According to Christie the effective temperature and radius of the *M*-star is 2500 K and 2400 $R_\odot$ whereas the corresponding quantities, for the B-star are 11000 K and 240 $R_\odot$. Their orbital inclination has been calculated to be 72°. The M-type star is a supergiant of enormous size and both the components are quite massive.

7.5   MULTIPLE STARS

William Herschel observed a peculiar phenomenon while studying the brighter component of the visual binary ζ Cancri, in the year 1781. The brighter one is itself a double star of identical components. After a long interval of time, Aitken announced that about 4 or 5 per cent of binary stars are multiple-bodied systems and he catalogued about 150 of such systems. In general, a triple system consists of a closer pair and a third star widely separated from them. Powerful telescope or spectroscopic observations are necessary to detect the closer pair separately.

The eclipsing binary *Algol* is already familiar to us. It has a third component which is fainter than the brighter component of the close pair but brighter than the fainter companion. The distance of the third body from the close pair is about two and a half astronomical units and its period of revolution is 1.8 years.

The visual binary *α* Centauri with a period of revolution of 80 years has a very distant companion Proxima Centauri more than 0.2 light years away, *α* Centauri consists of a Sun-like star (G0 spectral class and +4.73 absolute magnitude) as a brighter component and a fainter *K*-type star of absolute magnitude +6.10. The third star proxima is physically connected with these brighter stars. Proxima is a very faint star having the absolute magnitude +15.0. This is the multiple system nearest to the Sun.

Castor is a triple star system having a close pair 86 a.u. apart, revolving about their common centre in the period of 380 years. The third component is a *M*-type star at a distance 1000 a.u. apart. Each of these three stars is itself a spectroscopic binary. So ultimately we find that Caster has six components.

"*Trapezium*" is another multiple star system found in the Orion Nebula. Four brighter and two fainter components together form a compact six-body system which is unstable.

According to some authors all triple and multiple systems are unstable. The lightest star usually escapes from the system due to perturbation produced by field stars passing by.

7.6 ORIGIN OF BINARY STARS

Since about 50 per cent of all detectable stars in the Galaxy belong to the binary systems, it is assumed that there should exist some easy physical process which lead to their formation. Many theories have been proposed to explain the formation of binary systems. We shall discuss here some of these theories.

In 1867, John Stoney proposed the *capture theory* for the formation of binaries. According to this theory, two independent stars approach each other under certain circumstances and fall under the influence of their mutual gravitational attraction. As a result they begin to revolve about their common centre of gravity. This theory, however, has been long discarded, as subsequent investigations have proved that such close approach of two stars is a very rare phenomenon in the history of the Galaxy. But the importance of this theory has been kept alive by proposing it from time to time in various modified forms.

According to MacMillan, *binary systems have originated from planetary systems*. Since stars with their planets pass through the interstellar medium, they accumulate interstellar materials and grow in mass. Some amount of mass gained by the star is re-ejected but a planet stores almost all the material it has accumulated. As a result, a large planet grows more rapidly than its parent star. As they increase in mass, their mutual attraction increases and the distance between them decreases, and ultimately the large planet is converted into a small faint star; other smaller planets lose their individual existence being absorbed either by the main star or by the largest planet. Hence the planetary system reduces to a close binary system. But this hypothesis faces a severe contradiction. In this case, the mass of the faint star must be much less than the parent body. But the observed mass-ratio in the binary systems is not much less than half. Hence this theory has not been accepted.

It is the *fission theory* which can explain possibly all aspects of the origin of binary stars. When a gaseous body is rotating slowly, it takes up oblate spheroidal shape, whatever its internal composition may be. For gaseous body, the rotating mass has a tendency to concentrate at the centre with the corresponding increase in the rotational speed. The gaseous bulge rotating with high speed elongates gradually, and as motion continues concentration of mass takes place not at the centre but at two particular points on its longest axis, and ultimately the gaseous body is split up into two separate parts. This also corresponds to the rotational theory of homogeneous incompressible fluid as developed by C.G.J. Jacobi and C. Maclaurin. Sir James Jeans applied this theory to origin of binary stars. According to his view, some stars while rotating with high speed in its newborn nebular stage are split up into two components, thus producing binary stars.

Roxbourgh has proposed that the fission process starts when a star is in its premain sequence contraction stage. Mass of the star should lie in the range 0.8 to 4 $M_{\odot}$ and the star should possess an initial angular momentum.

According to Vant Veer about fifty per cent of contact binaries are formed by the fission process. He divided binary stars into two classes, F-G or solar type and A-F or non-solar type. The solar type binaries are formed by fission process in their premain sequence stage or when they just begin their main sequence lives. They are comparatively young stars having the age limit of the order of $5 \times 10^7 - 10^8$ years.

A-F type binaries are usually born by capture process. Vant Veer has proposed two types of capture processes. In one process, outflow of mass between two neighbouring stars takes place and as a result of continuous mass exchange their mutual gravitational pull increases and two stars approach towards each other ultimately producing a binary star.

The second process occurs when two evolved stars expand to their giant or supergiant stages and the distance between them decreases and finally behave as binary components. These non-solar type binaries are originated from old evolved stars, their age being of the order of 10^9 years.

7.7 STELLAR MASSES AND MASS-LUMINOSITY RELATION

Before the discovery of binary stars the Sun was the only star whose mass could be determined from the motion of its planets with the help of Kepler's third law. This law has the form:

$$(m_1 + m_2)\tau^2 = a^3 \tag{7.7}$$

where m_1 and m_2 are the masses of the gravitating bodies measured in solar mass unit, a is the semi-major axis of the orbit measured in a.u. and τ is the period of describing the orbit measured in years. The linear semi-major axis a is related to the angular semi-diameter a'', a measurable quantity, by the relation:

$$a = a''/p'' \tag{7.8}$$

where p'' is the parallax. So Eq. (7.7) can be expressed in terms of the observable quantities as:

$$(m_1 + m_2)\tau^2 = (a''/p'')^3 \tag{7.9}$$

If the parallax p'' of the binary system can be measured, then the sum of the masses of the system, $m_1 + m_2$, can be determined from Eq. (7.9). The individual masses can be determined only if the relative distance of each star from their common centre of mass is known. For then,

$$m_1 a_1 = m_2 a_2$$

where

$$a = a_1 + a_2 \tag{7.10}$$

and the individual masses are determined from Eqs. (7.8), (7.9) and (7.10). It is to be noted that the true semi-major axis a has to be determined from the apparent orbit by measuring the displacement of the primary from the apparent focus.

Thus, the binary systems yield clues to the determination of masses of stars in general. Since the knowledge of the mass of any system requires the already known parallax of the system, its apparent magnitude yields the luminosity also. Thus the mass and luminosity of many binary systems have been known. When these data are plotted in a graph with logarithm (base 10) of the masses in solar mass unit along the abscissa and the absolute bolometric magnitudes along the ordinate, we obtain what is known as the *mass-luminosity relation* for common stars. Such a relation is shown in Fig. 7.5.

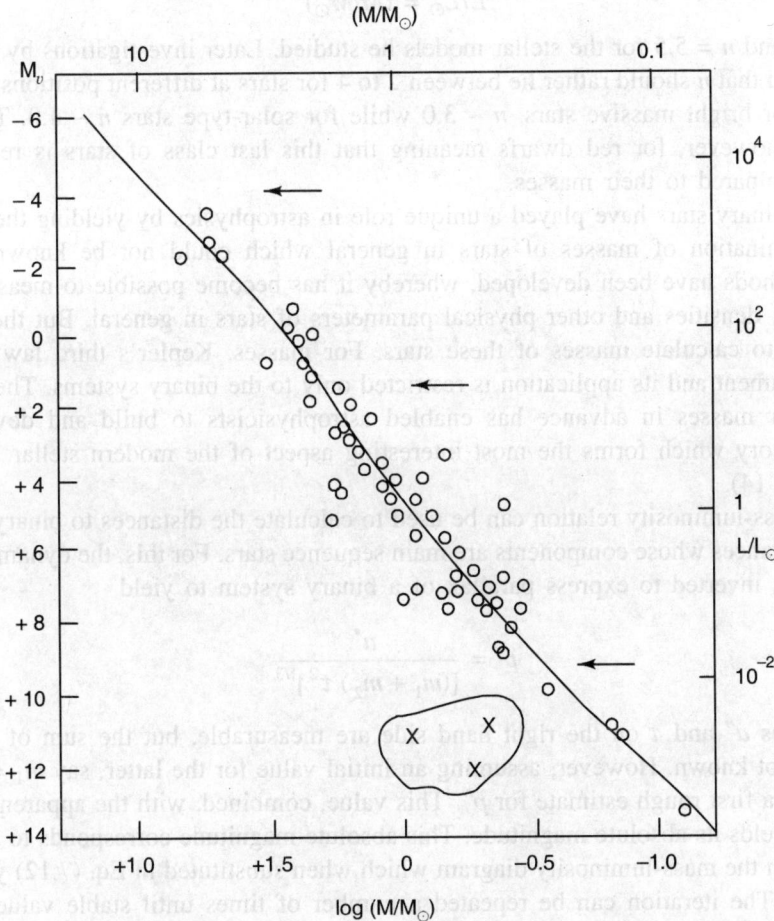

FIGURE 7.5 The mass-luminosity relation.

When the mass-luminosity relation is carefully analyzed, several important features are revealed. First, only the main sequence stars are found to conform to this law. The evolved stars, such as the red giants and white dwarfs deviate from the law. Thus the mass-luminosity law is obeyed by the overwhelming majority of stars (~ 90%). Secondly, the masses of stars are found to lie in a relatively short range of 0.05 to 50 $M_\odot$ (covering a factor of 10^3), and it is found that brighter stars are in general more massive (among main sequence stars). We know that the luminosity of stars has a wide range. Stars having absolute bolometric magnitudes from about +15.0 to –10.0 are observed which correspond to a luminosity range from 10^{-4} to 10^6 $L_\odot$ (covering a factor of 10^{10}). This at once suggests that the mass-luminosity relation among stars is non-linear. Such a non-linear relation was proposed as early as in 1924 by Sir A.S. Eddington. He suggested that the mass and luminosity of normal main sequence stars should conform to a relation of the form

$$L/L_\odot = (M/M_\odot)^n \tag{7.11}$$

where he found $n = 5.5$ for the stellar models he studied. Later investigations by astronomers have revealed that n should rather lie between 2 to 4 for stars at different positions on the main sequence. For bright massive stars, $n \sim 3.0$ while for solar-type stars $n \sim 4.0$. The value of n is lower, however, for red dwarfs meaning that this last class of stars is relatively less luminous compared to their masses.

Thus, binary stars have played a unique role in astrophysics by yielding the clue to the direct determination of masses of stars in general which could not be known otherwise. Physical methods have been developed, whereby it has become possible to measure colours, temperatures, densities and other physical parameters of stars in general. But these methods do not help to calculate masses of these stars. For masses, Kepler's third law is the only helpful instrument and its application is restricted only to the binary systems. The knowledge of the stellar masses in advance has enabled astrophysicists to build and develop stellar evolution theory which forms the most interesting aspect of the modern stellar astrophysics (see Chapter 14).

The mass-luminosity relation can be used to calculate the distances to binary systems of unknown distances whose components are main sequence stars. For this, the dynamical relation (Eq. (7.9)) is inverted to express parallax of a binary system to yield

$$p'' = \frac{a''}{[(m_1 + m_2)\,\tau^2]^{1/3}} \tag{7.12}$$

The quantities a'' and τ on the right hand side are measurable, but the sum of the masses, $m_1 + m_2$ is not known. However, assuming an initial value for the latter, say $m_1 = m_2 = M_\odot$, one can find a first rough estimate for p''. This value, combined, with the apparent magnitude of the star, yields its absolute magnitude. This absolute magnitude corresponds to a new value of the mass in the mass-luminosity diagram which when substituted in Eq. (7.12) yields a new value of p''. The iteration can be repeated a number of times until stable values of p'' are obtained. The parallax of the binary system thus obtained is known as the *dynamical parallax,* the name being derived from the fact that it is obtained from the dynamical relation Eq. (7.9) or Eq. (7.12). The method however is limited to only relatively nearby binary systems (distances $\leq$ 50 pc) since only for such systems a'' can be measured with reliable accuracy.

7.8 MASS TRANSFER IN CLOSE BINARY SYSTEMS

Some close binary systems, during certain stage of their evolution, undergo mass transfer. The mass transfer between the components may be attributed mainly to the following two reasons:

1. One of the member stars in the binary system may evolve to a large radius or the mutual separation between the pair may shrink due to the loss of orbital angular momentum. The latter may occur due to the effect of stellar wind mass loss or gravitational radiation. In any case, under such situation the outer layers of the envelope of a star may be gravitationally pulled off by the companion.
2. One of the stars in the system may, at some point of evolution, eject much of its mass from the surface layer in the form of a stellar wind. A part of this ejected matter may be attracted gravitationally by the companion.

The situation described in (1) in which the surface matter of one star is drawn by its companion, gives rise to the Roche problem. The Roche problem assumes that the Keplerian orbits of the two stars about each other in a plane are each *circular*. It further assumes that the stars are *centrally condensed*. The detached mass gradually fills regions known as *Roche lobes* surrounding each star and defined by surfaces of equipotential as shown in Fig. 7.6. The saddle point L_1 is the *Inner Lagrangian Point* which acts as a *pass* between the two Roche lobes surrounding the member stars of masses M_1 and M_2. This means that the material inside one of the lobes finds it much easier to move to the other lobe through the point L_1 than to escape the critical surface altogether.

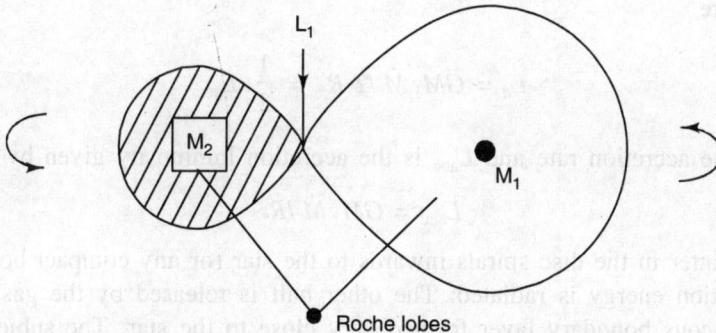

FIGURE 7.6 The secondary filling the Roche lobe in an interacting binary system. L_1 is the inner Lagrangian point.

Suppose the mass pulled off from the surface of one of the stars eventually fills its Roche lobe. We call this star as the *secondary* and denote its mass by M_2. The other star is called *primary* and its mass is M_1, as shown in the Figure. Here the primary is a compact star. The binary system is said to be *detached* so long as no matter is pulled off from any member by the gravitational attraction of the other. In such a case, the mass transfer can take place only through the stellar wind. But when the envelope of the secondary fills either completely or partially its Roche lobe, its material can easily be pushed to the point L_1 either naturally or by perturbation. This material will then be moving into the Roche lobe of the primary and will ultimately be captured by this star. Mass can thus be efficiently transferred from the secondary to primary as long as the former fills the Roche lobe. Such a system is called *semi-detached* and the mass transfer in this case is called Roche lobe overflow. In some cases, it may so happen that both stars fill their Roche lobes simultaneously. Such a system is called *contact binaries*. The mass thus flown from the secondary to the primary cannot however easily land on the latter, because the gas initially possesses a high specific angular momentum. At various stages and by various mechanisms such as external and internal torques, the gas may be relieved of its angular momentum before settling in a disc around the primary in the binary orbital plane. This disk is known as the *accretion disc*. The amount of gas in the accretion disc is very small compared to the mass of the primary. The gas in the disc therefore revolves around the primary in Keplerian circular orbits with angular velocity

$$\omega_d(R) = \left(\frac{GM_1}{R_d^3} \right)^{1/2} \tag{7.13}$$

R_d being the radius of the disc. Simultaneously, the gas in the disc spirals in toward the primary by the latter's gravitational pull releasing its binding energy. The accretion disc thus becomes luminous by the radiation of the gravitational energy released by the infalling gas. The initial binding energy of a gas element m when it is far away from the primary is very small. But when it reaches a grazing Keplerian orbit around the primary, its binding energy becomes $GmM_1/2R_*$, R_*, being the radius of the primary. The total disc luminosity in a steady state is therefore

$$L_d = GM_1 \dot{M} /2 R_* = \frac{1}{2} L_{acc} \tag{7.14}$$

where $\dot{M}$ is the accretion rate and L_{acc} is the accretion luminosity given by

$$L_{acc} = GM_1 \dot{M} /R_* \tag{7.15}$$

Thus, as the matter in the disc spirals inwards to the star (or any compact body) half of the available accretion energy is radiated. The other half is released by the gas in penetrating through the viscous boundary layer formed very close to the star. The subject being fairly difficult and lengthy is beyond the scope of our present discussion. The interested reader may consult the relevant references given at the end of the chapter.

EXERCISES

1. Most of the visual binaries have relatively long periods and most spectroscopic binaries have relatively short periods. What is the reason for it?

2. A few stars are both visual binaries *and* spectroscopic binaries. Why such binaries are rare?

3. The total mass of a binary system is 5 $M_\odot$. The star A is twice as far from the centre of motion as star B. Compute the mass of each star.

4. The period of a visual binary is 100 years. Its parallax is $0''.15$ and its semi-major axis is $5''.0$. Compute the total mass of the system.

5. Describe how the (i) separation of stars, (ii) periods, and (iii) eccentricities of orbits vary among visual, spectroscopic, and eclipsing binaries?

6. Distinguish between a spectroscopic binary and a spectrum binary. Draw a schematic diagram of the velocity curves of a double-line spectroscopic binary with amplitudes of 20 km s^{-1} and 30 km s^{-1} and a period of 100 days. Assume that the centre of motion recedes with a constant velocity of 25 km s^{-1}.

7. A binary star system is composed of solar mass stars having an orbital period of 15 days. Compute the angular separation between the components and their orbital velocity (assume circular orbits). If the distance of the system be 30 parsecs, find the linear separation of the stars.

8. Compute the luminosity relative to that of the Sun for a main sequence star of (i) least mass ($M = 0.08\ M_\odot$), (ii) of highest mass ($M = 60\ M_\odot$), and (iii) of a star having a mass of 2.5 $M_\odot$. Use appropriate values of n in each case.

9. Why cannot we use the mass-luminosity relation for computing the masses of giant and supergiant stars?

10. A star has a luminosity equal to that of the Sun. Its surface temperature is 2500 K. Compute the radius of the star in terms of the radius of the Sun.

11. The radius of a red supergiant star is 100 $R_\odot$, its temperature is 3000 K, find its luminosity relative to that of the Sun. If a blue star of temperature 30,000 K has the same luminosity as the above star, find its radius in $R_\odot$.

12. If the time interval between the primary and secondary eclipses of a binary is exactly the same, does it follow that their orbits are circles? Explain.

SUGGESTED READING

1. Aitken, Robert G., *The Binary Stars,* Dover Publications, New York, 1964.

2. Baker, R.H. and Fredrick, L.W., *Astronomy,* 9th ed., Van Nostrand Rinehold Company, New York, 1971.

3. Kopal, Zdenêk, *Close Binary Systems,* John Wiley & Sons, New York, 1959.

4. Kurganoff, V., *Introduction to Advanced Astrophysics,* D. Reidel Publishing Company, Dordrecht, Holland, 1980.

5. Strand, K.Aa. (Ed.), *Basic Astronomical Date,* University of Chicago Press, Chicago, 1963.

6. Swihart, Thomas L., *Astrophysics and Stellar Astronomy,* John Willey & Sons, New York, 1968.

7. Zeilik, Michael, *Astronomy—The Evolving Universe,* John Wiley & Sons, New York, 1994.

8

Variable Stars

8.1 CLASSIFICATION OF VARIABLE STARS

A variable star is broadly defined as any star whose apparent brightness changes with time. Variable stars may be divided into two broad groups: (a) extrinsic variables, (b) intrinsic variables.

Those stars whose variations are only apparent and are caused by obscuration of their light by stars (or dust) have been called *extrinsic variables*. The eclipsing variables belong to this group.

Stars, in which changes of brightness are the results of the physical changes occurring inside are known as *intrinsic variables*.

Intrinsic variables can further be divided into two classes:

1. Pulsating variables
2. Exploding and eruptive variables.

In this chapter we shall be mainly concerned with the *intrinsic pulsating variables,* in which changes of luminosity are mostly regular and periodic and also consider some types of eruptive variables. We shall discuss the exploding variables (novae and supemovae) in Chapter 9.

Pulsating variables can be classified on the basis of the length of their periods, nature of variation of their light and their spectral characteristics. On the basis of these physical characteristics they can be divided into the following classes:

1. The Cepheid group of stars
   (i) the classical Cepheids (Type I Cepheids),
   (ii) the RR Lyrae stars (also called the cluster variables),
   (iii) the W Virginis stars (Type II Cepheids),
2. The RV Tauri stars
3. The red semi-regular and irregular variables
4. The long-period regular variables (Mira-type variables)
5. The Beta Canis Majoris variables.

Besides these, we shall discuss the following types of eruptive variables:

1. The U Geminorum type variables (or SS Cygni or dwarf novae)
2. The flare stars or flash stars.

8.2 THE CEPHEID GROUP OF VARIABLES

Classical Cepheids of Type I

The Cepheid variables are named after one of the earliest recognized members of the group, the Delta Cephei. More than 600 of such Cepheid variables are now known in our Galaxy. Their periods of variation of light range generally from 2 to 45 days and more usually about 5 to 10 days. They are yellow supergiants with spectra varying between F and K. Many of them are visible to the naked eye. Some of the brightest known stars among these are Polaris, Delta Cephei, Eta Aquilae, Zeta Geminorum and Beta Doradus. The Cepheids are highly luminous stars, with absolute magnitudes ranging from about −1.5 to about − 6. This means that the brightest of these are about 10^4 times more luminous than the Sun. The Type I Cepheids are strongly concentrated near the equatorial plane of the Galaxy and are supposed to belong to the extreme Population I (see Chapter 11). If one plots the change in apparent magnitudes of a variable star against the time of measurement through a complete period, one obtains what is called the *light curve* of the variable. In one complete period, this curve changes through one complete cycle, i.e., from crest to crest or from trough to trough. Different types of light curves are exhibited by Type I classical Cepheids, depending mainly on the length of their periods.

1. Stars with periods between 2 to 6 days (e.g. Delta Cephei) brighten more rapidly than they fade, but have smooth curves as shown in Fig. 8.1.

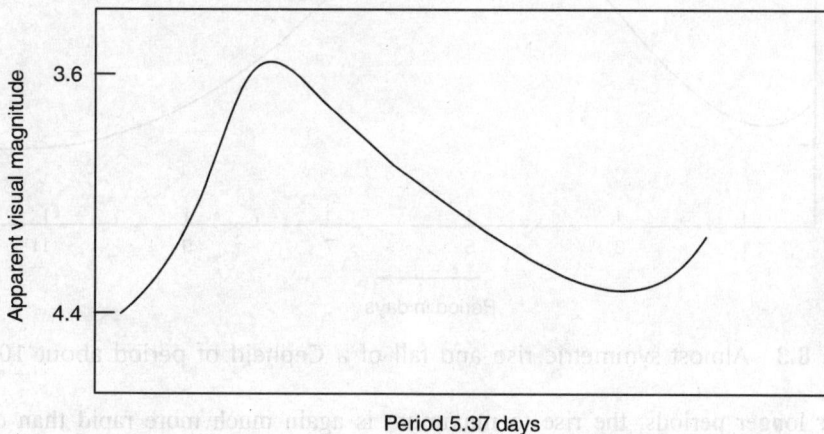

FIGURE 8.1 The light curve of δ Cephei with period of 5.37 days.

2. Stars with periods in the range of 7 to 8 days have a pronounced hump on the descending branch of the light curves as shown in Fig. 8.2. Eta Aquilae provides an example of this type.

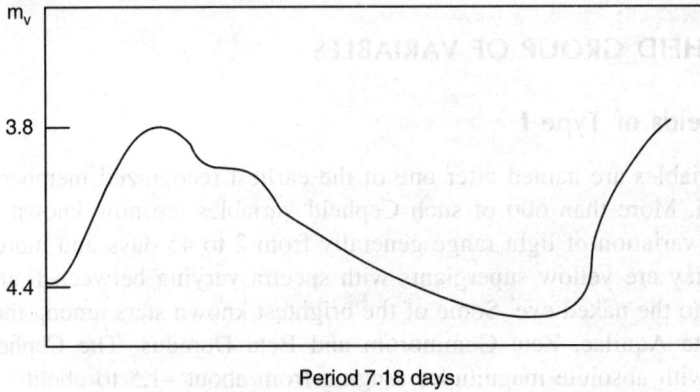

FIGURE 8.2   The light curve of Eta Aquilae.

3. For the stars with periods of 10 days, the curve becomes very symmetrical, rising and falling at almost equal rates as shown in Fig. 8.3.

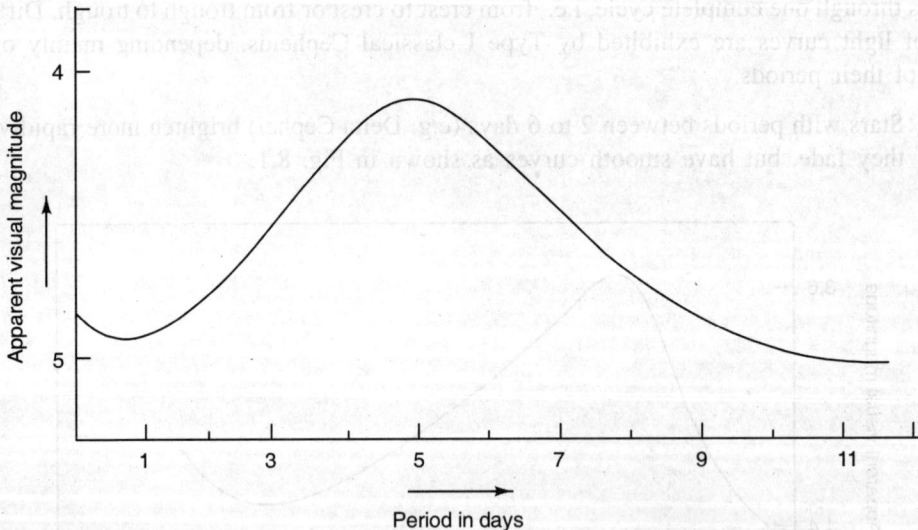

FIGURE 8.3   Almost symmetric rise and fall of a Cepheid of period about 10 days.

4. For longer periods, the rise to maximum is again much more rapid than decline to minimum.

Observation of radial velocity variations in these stars yields some clue to the understanding of the nature of pulsation undergone by them. In 1894, Belopolsky discovered that the radial velocity of Delta Cephei measured from the Doppler shifts of absorption lines varies with period. The radial velocity variation is found to be strongly correlated with the variations of light curve. The total range of velocity variations is about 40 km s^{-1}. Figure 8.4 shows the variation of light and radial velocities with respect to the period for Delta Cephei. Conventionally, the velocity of approach is considered as negative radial velocity. The velocity curve is very much like the light curve reflected on a horizontal mirror. The maximum light occurs where the velocity curve attains a minimum. This corresponds to the phase at which the surface of the star is approaching most rapidly towards the observer. Similarly, the light minimum coincides with the velocity maximum.

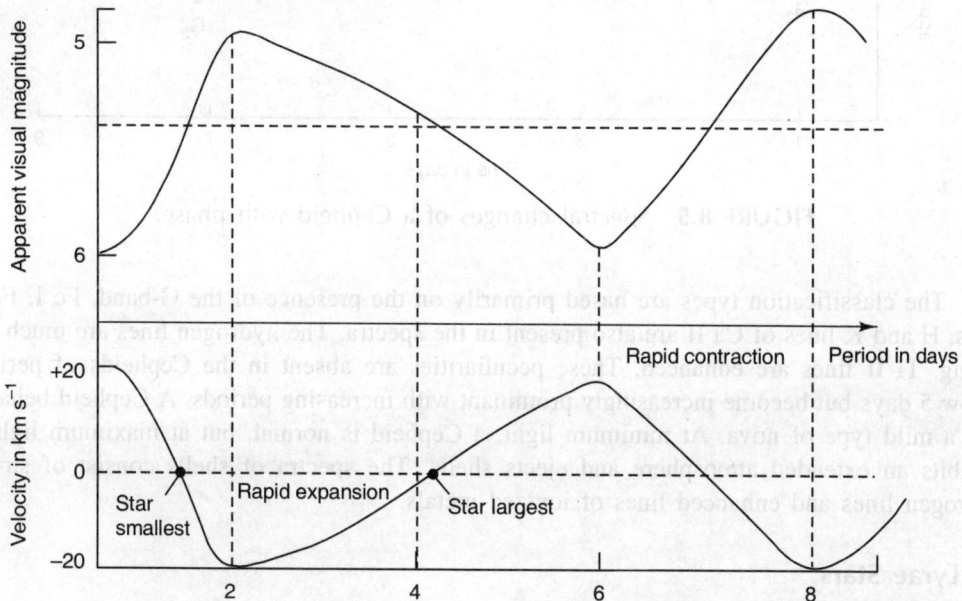

FIGURE 8.4   The upper curve represents variation of light with period and the lower one indicates the variation of radial velocity with period.

Of course, the exact mirror-image relation is not true of all Cepheids. For Cepheids of longer periods, light maximum is attained distinctly before the velocity minimum is reached. Further discussion of the relation between light and velocity curves will be made in section 8.11, where we shall briefly discuss the basic features of the Pulsation Theory.

Analysis of Cepheid spectrum.   During the variation cycle of Cepheid light, the spectrum changes continuously and this change amounts to a whole spectral class or sometimes even more. This is illustrated in Fig. 8.5. On analysis of the Cepheid spectra, one can notice the following broad characteristics.

1. At minimum light the Cepheids exhibit nearly normal F-G-K-class spectra which change with periods.
2. The spectral type at maximum light usually ranges from F_5 to F_8.
3. The spectra at maximum light (or perhaps a short interval before maximum) are not normal.

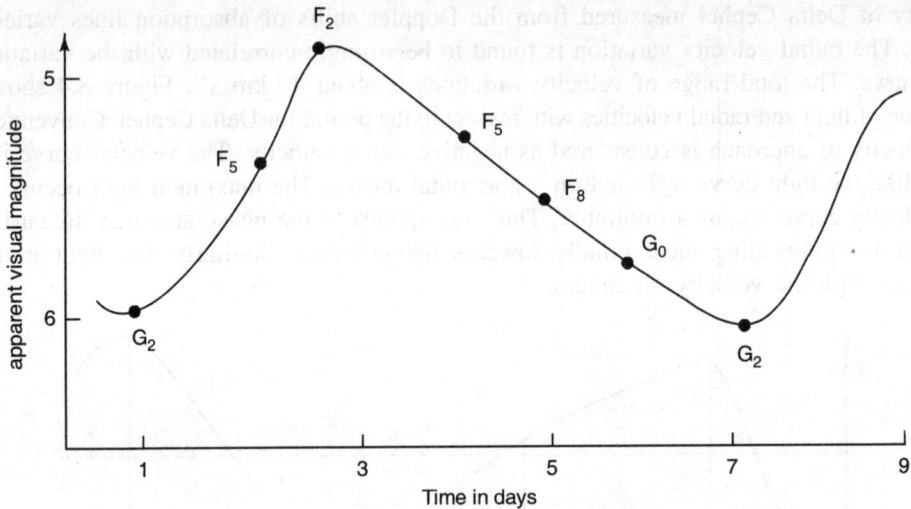

FIGURE 8.5   Spectral changes of a Cepheid with phase.

The classification types are based primarily on the presence of the G-band, Fe I, Fe II lines; H and K lines of Ca II are also present in the spectra. The hydrogen lines are much too strong. Ti II lines are enhanced. These peculiarities are absent in the Cepheids of periods below 5 days but become increasingly prominant with increasing periods. A Cepheid behaves like a mild type of nova. At minimum light, a Cepheid is normal, but at maximum light it exhibits an extended atmosphere and ejects shells. The spectra of shells consist of strong hydrogen lines and enhanced lines of ionized metals.

RR Lyrae Stars

These are called cluster Cepheids (or cluster variables) because they are found in abundance in globular clusters. They are also found in sufficient abundance in general field of the Galaxy. Their distribution and motion in space distinctly identify these to be the members of Population II. Periods of RR Lyrae stars range from less than two hours to 24 hours.

The long-period RR Lyraes with $P > 0.^d9$ that belong to the old globular clusters having only 1% or less of the normal metal abundance, are believed to link the RR Lyrae stars with Type II Cepheids. These are more luminous than other RR Lyrae stars and have large amplitudes of variation. However, the field RR Lyraes having periods $> 0.^d9$ are difficult to recognize because other types of variables of similar periods are found mixed in the field. Thus, the star BX Del having a period of $\simeq 1.^d09$ which was originally classified as an RR Lyrae star has later been recognized as a Type I Cepheid variable of shortest known period. On the other hand, some astronomers like F.G. Smith and L. Woltjer have suggested that the variables having $P < 0.^d2$ probably belong to a different class of objects which have been called *dwarf Cepheids*. Their galactic distribution also differs from the normal RR Lyrae stars and are alike the strong-lined RR Lyraes whose metal abundance differs little from the normal solar abundance. Thus at both the period boundaries the general criteria for the recognition of the RR Lyrae characteristics may be rendered ambiguous, posing classification problems. Two types of light curves have been recognized in RR Lyrae stars. The majority of stars show a rapid increase of light, a pointed maximum, moderately rapid decline and a broad and rather flat minimum. The range from maximum to minimum light variation is about one magnitude. Figure 8.6 shows the light variation characteristics in most of the RR Lyrae stars. The other type has smaller range of brightness and more symmetrical and smoother curve.

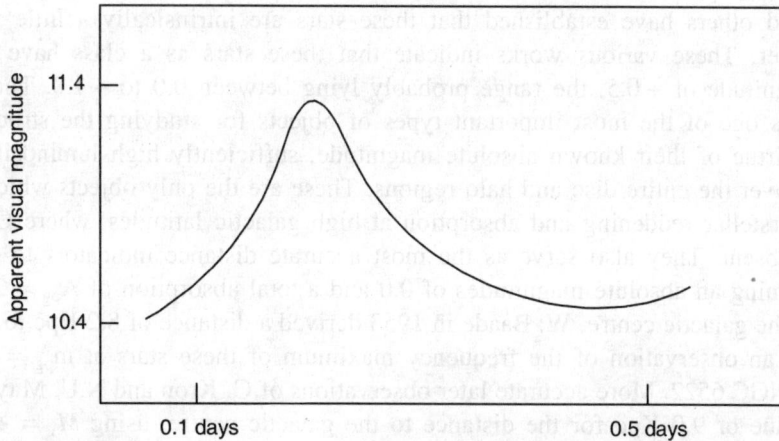

FIGURE 8.6 Light curve of a typical RR Lyrae star.

Several interesting characteristics are revealed by spectroscopic studies of RR Lyrae stars. In general, these variables belong to spectral classes A to F, the spectral class of any particular star varying with phase, H and Ca II lines are found in emission during rising light. These emission lines are observed at different strengths in different stars while some stars such as SV Ari ($P = 0.^{d}714$), shows no emission at all. Strongest H emission is observed in the spectrum of X Ari ($P = 0.^{d}651$). The emission lines are often observed to be flanked by an absorption feature into components, one shortward and the other longward of the natural wavelengths of lines. One probable interpretation of this phenomenon is that while the emission feature is produced in the atmosphere of the star, the superimposed absorption feature probably originate in the overlying infalling cooler material. However, other interpretations are also possible.

From a spectroscopic study of a large number of field RR Lyrae stars, G.W. Preston observed that these stars comprise mixed population having varied amount of the metal content. He introduced the parameter $\Delta S = 10\,[\text{Sp(H)} - \text{Sp (Ca}^{+})]$ to distinguish between these populations in which smaller values of ΔS means stronger metallic-lines while the larger ΔS values indicate weaker metallic lines. Preston's study revealed that strong-lined stars having $\Delta S < 3$ have shorter periods (average $P = 0.^{d}43$), were more concentrated about the galactic disc and participate more strongly in the galactic rotation compared to the weak-line group having $\Delta S > 5$. Preston's study further revealed that for a single star, the spectral class assigned by the hydrogen lines and that assigned by the metallic lines may differ by the full range of an entire spectral class which corresponds to a value of the Preston's parameters $\Delta S = 10$.

The velocity curves of RR Lyrae stars have approximately a mirror image relationship with the light curve as in the case of the classical Cepheids; but velocity minimum takes place a little after the light maximum is attained.

For a long time, the RR Lyrae stars were assumed to have an absolute magnitude zero, regardless of their periods. But later works of such astronomers as A.R. Sandage, O.G. Eggen, H.C. Arp and others have established that these stars are intrinsically a little fainter than thought earlier. These various works indicate that these stars as a class have an average absolute magnitude of $+ 0.5$, the range probably lying between 0.0 to $+ 1.0$. The RR Lyrae stars serve as one of the most important types of objects for studying the structure of our Galaxy by virtue of their known absolute magnitude, sufficiently high luminosity and their distribution over the entire disc and halo regions. These are the only objects which allow the study of interstellar reddening and absorption at high galactic latitudes, where OB stars are completely absent. They also serve as the most accurate distance indicators to the galactic centre. Assuming an absolute magnitudes of 0.0 and a total absorption of $A_{pg} = 2.^m75$ in the direction of the galactic centre, W. Baade in 1953 derived a distance of 8.2 Kpc to the galactic centre, from an observation of the frequency maximum of these stars at $m_{pg} = 17.5$ in the direction of NGC 6522. More accurate later observations of G. Kron and N.U. May all in 1960 yielded a value of 9.0 Kpc for the distance to the galactic centre, using $M_v = +0.3$ for RR Lyrae stars. With their average absolute magnitude of $+0.5$, they are about fifty times more luminous than the Sun.

Type II Cepheids (or W Virginis stars)

The W Virginis variables or Type II Cepheids are frequently recognized in the globular clusters and near the centre of our Galaxy. Their periods are mostly from 10 to 30 days with absolute magnitudes from 0.0 to about -3.5. Their light curves are similar to those of Type I Cepheids except that their decending branches fall less stiffly. For the same period, Type II Cepheids are about 1.5 magnitudes fainter than classical Cepheids of Type I. But the former follow a period-luminosity relation very similar to that of the latter (Fig. 8.7).

The class as a whole has been named after their representative W Virginis star. They are high velocity and metal-poor stars. These characteristics assign to them Population II whereas the classical Cepheids of Type I are members of Population I (see Chapter 11). The colours of W Virginis class are bluer than those of Type I Cepheids.

The stars belong to spectral classes F to G, the spectrum of any particular star varying with phase as in the case of Type I Cepheids and RR Lyrae stars. Their spectrum contains well-marked bright lines of hydrogen, specially during the phase when the brightness of the star increases. But the most remarkable feature lies in the doubling of H lines at maximum. The violet component of the Ca II doublet appears at one maximum and the red component at the next maximum. This is interpreted as due to the presence of two concentric shells, alternately rising and falling above the photosphere of the star. Shortly after these two have collided bright lines of hydrogen are observed.

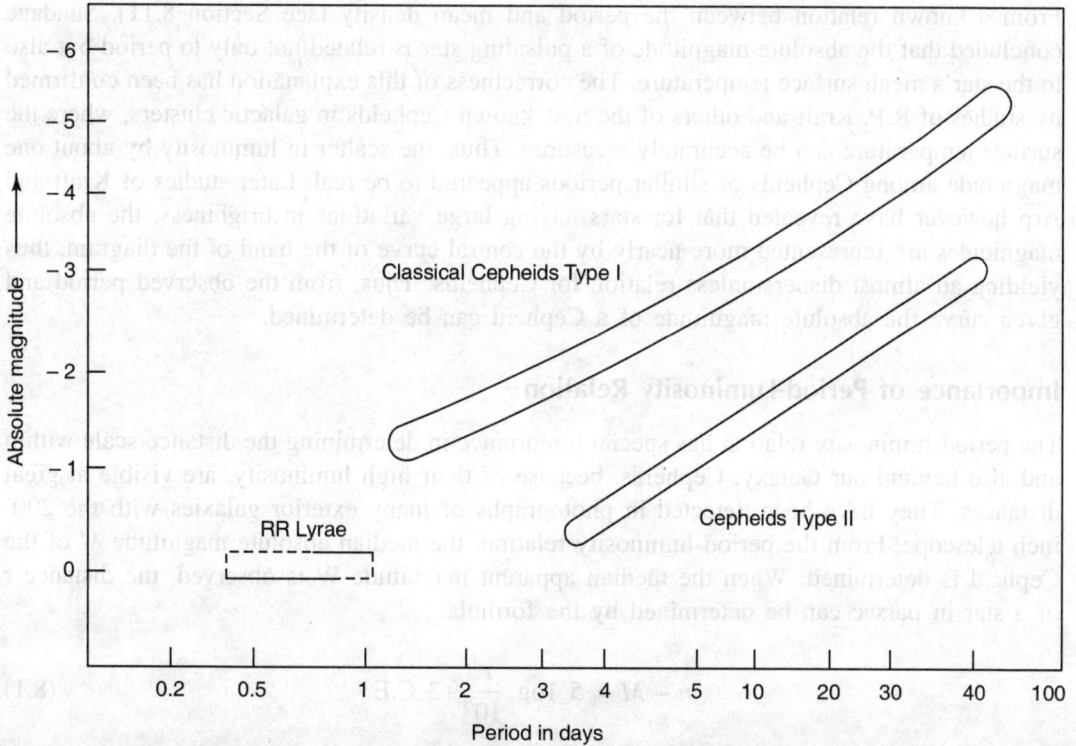

FIGURE 8.7 Period-luminosity relation of Cepheids.

8.3 PERIOD-LUMINOSITY RELATIONS OF CEPHEID GROUP OF VARIABLES

During her study of variable stars in the Small Magellanic Clouds in 1911, H.S. Leavitt at the Harvard University Observatory found that the length of the period progressed with the average brightness of these variables. The stars of shortest period (2 or 3 days) are nearly 3 magnitudes fainter than those with periods around 30 days. Stars of period 2 days are 700 times as bright as the Sun whereas those of period 30 days are 10,000 times brighter than the Sun.

Later works of H. Shapley, E.P Hubble and others have confirmed this relationship of period to the apparent magnitude among the Cepheid variables in other systems, e.g. the Large Magellanic Cloud, the Andromeda spiral, M33 etc. In each of these systems, all the variables are at nearly the same distance from us; hence the relation between the apparent magnitude and period is actually the relation between the absolute magnitude and the period.

H. Shapley in 1917 considered the P-L curve showing how logarithm of the period increases as the median absolute magnitude (i.e. the average between the magnitude at maximum and minimum brightness) increases. In Fig. 8.7, the period-luminosity relation of both types of Cepheids and of RR Lyrae stars is represented by bands at least one magnitude wide rather than by a single curve. This scatter in absolute magnitudes was suggested from H.C. Arp's studies of Cepheids in the Small Magellanic Cloud and was later explained by A.R. Sandage.

From a known relation between the period and mean density (see Section 8.11), Sandage concluded that the absolute magnitude of a pulsating star is related not only to period but also to the star's mean surface temperature. The correctness of this explanation has been confirmed by studies of R.P. Kraft and others of the new known Cepheids in galactic clusters, where the surface temperature can be accurately measured. Thus, the scatter in luminosity by about one magnitude among Cepheids of similar periods appeared to be real. Later studies of Kraft and Arp however have revealed that for stars having large variations in brightness, the absolute magnitudes are represented more nearly by the central curve of the band of the diagram, thus yielding an almost dispersionless relation for Cepheids. Thus, from the observed period and given curve the absolute magnitude of a Cepheid can be determined.

Importance of Period-Luminosity Relation

The period-luminosity relation has special importance in determining the distance scale within and also beyond our Galaxy, Cepheids, because of their high luminosity, are visible at great distances. They have been detected in photographs of many exterior galaxies with the 200-inch telescope. From the period-luminosity relation, the median absolute magnitude M of the Cepheid is determined. When the median apparent magnitude W is observed, the distance r of a star in parsec can be determined by the formula:

$$m - M = 5 \log \frac{r}{10} + 3 \text{ C.E.} \qquad (8.1)$$

where the last term (3 C.E.) is the correction term to be introduced for any dimming of starlight by intervening cosmic dust. C.E. means colour excess which is the difference between the observed colour index and the colour index the star would have in the absence of obscuration and reddening. RR Lyrae stars are useful for determining distances within our Galaxy. They are not bright enough to be observed with present telescope beyond the nearest exterior galaxies LMC and SMC. But the Type I classical Cepheids are luminous enough to be detected in external galaxies situated much farther. By observing classical Cepheids, in 1924, it was first shown by Hubble that the Andromeda spiral was another galaxy. Thus Cepheids of Type I and RR Lyrae stars are important extra-galactic distance indicators.

8.4 RV TAURI VARIABLES

These are yellow to red bright giants of the spectral classes G to K showing semi-regular variations of light within periods between 30 to 150 days. The typical star is RV Tauri which varies by about one magnitude in a period of around 79 days. Its light curve has two alternate maxima of nearly equal amplitudes, but the corresponding minima are unequal (see Fig. 8.8). Like a Cepheid, the star appears *bluer* at the maximum light than at the minimum, the variations in spectral classes being from K5 to G4.

The 1958 edition of the Soviet General Catalogue of variable stars lists 92 RV Tauri stars in our Galaxy. Most of these stars are observed in the halo and nuclear regions of the Milky Way. The light curves of these stars usually have alternate large and small maxima

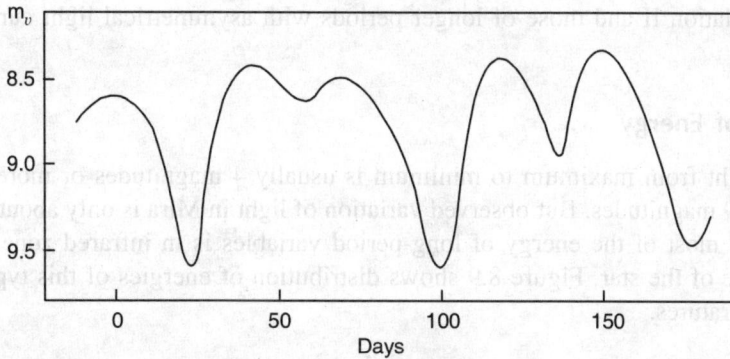

FIGURE 8.8 Light curve of an RV Tauri variable, V Vulpeculae.

whose amplitudes may range up to 3 in magnitudes, the absolute magnitudes in median lights being –2 to –3.

The semi-regular variations of brightness of the RV Tauri variables are thought to be partly due to pulsation and partly due to other intrinsic causes. The study of their light variation, spectrum and velocity is accordingly more difficult.

On detailed analysis three outstanding features of RV Tauri stars appear. These are:

1. Strong bright lines of hydrogen are observed during increasing light; where a light curve has a secondary maximum, weaker Hydrogen emission may appear there also.
2. Although the spectrum of the star varies within G and K classes during the cycle, at minimum titanium oxide band is often observed which is a characteristic of class M.
3. The velocity curves may sometimes be discontinuous. Analysis of all these characteristics indicate that these stars belong to Population II. The bright lines of hydrogen and discontinuous velocity curves link the RV Tauri stars with the RR Lyrae stars and W Virginis stars on the one hand, and on the other, titanium oxide bands at minimum and hydrogen emission suggest a link with long-period variables.

8.5 LONG-PERIOD VARIABLES (MIRA-TYPE VARIABLES)

The largest number of observed variable stars belong to this class. The General Catalogue of variable stars (1958) lists 3657 such stars. Periods of long-period variables range from 100 to 1,000 days, though most of these are between 150 to 500 days.

Examples are, ω Centauri which has a period of 84 days, Mira (Omicron Ceti) with a period of 330 days and Messier 22 with a period of 200 days. A few long-period variables are observed in globular clusters. Three of them with periods around 200 days lie in the cluster 47 Tuc.

The median absolute magnitudes are in the range +2 to –2, variation of amplitude from minimum to maximum being on the average about 3 magnitudes. With their spectra belonging to the class M, long-period variables therefore belong to the red giant class of stars. Their radii are certainly a few hundred times that of the Sun. Radius of Mira is equal to about 460 solar radius. Long-period variables with shorter periods ($P < 150^d$) having symmetrical light curves belong to Population II and those of longer periods with asymmetrical light curves belong to Population I.

Distribution of Energy

Variation of light from maximum to minimum is usually 4 magnitudes or more. A few stars have ranges of 9 magnitudes. But observed variation of light in Mira is only about 1 magnitude. This is because most of the energy of long-period variables is in infrared zone owing to the low temperature of the star. Figure 8.9 shows distribution of energies of this type of stars for different temperatures.

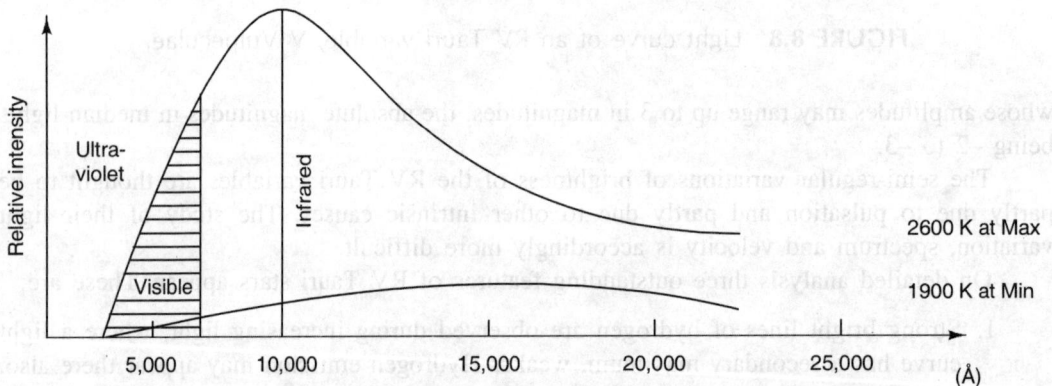

FIGURE 8.9   Energy curves of Mira at maximum and minimum light.

Period-luminosity Relation

Variation of light is roughly correlated with the length of the period; the stars of longest periods having the greatest change of brightness. In general, stars of shorter periods have fairly symmetrical curves; for those of longer periods the rise is steeper than the decline, having a decided pause on the rising branch.

Period Spectral Relation

From spectra of long-period variable stars, we find that longer the period of the star, the 'cooler' is its spectral class. Eighty-five per cent of these stars have M-type spectra with bright lines of hydrogen and a few other elements. The remaining 15 per cent have spectral classes R, N, S—not always with bright lines of hydrogen. The absorption spectra vary in the same direction as in Cepheids, i.e. hotter at maximum light than at minimum. The typical star Mira varies from M9 at minimum to M6 at maximum. Bright lines of hydrogen are absent at minimum, but during the increase in luminosity the lines become gradually brighter and show with great strength after maximum light. Typical long-period variable stars are of spectral types M, S, N. Carbon is in abundance in N-type stars. Traces of Sr, Zr, Ba, La, Ca, TiO, Mg, Fe, etc. are present in the spectra of long-period variables at lower temperatures. Formation and dissociation of molecules take place simultaneously in these stars.

Radial Velocity

In long-period variable stars, there is a marked difference in velocity obtained from emission and absorption lines. The absorption spectra of Mira has been studied to get detailed information about variation of velocity curves. In general, the maximum velocity occurs at the time of light maximum which is just opposite to the relation as far as the Cepheids are concerned. The radial velocity of bright lines differs greatly from those of dark lines of the spectrum. The velocity of approach of bright lines is greater than that of dark lines by 15 to 20 km s^{-1} in stars like Mira and is comparatively less for those of smaller periods.

8.6 RED IRREGULAR AND SEMI-REGULAR VARIABLES

There is a large group of red variable stars in which variation in brightness in certain aspects is related to the long-period variables and in certain others to the semi-regulars. The stars which are related to long-period variables have a range of brightness variation of over a quarter of a magnitude. These stars are of high luminosity or of advanced spectral class (M-class) or of both. They correspond to cycles of around 100 days. Another type (semi-regulars) belonging to the subclass N, have cycles over 350 days.

Betelgeuse (Alpha Orion) and Mu Cephei are examples of this group of red variable stars. Betelgeuse has a diameter of about 900 to 1000 times that of the Sun.

8.7 BETA CANIS MAJORIS VARIABLES (β CEPHEI STARS)

Beta Canis Majoris stars are blue giants with spectral class belonging to early B and have an average absolute visual magnitude of –3. The range for individual stars varies from –2 to –4. These stars usually have very small range of light variations as also similar variations in spectral classes. In a typical star, like BW Vulpeculae (Petrie's star) the amplitude of brightness fluctuations appears to be greatest with 0.2 magnitude. The other stars in this group usually have amplitude less than 0.1 magnitude which may vary up to a few hundredths of a magnitude at their minima. Due to such small fluctuations in light, these stars are hardly detectable in the sky and have only been observed among the bright stars. Only 11 of these are listed in the 1958 Soviet catalogue. Since such small variations in light are difficult to record photographically, the only way these objects could be traced is by means of improved photoelectric devices.

Periods of these variables range between 3 and 8 hours. The prototype β Canis Majoris has a period of 6^h which sometimes varies to 6^{h}1^m from cycle to cycle with corresponding changes in the amplitudes. These two different modes of variation have been observed to form a conspicuous beat phenomenon. There is an interesting phenomenon associated with the variations of brightness of the variable γ Böòtis. This star shows small fluctuations in brightness for definite periods that continue for quite a number of years. But for the next few years the star does not show any variation. Its brightness remains constant throughout this interval. After a prolonged halt at the same luminosity, the star suddenly begins to pulsate with a definite period as before.

Like classical Cepheids, β Canis Majoris stars also show variations of radial velocities from a few kilometres per second to several tens of kilometres per sec. Their radial velocity curves are not always uniform, but show irregularities in forms and amplitudes. Study of these curves also suggests the small but stable variations in these stars. Due to the similarities with the Cepheids in many respects, these stars are also called β *Cephei stars or dwarf Cepheids.*

The position of β Cephei stars in the H-R diagram is quite interesting. These stars have emerged from the main sequence with γ Pegasi (at spectral class B3) and extend to the left of the diagram above the main sequence upto the star β Cephei at class Bl. A strong correlation exists between the average periods of the stars and their positions in the H-R diagram. Thus, γ Pegasi is a star of class B3 with period $3^h.4$ while β Cephei of class Bl has an average period of $6^h.0$, that is, the period of pulsation increases as the stars become brighter, which is much like the period-luminosity relation of the classical Cepheids. Calculations show that β Cephei stars at the bottom of the H-R diagram (such as γ Pegasi) have radii 5 times that of the Sun while those at the top (such as β Canis Majoris) have radii 10 times the solar radius. Probably these stars have an extensive semi-convective zone in the envelope which destabilizes their superficial layers thereby initiating the mild pulsation in them. This semi-convective zone is not developed in stars much fainter than $M_v = -3.0$.

8.8 U GEMINORUM STARS (SS CYGNI OR DWARF NOVAE)

The 1958 Soviet catalogue of variable stars lists 112 stars belonging to this small group of variables. The variation is sudden, rising to maximum in 2 to 3 days and declining again to minimum in 10 to 20 days. The total amplitude of variation ranges from 2 to 6 magnitudes. U Geminorum, SS Cygni and AE Aquarii are examples of this group of stars.

They belong to the class of eruptive variables. The stars are of hot subdwarf type, the spectral class usually ranging from A to F. U Geminorum stars, possess several interesting spectral characteristics. Emission lines of hydrogen and helium are revealed. Some stars have G-type spectra at minimum, while some others possess only continuous spectra without any lines superposed on them. These continuous spectra have energy distribution like that of G class. After the outburst, strong continuous spectrum with energy distribution of G-class is observed and at maximum light bright lines of G-class spectra are submerged by the strong continuous spectrum. As the star becomes fainter the bright lines appear in the picture. The cause of outburst is unknown.

8.9 FLARE STARS

These are normal dwarf M stars of spectral types ranging from M3 to M6 but undergo sudden outbursts of light with amplitudes ranging from 1 to 6 magnitudes. The outbursts last from a few minutes to about half-an-hour and the rise to maximum is very rapid; it takes only a few seconds. These belong to the eruptive variable group of stars. About two dozen flare stars are known; important examples are Proxima Centauri, Kruger 60 and UV Ceti. During the outbursts, a continuous spectrum of hotter energy distribution obliterates the normal M spectrum and the colour associated with flares corresponds to A and B type stars. Calculations of temperatures of flares and the total luminosities of stars indicate that the flares must be localized phenomena. The flares probably are manifestations of outbursts of hot gases from below the photospheric level as are believed to be the case with solar flares. A convection zone below the photospheres of stars is presumably the underlying source of flares.

UV Ceti is probably the best known flare star. In this star outbursts occur at an average interval of $1^1/_2$ days. The amplitudes of variation in light generally ranges from 1 to 2 magnitudes; however in 1952 an increase of 6 magnitudes was observed. Spectra of these stars normally contain emission lines of hydrogen and ionized calcium; some bright lines of helium are also observed during the outbursts. Emission lines are normally strong and narrow. At flare times, bright lines of hydrogen are wider and stronger.

8.10 A SURVEY OF VARIABLE STARS AS A WHOLE

In this section we shall summarise the characteristic features of the principal types of variable stars (Fig. 8.10).

(a) *Period.* The range in periods varies from less than 2 hours for some RR Lyrae stars to almost thousand days for some long-period variables.

(b) *Spectral class.* Their colour runs from bluish white for β Canis Majoris (B–class), moderately blue for RR Lyrae stars (A–F class) to extremely red for long-period variables and red semi-regular or irregular variables (K–M class with effective temperature 2000 K).

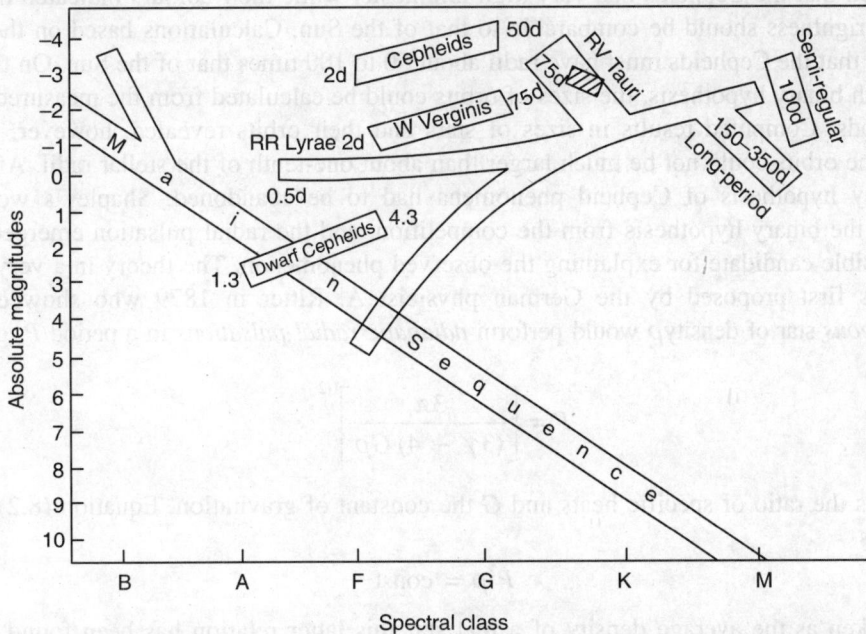

FIGURE 8.10 The H-R diagram showing the position of the intrinsic variables.

(c) *Luminosity.* The luminosity of variable stars is fairly high and extends over a range of 6 magnitudes or more.

(d) *Populations.* Classical Cepheids and β Canis Majoris belong to Population I. Type II Cepheids, RR Lyrae stars and RV Tauries belong to Population II. Long-period variables, semi-regular variables and red irregular variables lie in the domain between the two stellar populations.

For a discussion of stellar populations see Section 11.5.

8.11 THE PULSATION THEORY OF VARIABLE STARS

Before pulsation was accepted as the actual cause of the observed light and velocity variation of the intrinsic variable stars, various other hypotheses were put forward by early astronomers in order to explain the observed phenomena. In 1894, Belopolsky discovered that the radial velocity of δ Cephei varies in the same period as its light in the manner that the maximum radial velocity corresponds to its minimum light. This observation led many astronomers to believe that the Cepheid phenomena are associated with the binary nature of these stars. It was suggested that δ Cephei as also all such stars might be spectroscopic binaries. But the second spectrum of the less luminous companion was never found. During the following decade, astrophysicists like K. Schwarzchild, R.B. Roberts, H.D. Curtis, etc. proposed various models under the binary hypothesis to explain the observed Cepheid phenomena, but all these met with little success. The idea continued, however, until in 1914 the binary hypothesis received a death blow from H. Shapley's work. By that time the works of H.N. Russell and E. Hertzsprung established that the Cepheids had very high luminosity while their colours indicated that their surface brightness should be comparable to that of the Sun. Calculations based on these data indicated that the Cepheids must have radii about 50 to 100 times that of the Sun. On the other hand, with binary hypothesis, the sizes of orbits could be calculated from the measured speeds and periods. Computed results in sizes of stars and their orbits revealed, however, that the radii of the orbits could not be much larger than about one-tenth of the stellar radii. After this, the binary hypothesis of Cepheid phenomena had to be abandoned. Shapley's work thus removed the binary hypothesis from the competition, and the radial pulsation emerged as the only possible candidate for explaining the observed phenomenon. The theory in a very simple form was first proposed by the German physicist A. Ritter in 1879 who showed that a *homogeneous* star of density ρ would perform *adiabatic radial pulsations* in a period P, given by

$$P = \left[\frac{3\pi}{(3\gamma - 4) G\rho} \right]^{1/2} \tag{8.2}$$

where γ is the ratio of specific heats and G the constant of gravitation. Equation (8.2) can be written as

$$P^2 \rho = \text{const} \tag{8.3}$$

If ρ is taken as the average density of a real star this latter relation has been found to hold approximately in general, for all variable stars. Ritter's work was not pursued, however seriously by other astronomers and so the idea temporarily lost its importance.

After Shapley's work in 1914, the radial pulsation hypothesis regained its importance and was vigourously pursued by many astronomers; the most noteworthy among them was Sir Arther Eddington. Eddington considered the adiabatic pulsation of a star with certain density distribution and arrived at a period-density law not much different from Ritter's law. However, in view of the overwhelming abundance of hydrogen in the stellar material, if γ be taken equal to 5/3, the computed periods with Eddington's formula fall short of the observed periods, if normal main sequence stellar density is adopted. This led Eddington to the conclusion that density in Cepheids must be very much less compared to that of the main sequence stars.

According to the original form of the pulsation theory, the star as a whole expands and contracts alternately. This implies that it should be hottest and therefore brightest and bluest when its size is smallest, i.e., when the radial velocity is zero. In actual practice however, these two phenomena are observed at an interval of a quarter of a period. This phase-lag posed intriguing problems to the original over-simplified radial pulsation theory. Serious attempts were made by several physicists, but the problem remained obscure in spite of their efforts.

The most satisfactory answer about the true nature of pulsation that properly explains the observed phenomena of Cepheid variation came from Martin Schwarzchild in 1938. He suggested that the star as a whole does not pulsate. The interior of the star pulsates and in the process sends compressional waves to the photospheric layers of the star. According to this model, the maximum brightness does not occur when the star is smallest but at the time when the compressional waves are moving fastest, that is, when the velocity of approach is maximum. This is exactly what is observed from the light curve and velocity curve relation. The light minimum will occur when the rarefaction wave reaches the photosphere. Thus, Schwarzchild's modified pulsation hypothesis ultimately emerged as the only acceptable explanation which is in harmony with the observed Cepheid phenomena. The hypothesis was also supported later by spectroscopic studies of the behaviour of lines originating in different photospheric levels. It is known that some lines in the stellar spectrum originate in deeper layers than others. Studies of Cepheid spectra have revealed that the changes corresponding to the maximum compression are manifested slightly earlier by low-level lines than by the high-level ones. This fact has been accepted by most astronomers as a definite proof for the presence of the pulsation mechanism proposed by Schwarzchild.

It should be mentioned, however, that although the Cepheid phenomena have been satisfactorily understood in terms of Schwarzchild's pulsation hypothesis, the physical cause of pulsation is still unknown. The physical processes which initiate the disturbance in the internal structure of the star, and for how long does this disturbance continue, are some of the questions which have not yet been well understood. However, the Russian astrophysicist S.A. Zhevakin claims that the mechanical oscillations of Cepheids can be maintained by pumping energy generated in some critical zone of ionization of certain element. His investigations revealed that hydrogen ionization was ineffective in exciting the auto-oscillations in these giants and supergiants because of the highly nonadiabatic nature of the oscillations. On the other hand, he found that the critical zone of the He ionization would be efficient enough to initiate and maintain the mechanical oscillations in Cepheids provided the helium content in the envelope of these stars were around 15 per cent by number of atoms. In a series of works, Zhevakin demonstrated that if this abundance in the envelope of these variable stars was assured, then the He ionization zone could explain satisfactorily a large number of observed phenomena in these stars. The phase shifts between the variations in light and radius of the star, the relation between the amplitudes of light and radius variations, the observed asymmetry in the radial velocity curve, the observed period-luminosity relationship etc., are some of these observed phenomena. The problem thus appears to rest on the correct assessment of the helium abundance in the envelopes of the variable stars. This again is a problem whose solution does not appear in sight in near future, because these stars are cool and helium has very high ionization potential. Nevertheless, helium has been observed in emission spectra of some variable stars, such as the line λ 5876 of helium has been observed in emission behind the shock-front by Kraft in the spectrum of the Type II Cepheid W Vir. From an analysis of this observation, Wallerstein computed the ratio of hydrogen to helium by number of atoms to lie between 3 and 10 in W Vir. Although the large range demonstrates the inherent uncertainty in such computations, it includes the requirement of the helium abundance set by Zhevakin's theory. More work of this nature but of higher accuracy would be required before any definitive answer to the question of the helium abundance in these stars could be established. Zhevakin's theory although has emerged as the most plausible one has yet to establish itself on the firm basis of observations.

EXERCISES

1. Make a table of the different kinds of variable stars you have been familiar with, indicating for each of them the following properties:
   (a) the mean visual absolute magnitude;
   (b) the mean spectral class;
   (c) the range of variation in absolute magnitude;
   (d) the range of variation in spectral class;
   (e) the temperature at the mean absolute magnitude;
   (f) the mean period of pulsation;
   (g) the range of pulsation period; and
   (h) the population (See Chapter 11).

2. Discuss the importance of the period-luminosity relation of Cepheid variables. How it came to be known that Type II Cepheids form a parallel sequence as Type I Cepheids but are fainter by about 1.5 magnitude? Discuss the difference between the period-luminosity relations of Type I Cepheids and RR Lyrae stars. Explain the scatter in the period-luminosity relation of Type I Cepheids.

3. Explain the effect of the Preston parameter ΔS on the metallic content in RR Lyrae stars. Discuss how the values of ΔS influence the periods of these stars.

4. Using the period-luminosity relations of Type I and Type II Cepheids, calculate the absolute magnitudes of these stars having the periods of (i) 5 days, (ii) 10 days and (iii) 25 days.

5. Find the distances to the stellar systems in which Type I Cepheids are observed to have:
   (a) $P = 40$ days, $m = +20$
   (b) $P = 32$ days, $m = +15$
   (c) $P = 5$ days, $m = +18$

   where m denote the mean apparent magnitude.

   Find the corresponding distances if the above data represent observation of Type II Cepheids.

6. The distance of a stellar system is determined by observing a Cepheid with $P = 15$ days and $m = +20$. What error will be committed in distance if the star is considered as of Type I while actually it is of Type II.

7. An RR Lyrae star is observed to have an apparent magnitude +20 in a globular cluster. Calculate the distance of the cluster.

8. The apparent magnitude of a Type I Cepheid is +3 and its period is 30 days. Find the distance of the Cepheid. Find the corresponding distance if the above data represent a Type II Cepheid.

9. If a large telescope has a limiting magnitude of +22, find the maximum distance to which we can see the following stars.

   (a) RR Lyrae stars;
   (b) Classical Cepheids of Type I and Type II;
   (c) W Virginis stars;
   (d) Mira-type stars; and
   (e) β Canis Majoris stars.

10. A Type I Cepheid with a period of 10 days has an apparent magnitude of + 5. How bright would an RR Lyrae star be, if it were in a globular cluster near to the Cepheid?

11. A Cepheid variable in a hypothetical Galaxy is observed to have a period of 10 days and its mean apparent magnitude is +20. It is not known whether it is a Type I or a Type II Cepheid.

    (a) What are the two possible distances to the Galaxy?
    (b) What is the ratio of these distances?
    (c) How do the distances compare to the diameter of the Milky Way Galaxy?

12. Use Stefan's law to calculate the maximum ratio of the energy radiated per unit area of Mira's surface during its cycle.

SUGGESTED READING

1. Kourganoff, V., *Introduction to Advanced Astrophysics,* D. Reidel Publishing Company, Dordrecht, Holland, 1980.

2. Levy, David, H., *Observing Variable Stars,* Cambridge University Press, Cambridge, 1989.

3. McLaughlin, Dean B., *Introduction to Astronomy,* Houghton Mifflin Company, Boston, 1961.

4. Otto Struve and Velta, Zebergs, *Astronomy of the 20th Century,* Macmillan, New York, 1962.

5. Percy, John R., Mattei, Janet Akyuz and Sterken, Christian (Eds.), *Variable Star Research: An international perspective,* Cambridge University Press, Cambridge, 1991.

6. Shu, Frank H., *The Physical Universe, An Introduction to Astronomy,* University Science Books, Mill Valley, California, 1982.

7. Zeilik Michael, Gregory, Stephen A. and Smith Elske, V.P., *Introductory Astronomy and Astrophysics,* 3rd ed., Saunders College Publishing, Orlando, Florida, 1991.

9 Erupting and Exploding Stars

9.1 INTRODUCTION

In the previous chapter, our discussions were confined to pulsating variable stars. In this chapter, we shall discuss those types of variable stars which flare up suddenly and violently, and sometimes explode. Novae and Supernovae belong to these groups of stars. Let us first consider novae which undergo occasional outbursts. These are peculiar astrophysical objects, generally faint subdwarf stars, which suddenly flare up to very high luminosities.

At the time of outburst the luminosity of a nova becomes several hundred to more than a million times greater than that of the original star. The maximum luminosity continues for a few days for the fast novae and for months or even years for slow novae, but both ultimately decline to their original brightness. There are also some novae, known as *recurrent novae,* whose outbursts repeat after some years. For example, T Cr B was observed to burst out in 1866 and 1946. RS Oph flared up first in 1898 and then in 1933. T Pyx had four violent outbursts in 1891, 1902, 1920 and 1944. WZ Sge, U Sco and many others belong to the group of recurrent novae. Kukerkin and Parenago derived a formula connecting the amplitude A and time-interval P between two consecutive maxima of a recurrent nova which is given by

$$A = 0.80 + 1.667 \log P \qquad (9.1)$$

The first nova-like object was observed by the Chinese astronomers near ρ Tauri on July 4, 1054. In 1572, Tycho Brahe observed a nova which was visible at noon and was as bright as Venus. It became one magnitude fainter in three months and after about 18 months it was too faint to be observed with the naked eye. In 1600 another nova was observed which remained fairly bright for many years. Now it is a fifth magnitude star and is known as P Cygni. In 1604, Kepler observed a nova in Ophiuchus.

The nova observed in the year 1054, is now recognized as the supernova remnant named Crab Nebula. Novae observed in 1572 by Tycho Brahe and in 1604 by Kepler were also supernovae. The detailed discussion of these will be given later in this chapter.

The number of novae observed before the present century was not too large. They were mostly chance discoveries. But great advancement in instrumentation and observational skill has enabled astronomers to detect and study numerous novae in the present century in our own Galaxy and also in some nearby external galaxies.

As early as in the beginning of 1920s, Zinner and Lundmark worked on the distribution of novae in our Galaxy. It was observed that most of the novae were concentrated in Sagittarius region of the Galaxy. Hubble discovered many novae in M31. He also found a few novae in M33. Nine novae were found in the Megallanic Clouds by Heinze, Hoffleit and Nail. Novae in M81 were observed by Humason and Sandage. Payne-Gaposchkin gave a thorough investigation of the nature of light curves, period-luminosity relation and many other physical aspects of the novae. More novae have been observed in the subsequent years.

9.2 DISTRIBUTION OF NOVAE IN OUR GALAXY

Besides Zinner and Lundmark, Hubble, Kopylov, McLaughlin and others also have studied the distribution of novae in our Galaxy. It was observed that novae are, in general, concentrated near the galactic plane as well as near the galactic centre. A brief account of the longitudinal and latitudinal distribution of galactic novae is given below.

The study reveals that 63 per cent of known novae are situated in the direction of the galactic centre, 44 per cent are scattered in the range where longitude varies from 320° to 340°. The minimum number of novae are found in the neighbourhood of 100° longitude. It may be stated from statistical data, that slower novae lie near the galactic nucleus and fast novae are away from it. Let us next consider the distribution of novae in latitudes. Nearly 87 per cent of novae (including 84 per cent of slow novae and 78 per cent of fast ones) lie within 500 parsecs from the galactic plane. Hence their latitude varies generally from –10° to +10°.

Some novae, as for example, T Cr B, RR Pic, DQ Her etc. are very luminious. Since they are quite close to the Sun, their latitudes exceed 10°, but they are not at great distances from the plane of the Galaxy. There are however several novae whose distances from the galactic plane are greater than 5 kpc. Such novae are however very rare. VY Aqr, W Ari etc. are novae observed in the halo of the Galaxy. Thus, in spite of a few deviations, most of the novae lie not too far from the galactic plane.

9.3 DETERMINATION OF DISTANCE AND LUMINOSITY OF NOVAE

Novae are highly luminous objects, their absolute magnitudes being higher than –7 as deduced by Lundmark. The following methods are adopted to determine the distances of Novae:

1. The trigonometric parallax;
2. The rate of expansions of nebular disc;
3. Intensities of interstellar Ca lines in the spectrum of the nova concerned;
4. Measurement of galactic rotational effect on the radial velocities of these calcium lines.

Novae are so distant objects that we cannot measure their distances accurately by applying the method of trigonometric parallax. Thus, the first method is not very useful. However the method has been used to calculate the distances of a few novae.

The most accurate method of measuring the distance of a nova is by considering the angular rate of expansion of the nebulae around them. Here the measurement is based on the assumption that the expansion is spherically symmetric. This assumption probably involves some uncertainty in the computed results.

The third method is to consider the intensities of calcium lines in the nova spectrum as indicating the distances to them. This is possible because intensities of these lines increase steadily with distance. This method certainly yields a rough estimate of distances to novae. The same is true for the method based on the galactic rotational effect on the radial velocities of interstellar Ca II lines in the spectra of novae. Knowing distances of the individual novae by different methods and their apparent magnitudes, we can determine their mean absolute magnitudes. The value however, varies slightly with different methods. The mean absolute magnitudes yielded by the nebular expansion method at maximum and minimum are respectively –8 and +5.2; the interstellar Ca II line intensities give the corresponding quantities as –7.6 and +5.0; the mean absolute magnitudes obtained from galactic rotation method is –7.6 at maximum. It is not detected at minimum. Trigonometric parallaxes give $M_{max} = -7.3$ and $M_{min} = 4.4$. All these methods are far from high accuracy since errors are introduced due to interstellar absorption whose effect also depends on distances. It is difficult to take accurately this effect into account.

9.4 LIGHT VARIATION OF NOVAE

It is very interesting to study the light curve of a nova from its beginning to the postnova stage (Fig. 9.1). McLaughlin has divided a nova light curve into nine parts. Following McLaughlin, we discuss these different stages of light variation.

(a) *Pre-nova stage.* Here the luminosity of the star remains constant for many years. Sometimes their light fluctuates in a small range through 1 or 2 magnitudes.

(b) *Initial rise.* In general, most of the novae brighten from the minimum upto 2 magnitudes below maximum very rapidly. This phenomenon is known as nova outburst. Novae remain in this state for a short interval, not more than 2 or 3 days, even in the case of slow novae.

(c) *Premaximum halt.* After the initial rise, when novae are nearly 2 magnitudes below the maximum, their brightness remains unchanged for a few hours in the case of fast novae. In the case of slow novae, this steady state of luminosity continues for several days.

(d) *Rise to maximum.* At this stage the star attains its maximum brightness in one day for fast novae. Slow novae take several weeks to rise to maximum.

(e) *Maximum.* Fast novae remain at maximum for the period not exceeding one day. This period may be upto a few days for slow novae.

(f) *Decline after maximum.* The descending light curves of fast novae are rather smooth and flat. They fall through 3 of 4 magnitudes below the maximum steadily. The light variation curves of slow novae oscillate while declining from maximum.

(g) *Transition.* After descending through 3 or 4 magnitudes, light curves of novae begin to oscillate, which ceases after attaining luminosity 6 to 7 magnitudes below the

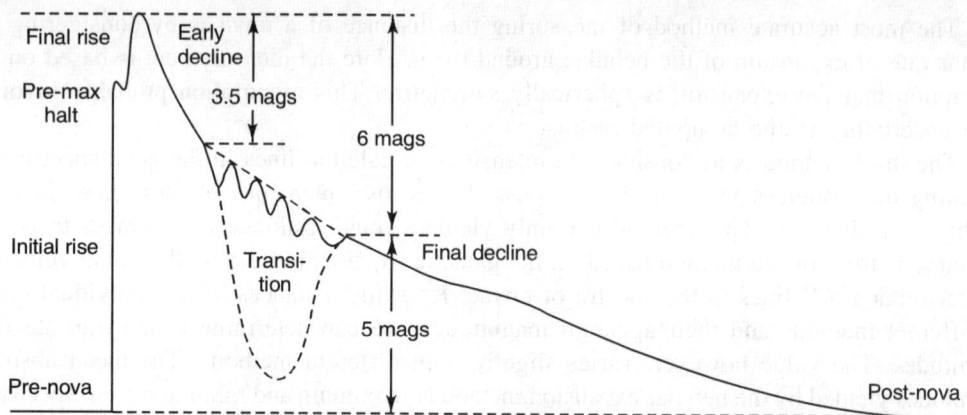

FIGURE 9.1 Schematic light curve of a Nova.

maximum. This oscillation continues for several weeks. Slow novae show prominent oscillations in their light curves.

(h) *Final decline.* After the transitional stage, there is a slow and steady fall of luminosity until reaching the minimum.

(i) *Post-nova.* After the final decline the stars attain the post-nova stage. At this stage some of them may remain constant, while some others may vary by 1 or 2 magnitudes. One must notice the differences in the nature of light curves of fast and slow novae while descending after maximum. The descending curves of fast novae are smooth. Slow novae however tend to oscillate in the early stage of decline.

The novae may be divided into three classes on the basis of the nature of their light variations.

1. Novae whose light curves remain constant after reaching minimum: V 603 Aql, T Aur, V 476 Cyg, DN Gem belong to this class.
2. Novae which oscillate slowly in their post-nova stage: DQ Her, T Cr B etc. are examples of this type of novae.
3. Novae having rapid variation of light after attaining minimum: Q Cyg, V 841 Oph, GK Per, etc. are members of this group.

This variability of the light curve of a nova after reaching minimum is most probably independent of any of its physical features. It does not depend on whether the nova is slow or fast. It is not related to its spectral property. As an illustration, let us notice that GK Per is variable in nature and contains strong emission lines in its spectrum, but V 841 Oph varies very rapidly and has continuous spectrum. So no correlation between the nature of light curve of a nova in its post-nova stage and its physical features can be established.

There is a striking relationship between *the luminosity of a nova and its rate of fading.* Luminosity of novae has been studied by many astronomers, like E.P. Hubble, H.C. Arp, M. Schmidt, C. Payne-Gaposchkin and others. The results obtained were almost similar and from these a linear relation between the absolute magnitude of a nova at maximum and its rate of fading could be deduced. This is given by

$$M_0 = -11.5 + 2.5 \log t \qquad (9.2)$$

where

M_0 = absolute magnitude of a nova at maximum, and

t = time (in days) taken by a nova to fade from maximum to 3 visual magnitudes below maximum.

Thus, the study of the light variation curve of a nova, in the first place, gives us information of its luminosity variation at nine different stages. Secondly, we can find the characteristic difference of the descending curves of fast and slow novae while declining from the maximum. Thirdly, we are informed that the light curve of a nova in its post-nova stage is not related to any of its physical characteristics. Finally, we can derive a relation between the luminosity of a nova at maximum with its speed of decline.

9.5 SPECTRA OF NOVAE

The study of a nova spectrum is very important because this yields clue to the physical cause of nova outburst. Much work has been done in this field by McLaughlin, Aller, Baker, Baade and other astronomers who have left a huge amount of observational information about the nova spectrum. Bowen, Grotrian, Swings, Schwarzschild and others helped in theoretical interpretation of the observational features. We shall now discuss observational aspects of nova spectrum and interpretation of nova phenomenon from such spectrum.

During the observation of a nova spectrum, the following sequences of phenomena can be recorded.

No nova has been spectroscopically observed during its initial phase of rapid rise prior to the pre-maximal halt.

When rising to maximum, the nova displays an early-type absorption spectrum in which lines are strongly displaced towards shorter wavelengths. This indicates that the envelope of the star is expanding rapidly, though the main stellar body does not expand. At this stage the continuum is weak and bright lines of hydrogen and other metallic elements with various intensities appear. In some novae, such as DQ Her, the bright lines observed are very strong.

Near the maximum, a well defined strong absorption spectrum which is strongly displaced shortward appears to be the principal feature. The initial absorption spectrum changes to later classes indicating that the surface temperature of the star is decreasing. But as the area of the radiating surface expands to a great extent, the luminosity of the star which is proportional to the square of its radius enormously increases. Expanding photosphere ejects hot gases which is indicated by the negative radial velocity of the absorption spectrum. The strong continuous background is due to the hot central star. The gaseous envelope between the star and the observer is the source of absorption lines. The two lobes of the envelope which are seen against the sky background emit bright lines. The bright lines formed at the near-side of the lobes are displaced towards shorter wavelength as the gas in these regions is approaching us, and those formed at the far-side of the lobes which is receding are displaced towards longer wavelength. There is no Doppler shift of bright lines originating across the line-of-sight (Fig. 9.2). The figure illustrates the nature of the Doppler shift in a nova-spectrum.

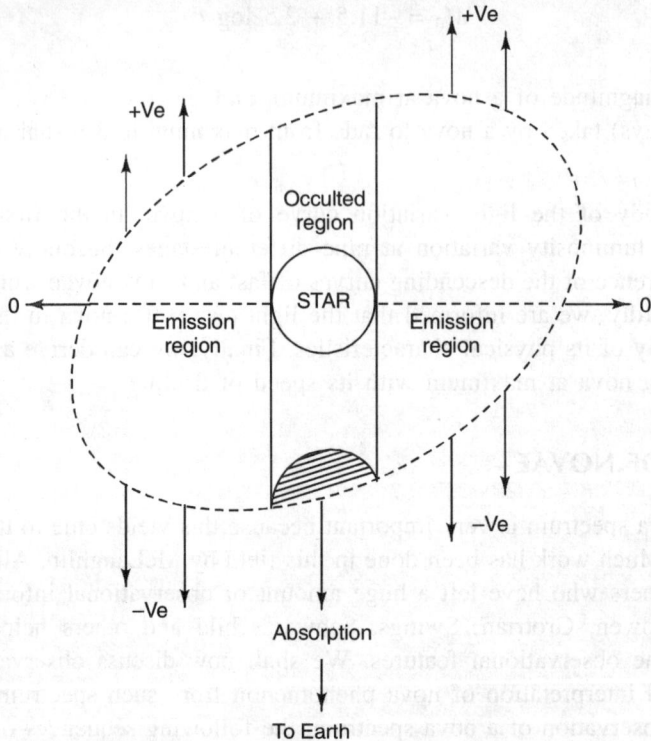

FIGURE 9.2 A schematic diagram explaining the Doppler shift of a Nova spectrum.

When the photosphere expands further, the opacity of the surrounding gases becomes almost insignificant. The continuum of the spectrum brightens to such an extent that the emission features are almost submerged, only the H_α line being occasionally visible.

The spectrum near the maximum has been observed to range from A0 to F8 or even to later classes. This information was provided by observations of several novae, such as GK Per (A0), RR Ric (F8), V 1148 Sgr (K) etc.

The spectrum of a nova when declining from maximum is very complicated. At this stage, a *typical nova spectrum* is revealed which contains bright lines widened symmetrically about their mean wavelength and is associated with dark lines on their shorter wavelength sides. Bright lines of hydrogen and metals and the forbidden lines of neutral oxygen and singly ionised nitrogen are prominent spectral features at this stage. When faded one magnitude below the maximum, a wide and diffuse absorption system is noticed which is recognized as diffuse *enhanced spectrum*. This spectrum is still farther displaced shortward and contains mainly hydrogen lines. Bright lines of Ca II, Na I, Fe II etc. are also found in diffuse enhanced spectrum. At two magnitudes below the maximum, this type of absorption system fades to insignificance and the nova displays an early B-Type absorption spectrum. This is called *Orion spectrum* which mainly contains N II and O II lines. Orion spectrum is very prominent in slow novae. In fast novae it is very difficult to distinguish Orion spectrum from diffuse enhanced spectrum.

When the luminosity fades to four magnitudes below maximum, the continuous spectrum fades more rapidly than bright lines and the nature of the bright line spectrum changes simultaneously. Bright lines of Ca II and Na I fade but those of He I, N III brighten. The forbidden lines of [O III] and [Ne III] are also observed.

It has been already mentioned that the light curves of slow novae and the novae of moderate rate oscillate when declining from 4 magnitudes below the maximum to further 3 magnitudes. This phenomenon is rather rare for fast novae. Oscillation curves of DN Gem and V 356 Aql were studied by Stratton who observed the following characteristics:

1. At minimum subsequent to the first maximum both diffuse enhanced and orion absorption lines are present.
2. At the second maximum orion absorption lines almost fade away while diffuse enhanced lines become most prominent.
3. Declining of light to second minimum is accompanied by the decay of diffuse enhanced lines and re-appearance of stronger orion lines. These latter lines again fade away at the epoch when the minimum is attained.

At this stage, emission lines also show some peculiarity. He I and Na I become prominent at minimum light. When luminosity rises, Na I remains strong whereas He I lines practically disappear. During the subsequent decline, the phenomena of He I and Na I lines are seen to reverse. Just at the time of oscillation, a peculiar phenomenon takes place which is known as *nitrogen flaring*. Wide and semi-luminous flaring bands of emission lines of N III are observed at wavelengths λ 4100 and λ 4640. These bands are bordered on the shortward side by strongly displaced absorption lines of N III. Nitrogen flaring takes place more or less in all the novae but is very important in slow novae. An increase in the intensity of N III accompanied by a decrease in that of [O III] and the great width of N III lines corresponding to a velocity of 3200 km s^{-1} has been suggested to be the principal factors behind the phenomenon of nitrogen flaring.

When the light is 7 magnitudes below the maximum, fluctuation ends and emission of [O III] becomes prominent. The spectrum is nearly the same as that of gaseous emission nebula.

When the star attains minimum light, the gaseous envelope is almost cleared away and the spectrum is a strong continuum which is brighter in the region of shorter wavelength, since the central star is still hot. Some novae, of course, still contain spectrum of weak lines of hydrogen and helium.

It should be mentioned here that the total mass ejected by the central star is negligible in comparison with the mass of the star itself.

9.6 CAUSE OF NOVA OUTBURST

For many years astronomers devoted their attention to solve the mystery of nova outburst. It was at first believed that collision of stars is the cause of nova outburst. But calculations reveal that such a collision is a rare event. It occurs once in a billion years in our Galaxy, whereas the nova outburst is a regular event which takes place quite frequently. Again, tremendous energy should be liberated as a result of a collision, but in case of nova outburst the energy liberated is much smaller than should be in a collision. Hence collision hypothesis is not satisfactory.

Another school of opinion suggested that nova outburst is caused when a star is pushed into a dark nebula. This will lead to a rise in temperature with increase in luminosity. However, the reverse is observed, it is found that the photosphere of a nova is coolest when the star brightens with a great brilliance.

It was Martin Schwarzschild who gave the proper explanation of nova outburst. We know that the nuclear processes at the centre of a star continuously convert hydrogen into helium and thus liberate energy in the interior of the star. A huge amount of helium is concentrated at the core in the process. When the helium concentration exceeds a critical limit, a huge amount of energy begins to be liberated from the stellar interior generating shock wave which propagates into the stellar surface shooting materials outwards. At this stage, the phenomenon is observed as a nova outburst.

Many astronomers, viz, W. Grotrian, E.R. Mustel, Z. Kopal and others investigated the nature, velocity and other physical features of the material ejected as a result of a nova outburst. According to Grotrian's view, the gas is ejected from the surface of the star as the photosphere expands, obeying the following relation

$$r_p = \text{const.} \sqrt{\frac{n}{v}} \qquad (9.3)$$

where n is the number of atoms ejected per square centimetre per second from the surface, r_p is the radius of the star upto the photosphere at the instant when the ejection velocity is v. It follows that the expansion of the photosphere can be realized either by the increase in the rate of ejection of atoms or by a decrease of the ejection velocity.

The view of Mustel is quite different. He believes that during the outburst the entire star expands until the maximum light is attained. At this stage, the outer layer is blown off in the form of shells. Then the body of the star contracts slowly but the ejection of shells continues for a longer time. Mustel's view however bears an apparent contradiction. We know that an enormous gravitational pull is necessary for contraction of a star. In order to supply such energy, the mass of the star should be of the order of hundred times that of the Sun. But there are several novae whose masses have been measured to be not very large. To overcome this difficulty Mustel suggested that if the contraction is caused by the effect of magnetic force, then large masses are not essential for novae.

Z. Kopal considers the expanding photosphere as the luminous shock-front moving through gaseous envelope of the star. Kopal's view is almost in agreement with that of Schwarzschild.

9.7 SUPERNOVAE

The evolution of stars is generally a slow and peaceful phenomenon. The centre of the star accumulates heat from gravitational contraction and when the central temperature is sufficient to start hydrogen burning process, the star becomes stable as a main sequence star. It remains in the main sequence for a long period and when hydrogen is exhausted at the core forming an inert helium core, the star leaves the main sequence. The evolution continues and the star passes from its dwarf stage to giant stage. It gradually burns all its fuel—both hydrogen and helium—and ultimately reduces to a white dwarf, occupying the lower left corner of the H-R diagram (see also Chapter 14).

But the fate of all stars is not the same. There are a few stars whose peaceful evolution is disturbed by sudden and violent outburst. These are called *exploding stars* or *supernovae*. Any star whose mass exceeds Chandrasekhar limit of 1.44 $M_\odot$ may have a chance to explode as a supernova. These massive stars experience several chemical changes in its innermost core. At first, hydrogen in the core is converted into helium. Subsequently, the core contracts; the temperature rises and the helium in the core is changed into carbon, oxygen and neon. Further contraction and increase in temperature in the core transform carbon group of materials in the core first into silicon group of metals and then into metals of iron group. At this stage, a peculiar change takes place; iron group of metals are suddenly changed back into helium and some neutrons are set free. This change requires huge amount of energy which is supplied by the severe gravitational contraction. The core hardly can bear such excessive contraction which therefore collapses. At the same time, the outer layer becomes so hot that thermonuclear reaction starts in that region liberating huge amount of energy. This energy cannot be radiated away as quickly as is necessary to hold the structure of the star. As a result, the star may suffer a violent explosion. The total energy liberated in a supernova explosion in nearly of the order of 10^{50} ergs. Supernovae are extraordinarily bright and dazzling objects having −17 as their average absolute magnitude. Some of them were reported to be visible even in the broad day light.

Long years of observation has enabled astronomers to distinguish between two types of supernovae, viz., Type I and Type II. Information gathered about them over long years of study are briefly described in the following (see also Table 9.1).

TABLE 9.1 The Characteristics of Two Types of Supernovae

Characteristics	Type I supernova	Type II supernova
Absolute photographic magnitudes at maximum	−18.9	−17.5
Energy liberated (in ergs)	10^{50}	3×10^{50}
Mass ejected (in $M_\odot$)	$10^{-1} - 1$	3–5 (Sometimes more)
Expansion velocity (km s^{-1})	~ 2,000	6,000–10,000
Number of supernovae per year in a galaxy	1/100	1/100
Population	II	I
Distribution	In spheroidal galaxies	In the spiral arms of galaxies
Light curve	Regular	Irregular
Hydrogen contents	Hydrogen-poor objects	Hydrogen in normal abundance
Probable mass (in $M_\odot$)	1.5–5	15 to 30

Masses of the Type I supernovae are generally small, of the order of a few solar mass. But Type II supernovae originate from massive stars. They may be as massive as 30 $M_\odot$ or more. A Type I supernova may be brighter than the combined brightness of all other stars in a Galaxy. The Type II supernovae are fainter by about 1.5 magnitudes. But the energy liberated in Type I is less than that in Type II. Type I supernovae look brighter since a larger fraction of the energy liberated in them is in the range of visible light. Type II supernovae on the other hand, emit a large part of their energy in material motion.

Type II supernovae eject more mass at much higher velocity. They eject almost 3 $M_\odot$ on the average, with velocity as high as 6000 km s^{-1} or even more. More recent calculations have shown that the mass ejected during a Type II supernova explosion in which a pulsar is formed may be larger than 5 $M_\odot$ and may be as large as 40 $M_\odot$. Type II supernovae originate from massive young stars, possibly of extreme Population I, whereas the Type I supernovae are supposed to originate from evolved Population II stars. Type I supernovae are generally observed in spheroidal galaxies and are not associated with spiral arms. On the contrary, supernovae of Type II are associated with spiral arms of galaxies and are almost absent in the spheroidal galaxies.

Hoyle and Fowler gave the main characteristic differences between the supernovae of the two types. For the Type II supernovae, because of their huge mass, explosion takes place when stars are almost near the main sequence in the H-R diagram. The core material of these stars remain non-degenerate. So hydrogen abundance is a characteristic property of a Type II supernova. At the time of explosion only a small fraction of hydrogen is consumed in the core of the star. For a Type I supernova the explosion takes place in the latest stage of its evolution. Hydrogen in the core is exhausted and the core material is in a degenerate state. So they are hydrogen-poor objects.

Light curves of Type I supernovae are rather regular in shape. It is rather difficult to obtain light-curves of Type II supernovae. Generally, they explode in such regions of our Galaxy where their light is obscured by about 5 to 8 magnitudes by interstellar dust particles when they are observed from the Earth. Light curves of Type II supernovae are irregular in comparison to those of Type I.

Various works on the spectra of Type I supernovae provide us reasonably good information about them. The spectrum consists of broad featureless emission bands. Hydrogen lines are not practically observed. Lines of He, ionized Na and Ca are also absent in the spectrum of Type I supernovae though forbidden lines of [Ca V] at λ 6086 Å and λ 5309 Å are sometimes seen.

The intensity of red bands ($\lambda > 5000$ Å) changes from day to day. The intensity of the bands at λ 6500 Å and λ 5865 Å attains great strengths 40 days after the light curve attains the peak. It declines afterwards and minimum intensity is attained in about 120 days. After that the intensity curve rises to its initial value in about another 20 days.

The development of the nature of blue bands proceeds in a quite different way. Changes in intensity as occurring in the red bands are not observed in blue bands ($\lambda < 5000$ Å) of the spectrum. The strongest blue band is observed at λ 4575 Å while weaker bands at λ 4300 Å, λ 4800 Å and λ 4900 Å are noticed. The most interesting feature of blue bands is that they shift in wavelengths towards red. Such a shift is not observed in red bands.

The nature of the spectrum of a Type II supernova is completely different from that of a Type I supernova. Before maximum light, the spectrum of a Type II supernova is characteristic of that with a very high temperature. Continuous spectrum is revealed after maximum light.

Normal abundance of hydrogen is a principal feature of the spectra of Type II supernovae. Lines of Balmer series can be identified. Other prominent lines observed are those of N III, Fe II, Fe VI and Mn II to Mn VI.

9.8   SN 1987 A

On the night of February 23, 1987, Ian Shelton of the University of Toronto, who had been working then at the Las Companas Observatory in Chile, took a long exposure photograph of the Large Magellanic Cloud with a small 10-inch astrograph. To his extreme surprise Shelton found that a new star had appeared since the previous night at the tip of the "Tarantula Nebula" which is a bright part of the Cloud. The event recorded for the first time in man's history that scientists could carefully observe and study a supernova when it was in explosion (really speaking, the star exploded some 160,000 years ago at the source but the signal of that explosion arrived on the Earth on 23 February, 1987). This was the supernova 1987A in the Large Magellanic Cloud, the first one observed in the year 1987.

Careful search in the field of explosion by groups of astronomers revealed that the exploded star was SK-69 202, a blue supergiant star in Nicholas Sanduleak's Catalogue of bright stars in the LMC. Spectra of the supernova showed strong lines of hydrogen Doppler shifted with speeds upto 40,000 km s^{-1}, indicating that the blast was of Type II. That a blue supergiant star would explode as a Type II supernova was a puzzle to the theoreticians because the progenitors of Type II supernovae were believed so far to be red supergiants. It may be that the star really was a red giant sometime past, but by the time the explosion occurred the surface layers of the red giant were ejected in space showing the star bare as a blue supergiant. If initially its mass was ~ 20 $M_\odot$, its mass at explosion was ~ 15 $M_\odot$, losing about 5 $M_\odot$ during the period of interim evolution.

Initially a star of magnitude 12 in Sanduleak's Catalogue, its brilliance rose swiftly to magnitude 4.2 within a week, and then rose more slowly to magnitude 2.9 in mid-May which was its brightest. SN 1987A was observed to be about two magnitude fainter than other Type II supernovae. This may be due to the fact that its progenitor being a blue supergiant instead of a red supergiant and so much smaller in size, more energy than normal had to be used in pushing the vast amount of debris away into the space. This will necessarily leave lesser energy for optical brilliance. The light curve of the object has been accurately traced for many months. The curve revealed that the light was falling from mid-June and the subsequent fall was exponential in nature. One of the most spectacular achievements of the astronomers this century was to detect the neutrinos released by the explosion of the SN 1987A. Supernovae were long believed by the physicists to be prolific sources of neutrinos. This was put to test and verified with positive results in the case of SN 1987A. The two largest neutrino detectors in the world—the Kamiokande II operated by a team of Japanese scientists and the IMB operated by a group of American physicists—were put to use for detection of neutrinos. Analyzing the data carefully in these two detectors, scientists came to the definite conclusion that nearly a day before Shelton's sighting of the big event, 11 neutrinos left their record in Kamiokande II, while in the IMB 8 neutrions left their marks. These observations were extremely important in nature due to the fact that the interior of an exploding star was seen for the first time. The observation put the theory of core collapse of Type II supernovae on a sound footing.

EXERCISES

1. What do you know of the distribution of novae in our Galaxy? From their distribution can you assign the Population characteristic of novae (see Chapter 11)?

2. What different methods can you apply to find the distance to a nova? Which of these methods appears to you to be the most accurate? Give reasons.

3. Deduce the following relations for a nova:

$$p'' = 4.74 \frac{\text{Angular rate of expansion (sec yr}^{-1})}{\text{Velocity from Doppler effect (km s}^{-1})}$$

and

$$d = \frac{1}{4.74} \cdot \frac{\text{Velocity (km s}^{-1})}{\text{Angular expansion (sec yr}^{-1})}$$

4. The nebular shell of a nova is observed to expand at a rate of 0″.25 per year. The velocity measured by the Doppler effect is 750 km s^{-1}. What is the distance to the nova (i) in parsecs and (ii) in light years. If the apparent magnitude measured at maximum is +4 and the interstellar absorption is 1.2 magnitude, calculate the absolute magnitude of the nova at maximum.

5. The outburst of Nova Aquilae (V603 Aql) occurred in June, 1918, when its apparent magnitude was –1.1 at maximum. Its spectra showed Doppler-shifted absorption lines corresponding to a velocity of 1700 km s^{-1}. In 1926, the star was surrounded by a faint shell of diameter 16″. Find the distance to Nova Aquilae in parsecs and its absolute magnitude at maximum (Assume no interstellar absorption).

6. Describe the main features in the pattern of variation of the light curve during the life cycle of a typical nova. What are the principal differences that can be noticed in slow and fast novae?

7. Trace the pattern of changes in the spectral feature during the life cycle of a typical nova. Discuss, in particular, the nature of lines as observed during the period of decline of the nova.

8. Explain your understanding of the cause of a nova outburst.

9. Discuss briefly the chain of events that causes some stars to explode.

10. What are the observational basis on which supernovae of Type I and Type II are distinguished?

11. Although Type I supernovae are, in general, brighter than Type II, the total energy involved in Type II events is much higher than that in Type I events—Explain.

12. The absolute photographic magnitudes of Type I and Type II supernovae are respectively −19 and −17.5. Compare the luminosities of these objects with that of the Sun.

13. A Type II supernova ejects 5 $M_\odot$ of material with a velocity of 10,000 km s^{-1}. Compute the kinetic energy of the ejected material and compare this energy with the total energy to be radiated by the Sun during its entire lifetime (10 billion years) at the present rate.

14. If the limiting magnitude of the currently available best telescope systems is 25, obtain the distances measurable by observing:

    (a) an ordinary nova with absolute magnitude −9;
    (b) a Type II supernova; and
    (c) a Type I supernova (Ignore the interstellar absorption).

15. Ignoring the interstellar absorption, calculate the apparent magnitude of a Type II supernova, if its distance be 1000 pc.

16. Ignoring the interstellar absorption, find at what distance a Type I supernova will be detectable with the unaided eye having a limiting magnitude of 6.

17. Discuss some of the important astrophysical information that were obtained by the event of 1987A supernova.

18. If the distance of the Large Magellanic Cloud, the seat of the 1987A supernova, be 60 kiloparsecs, find in which year B.C. the star exploded.

SUGGESTED READING

1. Birney, D. Scott, *Observational Astronomy,* Cambridge University Press, Cambridge, 1991.

2. Goldsmith, Donald, *Supernovae,* Oxford University Press, Oxford, 1989.

3. Marschall, L.A., *The Supernova Story,* Plenum Press, New York, 1988.

4. McCray, Richard (Ed.), *Supernovae and Supernova Remnants,* Cambridge University Press, Cambridge, 1994.

5. McLaughlin, Dean B., *Introduction to Astronomy,* Houghton Mifflin Company, Boston, 1961.

6. Murdin, Paul and Murdin, Lesley, *Supernovae,* Cambridge University Press, Cambridge, 1985.

7. Payne-Gaposchkin, C., *The Galactic Novae,* Dover Publication Inc., New York, 1964.

8. Shklovskii, I.S., *Supernovae,* John Wiley & Sons, New York, 1968.

9. Smith, Elske V.P. and Jacobs, K.C., *Introductory Astronomy and Astrophysics,* W.B. Saunders Company, Philadelphia, 1973.

More Stars of Interest

Section A   Stars with Extended Atmospheres

10.1   THE WOLF–RAYET STARS

A very rare type of objects in the Galaxy, the Wolf–Rayet stars bear the name of their discoverers—the French astronomers C. Wolf and G. Rayet who discovered three of them in 1867. Till to-day, only about 200 of them are known, about two-thirds of which are in the Southern Hemisphere. We shall refer to these as W stars in what follows.

The W stars are one of the several types of emission-line objects. Other emission-line objects are novae and supernovae, planetary nebulae, P-Cygni, Of and Be stars. These latter stars as well as W stars are all early-type objects. But the spectral features of W stars are so radically different from those of other emission-line objects that the former are easily recognized.

The most distinct feature of W spectra is the presence of very wide intense and numerous emission lines. The emission bands are in general, 50 to 100 Å wide with intensity 10 to 20 times that of the background continuum. So numerous are the emission lines that they contaminate the continuum almost over the entire visible region of the spectrum. Sometimes the emission bands are bordered by absorption bands on the violet edge. If Doppler broadening of the lines is assumed, the widths of bands correspond to ejection velocity as high as 3000 km s^{-1}.

A very unusual characteristic of W stars lies in the fact that they appear to be almost equally divided into two parallel sequences, WC and WN, the former showing predominantly strong and wide lines of carbon while the latter containing similar lines of nitrogen. Because of this predominance of two different elements exclusively of each other and because of the great range of ionization exhibited by individual stars, and their superficial similarity with other emission-line objects, great confusion prevailed earlier in classifying the W stars. The classification, however, of WC s into WC 6 to WC 8 and of WN s into WN 5 to WN 8 was devised by C.S. Beals which was adopted in the IAU meeting in 1938. The classification was mainly based on the intensity ratios of lines in successive stages of ionization. As a result of more intensive study this classification scheme has been modified during the subsequent years. At present, the adopted subdivisions of the two sequences are WN 3 to WN 8 for the nitrogen sequence and WC 5 to WC 9 for the carbon sequence. There are, however, a few characteristics which are usually common to all W stars. One such is the presence of strong broad emission of He II 4686 line with adjacent blend of N III in WN s and of C III or C IV in WC s. Absorption components are common on violet side of the lines of He I or else, of the strongest lines in the spectra.

The determination of luminosities of W stars is inherently difficult for the facts that no W stars lie within the galactic clusters and no accurate methods such as the trigonometric or spectroscopic parallax can be applied in this case. However, attack has been made on the problem from all possible directions with the result that a wide range values for luminosities of W stars has been derived. The results suggest that the W stars do really possess large scatter in intrinsic luminosity. A very convincing result comes from study of the W stars in the Large Magellanic Cloud. As a class, the average absolute magnitude of W stars is about $M = -5.0$. So they are highly luminous objects.

The temperature of W stars is also very high although large scatter seems to be present here also. Different methods used for the determination of temperatures have yielded different values, usually ranging between 30,000 to 100,000 K. Among others, H. Zanstra's method (Chapter 12) has been most extensively used for the purpose. The method of colour temperature T_c, becomes unreliable for W stars because almost every region of the continuum is highly contaminated by emission lines. Nevertheless, attempts have been made by various authors to determine the colour temperatures of these stars which have been found systematically much lower than those obtained by Zanstra's method. The colour temperatures obtained by various authors usually range from 10,000 to 30,000 K. These values, however, are unreliable for the reason we have already mentioned. The Wolf-Rayet stars are believed to be the hottest of all stars; and among these the WN stars as a class are somewhat hotter than the WC stars.

The fact that a large percentage of Wolf-Rayet stars are members of binary systems has presented great advantage for the determination of masses of these objects. The companion stars are generally O–B stars and, on the average, the mass of the W star is found to be about 1/3 to 1/4 of that of its O–B companion. Since the masses of the latter stars are known to be in the range 30 to 40 $M_\odot$, the average mass of a W star can be taken as of the order of 10 $M_\odot$. The radius R of a W star is usually in the range 2–3 $R_\odot$ and the size of the envelope usually satisfies $R_{\mathrm{env}} \not> 5\ R_\odot$. Since the envelopes of these stars are not very much extended, the radiation can never get very much diluted. Also, the electron density of the envelope is quite high, of the order of 10^{12} cm^{-3}. The absence of large dilution of radiation and the presence of high matter density there prevent the formation of forbidden lines in the envelopes of the W stars (Chapter 12).

The Wolf-Rayet stars are believed to be young objects of extreme Population I, although some of these having planetary nuclei may be older objects. Bart J. Bok described these stars as primary spiral tracers in M 31 and possibly also in our Galaxy. Bok estimated their age to lie between 10^5 to 10^6 years. The extremely young age of these stars is also supported by the fact that they are mostly distributed along and within a band of $\pm 5°$ of the galactic plane and that a large percentage of these is associated with stellar associations (Chapter 11).

The very wide emission bands bordered on the violet side by absorption clearly indicate large scale ejection of stellar matter with high velocities. The continuous ejection of matter creates the envelope and maintains it by replenishing the lost mass by newly ejected matter. In this manner, a permanent envelope is maintained around a W star, although mass is being continuously lost from the superficial layers of the envelope. The rate of mass loss has been variously estimated by authors using different values for the quantities involved in the formula

$$\frac{dM}{dt} = 4\pi r^2 \rho(r) \, v(r) \tag{10.1}$$

where r is the radius of the photosphere. V.I. Ambartsumyan gives a loss rate $\sim 10^{-5} \, M_\odot \, \text{yr}^{-1}$ which has been later supported by other authors. Such a high rate of mass loss has profound influence on the evolution of a W star. It further implies that the Wolf-Rayet phenomena cannot but be short-lived. Such a high loss rate cannot continue for a long time. This also explains the small number of W stars observed.

Quite a number of problems in W stars still await satisfactory solution. The most important of these is the presence of two parallel sequences with lines of carbon and nitrogen predominating. Whether this observed chemical difference is real or whether a fictitious difference is projected as real by some yet unknown excitation mechanisms prevailing in the two series, still remains a subject of controversy.

An extensive study of the distribution of W stars by M.S. Roberts has revealed that an overwhelming majority of these stars belonging to clusters and associations are of WN type; while among the field W stars the WC stars have the overwhelming majority. This has led Roberts to suggest that if we assume all W stars to have been formed in clusters and associations, it seems reasonable to conclude that the WN stars evolve into WC stars in a time-scale comparable to that for the evaporation of a cluster or an association in the general field. This view finds support in the observation of Nova Aquila (1918) and Nova RR Pictoris (1925) in which N III 4634–40 emission has been found to be replaced by C III 4650 emission over a period of 30 years. Over this period, therefore, the apparent abundance has been found to change in these novae. The same may happen in W stars over a much longer period of time owing to much milder conditions in them than in novae.

The most plausible explanation of this apparent abundance anomaly is believed to be the *selective fluorescence* as proposed by I.S. Bowen. If this mechanism is operating in the envelope, then a line may be radiated not only at the expense of the radiation from the star beyond the limit of the principal series of the atoms concerned, but also at the expense of radiation of the envelope in a different line. This will be more effective in the presence of a velocity gradient in the emitting layer. Another yet unsolved problem is the appearance of absorption components in certain lines but not in others. Absorption components generally appear in He I lines and some other very strong lines. Among these the lines of He I 3889 and N IV 3483 have lower levels metastable which helps the absorption components to appear in them. But some C III and N V lines arising from ordinary levels also have absorption components. The absorption on the violet side is best explained by the expanding model of the envelope. An equally unsolved problem is the asymmetry in the emission band profiles of some W stars. Although occultation effect has been suggested as a plausible explanation, but in that case all profiles should be asymmetric which is not observed.

Another interesting problem which still promises much scope for theoretical work is the variation in intensity of lines of some W stars which are the components of eclipsing binary systems. The eclipsing system HD 214419 (CQ Cep) was studied by W.A. Hiltner, by M.K.V. Bappu and S.D. Sinvhal and by others. Hiltner found that the intensity of He II 4686 line reaches a maximum during the primary and secondary minima and minimum at elongations. Similar behaviour was observed by Bappu and Sinvhal by the lines He II 6560, He I 5875, He II 5411 and N IV 4058 of the same system. They all show increased intensity at minima and lower intensity at elongations. Also, intense violet edges are developed immediately after primary and secondary minima. All He I and He II lines show this feature. On certain nights the emission intensity becomes much brighter above the average. This suggests that on such occasions possibly a general overall change in the excitation mechanism takes place. Bappu has suggested that the increase in intensity of emission lines at minima might be explained by a common envelope model. A common envelope may not be unrealistic in such an extremely close pair ($P = 1.64$ days).

The eclipsing binary system V 444 Cygni has been studied by many authors. The variation in intensity of lines has been studied among others by L.V. Kuhi. Kuhi observed that during the secondary minimum (O star in front of the W star) the intensity of all lines of He I and He II and of N III and N IV increases by 0.1 magnitude, but that of C IV 5808 line decreases by the same amount. The author suggests that the increase in intensity is probably due to heating effect. In secondary eclipse, the heated portion of gas is nearer to us, so there should be an increase in intensity. Other explanations are also possible. It may be asserted therefore, that there are still many unsolved problems in the Wolf-Rayet phenomena whose solution awaits more observational and theoretical works.

10.2 P CYGNI AND *A* CYGNI STARS

So far as the degree of excitation and ionization of atoms are concerned, the Wolf-Rayet stars appear to be more akin to the spectral class O. The prominent emission bands observed are those of the Balmer series of H and of He I, He II, C II to C IV, O III to O VI and N II to N V. Stars of P Cygni type (of which P Cygni itself is the prototype) have similar spectral features but the spectra in this case correspond to lower excitation and ionization compared to those of W stars. The prominent lines in P Cygni spectra are those of H, He I, N II, N III and Si III in emission. But the lines are much narrower in this case. While the widths of lines in W spectra correspond to ejection velocities ranging from 500 to 3000 km s^{-1}, those in P Cygni spectra correspond to velocities ranging between 100 to 300 km s^{-1} only. This suggests that the conditions prevailing in the atmospheres of P Cygni are very much milder compared to those in W stars. Spectroscopically, P Cygni stars belong to the spectral class early B supergiant.

The emission profiles of P Cygni are similar to but narrower than those of W stars. On the other hand, the absorption lines in P Cygni are much more intense. The absorption bands on the violet border of emission are also much more intense in these stars. The typical emission profiles of a P Cygni and a W star are shown schematically in Fig. 10.1. Any relatively sharp emission line bordered by absorption cores on the violet end is called by the general name as a P Cygni line whose profile will resemble that in Fig. 10.1(a).

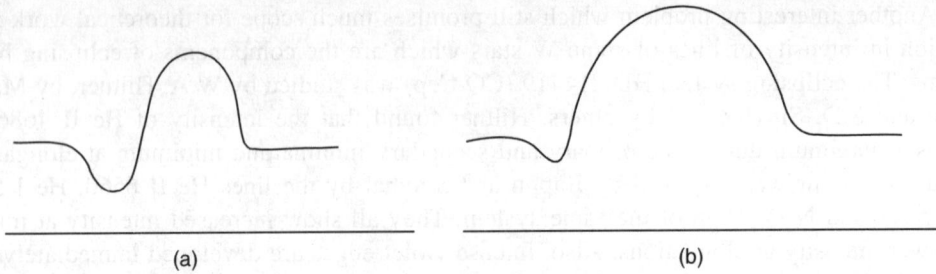

(a)   (b)

FIGURE 10.1   (a) Schematic emission profile of P Cygni star. Narrow and fainter emission bordered on the violet by strong absorption; (b) Schematic emission profile of a W star. Wider and stronger emission bordered on the violet by a weak absorption.

The P Cygni profile clearly indicates that these stars like W stars, are also surrounded by expanding envelopes of moderate density. The mass ejected by the star, according to Ambartsumyan, is $\sim 10^{-5}$ $M_\odot$ yr^{-1} like W stars. This implies that P Cygni phenomena are also short-lived like Wolf-Rayet phenomena.

α Cygni stars belong to spectral class A and have spectral features similar to those of P Cygni but of lower excitation. Spectra of the former contain H_α in emission and ionized metal lines in absorption. Lines are sharper in this case. All these characteristics suggest that α Cygni stars also possess expanding envelopes under much milder conditions.

As early as in 1940, C.S. Beals made a detailed study of the spectra of W, P Cygni and α Cygni stars and came to the conclusion that these stars form a sequence. They represent successive stages of a physical sequence differing only in temperature. Their emission profiles are all similar and form a progression of decreasing widths. The Wolf-Rayet stars correspond to O-type or earlier, P Cygni stars correspond to B-type while α Cygni stars correspond to A-type spectra in respect of excitation and velocity of ejection. A progression in excitation clearly emerges from the following fact. α Cygni have H in emission, He emission is absent. N II and He I are strongest, N III not strong in P Cygni, and He II is absent. In W 8 N II faint, He I weaker, He II moderate and N III is the strongest. So one can possibly draw the following sequence: W5 – W6 – W7 – W8 – P Cygni – α Cygni. Such a sequence, of course, does not explain the observed chemical difference in W stars.

10.3   BE STARS: SHELL STARS

A large number of stars belonging to spectral class B show strong emission lines of hydrogen, superposed on much broader and shallower absorption counterparts. Besides hydrogen, lines of some other elements also are occasionally seen in emission in some of these stars. The emission component is single in some cases, in others it is double, being flanked by a central absorption as shown schematically in Fig. 10.2. In some stars with single emission, the underlying absorption is also quite sharp. The stars having spectral lines of the type as in Fig. 10.2 are called B *emission* or Be stars. The spectra of these stars consist of the normal B

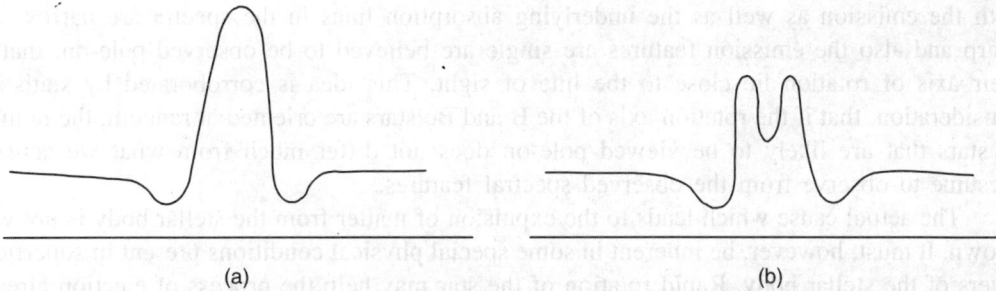

(a)   (b)

FIGURE 10.2 (a) Single emission component superposed on broad shallow absorption line; (b) Similar emission but flanked by a central sharp absorption line.

absorption spectra with superimposed hydrogen emission. Another sub-class of Be stars, called *shell stars,* have Be spectrum as stated above, with additional series of absorption lines. These latter lines, mainly of hydrogen and singly ionized metals, are sharp and resemble those in the spectrum of A supergiants like α Cygni. Around 4000 stars of spectral class B are currently known in whose spectra emission lines have been observed.

The strength of the hydrogen emission gradually decreases as the series advances. The emission is strongest at H_α and in most stars it fades away beyond recognition in case of some early member of the series. In some stars, however, such as in HD 32343 and II Cam, the Balmer emission can be recognized as far as H 22 which is much beyond even the limit upto which the underlying Balmer absorption series is recognizable. The intensity of emission lines varies greatly from one star to another. Again in any particular star, the emission intensity is found to vary with time. Sometimes the emission lines are found to disappear altogether and the Be spectrum changes to a normal absorption spectrum of a B star. The reverse has also been observed. Also, a normal B spectrum changing to a shell spectrum and vice versa has been observed. So whether a particular B star will appear as a normal B, a Be or a shell star depends on the presence or absence of some particular entity surrounding the star; that is the *amount of material* penetrated by the line of sight through the envelope of the star at the time of observation. If the envelope is thin so as to be unable to produce absorption lines, a Be spectrum will result. We shall observe a shell spectrum if there is enough material in the envelope to produce absorption lines. A normal absorption B-type spectrum will result if the envelope is totally blown off.

Observations of Be phenomena enable us to draw a few reliable conclusions: the presence of emission features suggests that like W and P Cygni stars, Be stars also have extended envelopes, but the amount of material in Be envelope is far too less compared to that in envelopes of W and P Cygni stars. The central star must be in very rapid rotation in order to produce the broad and shallow absorption lines as observed. The linear velocity of rotation of 200–300 km s^{-1} at the surface is very common among these stars. The material ejected in the envelope must also rotate to produce the widths of the emission lines, but the rotation speed in this case must be much slower in order to yield the smaller emission widths as observed. In fact, for the angular momentum to be conserved, the rate of rotation of the ejected material should decrease as it moves away from the parent body. Be stars in which both the emission as well as the underlying absorption lines in the spectra are narrow and sharp and also the emission features are single are believed to be observed pole-on, that is, their axis of rotation lie close to the line of sight. This idea is corroborated by statistical consideration, that if the rotation axis of the B and Be stars are oriented at random, the number of stars that are likely to be viewed pole-on does not differ much from what we actually presume to observe from the observed spectral features.

The actual cause which leads to the expulsion of matter from the stellar body is not well known. It must, however, be inherent in some special physical conditions present in superficial layers of the stellar body. Rapid rotation of the star may help the process of ejection already present, but it cannot be the cause itself. For, if rotation itself was the cause, ejection should be regular and systematic. But that the ejection is not regular is borne by the fact that Be spectrum often undergoes irregular changes. Also, the amount of material in an envelope or even a *shell* at any particular time is not significant to warrant a serious consideration from the point of view of mass loss, unlike the cases of W and P Cygni stars. Assuming a temperature of 12,000 K and an electron density of 10^{13} cm^{-3} in the shell, calculations indicate the mass in a shell to be 10^{-9} $M_\odot$, if the shell thickness be 0.1 R_* and its outer radius be equal to 10 R_*, R_* being the radius of the stellar body. If the outer radius of the shell be equal to 3 R_* and the same thickness is assumed, the computed mass of the shell becomes 10^{-8} $M_\odot$. Both these estimates are made on quite liberal basis. Thus, if we assume that the star loses one shell every year, on the average, then the rate of mass loss from a Be star is 10^{-8} to 10^{-9} $M_\odot$ yr^{-1}. This is several orders of magnitude less than the rate computed for W and P Cygni stars or for red giants and other objects which are known to lose mass. At this rate, the star should lose 1 $M_\odot$ in 10^8 to 10^9 years, a time-scale much larger than the stipulated lifetime of B stars. So mass loss from Be stars does not probably lead to serious evolutionary effects, as is the case for objects losing mass at much higher rates.

As an illustration of how shells are formed and destroyed over a period of a few years we cite the case of *Pleione*, a B5 star in Pleiades which has been most extensively observed by many astronomers. This star is observed nearly edge-on, that is, the observer's line-of-sight passes near the equatorial plane of the star. Spectroscopic observations over the years have confirmed that a shell of Pleione had formed and destroyed over the years 1938–54, a period of 16 years. Prior to 1938, no trace of a shell was observed. Then gradually the shell absorption lines grew in strength from 1938 on. The strength increased to a maximum in 1945 and started fading away subsequently. By 1951, the shell lines became very weak and by 1954 any trace of a shell vanished altogether. These observations confirm that during the year 1938–54, Pleione shed off a shell from its body which was finally diffused into the interstellar space. This fact has since been known as the 1938–54 *shell episode* of Pleione, It is now believed that all Be stars lose material in shells periodically as is observed in Pleione.

10.4 OF STARS

A significant fraction of O stars show emission lines of ions of several elements superposed on the absorption spectra of normal O-type stars. These stars have been designated as Of Stars. The spectra of most Of stars contain prominent emission lines of He II 4686 Å and of N III at 4634 Å, 4640 Å and 4641 Å. Weak emission at H_α and at 5696 Å of C III is usually observed. Also, lines of He I at 5876 Å, Si IV at 4088 Å and 4116 Å, N IV at 4057 Å and N V at 4603 Å and 4619 Å are observed in emission in spectra of some of the stars of very early type. The above statements strictly apply when low dispersion spectra are analyzed. In moderate and high dispersion spectra almost all O-type stars are observed to possess at least some weak emission of N III at λ 4640 Å. In view of this fact, the division of O stars into Of and normal O stars apparently becomes arbitrary. But when we talk of Of Stars, it is understood that we refer to those O-type stars which exhibit *stronger* emission lines and usually of more than one ion.

The observational results stated above definitely suggest that extended envelopes surround Of stars like W, P Cygni and Be-type stars. But Of envelopes appear to be a more permanent feature unlike those in Be stars. Also, the envelopes in Of stars probably do not extend too far from the stellar body to allow the radiation to be highly diluted. In particular, no forbidden lines are observed in any of the stars belonging to the classes, W, P Cygni, Be and Of. Whether the Of stars are more luminous than the normal O stars is not clearly understood. There has been suggestions, of course, that the Of stars are probably evolved O stars, and so more luminous than the latter. The matter, however, is still inconclusive.

Observations show that WN 7 spectra have close similarity with Of spectra. Lines of He II 4686 Å and the N III multiplet at 4640 Å are common. The important differences between the two, on the other hand, are: (a) the emission lines are much too broader in W stars and, (b) the counterpart of the background O absorption spectrum present in Of is absent in W stars. However, the close similarity of emission features of the two has led some astronomers to suggest that Of spectra might be taken as the transition between the pure absorption O spectra and WN spectra. According to this view point, the O stars, Of stars and WN stars appear to form an evolutionary sequence. This view, however, has not been seriously considered by many astronomers.

Section B   Some Cooler Stars of Interest

10.5 PECULIAR A STARS AND METALLIC-LINE A STARS

It has been known since early this century that many stars belonging to the spectral class A possess large-scale spectral peculiarities. More than 10 per cent of the brighter A stars have been observed to possess such abnormal spectra, and on the basis of extensive analyses of their spectra they have been now divided into two separate groups. The principal spectral features of the first group are: (a) normal abundance of hydrogen; (b) underabundance of helium by a very large factor and overabundance of He^3 relative to He^4 by a very large factor; (c) overabundance of certain elements, particularly of rare earth group, by very large factors; (d) overabundance by very large factors of the three elements Li, Be and B, which completely burn out at relatively lower temperatures and therefore are not generally observed in normal stars, (e) underabundance of C and O by large factors in some of these stars; (f) very large overabundance of Ph, Ga and Kr in some stars; and (g) Zeeman splitting of spectral lines in some of these stars indicating the presence of large magnetic fields ranging from a few thousand to as high as 34,000 gauss. The stars which exhibit some or all of the above peculiarities have been designated as *peculiar* A or Ap stars although subsequent investigations have revealed that some of these peculiar spectral properties may be possessed by stars belonging to spectral classes as early as B5. But such peculiar stars hotter than A0 are very rare. The second group of A stars showing abnormal spectra is called metallic-line A or Am stars which are recognized by the following unusual spectral features:

1. The metallic lines are, in general, much stronger; particularly, the lines of ionized metals and those of rare earths are very strong.
2. Spectral classes assigned on the basis of strengths of hydrogen lines are systematically later than those derived by the strengths of the K line of Ca II. While the group as a whole lies in the spectral range A1 to A5 on the basis of the K line strengths, the hydrogen line strengths will assign to them the range A8 to F0. Needless to mention that the Am stars as a group belong to later spectral classes than the group of Ap stars.

To summarize the observed peculiarities of Ap stars λ 4200—Si are the hottest members of them showing this high excitation line of Si II in great strength. Mn is overabundant in the next group of hotter members by a factor of 100 and are therefore called the maganese stars. In the next cooler ones Si is overabundant by more or less the same factor and so these are called the silicon stars. These are followed by the further cooler stars of Si-Eu-Cr group in which Si is moderately overabundant, Cr is overabundant by a factor of about 10 and Eu by a factor of 1000. The coolest of the Ap stars in which Sr is overabundant by a factor of about 100 are called the strontium stars. Besides, Sr, Zr and Y are overabundant by factors ranging between 10 to 100 throughout the entire group of Ap stars. Helium is, in general, underabundant in Ap stars by factors of 10 to 100 while the ratio $He^3 : He^4$ is very high. While the normal abundance of He^3 is only 10^{-6} to 10^{-7} times that of He^4, in the Ap star 3 Centauri A, the former (He^3) has been found to constitute about 80 per cent of the total helium. The three lighter elements Li, Be and B are overabundant by various factors in different stars. In some stars, overabundance of these elements, particularly of Be, is by factors as high as 100. Li^6 is overabundant with respect to Li^7. Abundance of hydrogen does not show practically any anomaly. The most surprising abundance anomaly is probably revealed by strong lines of P II, Kr II, Ke II, Ga I and Ga II in some stars, such as 3 Centauri A and K Cancri. While only weak P II lines are observed in some stars, no other line mentioned above has ever been observed in stellar spectra. In 3 Centauri A, overabundance of P, Kr and Ga is almost by factors of 100, 1000 and 6000 respectively. Some other rare elements such as Ce, Pr, Nd, Sm, Eu, Gd, Dy and Ho which are hardly observed in any normal stars, are observed in some of these stars in such great strengths that, on the average, these elements are overabundant in them by factors of 100 to 1000.

H.W. Babcock of the Mount Wilson and Palomar observatories undertook a project of searching for any possible large-scale magnetic fields in stars. The most obvious target objects for such fields were the rapidly rotating stars. Babcock was almost immediately successful in detecting Zeeman splitting of lines in spectra of sharp-lined Ap stars and deriving the strengths of fields present in these stars. It was subsequently found that air sharp-lined Ap stars possess magnetic fields of sufficient intensity, ranging from a few hundred gauss to as high as 34,000 gauss measured in HD 215441. Zeeman splittings in these stars can be measured because the lines are sharp in their spectra, indicating that they are either rotating slowly or else are viewed nearly pole-on. If the stars are rapidly rotating and also viewed nearly along their equatorial planes, then the Zeeman separations which are small, will be masked under the rotationally broadened lines in the spectra. In these stars therefore, magnetic fields will not be detected even if present in sufficient strengths. This is probably the reason why magnetic fields could not be detected in most of the Ap stars because lines are very broad in their spectra. But since all the Ap stars are characterized by similar large-scale spectral peculiarities, it is generally believed that all of them possess sufficiently strong magnetic fields. In fact, it is possible that all rapidly rotating A stars possess sufficiently strong magnetic fields. Of these, only a few which are suitably oriented allow their fields to be detected. Another important point to be noted in this connection is that large-scale magnetic fields have never been detected in any B stars which as a group also rotate rapidly. The most plausible explanation for this seems to be that the latter do not possess sufficiently deep convection zones in their envelopes. It has been suggested by H. Elsässer and by E.N. Parker that rapid rotation and strong convection can together work to create magnetic fields of sufficient strengths.

Babcock has also observed that field strengths in these sharp-lined Ap stars undergo variations. He has classified these stars into four different groups on the basis of the nature of their magnetic variations. His α-class, named after the star α^2 CVn (α^2 Canum Venaticorum), shows *regular periodic variation* in magnetic fields as well as in spectra; the period for both the variations is the same. Babcock found 7 stars belonging to this class. The stars of Babcock's β-class, numbering 8, represented by the member β Coronae Borealis, show *irregular* magnetic variations with *reversals of polarity,* while the γ-class, 6 in number, shows *irregular* magnetic variations with only one polarity. These last two classes of stars are characterized by *constant spectra.* The remaining of the magnetic stars are put by Babcock into δ-class which show no regularity or interrelation in their variations in magnetic fields and spectral properties. Subsequent observations by Steinitz and by G.W. Preston and C. Sturch have revealed that the star β Cor Bor itself belongs to Babcock's α-class with a period of 18.5 days.

Babcock's α-class of Ap stars in which the magnetic field and the line intensities in the spectra of a star change regularly in the same period is known as *spectrum variables.* The cause of these variations and any physical interconnections that may exist or not between the changes in magnetic fields and spectral line intensities are not yet clearly understood. It has however been suggested from studies of the phenomena in α^2 CVn that changes in line intensities could be explained by assuming some reasonable changes in temperature and electron pressure at the surface. In this star a change in temperature by about 500 K has actually been measured by measuring the change in its colour.

We have already summarized the distinguishing features of the Am stars. We shall now very briefly discuss some other observed characteristics of these stars. It has already been said that the Am stars spectroscopically are the cooler continuation of the Ap stars. The HD spectral classes of the former group lie in the range A5 to F0 corresponding to the colour range B–V = + 0.15 to + 0.30. The Am stars occur in clusters and moving groups which are older than those in which the Ap stars occur. The observed turnoff points from the main sequence of the former clusters and moving groups correspond to their ages in the range of a few times 10^8 to 10^9 years.

As a class the Am stars are slow rotators compared to the Ap stars. The mean value of V sin i for a group of Am stars has been measured to be about 40 km s^{-1}. On the other hand, the Ap stars are believed to be rapid rotators (with (V sin i)$_{av}$ ~ 100 km s^{-1}) seen pole-on. The slow rotation of Am stars is accompanied by their weak magnetic fields. In some of these stars, magnetic fields of only a few hundred gauss have been measured while in some others no fields at all could be measured. Another remarkable observed difference between Am and Ap stars, is the frequency of their occurrence in spectroscopic binary systems. The percentage of their occurrence in visual binaries is quite similar to the normal main sequence A stars. But while an overwhelmingly large percentage of Am stars occurs in spectroscopic binary systems, the occurrence in such systems of the Ap stars is disproportionately small. This fact seems to favour the conclusion that as a class Am stars are slow rotators than Ap stars.

The large-scale abundance anomalies, the great strengths of magnetic fields and their variations with accompanying changes in spectra and the various other spectral peculiarities exhibited by the twin groups of stars, Ap and Am, demand satisfactory explanations. But a proper understanding of the problem has eluded the astronomers for more than half-a-century, which only speaks of the great complexity of the problem. Nevertheless, theories have been proposed but none of these has emerged as fully acceptable. Any comprehensive theories should be able to satisfactorily answer the following questions: Are the observed anomalies of abundance real? If so, are the abundant elements produced in the interior of the stars or on their surfaces? What is the degree of correlation between the changes in spectral features and magnetic field intensities, and what physical processes actually bring about these correlated changes? How to explain the corresponding changes when they are not correlated as in δ-class of stars? These and many other questions have to be explained with the help of theories but this has not yet been satisfactorily done.

If the abundance anomalies be only apparent, then there must be some unusual excitation mechanism operative in the atmospheres of the Ap and Am stars which enhances the strengths of lines of certain elements in preference to the others. Such mechanism was suggested by I.S. Bowen to be operative in atmospheres of W stars whereby the predominance of carbon and nitrogen is manifested in the two sequences. It is possible that similar mechanism excites the lines of rare earth elements and some other metals in preference to the others in Ap and Am stars. A second mechanism by which lines of certain selected elements may be produced in great strengths even with their normal abundances in the star, has been suggested by Babcock. According to this mechanism the atoms, by virtue of their possessing magnetic moments, and being subjected to nonuniform magnetic fields (believed to be certainly present in Ap stars) may migrate to certain regions on the surface of a star and form abnormal concentrations of these elements. The rate of migration will depend on the magnitudes of the magnetic moment of the atom concerned and the gradient of the magnetic field present in the star. But detailed analyses indicate that this mechanism can at best explain the unusual strengths of lines of those atoms only which possess large magnetic moments. Only Mn, Cr and Eu belong to this category. Babcock's mechanism, therefore, cannot explain the unusual strengths of lines of many other atoms as observed in Ap stars. Also, for such large-scale isolation of certain elements as required for producing the observed effects would call for the presence of magnetic field gradients orders of magnitude larger than are actually found to exist in these stars.

For explanation of the observed periodic variations of the magnetic fields and spectral line intensities, Babcock has proposed that migration of elements in oscillating magnetic fields might manifest the observed effects. A.N. Deutsch, on the other hand, has suggested that the observed effects could be explained if one assumed that both the magnetic axis and the axis of rotation of the star (which are inclined to reach other) were both inclined to the line of sight. This is known as the *oblique rotator* model. As the star rotates, surface areas with different magnetic fields and chemical composition turn towards the observer, and produce the observed effects. This model, however, does not explain as to how areas of different chemical composition are actually formed on the surface of the star.

Many authors believe, however, that the excess abundances of certain elements as observed in Ap and Am stars are real. According to them, these elements are formed in large excesses in these stars by nuclear reactions; the mechanism however is not fully understood as yet. Authors like William Fowler, the Burbidges and Fred Hoyle believe that the excess elements are formed by reactions in the deep interior of stars. But others do not accept this view. They are of opinion that some processes confined to the stellar surface are responsible for the creation of abundance anomalies. These authors argue that large excesses of He^3, Be and Li cannot be the product of any interior reactions because these nuclei do not survive in high temperature existing in stellar interiors. Overabundance of He^3 has actually been observed in high energy particles ejected during solar flares which are nonthermal phenomena. This encouraging observation led the latter authors to believe that the abundance anomalies in Ap and Am stars might be produced by similar nonthermal nuclear reactions, restricted to the surface of these stars. However, matters concerning the abundance anomalies in these stars still remain unsettled. The problem is extremely complex and complete understanding of it has so far baffled the astronomers.

10.6   T TAURI STARS

The different emission line objects which we have discussed in Sections 10.1 to 10.4 are hot stars of very high luminosity and belonging to early spectral classes. There are, however, some stars of low luminosity belonging to later spectral classes which also display *emission lines* in their spectra. Among these stars, there is a conspicuous group of variable stars which has been called T Tauri stars after the name of the representative of the group, the T Tauri. These are dwarf variable stars mostly belonging to the spectral classes G to M.

The pioneering work of A.J. Joy led to the recognition of T Tauri (the name was given by him) stars as a distinct class of emission-line variables. He also noticed that these stars are always associated in groups with thick nebulosity such as the Orion complex. The peculiar distributional property of T Tauri stars was more intensively studied by Ambertsumyan, who found that these stars were invariably present in association with groups of hot O and B stars, all embedded in regions of extensive and thick nebulosity. This study led Ambertsumyan to propose that both these hotter and cooler groups of stars were born out of the nebular gas and that the two groups were coeval. He called these groups as O-associations and T-associations respectively (Chapter 11). Since the more massive O and B stars have settled on to the main sequence, Ambertsumyan thought it natural to conclude that the less massive T Tauri stars were yet to settle on it. It was proposed that these stars were still in the process of gravitational contraction and were evolving toward the main sequence. Thus, T Tauri stars are extremely young objects. Their location in the H-R diagram is usually 2 to 3 magnitudes above the main sequence with their absolute magnitudes usually lying within the range +2 to +5.

The most important physical features of T Tauri stars which distinguish them from other dwarf stars of the same spectral class are: (a) the presence of emission lines of hydrogen and the H and K lines of Ca II; (b) the ultraviolet and infrared excesses over stars of similar spectral types; and (c) irregular variation of light. Other lines that are usually observed in emission are those of Fe I and the forbidden lines of [S II] and [O I]. Strong Li I 6707 line is another important characteristic of some T Tauri stars. The work of L.V. Kuhi has indicated the existence of a positive correlation between the strengths of H_α line and that of the ultraviolet continuum in T Tauri stars. This has led to the interpretation that the emission lines and the ultraviolet continuum in these stars arise from a common region and that this region lies above the star's photosphere. It has been proposed that many of the observed features in T Tauri stars can be understood by a model with a hot envelope $(T \sim 20,000$ K) superposed on a late-type photosphere. It has also been shown that the contribution of the envelope to the total luminosity of the star is comparable to that of the photosphere in most T Tauri stars; and in some cases the former may even exceed. Such a model, however, has not yet been fully confirmed.

The presence of emission lines with violet-displaced absorption feature and other observed features in T Tauri stars leads to almost inevitable conclusion that these stars are ejecting material from their bodies by some unknown process. Attempts have been made by several authors to make reasonable estimates of the rate of mass loss. The amount of mass lost by an average T Tauri star has been computed to be about 10^{-7} $M_\odot$ yr^{-1}, but for the luminous T Tauri stars G. Herbig computed a mass loss rate of 10^{-5} $M_\odot$ yr^{-1}. From a study of six T Tauri stars Kuhi has computed rates lying in the range of 0.31 to 5.85 times 10^{-7} $M_\odot$ yr^{-1}. Kuhi also found a positive correlation between the emission-line intensities and rates of mass loss, in the sense that the stronger the emission spectrum, the higher the loss rate. T Tauri phenomena have thus generally been understood on the basis of the model of outflow of matter. All phenomena admit of satisfactory understanding with the ejection of matter hypothesis.

Many authors have, however, tried to explain the T Tauri phenomena with an assumption of accretion or inflow of matter from the surrounding on to the stellar surface. The earlier attempts in this direction were made by J.L. Greenstein, G. Herbig, and by others. This idea, however, soon gave way to the growing evidence, as it were, in favour of the ejection hypothesis which has dominated for nearly the last four decades. The inflow hypothesis has been revived again by Roger K, Ulrich who has attempted to show that all the observed T Tauri phenomena can equally be explained with inflow model in which the star accretes, on the average, a mass of 10^{-7} $M_\odot$ yr^{-1}. This has only shown that the matter which had so far been considered as settled is not so. One has to wait to see which of these opposite models ultimately emerge as the really operative one.

10.7   THE EMISSION-LINE RED DWARF (DME) STARS

Since the dwarf M(dM) stars have very long main sequence life-time (i.e., 10^{10} years) they contain a mixture of stars of various ages. It is therefore, necessary, to discover some criteria which may yield clues to separate these stars into new and old groups, or equivalently, between Population I and Population II M dwarfs. One way to do this is to classify them on the basis of kinematic characteristics. It is well known that Population II objects have larger space motions; particularly their orbits have larger inclinations with the galactic plane and thus, larger components of motion perpendicular to this plane. Population I objects, on the other hand, have smaller space motions and flatter distribution with respect to the galactic plane (Chapter 11). But this criterion can be applied for the purpose of separating these stars into two population groups only with limited success. This is because statistically, some members of the Population I group should possess large space motions while some of Population II group should have smaller space motions. It will be safer, therefore, if some intrinsic physical characteristic could be used as a criterion to discriminate between the old and new stars of this class. Fortunately, such a spectroscopic criterion has been available for the dM stars which enables us to divide these into population groups. It has been found that a large fraction of dM stars show emission lines of Ca II in their spectra. These have been called dMe stars. Besides the Ca II lines, a fraction of dMe stars exhibit also the hydrogen lines in emission. It is believed that among the dwarf M stars those which exhibit emission lines in their spectra are of more recent origin; they belong to Population I. Those which do not show any emission lines in their spectra are older and belong to Population II.

This view is supported by analyses of the kinematic behaviour of these two groups of stars. J. Delhaye analyzed space motions of a number of M stars within 5 pc. He found that the Z component of motion of 12 dMe stars averaged to 6 km s^{-1}, while the same for 15 dM stars having no emission lines averaged to 17 km s^{-1}. From an analysis of much larger sample of dwarf M stars, E.R. Dyer found a velocity dispersion of 18 km s^{-1} for 65 dMe stars, while the velocity dispersion for 240 M dwarfs showing no emission was found to be 30 km s^{-1}. These observations clearly show that dMe stars have smaller space motions and a flatter distribution, much like the Population I objects of spectral type A, while those M dwarfs without traces of emission have space motions and distributions very much like Population II objects. The dwarf M stars with emission lines are therefore, of more recent origin than those having no emission features.

10.8 R CORONAE BOREALIS (R COR BOR) STARS

Although the number of these stars discovered in the Galaxy is only about three dozens, by virtue of many peculiarities they exhibit, they have generated much interest among astronomers. These are variable stars belonging to spectral classes F to K, their periods ranging from 10 days to several hundred days. They are supergiants having an average absolute magnitude of –5. They usually undergo sudden and irregular variations. The luminosity drops suddenly by several magnitudes and then increases again slowly. As a group, they have a very large range of amplitude of variations, from 1 to as high as 9 magnitudes. Spectroscopic analyses show that these stars have lower hydrogen abundance and higher abundance of carbon than normal. It has sometimes been suggested that these stars are seen through envelopes composed of materials richer in carbon. But this has not been confirmed. The problem of the abundance anomalies of R Cor Bor stars still remains unsolved.

10.9 THE CARBON STARS (R AND N STARS)

Giant stars belonging to spectral classes G, K and M are usually divided into three classes on the basis of differing chemical composition. These are: (a) the *oxygen stars* which are of normal compositions characterized by bands of light-metal-oxides such as TiO, and VO. These are the normal red giants; (b) The *carbon stars* whose spectra are dominated by bands of carbon molecules such as C_2, CN and CH. These stars apparently possess-an overabundance of carbon, and metallic oxides are practically absent in them. Their Henry-Draper spectral classes are assigned by R and N; (c) The heavy-metal-oxide stars whose spectra are dominated by bands of heavy-metal-oxides such as ZrO, YO, BaO and LaO. Their Henry-Draper class is assigned by S. In the present section, we shall discuss the class (b) of late-type giants reserving the next section for the discussion of class (c) objects. It may be mentioned that stars belonging to both these classes are very *rare* objects.

Following the studies of P.C. Keenan and W.W. Morgan, carbon stars are subdivided into classes C0 to C9 which bear approximate relationship with HD classes R and N as follows: the classes C0 to C4 approximately cover the class R from R0 to R9 while the remaining classes C5 to C9 cover the HD class N from N0 to N9. The temperature of carbon stars range from 4600 K for C0 (corresponding to HD class G5) to 1500 K for C9 (corresponding to HD class M9–10). The visual absolute magnitudes of carbon stars have been computed by various authors, considering the average for R and N stars separately. Considering the results of various authors it appears to be justified to take $M_v \sim -0.4$ for R stars and $M_v \sim -2.0$ for N stars.

Most authors believe that a higher carbon abundance in these stars is real. Experiments have shown that the relative increase of C_2 and CN over TiO becomes rapid with the increase of the ratio C:O. The observed spectral properties of the C and M giants will be affected as a result of transition from normal C: O and C: N ratios to values a few times higher. Thus while the values of the ratios C:6 and C:N in normal later-type giants are ~ 0.25 and ~ 0.5 respectively, a change in the values of the ratios to ~ 2 and ~ 1 will ensure the manifestation of carbon star properties. Such experimental results have encouraged astronomers to believe that the C stars do actually possess higher abundance of carbon. The observed results are *not* due to some peculiar excitation mechanism which fictitiously displays an excess of carbon in stars with normal chemical composition.

The study of the distribution of carbon stars shows that they are in general, heavily concentrated to the galactic plane. There are practically very few carbon stars at latitudes beyond ±5°. Among these stars again, those belonging to the class N have apparently higher galactic concentration than stars of class R. Also, the N stars show a marked tendency to concentrate in regions of higher obscuration. The observed concentration of some bright N stars on the galactic plane in the direction of Monoceros is believed to be associated with the Orion Spur of the local spiral arm. Such observations apparently lend support to the view expressed by some authors that the dust particles may be formed in the atmospheres of N stars.

Among the carbon stars, the class R possesses higher velocity dispersion compared to the class N. The association of N stars with spiral arms and obscuration, their stronger affinity to the galactic plane and lesser velocity dispersion, all lead to the conclusion that N stars are younger objects than the R stars. All evidences support the view that the R stars have probably evolved from dwarf G and K stars and are older objects than the M giants. This is probably not true for N stars.

10.10 THE HEAVY-METAL OXIDE STARS (S STARS)

As has already been mentioned, the S stars are recognized by bands of heavy-metal oxides in their spectra. They are extremely rare objects and are relatively cooler than carbon stars. According to their classification criteria based on the relative strength of the oxide bands these stars are divided into sub-classes S3 to S10, their temperatures ranging from 3000 K to 1500 K. No star earlier than S3 has been recognized. The temperature of S3 corresponds to that of the class C5 of carbon stars. A large fraction of S stars are variables of which again many have long periods like Mira-type stars. Some exhibit irregular or semi-regular variations of small amplitudes while others do not show variation at all. The long-period variables among S stars are believed to be somewhat more luminous than other members of the group. For the group of S stars as a whole, there exists a dispersion of luminosities at least as large as 2 magnitudes. The average absolute magnitude $M_v \sim -2.0$ for the long-period variable S stars and that for the remaining members, $M_v \sim -1.0$. It has been suggested that like carbon stars the S stars also can be divided into two sub-groups on the basis of motion and galactic distribution. The first sub-group consists of the nonvariable and small-amplitude-variable S stars. This group has very flat galactic concentration similar to that of N stars and have also smaller velocity dispersion. The second group of S stars consists of those which are long-period variables. They have galactic concentration less flat than the former group and also larger velocity dispersion. This would assign them to disc population.

The most outstanding distinguishing feature of S stars is the strength of Zr O band in their spectra. Transition from TiO band implies the corresponding transition from normal M giants to the S stars. In both these groups, the abundance of oxygen is the same and normal. Great difference, however, is believed to exist in the relative abundance of heavy metals and Ti. For example, the ratio Zr:Ti is ~ 0.01 in normal late-type giants while in S stars this ratio may be as high as 10. This large relative abundance of heavy metals is manifested by strong heavy-metal bands in the spectra of these stars while bands of TiO are relatively suppressed. It is believed that the observed anomalies of abundance in these stars are real.

10.11 INFRARED STARS

It is to be mentioned at the outset that the name *infrared stars* carries no special significance except that these stars are the coolest that are seen and their detection is greatly facilitated by special *infrared devices*. The nomenclature is rather significant in the sense that due to their low temperature most part of their energy is radiated in the *infrared*. As such, these stars appear very faint in optical wavelengths. Most of these objects are probably variable giant stars of very large dimensions.

A large-scale systematic search for these stars first started in mid-1960s at the Mount Wilson and Palomar observatories by a group of astronomers at Caltech and also by another group at the Tonanzintla Schmidt telescope. The earliest survey of the very red stars was in the Yerkes Observatory with infrared sensitive emulsion. A list of 168 stars with high infrared index $m(5600$ Å$) - m(8500$ Å$)$ was prepared by Hetzler. Such observations are generally now made at two wavelength ranges in the infrared, one in 0.68–0.92 μ and the other in 2.0–2.4 μ, covering respectively the effective bandwidths of the colours I and K. The latter range falls within the atmospheric window between 2.0–2.5 μ. The first observations were carried on 10 cool stars by Neugebauer and his coworkers. For these stars the colour index I–K was found to be nearly 7.5. This study was extended to 14 more stars by R. Ulrich and co-workers who found for these objects I–K $\geq$ 6. Large values of the colour indices I–K for these stars indicate that they are extremely cool stars, the blackbody temperature for some of these may be as low as 1000 K. It has been confirmed that the extreme red colour is intrinsic to these stars. It is not caused by heavy interstellar absorption. A photometric study of a number of these stars by H.L. Johnson and his co-workers has confirmed that the temperature of these stars lies within the range 1000–2000 K, which is more like those of late M stars.

Systematic search during the subsequent years by different groups of workers has led to the discovery of more such objects of which several hundred are now known. Detailed study of the physical nature of such stars was first carried out by R. Wing, H. Spinrad and L.V. Kuhi who selected three stars from those discovered by Neugebauer and his co-workers. The stars chosen by these authors are the Cygnus object, the Taurus object and TK Cam. A detailed analysis of the measured strengths of molecular bands and energy distribution in the wavelength range 0.8–1.1 μ and a review of all other available information on the spectral features and light variations of infrared stars were carried out by these authors. They have concluded that most of the infrared stars were Mira variables with periods much longer than 1 year and observed near minimum. This was confirmed by subsequent works of other authors. The variations of many such stars have been studied in optical, infrared and microwave wavelengths.

The earliest spectroscopic study of these very red objects was made by C.F. Rust. He studied the brightest of the Hetzler objects and came to the conclusion that most of these objects were of late M type stars. This was confirmed by a more extensive study of 120 such stars by H. Albers and by others. Analyzing the strengths of TiO and VO bands in the spectra of these stars, Albers concludes that a great majority of them belong to the class late M. The same result was derived by Wing and his co-workers from the observations of the three objects they had chosen. Among other bands in the infrared, those of CO, H_2O and C_3 are sometimes prominent or at least traceable.

The most interesting spectroscopic feature of some of these infrared stars is that they show microwave emission of OH and H_2O. Microwave emission of OH in these stars has been studied among others by Wilson and his co-workers. Strong emission of OH at 1612 MHz and at 1665/1667 MHz and of H_2O at 22,235 MHz has been observed in some (about 15–20 per cent) of the infrared stars. But in none of these 1612 MHz OH/IR and 1665/1667 MHz OH/IR stars any emission of OH at 1720 MHz has been observed. These microwave emissions are believed to be the result of *maser effect*.

All available information leads to the conclusion that these infrared objects are highly luminous and of very large size. The representative luminosity of the OH/IR objects (to which belong the Taurus and Cygnus objects) is $\sim 5 \times 10^4 \, L_{\odot}$ and the diameter $\sim 10^4 \, D_{\odot}$. The galactic distribution of these OH/IR objects is less flat compared to those IR objects showing no OH emission. The latter group is probably distributed like extreme Population I stars.

10.12   SUBDWARFS

A fairly common group of stars are found to occupy positions in the H–R diagram about 1.5 magnitude below the main sequence. These stars have been called subdwarfs and are sometimes designated by the luminosity class VI. The nomenclature is more or less losely applicable except for the fact that for a given temperature these stars are subluminous compared to their *dwarf* counterparts. The subdwarfs are not a *homogeneous* group of stars. They can roughly be divided into several groups on the basis of difference in motions and chemical compositions and several other parameters. For convenience of the brief discussion presented here, however, we shall divide them into two groups on the basis of temperature: (a) the hot subdwarfs belonging to spectral classes O to B and rarely to early A, and (b) the cool subdwarfs extending through classes F to M. Presumably the hot subdwarfs are entirely different objects from cooler subdwarfs.

The earliest studies of hot subdwarfs were mainly conducted by J.L. Greenstein, G. Münch, Vorontsov-Velyaminov and others. These stars are extremely hot with their B–V colours lying in the range – 0.4 to – 0.2. The visual absolute magnitudes of these stars lie in the range of +2 to +5. The common spectral features of these objects are broad lines of H, He I and He II. At sufficiently high dispersion, some of these objects exhibit innumerable sharp lines of N II, N III, Si III, Si IV, Ne II, C III and some other elements. The fact that these lines arise from levels of very high excitation is indicative of high temperatures of these stars. The sharpness of the lines in these stars distinguish them from hot white dwarfs, the latter having very broad lines due to pressure effect. It is also known that the white dwarfs do not show H and He lines in the same spectrum. The sharpness of lines also indicates smaller rotation and turbulence in subdwarfs. The study of a typical hot subdwarf, HZ 44, by Münch to ascertain the physical characteristics of these stars reveals the following values: $T \sim 35,000$ K, $R \sim 0.27\ R_{\odot}$ and Y (= helium abundance) = 0.25. The hot subdwarfs are thus stars of very high temperatures but of very small size (hence low luminosity for the temperature) with probably normal helium abundance. It is believed that at least some of these stars are in the process of evolution from the Population II *horizontal branch* objects to the hot white dwarfs. The others may have evolved from different types of objects but heading very likely towards the same end. At least some of these latter objects are most probably Population I stars in the dying stages. It seems, in general, to be true as Vorontsov–Velyaminov observed that the extreme high-temperature edge of the spectral sequence is largely populated by dying stars such as the hot subdwarfs, the brightest white dwarfs and the planetary nuclei.

Cooler subdwarfs have been extensively studied by L.H. Aller, N. Roman, J.L. Greenstein, the Burbidges and others. Most of these stars belong to spectral classes F, G and early K and are characterized by weak metallic lines in various degrees. A-type subdwarfs are rare objects. Cooler subdwarfs are more difficult to recognize, for their spectra are heavily contaminated by molecular bands. The most important observable spectral feature of F and G subdwarfs is extremely weak metallic lines for the temperature assigned by their colours and level of excitation and ionization. In general, the spectral class determined on the basis of strengths of metallic lines will be a full one class earlier than that determined by the temperatures of these stars. Lines including those of hydrogen are very sharp, implying that both rotation and turbulence are insignificantly small.

In cooler stars of normal composition, numerous lines of neutral and ionized metals are present in the blue-ultraviolet region of the spectrum. These lines absorb a substantial portion of stellar flux emanated in this region, imparting the star somewhat redder colour for its temperature. In metal-poor stars like F–G subdwarfs, the metallic lines are extremely faint and most of the blue-ultraviolet flux emerges almost unhindered from the stars, imparting them bluer colours compared to normal stars of the same temperatures. Thus, subdwarfs have B–V colours bluer than normal stars of the same effective temperature. Also, they have U–B colour less for a given B–V colour than would be found in a normal star. The difference (U–B) = (U–B)$_{normal}$ (U–B)$_{sd}$ is the ultraviolet excess of the subdwarf and its value reaches as high as 0.^{m}25. The ultraviolet excess is a prominent parameter for subdwarfs which can be easily recognized by measuring it. Detailed study has revealed that a normal abundance of metals in a star of class A or F will produce a blanketing effect to redden it by 0.^{m}43 in U–B and 0.^{m}20 in B–V colours. Some authors have shown that if corrections are made in colours for the blanketing effects equivalent to those in stars like Sun, these subdwarfs will fall along the same line in the two-colour diagram as occupied by normal stars. The subdwarf phenomena, therefore, are essentially the effects of metal deficiency in these stars, so far as their positions in the H–R diagram are concerned. In still cooler subdwarfs of spectral classes later than G8, the problems are simpler because extreme metal deficiency is not observed in them. Small corrections for line weakening move them on to the main sequence.

10.13 BROWN DWARFS

The class of objects which are not capable of burning hydrogen in the core but maintain hydrostatic equilibrium against gravitational collapse by degenerate electron pressure, are termed 'brown dwarfs' (BDs). The most fundamental parameter of a brown dwarf is its mass which is not sufficient to ignite hydrogen in its core, so these objects remain at the bottom of and below *Main sequence*. Detailed evolutionary calculations give hydrogen burning minimum mass as m$_{HBMM}$ ~ 0.085 $M_\odot$. (Grossman et al. 1974). Generally the upper limit is between 75–80 Jupiter mass (M_J) and the lower limit is 13 M_J. It is somewhat surprising that the first discovery of a bonafide BD (Nakajima et al. 1995) and the discovery of the first extra solar planet (Mayor and Queloz, 1995) were announced simultaneously in 1995. Currently it is debatable that that what exact criterion differentiates a BD from an extra solar giant planet (EGP). For this a thorough knowledge of the dense objects regarding interior physics, mechanical and thermal properties, evolutionary theory, mass function etc. comprising the wide range from low mass stars (LMS) to sub stellar objects (SSO) is necessary. In the subsequent sections, a brief outline has been given regarding various theories which have come up for the above mentioned objects.

Equation of State

The ranges of central density (ρ_c), central temperature (T_c) and mass (m) for LMS to SSO are 10^3–10 g cm^{-3}, 10^7–10^4 K and 0.07 $M_\odot$ – 0.0005 $M_\odot$ with solar composition and at 5 Gyr. The electron degeneracy parameter $\psi \to \infty$ for Maxwell–Boltzman classical limit, $\to 0$ for complete degeneracy. ρ_c and T_c are also shown in Fig. 10.3 for LMS–SSO mass range. The value of $\psi \sim 2-0.05$ in the interior of LMS and BDs. Above $0.4\ M_\odot$ the structure evolves slowly from $n = 3/2$ towards $n = 3$ polytrope. This leads to increasing central pressure and densities for increasing mass. Below $0.3-0.4\ M_\odot$ the core becomes entirely convective and follows the behaviour of $n = 3/2$ polytrope. The gas is still in classical regime ($\psi > = 1$), $m \propto R$ and central density increases with decreasing mass, i.e. $\rho_c \propto m^{-3}$. Below $m = 0.085\ M_\odot$ electron degeneracy becomes dominant ($\psi < = 0.1$) so that it approaches $m \propto R^{-3}$ ($\psi = 0$) and $\rho_c \propto m^2$. So the gas behaves as non-monatomic from $0.4\ M_\odot$ down to H burning limit.

The thermonuclear reactions relevant to LMS and BD are PPI chain which is $p + p \rightarrow d + e^+ + \nu_e$; $p + d \rightarrow {}^3He + \gamma$; ${}^3He + {}^3He \rightarrow {}^4He + 2p$ (Burrows and Liebert 1993; Chabrier and Baraffe 1997).

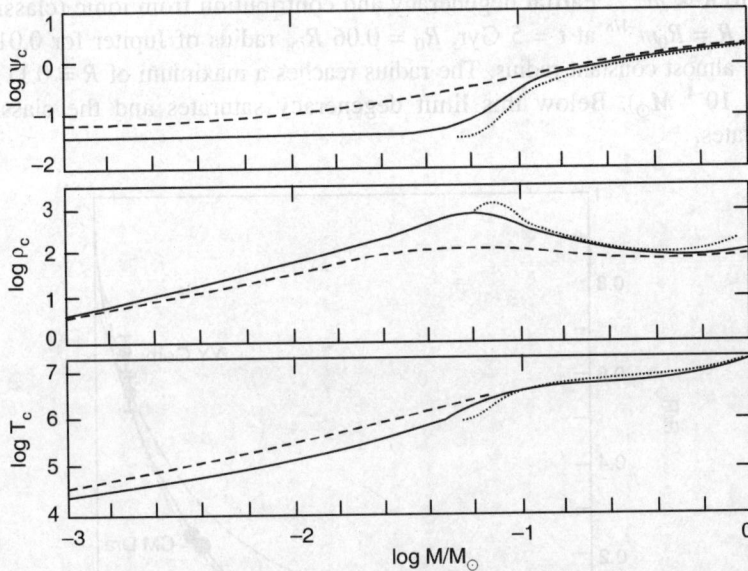

FIGURE 10.3 Central temperature (in K), density (in g cm^{-3}), and degeneracy parameter along the LMS–SSO mass range for objects with $Z = Z_\odot$ at 5 Gyr (*solid line*) and 10^8 yr (*dashed line*), and with a metallicity $Z = 10^{-2}\ Z_\odot$ at 5 Gyr (*dotted-line*) (Chabrier and Baraffe 2000).

Energy Transfer

Energy transfer for $M < 0.4\ M_\odot$ is mainly by convection especially when $M = 0.35\ M_\odot$ and $10^{-2} \lesssim Z/Z_\odot \lesssim 1$. But there are various factors like rotation, viscosity, density gradient and magnetic field which for their corresponding threshold values can inhibit convection. In the outer (molecular) layers having Rayleigh number $\sim 10^{15}$ mixing length theory (MLT) is no longer valid so a coupled hydrodynamic-radiation 3D simulationa (Demarque 1999) describes the the thermal structure to some extent.

Energy transport can occur by conduction also in the interior (Chabrier et al. 2000) of an object having high density and low temperature. Indeed below $m_{HBMM} \sim 0.07\ M_\odot$ the interior becomes degenerate enough so that conductive flux becomes larger than the convective flux. BD in the mass range 0.02–0.07 $M_\odot$ become degenerate enough to develop a conductive core which slows down the cooling.

Mass Radius Relation

The mass – radius behaviour shows the interior properties of LMS – SSO objects. It is clear from Fig. 10.4 that for $m \gtrsim 0.2 \ M_\odot$, $\psi > 1$ Fig. 10.3 for all ages with internal pressure, dominated by ion and electron pressure and the contribution from ion-electron interaction ($P = \rho kT/(\mu \ m_H)$; Debye contribution, $P_{DH} \propto -\rho^{3/2}/T^{1/2}$). When density is high so that $\psi < 1$ then the pressure is due to degenerate electron gas ($P \propto \rho^{5/3}$) for $m \sim 0.06 - 0.07 \ M_\odot$. Full degeneracy ($\psi \sim 0$) leads to $R \propto m^{-1/3}$. Partial degeneracy and contribution from ionic (classical) Coulomb pressure yield $R = R_0 m^{-1/8}$ at $t = 5$ Gyr, $R_0 = 0.06 \ R_\odot$, radius of Jupiter for $0.01 \ M_\odot <\ = m < \ = 0.07 \ M_\odot$, an almost constant radius. The radius reaches a maximum of $R = 0.11 \ R_\odot$ for $m \sim 4$. M_J ($M_J = 9.5 \ 10^{-4} \ M_\odot$). Below this limit degeneracy saturates and the classical Coulomb pressure saturates.

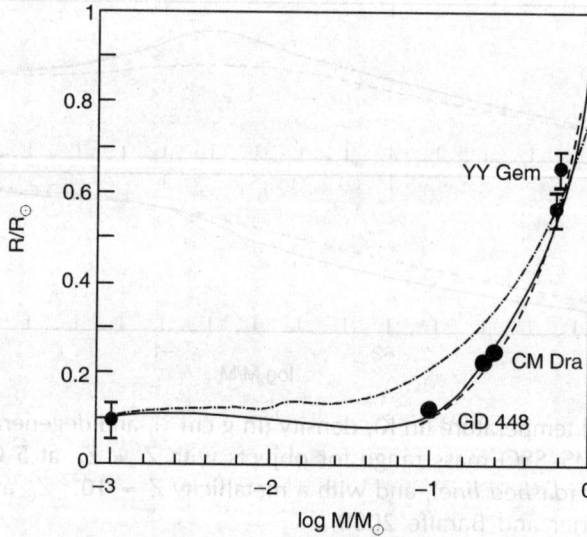

FIGURE 10.4   Mass-radius relationship for LMS and SSOs for two ages, $t = 6.10^7$ yr (*dotdash line*), 5×10^9 yr (*solid line*) for $Z = Z_\odot$, and $t = 5 \times 10^9$ yr for $Z = 10^{-2} \ Z_\odot$ (*dashed line*). The HBMM is 0.075 $M_\odot$ for $Z = Z_\odot$ and 0.083 $M_\odot$ for $Z = 10^{-2}$ $Z_\odot$. Also indicated are the observationally-determined radii of various objects (see text) and the position of Jupiter radius (J). The bump on the 6.10^7 yr isochrone illustrates the initial D-burning phase.

Thermal Properties

Figure 10.5 (Burrows et al. 1997) shows a representative central temperature and central density diagram for LMS and SSO. For LMS, T_c and ρ_c always increase with time until they

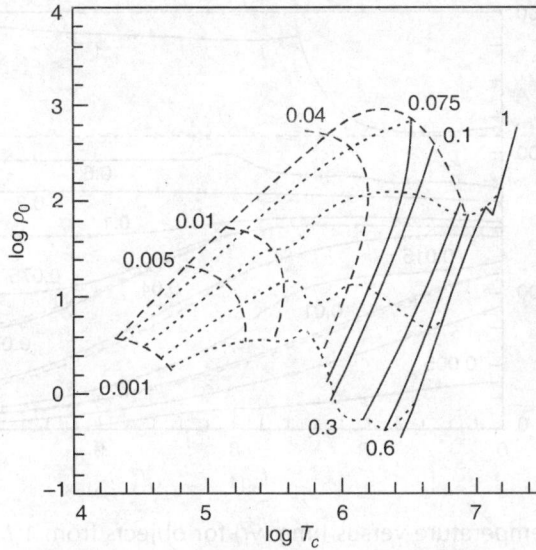

Figure 10.5 T_c–ρ_c relationship (in cgs) for LMS (*solid lines*) and SSOs *dashed lines* from 1 $M_\odot$ to 0.001 $M_\odot$ (masses in $M_\odot$ indicated on the curves). *Dotted lines* represent 10^6, 10^7, 10^8, 10^9 and 5.10^9 yr isochrones from bottom to top. The bumps on the 10^6–10^8 yr isochrones at log T_c ~ 5.4–5.8 correspond to the initial deuterium burning phase (Burrows et al. 1997).

reach the ZAMS. For BDs T_c first Increases for 10^7–10^9 yr for masses between 0.01 and 0.07 $M_\odot$, T_c reaches a maximum when degeneracy is dominant and then decreases. The bumps appearing on the isochrones between 10^6–10^8 yrs and log T_c ~ 5.4–5.8 result from initial deuterium burning.

Evolution

Figure 10.6 shows time evolution of LMS and SSO. When high mass BD are very young their luminosities are similar in nature with LMS since both objects release energy gained from gravitational contraction.

Eventually LMS settle onto MS which lasts for many Hubble times, BDs lack the required energy source and thus fade away.

Mass Function

In the beginning the mass function of BD have been determined using large telescopes in infrared. Spectroscopic determination of mass functions down to 20 M_J have been obtained in Taurus (Briceno et al. 2002; Luhman et al. 2003a), Ic348 (Luhman et al. 2003b) and Orion (Slesnick et al. 2004). Other surveys like 2MASS and DENIS have probed BD in solar neighbourhood. In LOCAL solar neighbourhood BD seem to be roughly as numerous as stars (Reid et al. 1999). Thus BD do not play significant role in making up Galactic missing mass.

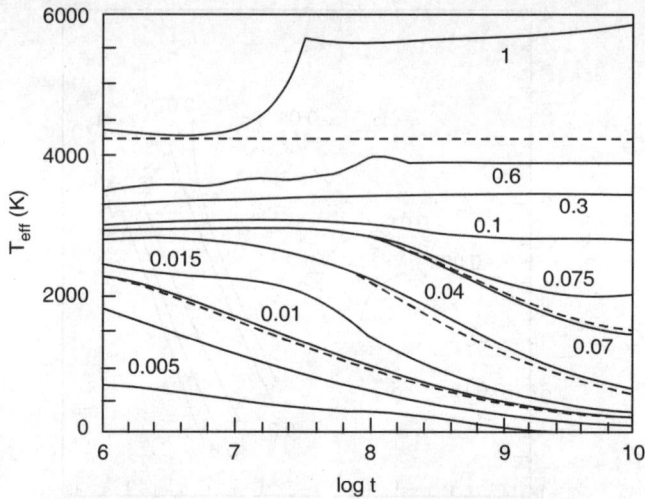

Figure 10.6 Effective temperature versus time (yr) for objects from 1 $M_\odot$ to 10^{-3} $M_\odot$ (masses indicated in $M_\odot$). (*Solid lines*): $Z = Z_\odot$, no dust opacity; (*dotted lines*): $Z = Z_\odot$, dust opacity included, shown for 0.01, 0.04, and 0.07 $M_\odot$; (*dashed line*): $Z = 10^{-2}$ $Z_\odot$ (only for 0.3 $M_\odot$) (Burrows et al. 1997).

But a proper census of the number of BD has significant implications for understanding of how stars and planets form. The parameterization of the IMF given by Chabrier (2003) is

$$dN/d \log M \propto 0.158 \exp [- (\log m - \log 0.079)^2/\{2.(0.069)^2\}], \quad m \leq 1\ M_\odot$$

$$\propto 0.044\ m^{-1.3}, \qquad\qquad\qquad\qquad m > 1\ M_\odot$$

and this is based on objects within 8 pc of the sun.

How to Distinguish between LMS, BD AND EGP

Lithium (Li) might be used as a distinguishing criterion between bonafide BD and LMS. Stars which achieve high temperature to fuse hydrogen in their core, Rapidly deplete Li. This occurs by a collision of ^{7}Li and a proton producing two ^{4}He. The temperature necessary for this reaction is just below the temperature necessary for hydrogen fusion. Therefore presence of Li line in a candidate BD's spectrum is a strong indicator that it is substellar. This is first proposed by Rebolo et al. (1992) to identify bonafide BD and is known as 'Lithium test'. However, Li is also seen in very young stars which have not yet had a chance to burn it off. Heavier stars like sun can also retain Li in their outer envelopes. These are distinguishable by mass from BD. On the contrary massive BD can be hot enough to deplete Li at their young ages. So the test is not perfect. A significant property of BD is that they all have roughly the same radius, the radius of Jupiter. For massive BD (60–90 M_J) the volume of a BD is governed by degenerate electron pressure whereas for less massive BD (1–10 M_J) it is governed by classical Coulomb pressure. So the radii of BD vary only by 10–15%. So it is difficult to distinguish a BD from a planet. In addition many BD undergo no fusion. So the alternative way is the 'radius'. So anything that size over 10 M_J is unlikely to be a planet. Some BD emit X rays or red or infrared until they cool to planet like temperatures (~ 1000 K). Currently IAU considers objects with masses above the limiting mass for deuterium burning (~ 13 M_J) to be BD whereas those objects under that mass are considered planets.

Classification of Brown Dwarfs

In the H–R diagram, the coolest stars are M stars whose optical spectra are dominated by absorption bands of TiO and VO. Objects like GD 165B, the cool companion to white dwarf GD165 has none of the above mentioned features instead their spectra contain strong metal hydride bands (FeH, CrH, MgH, CaH) and prominent alkali lines (NaI, KI, CsI, RbI). This fact led Kirpatrick and others to introduce a new spectral class, 'the L dwarfs', defined in the red optical region. As of 2005, over 400 L dwarfs have been identified., most by Two Micron All Sky Survey (2MASS), the Deep Near Infrared Survey of the Southern Sky (DENIS) and the Sloan Digital Sky Survey (SDSS).

After the discovery of Gliese 229B, the first observed BD, a new spectral class 'T' is introduced after 'L'. The near infrared spectra (NIR) of 'L' dwarfs show strong absorption bands of CH_4. FeH, CrH–characteristics of L dwarfs but abundant in broad absorption features from alkali metals Na and K. Theory suggests L dwarfs are mixture of LMS and SSO whereas T dwarfs are composed entirely of BD.

Brown Dwarf Catalogue (According to *Chris Gelino*)

Object	R.A. hh:mm:ss	Dec. dd:mm:ss	R	I	J	H	K	L	Spectral Type	Spectrum Yes or No	Lithium EW (Å)	H_α EW (Å)
Kelu-1 o	13:05:40.2	–25:41:06	19.2	16.8	13.17 UK	12.32 UK	11.79 UK	–	L2	Yes	1.7	1.9
GD 165B	14:22:12.0 (1950)	+09:30:45 (1950)	–	19.25	15.80 CIT	14.78 CIT	14.17 CIT	13.27 L_{CIT}	L4	Yes	<0.7	<0.8
DENIS–P J0020–4414	00:20:59.4	–44:14:43	–	18.32	14.97	–	13.6	–	M8	Yes	–	–
DENIS–P J0021–4244	00:21:05.7	–42:44:50	–	16.83	13.63	–	12.25	–	M9	Yes	–	–
DENIS–P J0205–1159 binary	02:05:29.0	–11:59:25	–	18.30	14.63	–	13.00	12.05	L7	Yes	<1.3	<2.7
DENIS–P J1058–1548	10:58:46.5	–15:48:00	–	17.80	14.08	–	12.71	12.00	L3	Yes	<0.3	1.6
DENIS–P J1228.2–1547 v o binary	12:28:13.8	–15:47:11	–	18.19	14.43	–	12.73	11.76	L5	Yes	3.1	<=0.6
DENIS–P J2040–3245	20:40:06.2	–32:45:24	–	17.86	14.89	–	–	–	M8	Yes	–	–
DENIS–P J2052–5515	20:52:55.0	–55:12:03	–	17.52	14.82	–	14.08	–	M8	Yes	–	–
DENIS–P J2146–2153	21:46:10.6	–21:53:09	–	18.40	15.42	–	–	–	M9	Yes	–	–
rho Oph162349.8–242601	16:23:49.8	–24:26:01	–	18.49	15.4	14.7	14.2	–	M8–M9	Yes	–	60
Roque Praesepe 1	8:38:2.9	+19:25:46.7	23.5	21.01 UK	17.7 UK	–	16.44 UK	–	M8.5	Yes	–	<3

(Contd.)

Object	R.A. hh:mm:ss	Dec. dd:mm:ss	R	I	J	H	K	L	Spectral Type	Spectrum Yes or No	Lithium EW (Å)	Hα EW (Å)
NPL 35(PPL 15)	3:48:04.8	+23:39:32.0	–	17.91 KC	15.43 CIT	–	14.48 CIT	–	–	–	–	–
NPL 36(Roque 12)	3:48:19.1	+24:25:15	–	18.66 KC	15.95 CIT	–	15.12 CIT	–	M7.5	Yes	–	10.0
NPL 37(Roque 11)	3:47:12.1	+24:28:31.4	–	19.06 KC	16.36 CIT	–	–	–	M8	Yes	–	–
NPL 38	3:47:50.4	+23:54:48.6	–	19.18 KC	16.30 CIT	–	–	–	M8	Yes	–	<6
NPL 39(Teide 1)	3:47:17.9	+24:22:31.9	–	19.26 KC	16.18 CIT	–	–	–	–	No	–	–
NPL 40(Roque 33)	3:48:49.1	+24:20:25	–	20.55 KC	17.15 CIT	–	–	–	M9	Yes	–	20
Roque Pleiades 4 o Finder Chart	3:43:53.5	+24:31:11	22.05	19.75	–	–	15.23	–	M9	Yes	–	<5
Roque Pleiades 5 o Finder Chart	3:44:22.4	+23:39:01	–	19.71 C	–	–	15.40	–	M9	Yes	<=8	<=8
Roque Pleiades 7 Finder Chart	3:43:40.3	+24:30:11	–	19.29 C	–	–	15.23	–	–	No	–	–
Roque Pleiades 9 Finder Chart	3:46:23.2	+24:20:37	–	18.99 C	–	–	14.96	–	–	No	–	–
Roque Pleiades 11 v o(NPL 37) Finder Chart	3:47:12.1	+24:28:32	21.41	18.73 C	–	–	15.10	–	M8	Yes	–	5.8
Roque Pleiades 12 o(NPL 36) Finder Chart	3:48:19.0	+24:25:12	–	18.47 C	–	–	15.10	–	M7.5	Yes	<=1.5	19.7
Roque Pleiades 13 v o Finder Chart	3:45:50.6	+24:09:03	20.57	18.25 C	–	–	14.60	–	M7.5	Yes	–	10.5
Roque Pleiades 16 o (CFHT–PL–11) Finder Chart	3:47:39.0	+24:36:22	20.03	17.79 C	–	–	14.61	–	M6	Yes	–	5.0
Roque Pleiades 17 o Finder Chart	3:47:23.9	+22:42:38	19.51	17.78 C	–	–	14.33	–	M6.5	Yes	–	15.0
Roque Pleiades 25 o Finder Chart	3:48:30.6	+22:44:50	23.92	21.17 C	–	–	16.27	–	L?(M)	Yes	<=5	<=5
G196–3B	10:04:21.7 (G196–3A)	+50:23:16.0 (G196–3A)	20.78 C	18.28 C	14.73 UK	–	12.49 UK	–	L1 (M)	Yes	5	>1.0
Calar 3 v o	3:51:26.0	+23:45:20	21.27	18.73 C	16.29 UK	15.45 UK	14.94 UK	–	M8	Yes	1.8	6.5
Teide 1 v o (NPL 39)	3:47:18.0	+24:22:31	21.54	18.80 C	16.37 UK	15.55 UK	15.11 UK	–	M8	Yes	1.0	4.5
Teide 2 v o (CFHT–PL–13)	3:52:06.7	+24:16:01	20.05	17.82 C	15.45 CIT	–	14.46 CIT	–	M6	Yes	0.77	4.7–9.5
LP 944–20 v o (BRI 0337–3535)	3:37:39.9 (J1950)	−35:35:30 (J1950)	–	14.16	10.70	10.03	9.58	–	L	Yes	0.53	1.2
CFHT–PL–11 (Roque 16)	3:47:39.0	+24:36:22.1	20.12	17.91 C	–	–	–	–	M7	Yes	0.5	5.1
CFHT–PL–12 v o	3:53:55.1	+23:23:37.4	20.47	18.00C	–	–	–	–	M8	Yes	0.8	17.2
CFHT–PL–13 (Teide 2)	3:52:06.7	+24:16:01.4	20.23	18.02C	–	–	–	–	M7	Yes	0.6	8.7
CFHT–PL–15 o	3:55:12.5	+23:17:38.0	20.96	18.62C	–	–	–	–	M7	Yes	0.5	7.0
CFHT–PL–18 o Visual binary	3:53:23.1	+23:19:19.5	21.38	18.81C	15.95 UK	15.23 UK	14.80 UK	–	–	Yes	–	–

(Contd.)

Object	R.A. hh:mm:ss	Dec. dd:mm:ss	R	I	J	H	K	L	Spectral Type	Spectrum Yes or No	Lithium EW (Å)	H_α EW (Å)
CFHT–PL–21 (Calar 3)	3:51:25.6	+23:45:20.6	21.5	19.00	–	–	–	–	–	No	–	–
CFHT–PL–24 (Roque 7)	3:43:40.2	+24:30:11.8	22.11	19.50	–	–	–	–	–	No	–	–
PPL 15 o spectroscopic binary (NPL 35)	3:45:06.4 (J1950)	+23:30.22 (J1950)	–	17.80	15.34	14.65	14.32	–	M6.5	Yes	0.5	7–48
PPL 1	3:42:43.1 1950	+23:44:52 1950	–	17.53	15.48	14.83	14.39	–	M7.5	Yes	2.4	6.5
PIZ 1	3:48:31.4	+24:34:37.7	–	19.64	–	–	15.5	–	M9	Yes	–	–
MHObd1	3:44:52.4	+24:36:50	–	17.95	15.54	–	14.50	–	M7	Yes	–	14.4
MHObd3	3:41:54.1	24:20:00	–	18.27	15.43	14.89	14.36	–	–	No	–	–
296A	1:22:36.1 (J1950)	–39:00:15.0 (J1950)	17.14 F	14.57 N	–	–	–	–	M6	Yes	0.5	8.5
2MASSW J0147+3453	1:47:33.4	+34:53:11	–	18.07 spec	14.94	14.16	13.34 short	–	L0.5	Yes	<1.0	<0.5
2MASSW J0326+2950	3:26:13.7	+29:50:15	–	19.17 spec	15.43	14.38	13.83 short	–	L3.5	Yes	<1.0	9.1
2MASSW J0345+2540 spectroscopic binary	3:45:43.2	+25:40:23	–	16.98 spec	14.03	13.23	12.68 short	–	L0	Yes	<0.5	<=0.3
2MASSW J0913+1841 v	9:13:03.2	+18:41:50	–	19.07 spec	15.92	14.84	14.20 short	–	L3	Yes	<1.0	<0.8
2MASSW J0918+2134	9:18:38.2	+21:34:06	–	18.68 spec	15.66	14.64	14.21 short	–	L2.5	Yes	<0.3	<0.3
2MASSW J1145+2317 v	11:45:57.2	+23:17:30	–	18.62 spec	15.51	14.44	13.87 short	–	L1.5	Yes	<0.4	4.2
2MASSW J1328+2114	13:28:55.0	+21:14:49	–	20.07 spec	16.04	14.88	14.25 short	–	L5	Yes	<3.0	<2.0
2MASSW J1334+1940	13:34:06.2	+19:40:34	–	18.76 spec	15.54	14.83	13.99 short	–	L1.5	Yes	<1.5	4.2
2MASSW J1439+1929	14:39:28.4	+19:29:15	–	16.12C	12.76	12.05	11.58 short	–	L1	Yes	<0.05	1.13
2MASSW J0242+1607	02:42:43.5	+16:07:39	–	19.01 spec	15.67	14.78	14.26 short	–	L1.5	Yes	<0.7	<0.5
2MASSW J0030+3139	00:30:43.8	+31:39:32	–	18.82 spec	15.49	14.58	13.99 short	–	L2	Yes	<1.0	4.4
2MASSW J1342+1751	13:42:23.6	+17:51:56	–	19.81 spec	16.06	15.12	14.59 short	–	L2.5	Yes	<=3.9	<2.2
2MASSW J1146+2230 v binary	11:46:34.5	+22:30:53	–	17.62 spec	14.23	13.24	12.63 short	–	L3	Yes	5.1	<=0.3
2MASSW J0355+2257	03:55:41.9	+22:57:02	–	19.49 spec	16.10	15.03	14.25 short	–	L3	Yes	<1.3	<0.5
2MASSW J0129+3517	01:29:12.2	+35:27:58	–	19.43 spec	16.74	15.29	14.68 short	–	L4	Yes	3.3	<0.5
2MASSW J1155+2307	11:55:00.9	+23:07:06	–	19.30 spec	15.98	14.78	14.30 short	–	L4	Yes	<0.5	<1.0
2MASSW 1553+2109	15:53:21.4	+21:09:07	–	20.79 spec	16.68	15.34	14.68 short	–	L5.5	Yes	18.5	<4.3
2MASSW J0850+1057	08:50:35.9	+10:57:16	–	20.33 spec	16.45	15.22	14.46 short	–	L6	Yes	15.2	<0.9

(Contd.)

Object	R.A. hh:mm:ss	Dec. dd:mm:ss	R	I	J	H	K	L	Spectral Type	Spectrum Yes or No	Lithium EW (Å)	H_α EW (Å)
2MASSW J1632+1904	16:32:29.1	+19:04:41	–	19.98 spec	15.86	14.59	13.98 short	–	L8	Yes	<=9.4	<=4.0
LHS 102B	00:04:32.8	–40:39:56	–	17.00	20.70	–	22.60 K	–	L4(M)	Yes	<2.5	<4.0
EROS–MP J0032–4405	00:32:55	–44:05:05	–	18.57	14.85	–	13.65 short	–	L0(M)	Yes	3.3	1.7
CRBR 14 o	16:23:17.3 (1950)	–24:19:25 (1950)	–	–	15.23	13.48	12.28	–	M7.5	Yes	–	–
GY 5	16:23:20.0 (1950)	–24:19:15 (1950)	–	–	12.70	11.57	10.91	–	M7	Yes	–	–
GY 10 o	16:23:20.8 (1950)	–24:17:08 (1950)	–	–	15.75	13.59	12.25	–	M8.5	Yes	–	–
GY 11 o	16:23:20.8 (1950)	–24:17:21 (1950)	–	–	16.52	15.37	14.15	–	M6.5	Yes	–	–
GY 37	16:23:26.5 (1950)	–24:19:58 (1950)	–	–	14.25	12.94	11.99	–	M6	Yes	–	–
GY 59	16:23:30.1 (1950)	–24:18:47 (1950)	–	–	14.75	12.89	11.68	–	M6	Yes	–	–
GY 64 o	16:23:31.1 (1950)	–24:19:52 (1950)	–	–	16.63	14.76	13.33	–	M8	Yes	–	–
GY 202 o	16:24:04.5 (1950)	–24:21:52 (1950)	–	–	16.76	14.67	12.97	–	M7	Yes	–	–
GY 310 o	16:24:36.9 (1950)	–24:31:58 (1950)	–	–	13.20	11.91	11.08	–	M8.5	Yes	–	–
Cha H_α 1 o	11:07:17.0	–77:35:54	–	16.4	13.3	–	12.3	–	M7.5	Yes	0.63	34.5
Cha H_α 7	11:07:38.4	–77:35:30	–	16.7	13.5	–	12.4	–	M8	Yes	0.80	35.0
Cha H_α 2	11:07:43.0	–77:33:59	–	15.3	12.1	–	10.6	–	M6.5	Yes	0.43	71.0
Cha H_α 8	11:07:47.8	–77:40:08	–	15.6	12.7	–	11.5	–	M6.5	Yes	0.49	8.4
Cha H_α 3	11:07:52.9	–77:36:56	–	15.0	12.3	–	11.1	–	M7	Yes	0.43	4.5
Cha H_α 4	11:08:19.6	–77:39:17	–	14.4	12.0	–	11.1	–	M6	Yes	0.48	9.7
Cha H_α 5	11:08:25.6	–77:41:46	–	14.7	12.0	–	10.7	–	M6	Yes	0.42	8.0
Cha H_α 6	11:08:40.2	–77:34:17	–	15.1	12.0	–	10.9	–	M7	Yes	0.43	61.7
2MASSW J0036+1821	00:36:15.9	+18:21:10	–	16.10 C	12.44	11.58	11.03 short	–	L3.5	Yes	<0.1	–
DENIS–P J0255–4700	02:55:03.3	–47:00:49	–	17.14	13.48	–	11.86	–	L6(M)	Yes	<1.0	<2.0
AP306	03:19:41.8	+50:30:42.0	20.74	18.40C	–	–	14.9	–	M8	Yes	–	10.5
AP326	03:38:55.2	+48:57:31.0	21.10	18.70C	–	–	15.09	–	M7.5	Yes	–	7
SDSSp J0330–0025	03:30:35.1	–00:25:34.5	22.21 r*	20.11 i*	15.29	14.42	13.83 short	–	L2	Yes	–	–
SDSSp J0413–0114	04:13:20.4	–01:14:24.9	22.48 r*	19.61 i*	15.33	14.66	14.14 short	–	L0	Yes	–	–
SDSSp J0539–0059	05:39:52.0	–00:59:02.0	21.49 r*	19.04 i*	14.00	13.08	12.55 short	–	L5	Yes	–	–
2MASSW J0746+2000	07:46:42.5	+20:00:32	–	15.11 C	11.74	11.00	10.49 short	–	L0.5	Yes	<0.2	1.38
DENIS–P J1047–1815	10:47:31.1	–18:15:58	–	17.75	14.24	–	12.88	–	L2.5(M)	Yes	–	–
DENIS–P J1159+0057	11:59:38.4	+00:57:27	–	17.32	14.25	–	12.67	–	L0(M)	Yes	–	–
SDSSp J1203+0015	12:03:58.2	+00:15:50.3	21.31 r*	18.88 i*	–	–	–	–	L3	Yes	–	–

(Contd.)

Object	R.A. hh:mm:ss	Dec. dd:mm:ss	R	I	J	H	K	L	Spectral Type	Spectrum Yes or No	Lithium EW (Å)	H_α EW (Å)
DENIS–P J1323–1806	13:23:35.9	–18:06:38	–	18.60	15.06	–	14.17	–	L0(M)	Yes	<1.0	1.1
SDSSp J1326–0038	13:26:29.8	–00:38:31.5	23.68 r*	21.69 i*	16.11	15.04	14.23	–	L8?	Yes	–	–
SDSSp J1440+0021	14:40:01.8	+00:21:45.8	22.63 r*	20.47 i*	–	–	–	–	L1	Yes	–	–
DENIS–P J1441–0945	14:41:37.3	–09:45:59	–	17.32	13.96	–	12.37	–	L1(M)	Yes	<0.5	2.1
2MASSW J1507–1627	15:07:47.6	–16:27:38	–	16.65 C	12.82	11.90	11.30 short	–	L5	Yes	<0.1	–
2MASSW J1523+3014	15:23:22.6	+30:14:56	–	–	16.32	15.00	14.24 short	–	L8/9	Yes	–	–
SDSSp J1636–0034	16:36:00.8	–00:34:52.6	21.30 r*	18.80 i*	14.59	13.93	13.41 short	–	L0	Yes	–	–
SOri 40	05:38:32.4	–02:41:57	20.27	17.93	–	–	–	–	M7	Yes	–	30.0
SOri 12	05:37:57.4	–02:38:45	18.22	16.82	–	–	–	–	M6	Yes	–	6.5
SOri 47	05:38:14.5	–02:40:16	22.9	20.5	17.2 UK	–	–	–	L1.5	Yes	4.3	<6
SOri 27	05:38:17.3	–02:40:24	19.20	17.07	–	–	–	–	M7	Yes	–	6.1
SOri 45	05:38:25.5	–02:48:36	22.47	19.59	–	–	–	–	M8.5	Yes	–	60.0
SOri 29	05:38:29.5	–02:25:17	19.21	17.18	–	–	–	–	M6	Yes	–	28.0
SOri 39	05:38:32.4	–02:29:58	20.16	17.82	–	–	–	–	M6.5	Yes	–	5.1
SOri 17	05:39:04.4	–02:38:35	18.82	16.80	–	–	–	–	M6	Yes	–	5.5
SOri 25	05:39:08.8	–02:39:58	19.33	17.04	–	–	–	–	M6.5	Yes	–	45.0
CTIO–061	IC 2391	IC 2391	19.45 C	17.31 C	–	–	–	–	M6	Yes	–	2.3
CTIO–113	IC 2391	IC 2391	19.42	17.28	–	–	–	–	M7	Yes	–	4.6

The following entries are methane dwarfs.

Object	R.A. (J2000) hh:mm:ss	Dec. (J2000) dd:mm:ss	R	I	J	H	K	L	Spectral Type	Spectrum Yes or No	Lithium Yes or No	H_α Yes or No
Gl 229B o	6:10:35.07 (Gl 229A)	–21:51:17.6 (Gl 229A)	–	–	14.32	14.35	14.42	12.18	T?	Yes	–	–
SDSS J1624+0029	16:24:14.37	+00:29:15.8	–	21.2 T–G	15.53 UK	15.57 UK	15.70 UK	–	T?	Yes	–	–
2MASSI J1047+2124	10:47:53.9	+21:24:23	–	–	15.82	15.79	>16.29	–	T?	Yes	–	–
2MASSI J1217–0311	12:17:11.1	–03:11:13	–	–	15.85	15.79	>15.91	–	T?	Yes	–	–
2MASSI J1225–2739	12:25:54.3	–27:39:47	–	–	15.23	15.10	15.06	–	T?	Yes	–	–
2MASSI J1237+6526	12:37:39.2	+65:26:15	–	–	15.90	15.87	>15.90	–	T?	Yes	–	–
2MASSI J1346–0031 SDSS J1346–0031	13:46:46.4	–00:31:50	–24.54 (r*)	–23.26 (i*)	15.86 15.82	16.05 15.85	>15.75 15.84	–	T?	Yes	–	–
Gl 570D	14:57:15.0	–21:21:48	–	–	15.33	15.28	15.27K$_s$	–	T?	Yes	–	–
NTTDF 1205–0744	12:05:20.2	–07:44:01	>26.7 Gunn	>26.3 Gunn	20.15	–	20.3	–	T?	Yes	–	–

EXERCISES

1. The Wolf-Rayet (W) stars present two parallel sequences with strong lines of carbon and nitrogen predominating. How would you like to explain this apparent chemical difference?

2. The radius R_* of a W star is given by $R_* = 3\,R_\odot$ and the radius of its envelope is given by $R_{\text{env}} = 5\,R_*$. The electron density measured in the envelope is 10^{12} cm^{-3}. Compute the total mass contained in the envelope. If the mass is ejected from the surface of the envelope with a velocity of 2000 km s^{-1}, compute the rate of mass loss per year. Compare this lost mass with the mass of the Sun.

3. Compare the points of similarity and variance between the spectral line properties of W and P Cygni stars.

4. Emission lines of different atoms and ions of different strengths are found to be a common feature in certain classes of early type stars. What would you suggest as the possible cause of this feature?

5. What observational features of T Tauri stars lead us to believe that they still are contracting towards the main sequence.

6. If $10^{-7}\,M_\odot$ yr^{-1} is ejected from a T Tauri star with a velocity of 200 km s^{-1}, compute the wind luminosity of the star and compare this with the luminosity of the Sun.

7. If an infrared star has a surface temperature of 1500 K and its luminosity is 1000 $L_\odot$, find the volume of the star and compare this volume with that of the Sun.

8. Describe the principal observed features of subdwarf stars. Why these stars are called subdwarfs?

SUGGESTED READING

1. Briceno C., Hartmann L., Stauffer J.R., and Mart´in E.L. *Astronomical Journal,* 115:2074, 1998.

2. Böhm-Vitense, Erika, *Introduction to Stellar Astrophysics,* Volume 3: Stellar Structure and Evolution. Cambridge University Press, Cambridge, 1992.

3. Burrows A., Marley M., Hubbard W.B., Lunine J.I., Guillot T., et al., *Astrophysical Journal,* 491, 856, 1997.

4. Chabrier G., Baraffe I., Allard F., and Hauschildt P.H. *Astrophysical Journal,* 2000, In press.

5. Demarque P., Guenther D.B., KimY-C. *Astrophysical Journal,* 510, 517, 1999.

6. Grossman A.S., Hays D., and Graboske H.C. *Astronomy Astrophysics,* 30, 1974.

7. Kutner, Marc L., *Astronomy,* Harper & Row Publishers, New York, 1987.

8. Luhman, K.L., Briceño, C., Stauffer, J.R., Hartmann, L., Barrado y Navascués, D., and Nelson, C. 2003a, *Astrophysical Journal*, 590, 348.

9. Luhman, K.L., Stauffer, J.R., Muench, A.A., G.H. Rieke, et al., *Astrophysical Journal*, 593, 1093, 2003b.

10. Mayor M., Queloz D., Udry S., in *Brown Dwarfs and Extrasolar Planets*, (Eds.) R. Rebolo, E.L. Martin, M.R. Zapatero-Osorio, *A.S.P. Conf. Ser.* 134:140, San Francisco: Astron. Soc. Pacific., 1995.

11. Nakajima, T., Oppenheimer, B.R., Kulkarni, S.R., Golimowski, D.A., Matthews K., Durrance ST., *Nature* 378:463, 1995.

12. Page, Thronton, and Page, Lau Williams, (Eds.), *The Evolution of Stars,* Macmillan, New York, 1968.

13. Rebolo R., Mart´in EL, Magazz'u A., *Astrophysical,* 389, 1992.

14. Reid I.N., *Annu. Rev. Astronomical Astrophysical* 37, 1999.

15. Slesnick, C.L., Capenter, J.L., and Hillebrand, L.A. Astronomical Journal, 131, 3016, 2004.

16. Smith, Elske, V.P. and Jacobs, Kenneth C, *Introductory Astronomy and Astrophysics,* W.B. Saunders, Philadelphia, 1973.

17. Swihart, Thomas L., *Astrophysics and Stellar Astronomy,* John Wiley & Sons, New York, 1968.

11
Clusters and Associations of Stars

11.1 INTRODUCTION

The vast stellar system, our Milky Way Galaxy to which the solar system belongs, contains about 10^{11} stars. Most of these stars are moving at random in space either alone, or with one or at most a couple of companions, as binary or multiple systems. But a small number of stars, may be about *one per cent* of the entire population or even less, is observed to be associated with groups of different sizes and shapes. These groups have been called *star clusters*. Their principal distinguishing features are: (a) stars in each group are *gravitationally bound together,* and (b) they possess a common motion which bears no relation whatsoever with the motion of the surrounding objects not belonging to the group. The number density of stars in clusters varies from 10 to 100 times that in the general field close to the plane of the Milky Way (the galactic plane).

On the basis of their size, shape, galactic distribution and the number as well as the physical characteristics of stellar content, *two* distinct types of clusters have been recognized, viz. the *galactic* or *open clusters* and the *globular clusters.* The galactic clusters are small in size, flattened in shape and are mostly distributed close to the galactic plane. The number of stars contained in the galactic cluster, usually lies between 10^2 to 10^3 and these stars have normal metal abundance like that in the Sun. So they are believed to have been formed out of the interstellar gas when the latter has been sufficiently enriched with heavy elements created inside the first generations of stars. This implies that the galactic clusters are relatively of later origin. The globular clusters, on the other hand, are normally slightly flattened spherical systems of stars which may contain a number of stars ranging between 10^5 to 10^7. They show no affinity to the galactic plane in their distribution. Rather they are distributed more or less spherically around the galactic centre. The individual stars in them are largely metal-deficient; the metal content usually lying between 1 to 10 per cent of that in the Sun. This observed fact led astronomers to believe that globular clusters were formed at some early phase of the Galaxy when sufficient amount of heavier elements could not yet mingle with the general interstellar gas out of which these clusters were created. Thus, the globular clusters are believed to be very old objects.

Besides the globular and galactic clusters, yet another type of stellar groups has been recognized in the Galaxy. They are the aggregates of hot, blue stars belonging to O and B spectral types. These groups are associated with thick concentration of gas and dust, and the T Tauri variables which are believed to be stars of very recent origin and still passing through their initial contracting phase. Unlike the star clusters, the number density of stars in these groups is extremely small, only about 10 to 50 per cent of that of the surrounding field. It is for this reason that they are difficult to be recognized. But their group characteristic is revealed by the common motion of the individual members of the groups and by their association with thick nebulosity. Such stellar aggregates have been called *O-associations* on account of the fact that the majority of the O stars are found to belong to these aggregates. They are of great astrophysical significance since they help us to understand the process of star formation and their evolution. It is now believed that stars are still being born *in groups* in regions of thick nebulosity of gas and dust which are then mingled gradually with the field stars by tidal disruption arising out of the differential galactic rotation. It is now believed that many or all of the hot B-type stars now observed in the general field were originally members of these aggregates of hot stars which have since been disrupted. So far as the galactic distribution is concerned, the O-associations are still more strongly concentrated towards the galactic plane as compared to the galactic clusters. They are believed to have been formed within the spiral arms of the Galaxy within a time-scale not exceeding 10^7 years. In subsequent sections, we shall discuss in more details the cases of the galactic and globular clusters and of O-associations.

11.2 GALACTIC CLUSTERS

About 1000 galactic clusters are known at present and reasonably accurate distances have been calculated for a large number of these clusters. When counts of the number of clusters at different distance-intervals are made they yield a number density of about 100 clusters projected per kpc^2 of the galactic plane. This would lead to an estimated total number of galactic clusters to be approximately between 10^4 to 10^5 in the Galaxy. The number of stars contained in an individual cluster varies greatly, from less than 10^2 to more than 10^3. The diameters of these clusters usually lie between a few parsecs to less than 10 parsecs. But for a very few clusters such as the double clusters h Persei and χ Persei, the diameters are as large as about 20 parsecs. Because the star density is nowhere very high in a galactic cluster, and the individual stars can be resolved even in the central part of the clusters, the name *open cluster* is alternatively used for a galactic cluster.

Galactic clusters are strongly concentrated near the galactic plane. Almost all the known galactic clusters lie within a thickness of about 1 kpc around the galactic plane. A galactic cluster with its distance greater than 500 pc from the galactic plane is an extremely rare object. This strong tendency of the galactic clusters to hover about the galactic plane is clearly demonstrated by Fig. 11.1. The figure shows a maximum concentration of galactic clusters at zero galactic latitude, that is, on the plane of the Galaxy. Another important characteristic of the space distribution of galactic clusters is that the relatively young clusters containing O and early B-type stars as members all lie within the spiral arms of the Galaxy while older clusters have moved out of the arms. The galactic clusters thus being associated with, the galactic disc and spiral arms partake into galactic rotation like Population I objects (Section 11.5).

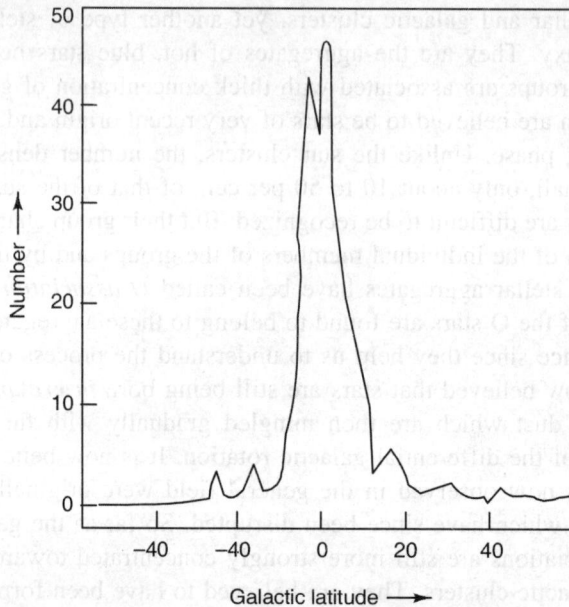

FIGURE 11.1 Distribution of open clusters in galactic latitude.

Accurate distance of a large number of galactic clusters have been known mainly from the photometric works of H.L. Johnson and his co-workers. It is found that the distances to galactic clusters in general, are large. The cluster nearest to us is the Hyades in Taurus with a distance of 40 pc. Other nearby clusters are Pleiades also in Taurus at a distance of 125 pc and the Coma Berenices at a distance of 90 pc. In these clusters a few brightest stars can be seen even with the naked eye. But the majority of the clusters are too far to be visible without the aid of telescopes. The Hyades cluster, by virtue of its closeness, spreads over a relatively large area of the sky and allows its convergent point to be determined accurately. This fact serves as a fundamental basis for calibration of the luminosity of the *Zero Age Main Sequence* (ZAMS), since the distance to Hyades can be computed by two independent methods, viz., the trigonometric parallax method and the moving cluster parallax method. By ZAMS we refer to those stars which have settled on the main sequence at the commencement of hydrogen burning in the interior, but which have not evolved further after exhausting a sufficient amount of hydrogen in the core.

The distances of individual members in a cluster are practically the same, owing to the large distance of the cluster itself. So a plot of the apparent magnitudes of the stars in a cluster and their spectral classes (or colours) is equivalent to the H–R diagram (or colour-magnitude diagram) of the cluster shifted vertically toward fainter magnitudes by the amount equal to the distance modulus of the cluster, provided that proper corrections have been applied for interstellar absorption. So *matching* the main sequence branch (where stars have not evolved) of a galactic cluster corrected for absorption, with the standard ZAMS by vertical shift, its distance modulus can be obtained accurately. In applying this method two basic assumptions are however made.

First, we assume that the calibration of luminosities of the standard ZAMS stars is error-free. This assumption appears to be closest to reality because it is based on luminosities of the Hyades stars, whose accurate distance can be obtained by two independent methods. We shall presently come back to this point for a more detailed discussion. The second assumption is that for stars of the age of the Hyades ($\approx 10^9$ years) and those younger, the ZAMS is the same, that is, they are identical in chemical composition and are unaffected by evolutionary effects.

For the construction of the standard ZAMS, the first basic data came from the H–R diagram of the main sequence stars of known distances nearest to the Sun. But these stars are not large in number and most of these are of lower intrinsic luminosity. A much wider and fuller base is provided by the main sequence members of the Hyades which contain a large number of stars and also extend to brighter magnitudes as shown in Fig. 11.2. The main sequence of the Hyades extends in colour index $(B–V)_0$ corrected for absorption from +1.0 to +0.2, but the upper part brighter than $(B–V)_0 = +0.4$ exhibits evolutionary effects. But the part between the values of $(B–V)_0$ from +0.4 to +1.0 coincides completely with the ZAMS obtained from the nearest stars, which runs down to $(B–V)_0 = +1.2$ at the fainter magnitude limit. The main sequence of the Pleiades cluster contains stars of brighter magnitudes than

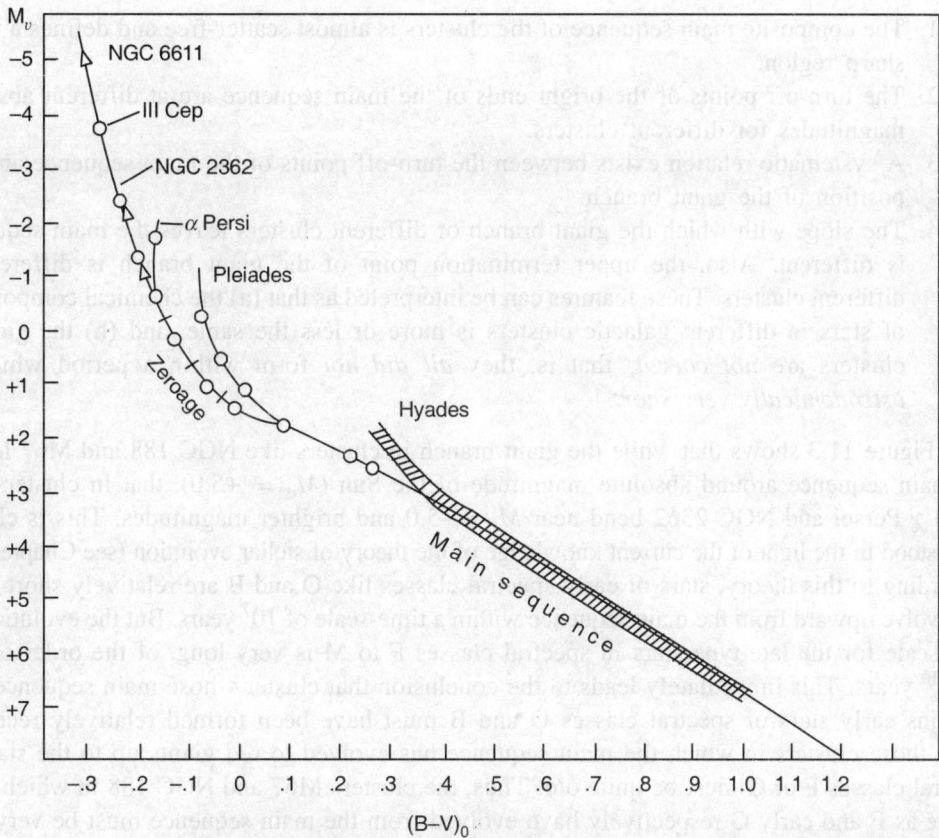

FIGURE 11.2 Calibration of ZAMS by matching the main sequence of galactic clusters.

those of the Hyades. So the next brighter part of the ZAMS is calibrated by matching the main sequence of the Pleiades with that of the Hyades. The Pleiades main sequence coincides with that of the Hyades in the lower part between +0.8 to +0.4 in $(B–V)_0$ values and extends upwards to $(B–V)_0 = +0.1$. The clusters that next come successively in the list for calibration of brighter and brighter magnitudes of the ZAMS are: the moving cluster α Persei extending through +0.6 to – 0.1; the cluster NGC 2362 extending through +0.5 to – 0.25; the cluster III Cep extending through – 0.25 to – 0.30; and the cluster NGC 6611 extending up to – 0.35 of the value of $(B–V)_0$. Thus, the calibration of the ZAMS is obtained from $M_v = +7.7$ at $(B–V)_0 = + 1.2$ to $M_v = –5.5$ at $(B–V)_0 = – 0.35$. The distance moduli obtained for these clusters by matching of their main sequences are: 5.55 for the Pleiades; 6.15 for α Persei; 10.8 for NGC 2362; 9.3 for III Cep and 12.6 for NGC 6611. The entire ZAMS thus calibrated is shown in Fig. 11.2.

When the standard ZAMS has thus been constructed, one can match the main sequence of the colour-magnitude diagram of any cluster with the standard ZAMS and derive the distance modulus for the cluster. Figure 11.3 shows the composite colour-magnitude diagram for eleven clusters corrected for absorption. When the figure is carefully analyzed, several interesting features are closely revealed:

1. The composite main sequence of the clusters is almost scatter-free and defines a fairly sharp region.
2. The turn-off points of the bright ends of the main sequence are at different absolute magnitudes for different clusters.
3. A systematic relation exists between the turn-off points of the main sequence and the position of the giant branch.
4. The slope with which the giant branch of different clusters leaves the main sequence is different. Also, the upper termination point of the giant branch is different in different clusters. These features can be interpreted as that (a) the chemical composition of stars in different galactic clusters is more or less the same, and (b) the galactic clusters are *not coeval;* that is, they *all did not* form within a period which is *astronomically very short.*

Figure 11.3 shows that while the giant branch in clusters like NGC 188 and M67 leaves the main sequence around absolute magnitude of the Sun ($M_v \simeq +5.0$), that in clusters like h and χ Persei and NGC 2362 bend near $M_v \approx –5.0$ and brighter magnitudes. This is clearly understood in the light of the current knowledge of the theory of stellar evolution (see Chapter 14). According to this theory, stars of early spectral classes like O and B are relatively short-lived and evolve upward from the main sequence within a time-scale of 10^7 years. But the evolutionary time-scale for the late-type stars in spectral classes F to M is very long, of the order of 10^9 to 10^{10} years. This immediately leads to the conclusion that cluster whose main sequence still contains early stars of spectral classes O and B must have been formed relatively recently; while those clusters in which the main sequence has evolved to red giants up to the stars of spectral classes F or G must be quite old. Thus, the clusters M67 and NGC 188 in which stars as late as F and early G respectively have evolved from the main sequence must be very old. These are extreme examples of old galactic clusters. The estimated age of the cluster M67 is of the order of 10^{10} years and that of NGC 188 is of the order of 1.6×10^{10} years. On the

Colour (B – V)$_0$

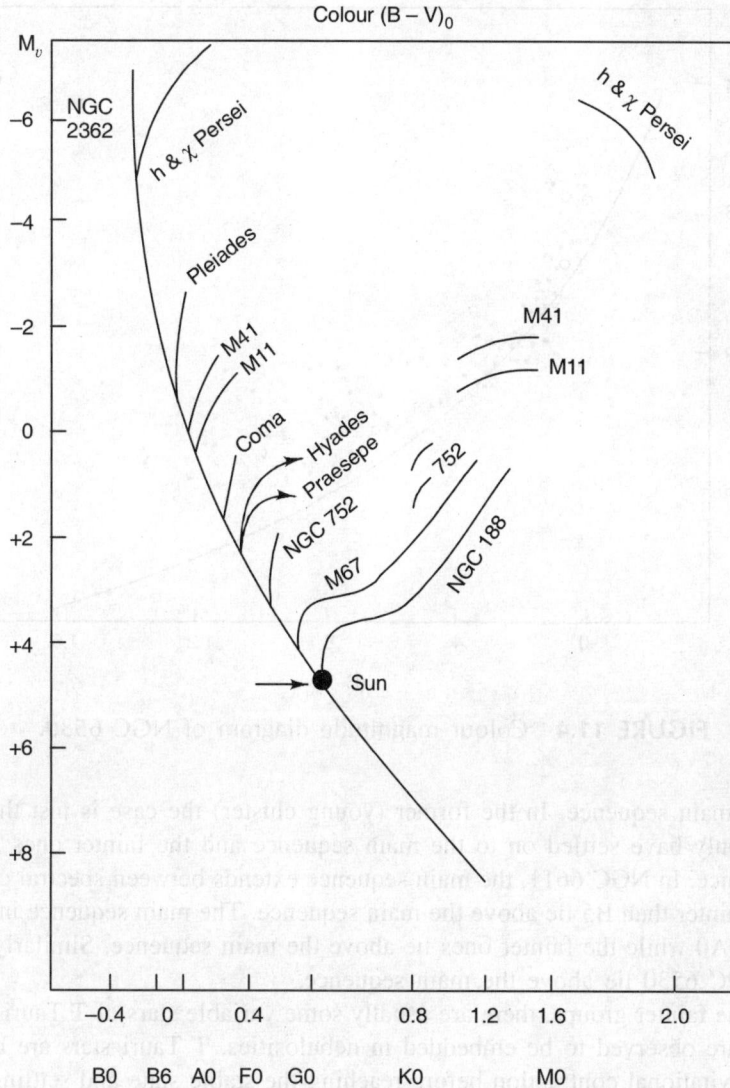

FIGURE 11.3   Spectral class: C–M diagram of some galactic clusters.

other hand, clusters like *h* and *χ* Persei, NGC 2362 and NGC 6611 which contain O and early B stars must be very young. They are probably not older than a few times 10^6 to a few times 10^7 years.

The extreme examples of young clusters are NGC 2264, NGC 6530, NGC 6611, IC 5146 and the Orion Nebula cluster. As typical examples of such clusters, we give in Fig. 11.4 the colour-magnitude diagram of NGC 6530. A clear difference between the colour-magnitude diagram of this cluster and those of older clusters as depicted in Fig. 11.3 is at once revealed. In the latter (older cluster) the brighter stars have evolved to giant sequence and the fainter

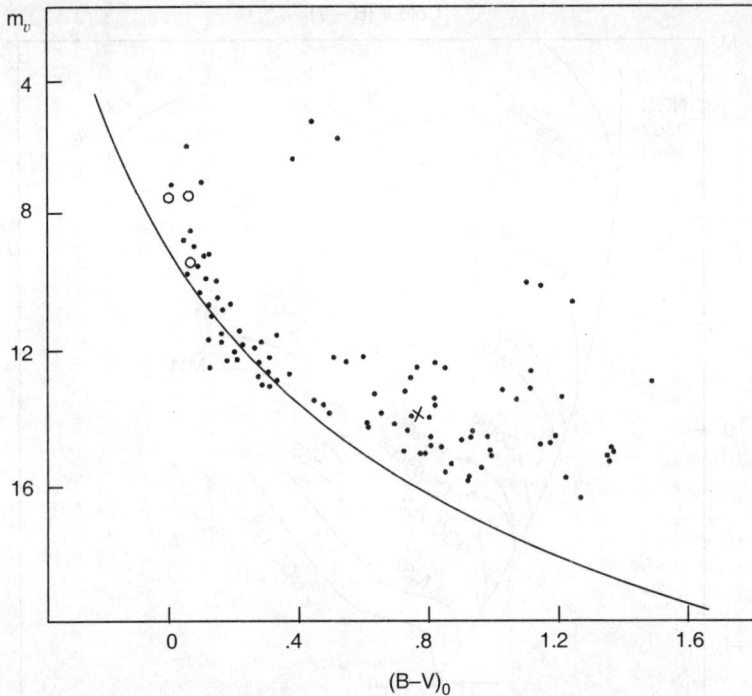

FIGURE 11.4   Colour magnitude diagram of NGC 6530.

ones form the main sequence. In the former (young cluster) the case is just the reverse; the brighter stars only have settled on to the main sequence and the fainter ones still lie above the main sequence. In NGC 6611, the main sequence extends between spectral classes O5 and B5 and those fainter than B5 lie above the main sequence. The main sequence in IC 5146 runs between B0 to A0 while the fainter ones lie above the main sequence. Similarly, stars fainter than A5 in NGC 6530 lie above the main sequence.

Among the fainter groups, there are usually some variable stars of T Tauri type. Further, these clusters are observed to be embedded in nebulosities. T Tauri stars are believed to be undergoing gravitational contraction before reaching the stable state and settling on the main sequence. All these observed facts lead to but one definite conclusion that the stars in these clusters have been born recently out of the nebulous matter associated with them, as only the more massive members have had enough time to settle on to the main sequence, but the fainter, less massive ones are still in the process of gravitational contraction and are yet to settle on the main sequence. These clusters are thus extremely young.

11.3   GLOBULAR CLUSTERS

As has been already observed, globular clusters form a spherical system of very large volume (Radius 20 kpc) around the galactic centre. This fact was first recognized by Harlow Shapley in 1920's, who investigated the distances to these clusters in order to deduce the distance to the galactic centre (see Chapter 16). Shapley, however, overestimated the distance because the phenomenon of interstellar absorption was unknown at that time and the correction necessary for it could not be applied by him.

At present about 125 globular clusters are known in the Galaxy. Since these clusters have larger density near the galactic bulge, which is an obscure region due to heavy interstellar absorption, a large number of clusters probably remain hidden from our view. The total number of globular clusters in the Galaxy may be as large as 500. Each cluster consists of a relatively bright central region (nucleus) surrounded by less bright regions. This is due to the high density of stars forming the nucleus, while in the outer region of low luminosity star density is much lower. Although the name "globular" was initially assigned to these clusters because of their geometrical appearance, later it was revealed from their colour-magnitude diagrams that these actually comprised a distinct group of stars having substantially different physical properties compared to the stars in galactic clusters and other common stars in the solar neighbourhood. It was further observed that the globular clusters do not form a homogeneous group and cannot be described by any single parameter. Thus, it has been suggested that for the detailed, accurate and meaningful classification of globular clusters one should use both *geometrical* and *physical* parameters. According to Arp, the geometrical parameters are: (a) shape, (b) density gradient of stars, and (c) numerical richness of stars. The physical parameters that should be used are: (a) chemical composition of stars, (b) age of stars, and (c) luminosity function.

The most important distinguishing parameter for globular clusters appears to be the chemical composition or, to be more specific, the metal abundance. This parameter decides the spectral classes or the colours $(B-V)_0$, corrected for absorption, of the member stars. Since most of the globular clusters are much too far away to yield colours and spectra of their individual stars, their integrated colours and integrated spectra are generally studied more conveniently. It was William W. Morgan who first analyzed the variation of intrinsic properties of globular clusters with galactic latitude. Integrated colours and spectral classes for 65 globular clusters in the Galaxy were measured by Kron and Mayall. These integrated spectral classes of globular clusters are plotted against galactic latitudes in Fig. 11.5. The figure clearly demonstrates that clusters with later-type integrated spectra are concentrated near the galactic plane and also concentrated near the galactic centre, that is, clusters with later spectral types are concentrated in denser regions of the Galaxy. This is to be understood in terms of metal abundances of stars in these clusters. Stars with higher metal abundance are intrinsically redder, indicating that clusters with later spectral classes contain higher abundance of metals. These observations therefore find logical explanation in the fact that in the denser inner regions of the Galaxy, formation and evolution of stars took place earlier. As a result, higher metal abundance was attained by material in these regions out of which these globular clusters were formed. This fact is further demonstrated in Fig. 11.6, obtained by plotting integrated spectra of globular clusters against their distances from the galactic centre. The spectral classification was based on the Morgan classification system, using I for the most weak metallic lines and VIII for the strongest metallic lines, the intermediate numbers indicating the corresponding intermediate strengths of metallic lines. The figure shows that the metal-weak clusters are situated far out from the galactic centre, while clusters with higher metal abundance lie closer to it.

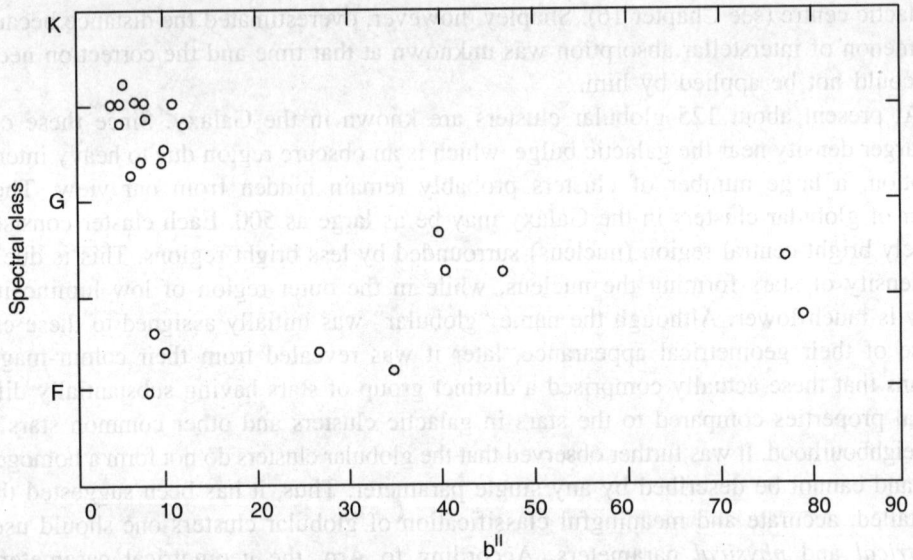

FIGURE 11.5 Spectrum of globular clusters against galactic latitude.

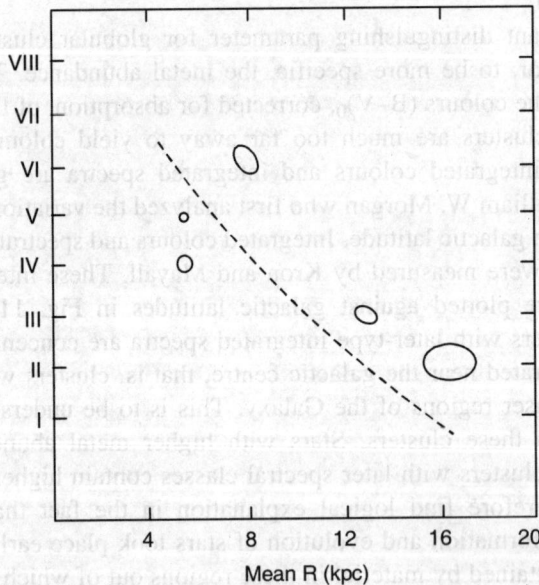

FIGURE 11.6 Variation of metal abundance in globular clusters with their distances from the galactic centre.

Radical differences are observed when one compares the colour-magnitude diagrams of globular clusters with those of galactic clusters. Intrinsically bright blue stars are totally absent here on the main sequence (Fig. 11.7). The entire group of brightest stars belong to the red giant class with $M_v \approx -3.0$, and this branch rises upward to the right, much more steeply than

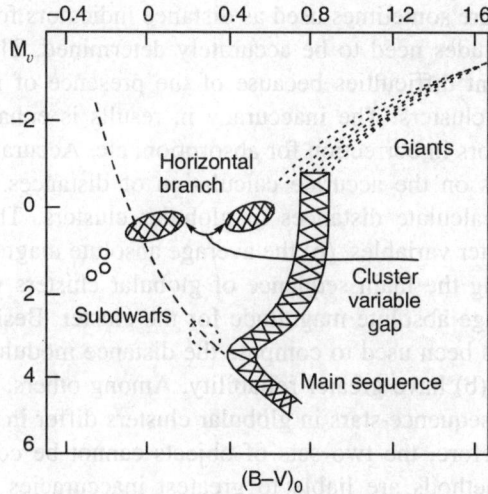

FIGURE 11.7   Schematic colour-magnitude diagram of a globular cluster.

does the giant branch in the vicinity of the Sun. The giant branch projects downward to join the main sequence through the intermediate group of yellow subgiants. The main sequence generally terminates at luminosities of the spectral class F, whence the subgiant branch rises to merge into the red giant group. At about the absolute magnitude –1 of the giant branch, another branch of stars, called the *horizontal branch,* moves to the left and slightly downward in the diagram. A portion of the intermediate range of this horizontal branch is filled only by the cluster variables (RR Lyraes) and no other star is found to intrude in this region. This portion has, therefore, been called the *cluster variable gap.*

In spite of the greatest care with which the photometric measurements may be made, the colour-magnitude diagrams for different globular clusters are found to differ. When the main sequence of different clusters is matched, it is observed that the giant sequence does not all coincide and conversely, although all of them exhibit similar appearances in broad features. This result is explained by the fact that globular clusters are not homogeneous in chemical composition. Spectroscopic studies have revealed that the metal abundance systematically varies from one globular cluster to another. The metal abundance between two clusters may differ by a factor as large as 10, and compared to the Sun, the metal content in these clusters is found to vary between 1–10 per cent. This metal deficiency in globular clusters is manifested in several important observable effects. First, in globular clusters the integrated spectra are earlier when spectral classes are assigned by strength of metal lines than when hydrogen lines are used. Secondly, relative to the horizontal branch, the giant branch is fainter by about one magnitude in metal-rich clusters. As a result, the cluster will appear too far away if the distance is determined by using the brightest star criterion. Thirdly, the number of RR Lyrae variables is convincingly smaller in metal-rich clusters than in metal-poor ones. The metal-rich clusters M12 and M71 are typical examples. In the former there is only one RR Lyrae variable and in the latter there are only four. On the other hand, there are hundreds of RR Lyrae stars in some metal-poor clusters.

As globular clusters are sometimes used as distance indicators for external galaxies, their integrated absolute magnitudes need to be accurately determined. This problem, however, is charged with some inherent difficulties because of the presence of inhomogeneous physical characteristics among the clusters. The inaccuracy in results is enhanced by such factors as inaccurate photometry, errors in correction for absorption, etc. Accurate calibration of absolute magnitudes primarily rests on the accurate calculation of distances. Four different methods can, in fact, be used to calculate distances to globular clusters. These are: (a) the known absolute magnitude of cluster variables, (b) the average absolute magnitudes of the 25 brightest stars in a cluster, (c) fitting the main sequence of globular clusters with the ZAMS, and (d) using some standard average absolute magnitude for the cluster. Besides these, the diameters of clusters have sometimes been used to compute the distance modulus. Of all these methods, those listed under (a) and (b) have greater reliability. Among others, the third one should not be used because the main sequence stars in globular clusters differ in physical properties from those of ZAMS, and therefore, the two sets of objects cannot be compared for any reliable accuracy. The last two methods are liable to greatest inaccuracies because large scatter is known to exist in absolute magnitudes and diameters of globular clusters.

Regarding the first method, the known absolute magnitudes of RR Lyrae variables have been frequently used. But different observers have obtained different values for absolute magnitudes of these variables, mainly because of photometric inaccuracies and use of different absorption values. Nevertheless, using the average of a number of determined values, a fairly reliable value can be obtained and used to compute distance moduli. H.C. Arp has used the value $M_v = +0.^m3$ for RR Lyrae variables to compute the distance to globular clusters. RR Lyrae variables yield the most reliable basis for computing distances to globular clusters. On this fundamental method is based the second method of calculating distances by using the average of 25 brightest stars. The average M_B for six clusters has been computed by Arp to be $-0.^m8$. The distance obtained by these two methods agree fairly well. With their distances known, absolute magnitude can be calculated using Eq. (3.5) after corrections have been made for interstellar absorption. Large scatter is found in the computed absolute magnitudes of globular clusters, the median of which is $\bar{M}_v = -8.4$.

Since the globular clusters form an almost spherically symmetric system, it is obvious that they will deviate largely from the rotational motion of the Galaxy, the latter being principally a disk phenomenon. They possess large radial motions and some may even possess *pendulum orbits* passing through the central region of the Galaxy. The mean rotational velocity of the globular cluster system is about 60 km s^{-1}. This imparts the globular cluster system as a whole large solar motions.

11.4 STELLAR ASSOCIATIONS

Stellar associations are aggregates of hot stars (O– and B-type) as well as of cooler T Tauri variable stars whose spectral classes vary in the range G to M. Most often these aggregates of two entirely different types of stars overlap and are together embedded in extensive thick nebulosities. The two types of aggregates are distinguished as *O-association* and *T-association,* but they are believed to have been formed out of the same nebulous matter and at the same time.

It is to be emphasized that stellar associations are different from star clusters. We have already mentioned that while the average star density in a cluster is much higher (10 to 100 times) than that in the surrounding field, the average density in an association is much lower in comparison. Associations are therefore recognized only by observing the space distribution of early-type stars. The *Catalogue of Clusters and Associations* by G. Alter and co-workers lists 50 O-associations and 25 T-associations. There may be as many as 10^4 O-associations present in our Galaxy; two of the most prominent members are the Orion association and the Scorpio-Centaurus association. The other important nearby associations are II Perseus, I Lacerta and I Cepheus. It is difficult to make an exact estimation of the number of T-associations in the Galaxy because these objects being much fainter cannot be seen at great distances. The number is likely to be comparable with that of O-associations. The dimensions of an association are, in general, very great. The linear diameters of the Orion association and the Scorpio-Centaurus association are about 130 pc and 300 pc respectively and their respective distances are 470 pc and 1300 pc.

For the determination of the space distribution of members of the associations, one has to calculate the distance to these stars. The distances can be determined by the method of spectroscopic parallax on the basis of two-dimensional spectral classification (that is, on spectrum and luminosity classes). Distances of a large number of O-associations have been determined. When these distances are projected on the galactic plane, the resulting distribution of the associations gives a spiral impression. The three spiral arms, the Orion, the Perseus and the Sagittarius arms, are clearly defined by associations.

The associations are very loose aggregates of stars so much so that they cannot withstand the tidal pull of the general gravitational field of the Galaxy against rapid disruption. And in the presence of the differential galactic rotation, the associations should undergo sufficient elongation before they are actually disrupted. But we do see many associations to exist at present and they do not appear very much elongated either. These facts led Ambertsumian to draw two very important conclusions regarding these objects. First, he asserts that these associations are *very young* objects, and secondly, their elongation is masked by the general expansion they are undergoing. The assertion that the associations are very young is supported by the presence of large number of O- and B-type stars in them. The stellar evolution theory predicts that these stars cannot live at the observed phase more than a few times 10^7 years. Also, the presence of T Tauri stars in them in large numbers confirms the extremely recent origin of these objects. T Tauri stars are pulsating and exhibit emission lines, proving that they are losing mass. These stars are believed to have been born very recently and are still undergoing gravitational contraction prior to their settlement on the main sequence. If the average lifetime of an association is assumed to be of the order of 3×10^7 years, since the age of our Galaxy is of the order of 10^{10} years, there would have been about 300 generations of stellar associations born so far in our Galaxy. If there are at present 10^4 associations, then about 3×10^6 associations have been born in the Galaxy during its currently estimated lifetime and they have contributed 3×10^8 to 3×10^9 stars to the Galaxy, if each association consists of between 100 and 1000 stars. Thus, a substantial part of the entire stellar community in our Galaxy may have originated through associations. In particular, *all,* the O- and B-type stars are believed to have originated in associations. This is corroborated by the observed fact that almost all the O stars are found in associations. There are hardly any O stars in the general stellar field uncorrelated with associations, but B stars are found in sufficient number in the general field. This finds the most plausible explanation in that the life of O stars ends before the association is disrupted, but the B stars outlive the disruption and mingle into the general field. The study of stellar associations thus indicates the very important fact that stars are being born continuously in our Galaxy.

If associations are expanding, as suggested by Ambertsumian, the member stars must have reached their present distribution moving with certain common speed from the initially convergent position of their origin. This position and the time during which the expansion has been going on may be deduced from a study of the spatial arrangement and space motion of the member stars. But since the motion, in general, is not very large, the three-dimensional arrangement is difficult to calculate. It can be reliably achieved only in the case of *runaway stars* (stars leaving the parent association with high velocities). Alternatively, the expansion is generally measured by calculating the proper motions and radial velocities of member stars. If the stars are linearly expanding from their centre of origin with certain distribution of speeds in arbitrary directions, then the present position of stars will be at distances proportional to their speeds. The measured proper motions and radial velocities should therefore reflect the speeds and positions due to linear expansion. If r is the stellar distance from the centre, v the velocity of the star and t the *expansion age* of the association, then $r = vt$, so that $v/r = t^{-1}$ is the rate of expansion. The fundamental work correlating the proper motions and radial velocities with linear expansion was done by A. Blaauw and others. In these calculations an expansion term r/t emerges in terms of other known quantities. Since the centre of expansion is known as the point of convergence of the proper motions, one can calculate the distance r of individual stars and thus determine the expansion age. The works of the astronomers such as Blaauw, Parenago and others have revealed that the rate of expansion (t^{-1}) is of the order of 10 km s^{-1} for most associations and the expansion age ranges from a few times 10^5 years to a few times 10^6 years. For Scorpio-Centaurus association Blaauw deduces an expansion age of 2×10^6 years and an age of 3×10^5 years has been derived for I Orion association by P.P. Parenago and by K.A. Strand. Similarly, the expansion ages of the associations IC 2602, I Gepheus and I Lacerta have been computed as 3×10^5, 4×10^6 and 7.4×10^6 years respectively. We thus see that most of the associations that are at present recognized are hardly older than 10^7 years. They are completely disrupted to be unrecognizable as associations within a time-scale of 2 or 3 times 10^7 years, after which their constituent stars mingle among the general field stars in the Galaxy.

11.5   STELLAR POPULATION CHARACTERISTICS

As early as in 1920s Harlow Shapley observed that all stars in the solar neighbourhood and those in galactic clusters imitate an H–R diagram similar to Russell's original diagram, whereas the H–R diagram of globular clusters were at variance with it in many important aspects. The real implication of Shapley's work however failed to attract astronomers for about two decades. The true interpretation was given by Walter Baade who introduced in 1944 the concept of two *stellar populations*. This was possible when Baade became able to resolve individual stars in the central bulge of M31 and in the elliptical galaxies NGC 221 (M32), NGC 205, NGC 147 and NGC 185, using red-sensitive plates at the focus of the 100-inch reflector at Mount Wilson. Baade found that the stars in the elliptical systems and in the central bulge of M31 were physically similar to those in globular clusters and their H–R diagrams are alike. The common observed feature in all of these systems is that the stars in red giant branch are much more luminous than the blue giant stars, unlike those which we observe for common stars in the solar neighbourhood. These observations led Baade to advance the hypothesis of the existence of two stellar Population types. The stars in the globular clusters and those similar to them were named Population II while those in the solar neighbourhood including the stars in galactic clusters were called Population I.

Baade's original concept of classification of all stars into two distinct population types was primarily based on *physical properties* and *space distribution* of objects belonging to these population types. Spectroscopically, the Population II objects are metal-deficient. Their metal content is only about 1 to 10 per cent of those of the Population I objects. As regards the luminosity, the brightest stars of Population II are *red giants* with $M_v \approx -3.0$, but in Population I the brightest stars are *red* and *blue supergiants* with $M_v \approx -7.0$ to -8.0 or even more. The high luminosity blue dwarfs $(M_v \approx -6.0$ or $-7.0)$ are exclusively the objects characteristic of Population I. Again, the space distribution of the objects belonging to the two population is markedly different. While the Population I objects are strongly concentrated to the galactic disk and spiral arms, those of Population II show a strong tendency to avoid these regions. The latter are distributed mostly away from the galactic disk. They are rather found in the spherical bulge in the central region of the Galaxy and in the halo, besides in globular clusters.

Subsequent investigations have revealed that the objects in the two populations differ not only in physical and positional features as discussed above, but the difference in their kinematical behaviour is also no less important. Population I objects being associated with the galactic disk share the rotational motion of the disk as the Sun does, and so the components of motion of these objects with respect to the Local Standard of Rest are usually small. On the other hand, the Population II objects being members of a more or less spherically symmetric system possess relatively large radial motions. They share little of the rotational motion of the galactic disc, and as such, possess large velocities with respect to the Local Standard of Rest.

On the basis of these differences in physical, distributional and dynamical properties, all objects and matter in the Galaxy can be roughly classified into the following two population types:

Population I	*Population II*
(i) Gas and dust	(i) Globular clusters
(ii) Hot stars and H II regions	(ii) RR Lyrae variables (Cluster-type variables)
(iii) Galactic clusters	(iii) Classical cepheids of Type II
(iv) Main sequence stars	(iv) Long-period variables with periods less than 150 days
(v) Normal giants and supergiants (non-variable)	(v) Subdwarfs
(vi) Classical cepheids of Type I	(vi) Planetary nebulae
(vii) Long-period variables with periods longer than 150 days	

The existence of these two entirely dissimilar types of objects in the Galaxy can be understood in the light of the evolution of the Galaxy through time, as different components were formed at different ages under different conditions prevailing in the Galaxy. Both physical and dynamical properties of Population II objects may be interpreted to bear evidence that these are *old* objects formed at an early phase of the Galaxy. At this phase, the Galaxy was more or less spherical and the matter possessed large radial components of motion. The rotational motion at this phase was not as large as it was at later phases. This explains the observational fact that Population II objects constitute a more or less spherical system in the Galaxy. But in the later phases as the gravitational contraction of the Galaxy proceeded, it attained more and more rotational velocity by virtue of the conservation of angular momentum, and became more and more a flattened system. It was at these later phases that the Population I objects were formed. This aptly explains their being restricted in a highly flattened system with large rotational but less random motions. The matter in the Galaxy out of which new stars have been formed had become *regimented,* as it were, being concentrated to the disk and the spiral arms. These objects therefore take part in the large rotational velocity of the galactic disc that developed in later phases of gradual contraction. They deviate very little from circular velocity. On the other hand, Population II objects possess great deviation from circular velocity of the galactic disc and share large radial components of motion.

The fact that Population II objects are old is also substantiated by spectroscopic evidences. They were formed in the early stage of the Galaxy when heavy elements were not formed in large quantities. In the later phases, larger amounts of heavier elements were ejected into the interstellar space from evolved and expired stars. This metal-enriched gas was then used as the raw material for the new generations of stars which form Population I. This clearly explains the observed spectroscopic feature of low metal abundance in Population II and high metal abundance in Population I objects. Thus Baade's concept of two stellar populations, which was advanced as a radical hypothesis at that time, subsequently gained acceptance on the grounds of physical and kinematical verifications. During the following decade the stellar population concept gave a tremendous impetus to astronomical researches both in observational as well as in theoretical fields of investigations.

However, knowledge was gained over the years that Baade's concept of the existence of two entirely dissimilar populations in the Galaxy was an oversimplification. The Galaxy, in fact, contains a mixture of objects formed almost continuously throughout its lifetime. Stellar associations provide unmistakable evidence that stars are being formed even now. The rate of formation was, of course, higher in early phases when more gas was available for star formation. This concept of continuous star formation and of the Galaxy containing a mixture of objects of all ages is now well accepted by astronomers. The globular clusters, subdwarfs and those RR Lyrae stars whose periods are longer than $0.^{\rm d}4$ are examples of what may be called *extreme Population II.* On the other hand, the gas, blue hot stars, members of stellar associations, Cepheid variables of Type I and the youngest galactic clusters form the *extreme Population I.* The other intermediate assemblages of objects link up smoothly these two extreme populations. These are *Intermediate Population II,* the *Disc Population* and *Intermediate Population I.* There is some controversy, however, as to the membership of the Sun in these population types. It is generally assumed that the Sun belongs to Population I. But the currently estimated age of at least 5×10^9 years would place the Sun rather in older population. On the other hand, its relatively high metal abundance would preclude it from being a member of a very old population. It will therefore be more reasonable to place it among the Disc Population objects. The present classification may undergo further changes in future. Availability of better techniques of observation, a better theoretical understanding of observed data coupled with newly discovered characteristics of various objects may lead to a far better knowledge of the subject.

11.6 STAR FORMATION

In the earlier sections, we have discussed the basic characteristics of the clusters and associations of stars. We have also seen how stars could be divided into different populations on the basis of the 'epochs' of their formation as the Galaxy evolved in time. In the present section we shall try to discuss our current understanding of as to how stars and clusters of stars are formed out of the interstellar matter.

The colour-magnitude diagram of the galactic clusters in Fig. 11.3 clearly demonstrates that the entire spread of age over which these clusters were formed is several billion years— from the very old cluster NGC 188 to rather recently formed clusters such as h and χ Persei and NGC 2362. Also, the colour-magnitude diagram of NGC 6530 in Fig. 11.4 is a clear proof that clusters are forming even in the present epoch. Star formation in our Galaxy (and also in other galaxies) thus appears to be a continuous process. Furthermore, the observation of some O-associations and T-associations embedded in thick and massive cloud complexes indicates that stars form most probably in groups in large and massive interstellar clouds, although the probability of formation of single stars cannot be ruled out at the moment.

All theories based on current observations indicate that stars are formed as a result of large-scale gravitational instability developed in the central region of the massive molecular clouds. Instability leads to collapse and breaking into pieces of the original cloud. Each subunit subsequently suffers further collapse and fragmentation leading to the birth of a group of protostars within the cloud, surrounded by the diffuse envelope. The collapse is isothermal, the compressional heat generated during collapse being radiated away in infrared and other low frequency radiation via heated grains and excited low lying levels of molecules.

The mathematical treatment of the problem of gravitational instability and collapse was first considered by Sir James Jeans in 1902. He considered the very simplified case of a gas in a static, uniform, equilibrium condition in which the velocity $v_0 = 0$ and the density and pressure ρ_0 and p_0 are constant. Assuming now a small perturbation in all the physical variables such as $\rho = \rho_0 + \rho_1$ and linearizing the relevant equations in the small perturbed quantities with subscript 1, the wave equation is derived. The relevant equations are the equation of continuity, Euler's equation of motion, Poisson's equation and the ideal gas equation. The linearized equations obtained are

$$\frac{\partial \rho_1}{\partial t} + \rho_0 \nabla . \mathbf{v}_1 = 0 \qquad (11.1)$$

$$\frac{\partial \mathbf{v}_1}{\partial t} = -\frac{1}{\rho_0} \nabla p_1 - \nabla \phi_1 \qquad (11.2)$$

$$\nabla^2 \phi_1 = 4\pi G \rho_1 \qquad (11.3)$$

and

$$p_1 = \frac{kT\rho_1}{\mu m_H} = C_s^2 \, \rho_1 \qquad (11.4)$$

where the isothermal condition of gas has been assumed, C_s is the isothermal speed of sound and the other symbols have their usual meaning. Equations (11.1) to (11.4) can be combined to derive the wave equation

$$\frac{\partial^2 \rho_1}{\partial t^2} = C_s^2 \nabla^2 \rho_1 + 4\pi G \rho_0 \rho_1 \qquad (11.5)$$

which admits of a plane wave solution

$$\rho_1 = A \, \exp[i(\kappa x - \omega t)], \qquad (11.6)$$

where κ is the wave number and ω the wave frequency. We have $\kappa = 2\pi/\lambda$ as the wavelength of perturbation. Substituting Eq. (11.6) in (11.5) we get the dispersion relation as

$$\omega^2 = C_s^2 \kappa^2 - 4\pi G \rho_0 \equiv C_s^2 \, (\kappa^2 - K_J^2) \quad \text{(say)} \qquad (11.7)$$

If $\kappa^2 < K_J^2$ then $\omega < 0$, so the disturbance grows exponentially in time. $\lambda_J = 2\pi/K_J$ is defined as the Jeans critical wavelength. Its value is given by

$$\lambda_J = 2\pi/K_J = \left(\frac{\pi kT}{\mu m_H G \rho_0}\right)^{1/2} = 6 \times 10^7 \left(\frac{T}{\mu\rho_0}\right)^{1/2} \text{cm} \qquad (11.8)$$

Thus, any fluctuation with wavelength $\lambda > \lambda_J$ will grow exponentially in time and will be unstable. In other words, a cloud of mass having a dimension larger than λ_J will be gravitationally unstable and will contract continuously under gravitational pull. The critical length λ_J defines the critical mass of a sphere with diameter λ_J which is given by

$$M_J = \frac{\pi}{6} \rho_0 \lambda_J^3 \simeq 10^{23} \left(\frac{T}{\mu}\right)^{3/2} \rho_0^{-1/2} \text{gm} \qquad (11.9)$$

M_J is defined as the Jeans mass. Any cloud of mass $M > M_J$ will be gravitationally unstable and will suffer a continuous collapse so long as the isothermal state is maintained by the collapsing cloud.

Jeans criterion of gravitational collapse of a static homogeneous and infinite medium under gravitational forces alone was however too simplified an assumption. Subsequently, more rigorous works of S. Chandrasekhar, E.N. Parker and others have established the fact, that the collapse cannot be averted even when the gas is subjected to other forces, say, arising out of uniform rotation, turbulence and uniform magnetic field, acting singly or even jointly.

Most of the calculations of collapse have been performed for spherical clouds. Once the gravitational instability sets in and the cloud starts collapsing, it will continue to contract with an acceleration so long as the cloud remains 'optically thin' to the compressional energy radiation, that is, so long as the process remains isothermal. Since the gravitational force per unit volume varies as M/R^5, whereas the pressure gradient force varies as MT_c/R^4 (where M and R are respectively the mass and radius of the cloud and T_c is the central temperature) the pressure force cannot resist the gravitational collapse if T_c remains constant. So the isothermal collapse proceeds with an acceleration without any resistance. Such a collapse has been called a *free-fall* collapse, and the time required for a mass element of a sphere of uniform density to reach from the surface to the centre is called the *free-fall* time given by

$$t_{\text{ff}} = \left(\frac{3\pi}{32 G \rho_0} \right)^{1/2} \text{sec} \qquad (11.10)$$

ρ_0 being the initial density of matter. The essential condition for the free-fall collapse to persist is that the compressional heat generated in the gas due to contraction is radiated away in a time-scale much shorter than the free-fall time, that is, $t_{\text{cool}} < t_{\text{ff}}$. Cooling is then said to be efficient. In a molecular cloud the important cooling agents are the dust grains and H_2 molecules. The grains are heated by the release of compressional energy which then radiates in infrared wavelengths. Similarly, H_2 rotational levels are collisionally excited, and downward transition radiates energy in infrared and other low frequency lines. Calculations show that cooling is very efficient and cooling time much shorter than the free-fall time in a standard molecular cloud of $T \sim 50$ K, $\rho \sim 10^{-21}$ gm and $\mu = 2.3$. Thus, the gas remains cold and isothermal and the collapse remains free-fall.

As the collapse starts, the initial density distribution within the cloud changes, whether it is assumed uniform or not. The collapse has been found to be highly nonhomologous irrespective of the initial density distribution. As the collapse advances, the density distribution becomes strongly centrally condensed—ultimately leading to a sharp central spike in the density distribution. The rapidly collapsing central core of the cloud shrinks in size and mass, while the extensive outer region of the cloud is left behind which collapses more slowly having its size and mean density not much different from the initial values. The central density of the gas continues to increase till it attains the critical density of $\sim 10^{-13}$ gm cm^{-3} at which the opacity starts building up in the cloud, terminating the phase of its isothermal collapse. The density distribution in the inner part of the Mailing cloud resembles that due to free-fall, that is, $\rho \propto r^{-3/2}$, while in the outer slowly collapsing region the density law remains as $\rho \propto r^{-2}$ resulting from the initial isothermal phase of the collapse. However, the region where the density law $\rho \propto r^{-3/2}$ applies, expands outwards as more and more gas falls into the centre, finally occupying almost the whole region of the infalling cloud.

As the gravitational instability sets in, not only the cloud contracts as a whole but also within the cloud condensations of subunits soon multiply. This phenomenon has been popularly called *fragmentation* of the cloud. In a contracting cloud, the process of fragmentation is accentuated for two reasons. First, so long as the density of gas in the cloud remains low, up to $\sim 10^{-21}$ gm cm^{-3}, the cloud is heated by penetrating high energy photons and low energy cosmic rays. But as the density exceeds 10^{-20} gm cm^{-3}, this heating becomes inappreciable below the surface layers of the cloud. The inner regions being cut-off from sources of heating and the gas density being higher there, the cooling becomes very efficient and the gas temperature can reduce to a value as low as 10 K. The high density and low temperature of the gas in the central region of the cloud reduce the Jeans critical masses M_J to lower and lower values. Thus, subunits of much lower masses form as the density increases by continuing contraction. This process of continued formation of subunits of smaller and smaller M_J in the inner region of the cloud has been aptly termed by Fred Hoyle as *hierarchical fragmentation* of the cloud. Detailed computational results indicate that this hierarchical fragmentation process continues within the cloud as long as the collapse remains isothermal, and the critical density at which the process terminates is ~10^{-13} gm cm^{-3}. At this density the gas within the fragment starts trapping its compressional heat and thus departs from the condition of isothermality. Any further fragmentation ceases at this stage and we get the protostars of minimum mass. Under favourable conditions, the minimum protostellar mass can be as low as 0.002 $M_\odot$ for T ~ 10 K, ρ ~ 10^{-13} gm cm^{-3} and μ = 2.3. But since subcondensations with different densities and temperatures generally form within the body of the cloud, protostars of different masses form in the central region of a contracting cloud. Thus gravitational collapse of a massive molecular cloud can give birth to a group of protostars having different masses that obey a mass distribution law of the form

$$N(m) = Am^{-\alpha} \tag{11.11}$$

where A and α are constants and $N(m)$ is the number of stars in the mass range m and $m + dm$. Calculations show that under a wide range of parameter values involved in the computation of protostellar masses, $2 < \alpha < 3$. The distribution of stellar masses as given by Eq. (11.11) is called the stellar mass function (SMF) and α represents the slope of SMF. The mass function produced just at the termination of the hierarchical fragmentation process is known as the Initial Mass Function (IMF). The IMF is then modified slowly due to the accretion of infalling gas on each individual protostar and also due to coalescence and disruption of protostars by mutual collisions.

The process described above has been popularly called as the *opacity limited fragmentation.* Subsequently, each protostellar body undergoes slow adiabatic contraction as increase in density inside leads to gradual increase in opacity. Finally, the protostellar body settles on the main sequence provided it has at least a mass of ~ 0.08 $M_\odot$ in order to be able to ignite the nuclear fuel at its centre. Calculations indicate that in a standard molecular cloud suffering gravitational instability and collapse, the hierarchical fragmentation process for formation of the first generation of stars is completed within less than ~ 10^5 years which is its free-fall time. Star formation may further continue in the regions of the cloud somewhat away from the centre. These new generations of stars are likely to form with higher masses because the density of gas is somewhat lower there and the temperature higher due to the radiation received from the central stars. The process of star birth can thus be continued till the cloud is dispersed by high energy radiation from massive hot stars formed within, and by the outburst of supernovae within the cloud. Calculations further show that if the star formation efficiency (SFE), that is, the ratio of the mass of the cloud converted to stars and the total mass of the initial cloud be at least fifty per cent, then the group of stars formed may hold together to form a bound star cluster. Otherwise, the group will be unable to hold together due to the action of the galactic disruptive forces and will gradually disperse and merge in the general field of the galaxy. The above picture of star formation is an extremely simplified one. The real problem is very much involved and complicated the full discussions of which are beyond the scope of this book (also see Chapter 14).

EXERCISES

1. Differentiate between a *group* and a *cluster* of stars. On what observational bases one can recognize a group of stars as a *cluster.*

2. Explain how the colour-magnitude diagrams are used to calibrate the galactic distance scale.

3. "The colour-magnitude diagrams of galactic clusters show clearly that these clusters are not coeval"—Explain.

4. Explain how the colour-magnitude diagrams of some galactic clusters prove that these clusters have been formed very recently.

5. Draw a schematic colour-magnitude diagram of a globular cluster and indicate the principal points of differences between this diagram and that of a galactic cluster. What, in your opinion, are the causes of this difference?

6. Describe how the galactic and globular clusters behave in the following respects:
   (a) Galactic distribution
   (b) Kinematic properties
   (c) Structure, size, and stellar number content
   (d) Chemical composition

7. Summarize your ideas of stellar population characteristics. In what respects do the Population I and Population II objects differ?

8. What are the important features of stellar associations? How do they differ from star clusters?

9. How do we know that stellar associations are expanding? Define *expansion age* of an association. How is the expansion age determined?

10. On what evidences can we conclude that stellar associations are extremely young objects?

11. Define Jeans length, λ_J, and Jeans mass, M_J. Derive expressions for them.

12. Explain Hoyle's concept of hierarchical fragmentation of a cloud, stating the conditions necessary for the progress of hierarchical fragmentation.

13. Calculate Jeans masses of clouds of pure H_2 molecules in which the following conditions prevail:
    (a) $T = 10\ K$, $n = 10\ cm^{-3}$;
    (b) $T = 100\ K$, $n = 10^3\ cm^{-3}$;

(c) $T = 100\ K$, $n = 10^6 \text{cm}^{-3}$;

(d) $T = 50\ K$, $n = 10^2\ \text{cm}^{-3}$;

obtain the corresponding Jeans masses if the clouds have cosmic abundance (assume 9 helium atoms for 100 hydrogen).

14. Find for $M_J = 1\ M_\odot$, T when $n = 10^8\ \text{cm}^{-3}$, and n when $T = 100$ K for a hydrogen cloud.

15. Find the radius of the protosun when its temperature $T = 10^3$ K and $n = 10^6\ \text{cm}^{-3}$.

16. Discuss how the minimum Jeans mass is obtained. Find the minimum Jeans mass of a hydrogen cloud at a temperature of 20 K. Compare this mass with the mass of the Sun.

SUGGESTED READING

1. Blaauw, A. and Schmidt, M. (Eds.), *Galactic Structure,* The University of Chicago Press, Chicago, 1965 (Chapters authored by: H. Arp, A. Blaauw, S. Shapless).

2. Cohen, Martin, *In Darkness Born: The story of star formation,* Cambridge University Press, Cambridge, 1988.

3. Goldberg, Howard S. and Scadron, Michael D., *Physics of Stellar Evolution and Cosmology,* Gordon and Breach Science Publishers, New York, 1986.

4. Goodman, J. and Hut, P. (Eds.), *Dynamics of Star Clusters,* D. Reidel Publishing Company, Dordrecht, Holland, 1985.

5. Hanes, D.A. and Madore, B.F. (Eds.), *Globular Clusters, Cambridge* University Press, Cambridge, 1980.

6. Hesser, J.E. (Ed.), *Star Clusters,* D. Reidel Publishing Company, Dordrecht, Holland, 1980.

7. Janes, K. (Ed.), Formation and Evolution of Star Clusters, *ASP Conference Series,* **13**, 1991.

8. Kutner, Marc L., *Astronomy,* Harper and Row Publishers, New York, 1987.

9. McLaughlin, Dean B., *Introduction to Astronomy,* Houghton Mifflin Company, Boston, 1961.

10. Spitzer, L. (Jr.), *Physical Processes in the Interstellar Medium,* John Wiley & Sons, New York, 1978.

11. Tenorio-Tagle, G., Prieto, M. and Sanchez, F. (Eds,), *Star Formation in Stellar Systems,* Cambridge University Press, Cambridge, 1993.

12

Galactic Nebulae

12.1 INTRODUCTION

Although most of the visible mass (nearly 90%) of the Galaxy is concentrated in stars, the vast regions of space between the stars is not completely devoid of matter. The interstellar matter which comprises only about 10% of the mass of the Galaxy has an overwhelming importance in maintaining the birth of new generations of stars and thus is responsible for the cosmic recycling process. Authentic evidences have now been gathered revealing that new stars are born in regions of large local concentrations of interstellar matter. Ionized and neutral gas, both atomic and molecular, fills with varying concentration, the entire interstellar space. Mingled with the gas, there are also minute solid particles of undetermined structure and composition which are more popularly called the *interstellar grains* or *interstellar dust* The gas and dust pervade the vast interstellar space. Also, there are cosmic rays bombarding the entire regions of interstellar space and a general magnetic field pervading this space. The interaction of cosmic rays with the galactic magnetic field and various physical and dynamical processes in ionized gas lead to many interesting but quite difficult problems, which make the study of the physics of interstellar space quite fascinating. While the former is responsible for most of the nonthermal galactic background radiation in the radio wavelengths, the thermal component of background radiation mostly originate in the latter. Interstellar grains play at least two vital roles. First, the polarization of starlight by them yields clues to the orientation of the galactic magnetic field. Secondly, the dimming and reddening of starlight by them yield clues to their size and structure.

In this chapter our discussions will be limited to the areas of dense concentration of interstellar matter. These isolated concentrations in the vast interstellar space have been known as nebulae for a long time. Some of these appear as dark obscuring clouds while others appear as bright clouds with various amounts of surface brightness. Dark nebulosities in Taurus, the Horsehead Nebula (Fig. 12.1) in Orion and Southern Coalsack are examples of the former, while the latter is best illustrated by examples like the nebulosity surrounding the stars of the Pleiades (Fig. 12.2), the Lagoon Nebula in Sagittarius (Fig. 12.3) and the Great Nebula in

Orion (Fig. 12.4), the nebula in Eta Carinae, the Ring Nebula in Lyra (Fig. 12.5), the Crab Nebula in Taurus (Fig. 12.6), etc.

The existence of the dark nebulae was not recognized by the astronomers until the turn of the present century. The works of E.E. Barnard and M. Wolf ushered in great advances in this field. Based on his photographic survey of the Milky Way, Barnard published a list of 349 dark nebulae in 1927. The catalogue gives position and size of each of these objects together with a brief description of each. Almost at the same time the Lundmark catalogue containing 1550 objects was published, giving the positions, sizes and description of all these objects. In 1934, the Ross-Calvert photographic *Atlas of the Milky Way* was published which subsequently served as the source work for two more catalogues of dark nebulae. The first of these was compiled by J. Khavtassi in 1955. Khavtassi also published an atlas, the *Atlas of Galactic Dark Nebulae* in 1960. This work was based on Ross-Calvert *Atlas of the Milky Way* and F.A. Hayden's Photographic *Atlas of the Southern Milky Way* published in 1952. Khavtassi's Catalogue lists 797 nebulae together with their galactic positions in different coordinate systems; their sizes in square degrees and an estimate of their opacity. The second catalogue was compiled by E. Schoenberg in 1964. The sources of this catalogue were the Ross-Calvert plates as well as the Lick photographs of the Milky Way. This catalogue gives the equatorial and galactic coordinates, sizes, shapes, and estimates of absorption of each of the 1456 dark nebulae listed therein. The most extensive catalogue of dark nebulae was compiled by B.T. Lynds in 1962. The material of this work was adopted from the Palomar sky survey red and blue photographs taken with the 48-inch Schmidt Camera at the Palomar 200-inch reflector. The catalogue lists 1802 dark nebulosities mostly contained within the general field of dark cloud complexes in Milky Way. The catalogue gives the equatorial and galactic coordinates, the surface area and a visual estimate of the opacity, for each of the clouds. An inherent difficulty is involved in the area determination of the dark nebulae, because the work is extremely sensitive to the limiting magnitude accessible as well as to the wavelength sensitivity of the photographic plates. Since the outer parts of dark nebulae are transparent to red light, the areas of these nebulae determined from red plates are systematically smaller than those determined from blue plates. The same effect is also exhibited by two blue plates having different limiting magnitudes, since very tenuous nebulae are not resolved except in high resolution photographs.

Great advances were made in observations of bright nebulae by the Herschels, William and his son John, by Charles Messier and by J.L.E. Dreyer during the later half of the eighteenth and the whole of the nineteenth centuries. As early as in 1781, Messier compiled a catalogue of 103 'nebulous-appearing' objects which included such diverse objects as star clusters, galaxies and true nebulae. The objects in Messier's catalogue are called by their number with prefix M before the number, for example, M31 stands for the 31st object in this catalogue which is the great spiral nebula in Andromeda. Several catalogues of nebulae were compiled by William Herschel during the last two decades of the eighteenth century, and in 1864 his son, John Herschel published the *General Catalogue of Nebulae*. This catalogue contains 5079 objects about 90% of which were discovered by him and his father. Herschel's *General Catalogue* was revised and enlarged to *New General Catalogue* by Dreyer in 1888 containing 7840 nebulae and clusters. This catalogue includes most of the bright nebulous objects in the sky, and most of the bright galaxies are now named by their numbers in this catalogue, such as NGC 1068. It includes all the objects already in Messier's catalogue and these common objects may be called by any of their names, such as NGC 224 or M31, both standing for the spiral galaxy in Andromeda. The new General Catalogue was subsequently enlarged to include nearly 15,000 nebulae by adding two supplements known as the *Index Catalogues;* the first supplement was published in 1895 and the second in 1908.

All these catalogues, however, list under the name *nebulae* such different objects as the true gaseous nebulae, planetary nebulae, star clusters, supernova remnants and remote galaxies. All these objects appear as unresolved, bright patches of gaseous matter when viewed visually through small and moderate-sized telescopes. As the true nature of these objects were unknown to early observers, these were all called by the common name *nebulae* as suggested by their appearance. But as we shall see in the next section, today the word *nebulae* is used by most of the astronomers to denote only the true gaseous nebulae belonging to our own Milky Way. These are clouds of gas and dust lying close to the equatorial plane of the Galaxy.

12.2 CLASSIFICATION AND GALACTIC DISTRIBUTION OF NEBULAE

The early observers could not distinguish between a luminous gaseous cloud belonging to our Milky Way and a far-off external stellar system like the Andromeda galaxy, because both appear alike when viewed through a telescope. The true nature of the latter objects was known mainly through the work of Edwin P. Hubble in 1920s when he discovered large red-shifts of the spectral lines in their spectrum. The large red-shifts at once revealed that these apparently nebulous objects must be hundreds of kiloparsecs or even many megapersecs away. Their bright appearance as observed at such great distances clearly indicates that these objects must be as luminous as an entire galaxy, Hubble further observed that the nebulae as one in Orion do not exhibit any such redshifts in their spectral lines. Hubble therefore called the former group of nebulae as the *extragalactic nebulae* while those like one in Orion were distinguished as the *galactic nebulae.* The extragalactic nebulae are each an individual galaxy while the galactic nebulae are local concentrations of galactic gas and dust mostly lying close to the equatorial plane of the Galaxy. A majority of extragalactic nebulae were found to possess regular shapes like projected spheroids or discs, a large number of them exhibiting spiral patterns, while a minority exhibiting no regular structure. These were later classified by Hubble as the elliptical, spiral and irregular galaxies.

In the present context, we shall limit our discussion to the galactic nebulae. These are luminous gas clouds as well as illuminated and dark dust clouds belonging to our own Galaxy. A schematic division of these nebulae into different types is given here.

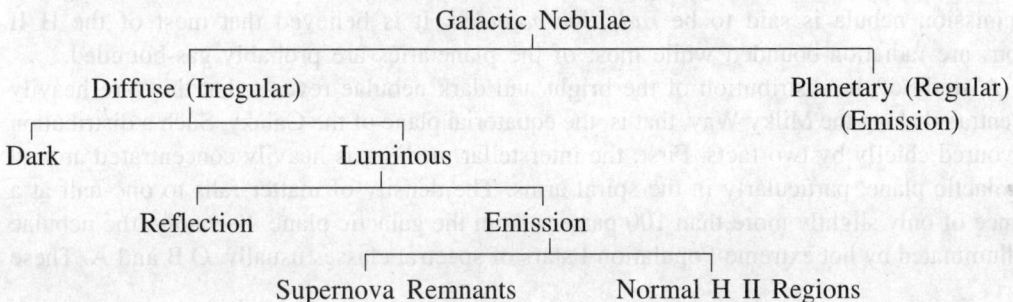

Galactic Nebulae

Diffuse (Irregular)   Planetary (Regular)
                                        (Emission)

Dark   Luminous

         Reflection   Emission

                Supernova Remnants   Normal H II Regions

From the external appearance the nebulae can be divided into *diffuse* and *planetary,* the latter being luminous patches with regular shapes, having a central hot star at the middle. The diffuse (irregular) nebulae are divided into *dark* and *luminous,* the latter being spectroscopically separated into two distinct classes, viz. *reflection* and *emission.* The diffuse emission nebulae are clearly divided into two distinct classes of entirely different types of origin. Of these, the *supernova remnants* created by exploding massive stars have been discussed in Section 12.8. The second type (Normal H II Regions) which are local concentrations of hot gas ionized by the high-energy radiation of hot stars associated with these concentrations will be first discussed in the following passages. Because of the overwhelming abundance of hydrogen in cosmic matter, these ionized concentrations are commonly known as H II regions.

The planetary nebulae are formed by the shells ejected by the central stars at certain critical stage of their evolution. Since the material is ejected symmetrically, these nebulae maintain regular shapes surrounding their central stars. They constitute a class of emission nebulae. The diffuse nebulae, on the other hand, are the results of random distributions of galactic gas and dust, and as such have no regular structure. The luminosity of a diffuse nebula is imparted by a nearby star or a group of stars lying close to or embedded in it. If there are no stars close enough to illuminate a nebula, it will appear as a dark nebula. Thus, dark and bright nebulae are compositionally the same; the difference lies in the presence or absence of an illuminating source in the neighbourhood of the nebula. Hubble's study at Mount Wilson Observatory in early 1920s led to the understanding that the bright nebulae could not be self-luminous and that they owe their brightness entirely to the stars close to or embedded in them. Hubble further concluded that if the illuminating star or stars were of spectral classes B0 or earlier, the associated nebula will be an emission nebula. It will be a reflection nebula, if the illuminating star or stars happen to be of spectral classes later than B0. The large flux of ionizing quanta from O and B0 stars into the surrounding nebula ionize all or a part of its gas, which acts as a source of the emission lines. The nebula in this case possesses a spectrum of its own, which is entirely different from the spectrum of the associated star or the group of stars. The intensity of flux of ionizing quanta strongly decreases with later spectral classes and for stars later than class B1, the flux is so small that it is insufficient to ionize even a minute fraction of the nebular gas. Under such condition the nebula is no longer recognizable as an emission nebula, but the radiation from the associated star or stars may still be sufficient to illuminate a large portion of the nebular gas and dust. Stellar light is then just reflected by the dust particles in the nebula, thus giving rise to reflection nebula. In this case, the nebular spectrum becomes identical with that of the associated star. An emission nebula is said to be *gas-bounded* if the associated hot star (or stars) ionizes all the gas around it. On the other hand, if the stellar ionizing flux is not sufficient to ionize all the surrounding gas, the emission nebula is said to be *radiation-bounded.* It is believed that most of the H II regions are radiation-bounded while most of the planetaries are probably gas-bounded.

A study of the distribution of the bright and dark nebulae reveals that they are heavily concentrated along the Milky Way, that is, the equatorial plane of the Galaxy. Such a distribution is favoured chiefly by two facts. First, the interstellar medium is heavily concentrated around the galactic plane, particularly in the spiral arms. The density of matter falls to one-half at a distance of only slightly more than 100 parsecs from the galactic plane. Secondly, the nebulae are illuminated by hot extreme Population I stars of spectral classes usually, O B and A. These stars, believed to have been born recently out of the thick interstellar medium within the spiral arms (which lie in the galactic plane) also lie close to this plane. These facts are consistent with observations that most of the conspicuous diffuse nebulae, both dark and luminous, lie within 200 pc of the galactic plane. Beyond this distance, both gas and hot stars drastically thin out. Thus the diffuse nebulae as well as the hot early-type stars, both constitute two almost entirely overlapping subsystems. This fact underlies the whole story of diffuse nebulae. Even if one of these subsystems extended far away from the galactic plane, the nebular phenomena would have been conspicuously absent due to the absence of the other. In fact, the diffuse nebulae are distributed in the spiral arms apparently in such close groups, that the entire length of the spiral arms may be considered as a large system of diffuse nebulae.

The story is, however, different with the planetary nebulae. While the diffuse nebulae hardly show any pronounced concentration in galactic longitudes, the planetaries do show a sharp maximum in their distribution near the galactic centre between longitudes 330° to 360°. Their preference to lie close to the galactic plane is much less pronounced compared to that of the diffuse nebulae. Both these distributional features of planetaries can be explained on the basis of the current understanding of the evolution of these objects. They are known to be old Population II objects and are much advanced towards the end of a stellar career. In fact, planetaries are believed to be precursors of white dwarfs which again are believed to be the end products of a stellar life. The old Population II stars are known to have a pronounced spherical distribution, contrary to the flat distribution of the extreme Population I objects. Further, the Population II stars are known to have a heavy concentration in the nuclear bulge of the Galaxy. It is therefore natural that the planetaries which are products of these old stars will have a heavy concentration toward the galactic centre and a less preference for the galactic plane. Another important point in this connection is that the appearance of planetaries is not conditional like that of diffuse nebulae. The latter always requires the coexistance of both hot stars and clouds of gas. The absence of any one of these will inhibit the birth of a diffuse nebula. But the planetaries are self-creative. The nebular gas in this case is indigenous to the central star. A planetary nebula can therefore be formed in any part of the Galaxy which is accessible to a Population II star having large non-circular motion.

12.3 OBSERVATIONAL TECHNIQUES

Direct photography is undoubtedly the first and the most useful method of nebular observation. Nebulae can initially be detected against the sky background only on photographic plates. In fact, all the nebular catalogues were compiled by direct counts of nebulae on the photographic records of different sky survey projects; the latest and most extensive of which is still the Palomar sky survey with 48-inch Schmidt at the 200-inch Hale reflector. But photographic observations have certain inherent limitations borne by the facts: (a) the survey yields information only of the gross superficial structure of the nebulae; (b) the method provides hardly any information about the physical conditions within the gas; and (c) there is no well defined correlation between the morphology and physical conditions within the gas of the nebulae.

The best information regarding the physical processes prevalent within the nebular gases comes from their spectroscopic studies. Spectroscopic studies only can reveal whether a particular bright nebula is of reflection or emission type. Also, analysis of spectral lines of an emission nebula can yield information regarding its composition, density, temperature, ionization structure, gas velocity etc. In fact, almost all the vital information regarding the physical as well as the dynamical conditions in bright nebulae is available from their spectral studies. Using the nebular spectrograph developed in Yerkes and Mc Donald Observatories in 1937–38 for the study of faint nebulosities, the work of Van Biesbroeck, L.G. Henyey and J.L. Greenstein ushered in a great advancement in the spectroscopic study of the nebulae after Hubble's pioneering work on these objects separating the galactic and extragalactic as well as the reflection and emission nebulae.

The radio observations of diffuse nebulae yield valuable information about the physical and dynamical conditions in them. We shall see in the last section of this chapter that the radio technique is essential for the understanding of the processes that are going on in nebulae formed by remnants of supernovae. Their masses and expansion velocities are also known from such observations. The same is true to some extent for other diffuse nebulae. The masses of the neutral gas in radiation-bounded nebulae and the velocities with which this gas is pushed outwards from the radiating stars are known from radio observations in 21-cm line. Such studies also give valuable information about the spectral energy distributions of ionizing stars as well as the ionized gas in the central parts of the nebulae. As early as in 1958, T.K. Menon observed the nebulae associated with the Orion aggregate and calculated the mass of the neutral hydrogen in the nebula to be 58,000 $M_\odot$ expanding with a velocity of 10 km s^{-1}. This yields a kinetic energy of expansion of the neutral gas to be 5.8×10^{49} erg which is achieved from the conversion of a fraction of the ultraviolet energy radiated by hot stars in the Orion aggregate. Similar observation was made by C.M. Wade (1958) on the nebulae associated with λ^1 Orionis which yielded a mass of $4.5 \times 10^4 \, M_\odot$ of neutral hydrogen, expanding with a velocity of 8 km s^{-1}. The associated kinetic energy is 2.88×10^{49} erg. Also, valuable information about the ionization structure, electron densities and temperature of the nebulae can be obtained by monitoring the radio emission produced by free-free transitions, and the centimetre radio lines arising out of transitions from very high excitation levels of hydrogen atoms, like those from $n = 110$ to $n = 109$, and $n = 105$ to $n = 104$. These lines and similar radio recombination lines of helium yield clues to the physical state of the nebula.

Lastly, with the advancement of the rocket flight technology, it is now possible to use extensively high-flying rockets for astronomical observations. This technique is particularly useful for studying the nebulae in the X-ray, ultraviolet and infrared regions of the electromagnetic spectrum which are inaccessible to the ground-based instruments. A wealth of information has been obtained by such studies about the radiation structures of the Crab Nebula, the Cygnus Loop, the Orion Nebula and several others. The former two are supernova remnants. The energy spectra of these objects greatly deviate from that of a blackbody. These nebulae strongly radiate in X-ray and ultraviolet by the same synchrotron mechanism with which they strongly radiate in radio wavelengths. Rocket observations of the dark dust clouds in the infrared can yield the best information about the physical conditions in them as well as their masses, the sizes and structure of the dust particles, etc., since these clouds are strong infrared emitters.

Thus, the galactic nebulae are now accessible to study in every important window of the electromagnetic spectrum, with various types of instruments so ingeniously built by astronomers.

So much information has thus been gathered about the physical and dynamical processes taking place in them that it can now safely be asserted that our current theoretical knowledge of the complex processes in them is quite sound.

12.4 DARK NEBULAE

The early astronomers of the eighteenth century noticed marked decrease or total absence of stars in several regions of the Milky Way. William Herschel in 1784 also drew attention to the fact that many of these regions were associated with luminous nebulae. The true nature of the dark regions, however, was unknown to these observers. The usual belief was that these regions were *actually devoid* of stars. Some observers thought that these regions were *holes* in space otherwise filled with stars with variable number density. The dark regions are spread along the Milky Way at low galactic latitudes. In the long region between the constellations Cygnus and Centaurus, which runs through Sagittarius, Aquila, Ophiuchus and Scutum, the Milky Way appears to be split into two parts by a wide dark lane. This is commonly known as the *Great Rift*. There are two regions, one in Cygnus (the Northern Cross) and the other near Crux (the Southern Cross), which appear almost completely devoid of stars. These regions have extremely black appearance and are thus named the Northern and Southern *Coalsack* respectively.

It was only through the photographic works of E.E. Barnard, M. Wolf, F.E. Ross and others early this century, that these dark markings were recognized as regions where the background stars have been obliterated by large nonluminous absorbing dust clouds. These are not holes but places where distant starlight was absorbed by dust. These absorbing dust clouds were named *dark nebulae*. The Great Rift was recognized to consist of a series of large dark clouds arranged in an almost continuous array. Photographs of the Milky Way belt also reveal many dark clouds in beautiful contrast against the background of bright nebulae. The Horsehead Nebula in Orion (Fig. 12.1) may be cited as one of the best examples of such objects. The dark nebulae associated with the bright nebulosities in M8 and NGC 6960 are two other examples.

Like bright nebulae, the dark nebulae are also almost entirely confined within galactic latitudes of ± 20°. The fact that the distances to the observable dark nebulae cannot be very large is substantiated by the argument that if distances are large, there will be too many foreground stars to indicate any impression of obscuration. So dark nebulae at sufficiently large distances will hardly be recognizable. In fact, all the important dark nebulae which have been studied in some details lie closer than 1 kpc and most of them are even nearer than 500 pc. This implies that the observed dark nebulae belong to the immediate neighbourhood of the Sun and the volume of space over which they are scattered hardly constitute even 1 per cent of the volume of the galactic disc. The observed number of these nebulae therefore indicate that the total number of dark nebulae in the Galaxy may well exceed 2.5×10^5.

The distances to the dark nebulae are however quite uncertain. The uncertainties are inherent in the methods that have been used in distance calculations. Two principal methods have been used in distance calculations. The first of these applies to those dark nebulae which

FIGURE 12.1 Horsehead nebula in Orion. [*Courtesy:* Indian Institute of Astrophysics, Bangalore]

are seen against the background of bright nebulae. In this case, the distance to the bright nebula is obtained from the spectroscopic parallax of the illuminating star (or stars) and the same distance is assumed for the associated dark nebula. The distance to a dark nebula determined in this manner becomes fairly accurate in cases where the dark and bright counterparts are physically associated with each other. But some of the dark nebulae seen projected on bright ones may not be physically associated; the dark counterparts may be much nearer but seen projected on the bright counterparts. This is particularly important in the case of small nebulae and *globules*. The globules being almost completely opaque small dark clouds of circular or oval shapes, are difficult to observe in the general stellar field. In such cases, the distances to the background bright nebulae are greater than those of the foreground dark counterparts. Thus in all cases, where the distances to the dark nebulae are determined from those of their associated bright nebulae, the calculated distances are taken as upper limits. The actual distances are equal or less than those calculated.

The above approach for determining distances of dark clouds is more suitably applicable to dense smaller clouds and globules which do not have many stars in their directions to yield a statistical study. For larger clouds which, in general, obscure the background stars over a sizable region of the sky, *the method of star counts* can yield not only the fairly reliable distances but also the depth and total extinction by these clouds. In this method, the number of stars per unit area (usually per square degree) of the obscured field is counted which lies between the apparent magnitude range $m \pm 1/2 \, dm$. A similar count per unit area of a nearby *unobscured, normal* stellar field is used as the comparison field. In choosing the comparison field one should be sure that this field has the same space distribution of stars as the obscured field. It has been found that this simple method of starcounts may yield the approximate values of the distance and total absorption of a dark cloud. The depth of the cloud can then be deduced if the density of particles in the cloud can be determined. This method however is associated with a few sources of uncertainty.

There is no well-defined accurate method for calculating the mass and density of a dark nebula. No precise dynamical analysis can be applied for the purpose; only approximate procedures have been devised which are based on many simplifying assumptions. At least some of these assumptions are likely to introduce significant uncertainty in the computed results. Nevertheless, acceptable values of masses and mean densities for many clouds have been derived. The basic observable data for the purpose again are starcounts and the surface areas of the nebulae. We have seen that analysis of starcounts yields approximate distance and total absorption of a cloud. These, together with the observed surface area of the cloud, can give at least the lower limit of the mass and mean density of the cloud. In Table 12.1 the calculated distance, the mass and mass-density for some of the principal nebulae have been listed.

TABLE 12.1   Calculated Parameters of Some Selected Dark Nebulae

Name (by constellation)	l^{II}	b^{II}	Distance (pc)	Absorption (mag)	Mass ($M_\odot$)	Density (gm cm^{-3})
Oph	1	+5	250	2.0		
Scu	28	−2	180–265	3.3		
Cyg	75	−2	600	1.5		
Cyg (Complex)	87	−2	600	1.0	700	1×10^{-25}
Cyg	92	+3	250, 600	1 + 1		
Cep	100	+2	200–500	0.9		
Cep	107	+2	200–600	0.8, 1.7		
Cas	120	+5	500	2.0		
Cas	130	0	500, 800	1.5, 3		
Cas	132	+5	300	1.8		
Heeschen K Tau, Ori,	135	−7	200	2.5	100	1×10^{-24}
Aur (Complex)	180	−6	150	1.0	80	1.25×10^{-25}
S Mon	201	+3	600	1.5		
Tau	204	−18	120	2.0		
Ori	206	−18	300	1.0		
Vela (Complex)	270	0	600	1.6	500	1.45×10^{-25}
Car	286	0	800	0.8		
Coalsack	304	0	170	1.8	15	2×10^{-24}
Oph	353	+17	200	4.0		

12.5   REFLECTION NEBULAE

The first reflection nebula to be discovered by V.M. Slipher was the extensive bright nebulosity associated with the Pleiades star cluster (M 45) (Fig. 12.2), particularly with its members Merope, Electra and Maia. Pleiades is a part of the extensive Taurus nebulosity which again belongs to the *Gould's Belt*. The latter is a local system of diameter about 1 kilo parsec and of mass about 10^8 $M_\odot$, consisting of many early type stars and diffuse dark and reflection nebulae. Pleiades is also one of the few best observed bright nebulae. In 1913, Slipher made a spectroscopic study of this nebula expecting emission lines to be observed. But to his surprise, instead of the expected emission lines, it showed a continuous spectrum superimposed by dark absorption lines. Slipher further observed that the absorption spectrum of the nebula was a true copy of that of the brighter Pleiades stars. The Pleiades nebula is one of the best examples of reflection nebulae. The bright nebulosities associated with NGC 7023, ρ Oph, σ Sco and Cep constitute some other examples.

FIGURE 12.2   The nebulosity of Pleiades star cluster. *[Courtesy:* Indian Institute of Astrophysics, Bangalore]

Hubble's early work revealed conclusively that if a bright nebula is illuminated by stars of spectral class B0 (surface temperature ~ 25,000 K) or earlier, the nebula manifests itself as an emission nebula. It will be a reflection nebula if the illuminating star is of spectral class later than B0. In some nebulae, of course, both emission and reflection components are present. This is revealed, conspicuously when a nebula is illuminated by a group of stars consisting of both hotter and cooler members than B0. The nebula IC 405, Orion nebula (NGC 1976), the Hubble variable nebula (NGC 2261), the Pelican nebula and the North America nebula (NGC 7000) are a few such examples. Hubble's investigations thus conclusively revealed that all bright nebulae owed their luminosity entirely to the associated stars, and that there were only two physical processes responsible for the observed brightness of those nebulae. These are *reflection and fluorescence*. The latter is the phenomenon produced when very hot stars having strong ultraviolet radiation beyond the head of the Lyman series at λ 912 Å, ionize the nebular hydrogen almost completely. In thermal equilibrium within the nebula, recombinations take place almost instantaneously. When the electrons recombine with protons at levels higher than the second and atoms subsequently *cascade* down, the entire Balmer series in the optical range may be produced in emission. The process actually consists in the degradation of the ultraviolet radiation to the optical radiation. Fluorescence will, dominate in those nebulae which surround hot stars whose ultraviolet energy output beyond the Lyman head is very strong. Thus according to Wien's Law, the energy intensity will be maximum at λ 912 Å if the star's temperature be 31,700 K. This corresponds to that of the O stars. The energy maximum shifts towards longer wavelengths and the output of ionizing photons decreases for cooler stars. This process continues until in stars cooler than B1, very little amount of ionizing photons beyond the Lyman head is emitted. Fluorescence is therefore, impossible in this latter case and consequently no emission lines are produced in the nebula. The nebula can then only *reflect* the light received from the associated stars and the reflection is produced by the dust particles associated with the nebula. This explains completely Hubble's finding about the relation between the emissive and reflective nature of bright nebulae and their illuminating stars. It appears quite certain that in the interstellar space in general, and in nebulae in particular, the gas and dust are thoroughly mixed. Which component will be manifested through spectrum depends entirely on the temperatures of the involved stars.

12.6 DIFFUSE EMISSION NEBULAE: THEORY OF EMISSION LINES

Diffuse emission nebulae are among the most spectacular objects in the sky. The Lagoon Nebula in Sagittarius (Fig. 12.3) and the Great Nebula in Orion (Fig. 12.4) are two of their best examples. If there is a star of spectral class earlier than B1 near the galactic plane, there is an associated diffuse emission nebula. The strong ultraviolet radiation of these very hot stars activate gaseous component of the nebula which then plays its key role through fluorescence. Stars of spectral classes B1 and later being poor producers of ultraviolet photons fail to activate the gas in the associated nebulae to fluorescence. So the gaseous component in such nebulae, due to the lack of sufficient excitations, remains mute and only the dust component plays its role by reflection of starlight. This does not mean however that the dust component has no role to play in emission nebulae. Like a reflection nebula, in emission nebulae also the

FIGURE 12.3   The Lagoon Nebula in Sagittarius. [*Courtesy:* Indian Institute of Astrophysics, Bangalore]

FIGURE 12.4   The Great Nebula in Orion. [*Courtesy:* Indian Institute of Astrophysics, Bangalore]

dust grains scatter and reflect light of the exciting star. But in this case, the additional glow imparted by fluorescence in gas far outshines the component reflected by grains. Otherwise, the composition of these two types of nebula is essentially the same. There are some observational evidences, of course, to suggest that at least in some localized regions the dust-to-gas ratio varies significantly from point to point. Thus O. Struve found that in the nebulosities surrounding σ Scorpii, ρ Ophiuchi and some other bright stars, emission components are conspicuous on one side while the reflection component on the other.

Emission nebulae are thus characterized by fluorescence. In the neighbourhood of the illuminating star almost all hydrogen is ionized by strong ionizing radiation beyond the Lyman head. Since hydrogen is by far the most dominant element of interstellar matter, these ionized hydrogen regions have been named by B. Stromgren as H II regions, the regions where hydrogen is neutral having been called H I regions. Almost all diffuse emission nebulae consist of an H II region surrounding the ionizing star and an H I region surrounding (or partially enclosing) the H II region. Almost all of the diffuse emission nebulae are probably radiation-bounded. There is no known example of such nebulae which are gas-bounded. The majority of the planetary nebulae are, however, believed to be gas-bounded. If the nebula contains sufficient amount of gas to absorb all the ultraviolet photons that are emitted by the central star (or stars), it is said to be *optically thick*. If, on the other hand, there is not enough gas in the nebula to absorb all the ultraviolet photons of the star so that some of the photons escape into space, the nebula is said to be *optically thin*. So most of the diffuse emission nebulae must be optically thick which is generally taken as a basic assumption for their physical studies. Only some planetary nebulae are probably optically thin.

The theory of the emission lines in nebulae now seems to be well understood, mainly through the early works of E.P. Hubble, D.H. Menzel, H. Zanstra, I.S. Bowen, L.H. Aller and others. The results based on theoretical predictions clearly explain the observed physical characteristics in these nebulae. Two different mechanisms are known by which the observed emission lines are formed.

The first is the photo-ionization of hydrogen and other elements by absorption of ultraviolet radiation from the embedded hot star, followed by recombination and cascading. Ultraviolet photons beyond the Lyman head at 912 Å which corresponds to a photon energy of 13.6 eV ionize hydrogen, the most abundant of the elements. The same radiation also ionizes most other elements in the nebula because their ionization potential is less than that of hydrogen. Only helium has an ionization potential much higher than that of hydrogen. Photons of energy 24.5 eV and higher, corresponding to the wavelength 506 Å or shorter are required for ionization of helium. But stars which emit sufficient number of ultraviolet photons beyond the Lyman head also emit photons of energy 24.5 eV and higher. One can therefore be sure that in H II regions, all other elements including helium are at least singly ionized. Many elements are also doubly ionized, but doubly ionized helium lines are never observed in diffuse emission nebulae. This is because the second ionization of helium requires photons of energy above 55 eV corresponding to wavelengths shorter than 225.5 Å. The illuminating stars are not so hot as to produce sufficient number of photons of such high energies, and consequently helium remains singly ionized. Since hydrogen is the overwhelmingly abundant element with a sufficiently high ionization potential, its fluorescence plays a major role in determining the physical processes in nebulae.

The kinetic energy imparted to the free electron when the ionizing photon of energy hv is absorbed by a hydrogen atom at the ground level is given by Eq. (2.12), where v_0 is the threshold frequency (corresponding to $\lambda\,912$ Å) and v is the velocity of the electron removed. Since hv may have any value greater than hv_0, so within the ionized gas free electrons move at random with all sorts of velocities that may arise out of the energy differences $hv - hv_0$. But these electrons do not remain free for a long time. The inverse process, recombination, soon takes place in which the free electrons are captured by protons to form neutral hydrogen atoms (the same thing happens also for other ions). Recombination may take place at any energy level of the atom. Each recombination is accompanied by the emission of a photon of energy corresponding to the difference between the kinetic energy of the free electron and the excitation energy of the level at which it is captured by the proton. When recombination takes place at a higher level, the electron subsequently cascades down the lower levels, emitting at each transition the corresponding line. In this way, all the series of hydrogen lines, viz. the Lyman series, the Balmer series, the Paschen series etc. can be produced in emission nebulae. But only the Balmer series of lines being in the visible range of the spectrum, are observed. Theoretical calculations show that of all the transitions involved in the Balmer series of lines, that from the third to the second, corresponding to the H_α line, has the highest probability. In excellent agreement with the theory, the H_α line is actually observed as the most intense among the Balmer series of lines in an emission nebula. Other lines of the Balmer series are also conspicuous. Similar processes of photoionization and recombination and subsequent cascading to lower levels occur in other atoms and ions also, giving rise to the corresponding emission lines many of which appear as very strong. These are the first type of emission lines observed in diffuse emission nebulae. Fluorescence is the basic physical mechanism that is responsible for the formation of these lines. They originate from *natural transitions*.

There is yet another mechanism which is operative in the formation of a second class of emission lines in these hot diffuse clouds. The free electrons released mainly by ionization of hydrogen, in general, possess kinetic energy of a few electron volts. These electrons suffer *inelastic* collisions with ions of trace elements like oxygen, nitrogen, neon, sulphur, etc. which possess low-lying excited levels. As a result of collisions the ions are excited to these levels which again come down to the ground level emitting lines of appropriate wavelengths. These lines therefore involve no absorption of radiation as in the case of fluorescence lines. The ions get their excitation energies at the expense of the kinetic energies of electrons. The most peculiar characteristic of these lines is that they originate from *forbidden transitions* and are very strong. In *natural transitions* the electrons remain in excited states only for a period of the order of 10^{-8} second. That is ordinarily the average lifetime of an electron in an excited level after which it is de-excited by the emission of a photon. But the low-lying levels to which the ions of these trace elements are excited by collisions are said to be metastable, since the excited ions remain in these states for a few hours before they get de-excited by emitting photons. The corresponding lines are therefore called forbidden since they are not formed under laboratory conditions. The essential conditions required for the formation of forbidden lines are that the medium must be highly ionized and extremely rarefied. These conditions are fulfilled in emission nebulae in which the electron density generally lies within the range 10^3–10^4 cm^{-3}. This is also the approximate range of gas density since most of the electrons are supplied by ionization of hydrogen which is the overwhelmingly abundant element. Even the best laboratory vacuum greatly falls short of such high rarefication. Gas density and therefore, gas pressure is much higher in laboratory vacuum or in stellar atmospheres. In such moderately high density gas the ions excited by collisions in the metastable levels will be soon de-excited by fresh collisions before they have time to undergo forbidden transitions by radiation of photons. As a consequence, the forbidden lines, even if they form, are extremely weak in stellar spectrum or in a spectrum produced under laboratory vacuum.

On the other hand, H II regions present an ideal medium for transitions from metastable levels and the consequent appearance of forbidden lines. High temperatures prevailing in these regions provide enough kinetic energies to electrons which take part in exciting the low-lying metastable levels of ions responsible for these lines. The extremely low density of the nebular gas further helps the condition for formation of these lines by greatly decreasing the probability of de-excitation. These two conditions make the situation so favourable that although the transitions from metastable levels are very slow, so many ions are excited to these levels at a time that the transition of even a fraction of these to the lower level will produce the forbidden lines in great strengths. The lines produced by natural transitions of any atom or ion will be comparatively weaker because in the extremely low density nebular gas, very many atoms or ions can never populate simultaneously a particular excited level (natural), the life time of an atom or ion in this level being 10 to 100 billion times less than that in a metastable level. Only the hydrogen lines, due to the overwhelming abundance of the element, appear in strengths at all comparable to the strengths of some forbidden lines. Thus H II regions present conditions that are just the opposite to what we are commonly familiar with.

This theory of the nebular emission lines is in perfect agreement with observations. The strongest emission lines in these nebulae are a doublet due to [O III] in the blue-green region of the spectrum at λ 4959 (N_2) and λ 5007 (N_1) and another ultraviolet doublet at λ 3126 and λ 3729 due to [O II], Among other prominent forbidden lines, the two ultraviolet lines at λ 3868 and λ 3967 due to [Ne III] and the two red lines at λ 6548 and λ 6584 due to [N II] appear in sufficiently great strengths. Since none of these forbidden lines could be produced in the laboratory, great difficulty arose in identifying these lines when they were first observed in spectra of planetary as well as of diffuse emission nebulae. They were therefore initially ascribed as due to an element 'Nebulium' which was unknown on Earth. The great strengths of these lines further suggested that the emission nebulae possessed a very high abundance of this element. The mystery of Nebulium, however, was solved by I.S. Bowen in 1927, who succeeded in showing that the lines ascribed to Nebulium were infact the forbidden lines of elements already known to Earth. The presence of these lines in the spectra of nebulae therefore indicate the existence of very peculiar conditions in them rather than the existence of unknown elements. Bowen's theoretical predictions about these lines were verified by many of their observed properties. For example, the observed frequencies of the lines were exactly identical with those predicted theoretically. Again, according to Bowen's theory, the blue-green doublet N_1 and N_2 of [O III] have a common upper level and their intensity ratio should be equal to 3. This is exactly what is found by measurement of intensities of these lines.

The physical reasons for which intense forbidden lines are produced in emission nebulae while no such lines can form in stellar atmospheres are now well understood. In nebulae the gas density is extremely low compared to that in stellar atmospheres. Moreover, the radiation emitted from an associated hot star is extremely diluted, the dilution coefficient at a point being measured by the ratio of the solid angle subtended by the radiating star at the point concerned to 4π. Quantitatively, the dilution coefficient at a distance x_n in the nebula from the centre of the star of radius x_s is given by

$$D = \frac{1}{2}\left[1 - \left\{1 - \left(\frac{x_s}{x_n}\right)^2\right\}^{1/2}\right] \qquad (12.1)$$

At a large distance from the illuminating star one can take $x_s/x_n \ll 1$; then Eq. (12.1) reduces to

$$D = \frac{1}{4}\,(x_s/x_n)^2 \qquad (12.2)$$

For an O star of radius, say $10\,R_\odot$, the value of D at a distance of 1 pc from the star equals $\sim 1.3 \times 10^{-14}$. Thus, stellar radiation density u_s is extremely diluted in the nebula and the value decreases with increasing distance from the star according to the equation

$$u_n = Du_s \qquad (12.3)$$

u_n being the radiation density where the dilution coefficient is D. The radiation density in the nebula is thus so small that the atoms occupying the metastable levels are not subjected to a strong radiation field in order to excite many of these to higher levels. Since the spontaneous transition probabilities from these levels are also very small, atoms are accumulated in large numbers to these levels. The process of accumulation is further helped by the extremely low density of matter, because not many atoms can undergo collisional de-excitation under such a condition. Thus, the nebular conditions greatly favour the accumulation of large number of atoms in their metastable states which when undergo spontaneous downward transitions, give rise to strong forbidden lines. Such favourable conditions are absent for natural line radiation in a nebula since the natural lifetimes of atoms in these levels are less by a factor of a billion and due to quick spontaneous transitions, accumulation of large number of atoms in these levels at any particular time is not favoured. So emission lines arising out of transitions from ordinary levels are generally weaker than those arising from metastable levels in a nebula. But the conditions prevailing in the stellar atmospheres are just the opposite. Here the lines originate in photospheres or in extended envelopes where the gas density is at least a million times higher than that in a nebula and the dilution coefficient is seldom less than 10^{-3} even for stars with very extended envelopes. Therefore, due to the presence of intense radiation fields and incomparably high density in stars, accumulation of atoms in the metastable levels can never reach such a stage as to give rise to a sufficiently strong forbidden lines in them. Consequently, even if forbidden lines are produced in stellar spectrum, they will be too faint to be recognized. On the other hand, the same conditions, viz., the presence of strong radiation fields and high matter densities in stars greatly favour the ordinary lines to be produced in sufficient strengths.

The entire physics of the nebular emission lines as observed is understood to be the manifestation of the process of transformation of the high frequency radiation to that of low frequencies. Ultraviolet radiation beyond the Lyman limit is absorbed by the gas which is then ionized. A part of the absorbed energy goes to the kinetic energy of the electrons, a fraction of which again is spent to excite low-lying metastable and other ordinary levels of atoms and ions. These atoms and ions then radiate lines in optical wavelengths by downward transitions. Again, recombinations take place between free electrons and ions, which by subsequent cascading to lower levels emit various optical lines. Thus ultraviolet radiation from hot stars is degraded to optical radiation in surrounding nebulae due to the peculiar physical conditions prevailing in them. Such degradation of radiation under favourable conditions has been well explained by a theorem due to Rosseland which may be stated as follows.

Consider the various ways of transitions that may occur between different energy levels of an atom. To be specific, we consider the upward and downward transitions between (say) the three lowest levels, 1, 2 and 3. Absorption of photons and subsequent reradiation may proceed in many different ways of which the two important cyclic converse processes are $1 \rightarrow 2 \rightarrow 3 \rightarrow 1$ and $1 \rightarrow 3 \rightarrow 2 \rightarrow 1$. The former process results in the emission of the high-frequency photon of energy $h\nu_{31}$ at the expense of two absorbed low-frequency photons of energies $h\nu_{12}$ and $h\nu_{23}$, while the latter results in the emission of two low-frequency photons of energies $h\nu_{32}$ and $h\nu_{21}$ at the expense of one absorbed high-frequency photon of energy $h\nu_{13}$. The former is therefore the process in which low-frequency radiation is converted to a high-frequency one. The reverse is the case with the latter. Rosseland's theorem asserts that in the presence of dilution of radiation the second process overwhelmingly gets the better of the first. In fact, the presence of very large dilution with coefficient D, Rosseland's analysis yields that in any volume of gas

$$\frac{\text{number of first process per unit time}}{\text{number of second process per unit time}} \approx D$$

As we have already seen, at a point in the nebula 1 pc away from the star, $D \sim 10^{-14}$. So, in a nebula the transformation of less energetic quanta to more energetic ones can be completely neglected in comparison to the reverse transformation. This explains quite clearly why the ultraviolet radiation received from the hot star is very efficiently converted to radiation at optical luminosity in the nebula than otherwise it would have been.

As has been already mentioned, the strongest lines in emission nebulae are those due to [O II] in the ultraviolet, [O III] in the blue-green and the hydrogen lines, particularly the H_α in the red. Since these lines lie at widely separated wavelengths, it should be possible by means of colour filter photography to trace the ionization structure at different regions of the nebula, in particular, with respect to the position of the ionizing star. With ordinary photographic plates, the efficiency of the process suffers seriously from the insensitivity of the standard colour emulsions at the blue-green region where the strongest lines due to [O III] are emitted. The dye transfer method developed by Lick Observatory astronomers to study the colour distribution in nebulae has made a great advancement in this respect. It is actually found, in agreement with the expectation, that the regions close to the illuminating stars are greenish in colour indicating that the radiation in these regions originate mainly from [0 III] lines. These ions requiring very high stimulus ($IP = 35.11$ eV) are naturally concentrated close to the radiating star. Away from the stars in the outer regions of the nebula are observed red rims and filamentary structures where H_α and [N II] lines are the chief sources of radiation. These lines requiring relatively low stimulus from the star dominate therefore the colour of the outer regions of the nebula. The ionization structure thus theoretically understood is also fully realized observationally.

We have not so far considered the role of dust grains in the physical processes in emission nebulae. This subject is of late gaining increasing attention of the astrophysicists. Previously it was thought that the physical processes in emission nebulae are totally dominated by the role of their gaseous component. The role of the dust component was hardly recognized and so neglected. Careful studies of the nebular spectra and planetary nebulae reveal some anomalous behaviour in respect of the observed strengths of certain lines. For example, the He I recombination lines have been found to be unusually weak in the giant H II regions toward the galactic centre. On the other hand, the strengths of forbidden lines of low excitation like those of [O II], [N II] and [S II] in planetary nebulae, have been observed to be greatly enhanced than predicted by theoretical calculations. To explain such anomalous behaviour, many authors have recently started looking to the dust component of the nebulae as the responsible agent. The preferential absorption of high energy photons by dust grains has been suggested to be the possible cause of the observable anomalies. Many authors have studied and analyzed the effects of dust grains on various observable features in emission nebulae. Many new interpretations and explanations regarding the function of dust in ionized regions have emerged as a result of these investigations. We mention here the more important conclusions arrived at by Balick in this respect from his analysis of a model nebula excited by a star of temperature 40,000 K. Balick finds that dust grains have only inconsiderable effect on the ionization of H and He. Absorption of ionizing photons by dust becomes very strong in the high density regions of the nebula and so in its central part, while in the outer part of the nebula absorption by gas is more pronounced. Dust therefore *softens* the intensity of the emerging flux from the illuminating star and can thus lower the electron temperature by as much as 15 per cent. If the absorption of such a high percentage of ionizing photons by dust in H II regions is real, it will be unreliable to take the radio continuum flux density as a measure of the stellar flux above the Lyman limit in dusty H II regions. A major part of the stellar ultraviolet luminosity in this case will be reradiated in the infrared by dust particles. The degrees of ionization of the trace elements like N, O, Ne and S are more sensitive to the amount and type of dust than is the case for H. The intensities of forbidden lines relative to H_α can be significantly affected even by a small dust content. Balick's study further revealed that the fraction of the ionizing photons absorbed by dust strongly increased with the temperature of the exciting star. This, evidently indicates that the far-ultraviolet radiation is more strongly absorbed by dust particles; a result which confirms the well-known absorption property of dust grains. At this stage, therefore, on the basis of more recent investigations on the role of dust component in emission nebulae, at least one conclusion can be drawn that the effect of dust grains on the thermal and ionization structures of these nebulae are quite complex, and so far have been only poorly understood. The investigations have only begun and when they would end, our present ideas of the physical processes in these nebulae might be significantly modified.

Thus, the diffuse emission nebulae (H II regions) are objects of great astrophysical importance. The determination of the temperatures of H II regions is a highly interesting and at the same time a complicated problem of astrophysics. These luminous objects serve as good spiral tracers in our Galaxy, and particularly, in distant spiral galaxies. The abundance of elements in interstellar medium, particularly the He/H ratio, can be evaluated from a spectral study of the H II regions. It was generally believed so far that the elemental abundance in the galactic H II regions does not differ much from the solar abundance. But recent studies have revealed that this identity of abundance does not quite hold. It has been observed that the spectral nature of H II regions systematically varies as one goes from the inner to the outer region of the Galaxy. The H II regions in the outermost arms are found to be deficient in nitrogen and oxygen compared to the solar abundance, while those in the innermost arms are comparatively richer in these two elements. The reason for such anomaly of abundance in H II regions along the galactic radius is not yet fully understood.

12.7 PLANETARY NEBULAE

Unlike the diffuse emission nebulae with complex irregular structure, planetary nebulae are emission nebulae of fairly regular shapes, each having an extremely hot blue star at the centre called the *nucleus of the nebula* (Fig. 12.5). In telescopic view, a planetary nebula appears as a circular or oval disc having finite size and well-defined boundary, thus imparting it the appearance of a planet. The name planetary nebula was derived from this similarity. But in detailed structure, planetary nebulae possess a wide range of differences. Many appear as rings, many as amorphos discs, while some possess quite complex structures including intersecting rings and wispy projections. The apparent angular sizes of planetaries range from about a quarter of a degree to a few seconds of arc and even less, down to star-like tiny spots, the latter being recognizable as planetaries only by their spectra. The average size is less than 1′, but the largest one, NGC 7293 in Aquarius has a diameter of 15′.

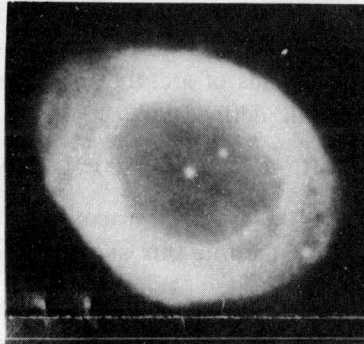

FIGURE 12.5 The Ring Nebula in Lyra. [*Courtesy:* ESO, Munich]

The early efforts of a systematic search of planetary nebulae were made by Rudolf Minkowski. It was mainly through his work that by 1951, 371 planetaries were discovered in our Galaxy. The systematic search was further continued by Minkowski, Abell, Parenago and others during the subsequent years which added a large number of these nebulae to their list. By mid-sixties somewhat less than 700 planetaries were known. The results of these and other surveys were compiled by L. Perek and L. Kohoutek (1967) in the *Catalogue of Galactic Planetary Nebulae.* This catalogue was improved by Kohoutek (1978) with results of subsequent surveys which added 226 new planetaries. The ESO surveys of the southern sky and new discoveries from the Palomar Sky Survey Plates added to the list many more planetaries. The results of all these surveys have been published in the Strasbourg Catalogue containing an exhaustive list of 1455 planetary nebulae compiled by Acker and his co-workers.

The absolute magnitudes of planetary nebulae are difficult to determine accurately as their distances are not precisely known. Planetary nebulae are distant objects and so the direct method of trigonometric parallax cannot be applied. Only in the case of NGC 7293, the nearest of the known planetaries, Van Maanen could measure parallax directly that puts the object at a distance of about 25 parsecs.

A few planetaries are known to be members of binary systems and therefore their distances can be derived from the spectroscopic parallaxes of the companion stars. The planetary NGC 246 is in a binary system having a G8-K0 star as companion, whose spectroscopic parallax has been computed by Minkowski to be 430 pc. NGC 1514, NGC 2346 and NGC 3132 are some other examples of planetaries whose central stars are members of binary systems. However, the number of such nebulae is not large enough to admit of any reliable statistical analysis. The usual method of determining statistical parallax of a group of objects with the help of their measured proper motions has been attempted earlier by a number of astronomers. The basic observational material for such works has been the proper motions of 21 planetaries measured by Van Maanen and those of 33 planetaries measured by Anderson. The results obtained by various observers on the basis of these data indicate that planetaries are in general, very distant objects.

One of the best earlier works on calculation of distance of planetary nebulae was done by O'Dell in 1962. He derived formulae for computing the mass, linear radius and distance of the nebulae using the recombination spectrum radiation of H_β. His formulae involve parameters such as the electron density N_e in cm^{-3}, the H_β flux at the earth $F(H_\beta)$ in erg $cm^{-2}s^{-1}$, the angular radius α of the nebula in seconds of arc, the filling factor ε and a function $f(T_e)$ of the electron temperature of the nebula defined by

$$f(T_e) = 10^6 \ T_e^{-3/2} \ \exp \ [9800/T_e - 0.98] \tag{12.4}$$

whose value becomes ~ 1 for $T_e = 10,000$ K, the usual value assumed for T_e in planetaries. $F(H_\beta)$ is related to the quantity $S(H_\beta)$, the true flux at the surface of the nebula in H_β line by

$$f(H_B) = S(H_\beta) \left(\frac{\alpha}{206265} \right)^2 \tag{12.5}$$

Using the value 0.7 for the average filling factor $\langle \varepsilon \rangle$, $T_e = 10,000$ K and an interstellar extinction of $1^m.1$ kpc^{-1} O'Dell computes $\langle M \rangle = 0.14 \ M_\odot$. With this average mass and filling factor, O'Dell derives a distance scale of the planetaries as

$$d = 75 \ \alpha^{3/5}[F(H_\beta)]^{-1/5} \tag{12.6}$$

where d is distance in parsecs. The formula gives fairly good representation of planetary distances.

The most extensively used method for determining distances to planetary nebulae is that originally proposed by I.S. Shklovsky. The basic assumption in this method is that at the fully ionized stage all nebulae have the same mass. The distance d in parsec is given by the formula

$$d = A\alpha^{-3/5} \, [F(H_\beta)]^{-1/5} \qquad (12.7)$$

where A is a constant involving parameters such as the absolute abundance of helium, the fractional ionization of singly to doubly ionized helium, the constant ionized mass of the nebula, the volume filling factor ε (that is, the fraction of the volume of a sphere that is actually occupied by nebular matter), and the effective recombination coefficient for the H_β line. In the above formula α is the angular radius of the nebula in seconds of arc and $F(H_\beta)$ is the flux in H_β line in ergs $cm^{-2}s^{-1}$ measured at the Earth. It must be remembered that this formula assumes that the nebula is optically thin. The computed distance with the above formula will be too large if this assumption is not fulfilled for a nebula.

The formula given in Eq. (12.7) has been calibrated by various authors using planetaries with known distances. After calibration is made, one can then fit the measured angular size and H_β flux of any planetary of known mass with the calibrated formula to derive its distance. Cahn and Kaler in 1971 derived the distances of a large number of planetaries combining an extensive set of optical and radio data on them and making use of Eq. (12.7). Another major work on planetary distances using Shklovsky's method was carried out by Milne and Aller in 1975, and later by Milne in 1979. They used radio observation data on 332 planetary nebulae and derived their distances.

In any calculation of distance of planetary nebulae, one has to assume a known ionized mass for the nebulae as the calibration mass. Thus while Cahn and Kaler used 0.18 $M_\odot$, E.A. Milne and L.H. Aller, and E.A. Milne used 0.16 $M_\odot$ for the calibration mass. Many of the earlier workers including R. Minkowski, L.H. Aller and I.S. Shklovsky used 0.2 $M_\odot$ as the calibration mass for calculating distances to planetary nebulae with Eq. (12.7). On the other hand, each calibration of Eq. (12.7) with planetaries of known distances will yield a value of their ionized mass when observed values of other parameters involved are used. Different calibrations have yielded different masses all of which lie in the range from 0.14 to 0.5 $M_\odot$. This appears to be a fairly good representative range of the actual planetary masses.

Works of S.R. Pottasch and his coworkers have, however, raised doubts about the validity of the use of Shklovsky method for determining distance to planetary nebulae. For a set of 28 planetary nebulae having precise distance determinations, Pottasch calculated the ionized masses of these planetaries. His calculated masses were found to vary by a factor as large as 200. Also, the ionized masses were found to vary linearly with the radii of the nebulae, indicating that the nebulae were optically thick. Thus, the fundamental assumption of Shklovsky method that the nebulae are optically thin is found to be violated, in general, according to the analyses of Pottasch. He concludes therefore, that Shklovsky method cannot be reliably applied for determining distances of a large sample of planetaries. A factor of 200 in masses introduces an error by a factor of 8.3 ($d \propto M^{0.4}$) in distance determination using the Shklovsky method.

When the distance scale is fixed with some degree of reliability, the density distribution of the planetary nebulae and their total number in the Galaxy can be computed. In his earlier work, O'Dell computed the linear radii of a large number of planetaries. Assuming the sizes of the nebulae to lie between the limit of identification and 0.7 pc, O'Dell calculates the total number of planetaries in the Galaxy to be 4.8×10^4. More recent determination by various authors point to the conclusion that the actual number should be comparatively less. Planetaries are believed to have heavy concentration around the galactic centre and a lesser but pronounced concentration in the disc of the Galaxy. Exponential surface density law of the form $\mu(r) = \mu_0 \exp[-\beta r]$ has been used to compute the total number of planetaries, assuming them to be projected on the plane of the Galaxy. Alternatively, the total number can also be calculated by computing the specific number of nebula, that is, the number of nebulae per unit mass, and then multiplying this number by the total mass of the Galaxy. It is assumed that the specific number used is the representative for the whole Galaxy. The total number of planetaries in the Galaxy has thus been variously determined by authors and these determinations reveal the fact that the number is probably limited to 30,000.

The uncertainties in the distance determination of the planetaries by any method being considered as tolerable, the general consensus that emerges from the various estimates is that these objects have absolute magnitudes in the range –1.0 to –3.0. There may be less luminous ones, but no planetary is probably more luminous than the absolute magnitude –3.0. The basic difficulty in determining distances, and so the absolute magnitudes, of planetaries lies in the fact that there is no single physical parameter which does not intrinsically vary from one planetary to the other. The above luminosity range of planetaries is also corroborated by observation of planetaries in the nearest external galaxies. As early as in 1955, Baade identified some planetary nebulae in M 31. With a distance of 675 kpc for this galaxy, he calculated the absolute magnitudes of these planetaries which lie between –2.0 to –2.5. The upper limit of absolute magnitudes of the planetaries identified in the Large and Small Magellanic Clouds was fixed at –3.0. This latter value seems to be an absolute upper limit for the luminosity of planetary nebulae in any galaxy. More recently, the surveys of the planetaries have been extended to other Local Group Galaxies. It has been found that the luminosity function of planetaries in these galaxies is very similar to that of the planetaries in the solar neighbourhood.

Opinion among the astronomers differs as regards the assignment of population to planetary nebulae. The main criteria on which this factor has to be judged are motions and galactic distribution of the nebulae and the evolution phases of their central stars. The distribution of planetaries is more or less analogous to that of RR Lyrae stars and the long-period variables of the disc population. Their concentration in the central region of the Galaxy is pronounced and the mean height above the galactic plane is considerably large. Their galactocentric motions differ significantly from circular, resulting in large value of the solar motion and velocity dispersions. These dynamical and distributional characteristics would definitely suggest that the planetaries belong between Intermediate and extreme Population II. The definite identification of a planetary nebula in the globular cluster M 15 confirms this suggestion. More recent works of some astronomers, however, have suggested the existence of two distinct types of planetary nebulae; viz., class B and C, classified on the basis of their kinematic properties. These difference in classes is revealed by the height distribution above the galactic plane and by the velocity dispersion perpendicular to this plane. The mean height above the plane, $\langle Z \rangle = 160$ pc, and the velocity dispersion perpendicular to the plane, $\sigma_z = 15$ km s^{-1} have been measured for the class B. For the class C, the corresponding values are $\langle Z \rangle = 320$ pc and $\sigma_z = 26$ km s^{-1}. These values indicate that the class B planetaries are evolved from rather younger population stars while the class C planetaries evolve from a fairly older population stars. More careful observations may reveal a continuous transition in population characteristics from moderately old to very old among the entire set of planetary nebulae. This suggestion is probably borne out by the presence of a planetary in each of the galactic cluster NGC 2818 and the globular cluster M 15.

As regards the evolutionary phase, it has been suggested by many astronomers that planetary nebulae are the precursors of the white dwarfs—the final product of a stellar life. So they must be very old objects passing through a highly unstable state in the evolutionary life of the central star. To cope with the structural instability, the central star has to shed off its layers of envelope with a velocity probably of the order of a few hundred km s^{-1}. By the time these ejected layers move through a fraction of a parsec and the object appears as a planetary nebula, the ejected gas is observed to be expanding with a velocity of the order of 10 to 20 km s^{-1}. Planetary nebulae are therefore expanding shells ejected from highly evolved stars, probably the end phase of red giants, the cores of these stars remaining as the hot, dense and small nuclei of the nebulae.

If the velocity of expansion is assumed to be uniform during the growth of the nebula from its initially recognizable size to the maximum size in which any nebula has so far been observed, then one can deduce the lifetime of a nebula. In this way, O'Dell deduced the lifetime to be 35,000 years, by assuming a maximum diameter of 0.7 pc and an expansion velocity of 20 km s^{-1}. But the more recent determinations prove the lifetime of nebulae to be substantially shorter. It was recognized that for calculation of lifetime only those nebulae should be considered which are optically thin, because the Shklovsky method is applicable to such nebulae only. According to Cahn and Kaler, the radii possessed by these nebulae is in the range of 0.08–0.40 pc. With an expansion velocity of 20 km s^{-1}, this yields a lifespan of only 16,000 years for the nebulae. However, the actual lifespan of nebulae is probably somewhere in between. Assuming a lifespan of 30,000 years and a total of 30,000 nebulae in the Galaxy, the birthrate of planetaries in the Galaxy turns out to be 1 nebula yr^{-1}. This may be an extreme lower limit and the more probable value of birthrate may be 50 per cent higher than this.

The physical state in a planetary nebula is somewhat similar to that in a diffuse emission nebula except that the former manifests a more extreme state of excitation. This indicates that the central stars are, in general, hotter than the illuminating stars of the diffuse emission nebulae. The conditions of extreme dilution of radiation ($D \sim 10^{-14}$) and of extremely low density ($n_e \sim 5 \times 10^3$ to 10^4 cm^{-3}) are present in planetaries as much as in diffuse emission nebulae. The electron temperature also has the similar range i.e. about 8000 K to 15,000 K, in both types of nebulae. This temperature is achieved by thermal balance between the energy received through the intense ultraviolet flux from the hot central stars and that radiated mainly through the forbidden lines excited collisionally by free electrons. Thermal energy is simply dissipated in forbidden lines which keeps the temperatures of the nebulae at such low values.

Like diffuse emission nebulae, the planetary nebulae also have very rich recombination and forbidden lines spectra. The basic mechanism for formation of these lines is the same as in diffuse nebulae, viz. fluorescence and collisional excitation. The excitation state is however more intense in the planetaries. The violet lines of [O II] at λ 3726 Å and λ 3729 Å, and the green "Nebulium" lines of [O III] are several times more intense than the Balmer lines of hydrogen in some of the planetaries. The He II line at λ 4686 Å is also very intense. Lines of highly ionized atoms such as C IV, O IV, O V, F IV, Ne IV, Ne V, Cl IV, A IV, A V, K IV K VI, Ca V, Ca VII, Mn V, Mn VI and Fe V Fe VII have been identified in many of the planetaries. On the other hand, lines of weakly ionized atoms such as C II, N II, O I, S I are also common. This suggests that there exists a gradation of excitation states among the planetaries which depends upon the temperature of the central star, the computed range for which lies between 30,000 to 100,000 K. The observed gradation of the ionization states confirms that this range in temperature of the central stars is probably real.

Degradation of intense ultraviolet flux of these extremely hot stars beyond the Lyman series limit to optical wavelengths causes the radiation of intense lines of various ions over the entire optical region of the spectrum. These lines dominate the planetary spectrum to such an extent that the continuum was completely neglected by the early observers. The continuum is very weak and can be intensively studied only in spectra of very long exposure time. Although the nature of line spectrum of planetaries has now been satisfactorily understood, some problems regarding the nature of their continuous spectrum still remain unsolved. One of the major problems is the observed fact that the continuum is almost constant over the entire optical wavelength range.

With the advancement in the infrared observation techniques, planetary nebulae have been extensively observed during the last two decades or so over a large range of infrared wavelengths. The best such observed nebula is NGC 7027. The most prominent feature in the infrared spectrum is a much stronger continuum in longer infrared wavelengths than can be produced by usual atomic processes under nebular conditions. In their pioneering work as early as in 1968, K.S. Krishnaswamy and C.R. O'Dell attributed this surplus IR strength to thermal emission by grains in the nebula. Several line features due to H, He and other ions, as well as those due to the quadrupole transitions of H_2 are revealed in the infrared spectrum. Several emission features have been observed in the spectrum of NGC 7027 in the wavelength range 8–13 μ, the origin of which has not yet been completely understood. Some of these features have been observed in other nebulae also. Fine structure lines of ions of S, Ne and O at different stages of ionization have also been observed in many nebulae including NGC 7027. These various infrared observations of planetary nebulae have been exploited to deduce information about abundance of various elements in them, the nature and origin of dust particles, dust-to-gas ratio, the dust temperature, the stratification of dust grain size within the nebulae, and so on.

The analysis of chemical abundances in planetaries has revealed that although there were marked differences in composition between the nebulae, their average composition presented no gross difference from that of the Sun. Because of the presence of numerous He I and He II lines in planetary spectra, the He to H ratio is most conveniently and accurately determined. The works of various authors have established an average He/H abundance of about 0.13 in the planetaries which does not differ much from cosmic-abundance. This value also appears to be almost constant in different planetaries. The abundance of other important elements appears to fluctuate somewhat from one nebula to the other. In some nebulae, particularly in the planetary in M 15, oxygen has been found to be deficient by nearly an order of magnitude. Planetaries with low surface brightness have been found to be hydrogen deficient. This fact has been interpreted by some authors as an evidence—that these nebulae are the ejected envelopes of highly evolved giant stars in which a significant fraction of hydrogen has been depleted by nuclear transmutation.

As we have already mentioned, the nuclei of planetary nebulae are the kernels of highly evolved red giants from which the surface layers have been as if, spilled off. They are therefore extremely hot objects, small in size, with their absolute magnitudes ranging from about +10 to −1. They are generally visible as small blue stars at the centres of the nebulae. But in some planetaries the nuclei are hardly visible against the background of much brighter nebulae and so special devices have to be employed in order to measure the luminosities and colours of the nuclei. Similarly difficulties are inherent in the spectroscopic studies of the nuclei, since the stellar lines are highly contaminated by nebular emission lines almost over the entire visible region of the spectrum. This is particularly true in the case of hydrogen and helium lines. The nebular emission lines of these two elements arise from natural transitions which therefore overlap the corresponding absorption lines of these elements in the spectra of the nuclei. This is also true to some extent for many lines of other elements. Such contamination of the spectra of central stars by those of the associated nebulae greatly complicates the problem of assigning spectral type to the planetary nuclei. Nevertheless, employment of ingenious techniques and persistent efforts by astronomers have yielded the satisfactory solution of the problem. A variety of spectral types all characterized by very high temperatures have been observed in planetary nuclei. All these types can be classified into the following three broad categories of spectra:

1. The first category of planetary nuclei are those with emission lines. These again consist of those with broad emission lines as in Wolf-Rayet stars, those with O-type spectra and lastly, those with very high excitation emission features such as O VI lines. Examples of such nuclei are those of NGC 40, NGC 6751, NGC 2392, NGC 246 etc.
2. The second class consists of nuclei with O-type pure absorption spectra having no trace of emission features. Examples are the nuclei of NGC 1535, IC II 2149, etc.
3. The remaining nuclei present purely continuous spectra having no observable absorption or emission features superposed on them. Examples of this category of nuclei are those of NGC 7009, IC II 4732, etc.

Thus, although the problem of the determination of temperatures of the planetary nuclei is quite complex, several methods have been employed for the purpose and all the results obtained point to the fact that temperature varies from one nucleus to another and covers the whole range between 30,000 to 100,000 K. Of the various methods, we shall here discuss very briefly the one which was developed independently by H. Zanstra and by D.H. Menzel both in 1926. This method rests on the principle that the nebula is a recorder of all the ultraviolet quanta beyond the Lyman series head that may be emitted from the hot central star. As already explained, almost each of these invisible ultraviolet quantum radiated by the central star reappears as one or more visible quantum in Balmer frequencies emitted by the nebula. Some ultraviolet quanta, of course, escape from the nebula after numerous scattering. Knowing the relative distribution of hydrogen atoms in various levels under the nebular conditions and using the recombination theory, it is possible to compute the total number of Balmer quanta by observing the number of quanta in any particular Balmer line from its measured strength. Thus, if the optical thickness of the nebula beyond the Lyman series limit is sufficiently higher than unity, then the total number of ultraviolet quanta beyond the Lyman series limit emitted by the central star will be at least equal to the number of Balmer quanta emitted by the nebula.

Now, the number of ultraviolet quanta beyond the Lyman series limit, N_L, emitted by a star of radius R_s and temperature T_s (if it radiates like a blackbody) is given by

$$N_L = 8\pi^2 \; \frac{R_S^2}{c^2} \int_{\nu_L}^{\infty} \frac{\nu^2 d\nu}{e^{h\nu/kT_S} - 1} \tag{12.8}$$

where ν_L is the frequency corresponding to the Lyman limit. The total number of Balmer quanta emitted by the nebula can be written as

$$N_B = \sum \frac{E_i}{h\nu_i} \tag{12.9}$$

where

$$E_i = A_i \nu_i \left(\frac{\partial E_S}{\partial \nu} \right)_i \tag{12.10}$$

is the total energy emitted in the ith Balmer line per unit time by the whole nebula, A_i, being a directly observed quantity, and

$$\left(\frac{\partial E_S}{\partial \nu} \right)_i = 8\pi^2 \; R_s^2 \frac{h\nu_i^3}{c^2} \; \frac{1}{e^{h\nu_i/kT_S} - 1} \tag{12.11}$$

is the total amount of energy emitted by the central star per unit time and unit frequency interval in the region of the same ith line of the spectrum. The inequality that must be satisfied by the frequency transformation process between the star and the nebula is $N_B \leq N_L$, that is

$$\sum \frac{\nu_i^3 A_i}{e^{h\nu_i/kT_S} - 1} \leq \int_{\nu_L}^{\infty} \frac{\nu^2 d\nu}{e^{h\nu/kT_S} - 1} \tag{12.12}$$

which is obtained by using Eqs. (12.8)–(12.11).

The inequality, as given in relation (12.12) changes to

$$\sum \frac{x_i^3 A_i}{e^{x_i} - 1} \leq \int_{x_L}^{\infty} \frac{x^2 dx}{e^x - 1} \tag{12.13}$$

by substitutions

$$\frac{h\nu_i}{kT_S} = x_i, \quad \frac{h\nu_L}{kT_S} = x_L \text{ and } \frac{h\nu}{kT_S} = x$$

The summation on the left-hand side of the inequality extends over all the Balmer lines including the Balmer continuum. In practice, various values of T_S are taken by trial and that value of T_S for which the values of x_i and x_L give an equality in (12.13) is taken to be the temperature of the nucleus. This method has been used first by Zanstra and subsequently by many other authors. The temperatures of the nuclei thus obtained usually range between

30,000 to 60,000 K and in some cases, the calculated temperature has been found to be as high as 100,000 K. A procedure based on similar principle has been employed to compute the temperatures of nuclei by using the intensities of the forbidden lines. The results thus obtained by Zanstra and others do not differ significantly from those obtained by using the hydrogen lines. The temperatures obtained are usually very high. These high temperatures have also been confirmed by other determinations such as those using the intensities of He lines. The temperatures of the nuclei determined on the basis of other principles such as the colour temperature and ionization temperature also lead to the same conclusion that the planetary nuclei are very hot, despite the fact that all such temperature calculations of the nuclei are beset with inherent uncertainties.

All the characteristics of planetary nebulae discussed above yield one definite conclusion that these objects are somewhat exceptional in the family of the Galaxy. Consequently, much efforts have been put by astronomers to ascertain the exact state of evolution of the nuclei. It has been generally suggested that the nebulae are the peaceful ejecta of the outer layers of highly evolved red giants whose nuclear fuel is almost exhausted. The expansion velocities of 10–20 km s^{-1} of the nebulae confirm the peaceful ejection as contrary to the catastrophic ejection from novae and supernovae with very high velocities. As the surface layers are ejected, the nucleus rapidly contracts and its temperature rises. The entire process results in the transformation of nuclei with higher luminosity and lower temperature to those, with lower luminosity and higher temperature, in a relatively short time-scale, i.e. only 30,000 years, although an upper limit of about 50,000 years has been suggested by some authors. In a computation with stellar models of masses 0.65 $M_{\odot}$ and 1.00 $M_{\odot}$ in 1975 R. Harm and M. Schwarzchild have found that the initial transition phase from a red giant to a blue nucleus lasts only for about 6000 years, prior to which the star continued to eject mass at the rate of about 10^{-4} $M_{\odot}$ yr^{-1}. The presence of only a small number of planetary nebulae in the Galaxy supports this relatively short time-scale for the evolutionary phase of these objects. By the time the star contracts to a temperature of 100,000 K, the degeneracy starts; its luminosity starts fading but temperature remaining almost constant at 100,000 K during this phase. The star then continues to cool down as its luminosity continues to decrease and it rapidly proceeds toward the white dwarf stage. Planetary nebulae thus fit with the theory that they are precursors of white dwarfs.

12.8 THE CRAB NEBULA: SUPERNOVA REMNANTS

In this Section, we shall discuss yet another type of emission nebulae whose origin and subsequent life follow quite different astrophysical processes. These are the nebulae formed by the gaseous debris ejected at the time of supernova explosions. These are called supernova remnants of which the Crab Nebula is the best example. This is also the nearest and the most extensively studied supernova remnant. Three different types of supernova remnants, each possessing completely different physical nature have so far been conjectured. These are: (a) the gaseous debris, (b) the pulsars, (c) the blackholes.

The *gaseous debris* consist of the blown off superficial layers of the exploding star; these subsequently form nebulosities which appear as radio sources. In fact, many of these remnants emit radiation in a very wide range of the electromagnetic spectrum, all way from X-ray to the decameter radio waves, with a pronounced flat distribution of radiation intensity unlike the blackbody radiation spectrum. The thrown-off debris sweeps away the surrounding interstellar medium suffering thereby a deceleration, mingling its rich abundance of heavy nuclei with the interacting medium. Supernova explosions are thus believed to be agents for enriching the interstellar medium with heavier elements (which were formed in their interiors), and this enriched mixture is subsequently used as the new material for the new generations of stars.

Pulsars which are believed to be rotating neutron stars formed from the innermost core of the exploding stars. If the original mass of the exploding star lies between certain limits, such as, between 1.5 to about 3 or 4 $M_\odot$, the core may reduce to a pulsar.

For exploding stars with higher original masses, the core may, reduce to objects of peculiar physical properties, namely, the *black holes.* We shall be concerned here only about the first type of remnants mentioned above, viz. the gaseous debris left after supernova explosions (henceforth called supernova remnants).

The best studied supernova remnant is the Crab Nebula (Fig. 12.6) which almost coincides with the radio source Taurus A, the first radio source to be discovered by J.G. Bolton in 1948 (other than the Sun, of course). The nebula was earlier identified by Hubble with the remnant of the supernova of 1054 A.D., about which detailed records were available from old Chinese sources. The light curve of this supernova that can be derived from the analysis of the observational records fits quite well with the light curve of a Type I supernova. But its calculated distance of 1170 pc by Baade from the expansion velocity and proper motion at the

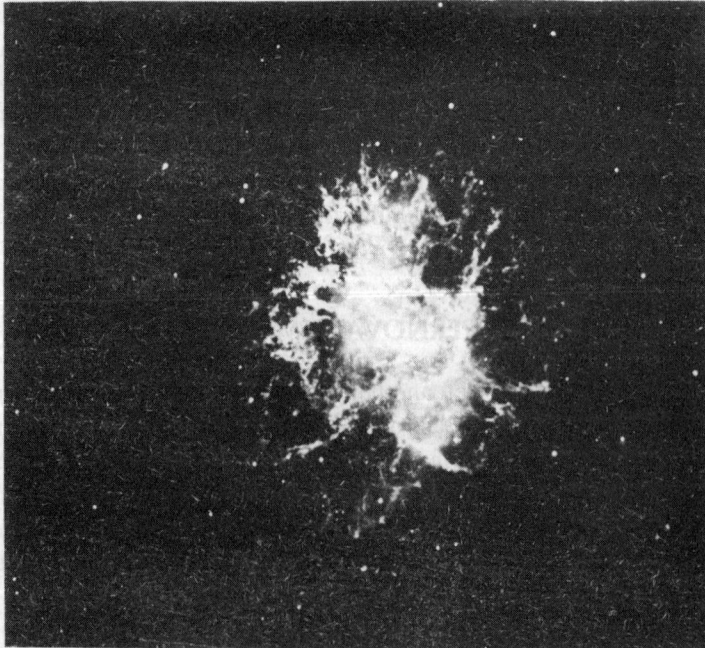

FIGURE 12.6   The Crab Nebula in Taurus. [*Courtesy:* Indian Institute of Astrophysics, Bangalore]

major axis, assuming it to be an oblate spheroid and its integrated apparent magnitude of – 6.25 with correction of 1.25 magnitude for interstellar absorption, gives it an absolute magnitude of –16.6. This contradicts the proposed membership of the Crab supernova among Type I. This difficulty is overcome, if one assumes the nebula to be a prolate spheroid. Then Baade's observations relate to the minor-axis of the spheroid, leading to a much larger distance of 1740 pc and absolute magnitude of –18.1. The prolate spheroid model therefore assigns Type I for the Crab supernova both in respect of light curve and absolute magnitude. This change in the spheroidal model from oblate to prolate also changes the height of the nebula above the galactic plane from 126 pc to 175 pc, which fits better with the concept of Population II stars as the origin of Type I supernovae. The volumes of the nebula for oblate and prolate models are respectively, 3.0×10^{55} cm^3 and 6.7×10^{55} cm^3. From a detailed analysis of the photometric data, O'Dell finds a mass of 0.64 $M_\odot$ for the nebula with Baade's distance and oblate spheroid model and with an electron temperature of 17,000 K. If the prolate spheroid model with its larger distance is assumed, the same observational data lead to a mass of 3.3 $M_\odot$ for the nebula. These masses reduce to 0.35 $M_\odot$ and 1.8 $M_\odot$ respectively with a reduction of electron temperature from 17,000 K to 10,000 K. It must be pointed out, however, that considering all the consequences of the two spheroidal models and comparing these with various types of observational results of the nebula, none of the models appear completely satisfactory, although the prolate spheroid model turns out to be a better choice.

The optical spectrum of the Crab Nebula consists of strong continuous spectrum crossed by emission lines of H, O II, O III, He I and other elements. The continuous spectrum arises out of the amorphous gas while the emission lines are due to luminous filaments. Earlier, Baade and Minkowski proposed individually that the optical continuum is radiated by the hot gas of the nebula with a temperature around 10^5 K. Such an explanation is, however, no longer accepted. It is now believed that synchrotron radiation which is responsible for the continuous radio frequency radiation from the nebula also can account for the observed optical continuum radiation. The nebula emits radio energy over a very wide frequency band from 19 to 10^4 MHz. The total power emitted below 10^4 MHz is 2.62×10^{33} erg s^{-1}.

The fact that both optical and radio continuum radiation from the nebula are due to relativistic electrons spiralling round the magnetic lines of force, is evidenced by the high degree of polarization of the radiation, both in optical and radio ranges. It is well known that synchrotron radiation is highly polarized. The energy of the relativistic electrons for the optical continuum radiation is required to be of the order of 10^6 MeV, while the electron energy less by about 3 to 4 orders of magnitude will be sufficient to radiate radio frequencies. The total energy of all the relativistic electrons calculated by Shklovsky is 2.7×10^{47} erg, while Woltjer derives the value of 7.0×10^{47} erg. The magnetic field strength is 5×10^{-4} gauss and the total magnetic energy lies between 10^{47} to 10^{48} erg. The field strength in the nebula is thus about 100 times that of the general galactic magnetic field. The present magnetic energy in the shell is 10^{46} erg and the energy of expansion of the shell is 10^{49} erg, if its mass is assumed to be 1 $M_\odot$. The energy of expansion of the shell may have been derived at the expense of the magnetic energy.

The Crab Nebula is also a strong source of X-ray emission. The fact that the X-ray emission comes not only from the central star, but also from the nebula itself, was demonstrated from rocket observation of the nebula during a lunar occulation of it. The spectral index of

X-ray intensity fits considerably well if the radiation is assumed to be due to synchrotron mechanism. The results of various calculations show that the total energy required by the Crab Nebula to maintain its radiation in all spectral ranges and also its velocity of expansion comes to the order of 10^{38} erg s^{-1}.

L. Woltjer made a detailed study of the various line intensity ratios of the nebula, in order to determine the relative abundance of elements in it. He used serveral electron temperatures, and comparing the observed decreasing intensity ratios of successive Balmer lines with the theoretical ratios, derived an electron temperature of 15,000 K as the best fit. Using three different electron temperatures, viz. 17,000, 10,000 and 8,000 K, Woltjer found that the H/He ratio in the nebula is only about 2, while the ratio is approximately 7 for normal cosmic abundance. Also, for the electron temperature as low as 8,000 K, the heavier elements N, O, Ne and S are overabundant. Such observation is consistent with the theory of supernova explosion—that the star explodes when the energy generation takes place in its superficial layers where heavy elements are produced before these layers are blown off.

Among other historically recorded supernova remnants, those observed in the years 1006, 1572 and 1604 are known. The first of these occurred in the constellation Lupas as has been identified from the Oriental chronicles. At maximum, the star was at least as bright as Venus and remained visible at night to the naked eye for a period of more than a year. The gaseous remnant of this event was detected as a radio source in 1965. The second remnant is the result of the supernova explosion in 1572 which was studied by Tycho Brahe till it was visible and has since been called Tycho's nova, otherwise known as Nova B Cassiopia. The remnant was first discovered in 1952 by Hanbury Brown and Hazard as a strong radio source. Subsequently, Minkowski found the optical remnant associated with the radio source. Tycho's nova is now known also as a strong source of X-ray emission. Analyzing the old records of this nova, Baade concluded that at maximum it had an apparent visual magnitude of -4, its light curve corresponding to that of a Type I supernova. The supernova observed in 1604 (SN Ophiuchi) was studied by Kepler and hence is known as Kepler's nova. Baade's analysis of old records indicated that this too was a supernova of Type I and maximum apparent visual magnitude attained by it was -2.25. Baade also discovered the optical nebula. The nebula was detected as a radio source by Hanbury Brown and Hazard in 1952.

There are yet a few supernova remnants in the Galaxy for which no records have been found. Among these, the one is Cassiopia, not far from Tycho's nova, probably provides with the most recent supernova explosion in the Galaxy. This was first discovered in 1948 by Ryle and Smith of Cambridge as the strongest radio source in the sky and was called Cassiopia A. The energy emitted in radio waves by Cassiopia A is 3.3×10^{34} erg s^{-1} which is higher than that emitted by the Crab Nebula by more than an order of magnitude. The optical nebula was later discovered by Baade and Minkowski, and from a study of this nebula the time of the event was found to be around 1700 A.D. The event could escape detection for two reasons: Firstly, the region of the sky where it occurred is heavily obscured by interstellar absorption. Cassiopia A is far away in the direction of the galactic centre where it suffers an absorption of at least 6 magnitudes by inter-stellar dust. And secondly, the supernova itself might be less brilliant than other recorded supernovae. This is plausible because the high velocity of the expanding gas (~ 7400 km s^{-1}) and large mass involved predict it to be a Type II supernova which is about 1.5 magnitude fainter intrinsically than Type I supernovae. In later years, Cassiopia A has been known as a very strong X-ray source.

Cygnus loop is a very old supernova remnant. It has also a very large dimension in the sky stretching over a region of about 3° in diameter. The study of its expanding shell indicates that the explosion of the original supernova took place at least about 50,000 years ago. The remnant is a very strong source of both radio and X-ray emissions. It appears that Cygnus loop is the remnant of a supernova of Type II. The radio energy emitted by Cygnus loop is $\sim 4 \times 10^{32}$ erg s^{-1}.

Other galactic supernovae that are mentioned in old Chinese and other eastern records, and which have been subsequently verified as radio sources, are those of A.D. 185, 393, 437, 827, 1181, 1203 and 1230. These were probably all supemovae of Type I. Recently, many tenuous nebulosities have been searched for detecting pulsars within them, but results were discouraging. These nebulosities are probably not remnants of previous supernova explosions, but are just local concentration of normal interstellar gas. For this reason, they have been named *Ghost Remnants.*

Although, due to their extremely high luminosities supernovae can be observed even in fairly distant external galaxies, remnants of previously exploded supemovae can hardly be detected except in the nearest extragalactic nebulae. D.S. Mathewson and J.R. Healy identified three supernova remnants—N49, N63A and N132D—in the large Magellanic cloud (LMC). Their identification was subsequently confirmed by D.S. Mathewson, K.G. Henize and B.E. Westerlund. Some more remnants have been subsequently discovered in the LMC. If the distance to this galaxy is assumed to be 55 kpc, then the mean surface brightness of its remnants appears to be higher by 25 per cent than that of remnants of the galactic supernovae. On the other hand, if the surface brightness be comparable among the remnants in these two galaxies, the distance to the LMC should be 46 kpc. D.S. Mathewson and G.W. Clarke have also reported the discovery of two supernova remnants in the Small Magellanic Cloud. The search is not yet complete and some more remnants are likely to be discovered in near future in these galaxies.

EXERCISES

1. Explain how dark nebulae are recognized in our Galaxy. What are the sources of our information about the dark nebulae?

2. Explain how galactic nebulae can be separated from extragalactic nebulae? What are the basic differences between these two classes of nebulae? Comment on the distribution of galactic nebulae within the Galaxy.

3. Give an ingeneous scheme of classification of galactic nebulae. Describe Hubble's work to classify bright nebulae into *reflection* and *emission* nebulae. What are the principal observable features of these two types of nebulae?

4. Name different types of emission nebulae. Explain the different causes that lead to their formation. Comment on the galactic distribution of these nebulae.

5. Explain the different physical processes that operate in the formation of emission lines in emission nebulae. Why strong forbidden lines are produced in emission nebulae while these lines are totally absent in the spectra of stars?

6. List the principal emission lines, both natural and forbidden, that are observed in emission nebulae.

7. Define *dilution factor*. What is its value in an average emission nebula?

8. Discuss how the ionizing photons beyond the Lyman limit are degraded to produce strong Balmer series of emission lines in diffuse emission nebulae.

9. Compare the points of similarity and contrast between planetary nebulae and diffuse emission nebulae.

10. Describe a method by which the temperature of the central star of a planetary is determined. What kind of temperatures are obtained for these stars?

11. What are the different methods that are used to determine the distances of planetary nebulae? Which of these methods, in your opinion, can be considered the most useful and accurate? How does the luminosity of a planetary nebula compare with that of the Sun?

12. Describe how the birth rate of planetaries in the Galaxy is calculated? How many planetaries do you think exist at present in the Galaxy?

13. What observational evidences lead to the conclusion that the planetaries are Population II objects?

14. The mass of a planetary is 0.2 $M_\odot$ which has been ejected during its lifetime of 30,000 years at a speed of 200 km s^{-1}. Calculate the rate of kinetic energy released by the event and compare this energy with that radiated by the nebula if its absolute magnitude is –3.0.

15. What are the distinguishing characteristics of synchrotron radiation?

16. A supernova explosion throws a shell of material of mass 2 $M_\odot$ at an initial speed of 5000 km s^{-1}. Calculate the kinetic energy and the initial momentum associated with the shell.
    (a) If this shell sweeps up interstellar material and slows down by conservation of momentum, how much mass will be swept up when the shell attains a velocity of 100 km s^{-1}?
    (b) What will be the radius of the shell when its velocity is 100 km s^{-1}, if the average density of the swept up material is 10 H cm^{-3}?

17. If the Crab nebula radiates at the rate equal to the Sun, how much energy it will lose by radiation till the end of this century? Compare this energy with the energy of the relativistic particles contained in the nebula.

SUGGESTED READING

1. Alter, L.H., *Gaseous Nebulae,* Chapman and Hall, London, 1956.
2. Aller, L.H., *Atoms Stars and Nebulae,* Cambridge University Press, Cambridge, 1991.
3. Ambertsumyan, V.A., *Theoretical Astrophysics,* Pergamon Press Ltd., New York, 1958.

4. Bailey, M.E. and Williams, D.A. (Eds.), *Dust in the Universe,* Cambridge University Press, Cambridge, 1989.

5. Heiles, C. Physical Conditions and Chemical Evolution of Dust Clouds, *Annual Review of Astronomy and Astrophysics,* **9**, p. 293, 1971.

6. Leitherer, C. Walborn, N., Heckman, T. and Colin, N. (Eds.), *Massive Stars in Starbursts,* Cambridge University Press, Cambridge, 1991.

7. Middlehurst, B.M. and Aller, L.H. (Eds.), *Nebulae and Interstellar Matter.* The University of Chicago Press, Chicago, 1968.

8. Minkowski, R,, Planetary Nebulae, *in Galactic Structures,* Blaauw, A. and Schmidt, M. (Eds.), The University of Chicago Press, Chicago, 1965.

9. Pikelner, S., *Physics of Interstellar Space,* Foreign Language Publishing House, Moscow, 1967.

10. Salpeter, E.E., Central Stars in Planetary Nebulae, *Annual Review of Astronomy and Astrophysics,* **9**, p. 127, 1971.

11. Sharpless, Stewart, Distribution of Associations, Emission Regions, Galactic Clusters and Supergiants, in *Galactic Structure,* Blaauw, A. and Schmidt, M. (Eds.), The University of Chicago Press, Chicago, 1965.

12. Tenorio-Tagle, G., Pietro, M. and Sanchez, F. (Eds.), *Star Formation in Stellar System,* Cambridge University Press, Cambridge, 1993.

Interstellar Matter

13.1 LARGE-SCALE DISTRIBUTION OF INTERSTELLAR MATTER

A little attention to the photographs presented in Chapter 12 will reveal the irregular and patchy distribution of interstellar matter in local regions of the Galaxy. The bright and dark nebulae are, in fact, regions of dense concentration of gas and dust. Such concentrations are also clearly revealed in photographs of external galaxies. In particular, spiral arms are defined by denser lanes of neutral and ionized gas and of dust. This is also true of the spiral arms in our own Galaxy. Irregularities are also common in the distribution of tenuous invisible component of interstellar matter. This is revealed in the dust component by the directional dependence of absorption by it. The multiple lines of Ca II observed in spectra of distant hot stars clearly indicate that the interstellar gas exists mostly in cloudy form, each cloud moving with its own characteristic velocity producing lines with different Doppler shifts. Attention was drawn to this irregular distribution of matter by early observers. R.J. Trumpler in his classical paper categorically emphasized that the general absorption was not uniform throughout the Galaxy and that the absorbing materials had local irregularities. Extensive photoelectric measurements of the selective extinction in O and B stars led J. Stebbins, C.M. Huffer and A.E. Whitford to conclude—"The absorption in the Galaxy is obviously so irregular and spotted that a constant coefficient cannot be used for any large region of space".

Attention was first drawn by C.S. Beals to the multiplicity and complex nature of the Ca II absorption lines. The subsequent work on radial velocities of interstellar lines by P.W. Merril and O.C. Wilson demonstrated conclusively that the lines originate in discrete clouds which possess sufficient random motions, over and above their motions due to galactic rotation. By the time W.S. Adams published his extensive work on observation of 300 O-B stars with high dispersion Coudé spectrograph, the cloudy distribution of interstellar matter was more or less accepted by astronomers. Detailed analyses of Adams' observational material however established this idea on a firmer basis. Also, quite a number of interesting features were revealed by these analyses regarding the structure of the interstellar lines and the deductions yielded thereby. In general, the H and K lines of Ca II or the D_1 and D_2 lines of Na I occur in the spectrum of any particular star as double, triple, quadruple or even with higher multiplicity, if sufficiently high dispersion is used. Of the 300 stars observed by Adams, 87 revealed double Ca II lines, 17 revealed triple lines, quadruple lines were measured in 4 and lines of greater complexity were observed in 40 stars. In general, one of these multiple components is stronger than all others. This stronger component shows the lowest Doppler shift with respect to the natural wavelength of the line. All other components are relatively weaker and of varying strengths, showing different Doppler shifts on one or the other side of the natural wavelength. When only a single interstellar line appears in the spectrum of a star, the line is usually strong having small Doppler shift. The radial velocity of the stronger component usually corresponds to that as would be produced by galactic rotation at a point midway between the star and the Sun. The analysis further indicates that neither the occurrence of multiple lines in the spectra of stars nor the Doppler shifts of the weaker component of lines are dependent on the direction in which they are observed. These observations immediately yield two definite deductions:

1. A uniform but relatively thinner distribution of interstellar gas which fully participates in galactic rotation is responsible for producing the stronger component of Ca and Na absorption lines.

2. Superimposed on this, there are gas clouds of smaller dimensions and higher densities which over and above the motion due to galactic rotation, possess sufficient random velocities of varying magnitudes. These clouds are responsible for the generation of the weaker components of absorption lines with different Doppler shifts.

The observational material has been subsequently enriched by works of other astronomers. Among these, mention must be made of the work of P. Mc R. Routly and L. Spitzer who observed and analyzed the Na I and Ca II lines in spectra of 23 stars. In addition to this, 112 stars were observed by G. Münch who analyzed the intensity and velocity shifts of interstellar lines of Ca II and Na I in the spectra of these stars. These data were further supplemented by Münch and others in subsequent years in which more distant stars have been included and higher dispersions were used. Among these, works of G. Münch and H. Zirin in 1961 are considered to be of great significance. These authors observed interstellar lines in the spectra of two dozen distant O and B stars situated at high galactic latitudes. The estimated distances of these stars from the galactic plane lie in the range between 240 to 2700 pc. The measured shifts of the interstellar lines correspond to velocities of the order of 50 km s^{-1} and in some cases velocities as high as 100 km s^{-1} have been measured. These observations therefore indicate the presence of high velocity clouds at large distances from the galactic plane. The analysis of the nature of the lines in these high velocity clouds at large galactic latitudes have revealed many peculiarities which subsequently created considerable astrophysical interest.

Side by side with these optical observations, the radio observations at 21-cm wavelength of hydrogen radiation were carried out which primarily began with the work of H.C. Van de Hulst, A.B. Muller and J.H. Oort in 1954. The large-scale spiral structure in our Galaxy was revealed for the first time in this radio diagram.

A more detailed study was carried out by M. Schmidt in 1957 which revealed large-scale local variations in gas density. Subsequent works of other authors at 21-cm have led to great advancement in the understanding of the general as well as the localized distribution of interstellar gas. The existence of clouds of smaller dimensions moving with random velocities, as deduced from earlier optical observations has been confirmed. Large cloud complexes in local regions, otherwise invisible, have also been observed by 21-cm radiation. On the basis of the above discussion, we can therefore summarize the principal distributional features of interstellar matter (gas and dust) as follows:

1. Clouds of matter of various dimensions swim across the vast ocean of a more tenuous and almost uniform distribution of matter pervading the entire region of the galactic plane.

2. The uniformly distributed component mainly participates in the galactic rotation while the clouds possess a sufficiently large range of random motions (a few kilometers to about 40 km s^{-1}) besides their share in the galactic rotation.

3. Some clouds extend much higher above the galactic plane. These clouds move with much larger velocities which can be as large as 100 km s^{-1}.

4. The gas and dust have, in general, identical distribution except probably in some exceptional regions where an over-abundance of dust is noticed.

5. A major part of the matter is concentrated along spiral arms of the Galaxy, the vast interarm regions having an extremely low density of matter.

In the following section, we shall discuss the optical and radio lines of interstellar origin that have so far been identified.

13.2   INTERSTELLAR LINES

The Optical Absorption Lines

Historically, the first interstellar lines to be detected were the H and K lines of Ca II in the spectrum of the spectroscopic binary δ Orionis by J. Hartmann in 1904. The stationary D$_1$ and D$_2$ lines of Na I were first observed fifteen years later in 1919 by M.L. Heger, following an early suggestion by V.M. Slipher in 1909. These lines were also first observed in the spectrum of δ Orionis and subsequently in other stars. Two years later, the ultraviolet Na I doublet were also discovered by Miss Heger. Valuable theoretical as well as observational works were carried out during the subsequent two decades which established the interstellar origin of the Ca II and Na I lines. These lines could be discovered with moderate dispersions used in earlier works because of their sufficiently great strengths. Higher resolving powers were achieved in subsequent years resulting in the discovery of fainter lines of other atoms, ions and molecules. Thus, lines due to K I, Ca I, Fe I and Ti II were discovered mainly through the efforts of W.S. Adams and T. Dunham Jr. on Mount Wilson during the late thirties and early forties. All these atomic and ionic lines have one thing in common, viz. all these are absorbed from the ground state and so the lines are sharp. Besides these sharp lines of atomic and ionic origin, some more interstellar lines were discovered during the same period. These lines, however, could not be associated with any atomic or ionic transitions. They were finally identified with those arising from the lowest rotational levels of the radicals CH, CH$^+$ and CN. Besides these atomic and molecular lines, some diffuse absorption features of interstellar origin have also been discovered. But in spite of ceaseless efforts of many astronomers these features have so far eluded identification. Many speculations have however been made about the exact source of these diffuse features. The strongest of such features is at λ 4430 Å and this has been found to bear a correlation with the colour excess. This fact has led some astronomers to speculate that the feature might originate in gases locked with the absorbing grains. Other suggestions have also been made, but the problem still remains where it was about four decades ago.

An analysis of the equivalent widths of Na I and Ca II lines reveals the interesting feature of different kinematical behaviour of interstellar Na and Ca. It is found that the velocity dispersion obtained from Ca II lines is systematically higher by about 50 per cent than that obtained from Na I lines. As a probable explanation, it has been suggested that Ca II is overabundant relative to Na I in clouds with higher velocities. It has also been found that about two-thirds of the higher-velocity clouds have negative velocities. In Adams' list of lines there are 82 components with $|v_r| \geq 15$ km s^{-1}. Of these, 55 components have negative and 27 only have positive velocities. This trend is faithfully followed by high-velocity lines in other observed materials. This preponderance of negative velocities' in clouds can be explained if many of these are seen through a shell accelerated by a nearby star. This is a likely phenomenon in view of the well-known clustering tendency of early type stars in whose spectra the lines are observed.

From an analysis of the equivalent widths of the lines of any element, one can calculate its abundance in interstellar matter. This has been done by various authors. The results of these studies have revealed that the abundances in interstellar matter do not grossly differ from cosmic abundances. On the other hand, if cosmic abundances are assumed to prevail in interstellar space, then from the calculated amount of any particular element in the line of sight from the lines it produces, one can compute the amount of any other element whose lines are not observed. Thus according to Routly and Spitzer, the mean equivalent width of Na I D$_2$ line for fast clouds is 40 mÅ. This corresponds to 2×10^{11} Na I atoms cm^{-2} if the lines are unsaturated. If the velocity dispersion in gas within the cloud be only 1 km s^{-1}, then an upper limit to the correction factor for saturation is about 1.5, giving 3×10^{11} Na I atoms cm^{-2}. But Na is mostly in the ionized state. The total number of Na atoms is I/n_e times the number of Na I where $I = n(\text{Na II})/n(\text{Na I})$, n_e being the electron density in the line of sight. A. Weigert has found $I = 0.68$ from photoionization calculations for $T = 100$ K. If we assume $n_H = 10$ cm^{-3} as the number density of hydrogen in cloud, then assuming a cosmic abundance and that all elements with ionization potentials less than that of hydrogen are fully ionized in space, one arrives at $n_e = 3.8 \times 10^{-4}$ cm^{-3}. When these values are used one gets 5.4×10^{13} Na atoms cm^{-2}. With cosmic abundances this yields 3.2×10^{19} H atoms cm^{-2}. Assuming some suitable model of the clouds, one can thus calculate their masses also. A wealth of information is thus possible to be derived from analyses of the properties of interstellar optical lines regarding the physical and dynamical characteristics of the interstellar medium.

Interstellar Radio Lines

Radio astronomy has ensured a great advancement in the knowledge of the physics and chemistry of interstellar space. 21-cm radio line of neutral hydrogen has played a very important role in enhancing our understanding of the structure of our Galaxy, and also of other galaxies. The centimetre radiation arising from transitions between very high excitation of hydrogen atoms such as, transitions between $n = 110$ to $n = 109$ and $n = 105$ to $n = 104$ levels, provide valuable information regarding the physical processes in H II regions. These are the only interstellar radio lines of atomic origin, and we shall not discuss these any further here. Many molecular lines, however, have recently been discovered in large cloud complexes, and almost all of these lines fall in millimetre and centimetre wavelengths. Since these lines are believed to be of great astrophysical importance, we shall discuss briefly some of the aspects relating to them.

The first interstellar molecular lines to be detected in the radio range were those of OH (hydroxyl). The possibility of observing these radio lines at nearly 18-cm in interstellar gas was first suggested by I.S. Shklovsky in 1953. The lines were actually observed in absorption ten years later in 1963 by S. Weinreb and his co-workers in the radio source Cassiopeia A, and soon after, in Sagittarius A in the galactic centre. The ground level of OH radical is separated in Λ-doublets. Each component of the doublet is split into two hyperfine levels leading to four possible transitions as shown in Fig. 13.1. The frequencies of the lines emitted (or absorbed) are 1720, 1667, 1665 and 1612 Mc s^{-1} which correspond closely in 18-cm wavelength. The theoretically predicted intensity ratios of these lines are as 1:9:5:1.

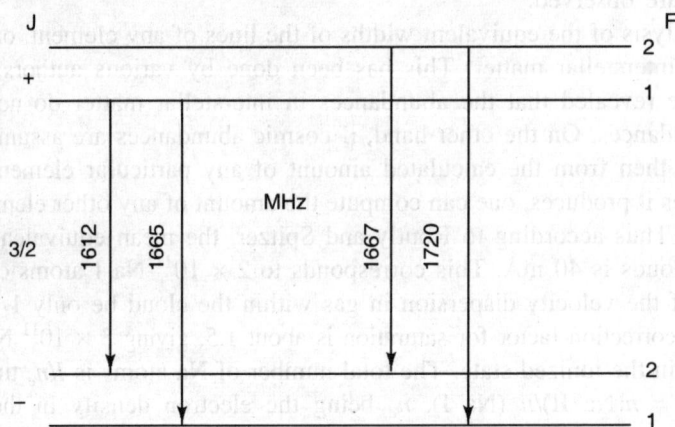

FIGURE 13.1 OH emission lines at 18-cm radio wavelength.

Thus, the strongest lines are those with frequencies of 1667 and 1665 Mc s^{-1} which were naturally observed first. Following the discovery of OH lines, it was first thought that the distribution and dynamical features of OH sources would be similar to those of neutral hydrogen. But subsequent OH observations revealed many surprises, one of which was that OH bore little relationship with neutral hydrogen in these respects. For example, 21-cm line of neutral hydrogen observed in Sagittarius A (the galactic centre) shows no Doppler shift as it should be since we are viewing clouds in this direction which are moving at right angles to the line of sight. But Doppler shifts corresponding to a velocity of approach as high as 120 km s^{-1} have been measured in OH lines in the same direction. Also, OH lines are of considerable width, indicating large internal motions in the source, the scale of which has never been observed in any of the hydrogen clouds. Many other peculiar features have been demonstrated by OH radiation. For example, the intensities of the four lines in Sagittarius A have been found to bear the ratio 1:2.7:2.2:1 instead of the theoretically predicted ratio 1:9:5:1. This anomaly in intensities has been attributed to saturation effect indicating very high concentration of OH molecules in the galactic centre region. Anomaly is also observed in relative intensities of different components of OH. For example, OH has been observed *in emission* in several sources, such as in W 49 (that is, the 49th object in G. Westerhout's catalogue of radio sources), where the 1665 line was found to be much stronger than the 1667 line. But theoretically, the former should be only 5/9 as strong as the latter. Saturation effect can not explain such anomalous result. It can only be explained if one is ready to accept the existence of a very peculiar excitation mechanism for OH molecules, and as such, *maser* action has been postulated. In this case, the conditions of thermal equilibrium breaks down completely and the Boltzmann distribution is violated. In the presence of thermal equilibrium, Boltzmann distribution,

$$n(\varepsilon) = n(0) \exp\left(-\frac{\varepsilon}{kT}\right) \tag{13.1}$$

where $n(\varepsilon)$ is the number density of particles in an excited energy state ε, and $n(0)$ is the number density in the ground level in a gas of temperature T, must hold. Since the exponential factor is always less than unity for positive T, $n(\varepsilon)$ is always less than $n(0)$ in thermal equilibrium. But if there exists an extraneous mechanism of pumping energy in the assemblage of particles in order to maintain a type of equilibrium with a larger number of particles populating the excited level than those populating the ground level, we find what is known as *population inversion* ($n(\varepsilon) > n(0)$), which is in complete violation of thermal equilibrium condition. The energy pumping mechanism maintains the higher population of particles in the excited level so that the corresponding transition gives rise to a line whose strength is *greatly in excess* of that predicted by the ordinary theory. This is what is known as the *maser action,* and *cosmic maser action* has been found to activate the OH and H_2O molecules. There are regions of interstellar space where certain energy states of these molecules are pumped up to maintain a population inversion.

NH_3 (ammonia) was the next molecule to be discovered in interstellar space in 1968. Subsequently, there was a breakthrough in the achievement of discovering new molecules in interstellar space. At present, around forty interstellar molecules have been detected, as given in Table 13.1. In view of the new discoveries that are likely to be made, the list in Table 13.1, should not be considered as complete and final. Many of these complex molecules have been detected in Sagittarius clouds in the galactic centre region and in Orion. In particular, the dense clouds at the region Sagittarius B2 have been found to be a storehouse of complex molecules. Since many of these molecules are known to be essential for the maintenance of living organisms, their abundance in the interstellar space assures a good reason for the astronomers to be encouraged in speculating the existence of life in the vast outer space.

All the molecules so far observed in radio frequencies radiate in the wavelength range from 2 mm to 40 cm. The shorter wavelength side tends to merge in the far infrared region of the spectrum. It may not be unlikely to detect infrared lines of atoms and molecules in near future.

TABLE 13.1   Radio Lines of Interstellar Molecules

Molecule	Chemical name	Spectrum	Wavelength (cm)
OH	Hydroxyl	Absorption	18 (four lines)
		Emission	18 (four lines)
$O^{18}H$	Hydroxyl isotope	Emission	18 (two lines)
CO	Carbon monoxide	Emission	0.26
$C^{13}O$	CO isotope	Emission	0.26
CN	Cyanogen	Emission	0.27 (two lines)
CS	Carbon monosulphide	Emission	0.20
NS	Nitrogen sulphide	Emission	0.26
SO	Sulphur monoxide	Emission	
HD	Deuterohydrogen	Emission	
SiO	Silicon monoxide	Emission	0.23, 0.34
H_2O	Water vapour	Emission	1.35
H_2S	Hydrogen sulphide	Emission	0.18
C_2H	Ethynyl	Emission	0.34 (four lines)
N_2H^+	Monoprotonated nitrogen	Emission	0.32
NH_3	Ammonia	Emission	1.2, 1.3 (five lines)
HCN	Hydrogen cyanide	Emission	0.35
$HC^{13}N$	HCN-isotope	Emission	0.35
HCO^+	Formyl cation		0.34
HNC	Hydrogen isocyanide	Emission	0.33
OCS	Carbonyl sulphide	Emission	0.27
DCN	Deuterium cyanide		
HC_3N	Cyanoacetylene	Emission	3.3
HDCO	Monodeutero formaldehyde	Emission	5.6
HCHO	Formaldehyne	Absorption	6.6, 6.2, 2.1, 1.0, 0.2
$HC^{13}HO$	HCHO-isotope	Absorption	6.5
HCHS	Thioformaldehyde		9.5
CH_3OH	Methanol	Emission	36, 1.2
CH_3CN	Methylcyanide	Emission	0.27
NH_3CN	Cyanamide	Emission	0.37, 0.3
HNCO	Isocyanic acid	Emission	1.36, 0.34
CH_2NH	Formaldimine	Emission	5.8
CH_3C_2H	Methylacetylene	Emission	0.35
CH_3NH_2	Methylamine		
HCOOH	Formic acid	Emission	18.3, 6.1
CH_3CHO	Acetaldehyde	Emission	28.1
NH_2CHO	Formamide	Emission	6.5
$HCOOCH_3$	Methyl formate	Emission	18.6
$HC^{13}CCN$			
$HCC^{13}CN$	Cyanoacetylene isotopes	Emission	3.3

13.3 INTERSTELLAR CLOUDS

Statistics

From what has been discussed so far, it has been clearly revealed that the interstellar matter is not uniformly distributed. Local concentration of cloud-like structure separated by low-density *intercloud medium* is the general distributional feature of the medium. The term *interstellar cloud* has been assigned to these local concentrations. The following observational facts clearly demonstrate such a cloud structure of the interstellar medium:

1. The photographs of bright and dark nebulae clearly reveal the cloudy structure of the local concentrations of the matter.
2. The longitudinal variations of the interstellar extinction and reddening along the galactic plane manifests the spotty distribution of matter in the plane.
3. The multiple components of interstellar absorption lines in spectra of distant bright stars indicate the existence of isolated clouds with random motions.
4. The cloud-like structure of the medium is suggested by the polarization measurements of star-light in different directions.
5. The 21-cm data clearly demonstrate the presence of local concentrations of varying sizes moving with different velocities.
6. The most recent observations of various molecules show that their radiation comes, in general, from highly localized concentrations. Various degrees of patchiness of the matter are suggested by these observations.

Analyzing the observational data, one can draw an idealized picture of the physical, dynamical and distributional properties of average individual clouds. An average cloud is about 7 pc in radius having a density of 10 H atoms cm^{-3}. If 10 per cent of He by volume is assumed, this yields the mass of an average cloud to be about 400 $M_\odot$. A line-of-sight in a random direction in the galactic plane intersects about 8 clouds kpc^{-1} and the number of clouds per cubic kpc is of the order of 50,000. The average line-of-sight motion of a cloud is about 8 km s^{-1} so that the average space motion is about 14 km s^{-1}. The total volume of space occupied by clouds is about 7 per cent. The clouds are almost entirely localized within the spiral arms in a small thickness of about 250 pc around the galactic plane. The vast intercloud region (about 93 per cent of the volume of the galactic plane) contains a very tenuous distribution of gas; 0.1 H atom cm^{-3} or less. An average cloud produces an absorption of light of about $0^m.3$. Besides these average clouds, observations show large cloud complexes of much higher masses and lower velocities.

Masses of these complexes may range from a few thousand to a few hundred thousand solar masses. Their linear dimensions may be between 30 to 100 pc and usually have much higher densities which may range from 20 to 10^3 H atoms cm^{-3}. These large clouds invariably have lower space velocities. These large complexes appear to be the seat of the birth of new stars. In many of these are embedded the associations of hot stars of spectral classes O and B. These hot stars, in turn, create large H II regions around themselves which push the surrounding H I gas with velocities of the order of 10 km s^{-1}.

Velocity Distribution of Clouds

As the observational data on cloud velocities began to pile up, astronomers became more and more interested in finding an appropriate velocity distribution law that would best fit the data. Most authors used Adams' observation of the H and K lines of Ca II in 300 O–B stars as the basic data. The most natural suggestion was the Gaussian distribution of radial velocities

$$f(V_r)\,dV_r = \frac{1}{\sigma_r\sqrt{\pi}}\,\exp\left[-\frac{V_r^2}{\sigma_r^2}\right]dV_r \tag{13.2}$$

which has been used by many authors. The radial velocity dispersion is determined from observations. Unfortunately, the observed radial velocities of clouds significantly deviate from Gaussian distribution chiefly for two reasons. Firstly, the observed number of higher-velocity clouds greatly exceeds that predicted by Gaussian distribution. Secondly, among these high-velocity clouds, two-thirds possess negative while only one third have positive velocities. We have already discussed how this preponderance of negative velocities among clouds with high velocities can be logically explained. Taking into account the effect of overlapping of line components, A. Blaauw has deduced a distribution law for the radial velocities of clouds which, according to him, fitted best with observational results. This law is:

$$f(V_r)\,dV_r = \frac{1}{2\eta}\,\exp\left[1 - \frac{|V_r^2|}{\eta}\right]dV_r \tag{13.3}$$

where η is the average random motion in the line-of-sight. From observed results, Blaauw computed the values of η ranging from 5 to 8 km s^{-1} which correspond to dispersions of 7 to 11 kms^{-1}. Blaauw's theoretical law is non-Maxwellian and simple in form. Also, it takes care of a larger proportion of high-velocity clouds.

Blaauw's simple law was however derived with the basic assumption of *identical clouds*. This assumption has been challenged by Schlüter, Schmidt and Stumpf. These authors have emphasized that the lines analyzed have been produced by *three distinct* classes of clouds whose physical differences have to be taken into account while formulating a general law. The three classes these authors have envisaged were:

1. Slow moving clouds which produce in the spectra of almost all stars the strong, relatively slightly displaced component; these clouds are believed to be more massive than the average clouds.
2. The fast clouds producing highly displaced line components; these are probably the average clouds.
3. The circumstellar clouds that are pushed by the expanding H II regions; these clouds are responsible for the lines with large negative radial velocities.

According to these authors, each of these three distinct types of clouds possesses its characteristic physical and dynamical properties which have to be given proper weightage. Su Shu Huang and Kaplan have suggested the law

$$f(V_r)\, dV_r = \text{Const.} \; \frac{dV_r}{V_r} \tag{13.4}$$

to represent the observed radial velocity distribution. This law is simpler than that of Blaauw and displays a better fit than the Gaussian law in respect of the observation of larger number of high velocity clouds. The law, however, does not take into account the physical differences among clouds.

One basic observational fact about the interstellar absorption lines that has been noticed from early days is that stronger lines have smaller velocity shifts. There is a strong inverse correlation between the intensity of the line and its Doppler shift—a higher intensity corresponds to a lower shift and vice versa. The relation that emerges from the analysis of materials obtained by Adams' observations has been found to fit approximately the formula:

$$I = \frac{\text{Constant}}{V_r} \tag{13.5}$$

This relation clearly suggests that the more massive clouds which produce stronger lines move with lower velocities, a fact that has been emphasized by many authors on both the theoretical as well as on observational grounds. If the density of gas in different clouds be assumed the same, these massive clouds must also be larger in size, so that an inverse correlation between the cloud size and its velocity exists.

13.4 H I AND H II REGIONS: STRÖMGREN'S SPHERES

The recent discovery of the various types of molecules particularly of H_2 in interstellar space suggests that the mass of the gaseous component of the Galaxy is likely to be much higher than was previously predicted. About 2–5 per cent of the total mass of the Galaxy was so long assigned to gas on the basis of observation of 21-cm hydrogen emission, although the problem of an unobserved component of mass was there. H_2 molecule was suspected as a likely constituent which eluded observational evidence for many years. But the various microwave observations and the ultraviolet observation from Copernicus have revealed that a large variety of molecules, and in particular H_2, is present in high abundance throughout the interstellar space. Although the exact amount of molecular mass could not yet be ascertained, it is likely that when better information would be available, the total amount of the gaseous mass would have to be substantially revised upwards. It may be as high as 20% of the total galactic mass, or even more. Thus the total mass of the interstellar matter is probably at least about 4×10^{10} $M_\odot$ (assuming 2×10^{10} $M_\odot$ as the mass of the Galaxy). Of this, the dust constitutes hardly 1% by mass.

The most detailed observations so far carried out have been on atomic hydrogen, both neutral and ionized. Due to the overwhelming abundance of hydrogen in cosmic matter those regions of interstellar space where hydrogen is neutral, have been called the H I regions. The regions containing ionized hydrogen have been called the H II regions. Every emission nebula is an H II region, while a dark or a reflection nebula is an H I region. Only a small fraction of volume of interstellar space belongs to H II regions, since the O and early B stars around which they originate are relatively rare objects. Since these hot stars are sprinkled along the spiral arms, H II regions probably fill only about 10 per cent of spiral arms. The remaining 90 per cent of the volume within spiral arms consists of H I regions. The exact physical nature of the vast interarm regions is not clearly known. Opinion varies as to whether these regions are occupied by tenuous hot ionized gas or by cold neutral gas.

In H I regions all the elements whose ionization potential is less than that of hydrogen are at least singly ionized. Among these ionized elements, the more abundant ones are C, Mg, Si, S and Fe. Each atom of these elements therefore enrich the medium with one electron. The cosmic abundance of these elements is C = 400, Mg = 25, Si = 32, S = 22, and Fe = 8 for H = 10^6. Taking also into account the contribution by other elements the electron density is 5×10^{-3} cm^{-3} in an H I cloud of density 10 H cm^{-3}. If we assume the average hydrogen density in the galactic plane to be 1 H atom cm^{-3}, then the average electron density in the H I region is about 5.0×10^{-4}. If low energy cosmic ray ionization of H atoms is taken into account, then the ratio n_e : n(H I) would be much higher, may be as high as 10^{-2} or even more. But since the spectrum of cosmic-rays with energies less than 10^{10} eV is only poorly known, and these low-energy particles are by far more efficient in producing ionization, cosmic-ray ionization of hydrogen must be considered with caution. The picture is however quite different in H II regions. Here hydrogen is fully ionized and so are also all the abundant elements including He. So the electron density n_e (and ion density also) exceeds that in an H I region by a factor of about 2000. The corresponding physical processes are therefore quite different in the two regions.

A very important theoretical work was done by B. Strömgren on the initial growth of H II regions around young hot stars that are born in dense interstellar medium. He made the simplifying assumptions that (a) the star was switched to its full luminosity instantaneously, and (b) the surrounding medium was spherical and homogeneous throughout.

Since the opacity of the initial H I gas surrounding the hot star is very high for the ionizing ultraviolet photons beyond the Lyman series limit, every such photon will be absorbed by the gas which will be ionized as a result. This effect can be clearly seen by quantitative analysis. The ionization cross-section σ_u for the ultraviolet photons is of the order ~ 10^{-17} cm^2. For a gas density n_H cm^{-3}, the mean-free path of the photon is of the order $l \simeq (n_H \sigma_u)^{-1} \approx 0.033/n_H$ pc. For a typical gas density n_H ~ 10^3 cm^{-3} initially around a hot star, l ~ 3.3×10^{-5} pc. When this length is compared with the observed sizes of H II regions which lie in the range from 0.1 to 40 pc, it will be apparent that no ionizing photon can escape unabsorbed through the gas. The gas is thus suddenly ionized, first in the immediate neighbourhood of the star. An ionization front then moves outward through the gas with supersonic speed, the gas itself remaining practically motionless. However, as the radius R_i of the ionized region grows, the speed of the ionization front quickly slows down for three effects:

1. The process of recombination stars immediately within the ionized gas. In fact, each recombination in a level higher than $n = 1$ will use up one ionizing photon within the ionized gas.
2. The number of neutral atoms encountered by the front at a distance R_i from the star increases as R_i^2, The energy density of photons beyond the Lyman limit therefore decreases as R_i^2 and the number of recombination increases as the volume of the ionized region.

3. The opacity of H I increases exponentially as $e^{-\tau}$, where τ is the optical depth at the Lyman limit. Ultimately therefore a stage is attained when the number of recombinations within the ionized volume becomes equal to the number of ionizations there. The initial growth of the ionized region ceases at this stage when the radius of the ionized gas sphere is R_s, say. This sphere is known as the *Strömgren sphere*. Within this sphere hydrogen is at least 90 per cent ionized, while outside this hydrogen is almost entirely neutral. The former is the H II while the latter the H I region. The transition zone between the H II and H I regions is very thin and is of the order of the mean-free path l.

Assuming that the gas density is constant with distance from the ionizing star and that the star itself is producing the ionizing photon flux at a constant rate, it is very easy to compute the radius of the initial Stromgren spheres for different classes of hot stars. The calculation follows from the condition that the rate of production of the energetic photons beyond the Lyman limit, $N(L_c)$, by the star is equal to the rate of recombination within the sphere of radius R_s. This is given by

$$N(L_c) = \frac{4}{3} \pi n_e n_i \, \alpha(T) \, R_s^3 \tag{13.6}$$

where $\alpha(T)$ is the total hydrogen recombination coefficient. It is a slowly varying function of temperature of the gas. For fully ionized hydrogen gas, $n_e = n_i = n_H$. Thus

$$N(L_c) = \frac{4}{3} \pi n_H^2 \, \alpha(T) \, R_s^3 \tag{13.7}$$

yielding

$$R_s = \left(\frac{3N(L_c)}{4\pi \, \alpha(T)} \right)^{1/3} n_H^{-2/3} \tag{13.8}$$

or,

$$n_H^{2/3} \, R_s = \left(\frac{3N(L_c)}{4\pi \, \alpha(T)} \right)^{1/3}$$

For a kinetic temperature of 10^4 K which is widely used for a H II region, the value of $\alpha(T)$ equals 2.6×10^{-13} cm^3 s^{-1}. Knowing therefore $N(L_c)$ for different spectral classes of stars, one can compute the values of $n_H^{2/3} R_s$ for these stars, thus yielding their Strömgren radii as functions of the hydrogen density. These are given in Table 13.2.

The table shows that R_s has a very strong dependence on the hydrogen density as well as on the spectral type of the central star. Stars of later spectral classes are poor producers of ionizing photons and as such, their Strömgren radii are small. For Sun, $R_s \sim 10^{-4}$ pc for $n_H = 10$ cm^{-3}. H II regions are important only for stars at least as hot as B1.

It must be mentioned that the value of R_s obtained by using Eq. (13.8) is a lower limit. In deducing the above formula, it has been assumed that every ionization utilizes a *virgin*

TABLE 13.2 The Strömgren Radii of Stars

Spectral type	$T_C(\text{K})$	$N(L_c)$ (s^{-1})	$R_s\, n_{\text{H}}^{2/3}$ (pc cm^{-2})	$R_s(n_{\text{H}} = 10\ \text{cm}^{-3})$ (pc)	$R_s(n_{\text{H}} = 1\ \text{cm}^{-3})$ (pc)
O5	56,000	31×10^{48}	100	22	100
6	44,000	9.0	66	14.4	66
7	36,000	2.7	44	10	44
8	30,000	0.77	29	6.3	29
9	25,000	0.19	18	4	18
B0	21,000	0.041	11	2.4	11
1	18,000	0.008	6.4	1.4	6.4

energetic photon from the central star. The actual condition is different from it, To explain this, we visualize two kinds of recombinations, those in an excited level and those in the ground level of hydrogen. Every recombination of the first kind will give rise to a Lyman photon, besides other photons of longer wavelengths. Since the H II region is optically thick in Lyman radiation, every Lyman photon will be reabsorbed immediately. The Lyman photons other than L_α will be soon degraded in the process by fluorescence into L_α, a Balmer photon and other photons possibly. The Balmer photons and other photons of longer wavelengths immediately escape, since the region is optically thin in these radiations. The L_α photons have very large optical depth for scattering so that each of these will be absorbed and reemitted many times in the regions before they finally escape from it.

Let us now consider the second type of recombinations where the free electron descends directly on to the ground level of the atom. Each such recombination is accompanied by the emission of a photon beyond the Lyman limit which is capable of further ionization. These ionizing photons are produced within the gas. They are not virgin photons emitted by the central star. Thus, a large fraction of ionizing photons are produced by recombinations within the H II region. Every ionization does not utilize a virgin photon emitted by the star. These excess photons will therefore push the equilibrium state given by Eq. (13.8) to larger values of R_s. The actual Strömgren radii will thus be larger than computed on the basis of the ideal case as given by Eq. (13.8). The effect is small in tenuous gases since the recombination rate is proportional to the square of the gas density. But for dense gases the effect becomes very important. The values of R_s will be 2 to 3 times larger than those given by Eq. (13.8) for densities of hydrogen in the range 10^4 to 10^5 cm^{-3}, which are usually observed around hot stars.

In Strömgren's theory as pictured above, however, the gas subsequent to ionization does not move. But in actual case, the ionized gas will be at much higher temperature and pressure than the surrounding neutral gas, and will tend to expand. Internal motions will therefore be set up in the expanding ionized gas. In Strömgren's theory these motions of gas were completely neglected and as such, his theory describes the static evolution of H II region. The actual evolution is, however, dynamic and thus in any realistic theory of the formation of H II regions, dynamical effects on the physical variables both in the neutral as well as in ionized gas, will have to be considered. One basic concept of Strömgren's theory, that the ionized and neutral hydrogen is separated by a relatively thin transition region, however, still remains valid in the dynamical theory of the evolution of H II regions.

The ionization front moves into the neutral gas very rapidly *in the initial phase*. Its velocity then is so much more than the local sound velocity in the medium that the latter does not feel any effect on it. Consequently, although the pressure in the hot ionized gas is about 200 times more than that in the cool neutral gas, due to a hundred times increase in temperature and a two-fold increase in particle density in the former, no tendency will be shown by the ionized gas to attain a pressure equilibrium with the surrounding neutral gas at this initial phase. But as the volume of the ionized gas increases, the number of recombination also increases, which will utilize more and more of the ionizing photons. Fewer ionizing photons will now reach the neutral gas in the outer region. The speed of the IF therefore suffers gradual retardation when the gas in the ionized region begins to expand outwards. This may be considered as the *second phase* of the growth of the H II region. At this phase, due to expansion the density of the ionized gas decreases. Since the number of recombinations is proportional to the square of the hydrogen density, the low density gas will have fewer recombinations. Consequently, relatively greater number of ionizing photons will have now chance to reach the boundary so that more atoms are ionized. The fall in density therefore increases the mass of the ionized gas that can be present in radiative equilibrium.

It is a common experience in terrestrial explosions that the hot expanding gas generates a shock wave which advances in cool gas compressing and thereby heating it. The same thing happens in the second phase of the growth of the H II region. A shock front (SF) moves out into the neutral gas ahead of the IF. The IF therefore now encounters a compressed high density neutral gas ahead and a hot, expanding gas with progressively falling density and pressure near the star. These various dynamical effects are likely to modify the structures and strengths of both the IF and SF. By the time the SF has moved sufficiently ahead of the IF, both the fronts have slowed down. Complex dynamical processes follow this last phase of the development of the H II regions.

The thickness Δx_1 of the SF in an HI gas is of the order of the mean free path of the atoms. Quantitatively, it is equal to $(4\sqrt{2} \ \sigma_c n_H)^{-1}$, where σ_c is the collision cross-section, and n_H the number density of H atoms ahead of the front. Since $\sigma_c \sim 10^{-15}$ cm^2, if we take $n_H = 10$ cm^{-3} then $\Delta x_1 \sim 6 \times 10^{-6}$ pc. This width is very small compared to the size of the gaseous mass through which the shock propagates and as such, the SF can be treated as a surface of sharp discontinuity. In H II regions, if the gas be regarded as fully ionized, the width of the shock Δx_2 is proportional to u_1^4, where u_1 is the velocity of the shock. This width is thus strongly dependent on the shock velocity. For very strong shock, say with velocity of the order of 1000 km s^{-1}, the width is of the order of 0.006 pc. This width strongly increases with higher velocity and strongly declines for lower velocity shocks. But in H II regions, the gas is not fully ionized. In equilibrium state, the density of neutral H atoms in the ionized region equals the density of ionizing photons there. This neutral component of atoms maintains the width of the shock to low values even in an H II region. Thus whether in H I or H II region, a SF can be treated as a thin surface of discontinuity.

Although both the SF and IF are surfaces of quite small widths, their dynamical effects on the nature and structure of H II regions are very great. Singly, and at certain stages jointly, they profoundly influence the entire dynamical processes of the growth of these nebulae. The discussion of these phenomena is beyond the scope of this book, but because of very frequent occurrence of shocks in the interstellar medium under different situations, we shall discuss the phenomena of interstellar shock waves with appropriate mathematical treatment.

13.5 INTERSTELLAR SHOCK WAVES

Any finite pressure change in a compressible fluid over a short distance will give rise to a shock. The changes in thermodynamic variables or the flow properties of the gas, such as pressure, density, temperature and velocity, are in this case *abrupt* across the transition surface, which is called the *Shock Wave* (SW) or the *Shock Front* (SF). Shock waves are generated in a gas, in general, if supersonic motions are present. Since interstellar gas motions are in general supersonic, shock waves are quite common in interstellar space. The flow properties of the fluid are discontinuous across the SF which is therefore treated as a surface of discontinuity. The dynamical phenomena that prevail *within* the SF itself is very complex and we shall not consider it here. We shall discuss the flow properties and relationships between them on the two sides of the SF.

To simplify the analysis, we shall consider a plane, stationary, normal shock wave in which the direction of flow is along, say, x axis. We also consider the SF to be fixed in space and the fluid flows steadily across this front from side 1 to side 2. The subscripts 1 and 2 thus denote quantities upstream (unshocked) and downstream (shocked), as shown in Fig. 13.2. As the gas moves across the SF, the physical principles of conservation of mass, momentum and energy must remain valid across the front. The first conservation law gives

$$\rho_1 u_1 = \rho_2 u_2 = J \tag{13.9}$$

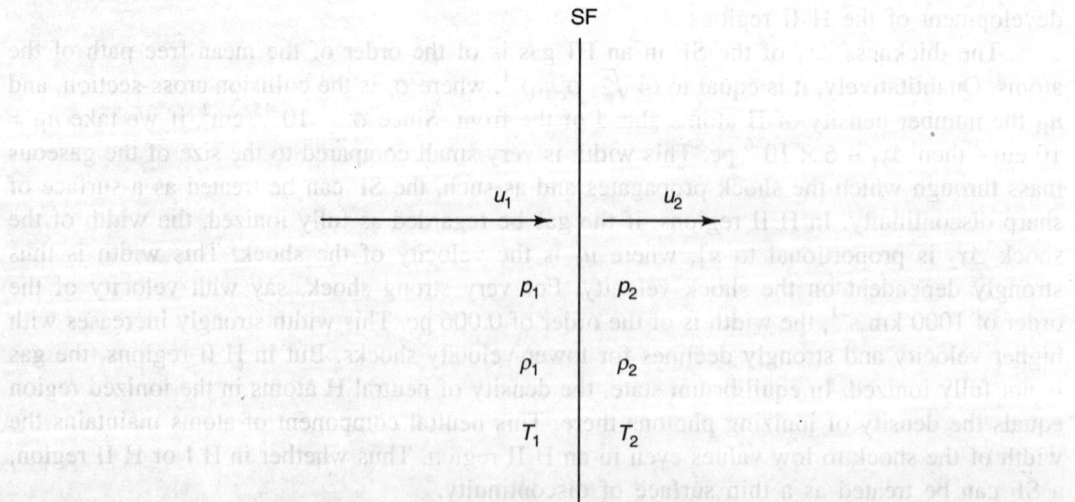

FIGURE 13.2 Normal stationary shock.

where J is the rate of mass flow across the front. The second conservation law is

$$p_1 + \rho_1 u_1^2 = p_2 + \rho_2 u_2^2 = K \tag{13.10}$$

or, since

$$p_1 = \rho_1 RT_1/\mu_1$$

and

$$p_2 = \rho_2 RT_2/\mu_2$$

for an ideal gas, where μ_1 and μ_2 are mean molecular weights of materials on sides 1 and 2 respectively, and R is the universal gas constant, the other symbols having usual meanings, we can write

$$K = \rho_1 \{RT_1/\mu_1 + u_1^2\} = \rho_2 \{RT_2/\mu_2 + u_2^2\} \tag{13.11}$$

where K is the rate of momentum flow. If the shock does not change the ionization structure of the gas, then $\mu_1 = \mu_2$. If c_1 and c_2 be the local sound speeds on sides 1 and 2 respectively, and M_1 and M_2 be the corresponding Mach numbers, then

$$\left.\begin{array}{l} c_1^2 = \gamma p_1/\rho_1 = \gamma RT_1/\mu_1 \\ c_2^2 = \gamma p_2/\rho_2 = \gamma RT_2/\mu_2 \end{array}\right\} \tag{13.12}$$

and

$$M_1 = \frac{u_1}{c_1}, \quad M_2 = \frac{u_2}{c_2} \tag{13.13}$$

by definition, γ being the ratio of specific heats. Interstellar gas is mostly monatomic for which $\gamma = 5/3$. Using Eqs. (13.12) and (13.13), Eq. (13.10) can be written as

$$p_1(1 + \gamma M_1^2) = p_2(1 + \gamma M_2^2)$$

whence

$$\frac{p_2}{p_1} = \frac{1 + \gamma M_1^2}{1 + \gamma M_2^2} \tag{13.14}$$

We consider first the adiabatic case in which no heat is added to or lost from the gas. Then the third conservation law is that of adiabatic energy equation of a perfect gas, which is

$$C_p T_1 + \frac{1}{2} u_1^2 = C_p T_2 + \frac{1}{2} u_2^2 = E \tag{13.15}$$

where E is the total energy of gas per unit mass and $C_p T$ is the enthalpy per unit mass. If C_p remains constant across the shock front, then using Eqs. (13.12) and (13.13), Eq. (13.15) can be written as

$$C_p T_1 + \frac{1}{2} M_1^2 \gamma RT_1 = C_p T_2 + \frac{1}{2} M_2^2 \gamma RT_2 \tag{13.16}$$

whence

$$\frac{T_2}{T_1} = \frac{C_p + \frac{1}{2}\,\gamma R M_1^2}{C_p + \frac{1}{2}\,\gamma R M_2^2} \tag{13.17}$$

Since $C_p = C_v\gamma$ and $C_p - C_v = R$, $\gamma R = C_p(\gamma - 1)$, and Eq. (13.17) thus reduces to

$$\frac{T_2}{T_1} = \frac{1 + \frac{1}{2}\,(\gamma - 1)\,M_1^2}{1 + \frac{1}{2}\,(\gamma - 1)\,M_2^2} \tag{13.18}$$

Again, using the equation of state for a perfect gas and the relation between specific heats, it is deduced that

$$C_p T = \frac{\gamma}{\gamma - 1} \cdot \frac{p}{\rho} \tag{13.19}$$

Thus Eq. (13.15) can also be written as

$$\frac{\gamma}{\gamma - 1}\,p_1/\rho_1 + \frac{1}{2}u_1^2 = \frac{\gamma}{\gamma - 1}\,p_2/\rho_2 + \frac{1}{2}u_2^2 \tag{13.20}$$

which is the usual form of the energy conservation equation for a steady one-dimensional shock.

We have here four basic equations, viz. the three conservation relations and the equation of state, in which eight quantities are involved, viz. the pressure, density, temperature and velocity on sides 1 and 2 of the shock front. Of these p_1, T_1 and the ratio p_2/p_1 are usually considered as known quantities. The remaining four unknown, viz. u_1, u_2, ρ_2 and T_2 can then be determined from shock relations.

Eliminating u_1^2 and u_2^2 from the three conservation equations (13.9), (13.10) and (13.20), the relation

$$\frac{\rho_2}{\rho_1} = \frac{(\gamma - 1)p_1 + (\gamma + 1)p_2}{(\gamma + 1)p_1 + (\gamma - 1)p_2} = \frac{p_1 + 4p_2}{4p_1 + p_2} \tag{13.21}$$

is obtained, where $\gamma = 5/3$ has been used. This fundamental relation between the pressure and density in an adiabatic shock is known as the *Rankine-Hugoniot relation*. It shows that in the limit of very large compression when $p_2/p_1 \rightarrow \infty$, the density jump is given by

$$\rho_2/\rho_1 = 4 \tag{13.22}$$

Thus, in an adiabatic shock, the density of the shocked material can not exceed four times the unshocked density. The equation of state can be used to obtain temperature jump from the relation

$$\frac{T_2}{T_1} = \frac{p_2 \mu_2 \rho_1}{p_1 \mu_1 \rho_2} = \frac{p_2 \mu_2 (4 p_1 + p_2)}{p_1 \mu_1 (p_1 + 4 p_2)} \tag{13.23}$$

When $p_2/p_1 \to \infty$, the temperature ratio reduces to the form

$$\frac{T_2}{T_1} = \frac{p_2 \mu_2}{4 p_1 \mu_1} \tag{13.24}$$

Thus both pressure and temperature in the shocked material increase in the same manner.

The gas velocities, u_1 and u_2, with respect to the shock front can be obtained using Eqs. (13.9), (13.10) and (13.21) as

$$u_1^2 = \frac{1}{3 \rho_1} (p_1 + 4 p_2) \tag{13.25}$$

and

$$u_2^2 = \frac{1}{3 \rho_1} \frac{(4 p_1 + p_2)^2}{p_1 + 4 p_2} \tag{13.26}$$

For strong shock therefore where p_2/p_1 is very large, p_2 and so T_2 increase as u_1^2. Also, in this case $u_1^2 \gg \gamma p_1/\rho_1$, the square of the local sound speed in the unshocked material.

We shall now discuss two basic thermodynamic properties which a shock always satisfies. First, the Rankine-Hugoniot relation Eq. (13.21) basically differs in form from the adiabatic *reversible* process

$$p = \text{constant} \cdot \rho^{\gamma} \tag{13.27}$$

This physically implies that the *changes produced by a shock are not reversible and are therefore not isentropic*. The deviation from isentropic character increases with the increase of compression p_2/p_1.

Secondly, we consider the change in entropy per unit mass as the gas moves from side 1 to side 2. This is given by

$$S_2 - S_1 = \int_1^2 \frac{dQ}{T} \tag{13.28}$$

But

$$dQ = C_v dT + p d (1/\rho) \tag{13.29}$$

Thus,

$$S_2 - S_1 = C_v \ln (T_2/T_1) - R \ln (\rho_2/\rho_1) \tag{13.30}$$

Substituting for T_2/T_1 and ρ_2/ρ_1, using the Mach numbers and simple algebra Eq. (13.30) can be put in the form

$$S_2 - S_1 = C_v \ln \left[\frac{2 \gamma M_1^2 + 1 - \gamma}{1 + \gamma} \left\{ \frac{(\gamma + 1) M_1^2 + 2}{(\gamma + 1) M_1^2} \right\}^{\gamma} \right] \tag{13.31}$$

The right-hand side of Eq. (13.31) is positive if $M_1 > 1$ and negative if $M_1 < 1$. Thus, $S_2 > S_1$ if $M_1 > 1$ and $S_2 < S_1$ if $M_1 < 1$. Since a decrease of entropy in nature is impossible, so $S_2 > S_1$ always, and so $M_1 > 1$ always. Thus, the velocity of the upstream gas is always *supersonic* with respect to the front. It can be shown easily that the velocity downstream is *subsonic*. This follows from the relation

$$M_1^2 = \frac{1 + \frac{1}{2}(\gamma - 1)M_2^2}{\gamma M_2^2 - \frac{1}{2}(\gamma - 1)} \qquad (13.32)$$

satisfied by the Mach numbers. It is easily shown that if $M_1 > 1$, then $M_2 < 1$. It follows therefore that $p_2 > p_1$ from Eq. (13.14); also $\rho_2 > \rho_1$ and $u_1 > u_2$. Thus, *a shock always compresses a gas.*

For strong shocks, $p_2 >> 4p_1$. Then the gas velocities upstream and downstream relative to the front are obtained from Eqs. (13.25) and (13.26) to be

$$u_1 = \left(\frac{4}{3}\,\rho_2/\rho_1\right)^{1/2} = 4\left(\frac{RT_2}{3\mu_2}\right)^{1/2} \qquad (13.33)$$

and

$$u_2 = \left(\frac{\rho_2}{12\rho_1}\right)^{1/2} = \left(\frac{RT_2}{3\mu_2}\right)^{1/2} \qquad (13.34)$$

where we have used $\rho_2/\rho_1 = 4$. Equations (13.33) and (13.34) yield

$$u_1 - u_2 = (3/4)\,u_1 \qquad (13.35)$$

and

$$u_1/u_2 = 4 \qquad (13.36)$$

The last relation can of course, be directly obtained by using Eqs. (13.9) and (13.22).

The above simple model of the shock cannot, however, be directly applied to interstellar space. The shock-heated H I gas behind the front cools down by infrared radiation and attains its preshock temperature within a cooling time given by $t_{cool} = 1.5\ kT/\Lambda n_2$ (Spitzer, 1978), where Λ is the cooling rate and n_2 is the number density of gas just behind the shock. This period is about a few times 10^3 years. An upper limit of this period may be taken as around 10^4 years. The gas moving upstream enters the shock at the plane, say, at x_1 and before attaining the preshock temperature moves to, say x_2 (Fig. 13.3). Then the region between the planes at x_1 and x_2 is the transition region within which the shock-heated gas has radiated all its excess energy in infrared radiation. In the absence of any magnetic field, the density increases very steeply while the pressure increase is slow, during the postshock cooling of the gas. If the gas velocity remains practically the same within the transition region, which we shall consider as the SF in this case, its thickness will be a few hundredths of a parsec, if the original upstream velocity was 10 km s^{-1}. The thickness of the shock behind which radiative cooling takes place is, therefore, much greater than that of the adiabatic shock we have already considered. If the shock does not change the ionization structure of the gas, the temperatures at x_1 and x_2 will essentially be the same, so that it will be appropriate to analyze the situation as in the case of an *isothermal shock*. In this case

$$\mu_1 = \mu_2, \qquad p/\rho = c^2 = \frac{kT}{\mu m_H} = \frac{RT}{\mu} \qquad (13.37)$$

With Eq. (13.37), Eqs. (13.9) and (13.10) together yield the density ratio in the very simple form

$$\rho_2/\rho_1 = u_1^2/c^2 \qquad (13.38)$$

where we have used $p_1/\rho_1 = p_2/\rho_2 = c^2$, since μ and T are equal on both sides of the isothermal shock which produces no new ionization. The density ratio in an isothermal shock is therefore proportional to the square of the Mach number of the upstream gas flow. For strong shocks therefore, where the Mach number is greater than 10, large compressions by factors of 100 and more are generally available. Due to large dimensions of interstellar gas clouds moving with supersonic speeds, usually with Mach numbers around 10 or more, interstellar shocks are often strong and are nearly isothermal. Thus, large compressions by shocks appear to be a usual feature of interstellar medium. This theoretical prediction is substantiated by observations of high density clouds which is a common characteristic of the distribution of the medium in space. An isothermal shock with density and temperature variations within the transition region and behind, is schematically shown in Fig. 13.3.

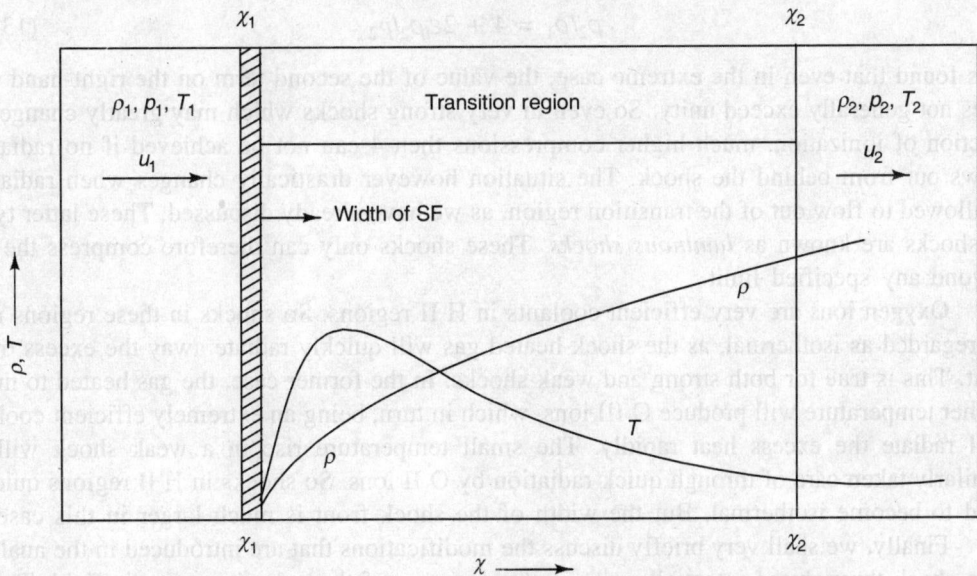

FIGURE 13.3 Interstellar shock showing the density and temperature profiles in the transition region.

If the ionization structure of the gas is changed by the shock, then $\mu_1 \neq \mu_2$ and $T_1 \neq T_2$, and the shock will not be isothermal. But in the presence of cooling by radiation behind a *strong shock*, Eq. (13.38) will approximately hold. Large compressions are therefore possible in this case.

If only the fraction of ionization in the hydrogen gas changes while passing through the SF, without any cooling by radiation behind the front, then compressions slightly higher than the adiabatic case only will be achieved. If the change in fractional ionization is $\Delta f = f_2 - f_1$ after passing through the shock, then the amount of energy of the shock used up for ionization is

$$\varepsilon_i = \frac{(\Delta f)(h\nu_L)}{m_H} \tag{13.39}$$

where ν_L is the frequency corresponding to the Lyman limit and m_H is the mass of hydrogen. The energy conservation equation, i.e. Eq. (13.20) has now to be modified to the form

$$\frac{\gamma}{\gamma - 1} \frac{p_1}{\rho_1} + \frac{1}{2}u_1^2 = \frac{\gamma}{\gamma - 1} \frac{p_2}{\rho_2} + \frac{1}{2}u_2^2 - \varepsilon_i \tag{13.40}$$

and the Rankine-Hugoniot relation will be modified to

$$\frac{\rho_2}{\rho_1} = \frac{p_1 + 4p_2}{4p_1 + p_2} + \frac{2\rho_2\varepsilon_i}{4p_1 + p_2} \tag{13.41}$$

For strong shocks where $p_2 \gg 4p_1$, Eq. (13.41) reduces to

$$\rho_2/\rho_1 = 4 + 2\varepsilon_i\rho_2/p_2 \tag{13.42}$$

It is found that even in the extreme case, the value of the second term on the right-hand side does not generally exceed unity. So even in very strong shocks which may greatly change the fraction of ionization, much higher compressions then 4 can not be achieved if no radiation flows out from behind the shock. The situation however drastically changes when radiation is allowed to flow out of the transition region, as we have already discussed. These latter types of shocks are known as *luminous shocks*. These shocks only can therefore compress the gas beyond any specified limit.

Oxygen ions are very efficient coolants in H II regions. So shocks in these regions may be regarded as isothermal, as the shock-heated gas will quickly radiate away the excess of its heat. This is true for both strong and weak shocks. In the former case, the gas heated to much higher temperature will produce O III ions, which in turn, being an extremely efficient coolant, will radiate the excess heat rapidly. The small temperature rise in a weak shock will be similarly taken care of through quick radiation by O II ions. So shocks in H II regions quickly tend to become isothermal. But the width of the shock front is much larger in this case.

Finally, we shall very briefly discuss the modifications that are introduced in the analysis of a shock through the interstellar gas in the presence of the galactic magnetic field. This is the domain of *hydromagnetic shocks*. The basic principle of analysis of hydromagnetic shocks does not substantially differ from that of ordinary shocks which we have already discussed.

Only the algebra is a bit more involved in the former. The mass conservation law remains as in Eq. (13.9). But the momentum and energy conservation laws are modified respectively in the forms

$$p_1 + \rho_1 u_1^2 + \frac{B_1^2}{8\pi} = p_2 + \rho_2 u_2^2 + \frac{B_2^2}{8\pi} \tag{13.43}$$

and

$$\frac{\gamma}{\gamma - 1}\frac{p_1}{\rho_1} + \frac{1}{2}u_1^2 + \frac{B_1^2}{4\pi\rho_1} = \frac{\gamma}{\gamma - 1}\frac{p_2}{\rho_2} + \frac{1}{2}u_2^2 + \frac{B_2^2}{4\pi\rho_2} \tag{13.44}$$

where $B^2/8\pi$ is the magnetic pressure and $B^2/4\pi\rho$ the magnetic energy per unit mass of the gas. The latter is related to Alfvén velocity V_A by the relation

$$V_A^2 = B^2/4\pi\rho \tag{13.45}$$

Here, it has been assumed that the magnetic lines of force are straight and perpendicular to the direction of propagation of the shock. Since a new variable, viz. B_2 (B_1 is assumed to be known) has been introduced in the above equations, we need an additional relation for solution of the problem. This is obtained for the one-dimensional shock as

$$B_1/\rho_1 = B_2/\rho_2 \quad \text{or} \quad \rho_2/\rho_1 = B_2/B_1 \tag{13.46}$$

which is the statement of the physical principle that through any circuit moving with a conducting fluid, the magnetic flux remains constant.

By substituting the values of ρ_1 and ρ_2 (see Eq. (13.9)), Eq. (13.46) can be written as

$$B_1 u_1 = B_2 u_2 \tag{13.47}$$

This relation is important which shows the inverse dependence of velocities on the magnetic field strengths on two sides of the front.

Using these equations, one can calculate the various parameters of a hydromagnetic shock and interpret the results physically under various assumptions relevant to the interstellar space. The detailed discussion of these is beyond the scope of this book. The interested reader may consult the work of S.A. Kaplan (1966) and L. Spitzer (1978). We shall conclude this discussion by mentioning some of the principal points that emerge from analysis of an interstellar hydromagnetic shock whose plane is parallel to the magnetic lines of force.

Firstly, the compression (the value of ρ_2/ρ_1) achieved in a hydromagnetic shock of a given strength is less than that achieved in an ordinary shock of the same strength. For strong *adiabatic* hydromagnetic shock $\rho_2/\rho_1 \leq 4$. In such a shock the influence of a magnetic field is found to be negligible. Secondly, for a strong *isothermal* hydromagnetic shock, the compression is given by the formula

$$\rho_2/\rho_1 = \sqrt{2}\, u_1/V_A \tag{13.48}$$

In interstellar conditions, the Alfvén velocity V_A is usually greater than the isothermal sound speed c. Moreover, the ratio of upstream velocity u_1 to the Alfvén velocity V_A now appears only with first power. Since for strong shocks this ratio >> 1, it follows that the compression achieved by Eq. (13.48) is much lower than that achieved by Eq. (13.38). Magnetic field therefore effectively inhibits the compression in an isothermal shock. Thirdly, in the case of a strong adiabatic hydromagnetic shock, the magnetic field has practically no influence on the dissipation of the energy of the wave. But if the shock energy is comparable to the magnetic energy, the magnetic field greatly inhibits the energy dissipation. In other words, hydromagnetic shock of a given strength survives much longer than an ordinary shock of comparable strength.

13.6 INTERSTELLAR CLOUD COLLISIONS

We have seen that the interstellar H I clouds move at random in space with an average velocity of 8 km s^{-1} in one direction. There are about 50,000 such clouds per cubic kpc. So one can be sure that these clouds will suffer occasional collisions. Since the average internal thermal velocity in clouds with average temperatures around 100 K is of the order of 1 km s^{-1}, and the average relative velocities between the colliding clouds are usually in the range 5 to 15 km s^{-1}, the collisions are highly supersonic with the Mach number M usually lying in the range 5 to 15. Such collisions will therefore be highly compressible and will give rise to shock waves which will heat the gas to a few thousand degrees Kelvin and compress it behind the shock. The shocked gas will subsequently cool down in infrared radiation to the preshock temperature over a period of a few thousand years. The kinetic energy of the cloud motion is thus dissipated in infrared lines.

The theory of the heating of interstellar gas by cloud collisions was first proposed by F.D. Kahn (1955) to explain the large discrepancy between the observed mean harmonic temperature of 125 K and the theoretically calculated equilibrium temperature of H I clouds which is usually less than 50 K. In H I clouds where other sources of energy input are relatively weak, collisions are likely to be the most dominant mechanism for heating the gas. According to calculations by Kahn every element of volume of a H I cloud will suffer a collision within a time-scale of 6.8×10^6 years and thereby lose a fraction of its kinetic energy. For head-on collisions between two clouds of equal masses with a relative velocity V along the line of centres, each cloud loses kinetic energy corresponding to a velocity 1/2 V. So the kinetic energy lost per unit mass is (1/8) V^2. This is therefore the gain in the heat energy per unit mass of the gas. Kahn has also shown that as a result of collisions between two clouds, one, two or three clouds may be formed, and a particular cloud can gain or lose mass due to collisions.

Calculations reveal that the observed mean harmonic temperature of 125 K can be realized in the light of Kahn's heating mechanism, provided cooling by electron-ion collision only is important. But the presence of H_2 molecules will turn the balance altogether. If H_2 exceeds 1% of H, then cooling will be so rapid that even Kahn's collisional heating mechanism will be inadequate to explain the observed temperature. The shock-heated gas in this case will cool down to the preshock temperature within a time-scale of only a few thousand years if all the cooling processes are operative. Of all the cooling processes, that by O–H collisions is the most dominant as long as H_2 does not exceed H by more than one per cent. Cooling by H_2 begins to dominate when its amount exceeds that value. Since the amount of H_2 is likely to be more than 1 per cent of H, as the available data suggest, its dominant cooling influence seems to be present in H I regions. Then Kahn's heating mechanism is likely to be inadequate to explain the observed temperature and other heating processes must be looked for. As we have already seen, heating due to low-energy cosmic-rays and soft X-rays are two likely candidates. But even then, for ideal cloud model which has usually been used, a balance between the observed and theoretically computed temperature is difficult to reach.

13.7 ENERGY BALANCE IN INTERSTELLAR GAS

We have thus seen that interstellar clouds, moving at random in space, are involved in occasional collisions and thus dissipate a part of their kinetic energy. Roughly speaking, we find that a gram of interstellar matter which loses one component of its random motion (= 8 km s^{-1}) in collisions in a time-scale of about 6×10^6 years, dissipates energy at the rate of $1/2 \, v_x^2 = 3.2 \times 10^{11}$ erg gm^{-1}. With the above rate of collisions, this amounts to a dissipation rate of 1.8×10^{-3} erg gm^{-1} s^{-1}. Assuming an average density of 1 H atom cm^{-3} together with solar abundances, the average density of matter in interstellar space turns out to be 2.5×10^{-24} gm cm^{-3}. The dissipation rate therefore becomes 4.5×10^{-27} erg cm^{-3} s^{-1}. If the gas is supposed to be confined within a cylindrical volume of radius 10 kpc and height 300 pc, then the total dissipation rate becomes approximately 1.1×10^{40} erg s^{-1}. This rate of dissipation will be achieved provided the interstellar magnetic field is weak and the collisions are perfectly inelastic. The latter condition is fulfilled if the kinetic energy density is much higher than the magnetic energy density in the interstellar space. With a mass density given above and a root-mean-square gas velocity of 14 km s^{-1}, the kinetic energy density in the interstellar space will be approximately 2.5×10^{-12} erg cm^{-3}. On the other'hand, the magnetic energy densities are 1.6×10^{-11} and 3.6×10^{-13} erg cm^{-3} depending on whether we assume a strong field of 2×10^{-5} or a weak field of 3×10^{-6} gauss in the interstellar space. Therefore, collisions of clouds will be perfectly inelastic only in the case of weak field, but in the other case, where the magnetic energy density far exceeds the kinetic energy density in space, only the energy corresponding to the component of motion parallel to the lines of force will be dissipated. The rate of dissipation will therefore be reduced in the latter case by a factor of $3^{3/2}$ which is nearly equal to 5. Thus if the interstellar magnetic field is strong, the rate of dissipation of kinetic energy by cloud collisions reduces to about 9×10^{-28} erg cm^{-3} s^{-1} and the total dissipation rate in the entire Galaxy becomes 2.2×10^{39} erg s^{-1}.

Since many theoretical and observational difficulties arise when strong magnetic field is assumed, we shall take into consideration the case of a weak magnetic field in the interstellar space. Thus, the rate of dissipation will be taken as 1.1×10^{40} erg s^{-1}. Then the question arises as to how is this energy replenished in order to maintain the cloud velocity. We shall discuss different sources of kinetic energy that can be considered as means for replenishment of the lost energy.

Cosmic rays have long been regarded as an efficient supplier of kinetic energy to the interstellar gas. The energy supplied to the interstellar gas of average density 1 H atom cm^{-3} by cosmic rays, according to the reasonable data is 6.4×10^{-27} erg cm^{-3} s^{-1}, and only 9 per cent of this energy becomes available to the interstellar gas as the kinetic energy. Thus, kinetic energy available to the interstellar space from cosmic rays is 5.76×10^{-28} erg cm^{-3} s^{-1}. This is less than the calculated rate of dissipation by about an order of magnitude. Moreover, the spectrum of the low-energy cosmic rays (energy $< 10^{10}$ eV) which has been *assumed for* such computations as the ionizing agents of H atoms is *not observed*. One has to assume a law of energy spectrum for such particles which may lead to significant errors. Thus, cosmic rays do not appear to be an effective agent for replenishment of the dissipated energy of the interstellar gas. A similar conclusion can be drawn for soft X-rays.

A rotational velocity of 250 km s^{-1} and an average matter density of 2.5×10^{-24} gm cm^{-3} in the solar neighbourhood will yield a kinetic energy density of about 7.8×10^{-10} erg cm^{-3} due to rotation of the Galaxy. At the computed rate of dissipation this energy will all be spent in a time-scale of 5.5×10^9 years. This means that if no other sources of energy except rotation were present in the Galaxy, then within a time-scale of about five billion years the rotation of the Galaxy would have ceased. But the age of the Galaxy is believed to be at least 10^{10} years, and thus, the rotation appears to be an inadequate mechanism to maintain the velocity of clouds as observed.

Exploded supernova shells have often been considered as a possible source of the required amount of kinetic energy. Of the two types of supernovae, Type II eject much more material with a much higher velocity than Type I. One solar mass or more is usually ejected in a Type II supernova explosion. The ejection velocity may be as high as 7000 km s^{-1} and the frequency of such occurrence may be roughly taken to be 1/300 per year. If we assume that roughly one solar mass of gas is injected by supernova explosions every 300 year with a velocity of 7000 km s^{-1}, the rate of kinetic energy supply to the interstellar space bounded in a cylindrical volume of radius 10 kpc and height 300 pc by this process is about 2.2×10^{-26} erg cm^{-3} s^{-1}. This is about five times the rate of dissipation of energy. Thus, supernova explosions may be a sufficient source of kinetic energy to drive the interstellar gas with observed velocity. But the statistical information about the galactic as well as other supernovae in general, is so inadequate that too much reliance should not be placed on this process. Also, how these extremely localized and sporadic events can distribute kinetic energy so uniformly and continuously throughout the region occupied by gas remains difficult to be explained. One should therefore look for a source which is known to fulfill these conditions of continuous and uniform energy supply to the space. It is natural therefore that for such a source astronomers look towards the hot stars.

The hot stars can be regarded as a more or less permanent feature of the Galaxy with their distribution almost coinciding with that of the interstellar clouds. So the problem of transfer of energy from one region to the other is automatically solved. Also, systematic observations over many years have left much more satisfactory data on the ultraviolet energy radiated by these stars and the consequent driving out of the surrounding cool gas by the expanding hot gas. The mechanism of driving the cool gas surrounding an O star was first investigated by Oort. Spitzer has shown that the total available power from early-type stars for heating the gas in a cylindrical volume of radius 1 kpc and thickness 300 pc is about 1.59×10^{39} erg s^{-1}. Assuming that these stars are uniformly distributed over the cylindrical volume of radius 10 kpc and height 300 pc (as we have considered for the galactic disk) the total ultraviolet power available in the disk is 1.59×10^{41} erg s^{-1}. This is larger than the computed rate of dissipation by more than an order of magnitude. In fact, the dissipation rate is only about 7 per cent of the available ultraviolet power. But all the ultraviolet power is not available as kinetic energy of the gas. In fact, the available kinetic energy will depend on the efficiency factor ε with which the interstellar gas can convert thermal energy into kinetic energy. The above calculations indicate that if the value of ε is only 0.07, that is 7 per cent, then the dissipated energy of the clouds will be replenished by the ultraviolet radiation from hot stars. The problem therefore rests on correct evaluation of efficiency factor ε.

Spitzer and his co-workers reasoned that an efficiency factor as high as 10 per cent is not unlikely. In that case, the ultraviolet radiation from hot stars appears to be a source which is more than sufficient to replenish the dissipated energy of clouds. But several other authors have argued against such a high value of ε. Schlüter and Biermann have suggested that the value of ε may be as low as 10^{-3}. Kahn has argued that the value of the efficiency factor ε may be 10^{-2} or even lower, because of two effects. First, hydrogen ionization consumes a large part of the ultraviolet energy output. Secondly, a substantial part is lost in heating the expanding H II region and subsequent cooling mostly by collisionally excited O II ions. After all such expenditure is met, hardly 1 per cent of the original ultraviolet energy is left in the gas in the form of kinetic energy. By considering various details of the problem, B.M. Lasker arrived at the conclusion that ε is of the order of 1 per cent of the total ultraviolet energy radiated from hot stars. The value of ε was recalculated by Spitzer with a somewhat higher rate of ultraviolet flux from hot stars. He calculated the values of the conversion factor ε ranging from 0.6 per cent for a star of type O7 to 2 per cent for an O9 star.

All these computations thus lead to the conclusion that the value of ε probably does not much differ from 0.01, implying that the available kinetic energy hardly exceeds 1 per cent of the total ultraviolet flux from hot stars.

These theoretical results can be compared with some observational data. We shall consider here two cases, viz. those of Orion Aggregate and λ^1 Orionis. Using the ultraviolet flux data and the number of O–B stars in the Orion Aggregate from the observations, one can calculate the total rate of ultraviolet power flux from hot stars in Orion Aggregate to be about 1.36×10^{38} erg s^{-1}. According to T.K. Menon, the age of expansion of the aggregate is 2.4×10^6 years. Thus the total power emitted by the hot stars in Orion Aggregate is about 102.54×10^{50} erg. Menon calculated that 58,000 $M_\odot$ of neutral hydrogen is associated with the Orion Aggregate. The expansion velocity of the gas measured by him is 10 km s^{-1}. Thus the kinetic energy associated with the expanding clouds in the aggregate is 58×10^{48} erg, implying that only about 0.5 per cent of the available ultraviolet power has been converted into kinetic energy.

λ^1 Orionis is an O8 star with a rate of ultraviolet power flux equal to 1.2×10^{37} erg s^{-1} according to Spitzer's data. The half-life of an O8 star is 2.2×10^{14} s according to Limber. Thus, the total power emitted by λ^1 Orionis to heat the surrounding gas is 2.64×10^{51} erg. Wade measured 4.5×10^4 $M_\odot$ of neutral hydrogen surrounding λ^1 Orionis expanding with a velocity of about 8 km s^{-1}. The kinetic energy of expansion is thus 2.88×10^{49} erg, yielding the converted power as about 1 per cent. It will therefore be quite justified to take 10^{-2} as the value of the conversion factor ε. If all our estimations are approximately correct, then kinetic energy available from ultraviolet radiation from hot star is about 1.59×10^{39} erg s^{-1} which is only about 1/7th of the dissipated kinetic energy of the interstellar gas, provided that the magnetic field is weak. Even if a strong magnetic field is assumed, the process seems hardly capable of replenishing the kinetic energy dissipated by cloud collisions.

Yet a fifth source of kinetic energy may be considered as important. This is the mass ejected with high velocities from surface of hot stars.* From rocket observations of ultraviolet spectra of blue giants and supergiants in Orion and of some other similar stars, D.C. Morton and his co-workers computed that these stars ejected mass at an average rate of 10^{-6} $M_\odot$ yr^{-1} with an average speed of 1400 km s^{-1}. Such high-velocity ejection of matter at a high rate from the hot stars has been confirmed very conclusively from the IUE data on various objects. High-velocity ejection of material appears to be a common feature of all hot evolved stars. The very broad emission lines of Wolf-Rayet stars indicate mass ejection from these stars with velocities as high as 3000 km s^{-1}. High-velocity mass ejection is also common to ordinary novae. Since the total number of the Wolf-Rayet stars and novae is not very large, we shall focus our attention to blue giants and supergiants. Assuming that 5 per cent of the blue stars are in giant and supergiant phases and that the number density of these stars is as given by Spitzer one finds that in the Galaxy of radius 10 kpc and height 300 pc there are about 1000 blue giant and supergiant stars each ejecting 1 $M_\odot$ in time T. The high-velocity mass ejection from these adds kinetic energy to the interstellar material at the rate given by:

$$K = 1/2 \ Mv^2/T \qquad (13.49)$$

where $M = 1000 \ M_\odot$, $v = 1400$ km s^{-1} and T is the time-scale over which this mass is ejected. Assuming $T = 6 \times 10^5$ years, we get $K \sim 10^{39}$ erg s^{-1}. This value is comparable to the kinetic energy obtained from ultraviolet radiation from hot stars, but amounts to only 9% of the total dissipation rate.

From what has been discussed above, we can draw the conclusion that although there are several sources of kinetic energy each of which can partly replenish the energy dissipated by cloud collisions, none of these sources except supernova explosions appears to be sufficient by itself. Supernova explosions may be a sufficient source, but the computation for this is based on very poor statistics. On the other hand, it may be that all the sources discussed above contribute their respective share to keep the clouds running in spite of energy dissipation. There may be yet other important sources of kinetic energy which are yet to be discovered. The final answer to the problem does not yet appear in sight.

13.8 THE INTERCLOUD MEDIUM

Most of the observational results in interstellar space describe the denser component of interstellar matter. Consequently, the theories constructed to fit these observations also relate to the denser medium. But we have seen that clouds occupy hardly 10 per cent of the galactic disk. The remaining 90 per cent of space remains *between the clouds* which has been called the *intercloud* regions. The very tenuous matter pervading this interstellar space is called the *intercloud medium.*

Since the density of matter in intercloud space is very low and the distances concerned are very great, nothing practically can be observed of the intercloud medium. But its existence

---

* The pioneering work was done by B. Basu in 1972. More work was done later by D.C. Abbott (1982), R. McCray (1983), D. Van Buren (1985) and others.

was originally proposed in order to explain the difference between the optical observations idealized with standard cloud model and the 21-cm observations which suggest a smoother gas distribution. The difference can be smoothed out if one assumes that a considerable fraction of the area under the 21-cm emission profile originates in the intercloud medium. Until some observational data have been gathered from the orbiting observatory *Copernicus,* the knowledge of the intercloud medium was based on hypotheses and indirect evidence. The physical picture of the intercloud medium that emerges at present on the basis of theoretical as well as observational evidences can be depicted as follows.

L. Spitzer and his co-workers originally proposed that many observational phenomena of the interstellar medium could be understood in terms of two-component model of the gas, viz. a tenuous hot intercloud medium in pressure equilibrium with a denser cool medium. The theoretical density and temperature of the rare medium have been suggested to be of the order of 0.1 to 0.2 H atom cm^{-3} and 10^4 K respectively. The denser medium consists of the standard clouds of density 10 H atoms cm^{-3} and of temperature not much higher than 100 K. The maximum kinetic pressure in such a medium is achieved at hydrogen density of nearly 0.2 atoms cm^{-3}. Because of the very low hydrogen density in the intercloud medium and because of the low-energy cosmic ray flux is more or less uniform over the entire galactic disk, the percentage ionization by energetic particles in the intercloud medium is much higher, may be as high as 10 per cent or more. On the other hand, the density of cooling ions being very low, the intercloud medium can maintain a temperature as high as 10^4 K.

Such a theoretical picture of the intercloud medium is consistent with the observational data gathered from the orbiting observatory, Copernicus. From the spectral lines study of four relatively nearly unreddened stars, Rogerson and his co-workers derived hydrogen densities in the intercloud medium ranging from 0.02 to 0.22 cm^{-3} on the assumption that nitrogen has its solar abundance relative to hydrogen in the intercloud medium. Highly ionized species such as N V, O VI, Si IV and S IV have been detected but their relative quantities are quite small compared to the singly ionized species. The elements whose ionization potentials are greater than that of H are concentrated in the lowest stage of ionization. This leads to the conclusion that processes which produce highly ionized species are not dominant in the intercloud medium. The observations are more or less consistent with a model of the heating mechanism in which the heating is due to the observed X-ray background with energies greater than 150 eV, together with a cosmic-ray spike at 2 MeV. Such a model gives approximate ratio of N I/N II which is consistent with observations. Thus, both X-ray as well as cosmic-ray heating appears to be important in the intercloud medium.

The abundances of elements whose lines have been observed are not very much discordant with the corresponding solar abundances. The differences do not exceed the limits of observational uncertainty.

13.9 INTERSTELLAR GRAINS

We have already had several occasions to discuss the various roles that the interstellar grains (or dust) are believed to play and also tried to point out the astrophysical importance of this component of the interstellar matter. It will be worthwhile to summarize here these various aspects.

1. The grains absorb and scatter starlight and thus produce a general extinction of light from distant stars. The absorption of light in the galactic plane region is, on the average, 1 magnitude kpc^{-1}.

2. Grains in clouds reflect light from the associated bright stars. For stars of spectral class later than B0, this gives rise to reflection nebulae. Also, grains soften the intensity of ultraviolet flux close to the ionizing star in an H II region.

3. Polarization measurements of light from distant stars lead to the conclusion that the grains are elongated particles oriented in some particular fashion by the galactic magnetic field. This study also leads to the very important clue to the orientation of the magnetic field itself. The polarization measurements indicate that the lines of force are oriented more or less along the spiral arms in the arm region.

4. Variable absorption of light in different galactic longitudes indicates that the distribution of dust is not uniform but spotty.

5. The total mass of the dust component is only about 1 per cent of the gas and the average size of the particles is less than 1 μ.

Absorption of light from distant stars by dust particles has been found to be wavelength-dependent so as to produce also a general *reddening* of stars. Let the absorption produced by dust be k_a magnitude pc^{-1} then the general distance formula (Eq. (3.5)) must be changed to

$$m - M = 5 \log r - 5 + k_a r \qquad (13.50)$$

so that the true distance modulus is, in this case, equal to $(m - M - k_a r)$. The actual distance modulus is thus overestimated if the correction for k_a is not applied suitably. The value of K_a is observed to vary both in galactic longitudes as well as in latitudes. The variation in longitude is somewhat irregular resulting from the irregular distribution of dust clouds in galactic longitudes. But the variation in latitudes follows a more or less regular pattern. Far away from the galactic plane, the absorption is negligible as the space is almost dust-free in that region. This is demonstrated by the constant number of external galaxies brighter than a given limiting magnitude discernible per square degree of the sky in higher galactic latitudes. Starting from about 40° galactic latitude downward, the number falls off systematically, until near the galactic plane the thick dust layer completely veils the distant galaxies. Such observations led Hubble to call the galactic plane region as the zone of avoidance.

Dust particles not only absorb light from distant objects but also redden these objects. This latter phenomenon is due to the existence of some particular size distribution among the particles. Spectrophotometric measurements have shown that almost over the entire optical wavelength region the absorption by dust, that is, the value of k_a varies inversely with λ. The law $k_a \propto \lambda^{-1}$ thus holds. This indicates that the light of shorter wavelengths will be absorbed more than that of longer wavelengths. An object will thus appear redder due to interstellar absorption. This phenomenon has been described by R.J. Trumpler as the *interstellar reddening*. The observed colour-index of an object will increase due to interstellar reddening. (B–V)$_{\text{obs}}$ – (B–V)$_0$ is called the colour-excess which actually gives a measure of the amount of reddening due to interstellar absorption. Numerous spectrophotometric observations have shown that the total absorption of light in visual magnitude, A_V, is nearly three times the colour-excess $E_{\text{B–V}}$. Thus the relation

$$A_V = (3.0 \pm 0.2) \, E_{\text{B–V}} \qquad (13.51)$$

almost systematically holds. This experimental relation is very important in stellar photometry, because it reduces the more complex problem of the measurement of total absorption to the simpler problem of measuring the colour-excesses.

The observed law $k_a \propto \lambda^{-1}$ yields a clue to the size of the interstellar particles. It is known that absorption and scattering of light by particles much larger than the wavelengths scattered are independent of these wavelengths. Such particles will just block the light depending on their geometrical cross-section. On the other hand, scattering by much smaller particles, say the molecular scattering, follows Rayleigh's law of λ^{-4}. The λ^{-1} law of interstellar absorption therefore indicates that the sizes of the particles are of the same order as the optical wavelengths of light. The size distribution between 0.3 μ to 0.7 μ is therefore very likely with an average mass of the order of 10^{-13} gm per particle.

The interstellar polarization discovered by Hall and Hiltner gives clues to the understanding of the shape and orientation of the dust grains. Polarization cannot be produced by spherical particles. Furthermore, even nonspherical particles cannot produce polarization if they are oriented at random. So the observed polarization points to the unavoidable conclusion that the interstellar grains are nonspherical particles oriented in some definite fashion. R.D. Davies and J.L. Greenstein have suggested that the observed polarization could be explained by elongated particles, oriented in more or less perpendicular manner to the galactic plane by general magnetic field of the Galaxy, with its lines of force running nearly along the spiral arms.

Much controversy still prevails among astronomers as regards the chemical composition and the formation process of the dust grains. The absorption and scattering properties of particles depend both on their size as well as on their composition. Both dielectric as well as metallic particles have been suggested for grains. Both of these, under suitable conditions, can give rise to extinction obeying λ^{-1} law. But while the dielectric particles mainly scatter light isotropically, the metallic particles chiefly absorb radiation converting it into heat. Experiments have shown that while the average size of the dielectric particles has to be comparable to the wavelengths of light to yield the observed extinction law, the average size has to be ten times smaller if the particles are of metallic origin. Thus, the observed extinction law by itself is not sufficient to definitely predict the true nature of the grains. Additional observational data are necessary for the purpose.

This was supplied by the observation of *scattered light* among dust particles in dark nebulae. These observations led Van de Hulst to suggest that dust particles do not absorb but scatter light and as such, they are dielectric particles. The average size of the particles was estimated by him to be about 0.8 μ. These observational results are further supported by theoretical considerations that if the grains were metallic, the metal content of the interstellar medium would be much higher than that in stars and other astrophysical bodies. Such an overabundance of metals in interstellar matter is therefore difficult to accept on logical grounds.

Even with all such experimental results and theoretical speculations, however, the controversy regarding the nature and composition of interstellar grains still persists. The observational results can be simulated by any one or a suitable combination of a few of the following likely candidates:

1. Elongated or flat particles of graphite (Carbon);
2. Particles with graphite cores and icy mantles;

3. Elongated probably needle-shaped or cylindrical dirty-ice grains;
4. Silicate particles; and
5. Diamonds or other forms of carbon.

Among these, opinion rests heavily on graphite or graphite-pius-ice models of the grains. Silicate particles also have achieved substantial popularity.

The model of particles ultimately accepted, should however explain the *formation* of these particles. In this respect, graphite particle model is the most acceptable of all others. Many authors believe that graphite grains (pure or mixed) may be formed on surfaces of cool carbon stars of R and N types. These are pulsating variable stars whose atmospheres undergo sufficient temperature fluctuations over a period. Carbon may condense in the cooling surface which may subsequently grow into graphite particles. Radiation pressure on the particles may then drive them into the interstellar space which ultimately becomes a part of the interstellar medium. These graphite particles can then act as nuclei which other atoms and molecules may adhere to giving icy mantles over graphite nuclei. Many authors have worked out the fitting of the observed extinction curve with pure graphite particles with fairly good success. Fitting of the observed data with graphite particles surrounded by dirty ice mantles has also been worked out by many authors with good success. The problem, however, still remains unresolved.

EXERCISES

1. What optical observations led astronomers to believe in the existence of an all-pervasive and uniform, but relatively thin distribution of the interstellar matter participating in the galactic rotation?

   How did the astronomers came to know that superimposed on this uniform sparce component of matter, there are gas clouds of smaller dimensions and higher densities having in addition, random velocities of varying magnitudes?

2. What is 21-cm radiation? How does it originate? What information about the distribution, density and velocity of the interstellar gas was obtained by 21-cm observations?

3. List the important optical absorption lines arising out of interstellar matter. What information do we get from these lines about
   (a) the velocity dispersion of the clouds; and
   (b) the abundance of interstellar matter?

4. List the important atoms and molecules producing interstellar lines in radio wavelengths, giving also the corresponding wavelengths.

5. Discuss the structure of the OH lines showing the transitions that give rise to these lines. Discuss the anomalous behaviour of the component lines of OH and try to give a physical explanation for the observed anomaly. What is *cosmic maser*?

6. The behaviour of interstellar matter as revealed by 21-cm study greatly differs in some respects from that as revealed by OH observations—Explain.

7. What observations indicate that the interstellar gas and dust possess *cloudy* structure? Describe the average properties of these clouds that have been known from 21-cm observations.

8. Discuss the present understanding of the velocity distribution of the interstellar clouds. Give the various distribution laws and comment on their merits and demerits.

9. Define H I and H II regions. Compare the physical parameters present in these regions.

10. Define a *Strömgren's sphere*. Describe the stages leading to the initial growth of a Strömgren's sphere. What subsequent events lead to its further growth?

11. Using the formula (13.8) and Table 13.2 calculate the radii of Strömgren's spheres in the following cases:
   (a) O5 star, $n_H = 10, 10^3, 10^5$ cm^{-3};
   (b) O7 star, $n_H = 5, 10, 10^2$ cm^{-3};
   (c) B0 star, $n_H = 1, 5, 10, 10^4$ cm^{-3}.

12. Describe the evolution of an H II region, taking into account the dynamical effects produced in the gas during the evolution.

13. Describe the various astrophysical events that may give rise to interstellar shocks. Write down the equations which are appropriate for studying the propagation of a plane, normal and adiabatic shock. Deduce the Rankine-Hugoniot relation.
   Show that even in the strongest adiabatic shock the density jump cannot exceed 4.

14. In the case of a propagating shock, prove the following:
   (a) the changes produced by a shock are irreversible; and
   (b) a shock always compresses a gas.

15. Explain why interstellar shocks are more *nearly* isothermal than adiabatic. Show that in the case of an isothermal shock, the density jump is proportional to the square of the Mach number of flow.

16. Write down the full set of equations necessary for analyzing a hydromagnetic shock. Show that the compression achieved in a hydromagnetic shock is less than that in an ordinary shock.

17. Discuss Kahn's model of interstellar cloud collisions and calculate the rate of kinetic energy dissipated by the process.

18. What are the different sources of input of kinetic energy in the interstellar medium? Do you think that these inputs can completely balance the energy lost by cloud collisions?

19. Describe the various effects produced by grains on astronomical observations. How do you know about the distribution, size, shape and orientation of the grains?

20. Discuss the status of the present understanding about the composition and formation mechanism of dust grains.

SUGGESTED READING

1. Bergers, J.M. and Van de Hulst, H.C. (Eds.), *Gas Dynamics of Cosmic Clouds,* North Holland Publishing Co., Amsterdam, 1955.

2. Burton, W.B., Elmegreen, B.G. and Genzel, R., *The Galactic Interstellar Medium,* Springer-Verlag, Berlin, 1992.

3. Combes, F., Distribution of CO in the Milky Way, *Annual Review of Astronomy and Astrophysics,* **29,** p. 195, 1991.

4. Diercksen, G.H.F., Huebner, W.F. and Langhoff, P.W. (Eds.), *Molecular Astrophysics,* NATO ASI Series, D. Reidel Publishing Co., Dordrecht, Holland, 1985.

5. Hartquist, T.W. (Ed.), *Molecular Astrophysics,* Cambridge University Press, Cambridge, 1990.

6. Hollenbach, DJ. and Thronson, A. Jr. (Eds.), *Interstellar Processes,* D. Reidel Publishing Co., Dordrecht, Holland, 1987.

7. James, R.A. and Miller, T.J. (Eds.), *Molecular Clouds,* Cambridge University Press, Cambridge, 1991.

8. Kahn, F.D. (Ed.), *Cosmical Gas Dynamics,* VNU Science Press, Utrecht, The Netherlands, 1985.

9. Kaplan, S.A., *Interstellar Gas Dynamics,* Pergamon Press, Oxford, 1966.

10. Kerr, F.J. and Westerhout, G., Distribution of interstellar hydrogen *in Galactic Structure,* Blaauw, A. and Schmidt, M. (Eds.), The University of Chicago Press, Chicago, 1965.

11. Ling, K.R., *Astrophysical Data I,* Springer-Verlag, Berlin, 1992.

12. Middlehurst, B.M. and Aller, L.H. (Eds.), *Nebulae and Interstellar Matter,* The University of Chicago Press, Chicago, 1968.

13. Spitzen, L. (Jr.), *Physical Processes in the Interstellar Medium,* John Wiley & Sons, New York, 1978.

14. Waltjer, L., Dynamics of gas and magnetic field: Spiral structure *in Galactic Structure,* Blaauw, A. and Schmidt, M. (Eds.), The University of Chicago Press, Chicago, 1965.

15. Waltjer, L. (Ed.), *Interstellar Matter in Galaxies,* W.A. Benjamin, Inc., New York, 1963.

14 Structure and Evolution of Stars

14.1 INTRODUCTION

In the earlier chapters we have attempted to present the various physical characteristics as manifested by different types of stars. These characteristics, however, mostly belong to the superficial layers of the stars. Many of these are certainly the reflection of the processes going on in the deep interior of the stars—a mysterious region which we can never "see", but of which we can only make hypothetical assumptions on the basis of various observational materials collected so arduously by generations of astronomers. The theory of structure and evolution of stars has been an absorbing story of the modern astronomy. The understanding of the physical processes and their mathematical formulation has been developed in slow but steady steps until at present we can claim that most of the essential processes underlying the various stages of stellar life are known with a fair degree of accuracy.

Yet there are many questions about the birth, growth and death of stars for which we do not have definite answers. Even the answer to the basic question as to how stars are actually born is not known. However, it is definitely known that the stars are born out of the gas and dust in a galaxy, attain certain age and meet the end of their life. The time-scale over which this phenomena occur is very long. For hot O and B stars the lifetime is a few times 10^7 years while for stars like the Sun this time is of the order of 10^{10} years. One of the most pertinent questions which the theory of stellar interior must answer is how such a huge amount of energy is generated within the stars in order to maintain their luminosities for such long periods. As we know, the energy radiated by the Sun is about 4×10^{33} erg s^{-1}, and for an O star the rate of radiation is $> 10^{38}$ erg s^{-1}. Thus, the total energy spent by a star during its active life is 10^{52} erg or even more. What is the physical process which can manufacture such an enormous amount of energy? The early investigators were content with the belief that chemical reactions inside were responsible for the production of stellar energy. Lord Kelvin had suggested in the 19th century that the stars maintain their extravagant life at the expense of their gravitational energy released by contraction. Calculations reveal that the total gravitational energy possessed by the Sun is:

$$\Omega = -\frac{GM_\odot^2}{R_\odot} \sim 4 \times 10^{48} \text{ erg}$$

At the present rate of expenditure, this energy can feed the Sun for a period of $\sim 10^7$ years. This has since been known as the *Kelvin contraction time*. Things would have been all right if the life time of the Sun were really comparable with the Kelvin contraction time.

But evidences gathered from geological studies and radioactive decay phenomena, and also more recently, from analyses of lunar rocks, reveal that the lifetime of the Sun is not actually comparable with the Kelvin contraction time. All these evidences point to one definite conclusion, i.e. the solar system and the Sun itself for that matter is nearly 5×10^9 years old. Thus, neither chemical reactions nor gravitational contraction could be responsible for the production of the required amount of energy. These ideas led to researches in new directions. It was finally discovered by H. Bethe and Von Weizsäcker in 1938 that nuclear transmutation of hydrogen into helium and of helium into heavier elements in the core of the stars was responsible for generation of the enormous amount of energy required. The process of transmutation annihilates a fraction of the nuclear masses which is converted into radiant energy following Einstein's law $E = mc^2$. Calculations show that a total of about 0.008 part of the mass is annihilated during the process. If, for example, about one-half of the solar mass is involved in nuclear transmutation during the lifetime of the Sun, the total radiant energy liberated will be $(0.008) (10^{33}) (c^2) \sim 7 \times 10^{51}$ erg which is sufficient to maintain a life-span of about 70 billion years of the Sun at its present rate of expenditure. Nuclear energy production in the stellar interior, therefore, has come out as the most satisfactory explanation for maintaining the long life-span the stars in general are believed to possess.

The theory of stellar interior must also explain satisfactorily as to how the energy generated in the deep interior of stars is transported through the stellar body to their surfaces. The interaction of radiation with matter inside stars gives rise to interesting physical processes. The nature of interaction and the resulting processes vary in different stars and also in different regions of a single star. These differences are likely to influence the structure of a particular star and the effects of such influence should be subjected to observational verification. In order to achieve this experimental justification of the theoretical inferences, stellar models are constructed with various initial parameters and checked how these models evolve in time along die H–R diagram. Valid inferences can be drawn by comparing the theoretical models with the actually known stars in the H–R diagram. The structure and evolution of stars thus provides a field of modern astrophysics where every step must be traversed by theory and observation together.

14.2 THE OBSERVATIONAL BASIS

In this section, we shall try to summarize some of the most important information about stars that can be gathered by various types of observations. Two of the basic observable parameters in stars are: The luminosity L (or absolute magnitude M), and the effective temperature T_e (or spectrum or the colour B–V). These basic quantities lead to the construction of the H–R (or C–M) diagram for the stars. These diagrams serve as fundamental guides to the theoretical construction of stellar models. In the H–R diagram, stars occupy certain well-defined regions and there are large regions in the diagram which is devoid of stars. Again, certain regions of the diagram are found to be heavily populated by stars while certain others are characterized by sparse population. The most obvious interpretation of this observed fact is that stars spend greater parts of their life-span in thickly populated regions (such as the main sequence, the giant branch and the region occupied by white dwarfs) of the diagram. A good model star should faithfully imitate this trend. For example, if a model star in course of its evolution is found to spend a large portion of its life in a region of the H–R diagram which is sparsely populated or entirely devoid of stars, such a model has to be rejected. The observed H–R diagram thus serves as a basic guide to the construction of stellar models.

Mass is a fundamental parameter in the theory of stellar structure and evolution. The mass of a star is related to its luminosity and radius at any particular phase of evolution. The general form of the mass-luminosity relation for main sequence stars can be written as:

$$L = \text{Constant } \frac{M^x}{R^y} \mu^z \qquad (14.1)$$

where x lies between 3.5 to 4.0, $y \sim 0.5$ and $z \sim 7.5$. Here M is the mass, R the radius and μ the mean molecular weight depending solely on the chemical composition and the ionization state of the composing material. The luminosity of a main sequence star steeply depends on its mass and the chemical composition while it is only weakly dependent on the radius of the star. In fact, in constructing a model star one has to fix at the outset its initial mass and chemical composition which are the two most fundamental quantities in the theory of stellar structure. There is a theorem put forth by H. Vogt and H.N. Russell which states that "given the initial mass and chemical composition, the structure of a star is fixed for all subsequent stages of its evolution". Observations of binary stars give clues to the determination of stellar masses and radii while the chemical compositions of stars are derived from the analyses of curves of growth. At present we know with fair degree of accuracy the abundance of elements in different kinds of stars. In stars like the Sun which we call to be of normal composition the approximate abundance is: 73 per cent of hydrogen (by weight) (X), 24 per cent of helium (Y) and 3 per cent of all other heavy elements (Z). Thus observations yield two of the most fundamental parameters required to construct stellar models. In considering the abundances in stars, it must be mentioned here of the fundamental assumption that is always made, viz. the chemical composition which is spectroscopically observed in the superficial layers of stars is representative also of their deeper layers. There is no valid reason to assume that abundances in deeper layers of stars differ significantly from those of the superficial layers. So in constructing stellar models, the initial composition is taken to be homogeneous throughout the star. As the time passes, the hydrogen is depleted in the core building up helium and the star *evolves*.

Finally, in considering the evolution of a model star with age, one must definitely assign its population characteristics. We have seen in Chapter 11 that significant differences exist in the basic features of the H–R diagrams of the two stellar populations. While the brightest stars of the Population I are blue, those of Population II are red. Also, the curves followed by giant branches of the two populations are quite different. The principal physical difference among the stars of the two populations as revealed in spectroscopic studies is the large difference in their metal content. Metal is under-abundant in Population II stars by factors ranging from 10 to 100, and even more in some cases. This makes the colours of Population II stars bluer for the same effective temperatures. Thus, the percentage of heavy elements (Z) assumed for the initial composition of the model star will determine whether the evolutionary track characteristic of Population I or that characteristic of Population II will be followed by the star. In other words, the percentage of heavy elements assigned in the initial composition of the model star will profoundly influence its subsequent evolution in time, and we shall be able to predict in advance whether it will evolve in the H–R diagram along the Population I or Population II sequence.

The theoretical study of the structure and evolution of stars is therefore no longer a futile exercise. Rather it has now entered in a very advanced stage. And we have now enough observational material and theoretical tools at our disposal to guide us through the complexities of various confounding physical concepts and mathematical obstacles that are generally encountered as one follows a star from its birth to death. Also, simultaneously a careful watch on the incidents occurring within the star through the layers beginning from the core to the surface is kept. The great advances in the fields of both theory and observation, together with the remarkable capacity of modern digital computers to handle rigorously large computational materials, have made possible rather satisfactory understanding of the problems of the structure and evolution of stars.

14.3   THE EQUATION OF STATE FOR STELLAR INTERIOR

The equation of state given in the form

$$p = \frac{\rho RT}{\mu} \tag{14.2}$$

where μ is the mean molecular weight of the gas holds precisely for a perfect gas at relatively low densities.

Apparently, the law should not hold in the stellar interior where the density of gas is very high. Fortunately, however, there is one favourable circumstance present in the stellar interior which renders the equation of state (Eq. 14.2) valid also in the stellar interior with a fair degree of accuracy. This is the existence of extremely high temperature which renders the matter in the stellar interior almost completely ionized. In the completely ionized gas the dimensions of individual particles reduce to those of minute nuclei rather than comparatively much larger dimensions of atoms. Under these circumstances, even the very high density gas in the stellar interior approximates to a perfect gas and the equation of state given by Eq. (14.2) holds with a fair degree of accuracy.

To apply Eq. (14.2) in the stellar interior, however, we must write the value of μ explicitly in terms of the prevailing composition there. Suppose the matter is composed of hydrogen, helium and minute fraction of heavy elements all considered together. Let X be the fraction of hydrogen (by weight), Y the fraction of helium and Z the fraction of all other elements considered together. Then,

$$X + Y + Z = 1 \tag{14.3}$$

It can be shown then that taking the proton mass as unity,

$$\frac{1}{\mu} = 2X + \frac{3}{4} Y + \frac{1}{2} Z \qquad (14.4)$$

Equations (14.2) and (14.4) together give the equation of state for stellar interiors as we have envisaged.

The above argument, however, is valid so long as we can neglect the radiation pressure in the interior and the entire pressure is due to the gas. Such a situation is approximately realized in the interior of stars like the Sun, the main sequence dwarf stars of low masses. In such stars, the radiation pressure is considered negligible compared to the overwhelmingly large gas pressure. The density of gas in the interior of these stars is very high. But the interior of hot massive stars is characterized by high temperature and lower gas density. The radiation pressure which is $\sim T^4$ therefore becomes significant compared to the gas pressure in the interior of these massive stars. So the pressure p in the equation of state in this case will be the sum of the gas pressure p_g and the radiation pressure p_r and the equation of state suitable in this case is

$$p = p_g + P_r = \frac{\rho RT}{\mu} + \frac{1}{3} \sigma' T^4 \qquad (14.5)$$

where μ is given by Eq. (14.4).

Our discussion of the equation of state remains incomplete so long as we do not consider its applicability to degenerate gases. Degeneracy of gas occurs at sufficiently high densities as a consequence of the *exclusion principle* enunciated by W. Pauli. Electron degeneracy is a consequence of *Pauli Exclusion Principle* which states that not more than two electrons having opposite spins can occupy an energy level in an atom. So in an atom with many electrons, most of the electrons being unable to find the lower energy states free to fill in, have to occupy the higher excited levels, with the result that the average excitation energy of the electrons in the atom is much in excess of the thermal energy kT. Such a gas in which all the lower energy states are filled up by electrons in accordance with the Pauli exclusion principle is called a *degenerate electron gas*. In a degenerate electron gas the average energies of electrons being much higher than that in an ordinary gas, these electrons also exert a much greater pressure than the pressure exerted by an ideal gas. This higher pressure is called the *degeneracy* pressure. Also, the average thermal energy being lower in such a gas, the temperature becomes unimportant in the pressure build-up, and the pressure depends only on the density of the gas as shown in Eq. (14.6). A degenerate gas is thus necessarily of much higher density than the density which prevails in an ideal gas.

The gas in white dwarf stars having average densities of the order of 10^6 gm cm^{-3} is degenerate. The red giant cores are also believed to be degenerate. Let us see how the state of degeneracy occurs in a gas. The Pauli exclusion principle states that at most two electrons of opposite spins can be contained in a cell of volume h^3 of the six-dimensional phase space. As the gas density increases the cells at lower momenta will be first filled by electrons. The degeneracy in the gas will start when the densities attain the range 10^2 to 10^4 gm cm^{-3} at temperatures normally found inside the stars. At this stage, some electrons will start occupying higher and higher quantum states as the gas density gradually increases. Finally, a stage is reached at a *critical* density when all cells having momenta smaller than some critical momentum P_{crit} are fully occupied while no electrons are found to occupy cells having momenta larger than P_{crit}. The distribution function of electrons which was initially Maxwellian is gradually distorted to such an extent that the influence of temperature on it becomes completely obliterated when the degeneracy is complete. The pressure p of the gas then depends only on the density p and the composition of the gas, and for complete degeneracy we have

$$p = 9.91 \times 10^{12} \, (\rho/\mu_e)^{5/3} \tag{14.6}$$

where μ_e represents the mass in atomic weight units per electron. With our fractional representation, we will have

$$\frac{1}{\mu_e} = X + \frac{1}{2} Y + \frac{1}{2} Z = \frac{1}{2} (1 + X) \tag{14.7}$$

the last equality being obtained by using Eq. (14.3). Thus, for pure hydrogen $\mu_e = 1$ and for pure helium $\mu_e = 2$. Equation (14.6) represents the equation of state for *non-relativistic degenerate gas* which is valid for gas densities of the order of 10^6 gm cm^{-3}.

At still higher densities the electrons are forced into cells of still larger momenta. The electrons then possess energies in excess of their rest energy $m_e c^2$. The non-relativistic analyses fail at this stage and we must look to the theory of special relativity for proper understanding of the situation. Pressure again is independent of the gas temperature. It depends on ρ and μ_e as before, and the equation of state in this case is given by:

$$p = 1.231 \times 10^{15} \, (\rho/\mu_e)^{4/3} \tag{14.8}$$

Equation (14.8) embodies the equation of state for a *relativistically degenerate gas*.

14.4 MECHANICAL AND THERMAL EQUILIBRIUM IN STARS

The set of observational data that forms the basis for formulation of the theory of the structure and evolution of stars has one property common to all, viz. the values remain constant over a long period of time. The basic observed quantities in stars, viz. the mass, the luminosity, the radius and the chemical composition of the superficial layers do not significantly vary even over long periods of time. This constancy of stellar structure, in fact, makes the theoretical study possible because we have to deal with equilibrium configurations only, where any changes with respect to time are neglected. This fact greatly simplifies the mathematical treatment of otherwise complicated set of equations.

The first basic equilibrium condition that must hold throughout the star is the *hydrostatic equilibrium*. If $\rho(r)$ be the density of matter at a distance r from the centre of the star, the mass of a shell at r and of thickness dr is $4\pi r^2 \rho(r) \, dr$. The mass of the sphere of radius r is therefore given by

$$M_r = \int_0^r 4\pi r^2 \rho(r) \, dr \quad \text{or} \quad \frac{dM_r}{dr} = 4\pi r^2 \rho(r) \tag{14.9}$$

Let us consider a cylindrical element of unit cross-section and length dr with its axial direction passing through the star's centre. Its mass is $\rho(r)dr$. So the force exerted by M_r on this mass is, according to Newton's law of gravitation, equal to $\rho(r)dr \cdot \dfrac{GM_r}{r^2}$, where G is the constant of gravitation. For hydrostatic equilibrium, this force must be counter balanced by the change in pressure, say dp, over the depth of the element considered. The condition of *hydrostatic equilibrium* is therefore expressed by

$$-dp = \rho(r)\,dr\,\frac{GM_r}{r^2} \quad \text{or} \quad \frac{dp}{dr} = -\frac{GM_r}{r^2}\rho(r) \tag{14.10}$$

It may be mentioned here that the pressure p in Eq. (14.10) represents the total pressure due to both gas and radiation. But in most stars except for the very hot ones the pressure due to radiation is insignificant as we have already mentioned in the previous section. So in the overwhelming majority of stars p differs little from p_g, the gas pressure.

The hydrostatic equilibrium equation in Eq. (14.10) can be utilized to calculate the order of magnitude of the pressure and temperature which usually prevails at the centre of stars. As an example, let us take the case of our most familiar star, the Sun. In Eq. (14.10), if we put $\rho(r) = \bar{\rho}_\odot = 1.4$ gm cm^{-3}; the average density for the Sun, take half the solar mass for M_r, the solar radius for dr and the difference between the central and surface pressures for dp, and neglect the surface pressure altogether, then the central pressure $p_{c\odot}$ turns out to be

$$p_{c\odot} \simeq 1/2\,\bar{\rho}_\odot\,\frac{GM_\odot}{R_\odot} = 1.5 \times 10^{15} \text{ dyne cm}^{-2} \tag{14.11}$$

The central temperature $T_{c\odot}$ can now be derived from the equation of state using relation (14.4) for μ. Using the composition we have already chosen, we have $1/\mu \approx 1.62$ in proton mass unit. Then the central temperature is given by

$$p_{c\odot} \simeq \frac{\bar{\rho}_\odot R T_{c\odot}}{\mu}, \quad \text{whence } T_{c\odot} = \frac{\mu p_{c\odot}}{R\bar{\rho}_\odot} \simeq 7.5 \times 10^6 \text{ K} \tag{14.12}$$

Equations (14.11) and (14.12) give a general idea about the magnitudes of the pressure and temperature that usually prevail at the centres of star. It is under conditions of such pressures and temperatures that nuclear transmutations occur there.

Mechanical stability is not the only condition that has to be satisfied by a constant star. The constancy will not be ensured unless the condition of *thermal equilibrium* prevails throughout the star. The thermal equilibrium required within the star is not to be confused with the concept of the perfect thermal equilibrium which means that every element of the system has the same temperature and no flow of energy between different parts of the system can take place. Such a condition cannot prevail inside a star because there is a temperature gradient from centre outward leading to a net flux of energy. The concept of thermal equilibrium in stellar interior means that the rate of energy lost by the star (that is, its luminosity) must be balanced by the production of energy inside it at an equal rate. This energy is produced by the process of nuclear transmutation. The other sources of energy are much too insufficient to balance the energy lost by the star. We have already seen that for the Sun, the gravitational energy can maintain its luminosity at the present rate for a time-scale of only the order of 10^7 years. The same is true for the case of the thermal energy of most stars. In fact, the calculation shows that under the conditions existing in the interior of stars, one-half of the total energy released by gravitational contraction goes to increase the internal energy of the star while the other half escapes from the star as radiation. It can be shown that the relation

$$2E_{\text{th}} + \Omega = 0 \qquad (14.13)$$

holds between the thermal energy E_{th} and the gravitational energy Ω. Therefore, like the gravitational energy the thermal energy of a star also turns out to be very much insufficient to feed the stellar luminosity for a sufficiently long time. Although both these forms of energy can play very important role in brief but in critical stages of stellar evolution, nuclear energy is the main source for maintaining the long and constant phases of stellar life. So thermal equilibrium in stars essentially means that the rate of loss of energy must be balanced by an equal rate of energy generation by nuclear processes. The condition of thermal equilibrium therefore can be stated by the relation

$$L = \int_0^R 4\pi r^2 \, dr \cdot \varepsilon\rho(r) \qquad (14.14)$$

where ε gives the rate of energy generated by nuclear processes per gram of stellar matter, e essentially depends on three parameters—density, temperature and chemical composition. Equation (14.14) is not the most convenient form for representing the condition of thermal equilibrium. For maintaining the equilibrium, balance in every small region within the star must be ensured and this concept is mathematically represented by the differential form

$$\frac{dL_r}{dr} = 4\pi\rho(r) \, r^2 \varepsilon \qquad (14.15)$$

where L_r stands for the total flux of energy across a spherical shell of unit thickness, and of radius r. Equation (14.15) represents the convenient form of the thermal equilibrium condition in stellar interior during the phases in which the star spends sufficiently long time. There occur critical phases of stellar life when other sources of energy, such as gravitational and thermal, play vital role in supplementing the function of the nuclear energy. During these phases the star adjusts its structure either by expansion or contraction, depending on its necessity in trying to attain the stability once again. These critical phases, however, are always shortlived and therefore the time derivative of the energy changes can no longer be ignored. During such critical phases, the condition for thermal equilibrium becomes

$$\frac{dL_r}{dr} = 4\pi r^2 \rho(r) \left[\varepsilon - \frac{3}{2} \rho^{2/3}(r) \frac{d}{dt} \left(\frac{p}{\rho^{5/3}(r)} \right) \right] \qquad (14.16)$$

where the second term within the brackets represents the rate of release of internal energy per gram of stellar matter. During the constant phase of stellar life this time rate of change is negligible and so the contribution from this term is practically zero. Equation (14.16) then changes to Eq. (14.15) which represents the thermal equilibrium condition for most part of a stellar life.

14.5 ENERGY TRANSPORT IN STELLAR INTERIOR

The thermal equilibrium in different regions of the stellar interior is maintained by a balance of the flux of radiation between these various parts. The magnitude of this flux essentially depends on the temperature gradient prevailing in the interior and the opacity of the material through which the energy is being carried out. The principal problem is to transport an enormous amount of energy (equal to the luminosity of the star) which is generated at the central region of the star, through the stellar body to its surface. There are three different mechanisms by which the energy may be transported, viz. conduction, convection and radiation. As we have already discussed in Chapter 6, among the processes of transport of energy through the superficial layers of stars, radiative transport of energy is the most important mechanism. This is also true for the deeper layers of stars. In the interior of stars, radiation plays the most important role in transferring the energy from central to the surface layers. Energy transport by conduction is effective under the condition in which the mean free paths of electrons and ions are sufficiently long. This condition is rarely satisfied inside normal stars where the gas density is very high. In degenerate gas, however, the electrons possess very long mean free path and therefore, the electron conduction is important in conditions of degeneracy. Transport of energy by electron conduction is, therefore, important in white dwarfs and in cores of red giants, where the gas is known to be degenerate. Energy transport by convection (carried by moving mass elements) is important in some limited zones of some types of stars. Under the conditions prevailing in these zones transport by radiation falls short of the required amount to be carried out in order to maintain the thermal equilibrium around these specified zones. This necessitates the setting up of convection currents which supplement the amount of energy required to maintain a thermal balance. The envelopes of stars like the Sun and the cores of the massive hot stars are known to be the regions where the convective transport of energy prevails. In the former case, the layer of the envelope where ionization of hydrogen takes place (known as the hydrogen ionization zone) consumes a large fraction of outgoing energy for the process of ionization. As a result, a large portion of energy which was being carried by radiation through the deeper layers in order to feed the stellar luminosity is obliterated on reaching the envelope. It is at this critical situation that convection comes to the rescue and helps transportation of the deficit energy. The convection prevailing in the cores of massive stars is due to somewhat different reason. Here, as the rate of energy generation by CNO cycle depends very steeply on temperature, a steep temperature gradient is formed at the stellar core. As a result, radiative process alone cannot cope with necessary amount of energy to be transported through the energy generating core. To cope with the situation, convection is called into play. But in the vast region outside the core of these stars energy transportation is carried out by radiative process.

It is thus clear that the transfer of energy by radiation is the principal mechanism operative in the stellar interior. We shall therefore first derive the condition for radiative equilibrium. To do this we recall Eqs. (6.8) and (6.7) and the equation of transfer given in Eq. (6.36). Equations (6.8) and (6.7) yield

$$\pi F = \int_0^\infty \pi F_\nu \, d\nu = 2\pi \int_0^\infty \int_0^\pi I_\nu(\theta) \cos\theta \sin\theta \, d\theta \, d\nu = \sigma T_e^4 \text{ [by Eq. (6.50)]} \quad (14.17)$$

which finally yields

$$\pi F = 2\pi \int_0^\pi I(\theta) \cos\theta \sin\theta \, d\theta \quad (14.18)$$

If we now integrate Eq. (6.36) over the frequency range, multiply by $2\pi \cos\theta \sin\theta$ and integrate over θ from 0 to π, we get

$$\pi F = 2\pi \int_0^\pi \frac{dI(\theta)}{d\tau} \cos^2\theta \sin\theta \, d\theta \quad (14.19)$$

the integral containing $B(T)$ vanishing since this function is independent of θ. If χ be the mass-absorption co-efficient then $d\tau = -\rho\chi \, dr$, where r is the distance from the stellar centre. With this, the relation (14.19) stands as

$$\pi F = -2\pi \int_0^\pi \frac{dI(\theta)}{\rho\chi \, dr} \cos^2\theta \sin\theta \, d\theta \quad (14.20)$$

Since the radiation in stellar interior is very nearly isotropic, $I(\theta) = I$ and θ integration may be separated and performed to yield 2/3. The relation (14.20) therefore can be expressed as

$$\pi F = -\frac{4\pi}{3\rho\chi} \frac{dI}{dr} \quad (14.21)$$

Also, for isotropic radiation we have

$$\pi I = \pi B = \sigma T^4 \quad (14.22)$$

using Stefan-Boltzmann law. Equations (14.21) and (14.22) together can be combined to yield

$$\pi F = -\frac{4\sigma}{3\rho\chi} \frac{d(T^4)}{dr} = -\frac{16\sigma T^3}{3\rho\chi} \frac{dT}{dr} \quad (14.23)$$

Since πF represents the flux of radiation per unit area of the surface, the total flux across a spherical surface of radius r is given by

$$L_r = 4\pi r^2 (\pi F) = -\frac{64\pi\sigma r^2 T^3}{3\rho\chi} \frac{dT}{dr} \quad (14.24)$$

where σ is Stefan's constant. Equation (14.24) represents the radiative equilibrium condition in stellar interior which relates the total flux of radiation with opacity and temperature gradient.

For radiative equilibrium condition in stellar interior the relation (14.24) must be satisfied for every value of r. At the surface of the star $r = R_*$, $L_{R*} = L$ is equal to the total luminosity of the star.

If at any point in the star Eq. (14.24) is not satisfied, the radiative equilibrium will be unstable and convection will set in. This occurs in the envelope of stars like Sun at the layers of hydrogen and helium ionization. L. Biermann has shown that the convection carried much more energy than radiation in layers where it is present. The energy is carried outward by hotter rising mass elements, while the cooler upper elements drift downward to take the positions left vacant by rising layers. A current of moving mass elements is thus set up whose main function is to transport energy through mass motion more quickly than the radiation is capable to do in the layer concerned. In this manner, the steep gradient of temperature which caused the instability in radiative equilibrium will be reduced by convection and the layer will soon be adjusted to equilibrium again. This new equilibrium that prevails in the convective layer is called the *convective equilibrium.* The pressure and temperature in a convective zone are connected by the adiabatic relation

$$T \sim p^{1-1/\gamma} \tag{14.25}$$

where $\gamma = C_p/C_v$. The gradients of temperature and pressure in a convective zone are therefore related by

$$-\frac{dT}{dr} = -(1 - 1/\gamma)\frac{T}{p}\frac{dp}{dr} \tag{14.26}$$

which is derived by differentiating logarithmically both sides of Eq. (14.25) with respect to r. In this equation, the quantity on the left-hand side represents the *actual temperature gradient* in the layer while the quantity on the right-hand side is defined as the *adiabatic temperature gradient.* If, therefore, in any layer inside the star the actual temperature gradient equals the adiabatic temperature gradient, we have to conclude that in the layer in question the radiative equilibrium condition has broken down and convection has set in. For radiative equilibrium to be stable the actual temperature gradient at every layer and at every moment must be less than the adiabatic temperature gradient. This latter condition is mathematically expressed by

$$-\frac{dT}{dr} < -(1 - 1/\gamma)\frac{T}{p}\frac{dp}{dr} \tag{14.27}$$

since the gradients dT/dr and dp/dr are essentially negative. In stellar model computation, one has to check for radiative equilibrium that the inequality relation (14.27) is satisfied at every point at every moment. If in any layer the above inequality does not hold, then one must at once consider the convective transport of energy in that layer. The thermal equilibrium is regained within the stars by the joint cooperation of radiative and convective transport of energy, which together carry outwards the exact amount of energy which is produced by nuclear transmutation within the stellar core.

The actual temperature gradient within a star depends on the opacity of stellar material. It is caused by the absorption of radiation by atoms and ions existing under various conditions within the star. The steep temperature gradient produced by absorption of a large fraction of outgoing radiation by hydrogen atoms in the hydrogen ionization zone in envelopes of the lower main sequence stars, leading to the onset of convection in the layer illustrates the influence of opacity. The opacity particularly depends upon the density and composition of the matter and on the frequency at which it is calculated. As we have already mentioned in Chapter 6, it is very difficult to compute the opacity in its fullest details. Various atomic processes contribute to the opacity. We have briefly discussed these processes in Chapter 6. For a rigorous treatment, one must consider the contribution due to each of these processes at different frequencies. But the problem is usually made simpler by taking some mean value of the absorption coefficient. This sacrifices certain amount of accuracy but the computational labour is greatly reduced in exchange. In stellar interior, the Rosseland mean absorption coefficient $\overline{K}_R$ defined by Eq. (6.66) has been used extensively.

Kramer's law of opacity which gives only an approximate value has also been often used because of its simple form. In the stellar interior hydrogen and helium are fully ionized. So bound-free absorption comes mainly from the heavy elements while the free-free absorption originates mainly from hydrogen and helium. The bound-free opacity law of Kramer is given by

$$K_{bf} = 4.34 \times 10^{28} \frac{\overline{g}_{bf}}{g_l} Z(1 + X) \frac{\rho}{T^{3.5}} \tag{14.28}$$

where $\overline{g}_{bf}$ is the average *Gaunt factor* for bound-free transition and g_l is another approximating factor called the *guillotine factor*. In stellar interior values of both of these correction factors are approximately equal to unity. The free-free opacity law due to Kramer is given by

$$K_{ff} = 3.68 \times 10^{22} \, \overline{g}_{ff} \, (X + Y) \, (1 + X) \, \frac{\rho}{T^{3.5}} \tag{14.29}$$

where $\overline{g}_{ff}$ is the average Gaunt factor for free-free transition. It may be noted that both of these opacity laws have the same dependence on density and temperature. Kramer's laws given above can be conveniently used in stars like the Sun. In more massive stars electron scattering is found to be an important source of opacity for which the appropriate law given by

$$K_e = 0.19(1 + X) \tag{14.30}$$

has to be used. In critical regions where more than one opacity laws are operative, all these laws have to be considered.

14.6 ENERGY GENERATION IN STARS

In this section, we shall discuss the various processes of the energy generation by transmutation of elements in the core of the star. The transmutation becomes effective in conditions of high density and temperature, the like of which are met within the core of stars. Under these conditions only, the nuclei can overcome the Coulomb barrier produced by the electrostatic potentials of like charges. The nuclei then come together to form a heavier nucleus while suffering a mass defect in the process. This lost mass appears in the form of energy following Einstein's law $E = mc^2$. The complete process of nuclear fusion can yield Fe^{56} using 56 H^1 (H^1 = proton) and producing 30 e^+ (e^+ = positron) and 30 v (v = neutrino); accordingly, we can write 56 $H^1 \rightarrow Fe^{56} + 30 \, e^+ + 30 \, v$. The process liberates an energy of 7.9×10^{18} erg gm^{-1}. Fusion cannot proceed any further and the above energy is the maximum which one can get through nuclear fusion. Of this, as much as 6.7×10^{18} erg gm^{-1} comes from the primary fusion process of hydrogen transmuting into helium, and this process takes place at relatively low temperatures. Transmutation of hydrogen into helium is therefore the most important source of nuclear energy inside a star.

At temperature around 10^7 K (10 million degree Kelvin), typical for stellar interior, the typical thermal energy of a particle is only about 1 KeV (= 3/2 kT) whereas the Coulomb potential barrier produced by the thermal repulsion of particles is about 1000 KeV (= $Z_1 Z_2 e^2 / R$) for particles of lowest charge (proton). For heavier particles the Coulomb potential is still higher. Although the Coulomb potential is about a thousand times higher than the typical particle velocity in the stellar core, depending on the magnitude of the barrier and the relative velocity of approach, a fraction of particles does penetrate through the barrier to give rise to collisions between pairs.

1. The probability $P_p(v)$ of penetrating the Coulomb barrier between the particles of charges $Z_1 e$ and $Z_2 e$ approaching each other with a relative velocity v has been calculated to be

$$P_p(v) \propto \exp \left[- \frac{4\pi^2 Z_1 Z_2 e^2}{hv} \right] \qquad (14.31)$$

   where h is Planck's constant.

2. Not all particles that cross the Coulomb barrier do involve themselves in nuclear interaction. The actual nuclear interaction is proportional to the probability P_N for nuclear interaction. The value of P_N is known from theory for the interaction of two protons. For other nuclei P_N has to be computed from experimental results.

3. Also the frequency of collisions between particles depends on their relative velocity v of approach towards each other.

4. The effective cross-section $\sigma(v)$ of collision. The latter is given by the square of the de Broglie wavelength, so that $\sigma(v) \propto 1/v^2$.

5. Due to the relative velocity dependence of several factors such as the collision frequency and penetration probability, a factor representing the velocity distribution of particles has, of necessity, to be introduced in the rate computation for nuclear reactions.

Since the interacting nuclei are non-degenerate, the Maxwellian velocity distribution is generally introduced for the purpose. This is given by

$$D(T, v) \propto \frac{v^2}{T^{3/2}} \exp \left[- \left(\frac{m_H A_{1,2} v^2}{2kT} \right) \right] \qquad (14.32)$$

where

$$A_{1,2} = \frac{A_1 A_2}{A_1 + A_2} \tag{14.33}$$

is the reduced atomic mass. This factor is very important in view of the fact that the high sensitivity of Maxwellian velocity distribution to temperature makes nuclear reactions highly sensitive to temperature.

Lastly, the reaction rate must be proportional to the product of the number densities of the different kinds of particles involved. If N_1 and N_2 be the number densities of the two kinds of particles whose abundances by weight are X_1 and X_2 respectively, then

$$N_1 = \frac{\rho X_1}{m_1}, \quad N_2 = \frac{\rho X_2}{m_2} \tag{14.34}$$

where m_1 and m_2 are the masses of the nuclei of Types 1 and 2. When all these various contributing factors are combined together, the reaction rate can be put in the form of the integral (over velocity) given by

$$r = \int_0^\infty N_1 N_2 v \, \sigma(v) \, P_N \, P_p(v) \, D(T, v) \, dv \tag{14.35}$$

Substituting the explicit expressions for all the factors occurring in the integral on the right-hand side Eq. (14.35) and integrating over the range of velocities for which the integral (it is significant over a narrow range of velocities in which the product $P_p(v) \, D(T, v)$) is high, the reaction rate r turns out in the form

$$r = C_1 \rho^2 X_1 X_2 T^{-2/3} \exp\left[-3\left(\frac{C_2 Z_1^2 Z_2^2}{kT}\right)^{1/3}\right] \tag{14.36}$$

where

$$C_2 = \frac{2\pi^4 e^4 m_H A_{1,2}}{h^2} \tag{14.37}$$

and C_1 includes all the proportionality factors together with m_1 and m_2. C_1 thus includes all the nuclear characteristics and so it varies from one reaction to another. Equation (14.36) shows that the reaction rate is sensitive to the density, composition and temperature.

Let us now consider the principal types of reactions that are believed to be responsible for generating energy in the stellar interior. After the star has been first formed from the interstellar medium, it gradually contracts—releasing gravitational energy. About one-half of this released energy is radiated from the surface, while the other half is preserved as thermal energy within the star thereby raising its internal temperature. When the internal temperature has been raised to about a million degrees, the first nuclear reactions start. In these reactions the lighter elements Li, Be and B are almost entirely burnt up except in the surface layers.

These lighter elements are thus converted into helium isotopes, and in the process energy is liberated as given in the following series of reactions. The numbers in the right column indicate the energy liberated in the corresponding reactions.

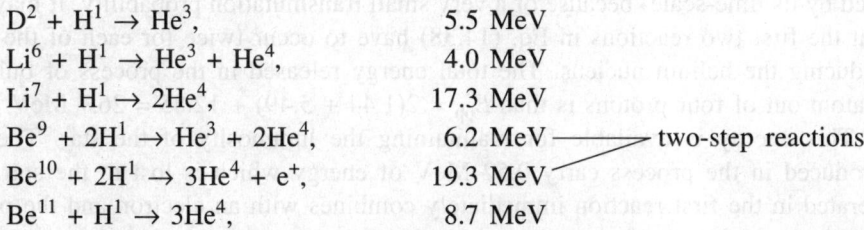

$$D^2 + H^1 \rightarrow He^3,$$ 5.5 MeV

$$Li^6 + H^1 \rightarrow He^3 + He^4,$$ 4.0 MeV

$$Li^7 + H^1 \rightarrow 2He^4,$$ 17.3 MeV

$$Be^9 + 2H^1 \rightarrow He^3 + 2He^4,$$ 6.2 MeV ——→ two-step reactions

$$Be^{10} + 2H^1 \rightarrow 3He^4 + e^+,$$ 19.3 MeV

$$Be^{11} + H^1 \rightarrow 3He^4,$$ 8.7 MeV

All these reactions take place at temperature lying in the range $(0.5–5) \times 10^6$ K. So these lighter elements are exhausted in the stellar interior except in uppermost layers of the envelope and this depletion takes place in early phases of stellar life when they are still contracting. This explains satisfactorily why these lighter elements are found to have extremely low abundance except in a few special types of stars. Some stars have been found to have an abnormally high lithium abundance for which no satisfactory explanation has yet been found. These have sometimes been called the *lithium stars*. Because of the low abundances of the above light elements and the relatively low output of energy in each of the above reactions (producing a helium isotope) they do not contribute much to the total energy output in the star required to maintain its luminosity.

As the star continues to contract, its central temperature gradually increases. When this temperature rises to the order of 10 million degrees, hydrogen burning sets in at the star's core. Because of overwhelming abundance of hydrogen in stars and the high energy output in building each helium atom out of four protons, transmutation of hydrogen into helium turns out to be the most prolific source of stellar energy for most part of their life. With the onset of hydrogen burning, no further contraction of the star is required for releasing energy. The star attains a mechanically and thermally stable configuration when it is said to settle on the main sequence. The star spends a greater and very important part of its life in this configuration. It is known, however, that the hydrogen burning in the stellar interior can proceed in two different sequences of reactions, viz. the proton-proton (p–p) chain and the carbon-nitrogen-oxygen (CNO) cycle (or bi-cycle). The former reactions will prevail in stars like the Sun while the latter will dominate in more massive stars because of the higher central temperatures attained in them. In reality, energy generation by both processes should simultaneously occur in most main sequence stars. But depending on the central temperature one or the other will dominate.

At around 10^7 K, the proton-proton reactions commence to build up He^4 out of $4H^1$. The final building of He^4 can be achieved through a variety of reactions. Below is given the most important chain of reactions, together with the amount of energy liberated in each reaction and the average time-scale over which the reactions take place at a temperature around 13×10^6 K.

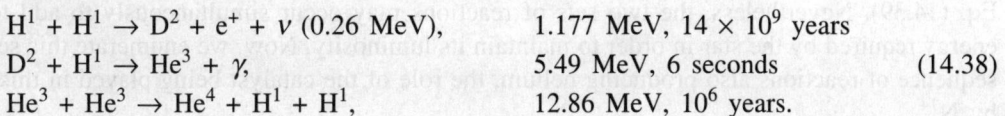

$$H^1 + H^1 \rightarrow D^2 + e^+ + \nu \ (0.26 \text{ MeV}),$$ 1.177 MeV, 14×10^9 years

$$D^2 + H^1 \rightarrow He^3 + \gamma,$$ 5.49 MeV, 6 seconds   (14.38)

$$He^3 + He^3 \rightarrow He^4 + H^1 + H^1,$$ 12.86 MeV, 10^6 years.

The first reaction produces a neutrino of energy 0.26 MeV. Due to the extremely low interaction cross-section of neutrinos ($\sim 10^{-44}$ cm^2), this neutrino escapes straight through the star and is lost to it. It may also be noticed that this proton-proton interaction has a very slow rate (indicated by its time-scale) because of a very small transmutation probability. It may also be noted that the first two reactions in Eq. (14.38) have to occur twice for each of the third reaction producing the helium nucleus. The total energy released in the process of building one helium atom out of four protons is thus $E_{pp} = 2(1.44 + 5.49) + 12.85 = 26.7$ MeV $= 4.2 \times 10^{-5}$ erg. This energy is available for maintaining the luminosity of the star. The two neutrinos produced in the process carry 0.52 MeV of energy which is lost to the star. The positron liberated in the first reaction immediately combines with an electron and the pair is annihilated with the emission of two gamma rays.

Hydrogen burning can occur with an entirely different sequence of reactions in which the C^{12} nucleus acts as a catalyst. In this process since the reactions proceed with the formation of nitrogen and oxygen isotopes, the entire cycle is called the CNO cycle (or sometimes bi-cycle). The ultimate produce, however, is the same as that in the proton-proton reactions, viz. a helium nucleus (α-particle) formed out of four protons. The set of six reactions completing the cycle, are given below together with the energy liberated in each reaction (in MeV) and the mean timescales over which these reactions take place at a temperature above 15×10^6 K.

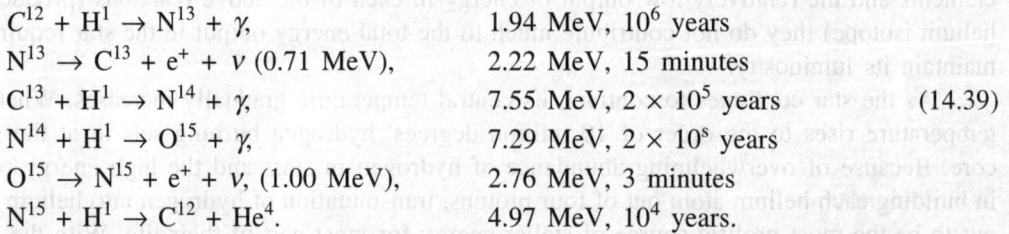

$$
\begin{aligned}
&C^{12} + H^1 \to N^{13} + \gamma, && \text{1.94 MeV, } 10^6 \text{ years} \\
&N^{13} \to C^{13} + e^+ + v \ (0.71 \text{ MeV}), && \text{2.22 MeV, 15 minutes} \\
&C^{13} + H^1 \to N^{14} + \gamma, && \text{7.55 MeV, } 2 \times 10^5 \text{ years} && (14.39)\\
&N^{14} + H^1 \to O^{15} + \gamma, && \text{7.29 MeV, } 2 \times 10^8 \text{ years} \\
&O^{15} \to N^{15} + e^+ + v, \ (1.00 \text{ MeV}), && \text{2.76 MeV, 3 minutes} \\
&N^{15} + H^1 \to C^{12} + He^4. && \text{4.97 MeV, } 10^4 \text{ years.}
\end{aligned}
$$

Thus the reactions which started with a C^{12} return it back ultimately, producing in the process one He^4. During the entire cycle two neutrinos are produced which again, as in the proton-proton case, escape from the star. These neutrinos, however, are more energetic than the former ones. The neutrino produced in the decay of N^{13} carries about 0.71 MeV while that produced in the decay of O^{15} carries about 1.00 MeV. The energy loss in this case is thus higher than that in the proton-proton reactions. The total energy available for radiation is $E_{cc} = 25.0 = 4.0 \times 10^{-5}$ erg. As in the previous case, the two positrons liberated in the process combine immediately with two electrons only to be annihilated with the emissions of two pairs of gamma rays. The mean time-scale for reactions given in the right-hand column of Eq. (14.39) are somewhat uncertain. These time-scales reduce sharply as the temperatures increase.

Starting from the last reaction given in Eq. (14.39), the helium building process may proceed along an entirely different sequence of reactions. But the probability that this latter sequence will be followed is only 4×10^{-4} times that of the sequence enumerated in Eq. (14.39). Nevertheless, the two sets of reactions may occur simultaneously to add to the energy required by the star in order to maintain its luminosity. Now, we enumerate this second sequence of reactions also producing helium, the role of the catalyst being played in this case by N^{14}.

$$N^{14} + H^1 \rightarrow O^{15} + \gamma, \qquad 7.29 \text{ MeV}, \ 2 \times 10^8 \text{ years}$$
$$O^{15} \rightarrow N^{15} + e^+ + v, \qquad 2.76 \text{ MeV}, \ 3 \text{ minutes}$$
$$N^{15} + H^1 \rightarrow O^{16} + \gamma, \qquad 1.21 \text{ MeV}, \ \text{occurring } 10^{-4} \text{ times the last reaction}$$
$$\text{in Eq. (14.39)}$$
$$O^{16} + H^1 \rightarrow F^{17} + \gamma, \qquad 0.60 \text{ MeV}, \ 2 \times 10^{10} \text{ years} \qquad (14.40)$$
$$F^{17} \rightarrow O^{17} + e^+ + v, \qquad 2.76 \text{ MeV}, \ 1.5 \text{ minutes}$$
$$O^{17} + H^1 \rightarrow N^{14} + He^4, \qquad 1.19 \text{ MeV}, \ 2 \times 10^{10} \text{ years}.$$

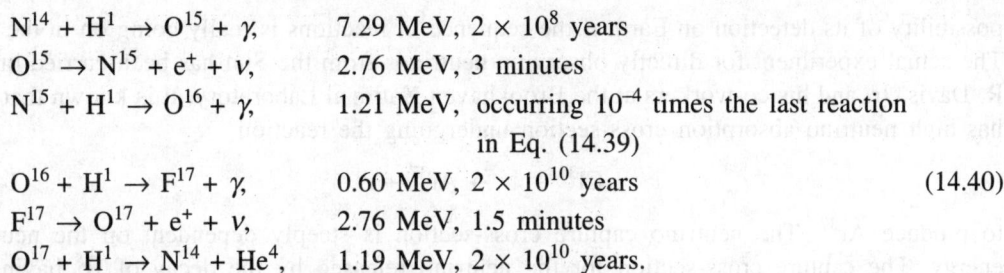

Not only that the sequence of reactions given in Eq. (14.40) occurs 4×10^{-4} times less frequent than that in Eq. (14.39), but also that the total energy released here is less than that in the latter. These reactions are therefore unimportant in the consideration of the total energy budget of the star.

Many other sequence of hydrogen-burning reactions are known. But their contribution to the total energy budget of the star is insignificant. To present a few of these processes, we start with the proton-proton chain. The reactions may alternatively follow either of the following chains, besides the most important one mentioned in Eq. (14.38).

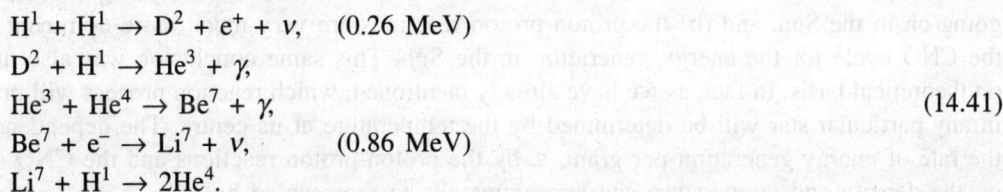

$$H^1 + H^1 \rightarrow D^2 + e^+ + v, \quad (0.26 \text{ MeV})$$
$$D^2 + H^1 \rightarrow He^3 + \gamma,$$
$$He^3 + He^4 \rightarrow Be^7 + \gamma, \qquad \qquad \qquad (14.41)$$
$$Be^7 + e^- \rightarrow Li^7 + v, \quad (0.86 \text{ MeV})$$
$$Li^7 + H^1 \rightarrow 2He^4.$$

Otherwise, after the formation of Be^7 the reactions may take the course,

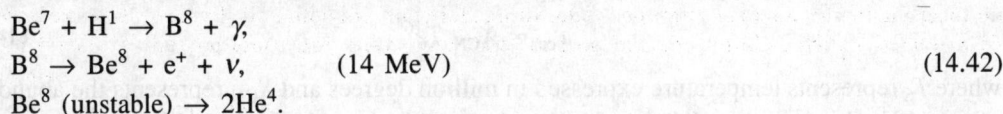

$$Be^7 + H^1 \rightarrow B^8 + \gamma,$$
$$B^8 \rightarrow Be^8 + e^+ + v, \qquad (14 \text{ MeV}) \qquad \qquad (14.42)$$
$$Be^8 \ (\text{unstable}) \rightarrow 2He^4.$$

Alternatively, hydrogen may be transmuted to helium through the Ne–Na sequence as,

$$Ne^{20} + H^1 \rightarrow Na^{21} + \gamma,$$
$$Na^{21} \rightarrow Ne^{21} + e^+ + v,$$
$$Ne^{21} + H^1 \rightarrow Na^{22} + \gamma,$$
$$Na^{22} \rightarrow Ne^{22} + e^+ + v, \qquad \qquad \qquad (14.43)$$
$$Ne^{22} + H^1 \rightarrow Na^{23} + \gamma,$$
$$Na^{23} + H^1 \rightarrow Ne^{20} + He^4.$$

This sequence requires high values of T but before attaining such high values hydrogen is used up.

Although the reactions enumerated in Eqs. (14.41)–(14.43) create He^4 from $4H^1$, these are only minor contributors to the total energy involved in stellar radiation. The reactions enumerated in Eq. (14.42), however, are important for a quite different reason. The neutrino emitted by the decay of B^8 is highly energetic (14 MeV) which, therefore, admit of the possibility of its detection on Earth if the sequence of reactions is really going on in the Sun. The actual experiment for directly observing neutrinos from the Sun has been carried out by R. Davis, Jr. and his co-workers at the Brookhaven National Laboratory. It is known that Cl^{37} has high neutrino absorption cross-section undergoing the reaction,

$$Cl^{37} + \nu \rightarrow Ar^{37} + e^- \tag{14.44}$$

to produce Ar^{37}. The neutrino capture cross-section is steeply dependent on the neutrino energy. The capture cross-section for the neutrino released by the decay of B^8 having an energy of 14 MeV is $\sim 1.35 \times 10^{-42}$ cm^2 while that for the neutrino released by Be^7 with energy 0.86 MeV is only $\sim 2.9 \times 10^{-45}$ cm^2. So the probability of capture by Cl^{37} is by far the higher for the neutrino from the boron decay than for the neutrino released by Be^7. The minimum neutrino energy required for the reaction (Eq. (14.44)) to occur is 0.81 MeV. According to the calculations of Davis and his co-workers, about 2 atoms of Ar^{37} should be produced per day from capture of solar neutrinos by Cl^{37} contained in 610 metric tons of C_2Cl_4. The actual observed rate of Ar^{37} was, however, much less. But since many uncertainties are involved in the experiment, no firm conclusions could be drawn. Tentatively, however, it has been possible to decide from the results of the experiment that, (a) hydrogen-burning is going on in the Sun, and (b) the proton-proton reactions are very much more dominant over the CNO cycle for the energy generation in the Sun. This same conclusion was also drawn on theoretical basis. In fact, as we have already mentioned, which reaction process will prevail in any particular star will be determined by the temperature at its centre. The dependence of the rate of energy generation per gram, ε, by the proton-proton reactions and the CNO cycle on the density and temperature can approximately be represented by,

$$\varepsilon_{pp} \propto X^2 T_6^4 \tag{14.45}$$

$$\varepsilon_{cc} \propto X_{CN} \, T_6^{16} \tag{14.46}$$

where T_6 represents temperature expressed in million degrees and X_{CN} represents the abundance of C and N taken together, X being the abundance of hydrogen. The above laws of temperature dependence hold fairly well for temperatures in the range 10–16 million degrees for ε_{pp} and 20–30 million degrees for ε_{cc}.

As the star advances in age on the main sequence, hydrogen at its core continues to be depleted and converted into helium. Eventually a helium core develops and continues to grow at the centre of the star, while the energy generation by hydrogen-burning becomes restricted in a shell surrounding the helium core. As the helium core grows, it continues to contract slowly releasing gravitational energy and slowly raising the core temperature. As the core continues to contract raising temperature constantly, the condition for the onset of helium burning at the core is attained. This happens when the central temperature reaches the order of 10^8 K. The core contraction then ceases and the star subsequently regains both thermal and hydrostatic balance. The helium-burning reactions, also called the *triple-alpha process,* because three α-particles are involved in the process, then proceed with the following two reactions:

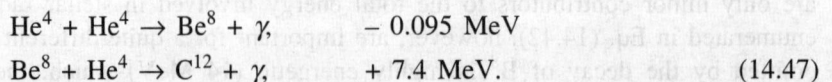

$$
\begin{aligned}
He^4 + He^4 &\rightarrow Be^8 + \gamma, & -\ 0.095 \text{ MeV} \\
Be^8 + He^4 &\rightarrow C^{12} + \gamma, & +\ 7.4 \text{ MeV.}
\end{aligned}
\tag{14.47}
$$

The first reaction forming Be^8 out of two a-particles is endothermic, absorbing an energy of 0.095 MeV in the process. The second reaction leading to the formation of C^{12} when Be^8 collides with He^4 releases 7.4 MeV of energy. It may be noted that the total energy available from three α-particles forming a C^{12} nucleus is small compared to that released by the sequence of hydrogen-burning reactions. The carbon forming reaction has a strong resonance at particle energy of 310 KeV. So most of the contribution to the reaction rate given in Eq. (14.35) comes from a narrow range of velocities centred around that corresponding to the resonance energy of 310 KeV. The total energy released in producing each C^{12} is $E_{3\alpha} \approx 7.3$ MeV = 1.17 $\times 10^{-5}$ erg, and the energy released per gram of α-particles is given roughly by,

$$\varepsilon_{3\alpha} = 10^{-8}\rho^2 Y^3 T_8^{30} \tag{14.48}$$

where T_8 represents temperature expressed in 10^8 K.

It may be noted that the energy generation by the triple-alpha process has extremely steep dependence on temperature. Also, unlike the energy generation by hydrogen-burning reactions which is proportional to ρ, that by the triple-alpha process is proportional to ρ^2. Thus, a high rate of energy generation by helium burning must be confined only within a relatively small region at the centre of the star where both ρ and T are highest. But the helium burning stage in a star cannot continue for a long time, since the total energy available from this process is comparatively small (only about 10 per cent of that available from hydrogen burning). Helium is soon depleted and a carbon core is formed. The helium burning continues in a smaller shell surrounding this core when both the core and the helium burning shell are surrounded by a hydrogen burning shell. If the temperature at the carbon core is sufficiently high, heavier elements such as O^{16}, Ne^{20} and Mg^{24} may be formed by successive capture of α-particles, releasing γ-radiation at each capture. This is the α-process designated by (α, γ).

Several other processes may operate in the stellar interior building heavier elements there. The chain must, however, terminate with the building of Fe^{56}. But in actual case, in most stars, the energy generation by nuclear transmutation probably terminates with the building of C^{12} by the triple alpha-process. Only the most massive stars are probably capable of attaining the central temperature exceeding 10^9 K which is required for the transmutation of C^{12} and heavier elements. At these temperatures, nuclear reactions may not proceed smoothly along the familiar sequences. Under certain conditions energy generation may be huge and abrupt leading to catastrophic explosion of the star. This is how supernova explosions take place. But all stars do not end in such explosions. Most of the stars are likely to meet their end smoothly as white dwarfs after using up their nuclear fuel of hydrogen and helium.

14.7 STELLAR EVOLUTION

In this section, we shall present a brief account of how Population I stars are born, evolve in age and finally come to the end of their luminous existence. Large clouds of interstellar gas at some point become gravitationally unstable, undergo fragmentation and some of these fragments gradually contract to form at first the pre-stellar bodies. The contraction continues, first at a rapid pace, but later as the internal pressure gradually builds up the contraction slows down. The pressure building is rather quickly achieved for two factors: firstly, as the contraction proceeds the density of matter increases; secondly, contraction releases gravitational energy about one-half of which is utilized to increase the internal temperature, the remaining being radiated away. This simultaneous increase of density and temperature causes a rather rapid increase in the gas pressure which checks to a great extent the initial rapid contraction. At this stage the Virial theorem (Eq. (14.13)) exactly holds in the star so that exactly half of the energy released by contraction is utilized in heating of the gas. As the pressure goes up, the stellar matter becomes opaque to radiation. Hydrogen and helium being the most abundant elements, a large fraction of the radiation is consumed in their ionization. Since an overwhelmingly large number of atoms are hydrogen, the internal temperature is restricted to below 10^4 K until hydrogen becomes mostly ionized. After hydrogen and helium ionization has become complete, the interior temperature further rises and the star proceeds towards achieving the hydrostatic equilibrium.

The currently accepted theory of the pre-main sequence contraction of stars was propounded by Japanese astrophysicist, C. Hayashi. He showed that the surface temperature of the star must be sufficiently high before it can achieve the state of hydrostatic equilibrium. Also its radius is still very large. According to Eq. (3.11) very large radius and sufficiently high surface temperature of the stars together require, the star to have a very high luminosity. At this stage the opacity of the stellar material is still very high which prevents the smooth transfer of radiation from the interior to the surface. As a result, a large-scale convection throughout the entire body of the star is called into play which carries the required amount of energy to the surface. The star is therefore, said to be in convective equilibrium. It has been shown by Hayashi that during the stage of convective equilibrium, the contraction of the star proceeds keeping its surface temperature almost constant. The luminosity of the star therefore systematically decreases and the star descends along the H–R diagram almost vertically. The initial descent is very rapid which slows down on arriving near the main sequence. A fully convective interior prevails from a few thousand to a few million years, depending on the initial mass of the star in the way that higher mass star has lower time-scale. This is also true for the total contraction time till the arrival on the main sequence. The dependence of the contraction time τ_c on the mass and luminosity of stars may be represented by the approximate formula,

$$\tau_c \simeq 8 \times 10^7 \frac{M/M_\odot}{L/L_\odot} \text{ years} \tag{14.49}$$

(Note that $M/M_\odot \ll L/L_\odot$ for $M \gg M_\odot$ and vice versa)

Throughout the contraction period the central temperature of the star continues to rise, and at a stage the complete convective equilibrium gives way to a radiative core which begins to grow. The subsequent stage of the pre-main sequence evolution of the star is characterized by the growth of the radiative core to larger and larger size until the convective region is pushed to the outer region either to restrict it in the envelope of the star or to remove it altogether out of the stellar body. The former situation is relevant to stars having masses like that of the Sun and lower, while the latter case is relevant to more massive stars. The growth of radiative core is accompanied by the corresponding increase in the central temperature which continues until it becomes high enough to ignite the nuclear fuel at the core. When the nuclear energy generation becomes sufficient to balance the stellar radiation, the gravitational contraction of the star halts for the first time, and the star is said to settle on the main sequence. In Fig. 14.1, the theoretical pre-main sequence contraction tracks in H–R diagram for several stars of different masses are given. These are correct subject to the condition that the masses remain constant as the stars evolve. The figure reveals that the more massive stars have smaller vertical drift down the Hyashi track as they achieve radiative equilibrium at the core very quickly. On the other hand, stars of low masses have long vertical drift along the Hyashi tracks. Very low mass stars ($< 0.5\ M_\odot$) probably remain almost completely convective throughout, and move down the Hyashi lines directly on the main sequence. The main sequence is terminated at the lower end by stars with the limiting mass which is just enough to generate nuclear energy sufficient to maintain the stellar radiation. This limiting mass is believed to be about 8 per cent of the mass of the Sun. At the upper end, the main sequence is terminated by stars which are massive enough to be gravitationally stable. These gravitationally stable masses are believed to lie within the range of 50–100 $M_\odot$.

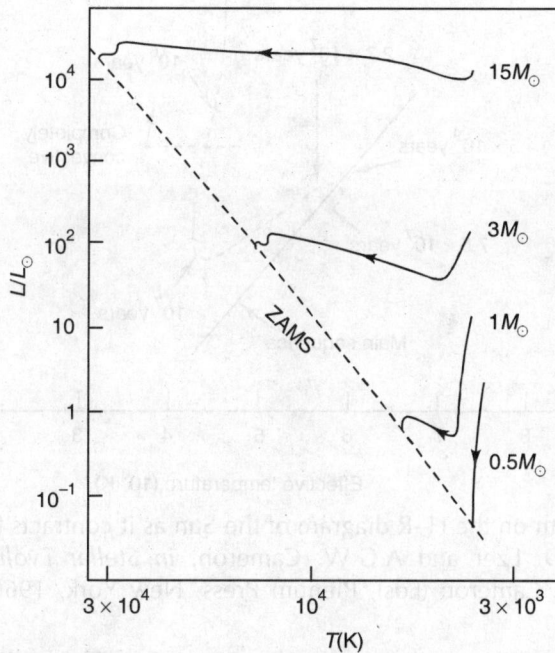

FIGURE 14.1 Evolutionary tracks of pre-main sequence stars of different masses on the H–R diagram.

Figure 14.1 also shows some enumerated points on the tracks of stars. The end points on the left of the tracks represent the position of stars as they just settle on the main sequence, and so define the theoretical ZAMS (Zero Age Main Sequence). Because of the enormous importance of the study of the evolution of the Sun, we give separately in Fig. 14.2 the theoretical pre-main sequence evolution of the Sun after D. Ezer and A.G.W. Cameron (1966). A feature common to all the tracks in these figures may be noticed in that with the development

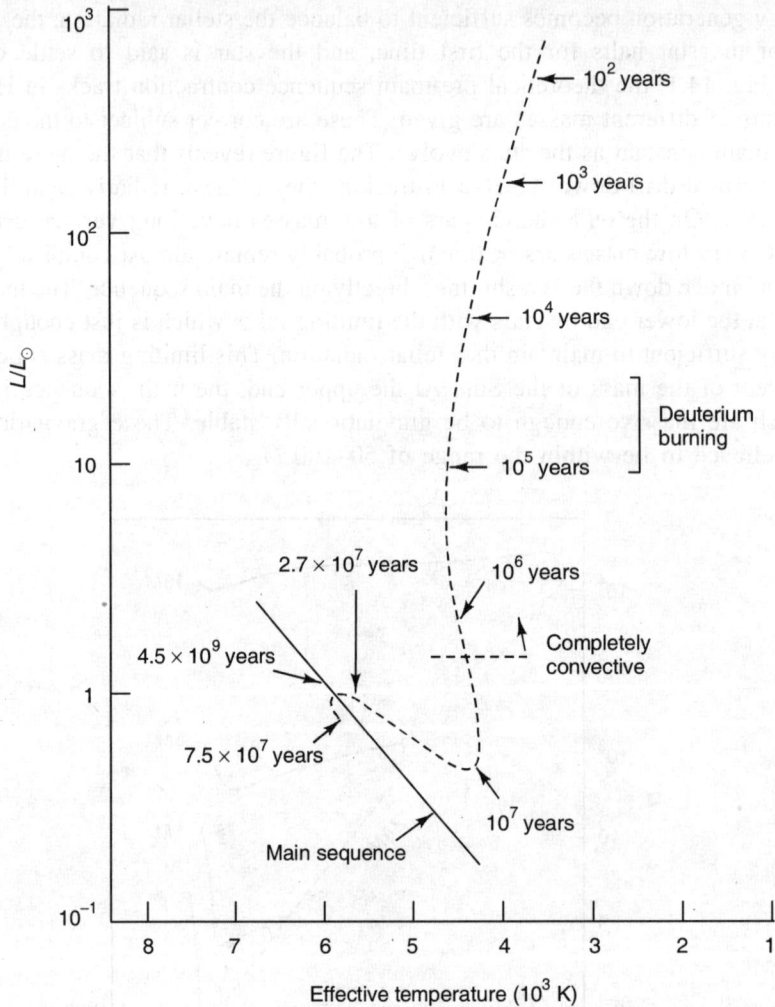

FIGURE 14.2 The path on the H–R diagram of the Sun as it contracts to the main sequence. (After D. Ezer and A.G.W. Cameron, *in Stellar Evolution*, R.F. Stein and A.G.W. Cameron (Eds), Plenum Press, New York, 1966.)

of the radiative core just after the end of the complete convective equilibrium, the stars move somewhat up and to the left before reaching the main sequence, implying that both luminosity and temperature of the stars increase at this phase. Just before settling on to the main sequence with the commencement of the nuclear energy generation, the star's luminosity and surface temperature slightly fall down as indicated by small hooks in the evolutionary tracks.

The star is now on the main sequence maintaining an almost static structure for a large part of its life. It is now smoothly fed by its nuclear energy source and no further contraction is required. As we have already mentioned, in massive stars ($M > 1.5\ M_\odot$) the chief mechanism of hydrogen burning is the CNO bicycle which is steeply temperature dependent. So a convective core is formed in these stars in order to carry energy quickly enough from the core outwards.

The surface layers of these stars, however, continue to maintain radiative equilibrium. Stars of masses similar to that of the Sun and lower than that, on the other hand, produce energy by proton-proton chain whose temperature dependence is relatively flat and so radiation alone is sufficient to carry necessary energy outwards from the core. But the initial convection still persists in the surface layers of these stars. Thus during their main sequence life the massive stars (upper main sequence) have convective core and radiative envelope, while the low mass stars (lower main sequence) have radiative core and convective envelope.

The main sequence life time of a star depends chiefly on two factors: the *luminosity* of the star and the *time* taken to convert upto a suitable fraction f of hydrogen into helium at the core. The star will not evolve away from the main sequence till the fraction of hydrogen depleted at the core exceeds f. If X be the fraction of hydrogen by weight in the initial composition, then the main sequence life time of a star of mass M and luminosity L is given by:

$$T_M = 1.1 \times 10^{11} \, fX \; \frac{M/M_\odot}{L/L_\odot} \; \text{years} \qquad (14.50)$$

where we have used the fact that an amount of 6.4×10^{18} erg of energy is obtained from burning of one gram of hydrogen. The value of f is, of course, riot exactly defined. It varies with the initial mass of the star. For stars of masses around that of the Sun, $f \sim 0.15$. Using this value, Eq. (14.50) becomes

$$T_M = 1.6 \times 10^{10} \, X \; \frac{M/M_\odot}{L/L_\odot} \; \text{years} \qquad (14.51)$$

whence the main sequence life time of the Sun is given by $16 \, X$ billion years. The actual value depends on X, the abundance of hydrogen in the initial composition.

While discussing the evolution away from the main sequence, we shall tacitly assume that we are considering only the upper main sequence stars, because no lower main sequence Population I star has been old enough to evolve from main sequence. The theoretical post-main sequence evolutionary tracks for some stars are given in Fig. 14.3. The structural and physical changes characterizing the basic features of this evolution are now believed to be well understood. The principles of hydrostatic equilibrium and the Virial theorem guide, in fact, the various stages of evolution.

As the hydrogen is depleted at the core with the build-up of helium and other heavier trace elements, the mean molecular weight μ increases. So the gas pressure is reduced in the hydrogen-depleted core. The core has therefore to contract slowly in order to hold the pressure of the overlying layers. Thus while the hydrostatic equilibrium is reinstated, the core heats up further. This heating of the core is accompanied by slow expansion of the star to increase its radius from the initial main sequence value. The star therefore begins to evolve from the zero-age main sequence. At first, the star remains confined for sometime within the main sequence band. As the hydrogen continues to be depleted, the nearly isothermal helium core grows in size, while the core contraction and its consequent heating continues. With the increase of the core temperature, the temperature in the hydrogen shell source also rises which therefore maintains the required amount of energy generation. The rise in internal temperature causes

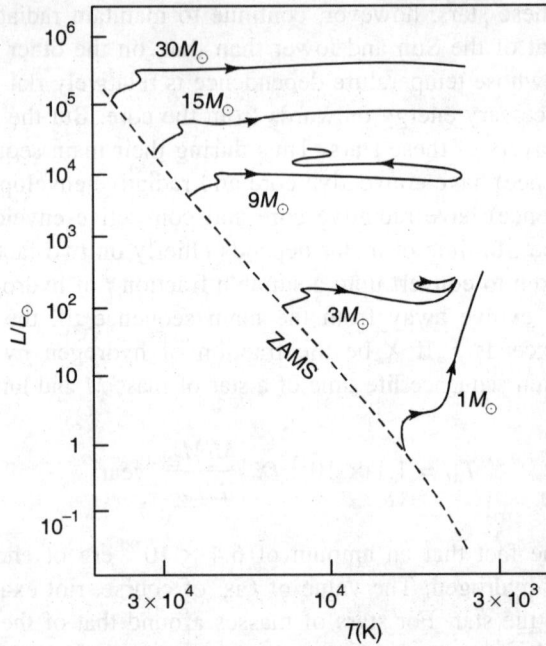

FIGURE 14.3 Evolutionary tracks of post-main sequence stars of different masses on the H–R diagram.

the radius of the star to increase rather rapidly in order to check the high value of the temperature gradient. The surface temperature of the star therefore falls. During the expansion the luminosity of the star first rises, moving it in the H–R diagram to the right and upwards. The star thus becomes a red *subgiant*. With further expansion, the surface layers of the star significantly cool down which causes the opacity mainly due to ionized metals to rise. Also, the rate of energy production may decrease for a certain period of this phase of evolution. The star's luminosity therefore may decrease for certain period of this expanding phase. This trend is however more than counter balanced soon due to the development of another new situation. With the capture of electrons by ionized metals the electron density decreases which greatly reduces the rate of formation of H⁻ ion which is the principal source of opacity in stars other than highly massive ones. The surface layers of these stars therefore become largely transparent causing their luminosity to rise sufficiently. As can be seen from Fig. 14.3, this rise is very substantial in stars whose masses do not greatly differ from that of the Sun, because H⁻ opacity is very high in these stars. The effect systematically diminishes with the increase of the mass of the star. At this phase both the radius and the luminosity of the star continue to rise. The star thus becomes a *red giant* or *red supergiant*. In order to feed the increased luminosity of these stars, the rate of energy transport from the interior must be high. Radiative energy transport becomes very much insufficient in the subphotospheric layers where opacity is still very high. Convection currents are therefore brought into play in these layers, whose thickness continues to increase with the increase of the radius of the star. Red giants, therefore, must have *deep convective layers* in the envelope.

The movement of the star to the right and upward is believed to continue till the onset of helium burning at the core by the 3-alpha process. The core had been contracting all the time with the consequent rise in its density and temperature. This continued rise in density and temperature ushers in the new phase of the star's evolution. When the temperature rises to the order of 10^8 K, helium burning commences. The star now produces energy at the core by helium burning, and in the shell surrounding the core the energy continues to be produced by hydrogen burning. In low-mass stars the central density becomes so high by the time as to produce a degenerate electron gas supporting the helium core. This causes the helium burning process to engulf the entire core within a few seconds. This phenomenon has been called *helium flash*. Because of the high temperature dependence of the energy generation by 3-alpha process, the core becomes again convective with the commencement of it when the star contracts again.

The subsequent evolutionary changes are somewhat poorly known. Depletion of helium at the core gives rise to a carbon core which continues to grow. This core also continues to contract for the same reason as that for the contraction of the helium core, and the star continues to expand for a second time. The expansion probably continues until the central temperature becomes sufficient for carbon burning. With the onset of carbon burning (most probably in a flash again) due to the presence of degenerate electron gas in carbon, the core becomes convective and the star undergoes contraction again. This may again be followed by oxygen burning with the repetition of the similar set of circumstances. But each subsequent energy generation process is characterized by lower amount of total energy produced, so that the time-scale during which the star can be fed with these processes are progressively lower. The star therefore has to repeatedly rely on gravitational energy in critical phases which is released by contraction. The star is therefore ultimately squeezed to a mass of degenerate gas, unable to produce any more energy either by nuclear burning or by gravitational contraction, leaving no other way for it but to die. The end product of active stellar life is the white dwarf which will be discussed in Section 14.8. The detailed changes undergone by a star during the last phases of its evolution are only poorly known. The star probably evolves through the horizontal branch, crosses the main sequence to the left and then moves downward to the white dwarf branch.

14.8 WHITE DWARFS

When all the nuclear fuel of a star has been exhausted by continual demand in course of its various stages of post-main sequence evolution, the only other source of energy left for it is the gravitational energy released by contraction. But it had already to contract repeatedly in order to be able to ignite the burning of elements heavier than hydrogen thereby raising its average density already to a high value. After the exhaustion of the last nuclear fuel, the star in a desperate attempt to 'live' as it were, further undergoes very high degree of contraction to release its gravitational energy. The collapse raises its temperature slowly but since the radius also decreases, no overall increase in its luminosity takes place. The density at the interior which was already high, increases further. The equation of state for perfect gas still holds upto a density of 10^2–10^3 gm cm^{-3}. With further increase in the density the degeneracy of the electron gas sets in. At densities higher than about 10^6 gm cm^{-3}, the degeneracy starts also in heavy nuclei and the relativistic effects are brought into play in the already degenerate electron gas. The amount of density that will prevail in the interior of the star is, however, determined by the mass of the star. Stars having average density in the range of 10^5–10^6 gm cm^{-3} have actually been observed. They must be in the final phase of a stellar life, having exhausted all their nuclear sources of energy, as well as their capacity to release gravitational energy by contraction. These are therefore the stars which are just radiating their thermal energy, and so are cooling to meet their final "death". These stars are called *white dwarfs*. The white dwarfs are therefore stars of average density in the range 10^5–10^6 gm cm^{-3} in which the electron gas is completely degenerate except in outer layers whose total thickness is very small compared to the radius of the star.

In such a degenerate configuration, the structure of the star is held in equilibrium by pressure of electrons. The pressure due to the nuclei, as well as the radiation pressure are negligible in comparison. Further, as has been already stated in Section 14.3, the temperature becomes irrelevant here in the equation of state, since under the prevailing conditions the transport of energy by electron conduction becomes highly efficient and plays the principal role in energy transfer. When electron conduction is the main energy transport mechanism, the role of temperature gradient becomes completely suppressed. The gas, therefore, quickly becomes isothermal. Most part of the interior of a white dwarf is degenerate and isothermal. Only in a thin outer layer, the gas is non-degenerate and energy is transported by radiation. Within the degenerate interior the equations of state given by Eqs. (14.6) and (14.8) with Eq. (14.7) hold in completely non-relativistic and completely relativistic cases respectively. In the transition region, the pressure and density are found to be given by.

$$p_e = \frac{8\pi m_e^4 c^5}{3h^3} f(x) \tag{14.52}$$

and

$$\rho_e = \frac{8\pi m_H m_e^3 c^3}{3h^3} x^3 \mu_e \tag{14.53}$$

where μ_e is given by Eq. (14.7) and $f(x)$ is the function, of

$$x = \frac{p_0}{m_e c} \tag{14.54}$$

defined by

$$f(x) = \frac{1}{8} [x(2x^2 - 3)(x^2 + 1)^{1/2} + 3 \sin h^{-1} x] \tag{14.55}$$

p_0 being the threshold momentum. Equations (14.52) and (14.53) together constitute the *equation of state for a partially relativistic degenerate gas*.

These equations can be utilized to investigate the equilibrium structure of white dwarfs. Since the equation of state is independent of temperature, the problems of the white dwarf structure can be considered in two separate parts, viz. the hydrostatic (mechanical) and thermodynamic (physical) ones.

The hydrostatic equilibrium conditions are still given by Eqs. (14.10) and (14.9) taken together. For a given composition μ_e is fixed. The most obvious choice is $X = 0$, $\mu_e = 2$, because hydrogen is supposed to be totally depleted in white dwarfs. One can now numerically integrate Eqs. (14.10) and (14.9) with the use of the equation of state defined by Eqs. (14.52) to (14.55) right from the centre of the star to its outer boundary where the pressure becomes zero, and calculate the values of M and R, the masses and radii, for different assumed central densities ρ_c. For any composition and central density, therefore, a unique solution can be obtained for the mass and radius of a white dwarf. In fact, this had been done long ago (early 1930s) by Chandrasekhar for $\mu_e = 2$, and various central densities. We enumerate his results in Table 14.1.

TABLE 14.1 Mass-Radius Relation for White Dwarfs Uniquely Determined by the Choice of Central Densities*

$\log p_c$	$M/M_\odot$	$\log R/R_\odot$
5.39	0.22	−1.70
6.03	0.40	−1.81
6.29	0.50	−1.86
6.56	0.61	−1.91
6.85	0.74	−1.96
7.20	0.88	−2.03
7.72	1.08	−2.15
8.20	1.22	−2.26
8.83	1.33	−2.41
9.29	1.38	−2.53
∞	1.44	∞

*After S. Chandrasekhar, *Stellar Structure*, p. 427, The University of Chicago Press, Chicago, 1938.

Two remarkable points are noticed in the Table. First, the complete degenerate configuration of a white dwarf is rigidly restricted by an upper limit to its mass. Even if the central density is infinite, the upper limit of the mass of the completely degenerate configuration cannot exceed

$$M_c = \frac{5.75}{\mu_e^2} M_\odot \tag{14.56}$$

which is known as the *Chandrasekhar mass-limit* for white dwarfs. The second remarkable point to be noticed is that the radius of a white dwarf decreases as its mass increases. For white dwarfs of masses around that of the Sun, the radius is about one per cent of the solar radius, while the radius of a white dwarf with its mass near the Chandrasekhar limit is vanishingly small. Since $\bar{\rho}_\odot \sim 1.4$ gm cm^{-3}, the average density of a white dwarf of one solar mass is $\sim 1.4 \times 10^6$ gm cm^{-3}.

The upper mass limit M_c obtained by Chandrasekhar about six decades ago, is however, only approximate, because in his computation a few points of correction have been neglected that should otherwise be applied to the white dwarf configuration for more accurate results. The first correction to be applied arises out of a little change that takes place in the number of quantum mechanical states which may be occupied by the electrons, under the condition of a prevailing electrostatic field generated by a kind of charge separation in the degenerate gas. The charge separation and a consequent electrostatic field is produced due to the fact that the gravitational force acts mainly on the nuclei whereas the reverse pressure force acts mainly on the electrons. The prevailing electrostatic field to a small extent, changes the number of quantum mechanical states mentioned above, a correction for which should be made. The second correction requires to be made in cases of high density. In extremely high densities (e.g. $\rho > 10^8$ gm cm^{-3}) some electrons may be forced on to the nuclei thereby increasing the value of μ_e. The effect of the neglect of these two corrections is to increase the mass of the white dwarf configuration by upto about 20 per cent. When these corrections are applied, the masses in the last few models in Table 14.1 will be reduced by some extent and the Chandrasekhar mass limit M_c (the last mass in Table 14.1) becomes about 1.2 $M_\odot$. Stars heavier than this limiting mass cannot have stable degenerate configuration. In such stars, the degenerate electron pressure will be unable to hold the onslaught of the gravitational force and the star will further collapse to form a neutron star as discussed in the next Chapter. The collapse will not be checked for a sufficiently larger mass even if the star possesses high amount of uniform rotation. The upper mass limit will, however, be much larger (≥ 4 $M_\odot$) only in the presence of high degree of differential rotation of the star. But the existence of such rapidly rotating white dwarfs has not been observationally confirmed. Although the number of white dwarfs with accurately known mass is small, the observational evidences gathered so far show that no white dwarf has so far been found to possess a mass in excess of the limiting value. In fact, the three white dwarfs, viz., Sirius B, 40 Eri B and Procyon B which are members of binary systems and whose masses are most accurately known, have all their masses below the Chandrasekhar limit. These masses are 0.45 $M_\odot$, 1.1 $M_\odot$ and 0.6 $M_\odot$ respectively. These observations confirm that the white dwarfs are examples of completely degenerate configuration in hydrostatic equilibrium.

From the fact that the white dwarfs are in stable configuration having a degenerate, nearly isothermal interior surrounded by a nondegenerate thin envelope in radiative equilibrium, and having no source of energy except the thermal energy of the nondegenerate nuclei, it is possible to calculate the temperature of the degenerate interior and the time required for *cooling* of the star of any given mass and luminosity. Since the interior is isothermal, the temperature of the degenerate interior is the same as at its boundary, which can also be treated as the transition layer. It can be shown that for a star of mass M with its chemical composition given by $Y = 0.9$, $Z = 0.1$ (so that $X = 0$), the relation between its luminosity, mass and temperature at the transition layer is given by

$$L = 2 \times 10^6 \ M/M_\odot \ T_{tr}^{3.5} \tag{14.57}$$

where an average guillotine factor of 10 has been used. The relation between the density and temperature at the transition layer is given by

$$\rho_{tr} = \frac{\pi \, \mu_e m_H}{3 \, h^3} \ (20 \, m_e k T_{tr})^{1.5} \tag{14.58}$$

Equations (14.57) and (14.58) can be used to compute the temperature and density at the transition layer of any star with known mass and luminosity. When the internal temperature is thus known, the cooling time for a star of mass M is approximately given by

$$t_{cool} = \left(\frac{3}{2} \frac{kMT_{tr}}{m_H \mu_A} \right) \Big/ (2.5\,L) \tag{14.59}$$

where μ_A is the mean molecular weight of the nuclei and has the value 4.44 for the composition chosen above. The cooling time calculated from the above formula is found to be so long that it is almost comparable to the total life time of the star upto its pre-white dwarf phase of evolution. The internal temperature and cooling times of white dwarfs having mass of 1 $M_\odot$ and a range of representative luminosities are shown in Table 14.2.

How do the white dwarfs move in the H–R diagram as they cool? In other words, what are their cooling curves like when depicted in the H–R diagram? The answer is rather simple. Since during cooling their radii do not change while their temperaturelind luminosity continue to decrease with time, they will move to the right and downward along the lines of constant radii as shown in Fig. 14.4.

TABLE 14.2 Internal Temperatures and Cooling Times of White Dwarfs

$(L/L_\odot)$	$T_{tr}(K)$	$\log \rho_{tr}$	t_{cool} (years)
10^{-2}	17×10^6	3.5	0.3×10^9
10^{-3}	9×10^6	3.1	1.6×10^9
10^{-4}	4×10^6	2.6	8×10^9

*Adapted from M. Schwarzschild's *Structure and Evolution of Stars,* 1958, Princeton University Press, p. 238.

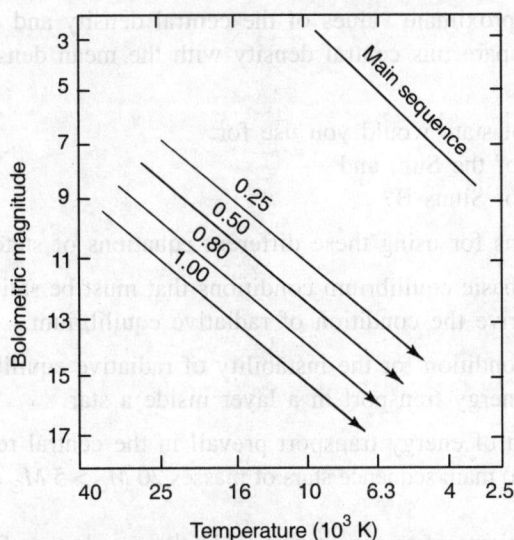

FIGURE 14.4 Evolutionary tracks of white dwarfs on H–R diagram. The numbers indicate the white dwarf masses.

The cooling curves of white dwarfs therefore, represent their evolutionary tracks on the H–R diagram. Along one such line a white dwarf will be destined to move as it cools, until it radiates all its thermal energy to become an invisible *black dwarf,* thereby ending its long luminous life.

EXERCISES

1. Assuming that 20 per cent of the mass of the Sun undergoes transmutation from H to He during its lifetime, calculate how long the Sun will shine maintaining its present luminosity. Compare this time with the *Kelvin Contraction time.*

2. For the Sun, $M_{\odot} = 2 \times 10^{33}$ gm and $L_{\odot} = 4 \times 10^{33}$ erg s^{-1}: How much hydrogen is converted to helium every second in the Sun? How much matter will be annihilated every second? What will be the corresponding values at the end of Sun's life (assume 12 billion years)? Compare these latter values with the mass of the Sun.

3. Consider the following (hypothetical) stars composed entirely of hydrogen:

   (a) $M = 100\ M_{\odot}$, $L = 10^6\ L_{\odot}$;
   (b) $M = 10\ M_{\odot}$, $L = 4 \times 10^3\ L_{\odot}$;
   (c) $M = 0.5\ M_{\odot}$, $L = 0.1\ L_{\odot}$.

   If these stars convert all the hydrogen into helium, calculate the periods over which these will shine.

4. The luminosity of a star is 10^{40} erg s^{-1} and its mass is $100\ M_{\odot}$. If 1 per cent of the mass is used up in nuclear transmutation, calculate the period over which the star will shine.

5. Calculate the approximate values of the central density and the central temperature of the Sun. Compare this central density with the mean density of the Sun if $R_{\odot} = 7 \times 10^5$ km.

6. What equation of state would you use for
   (a) the interior of the Sun; and
   (b) the interior of Sinus B?

   Give your reasons for using these different equations of state.

7. Write down the basic equilibrium conditions that must be satisfied by a stable stellar structure and derive the condition of radiative equilibrium.

8. Investigate the condition for the instability of radiative equilibrium and setting in of the convective energy transport in a layer inside a star.

9. What mechanism of energy transport prevail in the central regions of the following classes of stars; (a) main sequence stars of masses $20\ M_{\odot} > 5\ M_{\odot}$ and $0.5\ M_{\odot}$; (b) the Sun; (c) a red giant?

   What mechanisms of transport prevail in the envelopes of these stars? Give your answers with physical explanation in each case.

10. In which stars energy transport by conduction is important? Give reasons for your answer.

11. Describe the two kinds of nuclear reactions that transmute hydrogen into helium in the interior of a star. On what factors do these reaction rates depend?

12. Derive the nuclear reaction rate in the form

$$r = C_1 \rho^2 X_1 X_2 T^{-2/3} \exp\left[-3\left(\frac{C_2 Z_1^2 Z_2^2}{kT}\right)^{1/3}\right]$$

and write the expression for C_2.

13. Describe the reactions that transmute helium to carbon in the stellar interior and obtain the formula for the rate of energy generation in this process.

14. Compare the temperature sensitivity of the energy generation by pp, CNO and triple α processes. Describe in which stars these processes are important.

15. List the different nuclear reactions which may be present in the Sun. Discuss the solar neutrino puzzle.

16. Briefly trace the evolutionary history of the following stars from their birth from an interstellar cloud of gas and dust to their final demise:

    (a) $M = 10\ M_\odot$,
    (b) $M = 1\ M_\odot$,
    (c) $M = 0.1\ M_\odot$.

17. A white dwarf has an apparent visual magnitude equal to 10.5 and a parallax of 0".1. Find its absolute visual magnitude and compare it with that of the Sun.

    If the temperature of this white dwarf be 20,000 K, find its radius and compare this with that of the Earth.

18. Assuming $M = 1\ M_\odot$ and $L = 10^{-3}\ L_\odot$ for a white dwarf whose temperature is 30,000 K, find its radius and average density.

SUGGESTED READING

1. Böhm-Vitense, Erika, *Introduction to Stellar Astrophysics,* Vol. 3, *Stellar Structure and Evolution,* Cambridge University Press, Cambridge, 1992.

2. Chandrasekhar, S., *An Introduction to the Study of Stellar Structure,* Dover Publications, New York, 1957.

3. Clayton, Donald D., *Principles of Stellar Evolution and Nucleosynthesis,* McGraw-Hill Book Company, New York, 1968.

4. Cox, John P. and Guili, R. Thomas, *Principles of Stellar Structure,* Vols. 1 and 2, Gordon and Breach Science Publishers, New York, 1968.

5. Hansen, C.J. and Kawaler, S.D., *Stellar Interiors,* Springer-Verlag, Berlin, 1994.

6. Iben, Icko (Jr.), Stellar Evolution: Within and Off the Main Sequence, *Annual Review of Astronomy and Astrophysics,* 5, pp. 571–626, 1967.

7. Kippenhahn, R. and Weigert, A., *Stellar Structure and Evolution,* Springer-Verlag, Berlin, 1990.

8. Kurganoff, V., *Introduction to Advanced Astrophysics,* D. Reidel Publishing Company, Dordrecht, Holland, 1980.

9. Rose, William K., *Introduction to Astrophysics,* Holt, Rinehart and Winston, New York, 1973.

10. Schatzman, E.L. and Praderie, F. (Translated by A.R. King), *The Stars,* Springer-Verlag, Berlin, 1993.

11. Schwarzschild, Martin, *Structure and Evolution of the Stars,* Dover Publications, New York, 1965.

12. Ünsold, A. and Baschek, B., *The New Cosmos,* Springer-Verlag, Berlin, 1991.

15

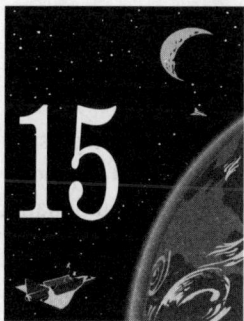

Neutron Stars and Black Holes

15.1 DISCOVERY OF PULSARS

The first *pulsar* was discovered in 1967, quite accidentally, by Anthony Hewish and his co-workers notably, Miss J. Bell, at Cambridge. Hewish and his fellow workers were engaged in studying interplanetary scintillation exhibited by radio sources of small angular size (usually less than 1" of arc) with a new radio telescope at Cambridge operating at a frequency of 81.5 MHz, that was set for the purpose. It is known that changes in interplanetary scintillation, when a radio source is observed at various angles with respect to the Sun, give information about the angular size of such a source. While carrying out the programme, Hewish and his co-workers came across an invisible object in the sky emitting sharp pulses of radio waves at exactly spaced intervals of time. Careful search revealed soon after, the presence of a few more such objects in the sky. They were subsequently given the name pulsar, of which the first to be discovered was Cp 1919 (meaning Cambridge pulsar at R.A. 19^h 19^m). So exact was the periodicity of emitted pulses from these objects that some astronomers could not resist the temptation of thinking that these pulses might possibly be transmitted into space by intelligent beings on other planetary systems. Such a fancy, however had to be soon given up, as more and more pulsars were discovered and observational data on them began to pour rapidly which enabled scientists to understand the basic nature of these mysterious objects. The discovery of these objects escaped so long because they emit strongly in longer wavelength range where radio observations are generally not favoured and because their pulse widths are much shorter than the integration time generally used for radio observations.

Earlier, pulsars were denoted by their provisional names based on a combination of their positions in right ascension and observatories where discovered. For example, the first pulsar discovered was named CP 1919 as explained above. Now, for the sake of uniformity the prefix PSR is used for all the pulsars followed by a four digit number indicating right ascension (in 1950.0 coordinates) together with a sign and two digits indicating declination. If required, a third digit is added to cover tenths of a degree in declination. Thus, for instances, the first pulsar CP 1919 is now denoted by PSR 1919+21. Two other very closely situated pulsars are denoted by PSR 1913+16 and PSR 1913+167. However, the provisional names are still very much in use. For example, the Crab pulsar PSR 0531+21 is still referred as NP 0532, its provisional name.

15.2 ROTATING NEUTRON STAR MODEL OF PULSARS

Various observational facts and theoretical consideration now lead us to believe that pulsars are *rapidly rotating neutron stars*. The current theory predicts that when a supernova explodes and its surface layers are blown off, the central core of mass of the order of 1 $M_\odot$ gravitationally contracts to a superdense body. It is not known whether this happens in every supernova explosion. In some explosions the original stellar bodies may be completely disintegrated leaving behind only gaseous debris. In some others, where the original stars are highly massive, the cores after explosions may shrink to black holes. Whatever may be the case, the present observational status definitely supports the theory that at least in some supernova explosions the cores contract to superdense neutron stars. Astronomers believe that these neutron stars have been observed as pulsars. They are objects with masses of the order of 1 $M_\odot$ and with radii of the order of 10 km, which yield a density of its material of the order of 10^{14} gm cm^{-3}.

The possibility of existence of neutron stars was theoretically predicted many years ago, independently by L. Landau and J.R. Oppenheimer. We have seen that any star whose mass exceeds the Chandrasekhar limit of 1.44 $M_\odot$ may explode as a supernova at a certain stage of its evolution. Stars whose masses are less than this limit, however, evolve to meet their end peacefully. In the final stage, by gradual contraction, the gas attains the state of degeneracy, and the star is called a *white dwarf*. The degenerate material of a white dwarf can exert sufficient pressure force to resist further gravitational collapse so that the white dwarf structure remains as a stable stellar structure. At this stage, the density attained by the material is very high. A star of mass 1 $M_\odot$ has, by gradual contraction, attained the size of the Earth leading to a matter density of 2×10^6 gm cm^{-3}. But the fate of the stars whose masses lie in the range 2 $M_\odot$ to about 10 $M_\odot$ or even greater, may be quite different. If such a star explodes, the surface layers are blown off in the interstellar space while, in the core, the gravitational pull may be so strong as to force the free electrons to stick on to protons thus producing neutrons. In this process, free electrons are depleted thereby decreasing the resisting pressure of free electrons to the gravitational pull. As a result, further gravitational contraction proceeds more or less smoothly. This contraction can be halted only when it is balanced by the opposing pressure of degenerate neutrons. At this stage, the matter density has reached the nuclear density which is 10^{14} gm cm^{-3}, as has been already mentioned. We now have a stable neutron star whose mass may lie somewhere between 0.7 to 2 $M_\odot$.

The intense gravitational contraction of the core necessarily leads to a corresponding increase in its rate of rotation and in the magnetic field. The material in the stellar cores is highly conducting and the theory says that in such material the magnetic lines of force become *frozen-in*. The magnetic lines of force therefore become packed within a small area as the star violently contracts to a small volume. The extreme contraction can therefore create a magnetic field of the order of 10^{12} gauss on the neutron star, even if the original magnetic field of the star before explosion be only a few gauss. Again, all normal stars possess at least some amount of rotational velocity (this velocity is 2 km s^{-1} for Sun and as high as 400–500 km s^{-1} in some O–B stars). Therefore, if a substantial part of the angular momentum possessed by the original star before explosion is retained by the contracting core, when its radius has decreased by about 10,000 times, its rotation in a few milliseconds is possible. Thus, if pulsars are actually such collapsed cores of supernovae, then the theory predicts that *pulsars are rapidly rotating neutron stars possessing an intense magnetic field.*

15.3 PERIOD DISTRIBUTION AND LOSS OF ROTATIONAL ENERGY

Pulsars emit radio pulses at extremely regular intervals of time. The periods in different pulsars range from 0.033 second for the pulsar NP 0532 (PSR 0531 + 21) (this is situated at the centre of the Crab Nebula, so called Crab pulsar) to 3.745 seconds for NP 0527 (PSR 0525 + 21) (the pulsar closest to NP 0532), corresponding to a factor of about 120. The radio pulses emerge in *trains* having duration for most pulsars lying in the range 10 to 50 milliseconds. There are fine structures (secondary pulses) within each pulse (primary pulse), a complete pulse being the envelope of these subpulses. The schematic diagrams of the primary and secondary pulses of a typical pulsar are shown in Fig. 15.1.

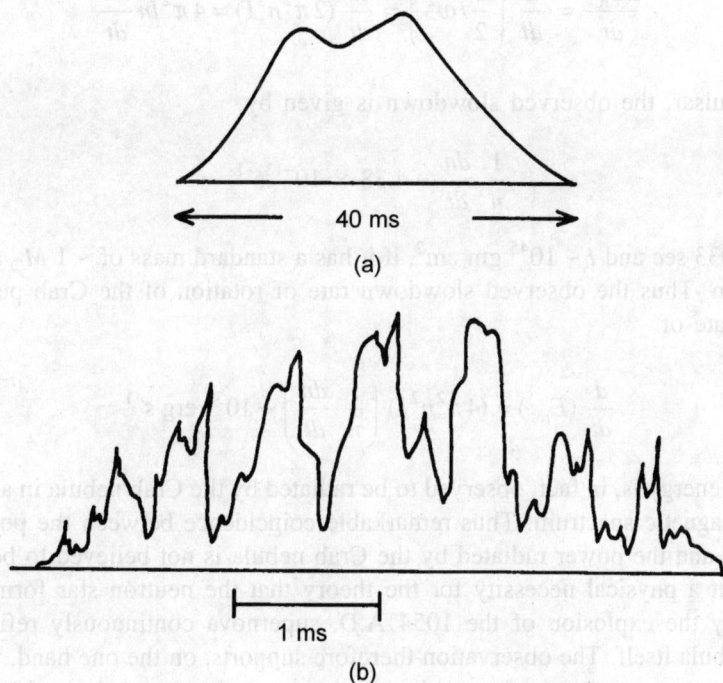

40 ms

(a)

1 ms

(b)

FIGURE 15.1 Schematic diagram of the primary pulse (a) and secondary pulses (b) of a typical pulsar.

A primary pulse may contain within it 20 to 30 or more subpulses each of a duration usually much less than a millisecond. The interval between any two consecutive primary pulses is repeated so exactly that it may be regarded as constant to 1 part in 10^9, although the amplitude may substantially vary from one pulse to another.

One of the most interesting observations on pulsars which has subsequently been proved to have great significance in understanding their, basic nature, was made by Frank Drake and his co-workers in Aricibo Observatory. They found that the rotation of the Crab pulsar was decreasing at the rate of nearly 1 in 2400 per year. Since the period of this pulsar is 0.033 sec, the observation implies that the period P of the pulsar was increasing at the rate of $\Delta P/P \sim 4.38 \times 10^{-13}$. Later investigations revealed that periods of all the pulsars are increasing with rates varying in the range 10^{-13} to 10^{-16}.

The observation of the rate of increase of pulsar periods has attributed great significance to the rotating neutron star hypothesis for pulsars. A body rotating with an angular speed ω (radians per sec) about an axis possesses a rotational energy of $(1/2)\omega^2$, where I is the moment of inertia of the body about the rotating axis. If n be the number of rotations per sec (frequency), then $\omega = 2\pi n$ so that, the rotational energy of the body in terms of the frequency becomes $2\pi^2 n^2 I$. For a neutron star with a period of rotation equal to 10 millisecond, a mass of 1 $M_\odot$ and a radius of 10 km, the rotational energy is $\sim 10^{50}$ erg, which is approximately equal to the energy liberated by a solar-type star during its entire lifetime. The rate of change of rotational energy of a neutron star is,

$$\frac{dE_{\text{rot}}}{dt} = \frac{d}{dt}\left(\frac{1}{2}I\omega^2\right) = \frac{d}{dt}(2\pi^2 n^2 I) = 4\pi^2 In\frac{dn}{dt} \tag{15.1}$$

For the Crab pulsar, the observed slowdown is given by

$$\frac{1}{n}\frac{dn}{dt} \sim 4.38 \times 10^{-13}\text{s}^{-1} \tag{15.2}$$

where $1/n = 0.033$ sec and $I \sim 10^{45}$ gm cm^2, if it has a standard mass of ~ 1 $M_\odot$ and a standard radius of 10 km. Thus the observed slowdown rate of rotation of the Crab pulsar generates energy at the rate of

$$\frac{d}{dt}(E_{\text{rot}}) = (4\pi^2 n^2 I)\left(\frac{1}{n}\frac{dn}{dt}\right) \sim 10^{38}\text{ erg s}^{-1} \tag{15.3}$$

This amount of energy is, in fact, observed to be radiated by the Crab nebula in all wavelengths of the electromagnetic spectrum. Thus remarkable coincidence between the power output by the Crab pulsar and the power radiated by the Crab nebula is not believed to be mere chance coincidence, but a physical necessity for the theory that the neutron star formed within the Crab Nebula by the explosion of the 1054 A.D. supernova continuously refills the energy spent by the nebula itself. The observation therefore supports, on the one hand, the theory that pulsars are rotating neutron stars formed by supernova explosions and, on the other, that the neutron stars thus formed continuously supply streams of relativistic particles to the surrounding nebulae which maintain the power flux from them at various wavelengths. The energy lost in radiation by the nebula may be replenished by the rotational energy of the pulsar, provided an efficient mechanism of conversion of rotational energy to high energy particles actually exists. It is believed that such conversion efficiency can be achieved in a *rapidly rotating intense magnetic field* as is believed to be present in pulsars. Rapid rotation and intense magnetic field may combine to create large-scale electric fields that accelerate charged particles to very high energy. These high energy particles stream into the nebula where they emit radiation in various wavelengths by synchrotron mechanism.

Unfortunately, the direct observational verification of the rotating neutron star model of pulsars can be applied only to the Crab pulsar, whose rotational energy loss gives a satisfactory measure of the electromagnetic energy radiated by the Crab Nebula. This is the only pulsar which has been optically identified with a small star-like object situated within the supernova remnant, namely, the Crab Nebula. Another pulsar that has been found to lie within a supernova remnant is the pulsar PSR 0833 within Vela X, but no optical identification has been made for it.

15.4 TEST OF ROTATING NEUTRON STAR MODEL OF PULSARS

Let us now investigate the soundness of the rotating neutron star theory for pulsars and consider what other models could possibly simulate the observed pulsar properties. The clocklike precision of pulse periods shows that the source of the pulse could not be a planetary object or a member of a binary system. For in that case, the periodicity would be changed due to Doppler shift caused by orbital motion. The periods will be P_1 and P_2 when the source of period P is moving toward or away from the observer respectively, with a relative velocity v, where

$$P_1 = P \left[\frac{1 - \dfrac{v}{c}}{1 + \dfrac{v}{c}} \right]^{1/2}$$
(15.4)

and

$$P_2 = P \left[\frac{1 + \dfrac{v}{c}}{1 - \dfrac{v}{c}} \right]^{1/2}$$
(15.5)

whence,

$$\frac{\Delta P}{P} = \frac{P_2 - P_1}{P} = \frac{2v}{c}, \quad \text{if } \frac{v}{c} \ll 1.$$
(15.6)

Since the measured values of $\Delta P/P$ for pulsars lie in the range 10^{-13} to 10^{-16}, $v/c \sim 10^{-13}$ to 10^{-16}, that is, if a pulsar is a member of a binary or a planetary system, its orbital speed would lie in the range 10^{-3} to 10^{-6} cm^{-1} s, which is absurd.

There are two other possibilities, namely, pulsars are *rotating* or *vibrating* stellar objects. The very short periods of pulsars compel us to accept that they must be very *compact objects* with extremely stable structure and, at the same time, capable of generating sufficient amount of energy. The only likely candidates are, therefore, white dwarfs and neutron stars. We can now ask the question: can the pulsation or rotation of a white dwarf produce the observed periodicities in pulsars? Calculations reveal that the rotation period of a white dwarf in stable configuration cannot be as short as the observed pulsar periods. If a white dwarf rotates in such a short period, the equatorial rotational velocity is required to exceed the critical velocity of mass escape from the surface of the white dwarf, and as such, the stable configuration of the white dwarf breaks down. To explain the observed pulsar periods in terms of the pulsation of a white dwarf is also a difficult task. The median period of pulsars is 0.66 sec. It can be shown that the period of fundamental mode of vibration of a white dwarf cannot be much less than 2 seconds. So the fundamental mode of pulsation of a white dwarf cannot satisfactorily explain pulsar periods. Higher overtone pulsation may occur in much shorter periods comparable to observed pulsar periods. But such higher overtones, which are excited under very special circumstances, are unlikely to stabilize, so as to yield extreme precision of periods as are observed in pulsars. If overtone pulsations were the mechanism for pulsar periods, not only that the drift in periods would have been much more rapid, but also that the periods would have gradually decreased contrary to observed increase, because the mean radius of the source would decrease slowly. Thus, on the basis of above arguments the pulsating or rotating white dwarf model for pulsars can be discarded.

We are, therefore, left finally with a pulsating or rotating neutron star as the only possible candidate to satisfy the observed properties of pulsars. Since neutron stars are much more compact objects than white dwarfs, their rotation or fundamental pulsation periods must be much shorter than those of white dwarfs. But it is well known that the fundamental pulsation period of a self-gravitating configuration is of the order of its free-fall time which is equal to $1/\sqrt{G\rho}$ second when cgs units are used. For a neutron star with $\rho \sim 10^{14}$ gm cm^{-3}, this is $< 10^{-3}$ second. Also, in a neutron star, the pulsational modes are damped out much more rapidly than indicated by changes in pulsar periods. Thus, the pulsation of neutron stars as the manifestation of the observed pulsar periods can be ruled out. Only the rotating neutron star model can explain satisfactorily all the observed characteristics of pulsar periods.

Such a model then must also be able to explain the observed radiation characteristics of pulsars. Observations have revealed that pulse duration is only about 2–4 per cent of the repetition periods of pulsar. The width of the radiating beam associated with the Crab pulsar is about 12°. If this width is typical of pulsars, in general, then only 1/30th of the total area is involved in radiation and by the same token, only one pulsar among 30 will sweep the solar system with its radiating beam. This probably explains why, even if the theory of the production of pulsars by supernovae explosions be accepted as true, only two pulsars have been definitely identified within supernova debris. However, it may not be unlikely, that all supernovae explosions do not leave behind them a pulsar and that pulsars are produced only under some special conditions.

15.5 GOLD'S MODEL OF PULSARS

It was T. Gold who first suggested a plausible mechanism by which the observed radio pulses with clock-like precision in periods could be radiated by a rotating neutron star. Many other theories have since been proposed by astronomers to explain the mechanism of radiation by pulsars. The discussion of these theories is beyond the scope of this book. We shall here make a brief discussion of Gold's model which, with a few modifications can yield a plausible model for pulsar radiation. In the intense magnetic field of a pulsar, any charged particle that may escape from its surface will be constrained to move only along the magnetic lines of force. The assemblage of particles is thus whirled around with the angular velocity of the neutron star. In this manner, a co-rotating magnetosphere (consisting of frozen in plasma) is formed around the star. The tangential velocity of these whirling particles gradually increases as they move farther and farther from the stellar surface. At a certain distance, depending upon the speed of rotation of the star, the velocity of particles approaches that of light. The co-rotation will cease at a distance where the speed of particles attains the speed of light. The circle described by particles at this distance is called the *velocity of light circle* as shown in Fig. 15.2. The relativistic plasma beam near this circle will radiate radio waves perpendicular to the beam, but at the velocity of light circle where the particles have attained the speed of light, they will break away from the magnetosphere and flow out into the surrounding regions of space. The observations of the Crab Nebula suggest that these high energy particles after leaving the influence of the parent neutron star stream into the nebula to supply the "perennial" energy source for radiation from the nebula over the entire spectral range from X-ray to radio waves by synchrotron mechanism. This appears to explain satisfactorily as to how the Crab Nebula could maintain its high intensity of X-ray flux for about a millennium, since particles emitting X-radiation are known to lose their energy in a relatively short time.

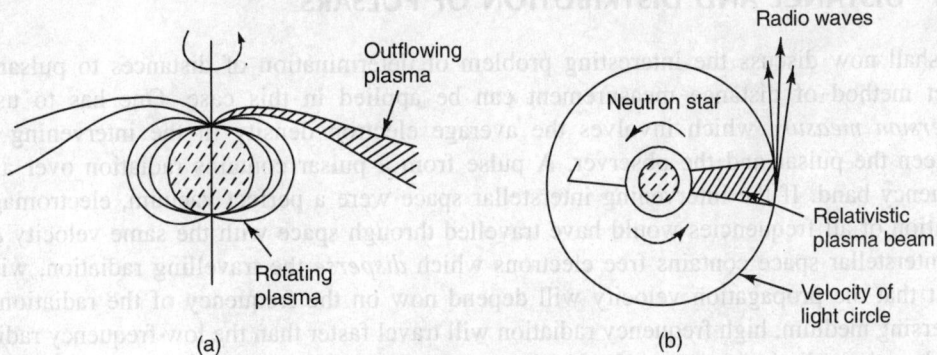

FIGURE 15.2 The rotating magnetosphere around a neutron star (a) and radio emission (b).

Turning back to Gold's model, it is deduced that the radius r of the velocity of light circle for the Crab pulsar NP 0532 with shortest known period (= 0.033 second) is ~1600 km, and that of the companion pulsar NP 0527 with the longest known period ($\approx$ 3.7 second) is ~ 160,000 km, as can be easily calculated from the simple formula

$$r = \frac{c}{\omega} = \frac{cP}{2\pi} \tag{15.7}$$

P being the period and c the speed of light. Near this circle the charged particles move along magnetic lines of force in helical paths producing synchrotron radiation. According to Gold magnetic activity will eject plasma from *only one* or *at most a few* places on the surface of the neutron star. The emitting particles are confined within a narrow cone. In fact, particles are confined in a narrow pitch angle $\theta \leq \tau/P$ in radian measure, where τ is the duration of a pulse and P is the pulsar period. Since the pulsars rotate with very high velocities, the radiation will be *strongly beamed* in a narrow cone in the forward tangential direction sweeping a particular region periodically, so that to a distant observer the pulsar behaves like a "light house" signalling radio waves.

According to this model, any asymmetry in the emission region may be responsible for the observed fine structure in pulses. Also, the observed fluctuations in pulse amplitudes may be due to the temporal changes in the interstellar medium through which the pulses propagate to great distances before reaching the Earth. It is found sometimes that the pulse structure remains fairly constant over a period of several pulses. This may be due to the fact that the interstellar medium remains unchanged during this period.

A modification of Gold's theory has been suggested by P. Goldreich and W.H. Julian in respect of the mechanism for particle acceleration to high velocities. These authors suggest that a rotating magnet will, due to unipolar induction, generate an electric field **E** in its vicinity which can cause charge separation and accelerate particles away from the surface to form a radiating magnetosphere. This theory predicts that acceleration of particles to energy as high as 10^{18} eV is possible in this way. Thus, whatever be the accelerating mechanism, such a mechanism is there, implying that pulsars may be a rich source of cosmic rays.

15.6 DISTANCE AND DISTRIBUTION OF PULSARS

We shall now discuss the interesting problem of determination of distances to pulsars. No direct method of distance measurement can be applied in this case. One has to use the *dispersion measure* which involves the average electron density in the intervening space between the pulsar and the observer. A pulse from a pulsar contains radiation over a large frequency band. If the intervening interstellar space were a perfect vacuum, electromagnetic radiation of all frequencies would have travelled through space with the same velocity c. But the interstellar space contains free electrons which *disperse* the travelling radiation, with the result that the propagation velocity will depend now on the frequency of the radiation. In a dispersing medium, high frequency radiation will travel faster than the low frequency radiation. So when the radiation from a pulsar is detected on Earth, receivers tuned to high frequencies will detect it earlier than those which are tuned to low frequencies. This time lag gives us clue to the distance to a pulsar when the average electron density over the intervening space is known. Solving the equation of motion of a free electron in the interstellar medium through which an one-dimensional electromagnetic wave of frequency ω and wave number κ propagates, one can derive the dispersion relation

$$\omega^2 = \omega_p^2 + \kappa^2 c^2 \qquad (15.8)$$

where

$$\omega_p = \left(\frac{4\pi e^2 n_e}{m_e} \right)^{1/2} = 5.6 \times 10^4 \sqrt{n_e} \text{ Hz} \qquad (15.9)$$

is the *plasma frequency* and n_e is the number density of electrons. If $\omega < \omega_p$, then κ becomes imaginary and the wave will not propagate. If $\omega > \omega_p$, then the electromagnetic wave propagates with group velocity

$$\frac{d\omega}{d\kappa} = V = \frac{c}{(1 + \omega_p^2/c^2\kappa^2)^{1/2}} \qquad (15.10)$$

If the pulse travels a distance R with this velocity, the arrival time is R/V. The frequency dependence of this arrival time is given by

$$\frac{d}{d\omega}\left(\frac{R}{V}\right) = \frac{R}{c}\frac{d}{d\omega}\left[1 + \frac{\omega_p^2}{\omega^2 - \omega_p^2}\right]^{1/2} \approx -\frac{R}{c}\frac{\omega_p^2}{\omega^3} \qquad (15.11)$$

where $\omega_p \ll \omega$.

We can now make a number of important deductions from Eqs. (15.9) and (15.11), together with the condition that $\omega > \omega_p$. First, by observing the cutoff frequency it is possible to find an upper limit to the average electron density $\langle n_e \rangle$ in the line of sight. Secondly, by substituting Eq. (15.9) in (15.11) we get

$$\frac{d}{d\omega}\left(\frac{R}{V}\right) \propto \frac{Rn_e}{\omega^3} \qquad (15.12)$$

where Rn_e is the total number of electrons within a column of 1 cm^2 cross-section in the line of sight between the source and the observer. Thus, the frequency dependence of the time lag of arrival of signals through the interstellar medium is *directly proportional* to the total number of electrons in a column of unit cross-section along the line of sight and *inversely* proportional to the cube of the frequency. The integrated number density of electrons in the line of sight, is called the *dispersion measure,* which is equal to

$$\int_0^R n_e\, dr = R\langle n_e \rangle, \qquad (15.13)$$

if $\langle n_e \rangle$ be taken as the average number density of electrons in the line of sight. Since the frequency dependence of the time lag is a measurable quantity, Eq. (15.11) can be used to determine the distance to a pulsar if the average electron density $\langle n_e \rangle$ in the line of sight is known. Conversely, if the distance to the pulsar can be known by some independent method, the value of $\langle n_e \rangle$ in the interstellar space can be derived. The mean value thus derived includes contributions both from electrons in the nebula, if any, surrounding the pulsar, as well as from those pervading the general intervening space. Results of extensive observation have indicated that the second contribution overwhelmingly predominates over the first. The value of $\langle n_e \rangle$, however, is not uniquely known and various values ranging from 0.01 cm^{-3} to 0.1 cm^{-3} have been used by authors for calculating the distances to pulsars, with the result that discordant results on pulsar distances abound in the literature. The problem of determining electron density in interstellar space is not easy. It depends on various factors that contribute to the ionization of the medium. Some of these factors are still only poorly known. Analysis of the available literature on the subject indicate that values of $\langle n_e \rangle$ between 0.02 to 0.03 cm^{-3} are most plausible.

How many pulsars are there in our Galaxy and how are they distributed? These are important questions which we would like to deal with briefly, at the end of this section. Uncertainties in the estimated average electron density in interstellar space have led to various assumptions by authors, thereby deriving distances to pulsars which differ by a factor of 4 to 5. However, if average of the estimated distances are used, it is observed that most of the pulsars lie within ~ 4 per cent of the total volume of the galactic disc. The currently observed number is ~ 400 and beaming of radiation puts a restriction to the effect that only one pulsar out every 30 is likely to sweep the earth. Thus, considering the very crude way of calculation, at present about 3×10^5 pulsars are likely to exist in our Galaxy. Approximately the same value is obtained if pulsars are assumed to originate in supernova explosions and supernovae statistics are used. If one supernova explosion every 30 to 40 years in the Galaxy is assumed, since the measured slowdown rate of rotation leads to a lifetime of ~10^7 years for a typical pulsar, about 3×10^5 pulsars are likely to exist in the Galaxy. Regarding the distribution of pulsars, if average of the estimated distances is used, pulsars are distributed with a tendency to concentrate near the galactic plane.

15.7   BINARY PULSARS

Since the first hundred pulsars discovered were single objects, astronomers were used to the idea that all pulsars would be single. So the discovery of the first binary pulsar PSR 1913+16 by R.A. Hulse and J.H. Taylor (1974) was received by astronomers with a great surprise. About a dozen binary pulsars are known to-date. Most of these are millisecond pulsars, the pulsar PSR 1855+09 having the shortest period of only 5.4 milliseconds. PSR 1913+16 has a period of about 59 milliseconds. This period was observed to vary by about one part in a thousand every 7^h45^m. This variation of the pulse periodicity is an indication of orbital velocities, thus confirming that the pulsar belongs to a binary system. The periodic variation of the pulse period is a very sensitive indicator of orbital motion. For most pulsars such variations have not been observed even with careful observations. Those, for which the variation has been observed are certainly binary systems.

The orbits of most of the binary pulsars are nearly circular, except those of PSR 1913+16 and PSR 2303+46 which are highly eccentric. The periods of the binary pulsars are decreasing very slowly indicating that they will enjoy long lifetime. In fact, for binary pulsars, transfer of angular momentum takes place from binary orbit to the rotating pulsar. These binary pulsars are a distinct class by themselves, their evolution running separately from general population of pulsars. In the P, $\dot{P}$ diagram, the binary pulsars occupy distinctly the lower left corner. The orbital velocity curve of the binary pulsar PSR 1913+16 is highly nonsinusoidal which indicates that the orbit is highly eccentric. The apsidal line (the line joining the periastron and apastron) of the orbit of this binary pulsar has been observed to rotate in space through a large angle of 4.2 degrees per year. This is in perfect conformity with the prediction of Einstein's general relativity. The prediction was made by Einstein himself in case of Mercury going round the Sun. He proposed that the line of apsides of an eccentric orbit will rotate slowly in space, the fact which he had termed "advance of the perihelion". The perihelion of Mercury, according to his prediction, should advance by 43 seconds of arc per century which was actually observed. The advance of perihelion could not be explained on the basis of Newtonian theory. The discrepancy between the theory and observation was thus removed by Einstein's theory.

In the case of the binary pulsar PSR 1913+16, the masses of both the components have been carefully deduced to be ~ 1.4 $M_\odot$ and most likely both the components are neutron stars. These heavy masses being very close to each other, the general relativistic effect should be much higher and the measured high value of 4.2 degrees per year is in perfect accordance with the theory. The observed decay of the orbital periods of binary pulsars demonstrates yet another relativistic effect. Einstein's general relativity theory predicts that two stars revolving about each other are subject to accelerations, which in turn will generate gravitational radiation. The gravitational radiation energy is derived at the expense of the energy and angular momentum of the orbital motion of the components. The orbital energy is thus dissipated leading to the gradual decay of the orbit. Since the decrease of periods is a common feature of binary pulsars, they serve to demonstrate the truth of the prediction of general relativity theory.

The origin of binary pulsars has not yet been well understood. Several theories have been proposed the discussion of which is beyond the scope of this presentation.

15.8 BLACK HOLES

Modern astrophysics is opening as a book of puzzles to the scientists. The *black hole* is one of these latest puzzles of astrophysics. Although the theory of general relativity and gravitation predicts the inevitable formation of black holes in the galaxy, they have not yet been observationally demonstrated. If, however, they do form according to the prediction of the theory, there may be as many as a billion or so of them in our Galaxy. As such, they should have a tremendous importance in the galactic phenomena, and fortunately, this aspect has not remained ignored by the theoretical physicists. What are then these black holes and how can they be formed? Why are they so called and what possibilities are there to actually "observe" them? These questions and the like we shall discuss briefly in the present section.

We have already seen that stars with masses less than about 1.4 $M_\odot$ invariably end their life as white dwarfs. These stars meet their demise peacefully. But any star having a mass greater than 1.5 $M_\odot$ has a chance to explode violently at a certain stage of its evolution, unless the catastrophe is prevented by rotation of the star or loss of a part of its mass before the catastrophic stage is reached. If the initial mass of the exploding star is not very high, say not greater than 10 $M_\odot$, then after the surface layers are blown off, the core will contract to form a stable neutron star in which the pressure of the degenerate neutrons counteracts the gravitational onslaught. The density of the stellar material at this stage is of the order of 10^{14-15} gm cm^{-3}. If, however, the original star is more massive, then after the explosion even the degenerate neutrons in the massive core may not be able to resist further gravitational contraction. Even if such a star does not explode for some favourable reasons, it will undergo the process of contraction. Because these massive stars quickly use up their nuclear fuel (their life-time is 10^7 years), they start cooling down. So the star will contract, and during this course some of its potential energy will be transformed to heat. This fresh quota of heat, however, is not enough to prevent the collapse, and therefore, the star continues to collapse with gathering speed. We shall assume that we are dealing all along with a spherical star. As the collapse starts, if the star cannot shed off most of its mass, the collapse continues with increasing speed, thereby producing more and more intense gravitation field and reducing the size of the star. In the process a stage is reached when the radius R_s, of the star of mass M is given by

$$R_s = \frac{2GM}{c^2} \qquad (15.14)$$

where c is the speed of light. R_s is called the *Schwarzschild radius.* When this radius has been attained, the object will have such a strong gravitational field that it will prevent even light from escaping out of its influence. In Newtonian mechanics also one finds the same result that at the radius R_s the velocity of escape from the object equals the velocity of light. Since photons cannot escape, no event occurring within the Schwarzschild radius can be communicated outside this radius. The object is therefore *lost* to the outside universe except that its strong gravitational field will be felt by other nearby objects. The Schwarzschild radius has therefore been called the *event horizon* (meaning that any event within it remains hidden from view as one below the horizon) and the resulting object as the *black hole.*

Even at this stage, not only that the collapse does not cease, but it continues with an appalling speed. In a *spherically collapsing star* all the material will simultaneously reach the centre within a period of about 1 μ sec after crossing the event horizon. The entire mass thus shrinks to a point of infinite density, which forms a *singularity.* The singularity will be formed even if the collapse is not radially symmetric. An understanding of the nature and physics of this singularity has of late become a stumbling block to physicists and relativists alike.

Equation (15.14) shows that the radius of the event horizon is proportional to the mass of the collapsed object. The Schwarzschild radius for the Sun is 2.67 km and the minimum density attained by the material at this stage is ~10^{16} gm cm^{-3}, if we assume that the percentage of mass converted into energy is not significant. For a star of 10 $M_\odot$ these quantities are respectively ~27 km and ~10^{14} gm cm^{-3}, while for a massive galactic nucleus of mass 10^9 $M_\odot$ they are respectively ~10^{-4} pc and 10^{-2} gm cm^3, the last value being only about 1 per cent of the average density of the Sun.

When the event horizon has formed, light from within cannot move out, while that from the outside neighbourhood of the horizon will propagate, being strongly redshifted by the intense gravitational pull of the black hole. Not only that, everything, both matter and light that may encounter the event horizon will be sucked in.

A clear understanding of the black hole physics and mechanics posed serious problems until a few years ago. But recently, the gateway to understanding these complicated problems has been opened to meaningful theoretical investigations, thanks to the very significant works in the field by Penrose, Hawking, Bardeen, Lynden-Bell, Chandrasekhar and several others. It is now understood that a black hole may capture mass from the surrounding space by its intense gravitational pull. The probability for such capture becomes almost certain, if the black hole be the member of a binary system consisting of a normal star. Owing to the intense gravitational pull the black hole will then draw gas from the upper atmosphere of the companion, this gas will form a rotating disc around the black hole by virtue of the conservation of angular momentum which the captured matter originally possessed. Frictional force will dissipate the kinetic energy of matter, particularly of that in the inner part of the disc, whereby the centrifugal balance will be upset and the matter will spiral towards the black hole to be sucked in by it ultimately, in the same manner as atmospheric friction brings down the artificial satellites to the surface of the Earth. The falling matter, before being sucked in, attains very high temperature so as to radiate intensely in very high frequencies, particularly, in X-rays. Calculations show that about 6 per cent or more of the total rest-mass energy of the infalling matter may be converted to radiation energy during the process. This is about 8 times more efficient than the energy generation process by nuclear transmutation. Thus gravitational energy generation process is much more efficient than any other known process, except that of complete annihilation by matter-antimatter interaction which is 100 per cent efficient.

The works of Penrose and Hawking have shown that a still more efficient process of gravitational radiation can be achieved when two black holes come in contact and merge together. In this case, the two event horizons also merge together to be enveloped by a common event horizon. When this happens, the theory predicts that the total surface area of the event horizon of the final black hole will increase. In an ideal case, let the two black holes be of equal mass, M, and the resulting black hole be of mass M'. Now, since the surface area of the event horizon is proportional to the square of the mass (the radius being proportional to the mass), we should have, according to the above theory,

$$M'^2 > 2\,M^2$$

whence,

$$M' > \sqrt{2}\,M \tag{15.15}$$

which corresponds to a maximum efficiency of $(1-1/\sqrt{2}\,) \approx 0.29$ for the conversion of original rest-mass energy to the radiation energy. This is the upper limit of conversion efficiency in the case of nonrotating black holes. It has been proved also that under certain conditions, if a mass is accreted by a rotating black hole, then before it is finally sucked in, about 43 per cent of its rest-mass energy may be converted to radiant energy; and this is the maximum limit of conversion efficiency so far known.

Such highly efficient mechanisms for energy generation are called for, in order to explain the observed radiation from radio galaxies, quasars and intense X-ray sources. So black holes, although not yet observed, may come out as the final answer to the energy problem that has been seriously intriguing the physicists for more than a decade.

Lastly, we can ask—What prospect is there for a black hole to be observed and where should we look for it? If the way we have described for the formation of black holes is correct, there may be at least 10^9 black holes in the Galaxy. But the most likely place to detect them is in the massive binary systems, from both the aspects of intense gravitational pull as well as high frequency radiation. We have already noticed that intense X-ray radiation results when mass is accreted by a black hole. So if in a binary system one of the components be a normal star while the other has become a black hole, the former will move in an orbit around, the latter which will be optically invisible, but will emit copious flux of X-rays. This condition is exactly fulfilled by the most intense X-ray source Cygnus X-l. It is now believed that Cygnus X-l belongs to a binary system having its companion as a normal BOIab supergiant. Reliable calculations have indicated that the mass of the supergiant lies in the range (20–30) $M_\odot$, while that of the X-ray source lies in the range (3.4–8) $M_\odot$. Such a high mass of the X-ray source lends belief that it is *not a neutron star but a black hole,* Such a belief is supported by the observation that the X-ray brightness of Cygnus X-l varies over a period of 0.1 second. This indicates that the radius of the source is smaller than 3×10^4 km, which is very small compared to the radius of a normal star ($R_\odot \sim 7 \times 10^5$ km). On the other hand, its estimated high mass excludes its membership among neutron stars or black dwarfs. Thus, the source is a compact object but belongs neither to neutron stars nor to black dwarfs. Its most likely classification is thus among the black holes. There are two other likely candidates similar to Cygnus X-l in our Galaxy, and one in the Small Magellanic Cloud. Each of these is a binary system with an X-ray source as one of the components. The two galactic systems with X-ray sources 2U 1700–37 and 2U 0900–4.0 respectively, have X-ray spectra similar to that of Cygnus X-l, but their masses have not been determined. The mass of the SMC source, SMC X-l, is estimated to be $\sim 10\ M_\odot$ from its measured X-ray energy flux.

The observed high radiation flux from radio galaxies, quasars and nuclei of some otherwise normal galaxies calls for its explanation some energy generation mechanism, whose efficiency should be comparable to that of total annihilation of mass. The only such mechanism as currently understood is the gravitational radiation in the intense gravitational field of black holes. So it is not unlikely that these objects contain at their centre black holes whose masses may range from 10^5 to $10^9\ M_\odot$.

EXERCISES

1. Consider stars of 1 $M_\odot$ in each case. Calculate the mean density and the luminosity L in the following cases:

   (a) the Sun, with $R_\odot = 7 \times 10^5$ km, $T = 6000$ K;
   (b) a white dwarf, with $R = 10^4$ km, $T = 20{,}000$ K;
   (c) a neutron star, with $R = 10$ km, $T = 10^5$ K.

2. Discuss the general nature of the *primary* and *secondary* pulses of a pulsar.

3. Calculate the rotational energy of a neutron star of 1 $M_\odot$ and 10 km radius, having a period of 100 milliseconds. If the observed slow-down rate of this pulsar is 10^{-13}, calculate the rate of rotational-energy loss by this pulsar.
   Calculate also the time for which this pulsar will remain alive.

4. It is believed that pulsars are rotating neutron stars. On what observational bases astronomers have been able to derive this conclusion?

5. Give a brief but precise description of Gold's model of pulsars. What is the "Velocity of light circle"? How does it depend on the periods of pulsars?

6. What do you know about the Galactic distribution of pulsars? Are all the observed pulsars Galactic? Give reasons to support your views.

7. Discuss how the distance to pulsars are determined. What are the sources of inaccuracy in the measured distances?

   How would you find approximately the total number of pulsars in the Galaxy?

8. Discuss the astrophysical importance of binary pulsars.

9. Define *Schwarzsckild radius* of an object Why this is also called the event horizon? Calculate the Schwarzschild radius and compute the minimum density of matter within this radius in each case, for the following objects:
   (a) the Sun;
   (b) a star of mass 10 $M_\odot$;
   (c) a star of mass 100 $M_\odot$;
   (d) a galactic nucleus of mass $10^9 M_\odot$;
   (e) the Milky Way Galaxy of mass $10^{12} M_\odot$.

10. On what observational bases some objects are believed to possess a black hole? List all such kinds of different objects.

11. Discuss the different theories that are known for the conversion of rest mass energy of matter into high energy radiation in the presence of black holes.

SUGGESTED READING

1. Kaufman, W.J., *Black Holes and Warped Spacetime*, Freeman, New York, 1979.

2. Kourganoff, V,, *Introduction to Advanced Astrophysics*, D. Reidel Publishing Company, Dordrecht, Holland, 1980.

3. Longair, M.S., *High Energy Astrophysics*, Vol. 1, Cambridge University Press, Cambridge, 1992.

4. Luminer, Jean-Pierre, *Black Holes*, Cambridge University Press, Cambridge, 1992.

5. Lyne, A. and Smith, EG., *Pulsar Astronomy*, Cambridge University Press, Cambridge, 1990.

6. Smith, F.G., *Pulsars*, Cambridge University Press, Cambridge, 1977.

7. Taylor, J.H. and Manchester, R.M., *Ann, Review Astronomy and Astrophysics*, **15**, p. 19, 1977.

16 Our Galaxy

16.1 INTRODUCTION

On a clear summer evening if one looks upon the vault of the sky, one may see the fascinating view of the milky band stretching from one corner of the vault to the other. The ancients attached the holy meaning to it of being the heavenly way along which the gods travel in their chariots. But we now know that this milky band is the manifestation of the diffused light of billions of stars that lie along the equatorial plane of the stellar system or the Galaxy to which our solar system belongs. This is our Galaxy, also called the Milky Way Galaxy. The Milky Way spans more or less over the entire heavens following the course of a great circle around the celestial sphere (Fig. 16.1).

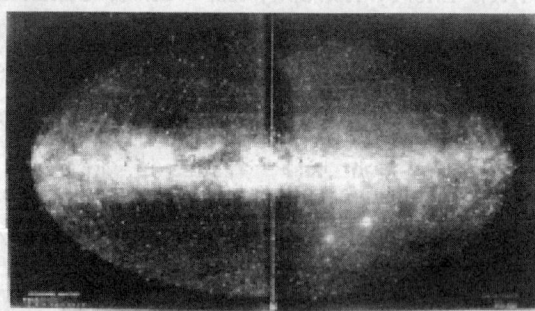

FIGURE 16.1 The mosaic of the Milky Way.

The brightest section of the Milky Way is its southern half lying between the two constellations Argo and Centaurus. The width of the Milky Way varies considerably. It is widest in the Scorpio and Sagittarius regions, while in the Taurus and Auriga regions both the width and the brightness are much reduced. Some regions of the Milky Way have thicker assemblage of gas and dust which obscure the stars behind them. When viewed through a telescope, the Milky Way is found to be composed of countless stars along its plane. The number of stars is, however, much less in a direction perpendicular to this plane. (This can be explained in terms of the existence of our Sun in a disc shaped system.) Years of observations and investigations on various components of the Galaxy, mainly its stellar components, have revealed that our Galaxy is a disc-shaped spiral of type Sb.

The study of the structure of our Galaxy and the physical and dynamical properties of its various constituents has assumed great importance in the current astrophysical research. Astronomers believe that such studies may ultimately reveal the mystery behind the process of formation and evolution of galaxies in general and the role played by cosmic rays, magnetic field, stellar evolution and nucleosynthesis etc., in particular. The galactic research is therefore, developing as a very promising branch of fundamental importance among the various current astrophysical topics. We shall discuss some aspects of these studies in the subsequent sections.

The main constituents of our Galaxy are the stars, the interstellar gas and dust, the magnetic field, the cosmic rays, and the unseen matter. The stars constitute the major portion of the total luminous* mass of the Galaxy and they are the principal 'actors' of the highly interesting 'galactic drama'. The contribution of the gas to the total luminous mass may fairly be taken to be approximately 5 to 10 per cent. The cosmic ray particles contribute only an insignificant mass but they have considerably large amount of energy, comparable to the kinetic energy of the interstellar gas. It is assumed that the cosmic ray particles are confined by the magnetic field of the Galaxy, otherwise they would have left the system within a period of about 10^5 years since the evolution of the Galaxy. The stars and the gas together constitute a somewhat complicated picture of the Galaxy. The stars, which are the chief luminous constituents of the Galaxy possess their own gravitational field. They revolve largely in circular orbits and also have velocity dispersions. The interstellar gas also has its own gravitational field and pressure. The magnetic field, the existence of which has definitely been established by the observation of polarization of starlight and nonthermal radiation, must also have its influence on the highly conducting interstellar medium.

Radio astronomical observations of the 21-cm line of hydrogen have revealed that the interstellar matter contributes not more than 10 per cent of the total luminous mass of the Galaxy. Oort's analysis has however revealed that in the solar neighbourhood the gas contributes at least 20 per cent of the total mass density. Interstellar matter is characterized by two types of absorption. The *continuous absorption* which is greater at shorter wavelength, is produced by minute solid particles, commonly known as interstellar grains. These are responsible for the observed *reddening effect* of the distant stars in the vicinity of the plane of the Galaxy where the bulk of the interstellar matter is concentrated. The grains constitute about 10^{-3} to 10^{-4} of the total luminous mass of the Galaxy. The *line absorption* is produced by the interstellar gas in the spectra of distant hot stars. Besides, there is (also) the line emission by atoms and molecules in various wavelength regions of the spectrum. The existence of the dense obscuring material close to the galactic equator is also revealed by the study of the distribution of external galaxies, in different regions of the sky. It is observed that these galaxies are almost uniformly distributed everywhere except in the region of the sky lying between +10° to -10° galactic latitudes, where the distant galaxies are almost completely veiled by the thick obscuring interstellar matter.

---

*The mass that can be detected through its emitted electromagnetic radiation.

As early as in the late eighteenth century, Sir William Herschel's investigation on the distribution of stars in the solar neighbourhood revealed that our stellar system is a very flattened one. This has been confirmed by the later investigations, both by optical and radio means. We now know that the shape of the Galaxy is that of a very flattened spheroid having a few open ring-like structures called spiral arms. The Sun lies very close to the equatorial plane of this system at an approximate distance of 10 kpc from the centre (Fig. 16.2). From a more recent study it has been calculated that the Sun's distance from the galactic centre is 8.5 kpc. In order to maintain stars and gas in their orbits the Galaxy should possess a very high rotation which develops the centrifugal force necessary to balance the gravitational attraction exerted by the large mass of the central core. A large number of globular clusters surround the Galaxy and are moving in elongated orbits around the galactic centre. Groups of stars and star clusters are born in the clouds of dark matter in this great conglomeration. These newly born stars lie mostly close to the galactic equator where the parent material out of which they form is concentrated. The life-spans of these stars are different and depend on the rate of using their fuel which again depends on their masses and chemical compositions. During their life time hydrogen is converted into heavier elements in their inner parts while a part of the matter from the surface layers is ejected into interstellar space in course of their evolution. Thus the aspect of the Galaxy is changing during a time-scale short compared to the galactic life time. Stars are born and they die continuously over this changing time-scale. In every region of the Galaxy, viz. in the solar neighbourhood, in the globular clusters, in the nucleus and in the spiral arms, stars which are less bright compared to the Sun are still in the main sequence. They have not yet reached the turning point of their existence. According to the theory of stellar evolution, these stars can maintain this phase of their life for about 5 to 10 billion years from their birth. The age of the Galaxy is also believed to be somewhat more than 10 billion years.

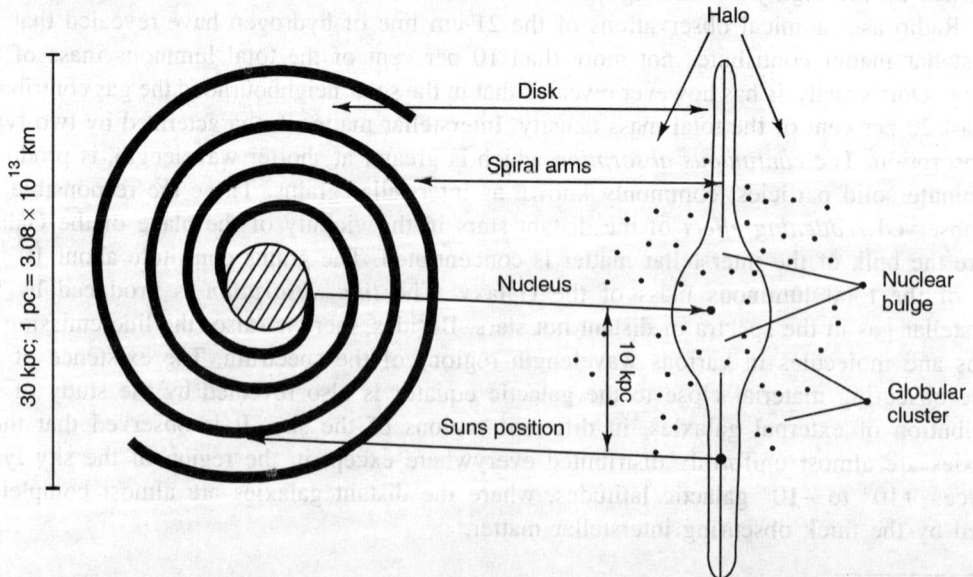

FIGURE 16.2   Schematic representation of the general structure of our Galaxy.

16.2 ROTATION OF THE GALAXY: DIFFERENTIAL ROTATION

The rotation of the Galaxy is entirely different from that of a solid body. In the case of a solid body rotation, all the points situated at different distances from the centre of rotation have the same angular velocity and the points located at larger distances from the centre are moving more rapidly than those situated inside. In the case of galactic rotation, the situation is almost the reverse. Stars away from the centre move more slowly in the same manner as the planets move away from the Sun. This is explained in Fig. 16.3. This type of rotation of the Galaxy has been referred to as the *differential galactic rotation* that produces a type of shearing motion in the plane of the Galaxy. As a result of this differential galactic rotation the radial

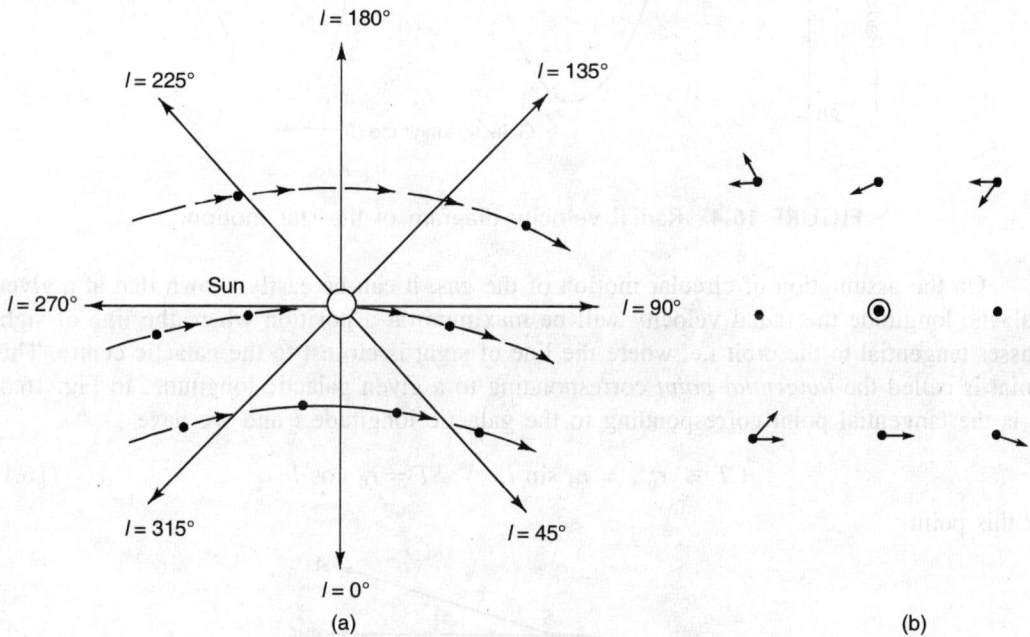

FIGURE 16.3 Differential rotation of the Galaxy.

velocities of the stars moving in circular orbits—concentric to that of the Sun (but viewed from different directions from the Sun in the galactic plane) are affected. Stars located in the direction of the galactic centre and in the direction just opposite to it (anticentre direction) have no relative motion toward or away from the Sun. These stars have thus zero radial velocities. Stars on orbits situated very close to that of the Sun but at galactic longitudes 90° and 270° respectively are also moving with approximately the same velocity as that of the Sun and have no radial velocities. Positive radial velocities are observed in stars at 45° and 225° galactic longitudes. In the first case, the stars are driven away from us owing to their faster velocities compared to that of the Sun. In the second case, we are being pulled away with much faster velocities compared to the velocities of those stars. Similarly, stars at 315° galactic longitudes are observed to be approaching towards us and we are approaching towards those at 135° galactic longitudes. These stars are assumed to possess negative radial velocities. A plot of the observed radial velocities of all stars lying on the galactic plane at a particular distance from the Sun against various galactic longitudes yield a curve as shown in Fig. 16.4.

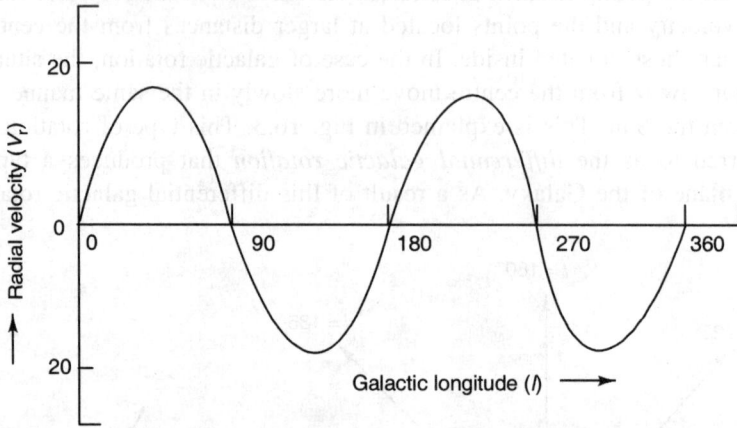

FIGURE 16.4 Radial velocity diagram of the star motion.

On the assumption of circular motion of the gas, it can be easily shown that at a given galactic longitude the radial velocity will be maximum at a position where the line of sight passes tangential to the orbit i.e. where the line of sight is closest to the galactic centre. This point is called the *tangential point* corresponding to a given galactic longitude. In Fig. 16.5, T is the tangential point corresponding to the galactic longitude l and we have

$$CT \cong r_{min} = r_0 \sin l, \qquad ST = r_0 \cos l \qquad (16.1)$$

at this point.

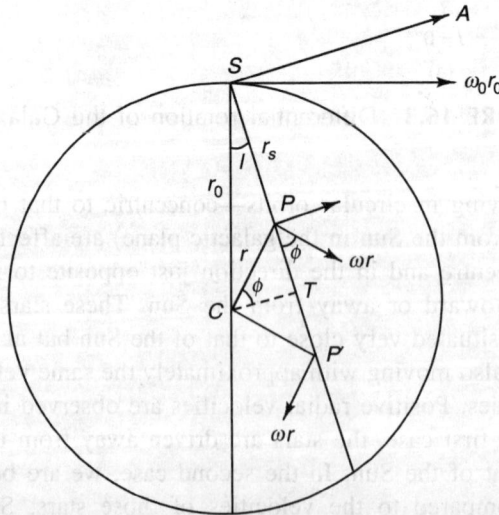

FIGURE 16.5 A schematic diagram of the galactic rotation.

The angular velocity ω is a decreasing function of r, the distance from the galactic centre. Let P be any other point on the galactic plane, whose distance from the centre is r and that from the Sun is r_s. Let the galactic longitude of the point be l. Assuming $\angle$ PCT as shown in the figure to be ϕ we have from triangle SCP

$$\frac{\sin l}{r} = \frac{\sin (90° + \phi)}{r_0} = \frac{\cos \phi}{r_0} \tag{16.2}$$

Thus assuming the average motion of the medium to be everywhere perpendicular to the radius vector from the centre and considering the angular velocity ω to be dependent solely on r, the radial velocity v_r of the material near P relative to that near the Sun may be shown to be

$$v_r = r_0 \left[\omega(r) - \omega_0 \right] \sin l \tag{16.3}$$

where $\omega(r)$ and ω_0 are angular velocities at a distance r from the centre and at Sun respectively. Here a perfect circular symmetry of motions has been assumed.

If the point P is not on the galactic plane but has latitude b, Eq. (16.3) will change to

$$v_r = r_0 \sin l \left\{ \omega(r) - \omega_0 \right\} \cos b \tag{16.4}$$

We shall however consider points on the galactic plane and take $b = 0$.

If the point P is in the neighbourhood of the Sun (say, within 2 kpc), then $r \approx r_0$ and $r_s/r_0 \ll 1$. We can expand $\omega(r)$ in Taylor's series in the neighbourhood of r_0 and write

$$\omega(r) = \omega_0 + (r - r_0) \left(\frac{d\omega}{dr} \right)_0 \tag{16.5}$$

Equation (16.3) then can be written as

$$v_r = r_0 \sin l \, (r - r_0) \left(\frac{d\omega}{dr} \right)_0 \tag{16.6}$$

to the first order. The quantities $r-r_0$ and $\left(\dfrac{d\omega}{dr} \right)_0 = \omega_0'$ represent local characteristics. Now the first Oort's constant A is defined by

$$A = -\frac{1}{2} r_0 \omega_0' = -\frac{1}{2} r_0 \left[\frac{d}{dr} \left(\frac{v}{r} \right) \right]_0 = \frac{1}{2} \left(\frac{v}{r} - \frac{dv}{dr} \right)_0 \tag{16.7}$$

Physically, A represents the rate of local shear.

Equation (16.6) now reduces to

$$v_r = 2A \, (r_0 - r) \sin l \tag{16.8}$$

This represents the *Camm approximation* for radial velocities. Also, we have

$$r^2 = r_0^2 + r_s^2 - 2r_0 r_s \cos l \tag{16.9}$$

Therefore,

$$\frac{r}{r_0} = \left[1 - 2\frac{r_s}{r_0}\cos l + \left(\frac{r_s}{r_0}\right)^2 \right]^{1/2}$$

$$= 1 - \frac{r_s}{r_0}\cos l + O\left(\frac{r_s}{r_0}\right)^2$$

Thus

$$r_0 - r = r_s \cos l \qquad (16.10)$$

to the first order. Equation (16.8) therefore reduces to

$$v_r = A r_s \sin 2l \qquad (16.11)$$

This is *Oort's formula* of double sine wave for radial velocities which has been depicted in Fig. 16.4. The radial velocity is thus a measure of the differential rotation of the Galaxy. The transverse velocity of the point P with respect to S is given by

$$v_T = v \sin \phi - v_0 \cos l \qquad (16.12)$$

v_T being assumed positive in the direction of increasing galactic longitude. Using the geometry of Fig. 16.5, Eqs. (16.5), (16.10) and $r_s/r_0 \ll 1$, and employing some algebraic simplifications*, one obtains

$$v_T = - r_s(r_0\omega_0' \cos^2 l + \omega_0) \qquad (16.13)$$

where

$$\omega_0' = \left(\frac{d\omega}{dr}\right)_0$$

On further simplification, the equation reduces to

$$v_T = r_s (A \cos 2l + A - \omega_0) \qquad (16.14)$$

Introducing the second Oort's constant $B = A - \omega_0$, so that $\omega_0 = A - B$, Eq. (16.14) changes to

$$v_T = r_s (A \cos 2l + B) \qquad (16.15)$$

---

* From Fig. 16.5, $r \sin \phi = r_0 \cos l - r_s$

$\therefore v_T = \omega_r \sin \phi - \omega_0 r_0 \cos l$

$\quad = \omega(r_0 \cos l - r_s) - \omega_0 r_0 \cos l$

$\quad = (\omega - \omega_0) r_0 \cos l - \omega r_s$

$\quad = (r - r_0)\omega_0' r_0 \cos l - \omega r_s$ [using 16.5]

$\quad = - r_s \cos^2 l \; \omega_0' r_0 - \omega r_s$ [using 16.10]

$\quad = - r_s \cos^2 l \; \omega_0' r_0 - \{\omega_0 + (r - r_0)\omega_0'\} r_s$ [using 16.5]

$\quad = - r_s \cos^2 l \; \omega_0' r_0 - \omega_0 r_s,$

when $r \approx r_0$

Since the proper motion μ is given by

$$\mu = \frac{v_T}{4.74 \, r_s} \qquad (16.16)$$

Combining Eqs. (16.15) and (16.16) we get

$$\mu = \frac{A}{4.74} \cos 2l + \frac{B}{4.74} \qquad (16.17)$$

Equation (16.17) shows that in proper motion, besides a sinusoidal component, there is a systematic effect as a result of galactic rotation, and this effect is independent of distances of stars. This relation can also be utilized to determine the values of the Oort's constants A and B, if proper motions for any group of stars are accurately measured. Now,

$$B = A - \omega_0 = \frac{1}{2}\left(\frac{v}{r} - \frac{dv}{dr}\right)_0 - \frac{v_0}{r_0}$$

$$= -\frac{1}{2}\left(\frac{v}{r} + \frac{dv}{dr}\right)_0 \qquad (16.18)$$

Therefore,

$$\left(\frac{dv}{dr}\right)_0 = -\omega_0 - 2B = -(A + B) \qquad (16.19)$$

Thus, the local rotational velocity gradient is determined by Oort's constants.

Again for purely circular motion the normal force per unit mass is given by

$$F = \frac{v^2}{r}$$

Therefore,

$$\left(\frac{dF}{dr}\right)_0 = \left(\frac{2v}{r}\frac{dv}{dr} - \frac{v^2}{r^2}\right)_0 = 2\omega_0\left(\frac{dv}{dr}\right)_0 - \omega_0^2 \qquad (16.20)$$

Therefore,

$$\left(\frac{dF}{dr}\right)_0 = -(A - B)(3A + B) \qquad (16.21)$$

by using the relation $\omega_0 = A - B$. Thus, if the local force law is of the form

$$F = Kr^n \qquad (16.22)$$

where K and n are constants, the index n is given by

$$\left(\frac{d \ln F}{d \ln r}\right)_0 = \left(\frac{dF}{dr} \bigg/ \frac{F}{r}\right)_0 = n = -\frac{3A + B}{A - B} \tag{16.23}$$

The values of the Oort's constants therefore give the local force law in the Galaxy. Observations suggest that the index n may vary from +1 close to the centre to –2 in the outer region of the Galaxy, as shown in Fig. 16.6.

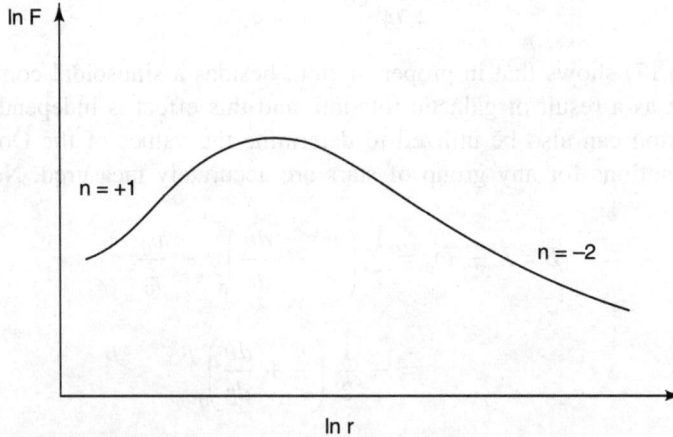

FIGURE 16.6 The variation of the force law in the galactic plane.

Again in the solar neighbourhood $r \approx r_0$. Also, $(v_r)_{max}$ corresponds to $r_{min} = r_0 \sin l$. We then have from Eq. (16.8)

$$(v_r)_{max} = -2A (r_{min} - r_0) \sin l$$

$$= 2Ar_0 (1 - \sin l) \sin l \tag{16.24}$$

Thus, if we observe the maximum radial velocity in the direction of longitude l for some type of objects in the vicinity of the Sun, we can calculate the product Ar_0 of the two basic quantities A and r_0 of the Galaxy. Sufficiently accurate values of the product Ar_0 can be determined by observations.

16.3 DETERMINATION OF THE ROTATION PARAMETERS IN THE SOLAR NEIGHBOURHOOD

A knowledge of the accurate values of the fundamental rotation parameters A, B, r_0 and ω_0 is essential in order to understand the complex structural and dynamical features of the Galaxy together with the large-scale mass distribution and the total mass in it. Many authors have attempted to calculate the values of these constants by observations of different types of objects. The results have not been identical but a workable average value for each of the parameters can be ascertained out of them. We shall discuss these here.

The Constant A

The type of objects that are suitable for the determination of A should fulfill several requirements, viz. (a) they should lie close to the galactic plane, (b) they should participate predominantly in the galactic rotation, and (c) they should be of high luminosity in order to be observable from a distance of 1 to 2 kpc. The O–B stars, the classical cepheids of Type I and the galactic clusters fulfill all these requirements; and therefore, these have been used by the authors for the purpose. The relevant formulas which can be used for the purpose are those in Eqs. (16.4), (16.7), (16.11) and (16.17). The first and third of these require observation of radial velocities while the fourth requires that of the proper motions of stars. To use the second, one has to compute ω against r for some type of object and calculate the slope of the curve at $r = r_0$ thus leading to the value of A when r_0 is known. A working value of $A = 15$ km s^{-1} kpc^{-1} has been used previously for many years. This value was adopted as an average of all previous determinations of the values of A. Later determinations of A have shown that this value should be modified somewhat. The straight mean and standard deviation of all the recently calculated values, using both radial and proper motions of different objects is given by $A = 14.4 \pm 1.2$ km s^{-1} kpc^{-1}. This value has been adopted by the IAU in its General Assembly held in New Delhi in 1985.

The Constant B

The constant B can be directly determined from observations of proper motions by using Eq. (16.17). This was done by Morgan and Oort who used proper motions of stars and derived the value $B = -7$ km s^{-1} kpc^{-1}. The value is however not very reliable, as it was based on a small number of stars and their proper motions containing some uncertainty. Using the proper motions from the General Catalogue, the value $B = -13$ km s^{-1} kpc^{-1} is obtained. The value of B can also be obtained from dynamical relation if the value of A is predetermined. In the ellipsoidal distribution of stellar velocity dispersions, the radial and tangential velocity dispersions in the solar neighbourhood are related to A and B by

$$\frac{\sigma_v^2}{\sigma_u^2} = \frac{-B}{A - B} \tag{16.25}$$

or

$$-\frac{B}{A} = \frac{1}{\dfrac{\sigma_v^2}{\sigma_u^2} - 1} \tag{16.26}$$

From dispersion analysis of proper motions of faint stars as well as from an analysis of dispersion of velocities of different types of stars, average value of the ratio A/B has been found to be near –1.5. Taking $A = 15$ km s^{-1} kpc^{-1} the results yield the average value of $B = -10$ km s^{-1} kpc^{-1}. This value has been extensively used. But more recently calculated values of B yield a straight mean and standard deviation of -12 ± 2.8 km s^{-1} kpc^{-1}. This value has been adopted by the IAU in its General Assembly held in New Delhi in 1985.

Sun's Distance r_0 from the Galactic Centre

For the determination of the scale of the Galaxy an accurate value of the solar distance r_0 from the galactic centre is essential. This has been variously determined by authors by direct observations of various types of objects. Three direct methods and several indirect methods have been used by the authors for the purpose. The first direct method was used by H. Shapley. The method consists in determining the centre of the system of globular clusters which are spherically distributed about the centre of the Galaxy. The second method was propounded by W. Baade. Observing a large number of RR Lyrae stars in the central bulge of the Galaxy and assuming the value 0.0 for the absolute magnitude of these stars, Baade obtained the value $r_0 = 8.2$ kpc. With the present value of +0.5 for the absolute magnitude of RR Lyrae stars, these data yield $r_0 \approx 10$ kpc. From his analysis of the globular cluster system and adopting $+ 0.5 \pm 0.1$ for the absolute magnitude of RR Lyrae stars, H. Arp finds $r_0 = 9.9 \pm 0.5$ kpc. Analyzing the radial velocities of southern B stars M.W. Feast and A.D. Thakeray calculated $r_0 = 8.9$ kpc. The third method uses the density peak of Mira variables in the galactic centre window. The indirect method consists in relating the kinematic model of the Galaxy to the distance r_0 of the centre. For the purpose OB stars have been extensively used. The results of early determination led to an average working value of $r_0 = 10$ kpc. This value has been used for a long time in the works on structure and dynamics of the Galaxy. But many later determinations of r_0, both by direct and indirect methods have indicated that the value $r_0 = 10$ kpc had been an overestimation. The more plausible value should rather be $r_0 = 8.5$ kpc. The New Delhi General Assembly of IAU has adopted the value $r_0 = 8.5 \pm 1.1$ kpc.

The Circular Velocity v_0 of the Sun

The circular velocity $v_0 = r_0\omega_0$ of the Sun around the galactic centre is a very useful parameter for the study of the structure and dynamics of the Galaxy. Its determination however is not an easy task. Direct determination of u_0 has been done by observing the solar motion with respect to the motion of some distant system of objects. Thus, T.D. Kinman has calculated solar motion with respect to the globular cluster system by measuring the radial velocities of the latter and found $v_0 \geq 167 \pm 30$ km s^{-1}. But since the average rotational velocity of the clusters about the galactic centre is unknown, the above determination cannot be treated as reliable. Similarly, the determination of the solar motion with respect to the Local Group of galaxies will involve uncertainties due to the unknown peculiar motions of these galaxies. Thus, any direct determination of the solar circular velocity is liable to be vitiated by the inherent uncertainties in the data used. An indirect determination therefore appears to be more reliable and this can be done from the relation $v_0 = r_0\omega_0 = r_0 (A - B)$. Using the values of A, B and r_0 as discussed above, we get $v_0 = 250$ km s^{-1}. This value has been extensively used for several decades. Now ω_0 is obtained to be 25 km s^{-1} kpc^{-1}. In angular measure this amounts to about $0".053$ yr^{-1}. With $r_0 = 10$ kpc and $u_0 = 250$ km s^{-1}, the period of revolution of the Sun around the galactic centre comes out to be about $T = 2.5 \times 10^8$ years. Table 16.1 lists the values of the parameters which have been used for many years till the modified values were accepted in the New Delhi General Assembly of the IAU. Newly adopted values for the parameters also are listed in the last column of the table.

16.4   RADIO OBSERVATION OF THE GALAXY AT 21-cm WAVE LENGTH

The energy levels of a hydrogen atom are each split into a number of sublevels of multiplicity $2n^2$ corresponding to the level n. This is called the *hyperfine splitting* of the energy levels. Accordingly, the ground level ($n = 1$) of hydrogen atom is split into *two* hyperfine levels corresponding to the *parallel* or *antiparallel* spins of the electron and the proton (Fig. 16.7).

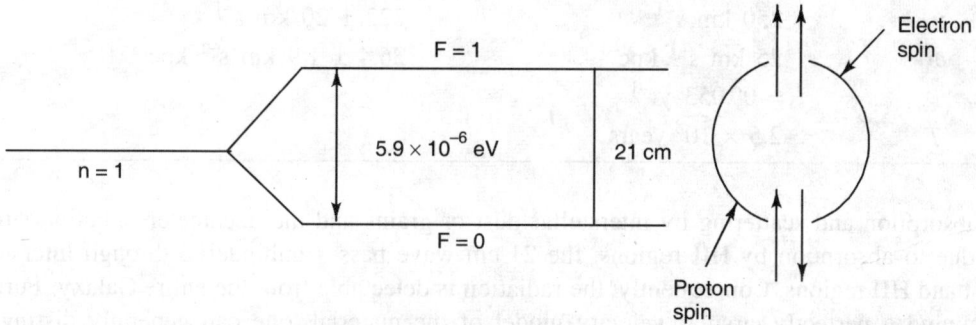

FIGURE 16.7   The hyperfine structure of the ground level of hydrogen atom.

In the former case, the total spin angular momentum quantum number $F = 1$ and in the latter, it is given by $F = 0$. The energy difference between these two levels is $\sim 5.9 \times 10^{-6}$ eV, and thus if transition from the upper ($F = 1$) to the lower ($F = 0$) level takes place, a quantum is radiated at a frequency of ~ 1420.4 MHz which corresponds to a wavelength of ~ 21.1 cm. If the interstellar hydrogen were perfectly cold ($T = 0$ K), all atoms would have remained in the $F = 0$ level. But the gas is not absolutely cold. An average temperature of about 100 K prevails there and so the atoms will be subjected to collisions. Collisions change the electron spin states leading to transitions of atoms both ways. But calculations show that in such random transitions $F = 1$ level is preferred to the level $F = 0$ in the ratio of about 3 to 1. Thus, in the process the majority of the hydrogen atoms seek their existence in the upper hyperfine state. An atom in the upper state may then be de-excited to the lower state either by another collision or by emitting a photon of 21.1-cm (we shall henceforth take it to be 21-cm for convenience) wavelength. The upper state however is highly metastable making the radiative transition highly forbidden. The average time for a transition is nearly 3.5×10^{14} sec ($> 10^7$ years) which can be compared with the normal transition time of the order of 10^{-8} second between permitted levels. Although collisional de-excitation rate is very much higher than this, yet a few atoms somehow escape the process and undergo radiative transition each emitting a 21-cm photon. Because of the enormous number of hydrogen atoms lying along the line of sight, the effect of even these rare events accumulate in any direction to give sufficiently strong radiation to be detected by just ordinary radio telescopic equipments.

The possibility of observing the 21-cm line of hydrogen in the Galaxy was first theoretically suggested by the Dutch astronomer H.C. Van de Hulst in 1944, and the line was actually observed first in 1951 by H.I. Ewen and E.M. Purcell of Harvard College Observatory. Since then the 21-cm line has served as the only method of tracing the large-scale structure of the entire Galaxy. While the optical radiation suffers attenuation on their way through space due

TABLE 16.1 The Rotation Parameters

Parameter	Previous values	Values adopted in 1985 IAU General Assembly
A	15 km s^{-1} kpc^{-1}	14.4 ± 1.2 km s^{-1} kpc^{-1}
B	-10 kms^{-1} kpc^{-1}	-12.0 ± 2.8 km s^{-1} kpc^{-1}
r_0	10 kpc	8.5 ± 1.1 kpc
v_0	250 km s^{-1}	222 ± 20 km s^{-1}
ω_0	25 km s^{-1} kpc^{-1}	26.4 ± 1.9 km s^{-1} kpc^{-1}
	$= 0''.053$ yr^{-1}	
T	2.5×10^8 years	

to absorption and scattering by interstellar dust or grains and the decameter radiation suffers so due to absorption by HII regions, the 21-cm wave passes unhindered through interstellar dust and HII regions. Consequently, the radiation is detectable from the entire Galaxy. Further, assuming a perfectly circular velocity model of the material, one can generally distinguish between the contributions from different distances through Doppler dispersion produced by galactic rotation.

Extensive surveys of 21-cm radiation from the Milky Way have been made by the Dutch (Leiden) and the Australian (Sydney) groups of radio astronomers, jointly covering a large part of the sky around the galactic equator at approximately the same resolution.

The principal information that has been provided by 21-cm observations may be summarized in the following:

1. The gross velocity structure of the gas over the entire Galaxy is found to be predominantly circular superimposed by small-scale random velocities. Large-scale radial flow of gas in the central region is observationally established.

2. The overall spatial and velocity distribution of gas clouds and random cloud velocity model generalized from the locally observed velocities have been pictured.

3. The large-scale density distribution over the entire galactic disc together with quantitative estimates of number density of neutral hydrogen at different parts of the disc have been found, and the spiral arms as relatively large density regions have been delineated.

4. The distribution of gas relative to the galactic plane and the thickness of the disc as defined by the distance between the half-density points have been known over the entire Galaxy.

5. The circular velocity of the gas at different radial distance from the galactic centre upto the Sun's distance has been measured, leading to the construction of the *Rotation Curve* of the Galaxy. This in turn has yielded the *mass distribution* over the entire Galaxy and its *total mass* therefrom.

6. The quantitative estimate of the average gas temperature in the entire disc has been derived.

Many more details have been provided by 21-cm observations. We now know that between 3 kpc and the Sun's distance from the centre the thickness of the gaseous disc is almost constant, that nowhere in this region the dispersion of this mean plane exceeds 25 parsecs and that outside the Sun's distance the dispersion increases gradually to very large values of 600 parsecs or more, upwards to the north and downwards to the south. This phenomenon is known as the *warping* of the disc, the cause of which is still not known satisfactorily. High-latitude observations indicate that there are clouds at sufficient heights above the galactic plane. Such observations, though not quite very conclusive, substantiate the hypothesis of the existence of a massive galactic halo. Information has been obtained about cloud structure, their average size, mass, density, internal velocity, their number density and distribution of velocities. The average density of the gas has been known to be about 1 hydrogen atom cm^{-3} in our neighbourhood and that the gas contributes to about 20% of the total mass in this region. These results were however derived using several simplifying assumptions, the most important ones of which are the following:

1. The velocity at each point is purely circular around the galactic centre and so the velocity distribution is known at each point. This gives the distance discrimination on the basis of Doppler shifts arising from differential galactic rotation.
2. The velocity dispersion of the gas clouds is the same over the entire disc of the Galaxy.
3. The gas temperature is the same everywhere in the region observed.

The detailed discussion of *all* the above mentioned aspects of the 21-cm observations and their reduction is beyond the scope of this book.

16.5 THE ROTATION CURVE OF THE GALAXY: THE GENERAL ROTATION LAW

According to the perfect circular velocity model, the radial velocity v_r of any point P in the plane of the Galaxy with respect to the Sun is given by formula (16.3). At the tangential point having longitude l, v_r is maximum and $r_0 \sin l = r_{\min}$. If the constants r_0 and ω_0 are known, observing $(v_r)_{\max}$ (after correction for random cloud velocities) one can find the relation $\omega(r)$ versus r and thus yields the relation $\omega(r) \times r = \theta_c$ (circular velocity) versus r, which is the rotation curve. In all 21-cm work, the values $r_0 = 8.2$ kpc and $\omega_0 = 26.4$ km s^{-1} kpc^{-1} have been used. The corresponding rotational velocity v_0 at the Sun is 216 km s^{-1}. These values are very close to those accepted by the IAU in 1985. It may be pointed out in this connection that incorrect values of r_0 and ω_0 will lead to a change in the scale and structure of the Galaxy derived from 21-cm data, but not serious changes in its general shape. It may also be mentioned that the method adopted for the reduction of the 21-cm data, viz., the use of tangential points, can yield rotational velocities, and so the rotation curve, for $r \leq r_0$. The rotation curve for the region $r > r_0$ must be derived either from some mass distribution model which fits the inner curve, or from optical observations.

The rotation curves for the Galaxy have been derived both from the northern observations of the Leiden group as well as from the southern observations of the Sydney group. The reduction of the data of the two groups was made with same or similar assumptions. The resulting smoothed out rotation curves from both Leiden and Sydney observations are shown in Fig. 16.8. In the actual curves two definite features are noticed. The first is the existence of a few dips in the curves, particularly in the northern curve. These dips can be conveniently

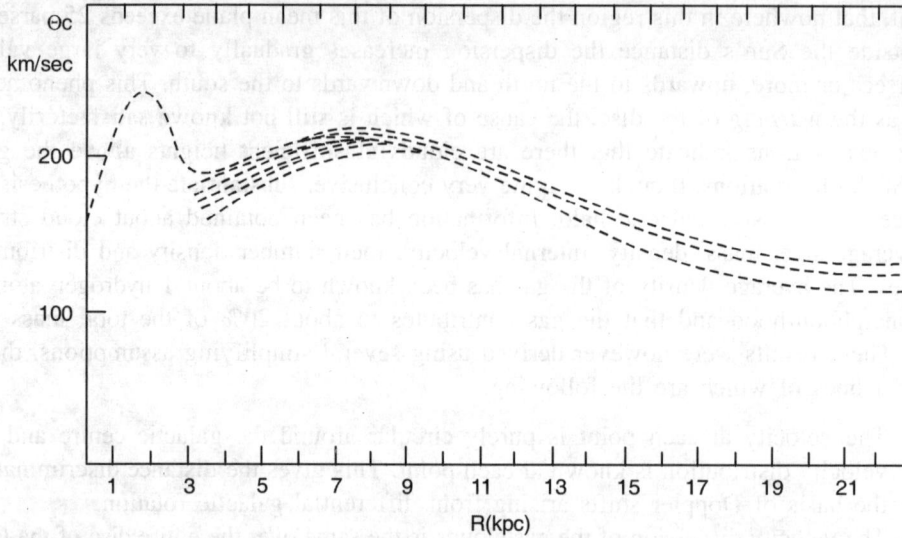

FIGURE 16.8 The northern and southern rotation curves derived from 21-cm line.

explained if one assumes that in the directions corresponding to the dips, the tangential points contain no hydrogen. The maximum radial velocity in the corresponding direction then must come from some other point more distant from the centre where the value of $\omega(r)$ is smaller. One thus gets a smaller rotational velocity, giving a dip. According to this assumption then the true rotation curve should run through the upper points, the second feature is the lower values of the circular velocity θ_c for the region between $r = 4.5$ and 7 kpc as have been derived from the southern data. Careful investigations have suggested that these *lower* values were probably real. The difference is well explained if one postulates the existence of an *outward motion of the local standard of rest* relative to which all the velocities have been measured. For an outward velocity component v_{LSR} of the LSR, the measured radial velocity v'_r at the tangential point at longitude l will be given by

$$v'_r = v_r + v_{LSR} \cos l \qquad (16.27)$$

where v_r is the true radial velocity at the tangential point. In this model, the measured radial velocities in the northern and southern hemispheres will differ by $2v_{LSR} \cos l$. The value $v_{LSR} = 7$ km s^{-1} fits well the observed data. The true rotation curve, according to this hypothesis, should lie midway between the observed curves. In fact, the observations are fairly satisfactorily explained by the general expansion velocity model of $470/r^2$ km s^{-1} for the gas in the region beyond $r = 3$ kpc, as has been proposed by F.J. Kerr, r being measured in kpc.

Thus, the rotation curve for the inner parts of the Galaxy ($r < r_0$) has been constructed from 21-cm data. Reliable rotational velocities *for four points* only have been given by K.K. Kwee, C.A. Muller and G. Westerhout. Besides these, the observation by G.W. Rougoor and J.H. Oort of the central region has provided rotational velocities at two more points. With $r_0 = 10$ kpc and $\omega_0 = 25$ km s^{-1} kpc^{-1} these results are given in Table 16.2.

TABLE 16.2 Circular Velocities in the Inner Part of the Galaxy

r(kpc)	θ_c km s^{-1}	
0.32	220.0	} Rougoor and Oort
0.67	265.0	
3.53	206.4	} Kwee, Muller and Westerhout
6.18	239.6	
7.74	248.5	
8.01	252.2	
10.00	250.0	revised values of r_0 and ω_0

As we have already mentioned, the 21-cm method fails to calculate the circular velocities in regions where $r > r_0$, since the maximum radial velocity cannot be measured in longitudes $90° \leq l \leq 270$. To get the circular velocities in the outer regions, therefore, one must rely on some mass model or optical observations. Observing the radial velocities of Cepheid variables with known absolute magnitudes from their period-luminosity relation, R.R. Kraft and M. Schmidt obtained a plot for $\theta_c(r)$ against r upto a few kpc beyond $r = r_0$. The data give the values of $\theta_c(r)$ = 244, 236 and 227 km s^{-1} at r = 11, 12 and 13 kpc respectively. Combining all the radio and optical data, one can therefore obtain the rotation curve of the Galaxy all the way from the centre upto 13 kpc. This is shown in Fig. 16.9 as has been given by G. Contopoulous and B. Stromgren. These authors have also derived the rotation law

$$\theta_c(r) = 67.76 + 50.06 \; r - 4.0448 \; r^2 + 0.0861 \; r^3 \tag{16.28}$$

which fits well with observed values in the region 3 kpc $\leq r \leq$ 13 kpc, r being measured in kpc.

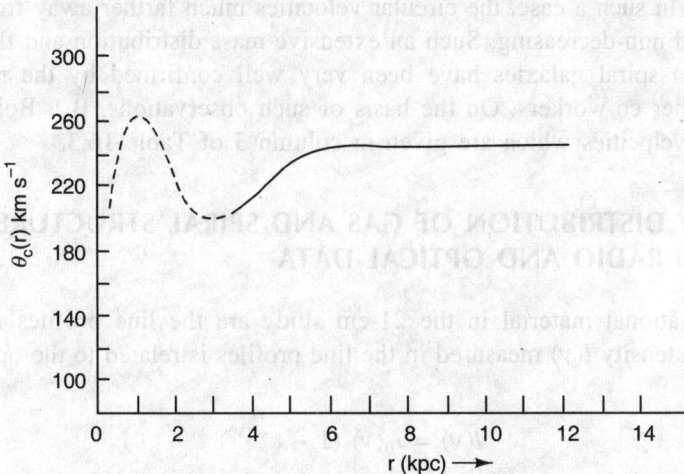

FIGURE 16.9 The rotation curve of the Galaxy.

The circular velocities of the Galaxy can be derived up to much larger distances from some suitable mass distribution model of the Galaxy. This can be done by assuming the Galaxy to consist of several components, which in combination will yield the observed circular velocities in the inner regions. For the various components, a central mass-point, inhomogeneous spheroids and outer shells have been considered. Models have been constructed by many authors, the pioneering works having been done independently by M. Schmidt, G. Burbidge with his co-workers, and L. Perek. Schmidt's model (1965) for the Galaxy, although rather simple, has many salient features. The circular velocities yielded by his model have been listed in column 2 of Table 16.3.

TABlE 16.3   Circular Velocities in Our Galaxy

Distance from centre (kpc)	θ_c (km s^{-1}) Schmidt's Model	θ_c (km s^{-1}) Bok's Model
1.0	200	240
2.5	190	210
5.0	227	230
7.5	250	230
10.0	250	230
12.5	235	250
15.0	218	260
17.5	205	300
20.0	193	300

Many authors have however, suggested from a dynamical point of view that the galactic disc can be stable only if the Galaxy possesses a very extensive (~ 100 kpc) and massive (~10^{12} $M_\odot$) halo. In such a case, the circular velocities much farther away from the Sun will be quite large and non-decreasing. Such an extensive mass distribution and flat or increasing rotation curves in spiral galaxies have been very well confirmed by the recent works of V.C. Rubin and her co-workers. On the basis of such observations, B.J. Bok calculated the galactic circular velocities which are given in column 3 of Table 16.3.

16.6   DENSITY DISTRIBUTION OF GAS AND SPIRAL STRUCTURE OF THE GALAXY: RADIO AND OPTICAL DATA

The basic observational material in the 21-cm study are the line profiles as illustrated in Fig. 16.10. The intensity $I(v)$ measured in the line profiles is related to the optical depth $\tau(v)$ by the formula

$$I(v) = I_0(v)\, (1 - e^{-\tau(v)}) \qquad (16.29)$$

where $I_0(v)$ is the intensity corrected for various effects such as the bandwidth and beamwidth and also for the continuous radiation at 21-cm wavelength. The optical depth is related to the kinetic temperature T of the gas and the number of hydrogen atoms in a velocity interval of 1 cm s^{-1} by the formula

$$N(V) = 1.835 \times 10^{13}\, T\tau(v) \qquad (16.30)$$

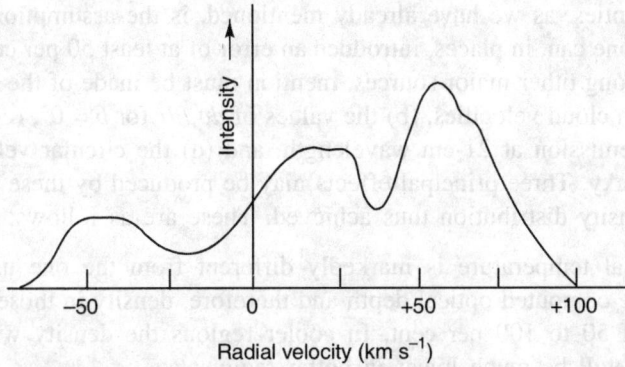

FIGURE 16.10 A typical line profile of neutral hydrogen at 21-cm wavelength.

Optical depth is affected by the random cloud motions, so one should correct it by assuming some model for cloud velocities. This has been done in the reduction of the 21-cm data. If $\tau'(v)$ is the corrected optical depth, then Eq. (16.30) changes to

$$N'(V) = 1.835 \times 10^{13} \, T\tau'(v) \tag{16.31}$$

In all 21-cm line work, $T = 125$ K has been used which probably contributes the largest single uncertainty in the 21-cm line reduction. $N'(V)$ is the number of hydrogen atoms in a velocity interval of 1 cm s^{-1} in the absence of any random motions. $N'(V)$ is related to the hydrogen number density $n_H(r)$ and the radial velocity gradient dv_r/dr, also called the widening factor, by the formula

$$n_H(r) = N'(V) \, dv_r/dr \tag{16.32}$$

When the gas temperature T is known (assumed to be 125 K in all reductions), $N'(V)$ is obtained from Eq. (16.31) by using the corrected optical depth $\tau'(v)$. dv_r/dr is known when the velocity-distance relation has been derived. In fact, dv_r/dr has been tabulated for all longitudes for the latitude $b = 0°$, that is, for the galactic plane. And it has been customary to use these values for other latitudes also which again introduces some uncertainty in the finally computed results. For example, for $b = 10°$ the values used may be in error by 10 per cent. When $N'(V)$ and dv_r/dr have been known as functions of r, the number density $n_H(r)$ of hydrogen is obtained as a function of r from Eq. (16.32). One can also express the number density in terms of the optical depth and the radial velocity gradient by the relation

$$n_H(r) = 0.0744 \, \tau'(v) \, dv_r/dr \tag{16.33}$$

where dv_r/dr is in km s^{-1} kpc^{-1} and $T = 125$ K has been used. Using suitable values on the right-hand side, one can compute the hydrogen density as a function of distance by using this formula.

Various simplifying assumptions had to be made at various stages of reduction of 21-cm line data in order to achieve the plausible density distribution. Due to this, large uncertainties may vitiate the finally computed density values. The most dangerous of these sources of uncertainties, as we have already mentioned, is the assumption of constant value $T = 125$ K. This alone can, in places, introduce an error of at least 50 per cent in the computed density values. Among other major sources, mention must be made of the use of (a) the same value of the random cloud velocities, (b) the values of dv_r/dr for $b = 0°$, (c) the discrimination of the continuous emission at 21-cm wavelength, and (d) the circular velocity model of the material in the Galaxy. Three principal effects may be produced by these uncertainties in the final picture of density distribution thus achieved. These are as follows:

1. If the actual temperature is markedly different from the one used, in some local regions, the computed optical depth and therefore, density in those regions may show an error of 50 to 100 per cent. In cooler regions the density will be much higher whereas it will be much lower in hotter regions.
2. The temperature differences will also change the radial width of the high density regions (spiral arms). The actual width will be larger if temperatures higher than actually used prevail and will be smaller for lower temperatures prevailing.
3. Deviation from the adopted circular motion of the material will *shift* the positions of spiral arms as a whole. This shift may be as much as 1 to 2 kpc from that measured on circular velocity model.

Considering various other factors that contribute to uncertainties in the computed density values, it would not be unreasonable to hold that the overall density and mass estimates obtained from 21-cm information may be low by a factor of two. Better quantitative estimates of density and total mass can be achieved only if much improved knowledge of temperature and velocity distribution can be obtained from higher resolution observations.

Whatever be the uncertainties, accepting the present picture as true, the distribution of neutral hydrogen in the galactic plane (based on the combined results of observations in the northern and southern hemispheres) indicates that the average density over the entire plane of the Galaxy is about 0.7 atoms cm^{-3} and the *density maxima corresponding to the spiral arms* are at $r = 4.5, 6.5, 8.5$ and 10.5 kpc (with $r_0 = 8.2$ kpc, $\omega_0 = 26.4$ km s^{-1} kpc^{-1}). With the revised values of $r_0 = 10$ kpc and $\omega_0 = 25$ km s^{-1} kpc^{-1}, these arms are placed approximately at $r = 6.0, 8.0, 10.0$ and 12.0 kpc corresponding to the inner arms of *Scutum, Sagittarius* and *Orion* and the outer arm of *Perseus* respectively. The arm spacings are thus approximately at intervals of 2 kpc. The northern, measurements show much more detail than southern, the counterpart of the Perseus arm being absent altogether in the southern hemisphere. The northern surveys show also two *intermediate arms* between Orion and Perseus arms and two more *outer arms* of lower densities than those of main arms. These are probably branches of Perseus arm. The Orion arm also branches off at $l = 45°$ and again at $80°$.

The greatest inaccuracy in the 21-cm work is introduced in the region of solar vicinity. On the other hand, the optical data are reliable only in this region. Therefore, the optical and radio observations are complementary to each other for depicting the entire Galaxy. The available optical data indicating spiral distribution have been summarized by S. Sharpless as given in Fig. 16.11. It is found that the optical arms and radio arms do not coincide. The optical arms cross the radio arms and sometimes pass through the regions where there is no hydrogen at all. The optical map depends on distances obtained from the apparent and absolute magnitudes of stars after correction for reddening and absorption of their light. The uncertainties

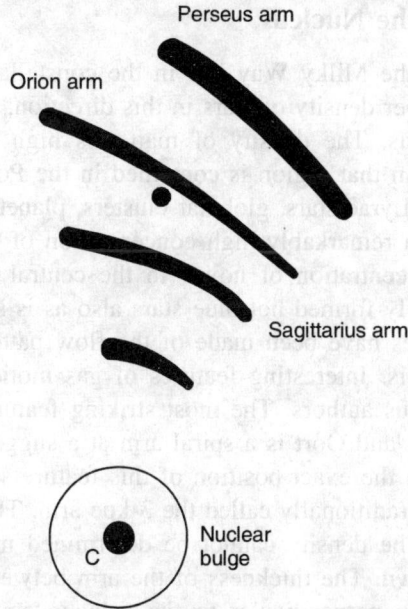

FIGURE 16.11   The nuclear bulge and the spiral arms of the Galaxy as obtained from optical data.

involved in radio reduction have already been mentioned. It has been suggested that if corrections could be properly made for both the data, almost perfect correlation between the gaseous and stellar arms could probably be achieved.

That the density distribution and spiral arm positions obtained for the Galaxy may be in substantial error is further suggested if we compare these with those of M31. These two galaxies are similar in many important respects. So one can expect that they will be similar also in hydrogen distribution and spiral pattern. The study of hydrogen distribution in M31 shows two outstanding differences from that in the Galaxy. First, at least four pronounced maxima have emerged in the Galaxy, while in M31 there are only two. Secondly, in the Galaxy the distribution abruptly falls off and there is hardly any hydrogen beyond 15 kpc, whereas in M31 the distribution is very flat and the gas layer extends about upto 25 kpc. Instead of these large differences being assumed real, one would like to believe that these are mostly due to the various simplifying assumptions made in the 21-cm line reduction in the Galaxy, particularly, due to the assumptions of constant temperature and circular velocity model. What has been derived for our Galaxy is thus just a working model.

16.7   THE GENERAL STRUCTURE OF THE GALAXY

The three principal parts into which our Galaxy (and all the normal spiral galaxies) can be divided are: (i) a dense central region in which the denser nucleus is embedded, (ii) the flat disc in which the spiral arms are embedded, and (iii) a large spherical halo.

The Central Region and the Nucleus

The brightest star cloud of the Milky Way lies in the constellation Sagittarius. Star counts indicate a much higher number density of stars in this direction, suggesting that the centre of the Galaxy lies in Sagittarius. The density of matter is high in the nuclear bulge and a significant part of the mass in that region is contained in the Population II objects. There is a great concentration of RR Lyrae stars, globular clusters, planetary nebulae etc., besides the old red stars. There is also a remarkably high concentration of novae in the Sagittarius star cloud, analogous to the concentration of novae in the central region of M31. The region appears to contain some newly formed hot blue stars also as is suggested by the presence of ionized gas. Extensive studies have been made of the flow pattern of the gas in the central region by 21-cm line analysis. Interesting features of gas motion in this region have been revealed by studies by various authors. The most striking feature of the region observed as early as in 1960 by Rougoor and Oort is a spiral arm at a suggested distance of 3-kpc from the galactic centre. Although the exact position of this feature seems to be more near 4 kpc from the centre, it has been traditionally called the 3-kpc arm. The average gas density in the arm is a few atoms cm^{-3}. The density cannot be determined more accurately as the radial width of the arm is not known. The thickness of the arm between half density points in the Z-direction (that is, direction perpendicular to the galactic equator) is 120 pc. The radial velocity measurements indicate that the arm is moving away from centre (in fact, moving towards us) with a speed of 53 km s^{-1}. The total gaseous mass (atomic and molecular) associated with this feature is of the order of 10^8 $M_\odot$. The observations of Rougoor and Oort further suggest the possible existence of an arm on the other side of the centre moving away from it with a much higher speed, probably in excess of 130 km s^{-1}. The total gaseous mass associated with this feature is of the order of one-half of that of 3-kpc arm. Inside 3-kpc arm, there appears to exist a vast region of low gas density. Further in, near the centre, the gas is concentrated in a rapidly rotating disc of radius 750 pc whose thickness is about 100 pc within 300 pc from the centre and about 250 pc in outer regions. The rotational velocities in the disc are very high, gradually decreasing inwards from 265 km s^{-1} at $r = 550$ pc to 180 km s^{-1} at $r = 70$ pc. The amount of hydrogen measured in the central region upto 750 pc is a few times 10^8 $M_\odot$ whereas, the total mass required to produce the observed circular velocity in this region is at least of the order of 10^{10} $M_\odot$. This mass is contributed mainly by stars and possibly also by a supermassive nonstellar body at the centre. Oort calculated the total mass to be 1.6×10^8 $M_\odot$ within a radius of 20 pc whereas, Rougoor and Oort computed the total density of 105 $M_\odot$ pc^{-3} at $r = 100$ pc to obtain the observed rotation of the nuclear disc. Superimposed on the rotational motion, explosive motions appear to be a very prominent feature in the central region. Gas is observed to be expelled radially with velocities as high as of the order of 200 km s^{-1}. These explosive phenomena are believed to be causally connected with high energy activities periodically undergoing in the deep central core. The mass of gas apparently flowing out of the central region seems to lie between 1 and 4 $M_\odot$ yr^{-1}. If this outward flow is real, then the lost gas must be replenished by some means in order to maintain this outflow over a sufficiently long time. One possible source of replenishment of the lost gas is probably the inner halo from where gas flows into the nuclear region of the Galaxy.

Completely veiled behind thick clouds of gas and dust, the centre of our Galaxy (the nucleus) remains hidden forever from human eyes. But radiation waves in radio, infrared, X-ray and γ-ray can penetrate through the layers of thick clouds and send message to us about the nature of the nucleus. All such information hints towards one conclusion that the galactic nucleus is a mysterious object.

All current observations lead to the conclusion that the galactic core lies at the strong radio source called Sagittarius A (Sgr A) or very close to it. Radio observations of Sgr A and its neighbourhood detect both thermal and nonthermal (synchrotron) radiation in sufficient strengths. The thermal radiation is emitted from ionized hydrogen clouds indicating that there must be many hot stars in the region. The fact that the hot stars are short lived further indicates that the star-formation process must be going on in the region. The detection of strong nonthermal radiation at the galactic centre region indicates the presence of motion of streams of relativistic electrons through the magnetic field in the region. So there should be high energy sources in the region that may accelerate electrons to relativistic speeds. Supernovae and pulsars are, among others, the likely candidates.

The central source Sgr A and its close neighbourhood also emits strongly in near infrared wavelengths. The analysis of infrared radiation shows some discrete infrared sources superimposed on background infrared radiation. The discrete sources may be dust clouds heated by strong radiation from hot stars, but most of the infrared luminosity is believed to originate in a fairly dense population of red giant stars. The galactic centre region thus appears to contain a rich population of hot blue as well as red giant stars. The region thus indicates a signature of brisk star forming activity.

Several orbiting space telescopes have peered through the galactic centre. The centre has been found to be a significantly strong source of X-rays (particularly, hard X-rays) and γ-rays. The sources of X-rays may be neutron stars, black holes in X-ray binaries and/or the local gas sucked in by a central massive black hole. Collisions of high-energy particles or the annihilation of the electron-positron pairs may produce the observed γ-ray intensity. The galactic centre, being so rich in both high-energy photons as well as high-energy electrons, is a likely seat where both the Compton effect as well as the inverse Compton effect play, important roles.

In close vicinity of Sgr A lies a giant molecular cloud, Sgr B2, known to possess the richest variety of interstellar molecules. This cloud is one of the best studied molecular clouds by the molecule hunting radio astronomers. The presence of molecular clouds in and around the galactic centre again ensures that the region is likely to be a seat of brisk star-forming activity.

The mass distribution and the total mass within a few parsecs from the galactic centre have been modeled by various authors on the basis of the observed velocities of gas clouds. Lacy and his co-workers (1980), and Serabyn and his co-workers (1985, 1988) have extensively measured the gas velocities in this region using the fine structure of Ne II. Velocities upto 400 km s^{-1} have been measured. Although circular velocity is predominant, turbulent velocities are also high. Based on the results of velocity measurements, attempts have been made by authors to draw a reliable picture of mass content within certain limits of the galactic centre. Analyzing the gaseous orbits Serabyn and his co-workers conclude that the dynamic centre of the gaseous motion is very close to Sgr A. The same conclusion prevails on the basis of the stellar measurements by Sellgren and his co-workers (1987) and by Ricke and Ricke (Townes, 1989). The various studies thus indicate that Sgr A is the most likely "candidate" for the galactic centre.

A very high concentration of mass at the centre has been worked out by most workers. Ricke and Ricke (Townes, 1989) computed the value of $10^7 \, M_\odot$ within 0.5 pc and $3.3 \times 10^7 \, M_\odot$ within 2 pc. Townes' (1989) value is $\sim 5 \times 10^6 \, M_\odot$ within 1.7 pc. A mass of a few times $10^6 \, M_\odot$ is predicted by M.A. Seeds (1990) within a radius of less than 10 a.u. Analyzing the motion of the molecular ring occupying the region outside 1.5 pc, Lacy (1989) computed the mass value of $4.2 \times 10^6 \, M_\odot$ inside 1.5 pc which increases linearly to $1.12 \times 10^7 \, M_\odot$ at ~ 4 pc. From models of ionized gas flow Serabyn and Lacy calculated the value to be greater than $2.8 \times 10^6 \, M_\odot$ inside 0.4 pc. The flow can also be modeled by a point-like mass $\sim 2 \times 10^6 \, M_\odot$ together with a star cluster containing about $2 \times 10^6 \, M_\odot \, pc^{-1}$ radius. Many other determinations of the core mass are available in the literature, but the discussion of these is beyond the scope of this book.

Analyzing the observed high-energy radiations and high-velocity gas motions, F.A. Shu (1982) prefers to predict the presence of a $10^6 \, M_\odot$ black hole at the centre of the Galaxy. Many astronomers do believe that the observed energetic phenomena in the nuclei of galaxies, in general, are most logically explained by assuming the presence of a massive black hole at the centres of the galaxies, and our own Galaxy is no exception. Dynamical evidence also supports the suggestion of a massive black hole at the galactic centre. However, the evidence is not compelling to accept the black hole hypothesis. On the other hand, the corroborating evidence is severely lacking. For example, the presence of a black hole at the centre would manifest the radiation emitted by the accreting gas as a single source, but no such source has yet been identified. The measured ionizing ultraviolet radiation also does not fit to that as may be produced by black hole accretion. In fact, such an accretion disc should be brighter by an order of magnitude than any of the observed sources. It therefore appears that the black hole does not provide the ionizing radiation. We conclude by suggesting that the central massive object remains mysterious and w.e badly need convincing evidence to establish its true nature. It may be a black hole or it may be not. Even if it is a black hole, it need not be as massive as $10^6 \, M_\odot$.

The Galactic Disc

Starting from beyond the nuclear bulge, a flat distribution of gas and stars extends upto about 15 kpc from the centre. The distribution of mass is very flat which forms as if a disc of matter, and so has been called the *galactic disc*. Most of the gas, both neutral and ionized, in the Galaxy is strongly concentrated along this disc. The spiral arms are all embedded in the gaseous disc. The distance between the half-density points hardly exceeds 250 pc anywhere upto about the Sun's distance. Beyond the solar distance, the gas layer becomes thicker, bending upward in the northern hemisphere and downward in the southern hemisphere. The distorted region exhibits an accumulation of a number of spiral features and gradually goes away from the plane as we move away from the centre. F.D. Kahn and L. Woltjer have tried to explain this distortion as the manifestation of the pressure exerted on the galactic halo as it moves through the intergalactic medium. The suggestion that the distortion has been caused by the tidal interaction of the Large Magellanic Cloud with the Galaxy has been made by various authors. But both F.J. Kerr and B.F. Burke have emphasized that mere gravitational effect of the LMC can produce only about one per cent of the observed phenomena. The bending or *warping* of the gaseous disc in the outer parts is probably caused by gravitational instability which originates within the Galaxy itself. This is corroborated by the fact that warping of the disc has been observed also in some of those galaxies which appear to be completely isolated from any possible gravitational influence of their neighbours.

The stellar disc is about 1 kpc thick. Among the stars the extreme Population I objects such as the blue and hot OB stars, H II regions, Cepheid variables, galactic clusters etc. are strongly concentrated along the gaseous disc. Stars of somewhat older origin such as the red variables and late-type giants and dwarfs have a much thicker distribution along the Z-direction and extends, on average, to about 500–600 pc on either side of the galactic plane. Also, some Population II objects such as the RR Lyrae and other blue variable stars, planetary nebulae and a few globular clusters are sprinkled along the disc. The Sun lies almost on the central plane at a distance of around 10 kpc from the centre. From a study of the stellar motion and their distribution perpendicular to the galactic plane $Z = 0$, Oort has computed the total mass density in the plane in the solar neighbourhood. He finds a total mass density of 0.148 $M_\odot$ pc^{-3} of which 20 per cent (= 0.03 $M_\odot$ pc^{-3}) belongs to observed gas, both ionized and neutral and 40 per cent (= 0.06 $M_\odot$ pc^{-3}) belongs to the observed stars. The remaining 40 per cent of the mass is not observed. This mass has remained hidden which may be in dead stars or in substellar objects like brown dwarfs and Jupiters. Another possibility is that the unobserved mass may exist partly in the gaseous and partly in the stellar form. The problem of the unobserved mass has not yet been settled with any definiteness.

The Galactic Halo

Although the main body of the Galaxy is confined to a relatively flat disc just described, the globular clusters and high velocity stars like subdwarfs and Population II cepheids (RR Lyraes and Type II cepheids) define a more or less spherical system superimposed upon the disc. These stars and globular clusters together with tenuous ionized gas and cosmic rays constitute what has been called the halo or the *Corona* of the Galaxy. The volume of the halo is vastly greater than that of the main disc of the Galaxy. The spatial density of stars and clusters in the halo increases as we go towards the galactic plane. On either side of the plane, towards Scorpius, Ophiuchus and Sagittarius, lie the greatest number of stars and clusters belonging to the halo. Individual RR Lyrae stars and globular clusters have been found as far away as 10 to 15 kpc on either side of the galactic plane. This indicates that the galactic halo must have a radial extent of at least 20 kpc; probably it is much more, may be as large as 50 kpc or even more.

High-velocity hydrogen clouds have been observed and analyzed by Munch and Zirin and by Muller, Oort and Raimond. Munch and Zirin observed the interstellar absorption lines of Ca II in stars more than 200 pc above the galactic plane while the latter authors observed in 21-cm radio line. Both observations indicate that neutral gas clouds extend in the halo to at least 1 kpc above the galactic plane. The extreme radial velocity of 175 km s^{-1} measured in one such cloud suggests that the region of their extension may be even much higher above the plane. The projected density of this cloud has been measured to be 2×10^{20} hydrogen atoms cm^{-2}. The mechanism by which such massive clouds can attain such high velocities is not well understood. All these observations point to one definite conclusion that some neutral hydrogen must be present in the halo. The available data suggest that the average density of neutral hydrogen in the halo is of the order of 10^{-3} atom cm^{-3}.

Much speculations have been made as to how these interstellar clouds maintain their equilibrium in the halo. If the halo was an empty space, these gas clouds would have quickly expanded and diffused into insignificance. In order to maintain their identity in the halo, there should be some kind of balance between the clouds and the halo material. Three different mechanisms have been proposed by which these clouds could be confined in the halo. According to the first mechanism suggested by L. Spitzer (Jr.), the clouds are assumed to be in equilibrium with a hot tenuous gas which is in hydrostatic balance with the galactic gravitational field. Spitzer's calculations indicated that the balance would be achieved between the moving clouds and a hot halo gas of temperature 10^6 K and particle density of 10^{-3} cm^{-3}. This model of the halo has generally been known as the *hot halo model* of the Galaxy. S.B. Pikelner and I. Shklovsky, on the other hand, have suggested a balance of the moving clouds with a much cooler high-density halo, with the temperature of 10^4 K and a particle density of 10^{-2} cm^{-3}. The halo proposed by these authors is supported by *hydromagnetic pressure waves* originating in the central regions of the Galaxy, and not by hydrostatic pressure as proposed by Spitzer. The halo proposed by Pikelner and Shklovsky has generally been known as the *cool halo model* of the Galaxy. Still a third halo model has been proposed by L. Woltjer according to which no gaseous medium in the halo is required to confine the clouds there. He has proposed that if there exists a magnetic field of strength about 6×10^{-6} gauss in the halo and the internal field of the halo clouds is of the same order as that in the disc clouds, then the clouds may be confined in the halo by achieving a balance between the internal and external magnetic stresses. The halo thus proposed by Woltjer has been generally known as *the magnetic halo* of the Galaxy.

It must be mentioned, however, that all the three halo models that have been proposed so far are subject to many criticisms. None of these models has been able to establish itself uniquely. However, a halo with magnetic field and cosmic rays is consistent with the observed continuous radio emission in the Galaxy. Such a halo must contain a tenuous distribution of matter, the density and temperature of which still remain as subjects of speculation.

16.8 THE MASS OF THE GALAXY

The subject of determination of the mass of the Galaxy has interested many astronomers since the last few decades. Prior to the construction of the rotation curve of the Galaxy in early fifties, the astronomers had to rely on approximate models of distributions of mass and velocity of the material in the Galaxy which finally could lead to its total mass. Various determinations of the mass of our Galaxy have thus been made. Camm (1938) obtained $M = 1.77 \times 10^{11}$ $M_\odot$, Lohmann (1953, 1956) obtained in two models $M = 2 \times 10^{11}$ and 2.5×10^{11} $M_\odot$ and the value obtained by Bucerious (1934) was 2.4×10^{11} $M_\odot$. Various other models such as those of Takase, Perek, Safronov etc. yielded masses of the Galaxy in the range $0.6 - 0.8 \times 10^{11}$ $M_\odot$. Many other earlier authors have constructed models to get the mass of the Galaxy. Using the rotation curve of the Galaxy, M. Schmidt constructed two models, one in 1956 and the other in 1965. His first model gave a mass of 0.71×10^{11} $M_\odot$ and the second model yielded 1.8×10^{11} $M_\odot$. Brandt's model yielded a mass of 1.8×10^{11} $M_\odot$.

More recent observations of the external galaxies by V.C. Rubin and her co-workers, however, indicate conclusively that galaxies are much more massive and are of much greater extension, in general. This mass, however, exists in unobserved form. Our Galaxy probably has a radius of 50 kpc or even more with a total mass of the order of 10^{12} $M_\odot$.

More recently, the galactic mass models have been computed by many authors on the basic concept that the total gravitational field in the Galaxy producing the observed rotation curve is the result of superposition of the individual gravitational fields of several components of the Galaxy. Of these various components, a massive dark halo or corona extending to very large distances from the centre emerges as a common feature. The existence of an extensive massive dark halo is suggested by recent observations of the extensive flat rotation curves in all types of spiral galaxies. Our Galaxy being a normal giant $Sb - Sc$ type spiral, it is also believed to possess a massive dark halo.

For flat rotation curves, we have θ_c = const, and for circular velocity balance, $M(r) \propto r$, so $\rho \propto r^2$. Thus shells of equal thickness contain equal masses. The halo therefore extends upto the distances where the density of matter in the halo is of the same order as the density of the intergalactic matter.

J.A.R. Caldwell and J.P. Ostriker (1981) compute a three-component model of the Galaxy— a modified exponential disc component, a spheroidal (bulge + nucleus) component and a corona. For local parameters they use $r_0 = 9.1 + 0.6$ kpc and $v_0 = 243 \pm 20$ km s^{-1}. The masses obtained by them for these components are $M_D = 0.0663$, $M_S = 0.0599$ and $M_c = 0.930$ in 10^{12} $M_\odot$ unit for a radius of 100 kpc for the Galaxy. The total mass computed for the model is therefore 1.0562×10^{12} $M_\odot$ upto a radius of 100 kpc of the Galaxy.

Also, a realistic three-component mass model of the Galaxy is given by K. Rohlfs and J. Krietschmann (1981). This model consists of a modified exponential disc, a spherical inner bulge and a massive and extensive outer corona. The model fits well with many observed features of the Galaxy within the solar distance such as, the steep rise of the circular velocity around $r = 2$ kpc and a reasonably constant rotation curve outside the solar circle. Also, the local density value $\rho_0 = 0.15$ $M_\odot$ pc^{-3} and values of k_z as given by Oort (1965) are well represented by the model. The values of the local parameters used are $r_0 = 8.5$ kpc and $v_0 = 225$ km s^{-1}, and the total mass derived for the Galaxy is approximately 5.826×10^{11} $M_\odot$ upto a distance of 100 kpc. J. Einasto's (1979) model of the Galaxy consists of six different components of mass distribution, viz., the nucleus, the bulge, the halo, the disc, the flat population and the corona. The values of the local parameters adopted by him are $r_0 = 8.5$ kpc and $v_0 = 225$ km s^{-1}. The total mass of Einasto's model approximately equals 1.2×10^{12} $M_\odot$ within a distance of 100 kpc from the centre. A more recent model of the Galaxy has been given by U. Haud, M. Joeveer, and J. Einasto (1985). These authors also consider the Galaxy to be composed of six different components, the same as considered by Einasto (1979). The values of the local parameters chosen here are $r_0 = 8.5$ kpc and $v_0 = 220$ km s^{-1}, but the truncated radius of the corona is considered to possess a very large value of 390 kpc. The total mass of their model is about 2.1×10^{12} $M_\odot$.

These and all other current mass models of the Galaxy indicate that the Galaxy consists of several components having different density distributions and that a massive and extensive dark corona envelops the visible matter in it. The total mass is $\sim 10^{12}$ $M_\odot$, about ten times of what was presumed a couple of decades ago.

16.9 MAGNETIC FIELD IN THE GALAXY

The existence of a large-scale magnetic field in the Galaxy was first proposed in 1949 by Enrico Fermi to explain the confinement of the cosmic rays in it over a long time-scale. Calculations show that these relativistic particles cannot be kept confined within the limits of the Galaxy for a long time unless the energy density of the magnetic field exceeds that of the cosmic rays which is of the order of 10^{-12} erg cm^{-3} in the solar neighbourhood. Almost simultaneously, it was observed by some astronomers (W.A. Hiltner, J.S. Hall, etc.) that light coming from distant stars, particularly from those lying close to the galactic plane, is polarized. These observations clearly demonstrated the existence of large-scale magnetic field in the Galaxy. Some astrophysicists came forward immediately to provide plausible interpretations regarding the structure and strength of the magnetic field. Subsequent years of studies have conclusively revealed that the Galaxy is pervaded by a large-scale magnetic field of strength of a few microgauss ($\mu G = 10^{-6}$ gauss) with a probable orientation parallel to the spiral arms near the galactic plane. The direct observable evidences for such conclusions come from the study of (a) the polarization of light from distant stars; (b) the Faraday rotation of the plane of polarization of linearly polarized radio radiation from discrete radio sources; (c) the Zeeman splitting of the interstellar 21-cm line; and (d) the magnetobremsstrahlung or synchrotron radiation which is non-thermal in nature.

It was recognized that the polarization of light of the distant stars was produced by some particular mode of orientation of the interstellar grains sprinkled tenuously throughout the intervening interstellar space. The amount of polarization was observed to vary with galactic longitude in the manner that when stars were observed in the direction perpendicular to the spiral arms, the polarization was greater and more regular. Lesser amount of polarization was observed, on the other hand, for the light from stars viewed obliquely to the spiral arms. These observations can be interpreted by assuming that the magnetic lines of force are parallel to the spiral arms and the elongated grains lie perpendicular to them. It has been proposed that if the grains were paramagnetic, a field strength of a few times 10^{-5} gauss could explain the observed results. On the other hand, a field strength of 10^{-6} gauss would be sufficient to explain the observations if the grains were ferromagnetic. But as the nature of the grains is unknown, no decisive conclusion regarding the field strength can be drawn. From a consideration of the dispersion of the polarization vectors around the mean, Chandrasekhar and Fermi estimated the field strength of the order of $(2–3) \times 10^{-5}$ gauss. If the mean square angular dispersion of the polarization vectors be α^2, ρ be the density of the gas, **B** be the magnetic field and σ_x be the velocity dispersion of the clouds in one direction, then we have

$$\alpha^2 = \frac{4\pi\rho\sigma_x^2}{B^2} = \left\langle \frac{b^2}{B^2} \right\rangle \qquad (16.34)$$

where **b** is the perturbation of the field.

By substituting plausible values for the quantities involved one would obtain the field strength of the order mentioned above. But this result also is not certain as the detailed structure of the field as well as the relative magnitudes of the seed field and the perturbed field are not known. Thus the polarization data although indicate a large-scale magnetic field in the

Galaxy, oriented probably parallel to the galactic plane as well as the spiral arms, do not conclusively yield the strength of the field.

We consider next the problem of estimation of the strength of the galactic magnetic field by measuring the *Faraday rotation*. When a polarized electromagnetic radiation traverses through a *dispersive* medium, its plane of polarization undergoes a rotation which is known as Faraday rotation. The polarized radiation emerges from some source outside the medium considered, the polarization being assumed to be independent of the frequency which traverses through the medium in which the refractive index and consequently, the speed of propagation of the two circularly polarized components slightly differ from each other. The right-hand circularly polarized component (which has counter-clockwise rotation when viewed into the plane of the sky) gives higher velocity. In a medium containing n_e free electrons per unit volume and pervaded by the magnetic field **B** the difference between the velocities is given by

$$V_R - V_L = \frac{n_e \lambda^3 e^3 B_{\parallel}}{2\pi^2 c^3 m_e^2} \qquad (16.35)$$

measured in the cgs units. Here λ is the wavelength of radiation, $B_{\parallel}$ is the component of the magnetic field in the direction of propagation, c is the speed of light in free space and e and m_e are the charge and mass of the electron, respectively. In consequence of this velocity difference of the two components, the plane of polarization of the plane-polarized wave undergoes a rotation to the right through an angle, which is expressed as

$$d\psi = \frac{\pi dl}{\lambda c} \, (V_R - V_L) = 2.62 \times 10^{-17} \lambda^2 n_e B_{\parallel} \, dl \qquad (16.36)$$

where dl is the path traversed by the wave measured in the cgs. unit and ψ is in radian. Integrating over the entire length, L of the path traversed by the radiation from the source to the observer, and expressing in units convenient for astronomical use we can write Eq. (16.36) as

$$\psi = 0.81 \, \lambda^2 \int_0^L n_e B_{\parallel} \, dl \qquad (16.37)$$

where ψ is in radians, λ is in metres, n_e is in cm^{-3}. L and dl are in parsecs and $B_{\parallel}$ is in μG. The quantity

$$RM = 0.81 \int_0^L n_e B_{\parallel} \, dl \qquad (16.38)$$

is called the *rotation measure* which is independent of the frequency of radiation and expressed in rad m^{-2}. Many authors have measured Faraday rotations for radiation coming from various sources. Faraday rotation of several radians has been measured and it is found to vary with galactic latitude. The method has yielded the strength of the galactic magnetic field of a few μG. There are however certain sources of uncertainty inherent in the method. In the first place, the distances of the radio sources are not exactly known. Secondly, the knowledge is far worse for the electron density in the intervening space. Only an average value is generally used in calculations which may be in error by a large factor. Thirdly, a part of the measured Faraday rotation may occur in the source itself. There is practically no way to separate completely the two parts, one occurring in the source itself and the other occurring in the intervening space.

The Zeeman effect on the 21-cm hydrogen line provides a direct evidence of the presence of the large-scale magnetic field in the interstellar space.

When exposed to a magnetic field of strength B, this line is split into three components whose frequencies are

$$v_0, v_0 \pm \frac{eB_\parallel}{4\pi m_e c}$$

where v_0 is the undisturbed frequency and $B_\parallel$ is the longitudinal component of the magnetic field measured in gauss. When viewed parallel to the magnetic field, the undisplaced component eB_n corresponding to the frequency v_0 is not observed while the two displaced components

$v_0 \pm \dfrac{eB_\parallel}{4\pi m_e c}$ are observed to be circularly polarized in opposite senses. J.G. Bolton and J.P. Wild first suggested that if this splitting was observed in the 21-cm line of hydrogen and could be measured, then the strength of the interstellar magnetic field could be inferred. Soon after, many astronomers set themselves to such observations but most of them were unsuccessful in detecting any Zeeman splitting in the line. Only R.D. Davies and his co-workers reported successful observations. They observed Zeeman spliting of one of the components of the 21-cm line in the direction of Taurus A corresponding to a field strength of 25 μG, while the other component showed null result. They also reported a smaller Zeeman splitting in the direction of Cassiopia A corresponding to a magnetic field strength of 10 μG. The subsequent works of many other authors however have demonstrated that a strong general field of the order of 25 μG in the Galaxy, reported by Davies and his co-workers differed from the actual results; a weaker general field of a few μG appears more plausible in view of the results obtained from various types of observations.

The last direct evidence of the existence of a large-scale general magnetic field in the Galaxy is supplied by the detection of the *synchrotron radiation* within it. This radiation is produced by the ultrarelativistic charged particles (mainly clectrons) as they spiral round the magnetic lines of force of the galactic field. The electrons move with velocity v in a field of strength B in helical paths with the gyrofrequency

$$\omega_B = \frac{eB}{m_e c \gamma}$$

where

$$\gamma = \left(1 - \frac{v^2}{c^2}\right)^{-1/2} \tag{16.39}$$

as shown in Fig. 16.12. The radius of gyration of the particle of energy E moving at an angle θ to the direction of the magnetic field of strength B is given by

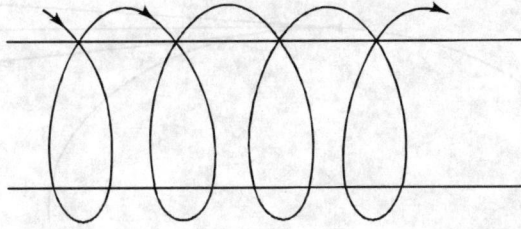

FIGURE 16.12 The helical path of the relativistic electron.

$$r_B = \frac{E \sin \theta}{eB} \tag{16.40}$$

This radiation is detected in radio frequencies from all directions of the sky with two notably strong components, one of which is supposed to originate in the disc and the other in the halo of the Galaxy. At 85 MHz, B.Y. Mills measured this radiation of about 1.5×10^{-39} erg s^{-1} cm^{-3} Hz^{-1} in the halo component whereas the value measured in the disc near the Sun was about 10 times larger than this. The spectral distribution of this radiation may be written as

$$J(v) \propto v^{-\alpha_s} \tag{16.41}$$

where $J(v)$ is the volume emissivity and α_s is the spectral index. The value of α_s has often been taken as constant and equal to 0.5, but it may depend on frequency. The radiation measured actually comes from the tenuously distributed ultrarelativistic electrons possessing a differential energy spectrum. This distribution with respect to the particle energy is represented by

$$n(\xi) = K\xi^{-\beta} \, d\xi \tag{16.42}$$

where ξ is the energy of the particle measured in BeV and β is the spectral index of particle energies, K being a constant to be evaluated by observing the flux of the particles near the Earth. With this distribution law of the radiating particles moving in a magnetic field **B**, the volume emissivity may be represented by

$$J(v) = K(10^5 \, B \sin \theta)^{(\beta+1)/2} \, (10^{-8} \, v)^{-(\beta-1)/2} \, F(\beta) \tag{16.43}$$

where B is measured in gauss and θ is the angle between, the direction of motion of the electron and the magnetic field **B**. The value $\beta = 2$ has often been used since this value corresponds to the value $\alpha_s = 0.5$ of the spectral index of the radio radiation. Also for this value of β, $F(\beta)$ does not depend heavily on β. The numerical value of $F(\beta)$ for $\beta = 2$ is given by

$$F(2) = 3.09 \times 10^{-27} \tag{16.44}$$

The electrons radiate in a narrow cone of nearly a minute of arc around the directions of their instantaneous motion (Fig. 16.13) and close to a critical frequency v_c given by

$$v_c = 1.61 \times 10^{13} \, B_\perp \, \xi^2 \text{ (BeV) C/s} \tag{16.45}$$

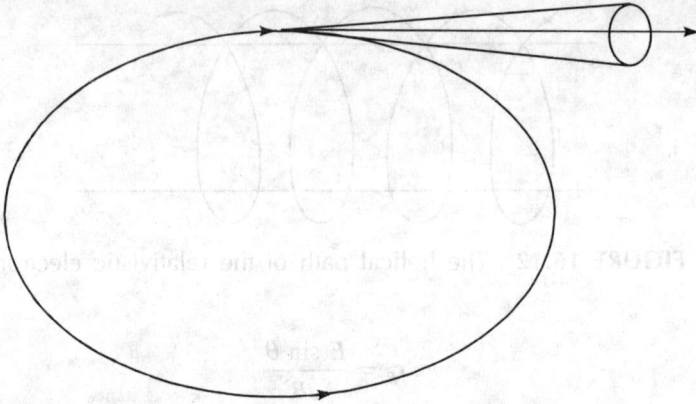

FIGURE 16.13   The radiation cone of the relativistic electron.

With some simplifying assumptions the observed volume emissivity near the Sun leads to the value

$$KB^{3/2} = 1.9 \times 10^{-10} \tag{16.46}$$

From observed flux of the particles near the Earth the value of K is found to be of the order of 10^{-12}. Allowing the uncertainty in the value of K, the values of $B = (17–35)\ \mu G$ are obtained in the spiral arm near the Sun. These values appear to be much higher than those obtained by most authors. The reason probably is that many uncertainties are involved in the measurements of the galactic magnetic field by using the synchrotron radiation data.

As we have seen so far, high values of the galactic magnetic field have been yielded only in a few cases. The works of majority of the authors however point to a galactic magnetic field of a few μG. The values between 5 to 7 μG seem to be plausible for the galactic disc. The value in the halo is probably less. In the nucleus of the Galaxy, however, due to the high density and large-scale turbulent motion of the gas, the magnetic field may be stronger by a factor of 100 than in the disc. The lower values in the disc are also supported by indirect evidences. The confinement of the cosmic rays in the Galaxy does not demand a much greater value than have been mentioned. Spitzer has argued that it would be difficult to explain the observed rate of star formation in the disc, if a large-scale uniform magnetic field of strength greater than a few μG did actually exist.

16.10 COSMIC RAYS

The atmosphere of the Earth is being continuously bombarded by swarms of relativistic and ultra-relativistic particles coming uniformly from all sides. These particles, originally discovered by V.F. Hess in 1912, have been named by him as *cosmic rays*. Particles possessing energies ranging from 10^9 eV to 10^{20} eV have been identified, although those having energies higher than 10^{16}–10^{17} eV are very rare. Those near the low energy end are overwhelmingly abundant and the mean energy per particle detected is about $(4–5) \times 10^9$ eV. Particles possessing energies lower than 10^9 eV (= 1 BeV) are probably still more abundant in the Galaxy, but due to their low energy they cannot penetrate through the atmosphere of the Earth and thus remain undetected. The energy spectrum of cosmic rays follows a power law of the form $n(E)\, dE = KE^{-\gamma}\, dE$, the spectral index having values near 2.5. The average number density of the particles detected at the Earth is about 10^{-9} cm^{-3} and the average energy density is about a few times 10^{-12} erg cm^{-3}. This is about the same as that of the turbulent motion of the interstellar gas. The magnetic energy density of the interstellar field is higher by an order of magnitude, if the field strength is a few times 10^{-5} gauss, whereas if the field strength is a few μG, the energy density is lower than that of cosmic rays by an order of magnitude.

The origin of cosmic rays is believed to be associated with violent events. Violent agitation of gas imbedded in magnetic field can probably generate these superenergetic particles. Thus, explosions in galactic nuclei, novae and supernovae explosions, stellar flares etc. are believed to be the sources of cosmic rays. The last mentioned event as the source of cosmic rays has some observational evidence in their enhancement during the solar flares. The gigantic supernova explosions are however believed to be the most prolific source of these superenergetic particles.

About 90% of the cosmic rays nuclei are protons. The rest are α-particles with smaller number of heavier nuclei. The electrons and positrons constitute only a few per cent of the cosmic rays. Relative to the overall cosmic abundance, the heavy nuclei with $Z \geq 20$ as also the light nuclei, Li, Be, B, are overabundant in cosmic rays by orders of magnitude. If the cosmic rays actually originate in the interiors of stars, then the overabundance of the heavy nuclei in them (almost by a factor of 10^3) finds a natural explanation, since the only places known where the heavy nuclei are overabundant are the stellar interiors. But the superabundance of the light nuclei, Li, Be, B, almost by a factor of 10^6 compared to the cosmic abundance, poses complicated problems for its explanation. There are no known objects in which these light nuclei are so overwhelmingly abundant. The most plausible explanation seems to be that these light nuclei are the result of spallation of the heavier nuclei in the interstellar space. Heavier nuclei disintegrate into lighter ones by collisions with other nuclei, the fragments retaining the velocity of the original nuclei. Assuming the validity of this assumption and observing the abundance of light nuclei relative to the heavy ones, one can estimate the amount of the interstellar material penetrated by these heavy nuclei before disintegration. Calculations show that the amount of matter required to be penetrated by the heavy nuclei is about 5 gm cm^{-2} which corresponds to a path length of $10^6\, n_{\mathrm{H}}^{-1}$ pc, n_{H} being the number density of hydrogen atoms cm^{-3}. A typical value in the galactic disc is $n_{\mathrm{H}} = 1$, giving a path length of 10^6 pc which corresponds to a lifetime of about 3×10^6 years if the cosmic rays are confined to the galactic disc. On the other hand, if they are confined in the galactic halo, where the typical gas density is 10^{-3} atoms cm^{-3}, the cosmic ray lifetime becomes about 3×10^9 years.

Assuming the confinement of the cosmic rays in the galactic disc of radius 15 kpc and of thickness 250 pc, yielding a total volume of about 4.8×10^{66} cm^3, the cosmic ray energy density of 1 to 2 times 10^{-12} erg cm^{-3}, and their lifetime of 3×10^6 years, the required energy generation rate of the cosmic rays is found to be about 5×10^{41} to 10^{42} erg s^{-1}.

It seems likely that the cosmic ray gas extends throughout the Galaxy in a quasi-steady state. If we agree that the cosmic ray flux in the solar neighbourhood at least has remained nearly constant during the last 10^9 years, then a uniform distribution of cosmic rays throughout the Galaxy seems most reasonable. During this period the Galaxy has revolved about its axis four or five times and the Sun has moved about 10 kpc relative to the surrounding stars. In spite of these facts, the marked isotropy of the cosmic rays suggests that they fill a region comparable to the entire volume of the Galaxy. Some of the highest energy cosmic ray particles may even be of extragalactic origin. In fact, three different theories on the quasi-steady extension on cosmic rays have so far been proposed:

1. Cosmic rays are confined to the galactic disc and halo.
2. They fill the region occupied by the Local Group of galaxies but do not extend beyond that region.
3. They fill the entire Universe.

Of these the galactic theory of the confinement of cosmic rays has been most widely discussed and generally held to be the most reasonable. The discussion of these different theories is beyond the scope of this text.

Cosmic rays play a major role in our galactic drama, the full implications of which are still not known. We have already seen that it is a dominant energy contributor in the interstellar region. The electron component of these particles are responsible for the entire non-thermal radio emission within the Galaxy. We have seen in the previous section how this radiation (synchrotron) enables to determine the strength and structure of the galactic magnetic field. These high energy particles are also believed to be major agents for heating the interstellar H I gas. In particular, the gas in the low-density region can be heated to a temperature of a few thousand degrees and also may be substantially ionized by the impact of these high energy particles. In that case, the generally accepted electron density (~ 0.1–0.01 electron-cm^{-3}) in the interstellar space has to be substantially revised. Unfortunately, the picture is not so clear due to the complete lack of knowledge of the cosmic ray energy spectrum on the low energy end ($< 10^9$ eV) which if extends smoothly, might be the most efficient agent for heating of the interstellar gas. The turbulent energy of the interstellar clouds which are periodically dissipated by cloud collisions may be partially replenished by the cosmic rays. These particles also play a vital role in the age determination of the meteorites.

Various other roles may be played by cosmic rays in the galactic phenomena which are probably still unknown. Nevertheless, physicists and astrophysicists are paying more and more attention to these relativistic space travellers.

16.11 CONTINUOUS RADIO EMISSION IN THE GALAXY

Cosmic radio emission was first detected and studied in the years 1930–35 by Karl Jansky, as a young physicist in the staff of the Bell Telephone Laboratories. In trying to locate the sources of noise that caused troubles in commercial radio telephony, he discovered that intense continuous radio noise was emitted from the entire region of the Milky Way band with a strong peak in the direction of Sagittarius which is now known to be the direction of the centre of our Galaxy. Jansky's work thus laid the foundation of modern *Radio Astronomy* and he earned the immortal glory of becoming the first radio astronomer. From a careful analysis of his long years of observations, Jansky concluded that the sources of the continuous radio emission were either the vast multitudes of stars or the interstellar gas sprinkled throughout the Milky Way. Subsequent studies by other astronomers have established that the stars cannot be the source of the intense radiation actually received on Earth and that the interstellar gas at various physical states is the real candidate. If the Sun is taken as a representative stellar source of radio emission, then the radiation intensity on Earth computed from the total number of stars in the Milky Way with their respective distances comes almost below the level of detection with the present-day instruments.

Observations have been carried out at various wavelengths ranging from a few centimetres to about 100 metres. Wavelengths shorter than the former are absorbed by various constituent gases of the Earth's atmosphere. Moreover, the intensity received at such wavelengths is very low. Wavelengths longer than the latter are almost completely reflected back by the ionosphere. The centimetre and meter wavelengths can pass freely throughout the vast span of interstellar space. These waves are not attenuated by interstellar dust particles which dim heavily the optical waves and travel unhindered the entire span of the Galaxy, On the other hand, the decametre waves are heavily absorbed in H II regions. So the suitable range of the radio spectrum to be used for the study of the galactic continuous radiation may be taken to lie between 0.1 to 20 metres. The observed intensity is high at longer wavelengths and falls off rapidly at shorter ones.

Radiation is received from all directions of the sky and careful analysis of the intensity distribution show mainly two components—(a) a background weaker component detectable in every direction of the sky, (b) another component superimposed on the weaker component which is many times more intense than the former, coming from localized regions of small angular sizes. These latter ones are called *discrete radio sources*. They may be both galactic and extra-galactic. Most of those sources which lie close to the galactic plane are galactic nebulae and have been named by Mills as Class I sources. They include the pulsars, supernova remnants, diffuse bright nebulae and the most intense source, i.e., the nucleus of the Galaxy. The discrete radio sources that are almost isotropically spread over the sky, away from the galactic plane, are almost all extragalactic. These have been called by Mills as Class II sources. They are *radio galaxies* and Quasi-stellar Radio Sources (QSS).

The background radiation also exhibits two principal features—a stronger component almost coinciding with the Milky Way belt with a sharp maximum in the direction of the galactic centre, and a general weaker component distributed almost isotropically but for a peak in the direction of the galactic centre. These observations were first explained by Shklovsky by proposing *two major subsystems* in which the galactic radio waves actually originate—an intensely radiating flat subsystem roughly coinciding with the galactic disc, and a nearly spherical subsystem of a vastly larger size than the former which he called the *Corona* of the Galaxy. Shklovsky thus recognized for the first time that our Galaxy possessed a Corona which was a source of continuous radio emission.

Years of investigations also revealed that the cosmic radio radiation as received might be produced by two distinct physical processes—*thermal* and *nonthermal*. Thermal radiation (first suggested by Reber in 1940) is emitted by the hot ionized gas due to free-free transition. The H II regions are the sources of this radiation. They have an average electron temperature T_e of 10^4 K and an electron density N_e of 10 cm^{-3}. If the radiation received both from the background as well as the discrete sources be interpreted as only due to the thermal radiation from the ionized gas regions and the corresponding brightness temperatures T_b are calculated, they generally lie in the range from 10^5 to 10^6 K for longer wavelengths of both the components. Since for thermal emission alone

$$T_b = T_e(1 - e^{-\tau}) \qquad (16.47)$$

where τ is the optical depth, $T_b \leq T_e$ always.

The observed radiation in the longer wavelengths thus cannot at all be reconciled with the possibility of thermal radiation alone. Also, the energy spectrum of the radiation at longer wavelengths differs greatly from that of thermal radiation. This is true for both the background and discrete sources. One is therefore forced to conclude that there must exist at least one more emission mechanism, which is largely responsible for the galactic continuous radio emission in the longer wavelengths. In fact, at these wavelengths thermal emission mechanism can explain only a minor fraction of the total observed radiation, although the observed emission at shorter wavelengths which is of greatly lower intensity than that at longer wavelengths can be explained in terms of the thermal emission alone.

Radiation by synchrotron mechanism has been accepted by the physicists as capable of explaining the high intensity of the cosmic radio radiation at longer wavelengths. This mechanism was first suggested by Alfvén and Herlofson to explain the observed radiation from discrete sources. The mechanism is nonthermal, the radiation being emitted by ultrarelativistic electrons (electron component of cosmic rays) spiraling round the magnetic lines of force of the large-scale galactic field as has already been explained in Section 16.9. Synchrotron radiation possesses two important properties: (a) the radiated power increases with wavelength until some critical wavelength is attained, and (b) the radiation is highly polarized. Both of these properties are verified for the high-intensity long-wavelength component of the cosmic radio emission. We have already seen that the analysis of the galactic synchrotron radiation imparts knowledge of the structure and strength of the galactic magnetic field and the large-scale distribution of the cosmic rays.

EXERCISES

1. Draw a neat sketch of the Milky Way Galaxy as seen from the Sun showing its different parts and indicate on it the distribution of:
   (a) gas and dust;
   (b) population I stars; and
   (c) population II stars including the globular clusters.

2. Discuss the observed effect of the differential rotation of the Galaxy on the radial velocity of stars lying in different longitude regions. Give a graphical representation of this effect.

3. Derive the formulae for the radial velocity, v_r and the tangential velocity, v_T, in terms of the Oort's constants A and B. Interpret physically A and B.

4. List the rotation parameters of the Galaxy. Discuss briefly how the value of each of these parameters is determined.

5. Describe the formation of the 21-cm radio line of neutral hydrogen. What information about the velocity and density structures of the Galactic gas have been obtained from 21-cm line observation? What assumptions were made for these observations?

6. Define *rotation curve* of a spiral galaxy. How is the rotation curve of our own Galaxy determined? Discuss both radio and optical observations.

7. The rotation curves of the Galaxy as obtained by observations from the northern and southern hemispheres are somewhat different. Discuss this difference and explain how this can be reconciled.

8. Draw a diagram of the rotation curve of the Galaxy and obtain a polynomial in the radial distance r that fits the rotation curve fairly well.

9. Read from your rotation curve the rotational velocities at $r = 1, 4, 6, 8, 10$ and 12 kpc, and calculate the periods of rotation of objects at each of these distances, assuming circular velocity.

10. How do we know that our Galaxy possesses spiral structure? How many spirals have been detected in the Galaxy? Draw a sketch of known spiral arms in the Galaxy relative to the position of the Sun in it.

11. Discuss how from the intensity of 21-cm line radiation the neutral hydrogen density is derived at different region of the Galaxy. What are the principal sources of uncertainty in thus deriving the gas density? What important information is obtained from this study about the structure of the Galaxy?

12. Describe the various methods that have been used for observing the Galactic centre. What physical and dynamical properties of the Galactic centre have been known on the bases of these observations?

13. Discuss briefly the important models of the halo of the Galaxy. Comment on the merits and demerits of each of these models. Which one of these models is best acceptable to you?

14. How can we know the mass of our Galaxy? What, in your opinion, appears to be the most reasonable value of the mass of the Galaxy? Assuming that the Sun revolves round the Galactic centre with a velocity 250 km s^{-1} at a distance of 10 kpc, and that the mass of the Galaxy is concentrated at its centre, calculate the mass of the Galaxy. Compare this mass with the most reasonable value.

15. Various observations indicate that the Galaxy possesses a large scale magnetic field: discuss these different observations and indicate how the strength of the magnetic field can be derived from them. What is the reasonable value for the strength of the field?

16. What are cosmic rays? How do they originate in the Galaxy? Comment on the composition and energy density of cosmic rays. What effects cosmic rays are supposed to produce on the interstellar gas?

17. What are the sources of the continuous radio emission that is measured in the Galaxy. Comment on the different components of this radiation.

SUGGESTED READING

1. Becker, W. and Contopoulos, G. (Eds.), *The Spiral Structure of Our Galaxy*, IAU Symposium No. 38, D. Reidel Publishing Company, Dordrecht, Holland, 1970.

2. Binney, J. and Tremaine, S., *Galactic Dynamics*, Princeton University Press, Princeton, New Jersey, 1987.

3. Blaauw, A. and Schmidt, M. (Eds.), *Galactic Structure*, The University of Chicago Press, Chicago, 1965.

4. Bok, B.J. and Bok, P.F., *The Milky Way*, 4th ed., Harvard University Press, Cambridge, 1973.

5. Burton, W.B. (Ed.), *The Large Scale Structure of Galaxies*, IAU Symposium No. 84, D. Reidel Publishing Company, Dordrecht, Holland, 1979.

6. Caldwell, J.A.R. and Ostriker, J.P., *Astrophysical Journal*, **251**, p. 61, 1981.

7. Lacy, J.H., IAU Symposium No. 136, p. 493, 1989.

8. Lacy, J.H., Townes, C.H., Gabelle, T.R. and Hollenback, D.J., *Astrophysical Journal*, **241**, p. 132, 1980.

9. Mavridis, L.N. (Ed.), *Structure and Evolution of the Galaxy*, D. Reidel Publishing Company, Dordrecht, Holland, 1971.

10. Mihalas, D. and Routiy, P.M., *Galactic Astronomy*, W.H. Freeman and Company, San Francisco, California, 1968.

11. Morris, M. (Ed.), *The Centre of the Galaxy*, IAU Symposium No. 136, Kluwer Academic Press, Dordrecht, Holland, 1989.

12. Rohlfs, K. and Kreitschmann, J., *Astrophysics and Space Science*, **79**, 289, 1981.

13. Seeds, M.A., *Foundations of Astronomy*, Wadworth Publishing Company, California, 1990.

14. Sellgren, K., Hall, D.N.B., Kleinman, S.G. and Scoville, N.Z., *Astrophysical Journal*, **317**, 1987.

15. Serabyn, E. and Lacy, J.H., *Astrophysical Journal*, **293**, p. 445, 1985.

16. Serabyn, E., Lacy, J.H., Townes, C.H. and Bharat, R., *Astrophysical Journal*, **326**, p. 171, 1988.

17. Shu, F.H., *The Physical Universe*, University Science Books, California, 1982.

18. Townes, C.H., IAU Symposium No. 136, p. 1, 1989.

19. Van Woerden, H., Allen, R.J. and Burton, W.B. (Eds.), *The Milky Way Galaxy*, IAU Symposium No. 106, D. Reidel Publishing Company, Dordrecht, Holland, 1985.

20. Zeilik, Michael, Gregory, Stephen, A. and Smith, Elske V.P., *Introductory Astronomy and Astrophysics*. Saunders College Publishing, New York, 1992.

17

External Galaxies

17.1 INTRODUCTION

On photographic plates, external galaxies appear as bright patches of nebulosity of different sizes and shapes. In photographs of those regions of the sky which are unobscured by the intervening dust and clouds, these nebulous objects appear in large number. Those with larger dimensions appear brighter, and the smaller ones gradually fade into the sky background. Earlier observers used to regard these objects as the nebulae belonging to our own Milky Way Galaxy. They could hardly form any idea of the actual distances and dimensions of these nebulae, although some philosophers like Emanuel Kant suggested that these nebulae were each a large stellar system similar to our Milky Way Galaxy. Better knowledge of these has however, gradually been developed. Observation of nova outburst in some of them first suggested that they were objects far beyond the limits of our own Galaxy, and the name *extragalactic nebulae* was given to them. Finally, E.P. Hubble's epoch-making work with the 100-inch Mount Wilson reflector in the early 1920s established beyond doubt that these extragalactic nebulae were each an extragalactic stellar system or an *island universe,* which has been known later as the external galaxies. Many of these galaxies belong to groups and many more to clusters. Many others again are observed to move through space in isolation or with a few companions. The clusters, in general, contain a large number of galaxies of various sizes and of different types. Certain rich clusters like *Virgo* and *Coma* are believed to be populated by several thousand galaxies. Our own Galaxy is a member of a group of more than 30 galaxies which constitute the *Local Group.* This group contains representatives of several types of galaxies. We shall discuss groups and clusters of galaxies separately in Chapter 18.

The galaxies were first classified in 1920s by Hubble into different types according to their shape and structural features. Later, Alan Sandage introduced some modifications in Hubble's classification and compiled a catalogue of them.

Much earlier than these workers, however, Charles Messier and J.L.E. Dreyer made an extensive observation of the galaxies (known to them as nebulae) and of globular clusters and made catalogues, the nomenclature of which are still in use. The celestial objects are represented by numbers in the catalogues made by them. An object in the Messier Catalogue is abbreviated by the letter M followed by a number. One such example is M31—the Andromeda nebula. Dreyer's catalogue which is a revised version of John Herschel's general catalogue of Nebulae is known as the *New General Catalogue* or simply NGC. The individual objects belonging to this catalogue are identified by a number. NGC 598 (the spiral galaxy in the Triangulam), NGC 205 (a companion of Andromeda spiral) are some examples of the members of this catalogue.

Two supplements of the New General Catalogue that include discoveries up to 1908 have been named as the Index Catalogues or simply IC. The majority of the NGC and IC objects are galaxies. The galaxies enlisted in the NGC catalogue are brighter than those in the IC catalogue. The observational techniques and the observing telescopes have substantially improved since 1908 and countless galaxies have since been discovered, but no extensive catalogues have been made to include these all in spite of the fact that much stress has been given to the studies of their distribution and characteristics.

17.2   CLASSIFICATION OF GALAXIES

Several classification schemes have been suggested for galaxies. The earliest and the simplest as well as the most popularly used today is that given by Hubble. His scheme consists of three regular classes—*ellipticals, spirals* and *barred spirals*. The irregular galaxies (Irr I and Irr II) form a fourth class of objects in his system of classification.

The elliptical galaxies are sub-classified by Hubble according to their degree of flattening or ellipticity. The spherical galaxies were termed by Hubble as E0 and the extremely flattened ellipticals as E7 as shown in Fig. 17.1. The galaxies having intermediate ellipticities were designated E1, E2, ..., E6. His classification of elliptical galaxies was based on the shapes of the images and not on their true shapes.

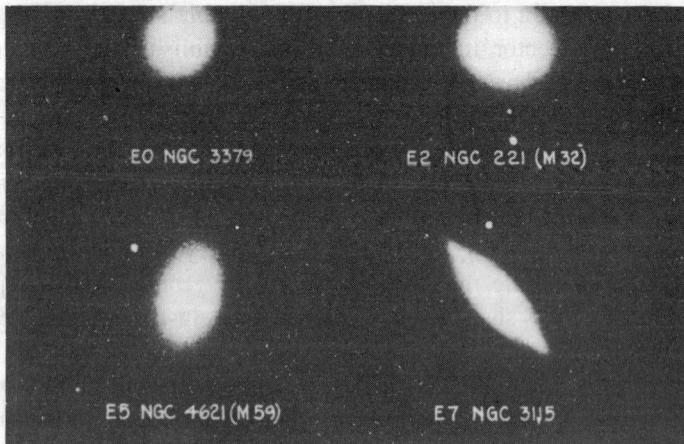

FIGURE 17.1 Elliptical galaxies with different ellipticities [*Courtesy:* Indian Institute of Astrophysics, Bangalore].

An E7 galaxy is in fact a very flat elliptical galaxy but it may be seen nearly edge-on, whereas an E0 galaxy may be of any degree of ellipticity but appears so as seen face-on. Each of the numbers 0 to 7 representing the flattening of the galaxies is defined in terms of the major and minor axes a and b by the relation $10(1 - (b/a))$, so that $b/a = 1$ for E0 and $b/a = 3/10$ for E7, the other values lying in between. The elliptical galaxies contain mostly Population II stars and consist of no spiral feature. They have much greater ranges in size, mass and luminosity as compared to the spirals. In clusters of galaxies nearly 80 per cent are ellipticals, but they form only 20–25 per cent of the field galaxies.

Our Galaxy, and M31 which is believed to be much alike ours, are examples of spiral galaxies (Fig. 17.2). M33 is another nearby spiral galaxy (Fig. 17.3). A typical spiral galaxy comprises a nucleus, a disc, a corona or halo and spiral arms. The arms of a spiral contain interstellar material and young stars including luminous supergiants. The spiral arms of a galaxy consist of the Population I stars whereas the nucleus and the corona contain mainly

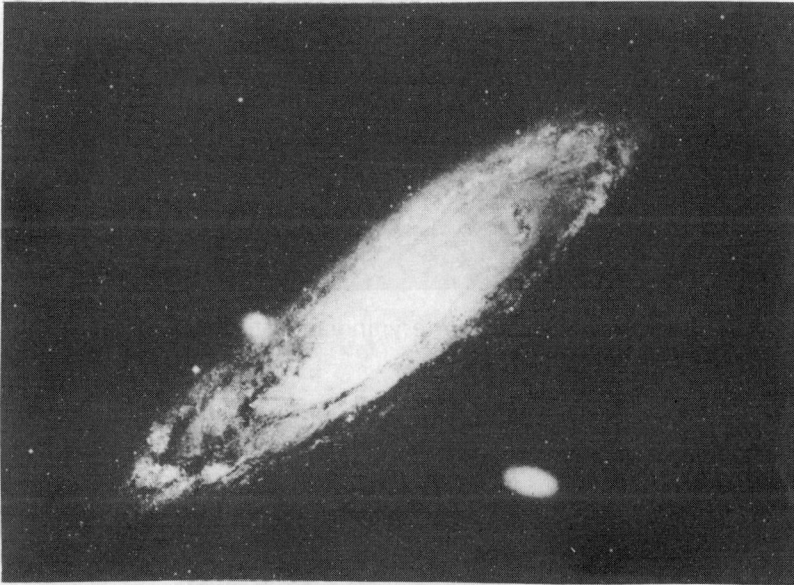

FIGURE 17.2   The spiral galaxy in Andromeda (M31) [*Courtesy:* Indian Institute of Astrophysics, Bangalore].

Population II stars. In general, the spiral galaxies have mixed stellar populations. A large number of spiral galaxies have "bars" passing through their nuclei. In the case of normal spirals, the spiral arms originate from two diametrically opposite sides of the nucleus, whereas in the case of 'barred' spirals, they begin from the ends of the bar. A famous example of a barred spiral is NGC 1300 (Fig. 17.4). The bar in a spiral is assumed to be the straight portion of a spiral arm and contains Population I stars and gas. Studies of rotation of different barred spirals have revealed that the inner parts of the galaxies are rotating as a solid wheel. The straight bars can persist, instead of winding up, in the absence of differential galactic rotation.

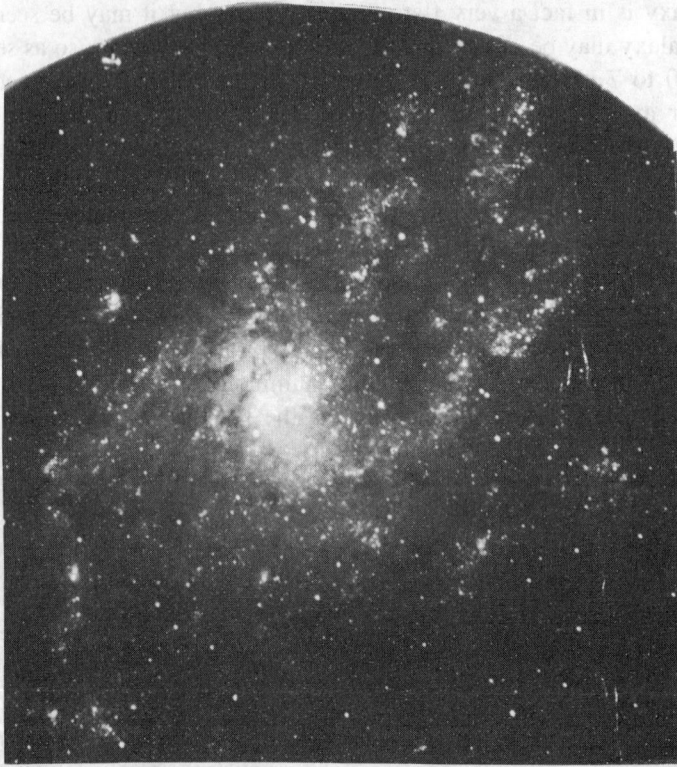

FIGURE 17.3   The spiral galaxy M33 [*Courtesy:* Indian Institute of Astrophysics, Bangalore].

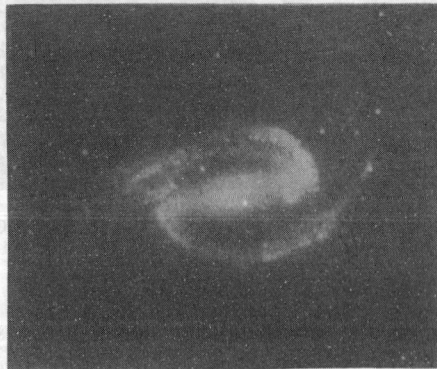

FIGURE 17.4   The barred spiral galaxy NGC 1300.

A gradual transition of morphology is found in normal as well as in barred spirals. At one extreme the nucleus is large, luminous, with arms small and tightly coiled, and bright emission nebulae and supergiants are practically absent, whereas at the other extreme the nucleus is small, the arms are loosely wound and contain luminous stars, star clusters and nebulae.

Hubble denoted the normal spirals by S and the barred spirals by SB. The sub-classes belonging to each are represented by the lower case letters a, b, c. Thus, Sa and SBa are spirals and barred spirals respectively at one extreme, having large luminous nucleus and arms tightly coiled. On the other hand, Sc and SBc are those in the other extreme with small nuclei and loose extended spirals. Nearly 70 per cent of the field galaxies are spirals. But among the members of the clusters of galaxies, spirals do not contribute more than 20 per cent. In some large clusters no spirals are seen at all. The Coma cluster of galaxies is such an example.

In some cases disc-shaped galaxies are observed with no trace of spiral arms. Hubble assumed these galaxies to be intermediate between spirals and ellipticals and designated them as S0 galaxies. Hubble's classification of galaxies is depicted in Fig, 17.5. This is known as Bubble's tuning fork diagram. Photograph of some spirals and barred spirals of various types are shown in Fig. 17.6(a–b).

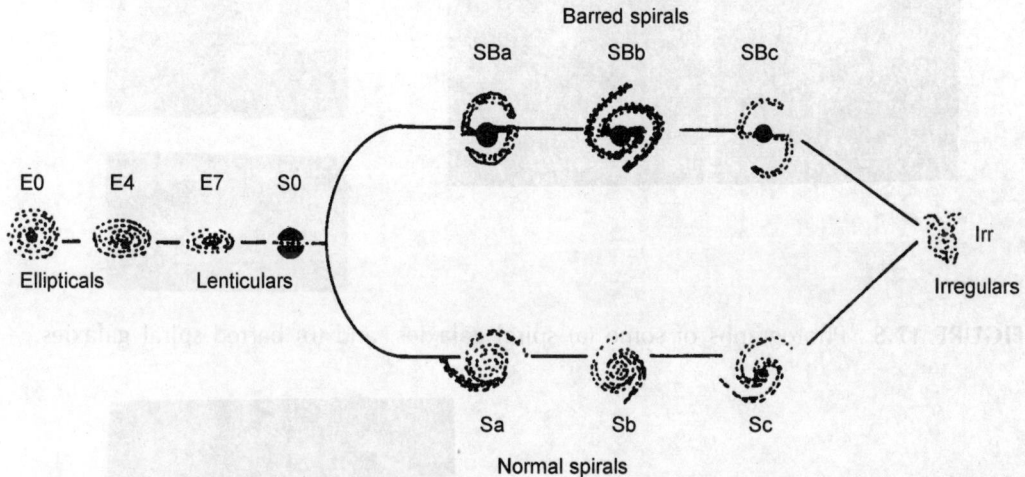

FIGURE 17.5 "Hubble's tuning fork diagram" showing four major classes of galaxies: 1. Ellipticals (E), 2. Lenticulars (S0), 3. Spirals (S and SB) and 4. Irregulars (Irr).

Among those galaxies so far found in the northern sky about 3 per cent are classed as irregulars, which show no trace of circular or rotational symmetry. Rather they have chaotic appearances. They are subdivided into two sub-classes. The first group that shows high percentage of stars with some emission nebulae, is denoted as Irr I galaxies. The most famous examples of this type are the Large and Small Magellanic Clouds, our nearest pair of galaxies, as shown in Fig. 17.7(a–b). A large number of star clusters, variables, supergiants and gaseous nebulae and also both Population I and II stars are found in them.

The second type of irregular galaxies designated as Irr II galaxies, are similar to the first type in regard to lack of symmetry but they show no resolution into stars or clusters. Examples of such galaxies are NGC 3034 (M82) (Fig. 17.8) and NGC 5195 (a companion to the spiral galaxy M51.

FIGURE 17.6 Photographs of some (a) spiral galaxies, and (b) barred spiral galaxies.

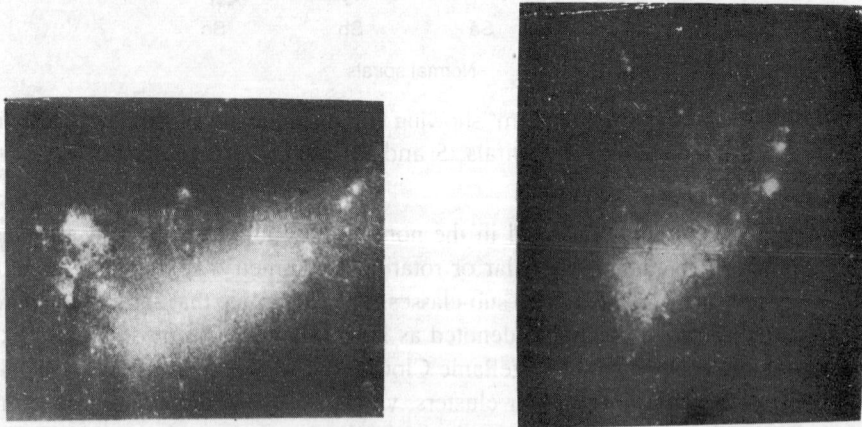

FIGURE 17.7 (a) The Large Magellanic Cloud; (b) Small Magellanic Cloud.

Partly because of our existence inside and partly because of the presence of dust acting as a type of fog lying along the plane of the Milky Way, it is not easy to have a clear idea of what our own Galaxy would look like, if it were possible to observe our Galaxy as a whole

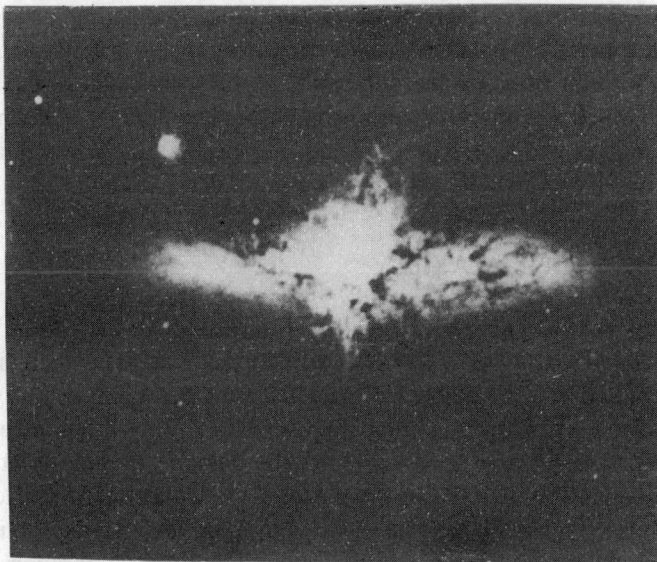

FIGURE 17.8 The irregular galaxy, M82.

from outside. But in recent years various painstaking and difficult researches have revealed that our Galaxy, if we could see it from outside, would have the general appearance exactly similar to that of M31 in Andromeda which belongs to the Sb type of galaxies, intermediate between the two extreme sub-classes of the spiral type.

Since the galaxy is a gigantic system consisting of innumerable stars and clusters and also of gas, dust and magnetic field, we have to study various aspects of it in order to understand it clearly.

Walter Baade found a close correlation between the shape of galaxies and the type of stellar populations in them. Population I found in the arms of spiral galaxies is recognized by the presence of blue stars and interstellar matter. These stars are very rare in elliptical and spherical galaxies without arm. These latter galaxies consist almost entirely of Population II stars and very little of interstellar matter. Population II stars are also found in the nuclei of spirals and ellipticals. Members of both the populations are thus found in our Galaxy and in general in all normal spiral galaxies.

Many important characteristics of the galaxies, viz. their population characteristics and their structural features are inherent in their classifications. Thus, the elliptical galaxies may be characterized by primarily the old stars, the spirals by both old (found in the disc and the halo) and the young stars (in the spiral arm) and the Irr I galaxies by a much higher proportion of young stars. The spectrum of a galaxy reveals the radiation contribution from the brightest hot stars and also from the fainter cool stars. W.W. Morgan suggested a classification scheme which is based on the relative contributions of light from population types to the total light. The central concentration of light is very important in this regard. Morgan's classification is correlated with that of Hubble. Some galaxies with extreme central concentrations of light have been discovered, that do not seem to fit themselves into any of these classification schemes. These galaxies are representatives of the most interesting extragalactic objects. These objects are the compact galaxies, some of which appear totally stellar (quasi-stellar objects), others may have brilliant nuclei with comparatively faint envelopes (Seyfert galaxies) and there are some which are more or less bright, small, galaxies with some peculiarities. There seems to have some evidences that gigantic explosions have taken place in the nuclei of many such galaxies (see Chapters 19 and 20).

17.3 DISTRIBUTION OF GALAXIES

In a schematic survey in 1920's with the 100-inch telescope on the Mount Wilson, Hubble photographed several hundred areas of the sky (actually 1283 areas) and counted, the galaxies in each of them. He found that the number of galaxies remained more or less constant all over the large galactic polar caps, but on approaching the Milky Way the number became fewer. This dependence of the number of galaxies upon the galactic latitude is now known to be simply the effect of obscuring material in the plane of the Galaxy. This galactic plane region was called accordingly by Hubble as the *Zone of avoidance*. Extragalactic objects suffer obscuration proportional to the lengths of the path of their light through these obscuring layers. When viewed towards the galactic pole ($b^{II} = 90°$), the obscuration is minimum and it increases with the decrease of the galactic latitude. According to Hubble's observation of galaxies as faint as 20th magnitude, the dependence of the number of galaxies per unit area brighter than a given limiting magnitude on the galactic latitude, is represented closely by a cosecant law given by:

$$\log N = \text{constant} - 0.15 \operatorname{cosec} b^{II} \tag{17.1}$$

We are surrounded by the obscuring material and thus in order to have a knowledge of the true distribution of galaxies, elimination of the effect of obscuration is essential. Without any serious error the sky may be roughly divided into two regions, the galactic belt between $b^{II} = -40°$ to $b^{II} = 40°$, and the polar caps between $b^{II} = \pm 40°$ to $b^{II} = \pm 90°$. The galactic belt that includes the zone of avoidance and its fringes of partial obscuration provides no current information about the galaxian distribution, but gives some information about the local obscuration. Information regarding the galaxian distribution may be obtained from the polar caps, which is free from local obscuring effect. However, owing to the relatively smaller area of the polar cap regions, information regarding the distribution of galaxies over the entire sky is not possible. As soon as the obscuring effect of the region of the galactic belt is eliminated, it is possible to have some information about that region following the general field of galaxies. A knowledge of the distribution of galaxies over a considerably large region of the sky (down to approximately about 15° galactic latitude) may thus be obtained. A schematic survey of the distribution reveals the following information:

1. No appreciable systematic variation in distribution of galaxies has been found in the general field.
2. The northern and the southern polar caps are found to be almost similar in regard to the distribution of galaxies.
3. Individual values of the logarithm of the number of galaxies per square degree are found to be distributed at random about the average values.

Figure 17.9 shows the distribution of galaxies as shown by the sample of Hubble's survey of 1283 regions. Each symbol represents one sample region and the size of the symbol indicates the number of galaxies counted in the region. The absence of any systematic variation in latitude and the similarity of the two polar caps is an indication of the fact that the large-scale distribution of galaxies over the sky is more or less uniform.

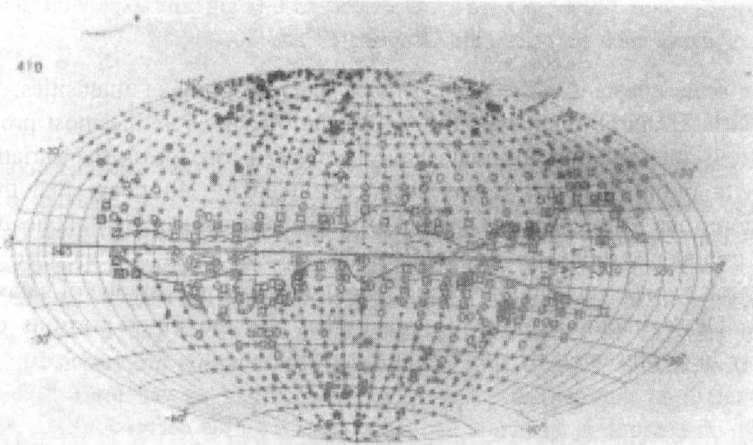

FIGURE 17.9 Distribution of galaxies, as shown by the sample of Hubble's survey of 1283 regions.

Galaxies are scattered at a mean interval of the order of 10^3 kpc, which is nearly 100 times their mean diameter. They often appear in groups and clusters. Analysis shows that a large number of systems consist of two or more galaxies, e.g. the spiral galaxy in Andromeda and its satellites or the Galaxy with the Magellanic Clouds. It is a common experience that almost all types of galaxies are found in such multiple systems. Groups as well as the double and multiple systems are almost a general feature of the galaxian distribution. It thus appears from Hubble's analysis and subsequent works of others that although the large-scale distribution of galaxies is more or less uniform in all directions when the obscuring effect is corrected, in small-scale analysis, the galaxies reveal clustering and clumping tendencies. We shall discuss this topic in greater details in Chapter 18.

17.4 LUMINOSITY DISTRIBUTION OF GALAXIES

Luminosity distribution of a galaxy is a basic obseravable quantity. It is one of the fundamental characteristics of a galaxy, because it can be taken to represent the spatial distribution of mass within a galaxy. The total apparent luminosity of a bright galaxy in a cluster is sometimes used as the distance indicator of the cluster. In 1930, Hubble investigated the luminosity distribution in elliptical galaxies and expressed the distribution law as

$$I(r) = \frac{I_0}{\left(1 + \dfrac{r}{a}\right)^2} \tag{17.2}$$

where I_0 refers to the central intensity and a is the radius at which the surface brightness has been reduced to one-fourth of the value at the centre. The latter may be conveniently termed as the *scale constant*.

De Vaucouleurs and Dennison, however, found that the intensity falls off more quickly in the outer region than the prediction of the Hubble's luminosity law. According to De Vaucouleurs, the variation of brightness in elliptical galaxies is best represented by

$$\log B = -ar^{1/4} + b \tag{17.3}$$

'a' being the scale constant while b determines the zero point of intensities.

In the case of spiral galaxies, the disc population serves as the most prominent source of light; various interpolation formulae have been used to express the variation of surface brightness with radial distance. It has been concluded by Holmberg that the spiral arms contribute more than 10 per cent of the total light in the case of Sc galaxies only. Separate and distinct treatments should be adopted for the nuclear bulge and the disc in the case of early-type spirals. This has been done by Van Houten for Sa and Sb type of galaxies. However, according to De Vaucouleurs, the luminosity law for the outer portions of spirals can approximately be represented by the formula

$$\log B = C_1 - C_2 r \tag{17.4a}$$

where C_1 refers to the constant for zero point and C_2 to that for the angular scale. More useful and accurate luminosity law now used for the disc of spiral galaxies is the exponential law given by

$$I(r) = I_0 \exp(-\alpha r) \tag{17.4b}$$

where α is a constant and I_0 measures the intensity at the centre. It has been shown by Holmberg that the surface brightness of inclined spirals is less than that of similar spirals when viewed face-on which has been assumed to be due to reddening within the galaxy concerned. The reddening effect has been estimated by Holmberg to be proportional to the cosecant of the inclination of the galaxy. For an increase of the cosecant value by unity, he deduced a reduction in the surface brightness by 0.43. The effect is larger in Sa and Sb systems and smaller in Sc galaxies. This is probably indicative of the larger dust content of the former galaxies. This is in agreement with other observations of the relative amount of dust in different types of galaxies.

The luminosity distribution of elliptical galaxies is an essential parameter for the determination of their masses. In general, the mass-luminosity ratio is assumed to be constant for any particular galaxy, in order to determine the mass distribution from the observed luminosity distribution. Also, to get the correct total apparent magnitude of galaxies, their luminosity distribution must be correctly expressed. The former is essential for the estimation of the correct distances of galaxies when their absolute magnitudes are known or otherwise, for the absolute magnitudes when their distances are known.

17.5 SPECTRA OF GALAXIES

The spectroscopic study of galaxies and assignment of a spectral class for each of these involves some inherent difficulties at least for three different reasons. Firstly, their spectral study is difficult because their surface brightness is in general, very low and consequently, the spectra are small and of very low dispersion. The study of any feature in details is therefore almost impossible. Again because of faintness, only the brightest central part (nuclear region) can be studied except for very close galaxies. Secondly, the individual stars cannot be studied, only the integrated spectrum with its nature varying according to wavelenghts is obtained. This spectrum is produced by the overall stellar population belonging to the central region of any particular galaxy. The contribution to the overall spectrum by stars in the outer parts of the galaxy is generally lacking or only poorly represented. The spectrum is dominated mostly by the contribution of the older population of stars in a galaxy and so tends to show a later type. Further, the interpretation of a composite spectrum by only the gross features of lines may often be erroneous. For example, the intensity of the resonance line at 4226 of Ca I increases along two sequences: first, in passing along early to late K dwarfs and secondly in passing along giant to dwarf K stars. It may not be easy to ascertain whether the observed intensity of the line is due to K giants or K dwarfs. This difficulty actually appeared in interpreting the blue light intensity in the spectrum of the inner part of M31. Initially, it was mistaken to be due to K dwarfs but later, detailed comparison and intensive study revealed that the bulk of the blue light is due to K giants. This is substantiated by the presence of CN bands in the same region of the spectrum.

Thirdly, the large red-shift of the entire spectrum of a distant galaxy presents problems for the correct assessment of the spectral features, particularly of the spectral energy distribution over the entire available range of wavelengths. This difficulty is enhanced by the large blanketing effect in the blue-violet regions of the spectrum due to the presence of numerous metallic lines in these regions. The entire effect produced by the above causes it to render quite difficult and uncertain the separation of the various components of the stellar material contributing the major part of light in different regions of the spectrum.

Nevertheless, the spectroscopic study of the galaxies has been carried out by many astronomers. Most of the galaxies exhibit composite nature of the spectrum as would be expected if it is produced by a mixture of stars belonging to all sorts of spectral classes and luminosities. However, some regular features are observed on the basis of which broad classification schemes can be adopted. Besides the prominent lines of H, Ca I (4226), Ca II (H and K lines), Na I (D lines), Fe I etc., and some bands like those of MgH, CN, TiO and G-band, emission lines of H, [O I], [O II], [O III], [N II], [N III], [S II], as well as the He I lines both in absorption and in emission are observed in spectra of galaxies. The emission lines evidently originate in the hot gaseous components of the nucleus. Analysis of the various absorption lines in variable strengths reveal three broad distinguishing features of the spectra of galaxies. These are in fact, interpreted as the manifestation of the relative contribution of the different types of stars to the light at different spectral regions, or to the overall spectrum. The first group of galaxies, those with He I lines, whether in emission or absorption, indicate that they contain a large relative number of early B-type stars. The second group exhibiting highly composite spectra may range from A to K in spectral class, depending on the predominance of the type of stars contributing the major portion of light in the blue-violet region. The third group has less complicated spectra of generally K-type in the violet region.

As we know, any predominant feature of the composite spectrum of a galaxy points to the corresponding predominance of the stellar type responsible for that feature. Utilizing this correspondence, one could deduce the principal stellar content in the nucleus of a galaxy. The analysis points to the fact that the nuclear regions of the latter two groups of galaxies contain mostly the late-type giant and dwarf stars. For example, the presence of the MgH band near λ 4800 in the spectrum of a galaxy indicates the predominance of late K and early M dwarfs in the nuclear region of that galaxy. This is because the band appears in considerable strength in the spectra of these dwarf stars. Similarly, those containing TiO bands contain large proportion of M-type stars in their nuclear regions. On the other hand, the galaxies showing bands of CN in the blue-violet region should contain larger proportions of late G and early K giants.

When the various difficulties and uncertainties mentioned above have been overcome to the best possible degree, one can assign the spectral class of a galaxy from the most predominant feature in its spectrum. When such a class has been assigned, it is to be understood that the assigned class is representative of the stellar population belonging to the nuclear region of the galaxy, since this brightest portion only can yield a good spectrum to be studied. The outer region of the galaxy is generally too faint to yield a good spectrum and can hardly be studied with any meaningful results. Moreover, the outer region is generally populated by different types of stars and the spectrum of that region is very likely to differ from that of the inner parts of a galaxy. This effect is more pronounced in the spiral galaxies in which the spiral arms are populated by extreme Population I stars while the nuclear regions contain mostly the older stars.

Of late, however, the astronomers are attaching more and more importance to the structure, contents and evolution of galactic nuclei. Spectroscopic studies by some authors of nuclei of some nearby galaxies indicate that there are enough hot stars and ionized gas in the nuclear regions of these galaxies. The equivalent widths of certain lines and the total continuum radiation from the M31 nucleus has been found to fit best with composite models comprising predominantly of dwarf enriched main sequence and hot stars. Also, the present studies indicate that the influence of the hot blue component increases with distance from the centre of M31. Similar studies of the deep central region of the Sb galaxy NGC 3031 also indicate from the appearance of various emission lines that there must be enough hot stars and ionized gas in the region. Also, the composite spectrum of the region has been found to fit best with models comprising of hot stars, stars with strong metallic lines and K and M-type giants. Thus, new findings and new interpretations are emerging in this field by new observations with newer efficient techniques and the conclusions drawn earlier or presently being drawn in this regard should not be accepted as final.

That the central regions of most galaxies and clusters of galaxies is rich with very hot gas is also borne by observations of the Einstein Observatory of strong X-ray radiation in the 2–10 KeV range from these regions. This will be discussed in greater details in Chapter 18.

17.6 THE LOCAL GROUP OF GALAXIES

We have already seen that in a large-scale view of the universe, the galaxies possess both large-scale clustering as well as small-scale regional grouping and clustering. The discovery of the conspicuous nearby galaxies such as M31, M32 and M33, the Large and Small Magellanic Clouds, etc. within a certain small range distance of less than 1 Mpc led to the concept that our own Galaxy also belongs to a local clumping. This has been called the Local Group of galaxies. The other nearby luminous galaxies such as the M51 and the M81 groups belong to a distance of more than 1.5 Mpc. The region between the Local Group and the nearby outsiders appears to be void except probably the presence of some faint dwarf galaxies which cannot be detected at such distances.

The number of galaxies in the Local Group is difficult to ascertain, partly because of the heavy obscuration in the direction of the galactic centre and partly due to the faintness of the smaller galaxies. At present more than forty galaxies are known in the Local Group, but this number is systematically increasing as a result of new identifications. Of the galaxies belonging to the Local Group, three are spirals, four are irregulars and others are ellipticals and dwarf ellipticals (*dE*) such as the sculptor system. The dwarf ellipticals have masses and luminosities of the same order as those of the galactic globular clusters. Their number in the Local Group supersedes all others taken together in the group. Whether this is a general characteristic of the galactic content in the universe, cannot be ascertained, because these faint objects cannot be identified at distances much beyond the Local Group. Even if such large abundance of these faint objects is true, the total mass and luminosity contributed by them will not be significant compared to the contribution in these respects of the luminous massive galaxies, since these faint objects are more than thousand times fainter and less massive by the same order than the large galaxies. For example, the two giant systems in the Local Group, the Milky Way and M31 are, taken together, many times more massive and luminous than all other members of the group put together. Both the Milky Way Galaxy and M31 are Sb spirals and they possess similar properties in many respects.

The Milky Way Galaxy has two satellites, the Large and Small Magellanic Clouds at distances a little over 50 kpc. Similarly, M31 has several satellites, the most conspicuous of which are the Sc spiral M33 and the E3 elliptical M32. The dwarf ellipticals may be the central remnants of Irregular galaxies which might have been torn apart by tidal interaction with large galaxies and their gaseous outer parts subsequently been swept away. Similar tidal interaction between our Galaxy and the LMC has disrupted the latter, giving birth to the Magellanic Stream. The Magellanic Stream which is an extensive and thin filament of gas originating at the Clouds and crossing almost to their antipodes, is believed to have been created and drawn by tearing the LMC by strong tidal force during a close passage of the clouds around the Milky Way a few times 10^8 years ago. If such close passage is repeated a few times, the Clouds may suffer a total disruption within several billion years leaving their nuclei as *dE* galaxies.

The Magellanic Clouds being our nearest neighbours have been the most well-observed galaxies. These galaxies are rich in stars and gas, and in particular, have a large population, of star clusters of a wide variety. The 30 Doradus or Tarantula Nebula in the LMC is a huge H II region containing a very large number of hot stars surrounded by extensive glowing gas.

The star SK-69202 which exploded as the SN 1987A belongs to this nebula. Extensive observations and careful analyses have shown that while the globular clusters in the Milky Way are all very old and metal-deficient, the Magellanic Clouds and also the spirals M31 and M33 contain globular clusters having different ages including those of more recent origin with significantly higher metal content. The reason probably is that while our Galaxy has long ceased to possess the source of agitation for high velocity gas motion, the latter galaxies still possess the sources for periodic agitation of their gas through tidal interaction.

The radial velocities of some of the galaxies of the Local Group have been calculated. It is observed that some of the galaxies are approaching us while some others are receding. For example, M31 is approaching us with a velocity of about 260 km s^{-1}, while the Large Magellanic Cloud is receding with a velocity of about 276 km s^{-1}. These valocities can be used to compute the circular velocity of the Sun round the centre of the Galaxy. The value obtained is about 250 km s^{-1}. This value of the solar motion about the centre of the Galaxy agrees very well with those determined by other methods.

17.7 DISTANCES OF GALAXIES

The galaxies are distant stellar systems and the calculation of accurate distances to them is one of the major astrophysical problems. One has to use various objects as distance indicators. Attempts have been made by various authors to calibrate the intrinsic luminosities of these distance indicators so that accurate distances were obtained. But unfortunately, in each of these indicators there is a certain amount of inherent uncertainty which is likely to vitiate the results. Nevertheless, calibration of the absolute magnitudes of various distance indicators has been made with utmost care and to the best possible accuracy, which have then been used to calculate the distances to the external galaxies and clusters of galaxies. The fundamental assumption underlying the method is that these distance indicators in the external galaxies have the same absolute magnitudes as those belonging to our own Galaxy. The method then consists in identifying in the external systems an object or a group of objects similar to the known objects in our Galaxy and having the same absolute magnitudes. The distance modulus *m-M* of a galaxy can then be obtained by comparing the apparent magnitudes of the objects with their calibrated absolute magnitudes. After correcting the distance modulus for absorption and reddening, the distance to the system can be obtained from the following formula

$$(m - M)_0 = 5 \log \frac{r}{10} \qquad (17.5)$$

where *r* is measured in parsecs and the subscript zero stands for the corrected distance modulus.

For the nearby galaxies (mainly the Local Group) the cepheid variables (including the RR Lyrae variables) serve as the most reliable distance indicators. Beyond the Local Group and upto the distances like that of the Virgo cluster, the brightest blue stars, the globular clusters, the ordinary novae and H II regions have been successfully used as distance indicators. Farther away, upto the distances like that of the Coma cluster and beyond, supernovae have sometimes served as the distance indicators. Beyond this range, the brightest galaxies in a cluster may be used to compute the distance to the cluster. But at these distances or even much before that (beyond the Virgo cluster) Hubble's redshift-distance relation (Eq. (17.8)) is commonly used to calculate the extragalactic distances. Accurate calibration of the Hubble's constant, however, has still remained as a formidable problem to the astrophysicists. In the following, we now discuss the calibration of each of the distance indicators and the use of these in calculating the distances of galaxies at various ranges of distance.

RR Lyrae Stars

Valuable works have been done by many astronomers to calibrate the absolute magnitude of the RR Lyrae stars. Different authors have used different and independent methods and the values of M_v obtained by them for these stars range from 0.0 to +1.0. Using the method of statistical parallax for a large number of field RR Lyrae stars, O.C. Wilson found $M_v = 0.0$. O.G. Eggen used the method of moving kinematic groups of stars for some groups and determined M_v for RR Lyrae stars in these groups whose average was +0.57. The method of main sequence fitting of the colour-magnitude diagram of several globular clusters has been used by H.C. Arp, W. A. Baum, A. Sandage, O.G. Eggen and others to calibrate the absolute magnitude of RR Lyrae stars. The average of these determinations give $M_v = + 0.85$. From these various determinations one can take $M_v = + 0.5$ for the RR Lyrae stars, although it is possible that there exists a real scatter in the absolute magnitudes of these stars. The photographic limit of the apparent magnitude for the 200-inch Hale telescope at Mount Palomar Observatery is about $M_v = + 22.0$. Thus, RR Lyrae stars can serve as distance indicators up to a distance modulus of 21.5 within which lie only our two nearest neighbours, the LMC and the SMC. The average observed value of $M_v = + 19.6$ for RR Lyrae stars in the SMC and the value $M_v = + 0.5$ for RR Lyraes correspond to a distance modulus of 19.1 for the SMC if the reddening is zero. The reddening for this galaxy has been measured to be about 0.2. This yields $(m - M)_0 = 18.9$ for the SMC.

The Cepheid Variables

The classical Cepheids of Type 1 are brighter than the RR Lyrae variables by 4 to 6 magnitudes and can therefore be detected at a distance of about 10 times more at which the RR Lyrae are seen. The works of various authors have led to the accurate calibration of the period-luminosity relation for these variables. R.P. Kraft deduced the following relations which have been subsequently used for many works:

$$M_v = - 1.67 - 2.54 \log P \tag{17.6}$$

and

$$M_B = - 1.33 - 2.25 \log P \tag{17.7}$$

where the magnitude averages have been calculated from the mean intensity during the variation cycle. If the Cepheids in die Local Group of galaxies are identical to those in our Galaxy and they can be observed in those galaxies, then their absolute magnitudes can be determined by using Eqs. (17.6) and (17.7), observing their periods and measuring their average apparent magnitudes. This has been done for most of the spiral and irregular galaxies of the Local Group and their distances have been determined accordingly. Distances of galaxies having distance moduli up to 27.0 or so can be determined by observing Cepheid variables. For galaxies beyond these distances and upto those having distance moduli as large as 31.0 or so, we can use four different types of distance indicators all of which have nearly the same order of absoulte magnitudes. These are (a) the brightest blue stars in a galaxy, (b) the normal novae, (c) the globular clusters and (d) the H II regions. It is important to mention that the Virgo cluster belongs to this range of distance and thus yields scope for determination of its distance by several independent methods.

The brightest blue stars.   Normally, the OB supergiants in our Galaxy have absolute magnitudes around −7 or so. But a few brighter stars such as 89 Her, HD 161796 and some others have been known to exist in our Galaxy. The absolute magnitudes of these stars range between − 9 and −10. Thus, such stars can serve as the distance indicators with the 200-inch telescope upto a distance modulus of 31.0. The underlying assumptions in using these stars as distance indicators are: (a) the brightest blue stars in the distant spiral systems have the same absolute magnitudes as those in our galaxy, and (b) we are actually observing the brightest stars in the system under study. The validity of the first assumption has been established by observing similar stars in the Magellanic Clouds and some other nearby galaxies. The brightest stars studied in these galaxies have absolute magnitudes around − 9. The validity of the second assumption however depends on the efficiency of the observer and the best check on this can be assured only by an expert observer.

Normal novae.   The extensive observational as well as the theoretical works by various astronomers on normal novae have proved that the brightest of these sporadic variables have average absolute magnitudes around − 9. The works on the two brightest galactic novae, Nova Pup 1942 and Nova Aql 1918, have established their absolute magnitudes to be $M_B = − 9.4$ and − 8.7, respectively, so that the average absolute magnitude for these two brightest novae is − 9.05. If the distance of M31 determined from the cepheid variables be accepted as correct, then the work of Arp on the two brightest novae observed in M31 yields an absolute magnitude of − 9 for both the novae. Normal novae can therefore, be used as distance indicators and the method can be used upto a distance modulus of 31.0—the same as with the brightest blue stars. The principal defect with this method however lies in the fact that the novae phenomena are sporadic, and the scatter of their absolute magnitudes may be as large as 4.0. Although the absolute magnitudes of the novae at their brightest are related to their lifetime, this relation has not been uniquely defined. Various authors have derived altogether different relations of novae. So one must be very cautious in deriving the absolute magnitudes of novae in distant galaxies before they are used to determine the distances.

The globular cluster.   Globular clusters in giant elliptical galaxies have sometimes been used to calculate the distances to these far-off systems. The absolute magnitudes of the globular clusters in our Galaxy are calibrated by observing the RR Lyraes in them. Once the absolute magnitude of the RR Lyrae stars has been calibrated, their apparent magnitudes can be observed in the galactic globular clusters and thus the distances to these clusters are known. Measuring now their total apparent magnitudes one will get their absolute magnitudes. More than a hundred globular clusters in our Galaxy have been studied by various authors. The studies reveal that their absolute magnitudes range from about -5.5 to about -9.5, a scatter too large to be explained as due to observational inaccuracy or due to the scatter in the absolute magnitudes of RR Lyrae stars. The large scatter in absolute magnitudes of globular clusters is thus real which makes these objects poor indicators of distances. They should not thus be used for a galaxy whose distance can be measured by any other more reliable method. Nevertheless, the method can be used as a check, and if one can be sure of having used the brightest of these objects in any galaxy, one can penetrate through a distance having a modulus of about 31.0, the same as that yielded by the brightest blue stars and normal novae.

H II regions. The average angular size of a few largest H II regions in a galaxy can be used as a statistical distance indicator for that galaxy. Extensive works have been done in this field by several authors who have calibrated the average angular size by studying a large number of H II regions in M33 and LMC, whose distances are known accurately by other independent methods. The method is powerful because large H II regions can be identified on plates of galaxies as far as the Virgo cluster in which the red-shift is significant. The accuracy of the method has been demonstrated by A.R. Sandage by evaluating the distances of some galaxies including some members of the Virgo cluster, by using the average angular size of the five largest H II regions in each galaxy. The results obtained are in very good agreement with those obtained by other independent methods.

The supernovae. The supernovae at their brightest are the most luminous individual objects in a galaxy. As we have seen already, two broad types of these objects are recognized, Type I and Type II having average absolute magnitudes of -18.9 and -17.5 respectively at their maximum. Thus, the explorable distance moduli with supernovae are of the order of 40.0 and the corresponding distances are of the order of 1000 Mpc. Supernovae can thus be observed at great depths of the universe. But there exists a large scatter in their absolute magnitudes on the one hand, and on the other, they are of extremely rare events, one supernova appearing in a galaxy on average over a period of a few hundred years. This makes their statistics extremely poor. Thus supemovae though allow us to penetrate through great depths of space, in respect of accuracy, they cannot be accepted as reliable distance indicators. At such distances, on the other hand, the red-shifts of galaxies are significant and Hubble's velocity-distance law can be more reliably used to calculate their distances. Nevertheless, the distances of many far-off galaxies have been determined by observing supernovae in them.

The brightest galaxies in a cluster. The distances to rich clusters of galaxies can be estimated by observing the apparent magnitudes of a few brightest members in the clusters. These giant ellipticals and spirals generally residing near the centre of the cluster have absolute magnitudes as bright as -23 which allows, at least theoretically, their detection with the 200-inch telescope at the photographic limit upto a distance modulus of about 45.0, corresponding to a distance of 10^4 Megaparsecs if one assumes the absence of any absorption and reddening throughout the intervening space. But absorption does take place in the Galaxy, in the intergalactic space and in the cluster observed and the theoretical distance modulus is thus somewhat reduced in practice in actual photometric work. Again at these great distances Hubble's law should be regarded as a more reliable method because the mean red-shifts of galaxies are quite large.

Hubble's constant. From the discussions we have so far made regarding the extragalactic distance measurements, it emerges that with increasing distances the applicability of other methods gradually dwindles away and Hubble's velocity-distance law (for very low redshift)

$$V = HD \qquad (17.8)$$

manifests itself with increasing usefulness. Here V is the speed of recession of the galaxy, D its distance in megaparsecs and H is Hubble's constant (see Section 21.2). The basic entity in this law is however the Hubble's constant H which has to be first calibrated accurately before the law can be used. The calibration of H however contains some inherent uncertainties in it. One has to derive by independent methods the distances to galaxies for which the red-shifts are significant. The recessional speed must largely supersede the random speed of the galaxy. For the purpose the Virgo cluster of galaxies has so far been considered the most suitably situated. It contains a large number of bright galaxies and at its distance ($m - M \approx 31$), the photometric method of distance measurement is applicable on the one hand and, on the other, red-shifts are significant. But unfortunately, the random velocities of the individual member galaxies of the Virgo cluster about the centre of mean recessional motion are of the same order as the mean recessional motion itself.

Since the mean recessional speed of the cluster is computed from the motions of these individual members, themselves having large random motions, large uncertainty may be introduced in the computation of the mean recessional speed. When this speed is used to compute the value of H, that value should be accepted with reservation. Nevertheless the problem being very crucial astronomers have devotedly worked for many decades for the correct evaluation of H. At present, the controversy rests between two groups of astronomers. Alan Sandage and his co-authors claim on the basis of their observation that the value of H should be around 50 km s^{-1} Mpc^{-1}. On the other hand, G. de Vaucauleurs and his co-authors claim that the value should be around 100 km s^{-1} Mpc^{-1}. The controversy persists while authors often work with some intermediate value of H. Much work has been done in sixties and seventies with $H = 75$ km s^{-1} Mpc^{-1}. Considering various aspects of the problem and inherent uncertainties in the determination, A. Dressier has suggested that $H = 70$ km s^{-1} Mpc^{-1} should be a better acceptable value. Many authors are however currently working with the value 50 km s^{-1} Mpc^{-1}.

17.8 NUCLEI OF GALAXIES

The central region of a galaxy is its densest part and the core of this region of maximum density is frequently termed as the nucleus of the galaxy. The nucleus is surrounded by a bright region of a much larger dimension. Some astronomers now believe that the overall evolution of galaxies including the formation of new generation of stars and morphological development is guided chiefly by the activities undergone into their nuclei. Among others, the Russian astrophysicist V.A. Ambertsumyan is a strong upholder of this theory, although there are many others who pay but little importance to it.

The nuclei of different galaxies have varying degree of luminosity depending upon the intensity of activities going on in them. From an analysis of the results of long years of investigations, Ambertsumyan has classified the nuclei of galaxies into four different types according to their activity. We can summarize in Table 17.1 their principal properties indicating the classification criteria in each case. The Table indicates that except probably some of the irregulars and most of the dwarf ellipticals of sculptor type, other galaxies do possess nuclei of varying sizes and luminosities. Ambertsumyan's classification system, although quite old, is still quite meaningful.

TABLE 17.1 Types of Nuclei in Galaxies

Type	Luminosity of the nucleus	Central condensation	Outflow of matter	Example
1.	Nucleus observed but luminosity very low: only about 2% or so of the total luminosity of the galaxy, a few emission lines present.	Not very marked but much higher than in those galaxies that exhibit no nuclei.	Definitely present but of quiet nature.	Our Galaxy, M31, M33
2.	Luminosity much higher than in Type 1; about 5 to 25 per cent of the total luminosity of the galaxy, emission lines present in larger number.	Sufficiently high	Quiet out-flow but with much higher speed than in Type 1.	NGC 4303 NGC 3162
3.	Bright nuclei with luminosity equal to a considerable portion of that of the entire galaxy; numerous emission lines present and starlike image shown.	Very high	Outflow with great speed producing large width or even splitting of lines.	Seyfert galaxies
4.	Starlike appearance, the entire luminosity of the galaxy concentrated in the nucleus.	Very dense and compact as a whole.	Supposed to possess flow of a violent nature with relativistic speed.	Quasars

Spectroscopic studies indicate that the major portion of the luminosity of the nuclei of Types 1 and 2 comes from their stellar population. The spectra are mostly continuous except that occasionally a few bright lines are observed superimposed. Besides, quiet outflow of gaseous matter is a common feature of these nuclei. The observed velocities range from 100–300 km s^{-1} and even more. Such outflow has been observed in M31 and in our Galaxy. Material flow with a speed of several thousand kilometers per second from the nuclei of Type 3 is indicated from the width and even splitting of lines originating in them. These nuclei therefore contain, besides the usual stellar population, large amount of gaseous materials, probably existing in the form of a supermassive non-stellar body the agitation in which is responsible for the high-speed gas ejection. The Type 4 nuclei mainly consist of supermassive non-stellar body of the form of a *Quasar,* this non-stellar body being the principal source of radiation from such objects. These bodies have been discussed in more details in Chapter 20.

It appears then that every nucleus contains stars, gas and one or more supermassive non-stellar body (bodies). The supermassive body is in a state of violent eruption in nuclei of Type 4, in a highly agitated state in the Seyfert nuclei (Type 3) and undergoes occasional agitation of a weak nature, a series of *little bangs*, as Fred Hoyle has put it, in nuclei of Types 1 and 2. There are ample evidences to suggest that the nucleus of our Galaxy suffered such *a little bang* several million years ago. Evidence of gigantic explosions in the nuclei of galaxies is provided by observations of M82, NGC 1275 and M87 phenomena.

The quiet nuclei (Type 1) have small dimensions having diameters of the order of 10–20 pc. A prototype of this class, the nucleus of M31, has an estimated diameter of 15 to 20 pc rotating with a speed of about 100 km s^{-1} and having a total mass concentration of the order of a few times 10^8 $M_\odot$ and a star density of $\sim 3 \times 10^5$ stars pc^{-3} at the deep core region. For our own Galaxy, Oort has given a mass of 1.6×10^8 $M_\odot$ within a radius of 20 pc from the centre. The observations of the [O II] λ 3727 doublet have revealed the existence of ionized gas in the nucleus of M31 and that about 1 $M_\odot$ of gas is flowing out of the nucleus at a speed of more than 50 km s^{-1}. The existence of ionized gas, of course, indicates the presence of hot stars in the nucleus along with other usual stellar components of G, K and M stars, both giants and dwarfs. The total infrared luminosity has been calculated to be of the order of 10^{41} erg s^{-1} in the nuclear region of M31 which is less than that at the centre of our Galaxy by a factor of 2 to 3. The spectrum of infrared radiation suggests that this radiation comes from the concentration of stars near the centre. The nucleus is also a source of nonthermal radio energy of the order of 10^{40} erg s^{-1}. This energy is certainly generated by the mild activity in the gaseous component of the nucleus. The investigations so far made of the nucleus of M81 indicate that this nucleus is analogous to that of M31 with regard to the properties of spectrum, colour, gas flow and radiation of energy. The Type 2 nuclei (e.g., NGC 4303) are also small but much brighter than those of Type 1. They are characterized by higher condensation, many more bright lines with higher intensity and higher velocities of gaseous outflow, but not to the extent of showing very large broadening or splitting of lines. The MK spectral class of around F5 in the violet of these nuclei indicates higher content of hot stars. However, with all such manifestations of active phenomena, the nuclei of Types 1 and 2 are considered as quiet and with mild activity. But those of Types 3 and 4 in which much larger scales of activity are observed have been specially named by astronomers as Active Galactic Nuclei (AGN). Recently, great importance has been attributed to this class of objects, because astronomers believe that these may hold clues to the understanding of the evolution of the large-scale structure of the Universe. Because *of their great importance,* we have discussed these objects (AGN) separately in Section 20.5.

17.9   THEORIES OF SPIRAL STRUCTURES OF DISC GALAXIES

The Prelude

Since Hubble photographed and classified the external galaxies, much interest grew among the astronomers in order to understand the magnificent manifestation of large-scale spiral pattern in some of these galaxies. These were called spiral galaxies and have been found to possess extremely flattened disc-shaped structure. When carefully scrutinized, two main observable characteristics of spiral patterns become apparent in the case of regular-shaped disc galaxies. First, the spiral pattern extends over the entire galaxy; so it is *not a local phenomenon*. This may be described as the *grand design* of the spiral phenomena. Secondly, the spiral pattern must persist for a very long time, because statistically, the majority of the galaxies belong to the spiral class. The spiral arms can be permanent or at least quasi-permanent feature of these galaxies. Both of these observable features are however difficult to explain theoretically in view of the presence of disruptive forces generated by the strong differential rotation, which are predominant in these disc galaxies. The effect of the differential rotation on the material in a galaxy is twofold. First, any local concentration will be drawn into a spiral pattern by the differential rotation. Secondly, the same agent will force the spiral patterns to be gradually stretched and wound up, and finally to be diffused into insignificance. This is illustrated in Fig. 17.10. So if the spiral features are permanent or at least quasi-permanent, they must be wound up in course of a few revolutions of the galaxy. Otherwise, they must be transient structures, periodically generated and destroyed by some system of forces present in the galaxy.

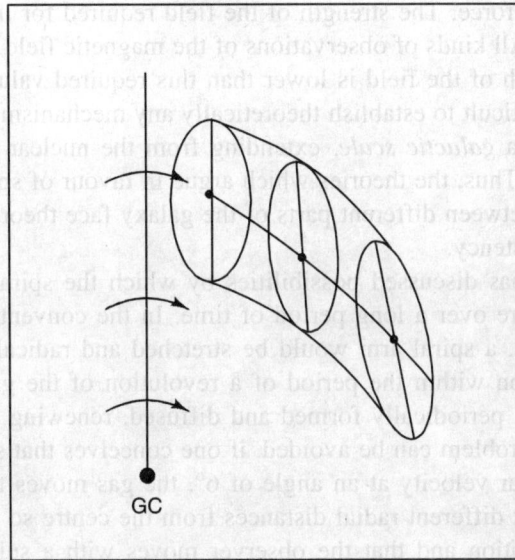

GC

FIGURE 17.10 Stretching of mass due to differential galactic rotation.

To explain the spiral phenomena, many theories have been formulated by astronomers and physicists, starting from 1920s. Most of these theories have since been abandoned and remain only as of historical interest. Of such theories, mention may be made of only a few. E. Brown in 1925 proposed that the observed spiral features could be interpreted as the envelopes of showers of individual orbits. About the same time James Jeans proposed that the origin of spiral arms might be attributed to the tidal phenomena combined with rotation. This idea had persisted among a group of astronomers for quite some time. The earlier theories of B. Lindblad that the spiral pattern may be considered as the manifestation of unstable quasi-asymptotic stellar orbits prevailing in highly flattened homogeneous ellipsoids, or in systems having high internal density gradients, also belong to the above category. The most fascinating theory belonging to this category is probably the *turbulence theory* of spiral structure proposed by W.K. Heisenberg and C.F. Von Weizsäcker. According to this theory, a hierarchy of turbulence elements will appear in the thin, slowly rotating interstellar gas. Depending on the physical condition of the gas, some turbulence elements of specific sizes may form clouds, which by differential rotation may be stretched into spiral arms. According to Von Weizsäcker, bar formation is also possible in this manner.

More plausible theories explaining the formation and maintenance of large-scale spiral pattern in disc galaxies have been developed in more recent years. The principal arguments put forward in these theories generally run along two entirely different lines. The first is to associate every spiral arm with a *given body of matter*. The arm is conceived as a *tube of gas* held together by the galactic magnetic field. The material concentration is the manifestation of certain pattern of circulation of gas between the disc and halo of a galaxy. To avoid disruption by the differential rotation of the galaxy, the arms should either be periodically generated and destroyed or they should be preserved as quasi-permanent structures by the strength of the magnetic force. The strength of the field required for the purpose is estimated to be as high as 50 μG. All kinds of observations of the magnetic field in our Galaxy indicate, however, that the strength of the field is lower than this required value by about an order of magnitude. Also, it is difficult to establish theoretically any mechanism which can periodically generate spiral arms on a *galactic scale,* extending from the nuclear region to as far as the periphery of the system. Thus, the theories which argue in favour of spiral arms being formed by material circulation between different parts of the galaxy face theoretical difficulty as well as observational inconsistency.

But Oort in 1962 has discussed possibilities by which the spiral arms as tubes of gas could retain their structure over a long period of time. In the conventional model of the gas moving in circular orbits, a spiral arm would be stretched and radically change its structure due to differential rotation within the period of a revolution of the galaxy. Spiral arms are, according to this picture, periodically formed and diffused, renewing many times during the age of the galaxy. This problem can be avoided, if one conceives that spiral arms are inclined to the direction of circular velocity at an angle of 6°, the gas moves through the tube of the arm at different speeds at different radial distances from the centre so as to correspond to the observed differential rotation and that the observer moves with a suitable chosen reference frame, so that the arms appear to have fixed structure. The motion of gas, in this picture, is chiefly guided by hydromagnetic forces. Alternatively, the permanent structure of spiral arms could also be maintained, if one supposes that the gas is transported by some non-magnetic agent from one arm to the next outer arm. In this manner, the stretching of the arm by differential rotation will be counterbalanced by systematic inward displacement of the arms. The transporting agents proposed by Oort are the supernovae shells which would push the gas from one arm to the other and also to all other directions. During one galactic revolution, about 10^6 supernovae are available in the Galaxy for the job which amounts to 10^3 supernovae per kpc arm length per revolution. A smooth transport over a long time-scale can be achieved in this manner, according to J.H. Oort.

The second class of theories regards the spiral structure as the manifestation of a *wave pattern* which *can remain* either stationary or quasi-stationary, in a frame of reference rotating around the centre of the Galaxy. According to this viewpoint, gravitational instability in the galactic disc is the basic agent for the formation of the large-scale spiral pattern. The disc would balance itself, in the presence of gravitational instability, with a *neutral density wave* extending over its entire domain. The idea of the density wave was first proposed by B. Lindblad who for many years investigated the mechanism by which density waves might be formed in the galactic disc and could grow into a large-scale spiral pattern through gravitational resonance. The same was later demonstrated by P.O. Lindblad and by R.W. Hockney and others by large-scale machine calculation of individual stellar orbits. Their works have indicated that under different conditions, both density waves as well as spiral structures do really show up.

The most successful demonstration that the spiral phenomena may be well understood in terms of density waves has been achieved by C.C. Lin and his collaborators (1964, 1966). Instead of considering the problem in terms of the individual stellar orbits as has been done by the Lindblads. Lin and his co-workers have worked out the average statistical behaviour of gas and stars by using the gas dynamical equations and stellar distribution function. They have shown that a quasi-permanent neutral density wave in spiral form may persist in the galactic disc, once it has been produced by a disturbance gravitational field of the spiral form. The theory has been shown to be capable of explaining many of the observed dynamical and structural features in our Galaxy.

About the same time Alar Toomre (1964) has demonstrated that transient spiral phenomena may be achieved in a thin stellar disc as the manifestation of gravitational instability, both axisymmetric and non-axisymmetic. After a few years, W.W. Roberts (1969) has attempted to show that the observed relation between the gaseous spiral arms of H I and optical spiral arms of OB stars and H II regions can be explained only if over and above the density wave in the disc a large-scale galactic hydrodynamical shock also is simultaneously assumed to be operative. Subsequently, the works of Piddington (1973) have attempted to disprove the very foundation of the density wave theory of Lin and his collaborators and to re-establish the hydromagnetic theory of spiral arms with new formulations. We shall *discuss some of these various aspects of the theories of spiral pattern in the following pages.*

Magnetic Field Theory of Spiral Arms

In the late forties and early fifties of this century, two very important discoveries were made regarding our Galaxy. The first was the existence of a large-scale magnetic field and the second was the large-scale spiral structure delineated by the 21-cm data. It was natural therefore for some astronomers to try to associate these two large-scale entities in our Galaxy. The protagonists of the magnetic field theory of spiral arms maintain that gravitational forces alone are insufficient to hold the spiral arms against the various disrupting forces, particularly that arising out of the differential rotation of the Galaxy. According to their viewpoint, the observed differences between the positions of gaseous spiral arms and of various types of stars which condensed in these arms, are due to differences in motions of gas and stars. Since gas and not stars, is largely influenced by the galactic magnetic field, these authors claim that magnetic force is probably the more important agent in creating and maintaining the gaseous spiral arms. This becomes almost obvious if the field strength is high, say, of the order of 50 µG. Even if the field is low, lower than the above value by an order of magnitude, the effect of the magnetic forces over large distances and long periods of time may predominate over that of the gravitational forces in shaping the large-scale structure of the Galaxy. This is particularly true when one considers the formation of spiral arms and their preservation over long time-scale against the disrupting forces of differential rotation. The observed regularity of gaseous spiral patterns is generally attributed to the predominant influence of the large-scale magnetic field on the gas. A large-scale magnetic field may also initiate and maintain a large-scale circulation pattern in the galactic gas. The spiral arms may be a manifestation of this circulation pattern. Whether such arms would be periodically generated and destroyed or whether they would be preserved against differential galactic rotation will be chiefly determined by the model of the magnetic field chosen and also on the strength of the field.

In order to explain the observed phenomena of spiral arms in terms of the magnetic forces, two models for the magnetic field in the arm have been postulated, viz., the cylindrical and the helical model. More popular of these two is the conventional *cylindrical* model. In this model, the lines of force are aligned along the arms but the strength of the field has been variously suggested. Equating the gravitational pressure in the spiral arm to the sum of the magnetic pressure and the pressure due to random cloud motion, S. Chandrasekhar and E. Fermi (1953) calculated the field in the spiral arm to be 6×10^{-6} gauss. They used the relation

$$p_{gr} = p_k + p_m \tag{17.9}$$

where

$$p_k = 1/3 \ \rho U^2 \tag{17.10}$$

p_k is the kinetic pressure for the random cloud motion.

$$p_m = \frac{H^2}{8\pi} \tag{17.11}$$

is the magnetic pressure, and

$$p_{gr} = \pi \ G\rho\rho_t r^2 \tag{17.12}$$

represents the total gravitational pressure both due to stars and gas, r being the radius of the cylindrical spiral arm assumed to be of uniform density. In the above expressions, ρ is the density of gas ρ_t is the total density including stars and U is the average velocity of the turbulent gas. Using plausible values of the parameters involved, the authors arrive at the above value of the magnetic field in the spiral arm. They suggested that the spiral arms could maintain themselves through this balance. These authors however neglected the effect of the differential rotation on the arms. The arms would be stretched by the differential rotation thus increasing the strength of the field in them, causing thereby the magnetic forces to predominate over self-gravitation.

In a detailed analysis, of differential rotation on a gaseous mass pervaded by a magnetic field, D.G. Wentzel (1960) found that after the mass had attained an elongated spiral form, the gas on the outer side of the arm-axis will gain angular momentum and would thus move outwards, while the reverse effect will be produced on the inner side of the axis of the arm. The spiral arm will thus be diffused gradually by differential rotation. The calculations of Wentzel indicate that the arm will be almost non-recognizable in this way after a period of about 10^9 years if a cylindrical model of the magnetic field is assumed. The typical rate at which the arm is dispersed within this period is of the order of 1 km s^{-1} in radial velocity, as obtained when the plausible values of the parameters involved are used.

In a classical treatment of the problem of spiral structure, F. Hoyle and J.G. Ireland (1961) discussed the possibility of the case in which a spiral arm, as a tube of gas held together by magnetic forces, is formed near the nucleus of the Galaxy. This gradually moves outward, reaching the periphery of the Galaxy of radius 10 kpc in a time-scale of 5×10^8 years. The velocity law followed by the outward moving spiral has been proposed by them to be of the form

$$u(r) = \frac{K}{r} \qquad (17.13)$$

r being the distance from the centre of the Galaxy and K is a constant. The material then flows out of the Galaxy of radius 10 kpc in a time of 5×10^8 years, provided $K \approx 3 \times 10^{28}$ cm^2 s^{-1}. The law yields an outward velocity of ~ 30 km s^{-1} at $r = 3$ kpc and ~ 10 km s^{-1} at $r = 8$ kpc. These velocities are actually of the observed order in these regions. According to these authors therefore, spiral arms are generated and destroyed during a time-scale of 5×10^8 years which is much shorter than the age of the Galaxy.

If a substantial amount of gas has to flow out of the Galaxy in a relatively short time of 5×10^8 years, one must seek a very large reservoir of gas as well as of angular momentum. Hoyle and Ireland take for this reservoir the halo of the Galaxy. They assume the halo to contain a mass of gas more than 10^{10} $M_\odot$. This is several times more than the mass of gas contained in the galactic plane. The model therefore gives an overwhelmingly large importance to the halo for the phenomena of spiral arms. The gas is supplied from the halo to the nuclear region and the angular momentum is supplied from the halo to the disc via the magnetic forces. The process of repeated generation of spiral arms and their loss from the galactic plane is thus maintained.

In order to give a much greater stability to the spiral arms, Hoyle and Ireland postulated a *helically twisted* lines of force around the axis of the spiral arms, instead of the *conventional model* of the lines of force running parallel to the arms. In such a twisted field, a force of the order of $H^2/4\pi a$ per unit volume, where a is the radius of the arm, is generated to oppose the shearing effect of the differential galactic rotation. With an arm field as low as 5×10^{-6} gauss, therefore, the arms can maintain their structure with a helical field, which cannot be achieved by the conventional cylindrical field.

Gravitational Theory of the Origin of Spiral Structure

Work of the Lindblads.   We shall now discuss the theories of the *origin* of spiral structure for which the gravitational forces play the predominant role and the magnetic forces are almost completely neglected. The pioneer in this thought was Bertil Lindblad. He, together with his collaborators, and later P.O. Lindblad, tried to explain the formation of spiral arms in disc galaxies as the manifestation of *density waves,* for which gravitational resonance was largely responsible.

To explain the theory of the Lindblads concerning the development of spiral structure in disc galaxies, we first need to define *dispersion orbits* in the central layer of such a galaxy. The entire body of such a galaxy is composed mainly of two different mass distributions—

those of Population II and Population I. Spiral arms are composed entirely of Population I material. Population II material comprises mostly older stars which have a much more spherical distribution with strong internal velocity dispersion and are heavily concentrated towards the centre. For the dynamical consideration of the entire system, one can assume that the bulk of the gravitational pull is provided by the Population II objects. Under the smooth gravitational field of these objects, the material in the plane moves in predominantly circular orbits with largely different angular velocities at different radial distances. This strong differential rotation imparts a sheering force on any condensation of material having a radial extension. A local cloud will therefore, be *drawn into a curve* before it finally loses its identity completely. This curve along which the cloud ultimately dissolves has been called by Bertil Lindblad a *dispersion curve*. One can now choose a coordinate system such that the orbit of an individual particle of the mass will coincide with this curve in the long run. The orbits thus formed are called *dispersion orbits*.

When a dispersion orbit has been formed by the dissolution of a stellar association or a cloud mass having a narrow range of velocity dispersion, the orbits of individual stars or fragments of the gas cloud will closely adhere to the dispersion orbit. The orbit of any other particle intersecting the dispersion orbit will have a tendency to be drawn along the dispersion orbit. A large complex of orbits will thus accumulate in the region of formation of the dispersion orbit which will then produce disturbing effects on the neighbouring orbits. The amount of disturbance will be different in different regions of influence and in the regions of maximum disturbance the nearly circular orbits will be drawn into considerably elongated orbits. The accumulated orbits will, in the long run of time, grow into a *diffuse elongated ring*. The ring will continue to have its share of disturbing forces on the neighbouring orbits: We can define a central orbit in the ring. The other orbits in the ring that happen to intersect this central orbit will accumulate around this orbit to give rise to a more *narrow ring*. The narrow elongated ring formed by the accumulation of dispersion orbits in the central layer of a galaxy as described above, has been called by B. Lindblad a *dispersion ring*. Dispersion rings are the products of differential rotation and gravitational disturbance on the matter lying in the central layers of the equatorial plane of disc galaxies. The rings are composed of *both stars and gas* and possess a high degree of stability. In the process of growth, it will draw in itself all the matter within the range of *r* in which it moves. A dispersion ring can therefore ultimately grow as a fairly massive one.

It is now natural to conceive that the gravitational perturbation experienced by a dispersion ring will give rise to a density fluctuation within the ring. This fluctuation will be propagated in wave motion along the body of the dispersion ring. These waves have been called *density waves*. Actually, two different modes of density waves will be, in general, propagated along the ring. One mode will move with angular speed faster and the other slower with respect to the circular velocity of the material in the ring at the corresponding distances. In the former case, the *maximum density* occurs at the minor-axis while in the latter it occurs at the major-axis of the elliptical dispersion ring. We are now close to the achievement of the spiral structure in disc galaxies produced by the sole effect of gravitational forces. It is found that a fairly stable dispersion ring along which steady bisymmetrical density waves are propagating will produce perturbing effects on the equatorial layer of the Galaxy. The perturbing effects will be strong in regions of gravitational resonance where the angular speed of circular motion equals the angular speed of the apsidal line of the disturbing ring. Spiral structure *will* develop in these regions of gravitational resonance which are often called the *Lindblad resonance.*

The formation of dispersion rings in the central region of a galaxy and the subsequent development of spiral structure by gravitational resonance effects has been demonstrated by P.O. Lindblad with the help of large-scale machine calculations. He considered elongated ring around the galactic centre, consisting of 48 mass points each of mass $16 \times 10^6 \, M_\odot$. The ring is subjected to the central force field of the Galaxy and the mass points act upon each other. His computations show that under the conditions even after a period as long as 1.168×10^9 years, the ring remains fairly stable, although a change in mass distribution has undergone inside the ring during this period. In a second computation, P.O. Lindblad considers a system of three rings, symmetrically situated around the galactic centre at distances of 2, 4 and 6 kpc respectively. Altogether 116 mass points in these rings, each of a mass of $64 \times 10^6 \, M_\odot$ are considered. A bisymmetrical pair of density waves is raised in the middle ring, one component of which moves with higher angular velocity than that of the masses forming the ring. These waves influence the surrounding matter, the perturbing effects being maximum in regions where the angular velocity of the matter is equal to the angular velocity of the apsidal line of the disturbing ring. Since the apsidal line moves much slower than the matter in the ring, owing to the smaller rotational velocities in the outer regions of the Galaxy, the region of maximum perturbation will occur farther out of the disturbing ring. This is demonstrated by P.O. Lindblad's experimental results in as much as the outer ring is greatly perturbed by the density waves propagating in the middle ring. Within 256×10^6 years from the start of the computation, the outer ring has been drawn into an elongated shape with some parts of it drawn closer to the middle ring and near about 400×10^6 years, the two rings join in their closest parts and break, thereby leading to a pair of spiral arms. The spiral arms are born as *leading,* but gradually in time by greater amount of mass exchange between the inner and outer regions, the arms transform to *trailing,* and also a *bar structure* is developed in the inner region. With the passage of more time, the prominent manifestation of spiral pattern deteriorates gradually by the differential rotation of the equatorial plane of the Galaxy, but the bar structure in the central region persists for a long time. This bar structure can again regenerate the spiral pattern in the outer regions in as much the same way as in the case of the first generation of the spiral arms. Thus a process of repeated formation and destruction of spiral arms can be carried on in the central layer of a disc galaxy under the joint influence of the bisymmetrical density waves and differential rotation. According to the experiment just described, one generation of spiral arms will appear and subsequently be liquidated within a time-scale shorter than 10^9 years.

In still another experiment with 5 rings and 192 point masses, P.O. Lindblad obtained similar results and arrived at the same general conclusion. The same type of experiments have been carried out by Hockney (1967) with ten times as many rodlike stars, which confirmed that under suitable conditions both density waves as well as spiral arms do really show up. Thus both the theoretical as well as experimental results indicate that the process of the formation of spiral arms in the equatorial layer of disc galaxies can largely be accounted for by the gravitational resonance effects.

Toomre's investigation. In his classical work, Toomre (1964) has investigated the problem of the gravitational stability of the galactic disc (assuming its thickness $\rightarrow 0$) and has arrived at the conclusion that massive material condensations (like those in spiral arms) in the galactic plane will be formed by gravitational instability which will then be destroyed by differential rotation of the galaxy in a time-scale short enough compared to the age of the Galaxy. In other words, Toomre favours the view of repeated formation and destruction of the spiral arms in disc galaxies.

Density-wave theory of Lin and Shu.   A quite different approach from that of Toomre regarding the formation and maintenance of spiral pattern in disc galaxies has been suggested by C.C. Lin and F.H. Shu and several other authors following them. This is the most widely discussed *density-wave theory of* the spiral structure of disc galaxies. In a series of papers, Lin and Shu and Lin first developed the basic logic and the mathematical formulation of the problem of development of *quasi-stationary, non-axisymmetric, self-sustained, neutral, spiral density-waves* as the manifestation of the *gravitational instability* in a thin material disc composed of stars and gas (Lin and Shu, 1964; 1966). Subsequently, these authors and several others attempted in another series of papers to explain many of the observed dynamical and structural features in our Galaxy in the light of the density-wave theory, thus exposing this theory to observational tests. We shall mention some of these features here: (a) the random gas motion throughout the galactic disc, (b) the migration of stars away from the gaseous arms, (c) the large-scale MHD shock formation, (d) the first gaseous arm to be observed at a distance of 3 kpc from the centre, (e) the large thickness of the gaseous disc in the outer parts, etc. For a comprehensive understanding the reader may consult some of the literature on this topic.

The *persistence* of spiral arms is more difficult to explain than their *formation*. Material condensation in local regions of the galactic disc in the form of spiral arms appears to be a natural consequence of gravitational clumping and differential rotation in the disc. But differential rotation will also disrupt this local clumping in the long run of time. This was well-illustrated by the works of the Lindblads. Although B. Lindblad had visualized a quasi-stationary spiral structure, the computation of P.O. Lindblad had demonstrated that spiral structure in a galactic disc would show up and subsequently be liquidated within a time-scale of less than 10^9 years. But as has been discussed by Oort (1962), the problem is to explain the formation of spiral structure over the *whole disc* of the Galaxy and its maintenance over a *very long period* of time as would be suggested by the statistically significant fact that about 70% of the galaxies were spirals. The former aspect of the spiral pattern has been described by Lin as the *grand design*. It is difficult to see how material arms of the galactic scale can repeatedly form and destroy themselves. Lin and Shu therefore try to explain this grand design of spiral arms as a *wave pattern*. The arms are manifestations of *density-waves* of the same general nature as has been suggested earlier by B. Lindblad. But while the works of the Lindblads depended heavily on the *individual stellar orbits,* Lin and Shu's theory of density-wave is based on the *collective behaviour* of the gravitational effects of the stars and of gas. Their approach is principally statistical, thus offering full scope for *quantitative analysis.* It is in this particular aspect that the theory of the Lindblads was severely lacking. The analysis of Lin and Shu shows that the stars and gas in the galactic disc can maintain a spiral density-wave through gravitational instability in the presence of the differential rotation of the disc. The spiral density-wave gives rise to a spiral component of the gravitational field superimposed upon the general axisymmetric field. This spiral gravitational field underlies the spiral arrangement of gas and young stars which can, in this way, be maintained over the entire scale of the disc. The density-waves conceived here remain either stationary or quasi-stationary to an observer, rotating around the centre of the galaxy with some proper angular velocity. This leads to the impression *of permanence of the grand design.*

The analysis of Lin and Shu has shown that non-axisymmetric disturbances of the spiral form can propagate around the *galactic disc without being sheared to destruction even in the presence of differential rotation.* This spiral pattern will therefore be *stationary* or *at least quasi-stationary* and thus the dilemma of winding and change of spacing is avoided. The simple theory is based on the assumption of *Zero dispersion* in velocities of gas and stars. It has been called by the authors as the quasi-stationary spiral structure hypothesis (QSSS-hypothesis).

17.10 DWARF GALAXIES

The galaxies with low luminosity, low metallicity having smaller size are termed *dwarf galaxies.* The distinguishing criteria between *giants* and *dwarfs* are generally absolute V magnitude ($M_v \leq -18$) or mass (~ 1/10 or 1/100 of $M_\odot$). But masses are not known for most of the dwarf galaxies, so luminosity criterion is widely used for differentiating *dwarfs* from *giant galaxies.* The study of dwarf galaxies is important from the aspect of evolution of early Universe as they are considered the counterparts of early galaxy forming fragments—especially the low mass dwarfs which have few star formation episodes compared to more massive systems. They provide insights into star formation and mass functions in low density environments, and give the impact effect of the neighbouring massive galaxies on star formation and find out pre conditions for star cluster formation.

Morphological Types

There are different types of dwarf galaxies depending upon their appearance, central surface brightness (μ_v), amount of H I mass, presence of H II regions, degree of compactness, etc. Their are: (i) dwarf spiral galaxies, (ii) blue compact dwarf galaxies (BCD), (iii) dwarf irregular galaxies (dIrr), (iv) dwarf spheroidal galaxies(dSph), and (v) tidal dwarf galaxies.

Dwarf spiral galaxies have morphological type S0, Sa, Sb, Sc and Sd with $\mu_v \geq 23$ mag arcsec^{-2} , HI mass $M_{HI} \leq 10 M_\odot$. They have rotation curves typical for rotationally supported exponential discs. Late type dwarf spirals exhibit solid body rotation. Blue compact dwarf galaxies are compact and zones of active star formation. They have $\mu_v \leq 23$ mag arcsec^{-2} and H I mass $\leq 10^9 M_\odot$. The inner part has solid body rotation but the outer part is decoupled from rotation.

Dwarf irregular galaxies are irregular in structure. They have clumpy pattern of H I clouds with traces of bright H II regions. In less massive dwarf irregulars, H I clouds are found off centred the optical region with little or no rotation. They have $\mu_v \leq 23$ mag arcsec$^{-2,}$ $M_{HI} \leq 10^9 M_\odot$. Massive dIrr show solid body rotation. They exhibit little concentration towards massive galaxies. The surface brightness profile fits with the exponential profile.

Dwarf elliptical galaxies have compact profile with spherical or elliptical structure with high central density and $M_v \leq -17$ mag, $\mu_v \leq 21$ mag arcsec^{-2}, $M_{HI} \leq 10^8\, M_\odot$ and are devoid of any rotation. The surface density profile follows the Sérsic generalization of deVaucouleurs $r^{1/4}$ law and the exponential profile (Jerjen et al. 2000).

Dwarf spheroidal galaxies are diffuse, gas deficient, with little central concentration with $M_v \geq -14$ mag, $\mu_v \geq 22$ mag arsec^{-2}, $M_{HI} \leq 10^5\, M_\odot$. There is no signature of rotation. The surface brightness fits with the King profile.

Tidal dwarf galaxies form from the debris of more massive galaxies during interactions and mergers. They do not have any dark matter and possess high metallicities (Duc & Mirabel 1998).

Local Group Dwarf Galaxies

Dwarf galaxies are found in large numbers in other environments particularly in nearby groups but Local Group (LG hereafter) dwarfs have the following aspects of consideration:

  (i) LG is primarily composed of dIrr and dSph/dE galaxies. The numbers are comparable. So it is a unique place to study the relationship between these two types of dwarf galaxies.
  (ii) LG dwarfs have the advantage of determining abundances from resolved stellar populations. Also the large luminosity range helps to study the variation of luminosity with other parameters like interstellar medium properties, dark matter (DM) content and star formation history.

Table 17.2 lists a short description of the LG dwarf galaxies. The spatial distribution (α, δ plot) of the LG galaxies shows that there are four subgroups of dwarf galaxies. The prominent group is located near the Milky Way. The second group surrounds M31. The third group forms an extended cloud populated by mostly dIrr galaxies. The fourth group is relatively isolated and contains NGC3109 as its most luminous member. The subgroup memberships are quoted in column 8 of Table 17.2.

Interstellar medium of Local Group Dwarf Galaxies

In studying the properties of interstellar medium in LG dwarf galaxies, HI maps are of importance in searching connections, if any, between dIrrs and dSphs/dEs. From Table 17.2 (column 9) it is clear that H I mass to total masses of dIrr galaxies range from 7% to over 50%. The spatial distribution is rather clumpy on scales of 100–300 parsecs and they are found in the neighbourhood of active star formation zone. Peak HI density is approximately 10^{21} cm^{-2} but for even such a high density in some dIrrs (Shostak and Skillman 1989; Young and Lo 1997) there is no evidence of current star formation. So it seems that a triggering mechanism (e.g. interstellar shocks etc.) is necessary for such event. Also H I emission is generally centred on the optical centroids of the galaxies. On the contrary, dSph galaxies are comparatively less enriched in H I clouds and these clouds are offset from the optical centre.

TABLE 17.2 Local Group Dwarf Galaxies [*Courtesy:* Mateo 1998]

Galaxy name	Other name	α_{2000}	δ_{2000}	l	b	Type	Subgroup	M_{HI}/M_{tot}
WLM	DDO 221	00 01 58	−15 27.8	75.9	−73.6	IrrIV–V	LGC	0.40
NGC 55		00 15.08	−39 13.2	332.7	−75.7	IrrIV	LGC	0.09
IC 10	UGC 192	00 20 25	+59 17.5	119.0	−3.3	dIrr	M31	0.10
NGC 147	DDO 3	00 33 12	+48 30.5	119.8	−14.3	dSph/dE5	M31	<0.001
And III		00 35 17	+36 30.5	119.3	−26.2	dSph	M31	—
NGC 185	UGC 396	00 38 58	+48 20.2	120.8	−14.5	dSph/dE3p	M31	0.001
NGC 205[j]	M110	00 40 22	+41 41.4	120.7	−21.1	E5p/dSph–N	M31	0.001
M32	NGC 221	00 42 42	+40 51.9	121.2	−22.0	E2	M31	<0.001
M31	NGC 224	00 42 44	+41 16.1	121.2	−21.6	SbI–II	M31	—
And I		00 45 43	+38 00.4	121.7	−24.9	dSph	M31	—
SMC	NGC 292	00 52 44	−72 49.7	302.8	−44.3	IrrIV–V	MW	0.004
Sculptor		01 00 09	−33 42.5	287.5	−83.2	dSph	MW	0.03
LGS 3	Pisces	01 03 53	+21 53.1	126.8	−40.9	dIrr/dSph	M31	0.07
IC 1613	DDO 8	01 04 54	+02 08.0	1.29.8	−60.6	IrrV	M31/LGC	—
And II[k]		01 16 27	+33 25.7	128.9	−29.2	dSph	M31	—
M33	NGC 598	01 33 51	+30 39.6	133.6	−31.3	ScII–III	M31	—
Phoenix		01 51 06	−44 26.7	272.2	−68.9	dIrr/dSph	MW/LGC	0.006
Fornax		02 39 59	−34 27.0	237.1	−65.7	dSph	MW	<0.001
EGB0427+63	UGCA 92	04 32 01	+63 36.4	144.7	+10.5	dIrr	M31	—
LMC		05 23 34	−69 45.4	280.5	−32.9	IrrIII–IV	MW	—
Carina		06 41 37	−50 58.0	260.1	−22.2	dSph	MW	<0.001
Leo A	DDO 69	09 59 24	+30 44.7	196.9	+52.4	dIrr	MW/N3109	0.72
Sextans B	DDO 70	10 00 00	+05 19.7	233.2	+43.8	dIrr	N3109	0.05
NGC 3109	DDO 236	10 03 07	−26 09.5	262.1	+23.1	IrrIV–V	N3109	0.11
Antlia[l]		10 04 04	−27 19.8	263.1	+22.3	dIrr/dSph	N3109	0.08
Leo I	DDO 74	10 08 27	+12 18.5	226.0	+49.1	dSph	MW	<0.001
Sextans A	DDO 75	10 11 06	−04 42.5	246.2	+39.9	dIrr	N3109	0.20
Sextans		10 13 03	−01 36.9	243.5	+42.3	dSph	MW	<0.001
Leo II	DDO 93	11 13 29	+22 09.2	220.2	+67.2	dSph	MW	<0.001
GR8	DDO 155	12 58 40	+14 13.0	310.7	+77.0	dIrr	GR8	0.59
Ursa Minor	DDO 199	15 09 11	+67 12.9	105.0	+44.8	dSph	MW	<0.002
Draco	DDO 208	17 20 19	+57 54.8	86.4	+3.4.7	dSph	MW	<0.001
Milky Way		17 45 40	−20 00.5	0.0	0.0	Sbc	MW	—
Sagittarius[j]		18 55 03	−30 28.7	5.6	−14.1	dSph–N	MW	<0.001
SagDIG	UKS1927–177	19 29 59	−17 40.7	21.1	−1.6.3	dIrr	LGC	9.2
NGC 6822	DDO 209	19 44 56	−14 48.1	25.3	−18.4	IrrIV–V	LGC	0.08

(*Contd.*)

TABLE 17.2 Local Group Dwarf Galaxies [*Courtesy:* Mateo 1998] (Contd.)

Galaxy name	Other name	α_{2000}	δ_{2000}	l	b	Type	Subgroup	M_{HI}/M_{tot}
DDO 210[m]	Aquarius	20 46 46	–12 51.0	34.0	–31.3	dIrr/dSph	LGC	0.35
IC 5152		22 02 42	–51 17.7	343.9	–50.2	dIrr	LGC	0.15
Tucana[n]		22 41 50	–64 25.2	322.9	–47.4	dSph	LGC	—
UKS2323–326	UGCA 438	23 26 27	–32 23.3	11.9	–70.9	dIrr	LGC	—
Pegasus	DDO 216	23 28 34	+1444.8	94.8	–43.5	dIrr/dSph	LGC	0.09

In dIrrs, dust is seen as compact absorption regions and these are seen near the cores of some dSphs. Also the dust extinction law differs significantly from the Galactic extinction law.

Almost all dIrrs contain H II regions while one dSph galaxy (NGC185) and one transition type (Antlia) contain H II regions. No diffuse X-ray emission has been detected in any LG dwarfs.

Chemical Abundances of Local Group dwarf galaxies

Photometric and spectroscopic techniques are used to calculate heavy element abundances of LG dwarf galaxies. Now the photometric technique is not very reliable for the determination of abundances of an individual element, rather it is suitable in the case of stellar populations (e.g. red giant branch – RGB population). Photometric abundances have now been measured in some LG dwarf galaxies where old/intermediate age RGB populations are detectable in deep field photometry. But unlike star clusters there is large dispersion in RGB colour and as a result very few galaxies have reliable abundance estimate which is a function of colour (V–I). Table 17.3 gives a list of photometric abundances and corresponding dispersions for dwarf galaxies. There is significant abundance dispersion for almost every galaxy (second column of Table 17.3).

Spectroscopy can be used for both individual star and emission nebula (e.g. H II regions and planetary nebulae). For dSph galaxies spectroscopic abundances [Fe/H] or [Ca/H] have been measured for individual stars (in Draco, Sextans, Carina, Sagittarius, Ursa Minor, etc.) and oxygen abundance has been measured for planetary nebulae (in Fornax, Sagittarius, NGC185, NGC205, NGC6822, etc.) and H II regions. Unlike [Fe/H] in early type dwarfs, there is no evidence for significant dispersion of the oxygen abundances in any LG dIrr in which multiple H II regions have been studied. Figure 17.11 shows a scatter diagram of mean abundances (corrected for reddening) vs absolute visual magnitudes of LG dwarf galaxies.

There is a clear segregation between dSph/dE and dIrr at $Mv \sim$ –13.5. So it remains controversial that whether dSph/dE are gas stripped dIrr or these two classes have different origin. Mayer et al. (2001) suggest that tidal shocks during perigalactic passage close to a massive galaxy produce tidal stripping of gas from dIrr and convert it to dSph. Thus in this situation dSph should always accompany a massive galaxy which is not the case always, e.g. dSph Tucana is found in distinct outskirts of LG.

TABLE 17.3 Heavy Element Abundances of Dwarf Galaxies in LG (*Courtesy:* Mateo 1998)

Galaxy name	[Fe/H] dex	σ[Fe/H] dex	12 + log[O/H][a] dex	[N/O][b] dex
WLM	−15±0.2	—	7.75±0.2	−1.46±0.15
NOC 55	—	—	8.32±01.5	−1.44±0.15
IC 10[6]	—	—	8.19±0.15	−1.37±0.12
NGC 147[h,i,j]	−1.1±0.2	0.4±0.1	—	—
And III	−2.0±0.2	≤0.2±0.04	—	—
NGC 185[j,k]	−1.22±0.15	0.4±0.1	8.2±0.2	—
NGC 205[j,k,l]	−0.8±0.1	0.5±0.1	8.6±0.2	—
M32[l]	−1.1±0.2	0.7±0.2	—	—
And I	−1.5±0.2	0.3±0.1	—	—
Sculptor	−1.8±0.1	0.3±0.05	—	—
LGS3	−1.8±0.3	0.3±0.2	—	—
IC 1613	−1.3±0.2	—	7.8±0.2	—
And II	−1.6±0.3	0.5±0.1	—	—
Phoenix	−1.9±0.1	0.5±0.1	—	—
Formax[j,k]	−1.3±0.2	0.6±0.1	7.98±0.4	—
EGB 0427+63	—	—	7.62±0.1	≤−1.5
Carina	−20±0.2	<0.1	—	—
Leo A[k]	—	—	73±0.1	—
Sextans B[m]	~−1.2	—	7.84±0.3	—
NGC 3109	−1.5±0.3	<0.3	8.06±0.2	—
Antlia	−1.8±0.25	0.3±0.1	—	—
Leo I	−1.5±0.4	0.3±0.1	—	—
Sextans A	−1.9±0.3	—	7.49±0.2	—
Sextans	−1.7±0.2	0.2±0.05	—	—
Leo II	−1.9±0.1	0.3±0.1	—	—
GR 8[m]	—	—	7.62±0.1	—
Ursa Minor	−2.2±0.1	≤0.2	—	—
Draco	−2.0±0.15	0.5±0.1	—	—
Sagitarius[j,k,n]	−1.0±0.2	0.5±0.1	8.30±0.08	−1.0±0.3
SagDIG	—	—	7.42±0.3	—
NGC 6822[k,m]	−1.2±0.2	0.5±0.1	8.2±0.2	−1.7±0.1
DDO 210[n]	<−1.0	—	—	—
IC 5152	—	—	8.36±0.2	—
Tucana	−1.7±0.15	0.3±0.2	—	—
UKS2323−326	—	—	—	—
Pegasus	−1.0±0.3	—	7.93±0.14	−1.24±0.15

[a] The oxygen abundance defined as 12 + log(O/H), where (O/H) is the number ratio of oxygen to hydrogen atoms.
[b] The nitrogen to oxygen ratio defined as [N/O] = log(N/O), where (N/O) is the number ratio of nitrogen to oxygen atoms.

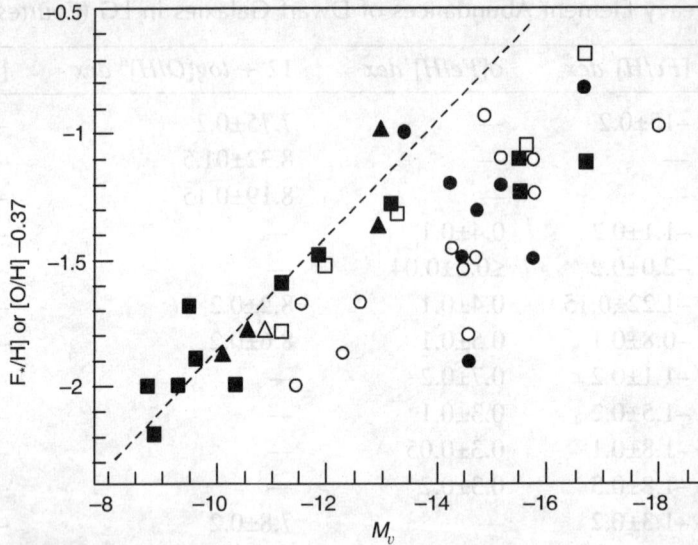

FIGURE 17.11   A plot of [Fe/H] (*filled squares*) or [O/H] [Anders and Grevesse (1989)] vs absolute V-band magnitude. The *dotted line* is a rough fit to the [Fe/H]–M_v relation for the dSph and transition objects. *Square symbols* refer to dSph or dE galaxies; *triangles* refer to transition galaxies (denoted dIrr/dSph in Table 17.2); *circles* refer to dIrr systems. *Filled symbols* correspond to [Fe/H] abundances determined from stars, while *open symbols* denote oxygen abundance estimates from analyses of H II regions and planetary nebulae.

Stellar Populations in Local Group Dwarf Galaxies

LG dwarfs vary widely in star formation history, mean metallicity and enrichment history, times of major star formation episodes, fractional distribution of ages and subpopulations, etc. The different features observed so far can be summarized as follows:

(i)   No two LG dwarfs have the same star formation histories.

(ii) Younger populations generally have the tendency to be centrally concentrated whereas older populations are more extended. This fact has been interpreted because tidal and ram pressures have removed most of the gas from dSphs whereas more distant dSphs are able to retain star forming material for longer periods of time (van den Bergh 1998). Though for MW this picture holds but unfortunately M31 dwarfs do not show presence of young populations in support of the above theory.

(iii) Almost all dwarf galaxies contain an old population (~ 10 Gyr) as is evident from the presence of RR Lyrae stars in these galaxies.

(iv) Oldest populations in majority cases of dIrrs and dSphs are comparable in age with the oldest globular clusters of Galactic halo and bulge (Grebel 2000). This indicates a common epoch of oldest measurable star formation episode.

(v) Nearby dwarf galaxies typically show either continuous or episodic star formation. In the case of episodic mode the reason for the periods of quiescence and the renewed onset of star formation are not yet known.

Kinematics and Dark Matter in Local Group Dwarf Galaxies

As dSph galaxies have very little or no rotation, so the velocity dispersion is measured on the basis of a pressure supported model which scales as (RcS0)1/2 (Richstone & Tremaine 1986) where Rc is the characteristic radial scale length and S0 is the central surface brightness in intensity units. Globular clusters have central velocity dispersions of 2–15 km s^{-1}. For dwarf galaxies they have radial scale length 10 times larger and surface brightness 10^3 times smaller and should have velocity dispersion ≤ 2 km s^{-1}. For non-rotating dwarfs the radial scale length is taken as King core radius or exponential scale length. But all low luminosity dwarfs have central velocity dispersion ≥ 7 km s^{-1}. Most of the dIrr galaxies have rotation. So the rotation velocity for this class of galaxies is used to estimate kinematics mass. For estimation of kinematics masses of dSph galaxies using velocity dispersion for these galaxies it is assumed that velocity dispersion is isotropic and mass follows the light distribution. Figure 17.12 shows a plot of derived *M/L* (integrated mass-to-light ratio in the V-band) ratios vs M_v for all LG dwarfs. In the figure dIrr and early type galaxies (dSph/dE) are clearly separated. Also *M/L* ratios are higher for dIrr galaxies showing that they possess extended halos. For dSph galaxies, mass follow light assumption is followed. So it is likely that *M/L* for the latter ones are under estimated.

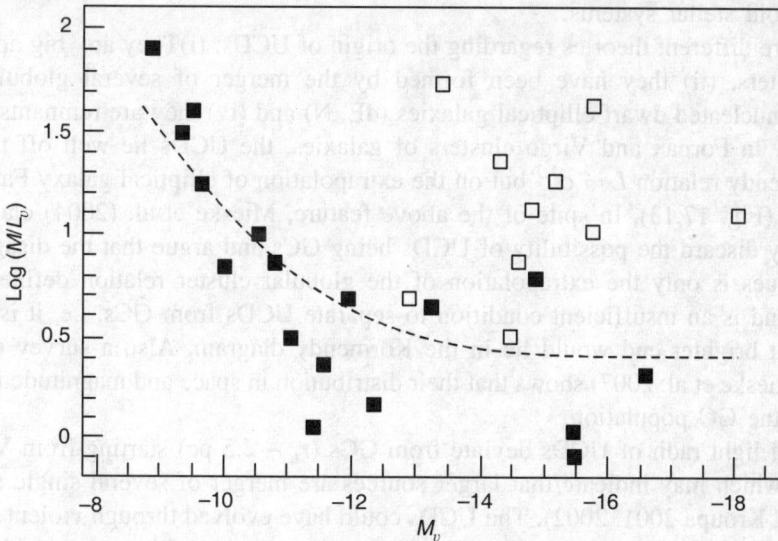

FIGURE 17.12 Kinematically determined mass-to-light ratios of local group dwarfs as a function of luminosity *Filled squares* are for dSph or dSph/Irr systems for which masses were determined from the central velocity dispersions, while the *open squares* represent Irr systems that have masses derived here from H I rotation curves Sagittarius is denoted as an *open circle*. The function log (M/L) = 2.5 + 10^7/(.L/L$_\odot$) as a *dashed line*.

There is an alternative theory for explaining the high velocity dispersion for dSph galaxies excluding the dark matter. Some authors explain that the observed high velocity for dSph galaxies might be due to tidal effects of some giant galaxy. But it is unlikely that the central part of a galaxy is affected by the tidal force until it is completely disrupted. Also Leo II type dwarfs are far away from any giant galaxy to experience such effect. So LG dwarfs are kinematically peculiar, whether dominated by dark matter or not.

These are the different aspects discussed in the present section but still there are many things to explore, e.g. are the dwarfs in the other groups similar? How do they differ? Whether the galaxy mass plays a dominant role in determining their property or whether the environment is the dominant factor? Why is there an ancient common epoch of star formation in all dwarfs irrespective of their morphology? What is the reason for episodic star formation in some dwarfs? These questions and many more require more extensive modeling and more observations beyond LG and therefore in the future surprising facts need to be explored from a rigorous study of dwarf galaxies.

17.11 ULTRA COMPACT DWARF GALAXIES (UCD)

Recently (Drinkwater et al. 2001; Phillips et al. 2001) discovered a new class of subluminous and extremely compact objects clusters and groups of galaxies. These objects are termed ultra compact dwarf galaxies (UCDs). They have intrinsic sizes ≤ 100 pc and absolute B magnitude in the range $-13 \leq M_B \leq -11$. In the luminosity regime they fall at the lower end of dwarf galaxy and at the upper end of globular clusters. The spectra show no traces of Balmer lines, i.e. they are old stellar systems.

There are different theories regarding the origin of UCDs: (i)They are big and luminous globular clusters, (ii) they have been formed by the merger of several globular clusters, (iii) they are nucleated dwarf elliptical galaxies (dE, N) and (iv) they are remnants of stripped disc galaxies. In Fornax and Virgo clusters of galaxies, the UCDs lie well off the globular cluster Kormendy relation $L \infty \sigma^{1.7}$ but on the extrapolation of elliptical galaxy Faber Jackson law $L \infty \sigma^4$ (Fig. 17.13). In spite of the above feature, Mieske et al. (2004) claim that one cannot readily discard the possibility of UCDs being GCs and argue that the disagreement of the UCD values is only the extrapolation of the globular cluster relation defined at fainter magnitudes and is an insufficient condition to separate UCDs from GCs, i.e. it is not known where GCs at brighter end would lie in the Kormendy diagram. Also a survey of UCDs in Centaurus (Mieske et al. 2007) shows that their distribution in space and magnitude is consistent with that of the GC population.

The half light radii of UCDs deviate from GCs ($r_h \sim 2.5$ pc) starting from Virgo UCDs (Fig. 17.14) which may indicate that larger sources are merger of several single star clusters (Felhauer and Kroupa 2001, 2002). The UCDs could have evolved through violent interactions among the GCs. Even for highly eccentric objects these merger objects are stable subject to strong tidal force and retain most of their mass even after 10 Gyr, e.g. Antanae galaxies, which produce knots of intense star formation result in massive young star clusters that have masses ranging from 10^7 to 10^8 $M_\odot$ and a dimension of few hundred pc. There are studies carried out to find resemblance between UCDs and dE, N galaxies. Figures 17.15 and 17.16 show a plot

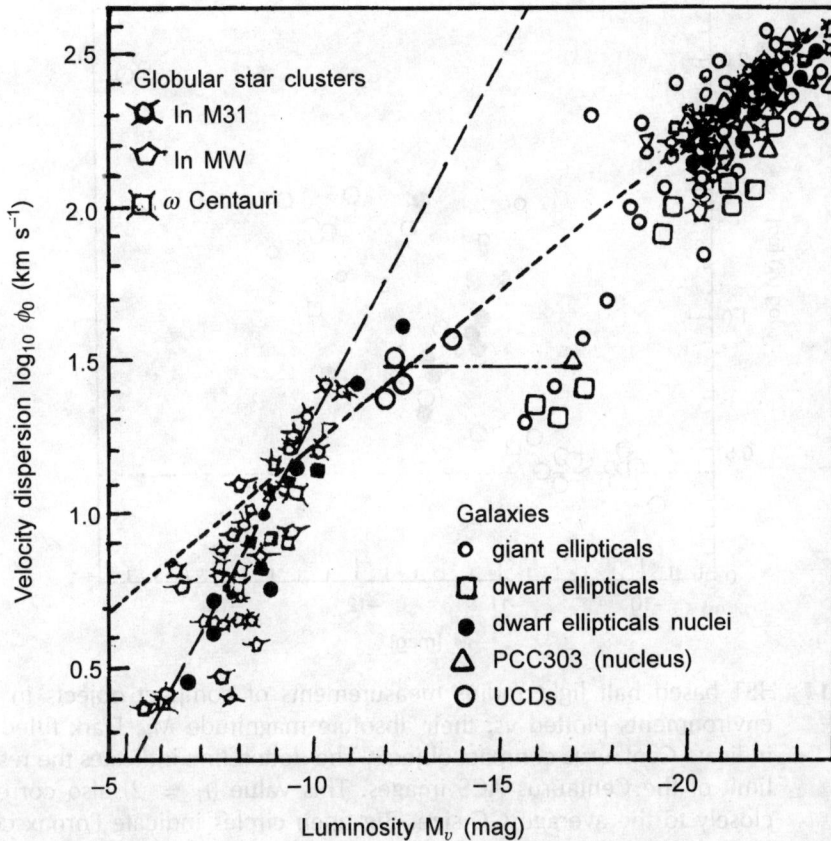

FIGURE 17.13 Kormendy diagram taken from Drinkwater et al. (2004).

of mean surface brightness vs absolute magnitude and surface brightness vs core luminosity (M_v) of UCDs and dE, N galaxies. There is a clear segregation between two classes. Also there is no sign of low luminosity envelopes around UCDs in Fornax and Virgo which correspond to galactic halo.

Another theory on UCD formation is that they are the remnant nuclei of stripped dwarf galaxies which have lost their outer envelope by tidal interaction with a massive galaxy. This phenomenon is known as *galaxy threshing* (Bekki et al. 2001), e.g. simulations with a dwarf galaxy and a massive galaxy NGC1399 show that after four passages the dwarf loses its envelope almost entirely. Since the central nucleus is compact, it survives. Varying the parameters of the simulation shows that not all nucleated dwarfs can transform into UCDs. The two possible outcomes are (i) stellar populations in UCDs are old as tidal stripping takes a long time and (ii) luminosity function of UCDs is not necessarily similar to that of the nuclei of nucleated dwarfs because tidal stripping is a selective process. Also repeated simulations (Bekki et al. 2003) result in a low surface brightness envelope around the nuclei of dwarf. Detection of such envelopes around UCDs is a possible support to this theory. In case of Fornax UCDs no such evidence is found. For Virgo, two UCDs show envelopes but the

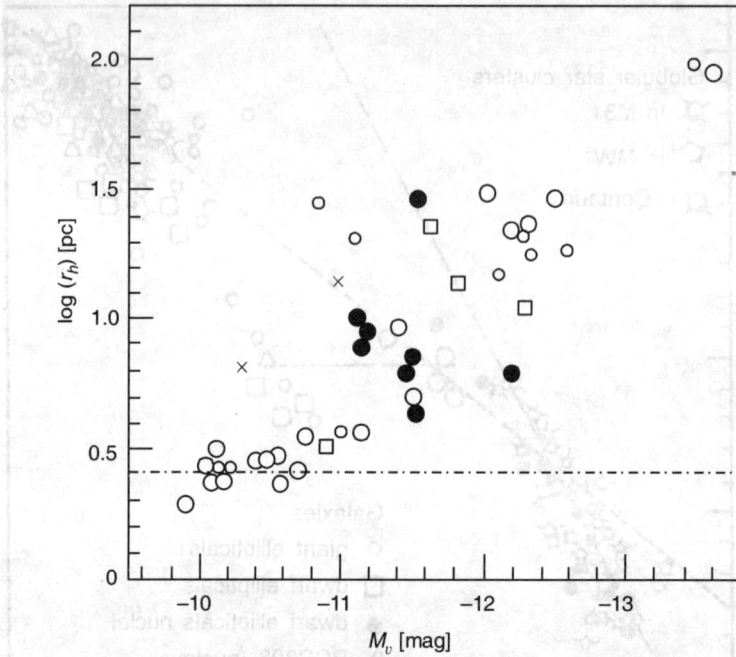

FIGURE 17.14 HST based half light radius measurements of compact objects in various environments plotted vs. their absolute magnitude M_v. Dark filled circles indicate Centaurus compact objects. The dotted line indicates the resolution limit of the Centaurus ACS images. This value ($r_h \simeq 2$) also corresponds closely to the average GC size. Big open circles indicate Fornax compact objects from Mieske et al. (2006a) and Evstigneeva et al. (2007). Small open circles are Virgo UCDs from Evstigneeva et al. (2007) and Haségan et al. (2005). Crosses are the Local Group compact objects ωCen (fainter) and G1 (brighter).

detection is highly uncertain. Also when dE,N transform to UCD, it moves along the magnitude axis in CMD diagram by 4.1 mag red ward but the nucleus has the same colour as the host galaxy and stellar population does not change during the threshing. On an average observations do not show this. The nuclei are found to vary about 0.07 mag bluer than the underlying host galaxy. Also, the UCDs of Abell 1689 are brighter and larger than those of the Fornax and Virgo (Mieske et al. 2004). Jones (2005) found that UCDs are always associated with a central massive galaxy supporting *galaxy threshing*.

17.12 COMPACT GROUPS OF GALAXIES

It is a very natural phenomenon that small systems (usually 3–4) of several galaxies configure very off and on within a very compact region of the sky. Naturally the question that arises is whether these systems are just the projection or transient effects or are they the sites of truly dynamical evolution consisting of a separate class called the *compact groups of galaxies*. To

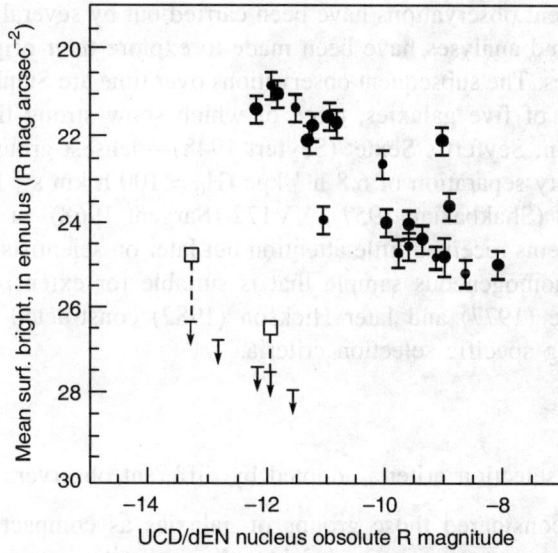

FIGURE 17.15 Magnitude vs surface brightness for dE,N and UCDs. 2.5σ upper limits used for the UCDs, shown as arrows. The two UCDs with formal detections are shown as squares. Virgo (large circles) and Fornax (small circles) dE,N galaxies are shown for comparison. Figure taken from Jones et al. 2005.

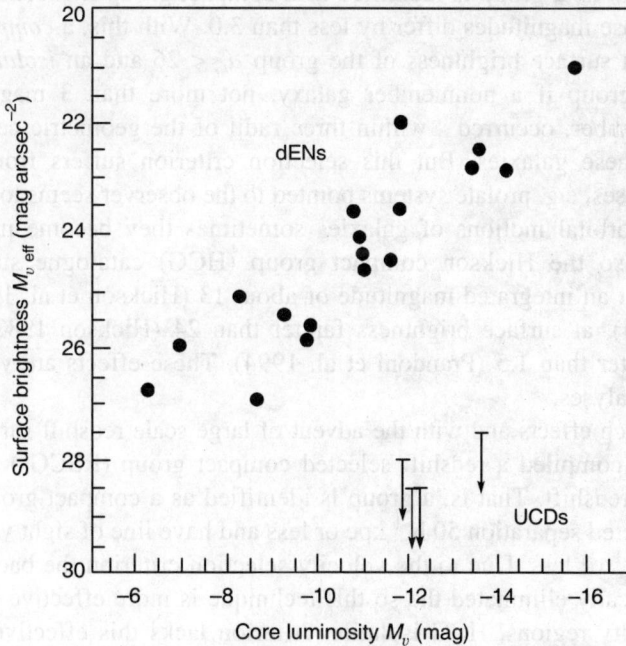

FIGURE 17.16 Plot of surface brightness vs core luminosity of UCDs and dE,N galaxies taken from Drinkwater et al. 2004.

find the answer, different observations have been carried out by several astrophysicists during the last few decades and analyses have been made to explore their origin, if any, using very sophisticated techniques. The subsequent observations over time are Stephan's Quintet (Stephan 1877)—a small group of five galaxies, three of which show strong tidal distortions due to gravitational interaction, Seyfert's Sextet (Seyfert 1948)—densest group of galaxies having a median projected galaxy separation of 6.8 h^{-1} kpc ($H_0 = 100$ h km s^{-1} Mpc^{-1}), a dense group of twelve red galaxies (Shakbazian 1957), VV172 (Sargent 1968)—a chain of five galaxies etc. Initially these systems received little attention but later on scientists became motivated for the preparation of a homogeneous sample that is suitable for extensive statistical analyses. With this desire, Rose (1977) and later Hickson (1982) constructed the first catalogue of compact groups having specific selection criteria.

Selection Criteria

Different quantitative selection criteria adopted by different observers are as follows:

1. Rose (1977) considered those groups of galaxies as compact groups if the groups contained three or more galaxies brighter than a limiting magnitude of 17.5 and had a projected surface density enhancement of a factor of 1000 compared to the surrounding galaxy density. But the fixed magnitude limit makes the sample susceptible to strong distance-dependent biases.

2. To reduce such effect, Hickson (1982, 1993, 1994) adopted a relative magnitude criterion. That is, a group is identified as a compact group if it contains four or more galaxies whose magnitudes differ by less than 3.0. With this, a *compactness* criterion, i.e. the mean surface brightness of the group $\mu_G < 26$ and an *isolation* criterion, i.e. to reject a group if a nonmember galaxy, not more than 3 mag fainter than the brightest member, occurred   within three radii of the geometric centre of the circle containing these galaxies. But this selection criterion suffers from projection and transition biases, e.g. prolate systems pointed to the observer seems to be more compact and due to orbital motions of galaxies sometimes they become momentarily more compact. Also the Hickson compact group (HCG) catalogue starts significantly incomplete at an integrated magnitude of about 13 (Hickson et al. 1989; Sulentic and Rabaca 1994), at surface brightness fainter than 24 (Hickson 1982) and magnitude interval greater than 1.5 (Prandoni et al. 1994). These effects are very important for statistical analyses.

3. To reduce such effects and with the advent of large scale redshift surveys Barton et al. (1996) have compiled a redshift selected compact group (RSCGs) on the basis of a criterion on redshift. That is, a group is identified as a compact group if the galaxies have a projected separation 50 h^{-1} kpc or less and have line of sight velocity difference of 100 km s^{-1} or less. Due to the velocity selection criterion the background galaxies are automatically eliminated the so this technique is more effective in finding groups at high density regions. HCG isolation criterion lacks this effectiveness.

Structural Properties

It is a matter of investigation that whether the compact groups are random projections or chance alignment objects (projection or transient configurations) or they constitute a different physical structure. In this regard the studies of luminosities and spatial distributions of the member galaxies are important. If they are chance alignment objects, then the luminosity function should be the same as that of the parent systems and the spatial distribution should be random in nature. Many authors (Arp 1973, Hickson et al. 1984; Malykh and Orlov 1986) have found *chains* or *elongated* structures which cannot be explained as random projections or chance crossings as this would erase any inherent ellipticity of a parent loose group. So if compact groups are not simply projection effects they might be expected to show centrally concentrated surface density profile as is seen in clusters of galaxies. By scaling and superimposing the HCGs Hickson et al. (1984) found evidence for central concentration. Similar results have been found by Mendes deOliveira and Giraud (1994) and Montoya et al. (1996). Their technique uses distributions of projected pair separations and thus avoids assumptions about the location of the group centre. The density profile found is consistent for all groups without scaling which implies that compact groups have a unique scale and they are not products of hierarchical clustering.

There have been several studies of the morphological types of galaxies in compact groups (Hickson 1982; Williams and Rood 1987; Sulentic 1987; Hickson et al. 1988). They have mainly two characteristics: (i) The fraction of late type galaxies is significantly less in compact groups ($f_s \sim 0.49$, HCG, $f_s \sim 0.59$ SCG) compared to field ($f_s \sim 0.82$, Gisler 1980, Nilson 1973). (ii) Also in any group the morphology of the galaxies are almost similar (Sulentic 1987; Hickson et al. 1988; Prandoni et al. 1994). This concordance of morphology might be the result of correlation between the morphology type and the other property of the group. The strongest such correlation is between velocity dispersion and morphology type (Hickson et al. 1988). Figure 17.17 shows that groups with higher velocity dispersion contain fewer late type (gas rich) galaxies. They are also more luminous. This suggests that the velocity dispersion is more fundamental in compact groups.

Also studies of rotation curves in 32 HCG galaxies (Rubin et al. 1991) show peculiar rotation curves in most cases. This asymmetry, irregularities in rotation curves and the above correlation all show evidences of strong gravitational interaction in compact groups. Also galaxies in compact groups should be the sites for starbursts and should harbour active galactic nuclei. In fact, many galaxies in HCGs show starbursts and harbour AGN (Ribeiro et al. 1996; Rubin et al. 1990). The degree of star formation activity in compact group galaxies can be determined from infrared observations. But the direct measurement of the ratio of far infrared to optical luminosity might induce error due to low spatial resolution (Sulentic and De Mello Rabaca 1993). To avoid such a discrepancy, the degree of star formation can be measured in terms of radio emission as it has a very good correlation with infrared flux (Hickson et al. 1989). Sometimes the infrared colours of the sources are used instead of the infrared/optical ratio. In compact group spirals the radio emission is primarily nuclear whereas in isolated spirals it originates in the disc. If this is also true for infrared flux there must be an enhancement of infrared/optical ratio in the nuclear region. This idea is supported by recent millimetre wavelength observations (Mamon et al. 1986) in which CO is detected in 55 of 70 IRAS

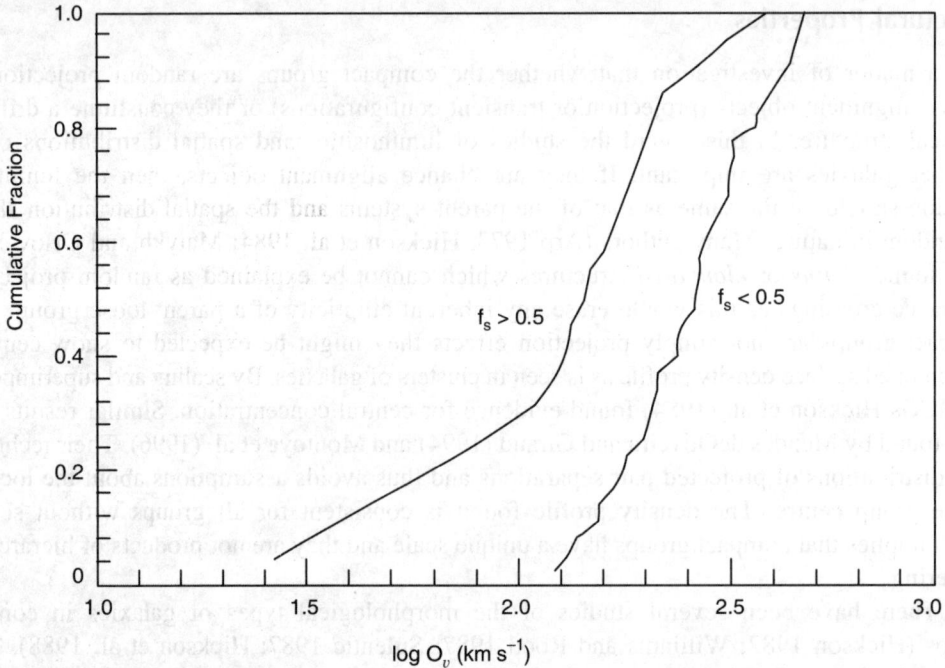

FIGURE 17.17 Morphology–velocity correlation for compact groups (White, 1990).

selected HCG galaxies. There are also evidences of tidal interaction in compact groups through diffuse light. As a result of tidal interaction, stripped-off stars should accumulate in the potential well of the group and may be detectable as diffuse light. Initially evidence of diffuse light was not satisfactory until 1981 when Berghvall et al. were successful in detecting ionized gas and a common halo around a compact quartet of interacting early type galaxies and evidence for a common halo in VV172 reported by Sulentic and Lorre (1983). Also Mamon (1986) indicated that the expected diffuse light would generally be very faint.

The luminosity function of compact groups was estimated by different authors leading to apparently conflicting results. One group (Heligman and Turner 1980; Hickson 1982; Kodaira et al. 1991; Mendes de Oliveira and Hickson 1991) found that the compact group contains relatively fewer faint galaxies compared to field galaxy sample while the other group (Sulentic and Rabaca 1994; Barton et al. 1996) found it to be consistent with field galaxies. Those conflicting results might be due to the groups with small magnitude range.

If there is tidal interaction among galaxies in a compact group, it is expected that the gas is removed from discs of spiral galaxies. This presumption is consistent with the observations (Menon 1995), i.e. compact groups of spiral galaxies are deficient in neutral hydrogen by about a factor of two compared to loose groups (William and Rood 1987). High resolution studies of individual groups also support the same picture. So the above results strongly indicate that these compact groups are physically dense systems. They are not chance alignment or transient configurations in loose groups.

The most extensive X-ray study of 85 HCGs (Ponman et al. 1996) has detected diffuse emission in almost 75% of the systems. X-ray emission is also detected in spiral reach systems but the surface brightness and the characteristic temperature are comparatively lower as is expected given the lower velocity dispersions of spiral rich compact groups. The diffuse X-ray luminosity is found to correlate with temperature, velocity dispersion and spiral fraction but not the optical luminosity. This leads to the conclusion that the gas is mostly primordial and not derived from the galaxies. The relatively low metallicuty (0.18 solar) in compact groups compared to rich clusters (0.3–0.4 solar) also supports the above conclusion.

Physical Properties

While studying the redshifts of member galaxies in compact groups, discordant redshifts have been measured for some galaxies. If the frequency of discordant galaxies is inconsistent with the statistics of chance projection, then we are in need of new theories for its explanation. Initially it was concluded that the number of discordant redshifts in HCG catalogue is too large to explain by chance. Later on, rigorous selection criteria found no strong statistical evidence for this. A question was raised that the latter result might be due to incompleteness issue. So this issue was addressed taking Monte Carlo simulations using the observational results from both HCG and SCG catalogues. It was ultimately found that except for Stephan's Quintet and Seyfert's Sextet the number of discordant redshifts is consistent with chance projections. For the above two, there is independent physical evidence that the galaxies at cosmological distances that correspond to their redshifts are therefore not group members (Kent 1981; Wu et al. 1994). There are questions, as discussed before regarding the projection effect of the compact groups. Since only three phase-space dimensions for galaxies can be measured, two components of position and one velocity the groups are subject to projection effects. Thus, what appear to be physically dense may not actually be the physically related systems. So different phenomena have been investigated to explore their physical origin.

(i) From the analysis of optical images 43% of HCG galaxies show morphological features which indicate interaction and /or merging. The high frequency of interactions observed in compact groups makes it difficult to accept the hypothesis of chance alignment.

(ii) A high fraction of HCGs show diffuse X-ray emissions. There is a correlation between the X-ray and optical properties. All these evidences reinforce the above conclusion.

So the fact that the compact groups are *physically dense systems* is well accepted. But it might happen that the degree of compactness is not as high as it may appear (Walke and Mamon 1989).

The next question that comes into play is that whether the compact groups are distinct entities or they are intermediate objects between loose groups and triplets, pairs and individual galaxies in the clustering hierarchy. Two mechanisms have been given for their formation.

(i) The first one depicts that (Diaferio et al. 1994) compact groups form continually from bound subsystems within loose groups. Observational supports are that most HCGs embedded in loose groups and Shakbazian groups contain early type galaxies and the number of blue (gas rich) galaxies seem to be very small.

(ii) A second mechanism proposes a model in which merging activity in compact groups is accompanied by infall of galaxies from the environment. So a mixture of morphological types occurs in such cases. Zepf (1993) estimated that roughly 7% of the galaxies in the compact groups are in the process of merging on the basis of (a) optical signature of merging (b) warm far-infrared colours and (c) sinusoidal rotation curves. This fraction may be as high as 25% if only one criterion is followed. Detailed studies of some of the individual groups like Seyfert Sextet, HCG31, HCG62, HCG94, and 95 show strong gravitational interaction while on the other hand low density groups, e.g. HCG44, are in a less advanced stage of evolution. Also merging allows compact groups to persist for a longer time and galaxy interaction should be associated with star burst galaxies or AGN. Recently, several groups have been found to be associated with luminous infrared galaxies, AGN and QSOs at z ~2.

So it now seems clear that in spite of contaminations by projections, a larger fraction of high surface brightness compact groups are physically dense. They form by gravitational relaxation process within the loose groups of galaxies, and the densest groups are in an advanced stage of evolution characterized by strong interactions, star bursts and AGN activities stripping off stellar and dark matter haloes and merging. They contain large amounts of dark matter and primordial X-ray emitting gas confined within the gravitational potential well.

EXERCISES

1. How was the concept that the galaxies are *island universes* gradually developed through observational knowledge of the astronomers? Why galaxies have been called *extragalactic nebulae?*

2. Describe Hubble's morphological classification of galaxies. What are the principal observable features that form the basis for this classification? What features distinguish the sub-classes?

3. "Hubble's morphological classes are correlated with the stellar Population characteristics" —establish this statement.

4. What do you know of the distribution of galaxies in space? Comment on the *zone of avoidance.*

5. Write down the luminosity distribution laws for different types of galaxies and comment on the merits and demerits of each of these laws.

6. Explain how spectral classes are assigned to galaxies. What are the inherent difficulties faced in such assignment?

7. Spectral classes of galaxies indicate the stellar content in them—explain.

8. List the prominent members of the Local Group of galaxies. Discuss the importance of these members in the study of more distant galaxies.

9. Discuss the use of various distance indicators leading to the determination of the distances to external galaxies.

10. What distance indicators you will use to determine distance of galaxies in:
    (a) the Local Group of galaxies;
    (b) the Virgo Cluster of galaxies;
    (c) the Coma Cluster of galaxies;
    (d) a very distant quasar?
    Explain your answers.

11. What are the principal difficulties that are faced in deriving the correct value of the Hubble constant *H*? Give a physical interpretation of the Hubble constant.

12. Discuss the observational steps leading to the establishment of Hubble's law *V = HD*. What, in your opinion, is the most reasonable value of *H*? Why?

13. What observational evidences are there to indicate that the nuclei of galaxies are *active*?

14. Describe galactic nuclei having different degree of activity. Discuss the effects of nuclear activity on the structure and evolution of a galaxy.

15. "The spiral arm is a tube of gas held together by the galactic magnetic field"—can you establish this statement on the basis of the theoretical works that are known to you?

16. Explain the terms: *density wave, dispersion ring, dispersion-orbit* and *Lindblad resonance.*

17. Discuss the works of B. Lindblad and P.O. Lindblad that attempt to establish that the spiral arms are manifestations of density-waves in the disc of the galaxies.

18. Discuss Lin and Shu's approach to the density-wave theory in order to explain the formation and maintenance of the spiral pattern.

SUGGESTED READING

1. Abell, G.O., *Exploration of the Universe,* Holt, Rhinehart and Winston Inc., New York, 1969.

2. Ambartsumian, V.A., *Structure and Evolution of Galaxies,* Interscience Publishers Inc., New York, 1965.

3. Arp, H., *Astrophysical Journal*, 185, 797, 1973.

4. Athanassula, E., *Internal Kinematics and Dynamics of Galaxies,* D. Reidel Publishing Company, Dordrecht, Holland, 1983.

5. Baade, W., *Evolution of Stars and Galaxies,* Harvard University Press, Cambridge, Mass., 1963.

6. Barton et al., *Astronomical Journal*, 112, 871, 1996.

7. Basu, B., Bhattacharyya, T. and Sen Gupta, A., *Bulletin Astron. Instts. Czech.*, **31**, p. 160, 1980.

8. Beckman, J.E. (Ed.), *Evolutionary Phenomena in Galaxies,* Cambridge University Press, Cambridge, 1989.

9. Bekki, K. et al., *Galaxy Threshing and the Origin of Ultra-Compact Dwarf Galaxies in the Fornax Cluster*, Astrophysics, 14 Aug, 2003.

10. Berkhuijsen, E.M. and Weilebinski, R. (Eds.), *Structure and Properties of Nearby Galaxies,* D. Reidel Publishing Company, Dordrecht, Holland, 1978.

11. Burbidge, G.R., The Nuclei of Galaxies, *Annual Review of Astronomy and Astrophysics,* **8**, p. 369, 1970.

12. Buserello, G., Capaccioli, M. and Longo, G. (Eds.), *Morphological and Physical Classification of Galaxies,* Kluwer Academic Press, Dordrecht, Holland, 1992.

13. Courvoisier, TJ.L. and Mayor, M. (Eds.), *Active Galactic Nuclei,* Springer-Verlag, New York, 1990.

14. De Sabbata, Venzo (Ed.), *The Origin and Evolution of Galaxies,* World Scientific, Singapore, 1982.

15. Diaferio, A. et al., *Astronomical Journal*, 107, 868, 1994.

16. Drinkwater, M. et al., *"Ultra-Compact Dwarf Galaxies: A new class of compact stellar system discovered in the fornax cluster"*, Astrophysics, 20 June, 2001.

17. Evstigneeva, et al., *Astronomical Journal*, 133, 1722, 2007.

18. Fall, S.M. and Lynden-Beli, D. (Eds.), *The Structure and Evolution of Normal Galaxies,* Cambridge University Press, Cambridge, 1981.

19. Franco, J., Ferrini, F. and Tenorio-Tagle, G. (Eds.), *Star Formation, Galaxies and the Interstellar Medium,* Cambridge University Press, Cambridge, 1993.

20. Genzel, R. and Harris, A.I., *The Nuclei of Normal Galaxies: Lessons from the galactic center,* Kluwer Academic Press, Dordrecht, Holland, 1994.

21. Grebel, E.K., 2000a, in 33rd ESLAB Symp., Star Formation from the Small to the Large Scale, ed. F. Favata, A.A. Kaas, and C. Wilson (Noordwijk: ESA), ESA SP-445.

22. Ha segan, M. et al., *Astrophysical Journal*, 627, 203, 2005.

23. Hazard, C. and Mitton, S. (Eds.), *Active Galactic Nuclei,* Cambridge University Press, Cambridge, 1979.

24. Heiligman, G.M. and Turner, E.L., *Astrophysical Journal*, 236, 745, 1980.

25. Hickson, P., *Astrophysical Journal*, 255, 382, 1982.

26. Hickson, P. et al., In Clusters and Groups of Galaxies, ed. F. Madirossian, G. Giuricin, M. Mezzetti, 367, Reidel, 1984.

27. Hickson, P. et al., *Astrophysical Journal*, 329, L65, 1988.

28. Hickson, P. et al., *Astrophysical Journal Supplement Series*, 70, 687, 1989.

29. Hickson, P., Astrophysical Letters Communication, 29, 1, 1993.

30. Hickson, P., *Atlas of Compact Groups of Galaxies*, Basel, Gordon and Breach, 1994.

31. Hubble, E.P., *The Realm of the Nebulae,* Dover Publications, New York, 1958.

32. Jerjen, et al., *The Astronomical Journal*, 119, 166, 2000.

33. Jones, et al., *Astrophysical Journal*, 620, 731, 2005.

34. Kent, S.M., *Pulication Astronomical Society of Pacific*, 93, 554, 1981.

35. Kodaira, et al., *Pulication Astronomical Society of Japan*, 43, 169, 1991.

36. Lequeux, J., *Structure and Evolution of Galaxies*, Gordon and Breach Science Publishers, New York, 1969.

37. Malykh, S.A. and Orlov, V.V., *Astrophyzika*, 24, 445, 1986.

38. Mayer et al., *Astrophysical Journal*, 559, 754, 2001.

39. Mamon, G.A., *Astrophysical Journal*, 307, 426, 1986.

40. Mateo, M., *Annual Review Astronomy and Astrophysica*, 36, 435, 1998.

41. McVittie, G.C. (Ed.), *Problems of Extragalactic Research*, The Macmillan Co., New York, 1962.

42. Mendes De Oliveira, C. et al., *Astrophysical Journal*, 437, L103, 1994.

43. Menon, T. K., *Astronomical Journal*, 110, 2605, 1995.

44. Mieske et al., *Astronomy and Astrophysics*, 418, 445, 2004.

45. Mieske et al., *Astronomy and Astrophysic*, 472, 111, 2007.

46. Mirabel et al., *Astronomy & Astrophysics*, 333, L1, 1998.

47. Mitton, Simon, *Exploring the Galaxies*, Faber and Faber, London, 1976.

48. Montoya, C.P. et al., *Astrophysical Journal*, 473, L83, 1996.

49. Nilson, P.N., *Uppsala General Catalogue of Galaxies*, Uppsala Obs, Ann 6, 1973.

50. Novikov, I.D., *Evolution of the Universe*, Cambridge University Press, Cambridge, 1983.

51. O'Connel, D.J.K. (Ed.), *Nuclei of Galaxies*, North Holland Publishing Co., Amsterdam, 1971.

52. Osterbrock, D.E. and Miller, J.S. (Eds.), *Active Galactic Nuclei*, Kluwer Academic Press, Dordrecht, Holland, 1989.

53. Palous, J., Burton, W.B. and Lindblad, P.O. (Eds.), *Evolution of Interstellar Matter and Dynamics of Galaxies*, Cambridge University Press, Cambridge, 1993.

54. Phillipps, S. et al., *Ultra-Compact Dwarf Galaxies in the Fornax Cluster*, Astrophysics, 21 June 2001.

55. Ponman et al., *Monthly Notices of Royal Astronomical Society*, 283, 690, 1996.

56. Prandoni et al., *Astronomical Journal*, 107, 1235, 1994.

57. Ribeiro et al., *Astrophysical Journal Letters*, 463, L5, 1996.

58. Richstone, D.O. and Tremaine, S., *Astronomical Journal*, 92, 72, 1986.

59. Rose, J.A., *Astrophysical Journal*, 211, 311, 1977.

60. Rubin et al., *Astrophysical Journal*, 365, 86, 1990.

61. Rubin et al., *Astrophysical Journal Supplement Series*, 76, 153, 1991.

62. Sepf., S.E., *Astrophysical Journal*, 407, 448, 1993.

63. Sersic, J.L., *Extragalactic Astronomy,* D. Reidel Publishing Company, Dordrecht, Holland, 1982.

64. Seyfert, C.K., *Astronomical Journal*, 53, 203, 1948.

65. Shapley, H., *Galaxies,* Harvard University Press, Cambridge, Mass., 1973.

66. Shostak, G.S. and Skillman, E.D., *Astronomy and Astrophysics*, 214, 33, 1989.

67. Sulentic, J.W. and Lorre, J.J., *Astronomy and Astrophysics*, 120, 36, 1983.

68. Sulentic, J.W., *Astrophysical Journal*, 322, 605, 1987.

69. Sulentic, J.W. and Rabaca, C.R., *Astrophysical Journal*, 429, 531, 1994.

70. Sulentic, J.W. and De Mello Rabaca, D.F., *Astrophysical Journal*, 410, 520, 1993.

71. Stephan, M.E., *Monthly Notices of Royal Astronomical Society*, 37, 334, 1877.

72. Tayler, Roger, J., *Galaxies: Structures and evolution,* Cambridge University Press, Cambridge, 1993.

73. Tinsely, B.M. and Larson, R.B. (Eds.), *The Evolution of Galaxies and Normal Populations,* Yale University Observatory, New Haven, Conn., 1977.

74. Tully, R. Brent and Fisher, J. Richard, *Catalogue of Nearby Galaxies,* Cambridge University Press, Cambridge, 1988.

75. Van den Berg, S., *Astrophysical Journal*, 505, L 127, 1998.

76. Walke, D.G. and Mammon, G.A., *Astronomy and Astrophysics*, 225, 291, 1989.

77. Williams, B.A. and Rood, H.J., *Astrophysical Journal Supplement Series*, 63, 265, 1987.

78. Wu, W. et al., *Bulletin of American Astronomical Society*, 26, 1494, 1994.

79. Young, L.M. and Lo, K.Y., *Astrophysical Journal*, 476, 127, 1997.

80. Zwicky, R, *Morphological Astronomy,* Springer-Verlag, New York, 1957.

18 Clusters of Galaxies

18.1 CLUSTERING NATURE OF GALAXIES

Hubble's investigation regarding the distribution of galaxies in space showed, as early as in 1920's that clustering tendency among galaxies is most common. Later, analysis of the Palomar Sky Survey plates and other surveys with large telescopes have revealed the presence of many large clusters and innumerable smaller groups of galaxies. The groups are rather loose aggregates consisting of number of galaxies varying from a few (even two or three) to a few dozen. They are far more common than larger clusters. The clusters are observed to be scattered all over the sky, barring the zone of avoidance where these are obscured by the galactic dust layer. The Local Group of galaxies containing about thirty members including our Milky Way is an example of a loose small group.

The larger clusters are richer and more compact aggregates of galaxies. As early as in 1933, Harlow Shapley catalogued 25 clusters. Many more were discovered in the subsequent years by F. Zwicky and his collaborators and by G. Abell. Zwicky proposed his famous model of distribution of galaxies, that all galaxies belong to clusters. The space can be considered to consist of *cluster cells;* the cells having diameters of the order of 50 Mpc (with the value of Hubble's constant $H = 70$ km s^{-1} Mpc^{-1}). Zwicky's catalogue of galaxies lists several thousand clusters. Abell's catalogue consists mainly of clusters located away from the plane of our Galaxy. His latest catalogue (1989) lists 4073 rich clusters of galaxies, each having at least thirty members within the magnitude range (m_3, m_{3+2}), m_3 being the magnitude of the third brightest cluster member. Many thousands more will be added to the list when the entire volume of the Palomar Sky Survey plates are scrutinized for clusters. The catalogues include from double or triple systems to rich compact clusters consisting of 10,000 galaxies or even more.

Analysis of the shape and structure (morphology) of clusters has led to the idea that they can broadly be divided into two categories: *irregular* and *regular,* although other bases of morphological divisions have been suggested (next section). Their structures differ principally in two respects, viz., their sphericity in shape and central concentration. The irregular clusters possess no well-defined regular shape and present more or less amorphous appearance. They have no well-recognized strong central concentration. This analogy in their shape and structure with those of the galactic clusters (open clusters) of stars has led some astronomers to call them as the open clusters of galaxies. The irregular clusters are much more common in space than the regular ones. The number of galaxies in irregular clusters range from a few to as many as several thousand. They contain galaxies of all morphological types—ellipticals, S0s, spirals and irregulars. Subsystems, such as double and triple galaxies are observed to be very common in them. The Local Group of galaxies is a typical example of small irregular clusters. Examples of rich irregular clusters are *Virgo* and *Hercules*.

The best observed among the clusters of galaxies (except the Local Group) is the Virgo cluster (so named because it lies in the direction of the constellation Virgo). It has a very complicated structure and contains many subsystems of double and triple galaxies, within it. The cloud extends over a region of the sky $12° \times 10°$, containing more than 1000 galaxies having many giant ellipticals and large spirals as the brightest members of the cluster. The brightest Virgo galaxies have $M_{pv} \simeq -23$. There are also many dwarf ellipticals. At the distance of Virgo, the O and B supergiants in large spirals are just separable with the 200-inch telescope which accounts for a distance modulus of about 30.7 corresponding to a distance of 14 Megaparsecs. This distance is confirmed also by the observed apparent magnitude of the globular clusters in the giant elliptical M87, a member of the Virgo cluster. At this distance the observed angular diameter of the cluster corresponds to a linear diameter of 3 Megaparsecs. The very great importance of the Virgo cluster lies in its unique position in space which allows an independent calibration of the value of the Hubble's constant H. The calibration depends upon the correct determination of the mean recessional velocity of the cluster as well as an independent determination of its correct distance. It has been done for this cluster. We have already seen, how the distance of 14 Megaparsecs for the cluster can be determined by observing the supergiant stars or the globular clusters in its giant members. The mean recessional velocity has been determined for the cluster to be +1136 by Sandage. Using these values in Hubble's formula (17.8), the value of $H = 81$ km s^{-1} Mpc^{-1} is obtained. This value may however be uncertain to a large extent, as the random velocities of the cluster members are comparable to the mean recessional velocity of the cluster. Unless the mean recessional velocity of a cluster supersedes the random velocities of its individual members greatly, Hubble's law cannot be reliably used. Selection effect may also affect the result. From observation of the globular clusters in M87, Sandage computed a distance modulus $m - M = 31.1$ for the Virgo cluster, leading to a value of H close to 50 km s^{-1} Mpc^{-1}.

As we have already noted, the structure of the Virgo cluster presents great complexity. The ellipticals and spirals belonging to the cluster exhibit appreciably different distributional as well as dynamical characters. The ellipticals are much more centrally concentrated than the spirals. The mean velocities of these two groups of galaxies have been obtained by G. de Vaucouleurs as +962 and +362 km s^{-1} respectively. Such significant differences in two of the most important cluster properties of the elliptical and spiral component of the cluster have led de Vaucouleurs to conclude that, what we call the Virgo cluster is actually two clusters; one predominantly consisting of spirals and the other of ellipticals, seen projected on each other. This suggestion, though plausible, has not yet been confirmed.

The regular clusters of galaxies are much rarer objects compared to the irregular clusters. They have more or less spherical symmetry and possess great central concentration. In these respects they are analogous to the globular star clusters and have therefore been called by some authors as the *globular clusters of galaxies*. Rich regular clusters contain thousands and even tens of thousand member galaxies. The brightest of these are generally found close to the centre of the system and are almost invariably giant ellipticals or S0 galaxies called cD galaxies. These giant ellipticals have been found to be metal-rich. Giant spirals are either completely absent or are very rare in regular clusters. Examples of rich regular clusters are the Coma cluster and the Perseus cluster.

No regular cluster is found reasonably close to our Galaxy. The nearest one is the Coma cluster at a distance of approximately 100 Megaparsecs. It is a very rich cluster containing 10,000 galaxies or more, and owing to its nearness it has become the best observed one among the regular clusters. The brightest galaxies in the cluster are the two giant ellipticals of absolute magnitudes –23 to –24 situated near the nucleus of the cluster. The mean recessional velocity of the cluster has been calculated to be +6850 km s^{-1}. With a value of the Hubble's constant equal to 70 km s^{-1} Mpc^{-1}, the above velocity corresponds to a distance of 100 Mpc and a distance modulus $m - M = 35.0$ for the cluster. At this distance its average angular diameter of 200′ corresponds to a linear diameter of 5.2 Megaparsecs.

Zwicky's study has shown that there are at least 3000 galaxies in the Coma cluster brighter than m_{pv} =18 which corresponds to $M_{pv} = - 17$. If the absolute magnitude distribution of the Coma cluster galaxies is similar to that of our Local Group members, then the above data indicate that there must be more than 12,000 galaxies in this cluster leading to a number density of about 160 galaxies Mpc^{-3}. Such high density of galaxies may lead to a sufficiently high frequency of collisions, during which the gaseous material of the colliding galaxies will be swept away to make them free of gas. This gas will then reside as the *intracluster gas*. This mechanism has been suggested by some authors as the possible explanation for the absence of any spiral galaxy in this cluster at present.

A careful study of the nature of the distribution of galaxies in space indicates that within a volume of space bounded by a radius of about 20 Mpc, there are several thousand galaxies. Beyond this distance, the distribution thins out until a much larger distance is reached. This observation has led some astronomers to suggest that just as the individual galaxies have a tendency to form clusters, so also the clusters of galaxies exhibit a clustering tendency. This *second-order clustering* nature form what may be called as the *'cluster of clusters'* of galaxies. The galaxies within about 20 Mpc as stated above are said to belong to the *Local second-order cluster* which has been called by G. de Vaucouleurs as the *Local supercluster*. The Local Group including our own Galaxy lies within the Local Supercluster dominated by the Virgo cluster at its centre. Studies also indicate that the Coma cluster forms part of a large Coma Supercluster and that the Perseus cluster is embedded in the Perseus Supercluster. A very large supercluster having a diameter of nearly 300 Mpc is the Corona Borealis Supercluster at a distance of nearly 500 Mpc. Careful investigations in recent years have revealed beyond doubt that superclustering is a common tendency of the clusters of galaxies. The superclusters have a very wide range of dimensions; usually lying between 30 to 300 Mpc, their average mass being of the order of 10^{16} $M_\odot$. Their geometrical boundaries are generally not well-defined and they usually have no pronounced concentration at any point inside including the centre. Only the Local Supercluster appears to have significant concentration at its centre containing the Virgo cluster. This lack of concentration and well-defined boundaries of the superclusters is best understood by the fact that they did not have sufficient time yet to *dynamically relax* and mix by the action of gravitational pull. This mixing actually depends on the crossing time of the supercluster by the individual clusters within it. For mixing, the crossing time must be sufficiently less than the age of the supercluster. Calculations however reveal that the average crossing time of a cluster is $\sim 3 \times 10^{11}$ years, a time that is an order of magnitude larger than the Hubble time.

Side by side with superclusters, the space is also full of *voids*. Voids are regions comparable to the sizes of superclusters, but matter-density in these regions is very low. Thus superclusters separated by voids appear to be a meticulous plan of nature in building the large-scale structure of the Universe.

The Great Wall: The Great Attractor

Several years of painstaking works by J. Huchra and M. Geller recently have revealed the existence of a gigantic structure of galaxy clustering. The structure actually consists of a thin sheet of galaxies, 160 Mpc in length, 60 Mpc in width and 5 Mpc in thickness at a distance of nearly 100 Mpc from us. Because of its enormous area and small thickness, the structure has been named by the astronomers as the *Great Wall*. This is our nearest wall of galaxies, but is believed not to be the only wall. The space may be divided into compartments by such walls of galaxies.

Again, analyzing the motions of galaxies it has been found that galaxies possess peculiar velocities superimposed on the Hubble flow. These peculiar velocities are acquired by the attraction of matter concentration in nearb'y groups, clusters and superclusters of galaxies. Observations show that the Local Group including the Milky Way and the Andromeda galaxies are moving with a velocity as large as 250 km s^{-1} with respect to the cosmic background radiation in the direction of Hydra-Centaurus Supercluster. Careful analysis of the motion of galaxies in the region around us has revealed that not only that the Local Group of galaxies is flowing towards the Hydra-Centaurus Supercluster with a large velocity, but also that the latter supercluster itself is moving in a direction in space with a much larger velocity. These observations have led the astronomers to believe that there is some very massive attracting object beyond the Hydra-Centaurus Supercluster that is attracting both the Local Group and the Hydra-Centaurus Supercluster. This massive object has been named by the astronomers as the *Great Attractor*. Its distance has been suggested to be at least twice the distance of the Hydra-Centaurus Supercluster. A concentration of tens of thousands of excess galaxies is required to pull the Local Group with a high velocity of 250 km s^{-1} from the great distance as calculated for the Great Attractor. A concentration of a very large number of galaxies is therefore predicted by astronomers in the region of the Great Attractor.

18.2 MORPHOLOGICAL CLASSIFICATION OF CLUSTERS

Various structural parameters have been used in classification systems for clusters of galaxies. But these systems are closely correlated to such an extent that all of them can be represented, though crudely, by a single system as *regular* and *irregular* clusters. We have already described in the earlier section the basic differences between these two types of clusters. We briefly summarize below the various other morphological classification schemes.

Zwicky and his co-workers have classified clusters as *compact, medium compact* and *open* depending on the nature and extent of concentration of galaxies within the cluster. A *compact* cluster has a single pronounced concentration of more than ten galaxies *in contact* in photographic plates. A *medium compact* cluster has either a single concentration of ten galaxies or several concentrations. But in these clusters the concentrated galaxies are not so close as to be in contact. *Open* clusters are devoid of any pronounced concentration.

Clusters of galaxies have been classified by W.W. Morgan and later by A. Oemler on the basis of the galactic content in them, that is, the fraction of spirals, SOs and ellipticals in a cluster. If the cluster contained a large number of spirals, Morgan classified it as type i and Morgan's type ii clusters contained only a few spirals. Later, Oemler modified Morgan's classification system defining three classes of clusters. Clusters having spirals as the most common galaxies were called by Oemler as *spiral-rich'*. The class of clusters containing SOs as the most common galaxies where spirals are relatively much less common was called by Oemler as *spiral-poor* class. Clusters which are dominated by a central cD galaxy (next section) and populated overwhelmingly by elliptical and SO galaxies were called by Oemler the cD *clusters*.

The degree of domination of a cluster by its brightest galaxies forms the basis of a classification system introduced by L.P. Bautz and W.W. Morgan. In this system, Type I clusters are dominated by a single cD galaxy at the centre. In Type II clusters, there is no cD galaxy but normal giant ellipticals dominate them. Type III clusters have no dominating galaxies. Type I–II and Type II–III are intermediates between the first three classes. The Rood-Sastry classification system is based on the nature and distribution of the ten brightest galaxies in a cluster. They define the following six classes in their scheme:

1. cD, in which the cluster is dominated by a central cD galaxy (example: A 2199).
2. B (binary), in which the cluster is dominated by a pair of luminous galaxies (example: Coma cluster).
3. L (line), where at least three of the brightest galaxies appear to lie in a straight line (*example:* Perseus cluster).
4. C (core), in which four or more of the ten brightest galaxies constitute the cluster core (example: Corona Borealis cluster).
5. F (flat), in which the brightest galaxies form a flattened distribution in the sky (example: Hercules cluster).
6. I (irregular), in which the distribution of the brightest galaxies is irregular (example: A 400).

The Rood-Sastry classification has more recently been modified by M.F. Struble and H.J. Rood by changing some of the definitions and adding some more subclasses.

As already emphasized, a careful analysis will reveal that the various classification systems as described above are closely correlated and can be understood in terms of Abell's classification system of the clusters as regular and irregular.

Galactic Content of Clusters

Observations easily reveal that the galactic content in regular compact clusters markedly differs from that in irregular clusters and in the field on several counts. First, many of the regular compact clusters are conspicuously dominated by a single central very luminous cD galaxy, or a pair of very bright giant elliptical galaxies. Secondly, the regular compact clusters are primarily dominated by elliptical and S0 galaxies while the spirals are much less important. In irregular clusters and in the field, just the opposite holds true. Such systematic variation in the galactic content in clusters of different types requires proper understanding and many theories have been proposed for the purpose. These theories, although varied, may broadly be divided into two classes. The first class of theories asserts that the galaxy morphology does not alter after being formed under the conditions prevailing in the region of their formation. This implies that more ellipticals will be formed in the region of formation of regular clusters; and so for the irregular clusters where formation of spirals is favoured. The theories under the second class predict that galaxies are formed initially with the same distribution of morphological types in different types of clusters, but later in course of time, under the influence of their physical environment, the population of galaxies alter their morphology. In particular, it has been suggested that in regular compact clusters, collisions being favoured due to higher density of galaxies, spirals are stripped of their gas and spiral arms due to collisions and alter into S0s or ellipticals, and S0s become ellipticals. That the gas in a galaxy can be removed by various mechanisms including a galaxy-galaxy collision, has been suggested by many authors. This will be further discussed in Section 18.4.

18.3 cD GALAXIES

cD galaxies are usually found at the centre of rich compact regular clusters of galaxies. About 20 per cent of all rich clusters contain cD galaxies. They possess a very luminous elliptical galaxy as the nucleus which is embedded in an extensive amorphous envelope of low surface brightness. cD galaxies are also seen in some poor clusters and groups of galaxies.

cD galaxies are extremely luminous having $\langle M_v \rangle \simeq -24.0$, showing a small dispersion of $\simeq 0.3$ magnitude only. Further, the magnitudes of cDs are almost independent of the richness of the clusters. Masses of cD galaxies have been determined by using two different methods. First, the cD mass can be calculated by measuring the velocity dispersion of stars populating its outer region. Secondly, many cDs have companion galaxies bound with them. The motion of the companions relative to cDs gives the masses of the latter. Both the methods yield very high mass for cDs, of the order of $\simeq 10^{13} M_\odot$.

Some cD galaxies are found to have double or multiple nuclei. Sometimes two cD galaxies are observed to be embedded in a common halo. These have been called *dumb-bell* galaxies. The multiple nuclei sometimes possess high velocities relative to the cD leading to the conclusion that these nuclei are nothing but chance projections. Many authors such as J.S. Gallagher and J.P. Ostriker and others have suggested that the halo of cD galaxies are formed by debris left from galaxy collisions. In rich clusters where cDs are most commonly formed, galaxy collisions are frequent. Collisions will sweep off the gas rich outer envelopes of the parent galaxies. This tidal debris will ultimately settle to the cluster centre which is a deep potential well.

Many speculations are there as to the mechanism of formation of cD galaxies. Of these, the hypothesis of *galactic cannibalism* proposed by J.P. Ostriker and M.A. Housman has been widely favoured. According to this hypothesis, cD galaxies are formed by the merger of massive galaxies at the core of a cluster. The dynamical friction within the cluster causes the orbits of galaxies to decay which then gradually spiral in towards the cluster centre, where they finally merge to form a very luminous and massive galaxy surrounded by the tidal debris. This is what we call a cD galaxy. Thus the cannibalism hypothesis as proposed by Ostriker and Housman for the formation of cD galaxies at the centre of rich clusters, although not above criticism, explains many observable characteristics of these galaxies.

18.4 INTERACTING GALAXIES AND GALAXY MERGERS

Since Hubble's work on the distribution of galaxies in space and their distances from us were known, it has been generally understood that the mutual separation of galaxies compared to their dimensions is smaller by orders of magnitude than that between the stars. It was therefore natural to assume that galaxies would be subjected to gravitational interaction. This point was first emphasized by J.G. Holmberg in early 1940s. He drew attention of astronomers to the importance of the tidal interaction among galaxies which could effect energy transfer between them and could even lead to mergers. In subsequent years, the observations of B.A. Vorontsov-Velyaminov (1959) and H. Arp (1966) led to the discovery of various unusual features like *bridges, tails* and other *appendages* associated with pairs and groups of galaxies. These features were described by Zwicky as "Clouds, filaments and jets of *stars* which are ejected massively from galaxies in Collision" and he attributed all these to "large-scale tidal effects". But such propositions by Zwicky and others did not gain ready acceptance by the astronomers, because many details could not be explained at that time on the basis of tidal hypothesis. But subsequent work of S.M. Alladin (1965), A. Toomre and J. Toomre (1972) and of many others clearly demonstrated that the observed peculiar features would be manifested due to tidal interaction between galaxies under the realistic assumptions about the various parameters. Assuming that the mass distributions of the galaxies resemble that of polytropes and that the intergalactic forces were given by poly tropic theory, Alladin showed that structures of the colliding galaxies would be considerably loosened and their translational velocity would be sufficiently reduced. The numerical calculations of Toomre and Toomre and of others showed decisively that the peculiar features like bridges, tails and other appendages would be produced by slow grazing encounters between galaxies under suitable choice of physical entities and galaxy orientations. Finally, the theoretical works of J.P. Ostriker, A. Toomre and others revealed that the discs of the spiral galaxies would be dynamically unstable unless these were enveloped by extensive and massive halos. This theoretical prediction was soon demonstrated to be true by observation of flat rotation curves up to very large distances in spirals of all Hubble types, by V.C. Rubin and her co-workers, and by others. It is now firmly believed that the halos of galaxies extend to very large distances of 100 kpc or even more, and thus the gravitational encounters between galaxies appear to be inevitable.

Numerical calculations of Toomre and Toomre, Wright and others have demonstrated that slow, grazing interactions between galaxies produce *bridges* and *tails* as are observed in many galaxies. Bridges often appear to link large galaxies to small companions. Prominent examples of the galaxy pairs connected by bridges are NGC 5194/95 (M51 and its smaller companion) and Arp 295. Tails are much more frequent than bridges. Prominent examples of galaxies having tails are Arp 242 which consists of two galaxies NGC 4676 A and NGC 4676 B, both having a long tail, and Arp 244 or NGC 4038/39 which is also known as 'The Antennae' galaxies. These galaxies have spectacular tails of unequal lengths. All these peculiar structures are strongly believed to be the results of tidal interaction between galaxies. The Magellanic Stream which is a long filament of H I extending from the Magellanic Clouds to the south galactic pole and beyond, is believed to have been formed by the tidal forces exerted by the Galaxy on the Magellanic Clouds.

Some galaxies appear to possess pronounced rings. These are called ring galaxies or cartwheel galaxies. Experiments have demonstrated that head-on on-axis collisions of disc galaxies can result in the formation of ring galaxies. It has also been suggested by many authors that the *warps* of the galactic discs as well as the spiral arms can originate as a result of tidal interaction between galaxies. J. Kormandy and C.A. Norman (1979) have shown that galaxies having spectacular spiral arms such as M51, M81 etc. have companions. It may be conjectured that the spiral density waves have been excited in these galaxies during close passage of their companions. Close passage of companions of S0 and elliptical galaxies can also produce fine structures such as *shells, ripples, plumes,* etc. as are observed in large fractions of S0s and ellipticals.

We have already seen that the formation of cD galaxies is quite well explained by the multiple merger hypothesis. Many astronomers believe that the S0s and elliptical galaxies may have been formed by the merger of spirals. The preponderance of S0s and ellipticals in the central parts of compact regular clusters of galaxies has been interpreted by some astronomers as the result of mergers of disc galaxies in these regions. Since the number density of galaxies is very high in the region, the collision frequency and so also the merger rate will be high, causing a higher fraction of galaxies to change their morphology.

Several categories of evidence have been cited in favour of the theory of formation of ellipticals by merger of spirals. The numerical simulation of a face-on collision between two axisymmetric spiral galaxies carried out by Toomre (1977) led to an elliptical galaxy with a central concentration as the merger product. Similar results were obtained by numerical simulation by R. Farouki and L. Shapiro (1982) and by J. Negroponte and S.D.M. White (1983). From an analysis of the currently observed merging pairs of galaxies, Toomre estimates that almost all the elliptical galaxies in the NGC catalogue might have been formed by merger of disc galaxies during the Hubble time of 10^{10} years. Observationally, Schweizer has identified Formax A, Centaurus A and Arp 226 as definite remnants of mergers. The elliptical galaxy NGC 7252 has also been identified by Schweizer after detailed observation as the prime candidate for a recent merger of two disc galaxies. From what has been discussed above, it is evident that galaxy merger hypothesis has gained much favour among astronomers on the basis of both experiments and observations, although some other astronomers such as S. Van den Bergh (1986) have argued against the hypothesis.

Galaxy collisions and tidal interaction between them lead to the tidal stripping of gas and stars from gas-rich disc galaxies in clusters. This hot and tenuous gas will be drawn in gradually in the deep potential well of the cluster. The hot gas at the cluster centre gives strong radiation in X-rays (see Section 18.5). Collision and merger of galaxies have also been suggested to be a very effective cause of formation of globular clusters in the merger products. An unusually high number of globular clusters per unit luminosity is observed in M87 (Virgo) and NGC 3311 (Hydra) (Van den Bergh 1983). These massive and luminous galaxies occupy the central positions in rich clusters and are likely to have been probably formed by merger of a number of galaxies. It is assumed that during the process, the gas in the merging members has suffered strong agitation and the clouds of gas have been torn off under highly agitative conditions favouring the formation of globular clusters.

Gravitational interaction between galaxies may in some way lead, also to the nuclear activity in them. This was first suggested by W. Baade and R. Minkowski in early 1950s and later theoretically investigated by Toomre and Toomre (1972). More recently, analyzing the observational data, both in optical and radio wavelengths of a large number of very strong radio galaxies, Heckman and his co-workers (1986) have suggested that a significant fraction of these elliptical galaxies is highly peculiar in optical morphology. The authors suggest that these latter galaxies are probably the results of merger of two disc galaxies or in which at least one of the merging pair must be a disc galaxy. This collision and merger of galaxies may lead to high energy activities in the resulting galaxies. Gravitational interaction between galaxies thus profoundly influences not only the general morphology as well as their fine structures, but also their internal dynamics and nuclear activities. The same phenomenon is therefore intimately related to the evolution of galaxies.

18.5   X-RAY EMISSION FROM GALAXIES AND FROM CLUSTERS OF GALAXIES: THE COOLING FLOW

The NASA devised the novel way of commemorating the birth centenary of Albert Einstein, by putting the Einstein X-ray Observatory in space in 1978 in order to peer through the X-ray emitting pointlike and extended sources in space. Prior to its launch, the only X-ray sources known were a few close binaries and some supernova remnants in the Galaxy, and some nearby galaxies like Magellanic Clouds and M31. The Einstein X-ray Observatory has detected X-ray emission from stars of all spectral classes of which the O stars are the most luminous. Some other X-ray emitting objects are the early type binaries, low luminosity binaries, Population II binaries, cataclysmic variables and some globular clusters. Also, the Galactic nucleus is an extended but weak source with an X-ray luminosity of 5×10^{35} erg s^{-1}. The X-ray emission has been detected from individual galaxies even beyond the Virgo cluster, and from the clusters of galaxies much farther away. Also, the detailed maps showing the distribution of point sources have been prepared of many nearby galaxies such as the Magellanic Clouds, M31, M33, M101 and others. We shall here principally restrict ourselves to the study of X-ray emission from clusters of galaxies, with an attempt to the discussion of the physical properties of this emission and its possible consequences on the environment.

The most luminous X-ray emitters in the Universe is the X-ray emitting hot gas lying in the central regions of rich clusters of galaxies. The X-ray luminosity of the regions has been found to span the values $L_x = 10^{43} - 10^{46}$ erg s^{-1}. The temperature range for the gas is $10^7 - 10^8$ K, having a low number density of $n \sim 10^{-3}$ cm^{-3}. The radiation mechanism has now been well understood to be the thermal bremsstrahlung (free-free emission). This is confirmed by the detection of strong X-ray line emission from clusters. The strong 7 keV Fe line emission observed from clusters is best understood with the model of thermal bremsstrahlung as the emission mechanism. The X-ray spectra of the clusters further confirm that the emitting gas is rich in heavy elements such as Fe, Ni, S and Ca. Much of this intracluster gas therefore appears to have been originated from the member galaxies in the cluster in which metal enrichment has been processed in the past through stars and supernovae. Several mechanisms have been suggested for transfer of the galactic gas into the intracluster medium. Among such mechanisms, galactic winds, evaporation and ram pressure stripping have attained keen attention. Whether any single or all of these mechanisms are actually operative is not yet clearly understood.

The actual amount of intracluster gas in a rich cluster is huge, i.e. of the same order as the total optically luminous mass in member galaxies, if it is assumed that eighty per cent of the total cluster mass belongs to the missing mass. The X-ray surface brightness profiles of the gas imply very high central gas densities and correspondingly short cooling times. Strong radiative cooling will drive the gas into smallar radii by the pressure of the surrounding gas. In a rich cluster, a large amount of gas, of the order of several hundred solar masses, may thus cool at the cluster centre. A cooling instability will thus set in, initiating a cooling flow in which the gas from outer region flows in to fill the vacuum created by the condensation of cooled gas at the centre. If this flow is sustained over the Hubble time, then the amount of gas involved in the cooling flow is 10^{12}–10^{13} $M_\odot$. This gas occupies a region at the centre of a typical rich cluster within a radius of nearly half a Megaparsec.

Cooling flows are common not only in clusters of galaxies but also in individual giant elliptical galaxies. In the latter, the gas involved is injected by stellar mass loss. This gas is heated by thermalizing the kinetic energy associated with the gas-losing stars. The hot intracluster gas and the hot gas in the individual galaxies emit not only the X-ray continuum and X-ray lines, but also very strong optical coronal lines all of which originate from forbidden transitions. The strongest of these lines are [SXII] at 7536Å, [CaXV] at 5496Å, [FeX] at 6374Å, [FeXI] at 7892Å, [FeXIII] at 10747Å and [FeXIV] at 5303Å. C.L. Sarazin and his co-workers (1987, 1989) have computed models of cooling flows on the basis of the observed intensities of the optical coronal lines. The total luminosities emitted in strong coronal lines have also been computed by these authors on the basis of the cooling flow models. The total luminosity emitted in a line will have a simple correlation with the assumed cooling rate $\dot{M}_c$, that is the mass of gas that flows in the central cooling region. Assuming a cooling rate of $\dot{M}_c = 300\ M_\odot$ yr^{-1} for the rich galaxy clusters, Sarazin and his co-workers calculated luminosities of lines of $\sim 3 \times 10^{39}$erg s^{-1}. For individual galaxies they assumed $\dot{M}_c = 1.19\ M_\odot$ yr^{-1} and computed the line luminosities of $\sim 10^{37}$ erg s^{-1}.

X-ray observations of the clusters of galaxies and individual galaxies have thus opened new gates to the field of research in astrophysics. The following facts have been revealed: (a) it has shown that the space is full of regions occupied by very hot gas, (b) there is much more luminous matter in the Universe than was previously thought, (c) the processes of chemical enrichment of the galactic and intracluster gas are evidently operative in these regions, and (d) the evolution of clusters as well as the individual galaxies is greatly influenced by the deeply penetrating cooling flows.

18.6 MASSES OF GALAXIES

The masses of galaxies can be obtained in several different methods, each of these being limited to galaxies of certain types and having particular theoretical and observational difficulties.

The Method Based on Velocity Dispersions in Nuclei of Elliptical Galaxies

Measurement of velocity dispersion of stars in the nuclear region with the help of high dispersion spectrograms serves the purpose of estimating the total mass in the elliptical galaxies. In this method all the stars are assumed to have the same mass and therefore the Virial theorem is applicable to them. In order to obtain the stellar velocity dispersion from the composite spectrum, one assumes a particular model of velocity dispersion law, e.g., the Gaussian law, and fits the observed line profiles with the experimental line profiles in the laboratory. The dispersion of particles with which the experimental profiles best fit with the observed profiles is considered to be the required dispersion of stars. The method is quite difficult and the results obtained may involve significant uncertainty. Nevertheless, when the radial velocity dispersion of stars has thus been obtained, one can multiply this value by some suitable numerical factor in order to realize the total velocity dispersion of stars in the galaxy. If the average of the space velocities of stars thus obtained be denoted by $\langle v^2 \rangle^{1/2}$, then the standard Virial theorem $2T + U = 0$ yields, assuming the same mass of the stars,

$$M\langle v^2 \rangle + U = 0 \tag{18.1}$$

where

$$U = -G \int_0^R \frac{M(r)}{r} \, dM \tag{18.2}$$

M being the total mass of the Galaxy, R the distance of its periphery from the centre and $\langle v^2 \rangle^{1/2}$ the average space velocity referred to its centre of mass. In order to evaluate U one must assume some mass distribution law for the galaxy.

Since the light distribution in the galaxy can be measured photometrically, it has been customary to assume the mass distribution to be identical with the measured light distribution and that the mass to light ratio is constant. Knowing thus $\langle v^2 \rangle$ and U, one can calculate the total mass of a Galaxy using Eq. (18.1). Following this method G. Burbidge and his co-workers have calculated the mass of M32 to be $3.6 \times 10^9 \, M_\odot$. Assuming 0.63 Mpc to be the distance of this galaxy, they found the mass-luminosity (photographic) ratio to be 27. These results are however considered to be only approximate due to various factors involved in the method that contribute to the inaccuracy.

The Average Mass for Cluster Members by Application of the Virial Theorem

The application of the Virial theorem to a cluster of galaxies yields the total mass of the cluster and so the *average mass* of an individual galaxy in the cluster. The validity of the application of the Virial theorem however demands that one must assume the cluster to consist of a *stable group of point masses,* having no invisible mass in the form of atomic or molecular gas or dust and no subgroupings like binary or multiple subsystems within the cluster. In other words, one must be content to accept that the individual point masses that are being observed only contribute to the *entire mass of the cluster* and these point masses are moving only under the smooth gravitational field of the cluster as a whole. Then the average mass of these individual members or a statistical distribution of mass among these can be obtained from the

Virial theorem. If $\langle m \rangle$ be the average mass of N galaxies observed in the cluster of mean radius R, then the Virial theorem for the cluster can be written as

$$\langle m \rangle N \langle v^2 \rangle = \frac{G \langle m^2 \rangle N^2}{R} \qquad (18.3)$$

where $\langle v^2 \rangle$ is the mean-square velocity for the cluster members obtained by multiplication of the observed radial velocities of the members by some suitable factor. For isotropic velocities $\langle v^2 \rangle = 3 \langle v_r^2 \rangle$, v_r being the radial velocity. To compute v_r for each individual galaxy with respect to the centre of mass of the cluster, one has to predetermine the radial velocity of the centre of mass itself from the observation of individual members.

Thus, the motion of the centre of mass of the cluster and the random motions of the member galaxies with respect to the centre of mass are *coupled* together. This may produce some uncertainty in the finally computed results. The potential energy of the cluster actually depends on the separation of the individual members from each other. But for distant clusters this is hardly possible to determine and the average radius of the cluster can be conveniently used. One can also compute the potential energy of the cluster by using Eq. (18.2). In that case, the light distribution of the galaxies within the cluster has to be measured and some law of mass to light ratio variation has to be assumed. In practice, only some giant bright galaxies are observed in the cluster making the assumption that these members contain most of the mass and light of the cluster.

When all this is done, the total mass of the cluster as well as the average mass of the individual cluster of galaxies are found to be unusually high. The total mass of the Virgo cluster is computed to be a few times $10^{14}\ M_\odot$ and that of the Coma cluster yields $\sim 3 \times 10^{15}$ $M_\odot$. The mass to light ratios for cluster of galaxies are also unusually higher compared to those determined by other methods. For the Virgo and Coma clusters, the latter values are of the order of 300. Also, these high values of mass and mass to light ratios of cluster of galaxies are found to increase progressively with the distances of the clusters. This fact has remained a puzzle to the astronomers. It may be that one or more assumptions on which the computations are based, may be incorrect. Some authors have expressed doubt about the stability of the cluster. If the clusters are expanding, the Virial theorem will yield too high an average mass. Also, the assumption of the constant mass to light ratios in computing the potential energy may introduce large errors. Another possible source of error has also been observed, viz., the coupling of the motions determined for the individual masses and the centre of mass of the cluster.

The above observational results opened up the gateway to the problem of the *missing mass* or *dark matter* in galaxies and clusters of galaxies. Though the situation has marginally improved by the existence of the hot intracluster medium, the problem still remains. The upper limit of the observed mass hardly exceeds 10 per cent, leaving 90 per cent for unobserved mass. All current theoretical considerations as well as observational tests definitely indicate that the existence of most of the matter in galaxies and clusters of galaxies as invisible matter is real (Section 18.8).

Double-Galaxy Method

Masses of galaxies can also be obtained by a statistical treatment of the orbits of binary galaxies, as in the case of the determination of stellar masses from binary stars. Many of the double galaxies are now known and in majority of these, the components have similar radial velocities so that they may be assumed to be at the same distance and to be gravitationally connected to each other. When the relative velocities of the two component galaxies with respect to the centre of mass can be determined, Kepler's third law can be applied to the system analogous to the case of binary stars. Generally, the measured difference in velocities is assumed to be equal to the projected circular velocity, and so statistical handling of the data demands the consideration of *projection effect*. The method requires a large number of assumptions, viz., the galaxies should act as point, masses, they are similar to the normal field galaxies, the orbits of the galaxies should be closed, the tidal effects between the galaxies may be ignored etc. These assumptions may lead to large uncertainties in computed values. Yet it may be expected that the uncertainties involved in these assumptions nearly or totally smooth out when a sufficiently large sample of galaxies is used. The masses of a number of such galaxies have been estimated by this method. The averages of these masses classwise are 4×10^{10} $M_\odot$ for spirals and irregulars and 7×10^{11} $M_\odot$ for ellipticals and S0s.

The Rotation Curve Method for Spiral Galaxies

The most reliable method for the determination of the mass of a spiral galaxy consists in determining the *rotation curve* of the galaxy and derive a density distribution model therefrom. This is a powerful and reliable method for determining the mass of a spiral galaxy which is not too far from us and is bright enough to yield good spectra at several points at different radial distances from the centre. The rotational velocities are derived from the Doppler shift and the tilt of spectral lines originating at different radial distances from the centre of the galaxy under consideration. The lines observed are usually the emission lines of H, [N II], [S II], [O I], [O II] etc. For our own Galaxy and the nearest spiral M31, the rotation curves have been constructed from the 21-cm hydrogen line observations. But this technique cannot be used for more distant galaxies mainly due to two reasons: (a) the 21-cm radiation is quite weak, and (b) the position measurements become uncertain due to the poor resolution of radio telescopes. Thus, except for the nearest galaxies, the optical emission lines are used to construct rotation curves of galaxies. The rotation curve is then converted to a model of mass distribution which, integrated over the entire dimension of the galaxy, yields its total mass. This has been done extensively by Burbidge and his co-workers for many nearby bright spirals and later by V.C. Rubin and her co-workers for more distant galaxies. The models used for mass distribution are inhomogeneous spheroids in which equidensity surfaces are concentric spheroidal shells with constant eccentricity.

In the model construction it is generally assumed that the gravitational force on any element of mass of the galaxy is equal to the centrifugal force on it due to rotation. The pressure arising out of the random motions of stars and gas is neglected. This implies, in fact, the perfect circular symmetry of the material motion around the centre of the galaxy. This assumption underestimates the gravitational force and introduces a consequent error in the calculated mass. But it has been found that the error introduced in this way is not significant.

The method followed by the Burbidges in analyzing the rotation curve to calculate the mass distribution and the total mass can be summarized as follows:

Let $\rho(a)$ be the density of a spheroidal shell of semi-major axis a and eccentricity $e =$

$\left(1 - \dfrac{c^2}{a^2}\right)^{1/2}$, being the other semi-axis. Let $\theta_c(r)$ be the circular velocity measured at a

distance r from the centre. Then these authors have shown that the density distribution $\rho(a)$ is related to the circular velocity $\theta_c(r)$ through the integral equation

$$\theta_c^2(r) = 4\pi G(1 - e^2)^{1/2} \int_0^r \frac{\rho(a)a^2\, da}{(r^2 - a^2 e^2)^{1/2}} \tag{18.4}$$

The equation is solved by substituting Taylor expansions

$$\theta_c^2(r) = r^2 \sum_{n=0}^{\infty} v_n r^n \tag{18.5}$$

and

$$\rho(a) = \sum_{n=0}^{\infty} \rho_n a^n \tag{18.6}$$

and equating powers of r. The total mass upto the farthest point a_t from the centre to which the rotation curve has been measured is finally obtained as

$$M_t = 4\pi(1 - e^2)^{1/2} \sum_{n=0}^{\infty} \rho_n \frac{a_t^{n+3}}{n+3} \tag{18.7}$$

where ρ_n's have already been evaluated from the series of linear equations obtained from Eqs. (18.5) and (18.6). Following this method, the masses of quite a few relatively nearby spiral galaxies have been obtained by Burbidges. Later, V.C. Rubin and her co-workers have computed the rotation curves of a large number of spiral galaxies and calculated the masses of many of these galaxies. The flat rotation curves of galaxies suggest that the mass distribution in them is spread to very large distances from the centre. Most of the mass in a galaxy thus resides in a dark halo extending far away from the centre of the galaxy.

Mass Determined by Density Wave Solution

B. Basu and collaborators used the density wave solution of C.C. Lin and co-workers to compute the masses of disc galaxies from their observed rotation curves. The basic formula adopted is

$$\mu F = \frac{K^2 - n^2(\Omega_p - \Omega)^2}{2\pi G\Phi'(r)} \tag{18.8}$$

which was obtained by Lin from the spiral-wave solutions of the general gas dynamical equations in two-dimensional cylindrical coordinates. Here μ is the surface density of matter, Ω and Ω_p are the angular velocities of matter and of the wave pattern respectively, and $\Phi'(r)$ is the radial wave number which depends on the radial spacings of the spiral arms. G is the universal constant of gravitation and n is a positive integer associated with the number of spiral arms in a galaxy, $n = 2$, has been used, in general. In our Galaxy and in many other galaxies the radial spacings of spiral arms is ≈ 2 kpc, so that $\Phi'(r) \approx \pi$ kpc^{-1}. K is the epicyclic frequency. The right-hand side of Eq. (18.8) represents the fraction of the basic mass density that *actually* participates in density perturbation. F, therefore is a reduction factor that cuts down the total basic mass to the mass "actually participating in the density perturbation". The value of F depends on several galactic parameters such as Ω, Ω_p, K and $\Phi'(r)$, and more sensitively on the average radial velocity dispersion of the material, the latter producing the smoothing effect on gravitational perturbation. Since velocity dispersions of stars and gas are different in galaxies of different Hubble types, the F-values will not be the same in these types of galaxies. The lower radial velocity dispersions ensure higher F-values and vice versa. Thus the F-value increases in the sequence Sa-Sb-Sc, and we *actually* see that the spiral patterns in Sc galaxies are more pronounced than in Sb and Sa galaxies. Toomre's analysis in 1964 has shown that the difference in the degree of manifestation of the spiral pattern could be explained by the actually existing difference in the magnitudes of the radial velocity dispersion of the material. For actual calculation of the distribution of mass density in the principal part of the discs, B. Basu and collaborators used $F = 0.10$–0.12 for Sa and Sb galaxies, and $F = 0.14$–0.16 for Sc galaxies.

The other important parameters to be evaluated in order to use Eq. (18.8) are Ω_p, Ω and K. Of these, the angular velocity of the wave pattern, Ω_p, is constant for any particular galaxy but varies from one galaxy to the other. Ω_p cannot be observed in any galaxy. Its value has to be determined from some plausible model. For their calculations, Basu and collaborators have used the model that the surface density μ vanishes at the periphery of the principal part of a galaxy which appeared quite plausible, with the arbitrariness of the assumed radial extension for any particular galaxy. The above assumption yields

$$\Omega_p = \Omega + K/n \tag{18.9}$$

at the periphery $r = R_0$, where R_0 is the radius of the principal part of the galaxy containing the grand design of the spiral pattern. Knowing the values of Ω and K at $r = R_0$, Ω_p is determined.

The basic observed parameter in this theory for computing the mass distribution is the same as that in any other theory, viz., the circular velocities $\theta_c(r)$ observed against r. The value $\Omega = \theta_c(r)/r$ is thus obtained as a function of r. K is the epicyclic frequency given by

$$K^2 = 2\Omega \left[2\Omega + r\frac{d\Omega}{dr} \right] \tag{18.10}$$

and is therefore related to the radial distance r from the centre of the galaxy. K is known at any point r when Ω and $d\Omega/dr$ are known at that point. As already mentioned Ω is known from the measured circular velocity at any point, and the derivatives of Ω can be evaluated by numerical differentiation using Newton's and Stirling's interpolation formulae.

When all the required quantities on the right hand side of Eq. (18.8) are thus evaluated at various distances r from the centre, Ω_p is known and the value of F is assigned, one gets the values of μ as a function of r. The surface density distribution is thus known over the principal part of the disc. These density values can now be fitted with suitable algebraic laws over different regions of the disc, excluding the core region, and the mass contained in the principal part of the disc can then be determined by using integrals

$$M_i = \int_{r_1}^{r_2} 2\pi\mu r \, dr \qquad (18.11)$$

where i refers to the particular region with r_1 and r_2 as the lower and upper boundaries.

We illustrate the method by applying it in evaluating the mass of our Galaxy. The Galaxy is of Sb type, so we choose the value $F = 0.1$. With $R_0 = 15$ kpc, the value of Ω_p is obtained as 22 km s^{-1} kpc^{-1}. The rotation curve has been taken from Schmidt's model of 1965. Fitting with the calculated values of μ, the density laws for different regions have been derived as follows:

$$\mu = -162.35 \times 10^{-6} \, r^2 + 1257.95 \times 10^{-3} \, r - 2668.5 + 2895.6 \times 10^3 \, r^{-1}, \; 1 \text{ kpc} \leq 4 \text{ kpc},$$

$$= -76.77 \times 10^{-6} \, r^2 + 1237 \times 10^{-3} \, r - 6512 + 13125 \times 10^3 \, r^{-1}, \; 4 \text{ kpc} \leq r \leq 7 \text{ kpc},$$

$$= 2768 \times 10^9 \, r^{-2.6}, \; 7 \text{ kpc} \leq r \leq 11 \text{ kpc},$$

$$= 475 \times 10^{16} \, r^{-4.15}, \; 11 \text{ kpc} \leq r \leq 13 \text{ kpc},$$

$$= 85.5 \times 10^{26} \, r^{-6.4}, \; 13 \text{ kpc} \leq r \leq 15 \text{ kpc}.$$

Applying formula (18.11) to each of these regions separately the corresponding sectorial masses are calculated to be 0.90×10^{11}, 0.35×10^{11}, 0.15×10^{11}, 0.05×10^{11} and 0.015×10^{11} solar masses respectively. Considering 15 per cent of the disc mass as the central mass within 1 kpc, the total mass of the Galaxy is evaluated to be $\sim 1.70 \times 10^{11} \, M_\odot$ which is a fairly reasonable value for the mass of the principal part of the Galaxy. Similar calculation with plausible values for the parameters of the galaxy M31 has yielded the mass $\sim 3.1 \times 10^{11} \, M_\odot$, again a reasonable value for the mass of M31 up to a distance of 25 kpc. For many other nearby disc galaxies, the method has been used to yield reasonable masses in each case (Basu et al., 1980). The method can therefore be reliably used to determine the masses of the principal part of disc galaxies.

18.7   EVOLUTION OF GALAXIES

Let us now have a brief discussion of the extremely important astrophysical and cosmological problems of the *evolution of galaxies*. Whether we can understand them or not, one point remains clear that the galaxies were born at some single or different epochs, have been evolving since then and would end their career at some other single or different epochs. It is therefore essential that to understand their nature, we know their course of evolution.

There were two early theories on the evolution of galaxies—the first, proposed by E.P. Hubble and the second, by H. Shapley. Hubble developed the classification scheme for galaxies, and suggested that the classification sequence in order represents the evolutionary sequence. He proposed that all galaxies begin their career as an E0 galaxy. Later due to rotation, gradual flattening takes place which thus pass through the stages from E0 to E7. Further flattening by rotation causes the development of spiral arms, at which stage material flows from the central nucleus and spreads into the arms. The nuclear material is thus gradually depleted and the spiral arms are spread widely when the galaxy reaches at the end of this sequence to the stage Sc. Finally, as the nuclear material is completely depleted the spiral arms get liquidated and geometric structure of the galaxy no longer exists—leading it to an irregular form. Thus, the evolutionary sequence is from elliptical to spiral and then eventually to irregular.

The second early theory of the evolutionary sequence of galaxies proposed by Harlow Shapley took just the reverse view. According to this theory, galaxies start their life as formless irregular masses. Form and structure gradually emerge in the process of evolution. Shapely reasoned that irregular galaxies, in course of time, turn into circular pattern of Sc type and gradually wind up to Sb and Sa types respectively. The galaxies then lose the spiral pattern and finally take the shape of symmetrical elliptical galaxies.

Both these early theories were, however, based on mere superficial structures of galaxies. There was no physical evidence behind their formulation. Also, observational material at that time was too insufficient to warrant any sound logical theory of the evolution of galaxies. At present most of the astrophysicists favour the rejection of both the theories as, none of these seems to be able to explain the various observed facts regarding the structure and evolution of galaxies. The currently accepted theory, based on the known facts of the stellar evolution theory, propounds that a galaxy does not evolve from one type to another, but the evolution of its contents, stars and gas, proceeds while retaining its fixed structure. An elliptical evolves as an elliptical, a spiral as a spiral, and so on. In other words, *chemical and not structural evolution takes place in galaxies.*

Before discussing the relative merits and demerits of these different theories, let us enumerate the basic observational facts about the galaxies:

1. The integrated colours of galaxies become progressively redder as we proceed along the sequence Irr-Sc-Sb-Sa-E. The gas content also decreases along this sequence.

2. The relative proportions of young (hot and blue) stars in galaxies increases along the sequence Sa-Sb-Sc-Irr. In ellipticals, these stars are almost or entirely absent. These blue stars are known to be of recent origin out of metal-enriched interstellar material. Thus, almost all stars in elliptical galaxies are of Population II type and more or less of the same age. In spirals, both populations of mixed age in sufficient numbers are seen, whereas in irregulars, Population I stars dominate although some old stars are also present in them.

3. Dwarf ellipticals are invariably metal-poor compared to the giants. Studies of the Coma cluster galaxies by Rood have shown that there is a direct correlation between higher abundance of metals and higher luminosity in elliptical galaxies. This correlation was also observed in earlier studies by W.W. Morgan, and N.U. Mayall and by H. Spinard.

4. In Sc galaxies, the spirals are well-developed and spread, and the stellar content in the nuclei is markedly different from that in the spiral arm region. The colour is much

bluer in spiral arms than in the nuclei. The bulk of the light in the spiral arms in Scs originate in blue hot O-B stars. In Sb galaxies these differences are noticed but with much less prominence. In these galaxies major part of the nuclear light comes from giant G, K, M stars while the light in the spiral arm region originates in a mixed population. In ellipticals and S0s as well as in spirals of Sa type, on the other hand, is observed a smooth distribution of light from centre to outer regions and the stars in the central region may be assumed to be representative of the entire galaxy.

5. The percentage of galaxies showing emission lines of [O II] at λ 3727 in the central regions progressively increases along the sequence E-S0-Sa-Sb-Sc, starting from about 20 per cent for Es to about 90% for Scs. For S0s the percentage is 50. Thus the ellipticals and S0s are not entirely devoid of either gas or hot stars, as is the common notion, although their percentage showing these characteristics is small compared to that of spirals. Some ellipticals such as NGC 4278 show very strong emission features.

6. The range of masses of ellipticals lies between 10^6 to 10^{13} $M_\odot$ while that for spirals lies between 10^9 to 10^{12} $M_\odot$. The absolute visual magnitudes for ellipticals range from –9 to –23 and those for spirals from –15 to –21.

7. The central condensation progressively increases along the sequence Sc-Sb-Sa-S0 giant E, the Irrs having no nucleus at all, and the activities in the nuclei also increase, in general, along this sequence.

8. Compact regular clusters of galaxies like the Coma cluster contains giant ellipticals but no spirals and the irregular compact clusters such as the Virgo contain both giant spirals and ellipticals. While about 75% of the field galaxies are spirals, in clusters their membership is only 20%.

9. The spirals possess much higher angular momentum per unit mass than the ellipticals.

10. The outflow of matter from the nuclei is a common feature of all galaxies. The speed of flow ranges from about 100 to several thousand kilometres per second.

In the light of these basic observed facts, we can now discuss the arguments in favour and against (if any) each of the three theories on the evolution of galaxies. The current theory, accepted in general, that the galaxies evolve, while each retaining the same structure, finds general support from the observed facts 1, 2 and 9. The structure of a galaxy may be uniquely determined by its initial mass and angular momentum, just as the structure of a star is now known to be uniquely determined by its initial mass and chemical composition. The angular momentum of a gaseous sphere has the property of preventing the collapse after a certain critical phase is reached. During the initial stages of condensation, the *protogalaxy* having smaller angular momentum per unit mass, will undergo a relatively rapid collapse. Rapid subcondensations will follow leading to star formation within a relatively short period. Most of the gas in the galaxy will thus be used up in the process of star formation within a relatively short period, thus preventing any major formation of subsequent generations of stars.

These stars thus formed, out of the initial metal-poor material will evolve through various stages of their life with the passage of time. They will therefore predominantly evolve as metal-poor red stars. This is what we observe of elliptical galaxies—spherical or spheroidal systems consisting of evolved, old, metal-poor, red stars and very little gas. Again the protogalaxies having higher angular momentum initially, will rotate faster as they collapse until at a stage, the collapse will be thwarted by the centrifugal force of rotation. Star formation will be slower in such systems. At the initial stage, some stars will be formed, but there still remains enough gas to initiate the formation of subsequent generations of stars. The gas, out of which these new generations of stars are formed has been enriched by now with heavy elements thrown out of the older stars. The cases of spirals and irregulars can be realized out of the picture just mentioned. In these systems we find old, middle-aged as well as young stars thereby pointing to the fact that star formation is a continuous process. The rate of star formation, however, is likely to depend on many factors, of which the gas density must be an important one. The exact density law can only be speculated and realized by observations for particular systems. The law can be represented in a general form as

$$\text{Rate of star formation} = \text{Constant} \cdot (\text{density})^n \qquad (18.12)$$

Values of n that have been used by different authors range from 1 to 3.

We now consider the theory initially proposed by Harlow Shapley and more recently supported by Alan Sandage in his revision of *The Hubble Atlas of Galaxies*. According to this theory, a galaxy begins as an Irr or Sc and subsequently evolves to earlier types. In view of the basic observed facts mentioned above, it is difficult to see how the galaxian evolution proceeds in this order. When a gaseous mass condenses, it assumes most probably either a spherical or spheroidal shape. It is difficult to visualize how it condenses first into a structureless form and after most of the gas has been consumed in star formation assumes subsequently smoother shape. Again, all observable evidences reveal that matter is being expelled from the central regions of all galaxies. The reverse phenomenon has not yet been detected. It is difficult to realize therefore how a small nucleus of an Sc galaxy, from which matter is continuously flowing out, grows to a larger and more compact nucleus as in an Sb or an Sa. Still another vital question has to be answered before accepting the theory. The masses of spirals range from 10^9 to $10^{12}\,M_\odot$ whereas those of ellipticals lie between 10^6 to $10^{13}\,M_\odot$. If the spirals evolve into ellipticals, in the process some have gradually to acquire while some have to lose mass. It is difficult to visualize as to how these mutually contradictory processes can be achieved during similar physical transformations.

Lastly, we discuss the theory of galaxian evolution, originally proposed by E.P. Hubble and later supported by V.A. Ambertsumyan. According to this theory, galaxies begin as ellipticals and subsequently evolve into later classes, Ambertsumyan proposed that this transformation is achieved by activities in the nuclei of galaxies. It is therefore natural to think that when a galaxy is first formed out of gas it will assume a spherical structure. Stars are formed in a relatively short time as already discussed. But these are only the first generation stars, the old stars as we call them, which constitute the elliptical galaxies. These stars evolve in time and during the process, a significant amount of mass is thrown into the intragalactic space. This mass will gradually be accreted in the central region of the galaxy, thereby forming a dense core of gas together with stars already existing there. The gravitational pull may gradually grow to reach a stage when a nonstellar superdense body will be formed at the nucleus. Some gas will also initiate the formation of next generations of stars—hotter, more luminous and metal-rich. This is exactly what is seen in the nuclei of giant ellipticals in which the process of evolution and formation of new generations of stars advances quite rapidly. In dwarf ellipticals, the rate of nuclear condensation may be too slow to initiate all these stages during the age of the Universe, analogous to the case of red dwarf stars. Thus, it may be argued that in elliptical galaxies we do not see large proportions of hot blue stars because they did not have enough time to go through the required processes.

Subsequent nuclear activities will drive enough gas from the central to the outer regions which may ultimately develop into spiral structures and form large number of hot, blue metal-rich stars. Thus, the nuclear activity may be the agent for transforming galaxies from one morphological type to another. This picture is in conformity with the observations of gradually depleting nuclei, as we pass along the sequence E-S0-Sa-Sb-Sc-Irr. This is also in conformity with the requirement that several generations of stars must pour their metal-rich products into interstellar space in order to achieve the solar abundance out of the abundance of extreme Population II. The observed fact enumerated in point (6) will be in perfect agreement with such a picture, if we are ready to accept the hypothesis that only large ellipticals have had sufficient time to evolve into the later morphological types.

The above picture of the galaxian evolution can also be understood in the light of the observed results stated in point (8). Many astronomers now believe that all galaxies are probably formed in clusters and groups just as the case with stars. The virial test when applied to these clusters, leads to the conclusion that almost all of these are unstable. The evidence of instability becomes unmistakable, particularly for large compact clusters like Coma and Virgo. We have but little choice other than to accept that the large clusters of galaxies are unstable and will disperse into space within some suitable cosmological time. By the time the members of a cluster merge into the general field, the larger ellipticals evolve into spirals through nuclear activities, which subsequently evolve into later classes through the same process.

The fact that the Coma cluster contains no spiral galaxies has been explained by W. Baade and L. Spitzer in 1951, reasoning that mutual collisions between galaxies have swept away the spiral arms and gas rendering all the spirals to S0s. Their calculations were based on the inaccurate distance to the cluster known at that time which yields the diameter of the cluster too small. This has resulted in a very high frequency of collisions. Thus, Baade and Spitzer found from their calculations that almost all galaxies in the cluster suffered at least one collision since it was formed. With the present estimate of the distance and size of the cluster, calculations reveal that only a small fraction of the members might have suffered a collision during the age of the Universe. This led G.O. Abell to suggest that spiral galaxies probably were never formed in Coma cluster or they all have evolved to other forms of galaxies. Alternatively, one can quite plausibly argue that no galaxy is initially formed as a spiral; it is formed as an elliptical. Evolution creates spirals. The Coma cluster is quite young and is yet to transform some ellipticals into spirals, and its globular shape favours this argument. By the time the cluster achieves a significant disruption, loses its regular shape and becomes irregular as the Virgo cluster, larger ellipticals will evolve as spirals. Further disruption with time will make the members lose identification with the cluster and we will see them as field galaxies. By that time many more larger ellipticals will evolve as spirals, but the dwarf ellipticals which may have a very large population will be too faint to be detected. This possibly explains the overwhelming majority of spirals among the field galaxies. In the process of disruption, the subgroups of double and multiple galaxies that are observed in sufficient number within the Virgo cluster will retain their double and multiple nature even when they merge with the general field. Thus the field galaxies seen now as single, double or multiple entities may be but the remnants of totally disrupted clusters.

Thus, we have a probable and plausible picture of the evolution of galaxies. All galaxies are first born *in clusters and as ellipticals—giants,* mediocres and dwarfs. Larger among these develop sufficient condensations in the central regions to initiate nuclear activities. These in turn get transformed gradually to the subsequent stages of spirals and then to irregulars. Eventually, any further transformation in form ceases due to the complete absence of nuclear activities. Such a sequence of evolution appears logical in view of the various observed characteristics of galaxies, *but it has not yet been proved.* The actual evolution may be in some different way. The final understanding of the problem does not yet appear in sight. As discussed in Section 18.4, merger of spiral galaxies may lead to the formation of some S0 and elliptical galaxies, but this may be a special rather than a general rule.

18.8 DARK MATTER IN GALAXIES

Hubble's observations in 1920s clearly demonstrated that every galaxy was receding away from every other galaxy with velocities proportional to the distances between them. This observational result was most suitably modelled by the concept of the Expanding Universe. Soon after, Abbe Lemaitre proposed the world model that all the matter in the present Universe was confined in a small volume at high temperature in the *Primeval Atom* as it was called, which exploded in a Big Bang throwing material at high speeds in all directions. Lemaitre's original concept, after some later modifications, developed into the present theory of Standard Big Bang for the universe which started some 10 to 20 billion years ago (depending on the value of H).

Calculations reveal that the expansion of the Universe will be halted if the mass density in the Universe be equal to the critical density, ρ_c given by

$$3H^2/8\pi G = 1 \times 10^{-29} \text{ gm cm}^{-3} \tag{18.13}$$

where $H = 70 \text{ km s}^{-1}\text{Mpc}^{-1}$ has been used. The ratio of the observed density, ρ_{obs}, and the critical density ρ_c is defined as

$$\Omega = \rho_{obs}/\rho_c \tag{18.14}$$

The expansion of the Universe will be halted if $\Omega = 1$, the Universe then being called flat. But all current observations indicate that $\Omega = 0.20$ is an extreme upper limit. This is the baryonic mass detected by its gravitational effect. The remaining 80 per cent of the matter is nonbaryonic, provided we like to cling to the concept of a flat Universe. The severe restriction on the amount of the baryonic matter is imposed by the present understanding of the Standard Big Bang Nucleosynthesis.

The limitation comes principally from the observed abundance of He^4 which is about 24 per cent, according to the best present determination. If all matter was baryonic, He^4 would have been about 30 per cent, a value absolutely precluded by present observations.

What kind of matter contributes to the 80 per cent nonbaryonic matter is not clearly known at present, particle physicists have prepared a long shopping list of particles in search of the plausible matter. The Big Bang may not have left just baryons and radiation, but other species as well which may contribute to Ω. In the Standard Big Bang model neutrinos are almost as abundant as microwave background photons (actually 9/11 times the background photons for three species of neutrinos) outnumbering baryons by a factor of about 2×10^8, if we use $H = 70$ km s^{-1} Mpc^{-1} and $\Omega = 0.20$. So if neutrions have masses as small as about tens of electron volts, they are able to supply enough mass for the flat Universe. All recent experiments in this direction indicate that the neutrions do possess such masses. Several other candidates have been proposed for the purpose by particle physicists. Light gravitinos, quark nuggets, axions and magnetic monopoles are a few such probables. But neutrinos are the most favoured candidate at the moment.

Again, of the observed baryonic mass the luminous matter is only 20 per cent. This luminous matter consists of bright stars and gas in different phases (molecules as well as neutral and ionized atoms) observed at various wavelengths of the electromagnetic spectrum including X-rays, radio waves and microwaves. In large clusters of galaxies about one-half of the luminous matter consists of the X-ray emitting high temperature ionized intracluster gas in the central regions. Thus, an overall 20 per cent of the detected matter is actually *seen.* The remaining 80 per cent remains *unseen* or invisible. Since this matter does not radiate in any wavelength of the electromagnetic spectrum it has been called *dark matter.*

How do we know that at least 80 per cent of this baryonic matter is dark? The answer comes from a comparison between the actually observed luminous matter and the gravitationally detected matter. The concept of hidden matter was first introduced by J.H. Oort as early as in 1932. Observing motions of stars perpendicular to the disc of our Galaxy in the solar neighbourhood, Oort estimated the surface density of gravitating matter in the galactic disc. He calculated a mass much in excess of that in the observed stars. At that time, methods for evaluation of gaseous mass were not known. Later, when this mass was calculated by 21-cm observation and by other methods, the total observed mass in gas and stars was still much in defect of the gravitational mass. Soon after, in early 1930s F. Zwicky and Sinclair Smith discovered much greater proportion of hidden mass in groups and clusters of galaxies. Assuming that a cluster is in equilibrium, we can use the Virial theorem to compute the gravitational mass of the cluster. Calculations show that the Virial mass of the cluster invariably exceeds the luminous mass by factors of more than one hundred. Finally, the flat rotation curves of spiral galaxies of all types clearly demonstrate that the total mass of a galaxy is at least ten times more than the luminous mass in it. In a galaxy the central region is very luminous and the luminosity falls off rapidly away from the centre. If all the matter was luminous, then the matter and light should have same distribution. So the matter density should fall equally rapidly as the luminosity, and therefore most of the matter should be concentrated in the central region where luminosity is very high. The velocity of rotation of a mass element in the galaxy should then fall off with distance following the Keplarian law which is derived from the balance and is given by

$$\frac{GM(r)}{r^2} = \frac{\theta_c^2(r)}{r} \tag{18.15}$$

where $M(r)$ is constant. But in spiral galaxies rather the circular velocity $\theta_c(r)$ is observed to the constant even at large distances from the centre and thus the same balance equation

(18.15) yields $M(r) \propto r$, which again yields $\rho \propto r^{-2}$. In spiral galaxies, therefore, the mass increases with r up to large distances from the centre, while the density falls as the inverse square of the distances. Since the volume of a spheroidal shell of radius r and thickness dr increases as r^2, therefore successive shells of equal thickness in the galactic body will contain equal masses. Calculations further show that the matter density even at great distances from the centre of the galaxy exceeds by orders of magnitude over that in the intergalactic medium. Thus, each galaxy contains an extensive distribution of matter in its halo out to far beyond its visible boundary. Each galaxy thus possesses a massive dark halo surrounding its visible disc. Calculations reveal that the actual mass of a galaxy is about ten times the mass of its luminous matter. Also, the mass luminosity ratio, M/L, measured in solar units increases systematically from the centre outwards, meaning that the relative amount of the dark matter increases outward in the galaxy.

Having thus been convinced that at least 80 per cent of the detected baryonic matter is dark, we should ask: In what form this dark matter exists? The possible forms are (a) black holes, (b) dead stars like white dwarfs and pulsars, and (c) substellar bodies like brown dwarfs and Jupiters. We discuss in the following the suitability of each of these types of objects for contributing to the dark matter as required.

Some astronomers believe that a massive black hole ($M > 10^6 \, M_\odot$) lies at the centre of each normal galaxy. Also, there may be a sparse distribution of smaller black holes in a galaxy which probably manifest themselves as isolated X-ray sources. But there cannot exist many massive black holes in a galaxy, sufficient to compensate for the dark mass by them, as their presence would manifest observable physical and dynamical effects on their environment, the like of which are not seen at all. Also, the existence of a distribution of innumerable low mass black holes that may provide a fair compensation for the required dark matter appears very improbable. It is difficult to see how most of the galactic mass can be converted into billions of tiny black holes. We therefore neglect the possibility of the presence of the galactic missing mass in the form of black holes.

We now consider the case for dead stars. From an extensive study of the distribution of pulsars in the local regions of the Galaxy, J.H. Taylor and R.N. Manchester have derived the birth rate of one pulsar in six years in the entire Galaxy assuming $\langle n_e \rangle = 0.03 \, \text{cm}^{-3}$. If $\langle n_e \rangle = 0.02 \, \text{cm}^{-3}$, the corresponding rate is one pulsar in forty years. We compromise with one pulsar in ten years, leading to the birth of 10^9 pulsars in the lifetime of the Galaxy. Assuming an average mass of $1 \, M_\odot$ for pulsars, the total mass dumped as pulsars in the Galaxy is $10^9 \, M_\odot$. A similar calculation can be made for the total mass in white dwarfs. Analyzing both theoretical and observational works on the formation rate of white dwarfs and planetary nebulae, C.E. Miller and J.M. Scalo have calculated the present birthrate of white dwarfs and planetary nebulae to be $(4\text{--}10) \times 10^{-9} \, \text{pc}^{-2} \, \text{yr}^{-1}$ in the Galaxy. We compromise with $7 \times 10^{-9} \, \text{pc}^{-2} \, \text{yr}^{-1}$ as the birthrate, take $1000 \, \text{kpc}^2$ as the effective area of the galactic disc and $0.7 \, M_\odot$ as the average mass of a white dwarf. This yields a mass $5 \times 10^{10} \, M_\odot$ in white dwarfs. Thus the total mass in dead stars in our Galaxy may be $\sim 5 \times 10^{10} \, M_\odot$. This is at most ten per cent of the total mass of the Galaxy. Almost an equal amount of mass is contained in luminous stars and gas and this is true for any normal spiral galaxy. The remaining 80 per cent of mass of the galaxy consists of some other form of dark matter.

Let us now examine the case for brown dwarfs and Jupiters. This refers to the theories of star formation process and stellar mass function.

The form of the SMF in Eq. (11.11) shows, that the number of stars steeply increases as mass decreases. There is no known theoretical basis to assign a cut-off mass at the low end and calculations have shown that the minimum Jeans mass can be as low as $0.005\ M_{\odot}$, or even lower. On the other hand, the stellar evolution theory predicts that unless a body has a minimum mass of $0.08\ M_{\odot}$, it cannot form star (nuclear fuel is not ignited at the centre) and will evolve as a substellar object, such as brown dwarf or Jupiter. Their, evolution time is 10^9 years. These objects will therefore populate the galaxy contributing to its dark matter. Various calculations have shown that under different parametric conditions these substellar objects can contain about twenty to eighty per cent of the total mass of the parent cloud. Thus, brown dwarfs and Jupiters appear to be the most favoured candidates at the moment as contributing to the major part of the dark matter in galaxies.

18.9 SUPERCLUSTERS AND VOIDS

Since 1980s large surveys using multi-object spectrograph, CCD detectors and some dedicated telescopes have revealed a very surprising picture of luminous matter in the Universe. Previously it was believed that galaxy clusters are floating in streams of field galaxies but present-day large scale observations show that galaxies are distributed along thin thread like structures called *walls and streamers* surrounding huge voids that appear largely empty. These enormous walls or sheets that can span a billion light years in length contain substructures of length 100 million light years. These substructures are called *superclusters*. Sometimes the existence of structures larger than super clusters are suggested, e.g. 'Sloan Great Wall and cfA2 Great Wall'. 'Sloan Great Wall' is a giant wall of galaxies and the largest known structure in the Universe. It was discovered by J. Richard Got III and Mario Juric' of Princeton University in 2003 based on the data from the sloan digital sky survey (SDSS). The wall is 1.37 billion light years in length and is at a distance of one billion light years from the earth. Figure 18.1 shows an image of the structure.

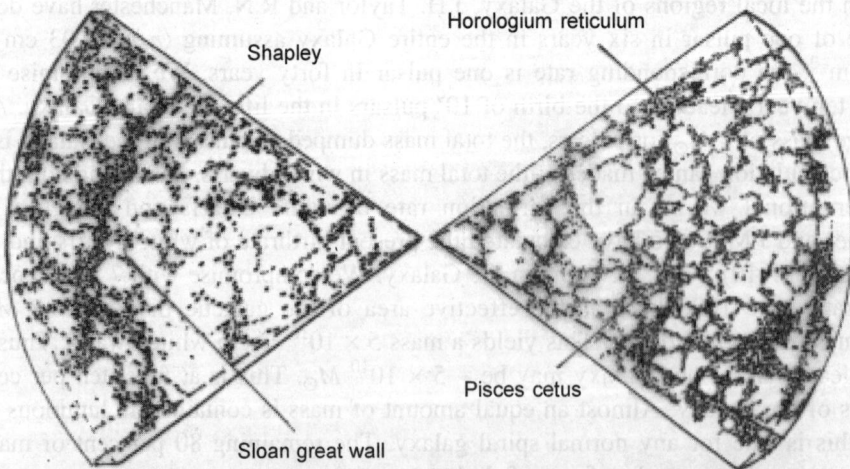

FIGURE 18.1 Sloan great wall based on data from SDSS.

On the other hand 'cfA2 Great Wall' is the second largest superstructure in the Universe. It was discovered by Margaret Geller and John Huchra based on red shift survey data of cfA red shift survey in 1989. This wall is 500 million light years in length and is at a distance of 200 million light years. There is the hypothesis that *dark matter* dominates the structure of the Universe. It gravitationally attracts normal matter and it is the normal matter that we see forming long thin walls of galaxies, known as *superslusters* (Fig. 18.2).

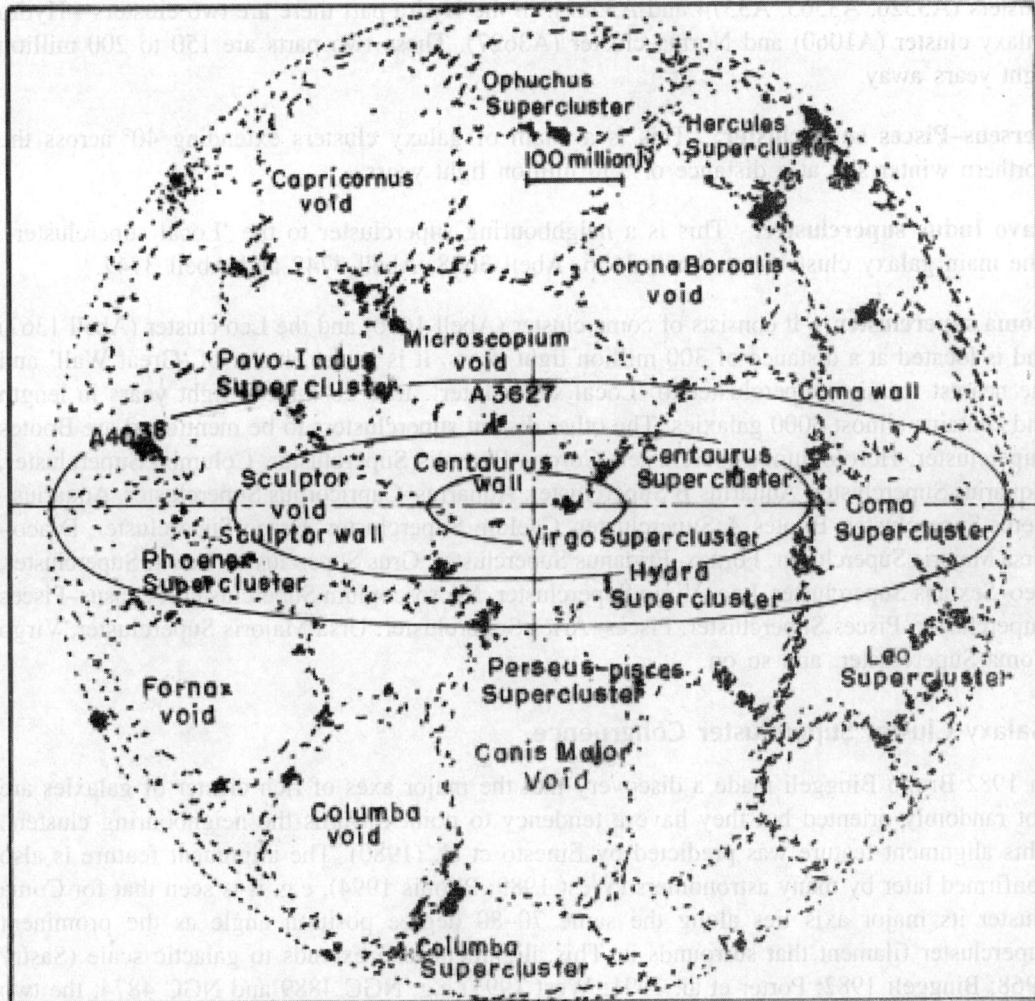

FIGURE 18.2   A schematic diagram of the nearby superclusters distributed along the 'Walls' as found by Got and Jumic.

The nearby superclusters as shown in Fig. 18.2 are:

Local supercluster.   It contains Local Group of galaxies and Virgo cluster along 100 more groups and galaxy clusters. It has two components, viz. disc and halo. The disc is flattened and pancake type in shape and contains 60 per cent of the galaxies and the halo consists of elongated objects and contains the remaining 40 per cent of the luminous galaxies. The diameter is 200 million light years. The mass is of the order of 10^{15} $M_\odot$. As the luminosity is very small it is thought that a large portion of the mass is dark matter. The entire supercluster is being pulled towards a structure called 'Great Attractor' located near the normal 'Norma cluster'.

Hydra supercluster. It consists of two parts. In Centaurus part there are four large galaxy clusters (A3526, A3565, A3574 and A3581). In the Hydra part there are two clusters—Hydra galaxy cluster (A1060) and Norma cluster (A3627). These two parts are 150 to 200 million light years away.

Perseus–Pisces supercluster. This is a chain of galaxy clusters extending 40° across the northern winter sky at a distance of 250 million light years.

Pavo Indus supercluster. This is a neighbouring supercluster to the 'Local supercluster'. The main galaxy clusters are Abell 3656, Abell 3698, Abell 3742 and Abell 3747.

Coma supercluster. It consists of coma cluster (Abell 1656) and the Leo cluster (Abell 1367) and is located at a distance of 300 million light years. It is in the cluster of 'Great Wall' and the nearest massive supercluster to 'Local superclutser'. It is 20 million light years in length and contains almost 3000 galaxies. The other distant superclusters to be mentioned are Bootes Supercluster, Horologium Supercluster, Corona Borealis Supercluster, Columba Supercluster, Aquarius Supercluster, Aquarius B Supercluster, Aquarius–Capricornus Supercluster, Aquarius–Cetus Supercluster, Bootes A Supercluster, Caelum Supercluster, Draco Supercluster, Draco–Ursa Majoris Supercluster, Fornax–Eridanus Supercluster, Grus Supercluster, Leo A Supercluster, Leo–Sextans Supercluster, Leo–Virgo Supercluster, Microscopium Supercluster, Pegasus–Pisces Supercluster, Pisces Supercluster, Pisces–Aries Supercluster, Ursa Majoris Supercluster, Virgo Coma Supercluster, and so on.

Galaxy Cluster Supercluster Congruence

In 1982 Bruno Binggeli made a discovery that the major axes of rich cluster of galaxies are not randomly oriented but they have a tendency to point towards the neighbouring clusters. This alignment feature was predicted by Einesto et al. (1980). The alignment feature is also confirmed later by many astronomers (West 1989; Plionis 1994), e.g. it is seen that for Coma cluster its major axis lies along the same 70–80 degree position angle as the prominent supercluster filament that surrounds it. This alignment also extends to galactic scale (Sastry 1968; Binggeli 1982; Porter et al. 1991; West 1994) e.g. NGC 4889 and NGC 4874, the two central galaxies in Coma cluster have the same 70–80 degree position angle. Also many brightest elliptical galaxies in Coma cluster share the same type of alignment scenario. Hence it appears that Coma cluster and its member galaxies have been greatly influenced by its supercluster surroundings. The above feature is not only limited to Coma cluster but also evident in Virgo cluster. Here the giant ellipticals lie along a line which has a position angle of 120°. Furthermore Virgo's principal axis points in the direction of Abell 1367, which is a rich cluster 50 Mpc away along a projected position angle 125°. This suggests the possibility that Virgo, Abell 1367 and Coma are all members of a common filamentary network. The alignment of Virgo and Abell 1367 is clear in Fig. 18.3 which plots the distribution of galaxy groups from the cfa redshift survey (Ramella et al. 1997).

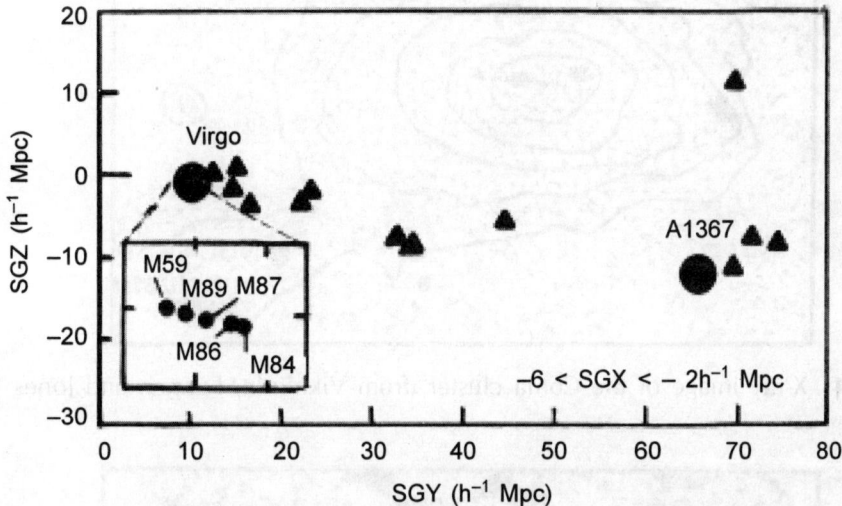

FIGURE 18.3 The distribution of groups of galaxies in the CfA survey (from Ramella et al. 1997). The positions are plotted in supergalactic coordinates SGY and SGZ. The insert shows a 10 Mpc by 10 Mpc region containing the brightest Virgo ellipticals.

Galaxy and Cluster Formation

The observational evidences of alignment phenomenon of galaxies, clusters of galaxies and superclusters led astronomers to think about how these objects have been formed. The most fascinating scenario is that of 'merger hypothesis'. According to this theory, elliptical galaxies and clusters are built by an anisotropic merger process. West et al. (1995) showed that the distribution of subclusters in clusters traces the surrounding filamentary distribution of matter on supercluster scale. This matter distribution on supercluster scales influences the properties of clusters. In Coma cluster the arrangement of its multiple subclusters is just the reflection of surrounding supercluster filament. Figure 18.4 shows the existence of a large subcluster associated with NGC4839 and this galaxy appears to be falling into the cluster along the direction of Coma A1367 filament. Other subclusters show similar trend (Mellier and Mathez 1987). This suggests that Coma has been built by mergers of subclusters which fall along the prominent filament in which it is embedded. Theoretical simulation on structure formation has been carried out by Bond et al. (1996). It shows that cluster alignment is an outcome of dynamical evolution of filaments resulting from some primordial matter distribution. Figure 18.5 shows a cosmological N body simulation of a region in a cold dark matter universe (Couchman 1997) which confirms the above scenario.

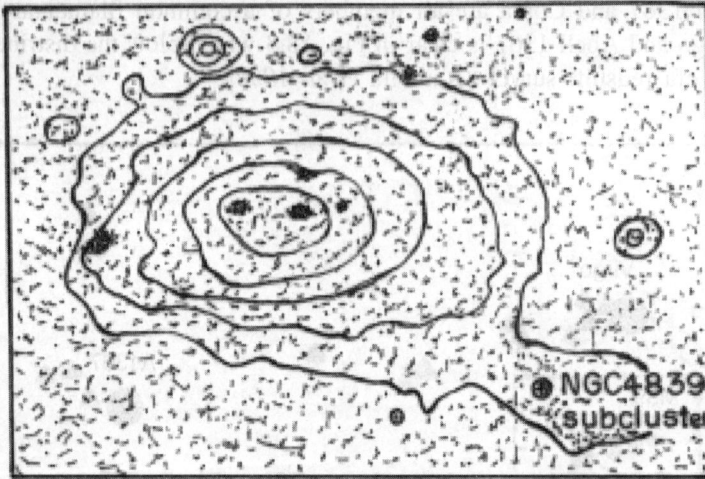

FIGURE 18.4   X-ray image of the Coma cluster (from Vikhlinin, Forman and Jones 1997).

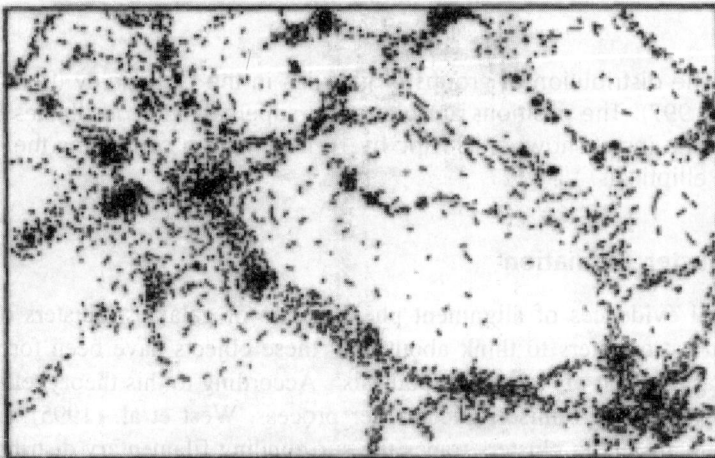

FIGURE 18.5   N body simulation of CDM universe.

EXERCISES

1. Do you agree with the concept that the galaxies have a clustering tendency? Give reasons for your answer.

2. "Superclusters separated by voids appear to be a careful plan of nature"—establish this statement on the basis of the current observations.

3. Write what you know of the *Great Wall* and the *Great Attractor*.

4. Discuss the various criteria which have been used by the authors for morphological classification of galaxy clusters. Which one of these appears to you as the best, and why?

5. Enumerate the principal differences between the regular and irregular clusters of galaxies. Give an example of each of them.

6. Write what you know of the Virgo Cluster of galaxies. Comment on the special astrophysical importance of this cluster.

7. Discuss the various theories that exist to explain the observed differences in the galactic content of regular and irregular clusters.

8. Describe the principal observable features of cD galaxies. What do you know of the mechanism of their formation?

9. Describe the observable features which indicate that galaxies mutually interact and often suffer merger. What are the possible effects of merger on the resulting galaxy?

10. "All elliptical galaxies are results of merger"—Do you support this statement? Give reasons.

11. Where in the Universe do we find strong radiation in X-rays? Describe the nature of the region of X-ray radiation in rich clusters of galaxies. What is the physical state of the gas emitting this radiation? What is the radiation mechanism?

12. What is a cooling flow? How is it related to the X-ray emission? How much mass is usually involved in a cooling flow into the central region of a rich cluster?

13. Explain how we can compute the masses of elliptical galaxies? What simplifying assumptions are made for such computations? What are the sources of uncertainty?

14. Discuss how the total mass of a cluster of galaxies can be calculated by application of the Virial theorem. Why does this calculated mass greatly exceed the actually observed mass of the cluster?

15. The mass-to-light ratio progressively increases as we consider (a) an individual galaxy, (b) a small group of galaxies, (c) a nearby cluster of galaxies, and (d) a distant cluster of galaxies—Explain why?

16. Describe how the rotation curve of a disc galaxy is determined. Show how the rotation curve of a galaxy can be used to determine the mass distribution in it and its total mass.

17. What observations lead us to conclude that the disc galaxies are enveloped by extensive and massive dark halos?

18. Describe how the density wave solution for spiral structure can be used to calculate the mass of a disc galaxy. What assumptions are made in this method? Comment on these assumptions.

19. Use Table 16.3, last column (Bok's model), for the rotational velocities in our Galaxy and calculate the mass of the Galaxy by applying the method of density wave solution.

20. Briefly narrate the different theories that have been proposed for understanding the evolution of galaxies. Which of these theories appeals to you most?

21. Various types of observations suggest that a very large fraction of matter remains hidden in individual galaxies, in clusters of galaxies and in the Universe. Discuss these observations and derive an estimate of the hidden matter.

    What are the possible states in which this hidden matter may lie?

SUGGESTED READING

1. Abell, G.O., Clustering of Galaxies, *Annual Review of Astronomy and Astrophysics,* **3**, p.1, 1965.

2. Abell, G.O. et at., A Catalogue of Rich Clusters of Galaxies, *Astrophysical Journal Supplement Series,* **70**, pp. 1–138, 1989.

3. Alladin, S.M. and Narasimhan, K.S.V.S., Gravitational Interactions between Galaxies, *Physics Report,* **92**, p. 339, 1982.

4. Barcons, X. and Fabian, A.C. (Eds.), *The X-Ray Background,* Cambridge University Press, 1993.

5. Barnes, J.E. and Hernquist, L., Dynamics of interacting galaxies, *Annual Review of Astronomy and Astrophysics,* **30**, p. 705, 1992.

6. Binggeli, B., *Astronomy and Astrophysics,* 107, 338, 1982.

7. Bond, J.R., Kofman, L., and Pogosyan, D., 1996, *Nature,* 380, 603, 1990.

8. Burgarella, D., Livio, M. and O'Dea, C.P. (Eds.), *Astrophysical Jets,* Cambridge University Press, Cambridge, 1994.

9. Casertano, S., Sackett, P. and Briggs, F. (Eds.), *Warped Discs and Inclined Rings Around Galaxies,* Cambridge University Press, Cambridge, 1991.

10. Couchman, H.M.P., in *12th Kingston Meeting: Computational Astrophysics,* D.A. Clarke and M.J. West, (Eds.), Astron, Soc. Pac., San Francisco, p. 340.

11. Dressler, A., The evolution of galaxies in clusters, *Annual Review of Astronomy and Astrophysics,* **22**, p. 185, 1984.

12. Elvis, Martin (Ed.), *Imaging X-Ray Astronomy,* Cambridge University Press, Cambridge, 1990.

13. Einasto, J., Joeveer, M., and Saar, E., *MNRAS,* 193, 353, 1980.

14. Fabian, A.C. (Ed.), *Cooling Flows in Galaxies and Clusters,* Cambridge University Press, Cambridge, 1988.

15. Giacconi, R. and Setti, G., *X-Ray Astronomy,* D. Reidel Publishing Company, Dordrecht, Holland, 1979.

16. Giacconi, R., *X-Ray Astronomy with the Einstein Satellite,* D. Reidel Publishing Company, Dordrecht, Holland, 1981.

17. Hirsh, Richard, F., *Glimpsing an Invisible Universe: The emergence of X-ray astronomy,* Cambridge University Press, Cambridge, 1985.

18. Hodge, P.W., *Galaxies and Cosmology,* McGraw-Hill Book Company, New York, 1966.

19. Hodge, P.W., *Galaxies,* Harvard University Press, Cambridge, Mass., 1986.

20. Kafatos, Minas (Ed.), *Supermassive Black Holes,* Cambridge University Press, Cambridge, 1988.

21. Longair, M.S. and Einasto, J., *The Large Scale Structure of the Universe,* D. Reidel Publishing Company, Dordrecht, Holland, 1978.

22. Mellier, Y., Mathez, G., *Astronomy and Astrophysics,* 175, 1, 1987.

23. Oegerle, W. (Ed.), with M. Fitchett and L. Danly, *Clusters of Galaxies,* Cambridge University Press, Cambridge, 1990.

24. Parker, Barry, *Colliding Galaxies,* Plenum Press, New York, 1990.

25. Peebles, P.J.E., *The Large Scale Structure of the Universe,* Princeton University Press, New Jersey, 1980.

26. Plionis, M., *ApJS,* 95, 401, 1994.

27. Porter, A.C., Schneider, D.P., and Hoessel, J.G., *A.J.,* 101, 1561, 1991.

28. Ramella, M., Pisani, A., and Geller, M.J., *AJ,* 113, 483, 1997.

29. Sarazin, C., *X-Ray Emission from Clusters of Galaxies,* Cambridge University Press, Cambridge, 1988.

30. Sastry, G.N., *PASP,* 80, 252, 1968.

31. Seitter, W.C., *Cosmological Aspects of X-Ray Clusters of Galaxies,* Kluwer Academic Press, Dordrecht, Holland, 1994.

32. Sulentic, J. (Ed.), *Paired and Interacting Galaxies,* IAU Colloquium No. 124, Kluwer Academic Press, Dordrecht, Holland, 1990.

33. Weilen, R. (Ed.), *Dynamics and Interactions of Galaxies,* Springer-Verlag, Berlin, 1990.

34. Woltjer, L. (Ed.), *Galaxies and the Universe,* Columbia University Press, New York, 1968.

35. West, M.J., *MNRAS,* 79, 268, 1994.

36. West, M.J., *Astrophysical Journal,* 347, 610, 1989.

37. West, M.J., Jones, C., and Forman, W., *Astrophysical Journal,* 451, L5, 1995.

38. Zeilik, Michael, Gregory, Stephen, A. and Smith, Elske V.P., *Introductory Astronomy and Astrophysics,* Saunders College Publishing, New York, 1992.

Radio Galaxies

19.1 INTRODUCTION

The first study of the extraterrestrial radio waves was done by Karl Jansky of the Bell Telephone Laboratories in the U.S.A. in early 1930's. In order to detect disturbances in the communication radio receivers, Jansky constructed a rotating aerial operating at 15 m wavelength. By the year 1933, he could draw the conclusion that the radio waves he received were of extraterrestrial origin. Jansky demonstrated that the extraterrestrial radio waves he measured originated from the Milky Way—the strongest intensity coming from the galactic centre. Jansky's work was subsequently pursued and extended by Grote Reber, a radio engineer, who installed a 30 feet parabolic reflector in his garden at his own cost. In order to achieve better resolution and higher power, Reber used 9 cm and 33 cm wavelengths for his radio observations. His attempts however, proved unsuccessful as the radiation intensity at these wavelengths was too feeble for his instrument. He succeeded, however, at the frequency of 160 MHz ($\lambda \approx 1.875$ m) and was able to draw the first radio map of the Milky Way Galaxy in 1940. He detected an intense radio source in the direction of Sagittarius, the constellation in which lies the centre of our Galaxy. He also located two other intense sources in the regions of Cassiopia and Cygnus respectively. Reber also suggested that the radiation was produced by thermal bremsstrahlung in the ionized interstellar gas.

In 1942, J.S. Hey of England, while working as a scientist in the British Army during the Second World War, found that the Sun was emitting intense radio waves at 4 to 8 m wavelengths. It was further known that this intense radiation was nonthermal and was associated with the solar activity which was at its maximum phase at that time. In the same year G.C. Southworth of Bell Telephone Laboratories detected the thermal radio emission at centimetre wavelengths from the Sun. The Sun was thus the first individual heavenly body known to be an emitter of radio waves.

The next important radio astronomical work was the detection of other discrete radio sources which was initiated by J.S. Hey, S.J. Parsons and J.W. Phillips immediately after the War, in 1946. They made a survey of radio noise at 64 MHz with the help of Yagi antenna of beam width 6° by 15°. From their contour map, it was found that there were at least two regions of intense radio emission—one in the direction of the galactic centre and the other in the direction of Cygnus. Rapid fluctuations were recorded by Hey and his co-workers in the radio intensity received from the direction of Cygnus as the antenna swept over the region. The only conclusion to which they could arrive from this observation was that the radio source Cygnus must be small and hence such objects were called *discrete radio sources.*

The discovery made by Hey and co-workers was soon confirmed by several radio astronomers using radio interferometric techniques. In Australia, J.L. Pawsey and his co-workers mounted an aerial system on a cliff at a seashore and used the sea surface as a reflector in the Lloyd's mirror interferometer, while the radio equivalent of Michelson's optical interferometer was used by Martin Ryle and Graham Smith of England. They determined θ, the angular width of a radio source in terms of λ, the wavelength of observation, and the distance d separating the aerial system, from the formula

$$\theta = \frac{\lambda}{d}$$
(19.1)

θ was measured by observing the fluctuations in intensity of the radio emission from the sources. In the case of Cygnus A (3C 405) the fluctuations were measured also by J.G. Bolton and G.J. Stanly by using similar interferometric techniques. Both the groups of Ryle and Bolton gave an upper limit of 8' of arc for the angular width of Cygnus A. Soon after, while studying the polarization of radio waves with a Michelson-type interferometer at 80 MHz, Ryle and Smith discovered the intense radio source in Cassiopia.

The pioneering work of Bolton as early as in 1948 in which he made a.radio survey of the sky with a sea interferometer at 100 MHz having a limiting sensitivity of 200 flux units (1 f.u. = 10^{-26} W m^{-2}Hz^{-1}) revealed at least six radio sources, in addition to Cygnus A. These include Taurus A (the Crab Nebula), Coma A, Hercules A and Centaurus A. Subsequent discoveries of the radio sources due to a concerted efforts of radio astronomers in different countries, notably of those in England and Australia, raised their number to thousands. By 1966, the survey conducted by the radio astronomers gave an account of about 10^4 sources detected to the sensitivity limit of 1 f.u. This number has greatly swelled in the subsequent years when important research centres were established in many countries, such as Australia, Holland, France, Canada, U.S.A., Russia, Japan and so on, and highly sophisticated radio telescopes were put into operation. With the progress in establishing the Giant Metrewave Radio Telescope (GMRT) near Pune, India's efforts to gain prominence in the field of radio astronomy is likely to succeed.

For the better understanding of the physical nature of these radio sources, the necessity for their optical identification and distance estimation becomes evident. In this respect, the pioneering workers are Ryle and Smith of Cambridge and Bolton and his co-workers of Sydney. The question whether these radio objects were within the solar system or beyond it had to be settled first, and it was demonstrated by Ryle and Smith that the sources Cygnus A and Cassiopia A definitely lie beyond the solar system. The first optical identification of the radio sources with known astronomical objects was made by Bolton and his co-workers, in 1948–49. They identified Taurus A, Virgo A and Centauras A, the three of the six sources already discovered by them. The known astronomical objects with which these sources were identified were the Crab Nebula, and the external galaxies NGC 4486 and NGC 5128 respectively. Subsequent observations by radio astronomers revealed, that the nearby bright normal galaxies such as M31 were also sources of radio waves.

During the early 1950's Walter Baade and Rudolf Minkowski took the task of optical identification of radio sources and many sources were thus identified by them. Smith found the accurate radio location of Cygnus A. Baade and Minkowski focussed the world's largest optical telescope in this direction and found that the radio source was also faintly visible optically. From the radio positions of the different objects as had already been found by Cambridge and Australian workers, Baade and Minkowski not only succeeded in detecting optically the objects such as Cassiopia A, Puppis A, NGC 1275 etc., but also could confirm the identification of the Crab Nebula, Virgo A and Centaurus A as had already been made by Bolton and his co-workers.

The discrete radio sources were divided by J.H. Oort and G. Westerhout and by B.Y. Mills into *two* distinct groups, viz., the *galactic radio sources* and the *extragalactic radio sources*. Baade and Minkowski further classified these objects into four specific types of radio sources. According to them the galactic radio sources can be divided into (a) *supernova remnants* such as Cassiopia A, Taurus A, Puppis A etc., and (b) *hot* H II regions. The extragalactic radio sources were divided by them into (a) the *normal galaxies* of which M31, M101 etc. are examples, and (b) *strong radio sources* located in elliptical galaxies such as NGC 5128, NGC 1316, NGC 1275, etc. The last class of objects, i.e. the strong radio sources, emit several orders of magnitude more radio energy than the third class of objects, viz., the normal galaxies of which our Milky Way Galaxy is one. In general, the former class radiates 10^2 to 10^6 times more energy in radio waves than the latter class. Further, the former class radiates more energy, in general, in radio waves then in optical wavelength while, in the latter class the reverse is, in general, true. The galaxies which radiate more energy in radio wavelengths than in optical wavelengths have been called the *radio galaxies*. The classical example of such galaxies is Cygnus A which is only second in radio brightness, the brightest source being the supernova remnant Cassiopia A.

Our Milky Way Galaxy, emitting only a few times 10^{38} erg s^{-1} of radiation in radio wavelengths gives out a few times 10^{43} erg s^{-1} in optical wave lengths. Thus according to the definition, our Galaxy is *not* a radio galaxy but just an intrinsically weak radio source.

19.2 TECHNIQUES OF IDENTIFICATION OF RADIO OBJECTS

It is an enigma to the astrophysicists to identify the radio galaxies with their optical counterparts. The problem involves the determination of the location and distance of the radio sources. The estimation of the distance D of the radio galaxies can be done with the help of optical redshifts from Hubble's Law in the form

$$D = \frac{cz}{H}, \quad z = \frac{\Delta\lambda}{\lambda} \qquad (19.2)$$

c being the speed of light and H the Hubble's constant. The radio galaxies are quite distant objects and appear as faint sources of light when viewed even through the world's largest telescope. Yet these objects exhibit very strikingly the strong and sharp emission lines in their optical spectra.

Due to their enormous distances in general, most of the radio galaxies are of very small angular size with low apparent radio power. For the accurate location of their radio position, the radio telescope to be employed should have high power of angular resolution and radio sensitivity. In practice, there is a remote possibility of achieving higher angular resolution upto the desired extent by any radio telescope with single antenna. The minimum size of a steerable parabolic antenna that might serve the purpose of the radio astronomers of the day in about 1.5 km in diameter. Even then, the huge antenna will not be as much sensitive as to trace the more distant radio objects to be studied in near future. Such a large dish is not technologically feasible either. The present largest radio telescope of single dish antenna of 1000 ft in diameter is installed at Arecibo in Puerto Rico. This telescope is however non-steerable.

In order to achieve high angular resolution, at present there are some successful alternative methods eventualy to compensate for the huge antenna systems. These are:

1. Radio Interferometer
2. Aperture Synthesis
3. Lunar Occultation.

Radio Interferometer

The interference properties of light waves were first applied in astronomical calculation by A.A. Michelson in early this century. To determine the angular diameter θ of Betelgeuse, a bright star in the constellation of Orion, from Eq. (19.1), Michelson used the 100-inch reflector at Mount Wilson. In order to make passage of incident light adequate for a few wavelengths, D was kept apart to the extent of a few metres (Fig. 19.1). Such an arrangement in the 100-inch reflector enabled Michelson to achieve angular resolution of $\theta > 0''.01$; but most of the stars have angular width beyond the resolution limit of $0''.01$.

Following the same principle of Michelson's optical interferometer, the radio interferometer (as shown in Fig. 19.2) is constructed. The radio signals are received by two antennae which are separated at a distance of $d = m\lambda$, where m is a factor depending on the incident wavelength λ. The received signals exhibit the interference pattern at the receiver communicated by both the antennae.

The line joining the two antennae is termed as the *base line*. Its effective length depends not only on the spacing of the antennae but also on their orientation. At different lengths of the base line, i.e. at different spacing and orientations of interferometer elements, the receiver enables recording of the radio output which is then integrated by Fourier transform.

The Fourier transform of the radio intensity gives valuable information about the diameter and physical nature of radio sources. By increasing the value of d is achieved not only the finer angular resolution $\theta \sim 1/m$ but also the higher degree of sensitivity. Radio astronomers now seek angular resolutions of the order of one millisecond of are $(0''.001)$ in order to analyze the structures of very distant radio sources. Such a high resolution although sounds amazing, has been made achievable by constructing Very Long Baseline Interferometer (VLBI) system spanning over countries and continents.

Aperture = D

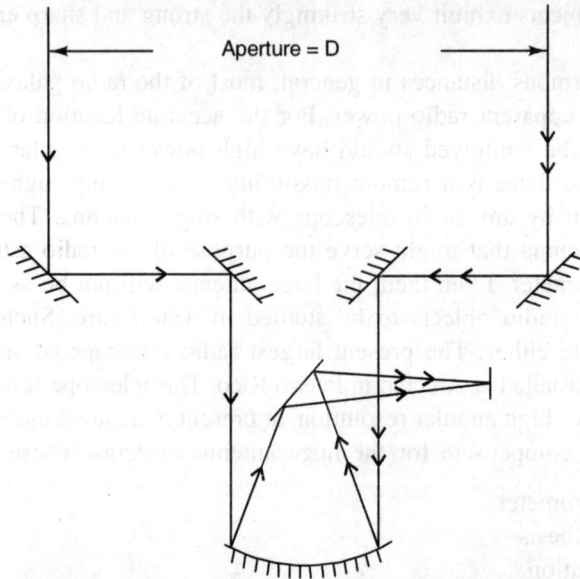

FIGURE 19.1 Michelson's optical interferometer.

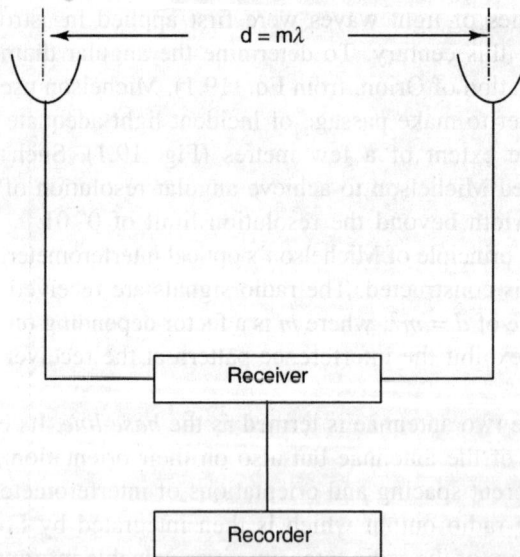

$d = m\lambda$

Receiver

Recorder

FIGURE 19.2 Radio interferometer.

The technique is used by the Jodrell Bank radio astronomy group. Their parabolic reflectors, one 250 ft in diameter is at Jodrell Bank and the other of 84 ft diameter at Royal Radar Establishment, Malvern are spaced 127 km apart. For such an arrangement $m = 2 \times 10^6$ for a wavelength of 6 cm giving a resolution of 0".1. The base line of NRAO (National Radio

Astronomy Observatory at Green Bank, West Virginia) system, the three-element radio interferometer is 8000 ft long which is workable at 3 cm, and the angular resolution achievable is about 2 seconds of arc. A highly sophisticated system of radio antennae is the Westerbork 12 element interferometer. It can achieve a resolution of 2.3 seconds of arc at 21 cm. Such a fine resolution yields radiographs which look almost like photographs of the radio galaxies.

Aperture Synthesis

Two or more variable spacing antennae of radio interferometer could be made to simulate a large aperture. If one could combine together the signals that are received by movable antennae at different positions, the effect produced would be equivalent to that of an aperture equal to the area covered by the movable antennae in question. This technique is known as *aperture synthesis.*

A sophisticated apearture synthesis system has been developed at Cambridge which is known as the "super-synthesis" radio telescope. It has three 18 m diameter paraboloid reflectors serving as the interferometer elements which are movable to a maximum distance of nearly 1.6 km. The system was subsequently further improved by adding a more extensive layout of aerials to provide better resolution and precision. The baseline was extended to 5 km which yields a resolution less than 1 arcsecond at 3 cm wavelength.

The two movable dishes out of its 12 antennae of the Westerbork Synthesis telescope while operating on the synthesis principle, need only five moves to complete a radio map of a source.

The most ambitious aperture synthesis system known as the Very Large Array (VLA) has been constructed by the National Radio Astronomy Observatory (NRAO) at New Mexico, U.S.A. The array is in the shape of Y and consists of 27 dishes each of 25 m diameter. Each of the three arms are 21 km long giving the synthesized telescope an effective diameter nearly 40 km. Such an orientation of the telescope will need no movement of its elements and will enable to have a complete map of a radio source in eight hours only since the positions overlap after 120° of the earth's rotation. This system detects objects with the resolution power of 1″ of arc at 21 cm and of 0″.05 at 1 cm. This is one of the best radio telescope synthesis system now available for accurate and detailed mapping. The Giant Meterwave Radio Telescope (GMRT), an ambitious interferometric system of 30 antennae, each of 45 metre diameter is in the process of completion in Pune, India, which holds a great promise for India's important contribution in this particular arena of scientific investigations.

Lunar Occultation

Sometimes an unwanted element also may serve a glorious purpose. In astrophysics, the Moon is such an object. It is considered as a disturbing element to the optical astronomers, but a very useful object for the radio astronomers.

The Moon in its journey round the Earth blocks the radio waves directed towards the Earth from radio objects like an opaque disc. By studying the change in intensity of the radio flux radiated from the source and the total blockade caused by its occultation by the Moon enables astronomers to determine the location and size of the source.

There are disadvantages too in this process.

1. Only those sources which lie on the lunar path can be studied.
2. The repetition of study of a particular source is not in general possible as it may be occulted only once in a year or more and that too for a very short period of time.
3. The uneven edge of the lunar disc may cause some uncertainty in calculations.
4. The Moon itself emits radio waves more intense than many weaker sources at short wavelengths.

19.3 STRUCTURES OF RADIO GALAXIES

The structures of radio galaxies are determined from the study of physical aspects observed in both the optical and radio wavelengths. The observations reveal that the radio structures of these galaxies are strikingly different from their optical counterparts. These structures are different irrespective of their optical forms. It has been already discussed how different radio techniques are employed to obtain the radio map or radiograph of a radio source, and also how the red-shift determines their distances. Now, we shall discuss as to how the polarization orientation and the wavelength dependence of the emitted radiation can be used to determine the structure of a radio source.

Polarization

The radio energy emitted from the radio galaxies originate from the synchrotron radiation process. It is known that synchrotron radiation is strongly polarized. It is believed that the condition prevailing in radio galaxies should produce as much as 70 per cent linearly polarized radiation from any region of uniform magnetic field. For such polarization, the maximum electric vector maintains the direction perpendicular to the projection of the magnetic field. Thus, the information about the magnetic field with respect to the structure of the galaxy is obtained from the polarized radiation emanated from the source. Also, the structure of the radio source can be derived from the polarization measurements. A high degree of polarization was observed by Miley, Wellington and Vander Laan in 1975. Observing the structure of the radio galaxy NGC 1265 at 5GHz, polarization distribution upto 60 per cent at some places was found by these authors.

Wavelength Dependence

It has been found that the appearances of the large double radio galaxies remain nearly unchanged when observations are made at different wavelengths. This implies that the distribution of relativistic electrons is similar throughout the various regions of these giant radio objects. For example, many authors have noticed that structure of the Centaurus A remains the same at intervals of 11 cm to 15 cm wavelengths although the total radio intensities vary with wavelengths of observation throughout the interval.

On the basis of interferometric observations made on 25 sources at 31 cm and 1.9 m, Moffet and Palmer came to the conclusion in 1965 that the radio brightness of a source is independent of the wavelengths at which the observations are made. But exceptions were seen in the cases of radio sources having core-and-halo type structure. Such objects give steeper spectrum from their halo regions than they do from their cores. The halos are more distinct at longer wavelengths. Similar observations were made earlier by Lequeux and Kellermann, These observers found that the sources having larger diameters have a tendency to exhibit steeper spectra than those of small diameters.

Structure

One of the earliest works on the discovery of the double component structure of radio galaxies was made by Jennison and Das Gupta (1953). They succeeded in resolving two radio components of the optical galaxy Cygnus A; each of these components are placed symmetrically on either side at a distance of 95″ of arc apart. The two-component structure theory gained momentum in 1962 when two teams, one consisting of Maltby and Moffet and the other of Allen, Brown and Palmer, found by analyzing 30 radio sources that most of the radio galaxies were double. Cygnus A or 3C 405 appears to be the strongest radio source whose estimated distance is about 320 Mpc. The radio-emitting regions are about 0.2 Mpc away on either side of the optical core. A schematic description of the double radio galaxy Cygnus A is shown in Fig. 19.3.

FIGURE 19.3 Diagram of the two-component radio galaxy Cygnus A or 3C 405 with centrally located optical core. (The radio lobes are each 200 kpc away from the optical core.)

Cygnus A provides with a typical example of radio-doubling with two identical components symmetrically placed about the optical component. Among other similar cases mention may be made of 3C 33, 3C 295, Fornax A etc. The second type of radio structure belongs to Centaurus A or NGC 5128.

With an estimated distance of about 5 Mpc, NGC 5128 emits a radio power of more than 10^{41} erg s^{-1}. The galaxy exhibits an interesting appearance both in respect of optical as well as radio wavelengths. Optically, the disc is divided by a dark absorbing band along its diameter, probably the band is constituted of dust particles. In radio observation, the galaxy has two large radio emission zones like Cygnus A. In addition to its two-component character, NGC 5128 is also spotted with two small intense radio sources located near the end of the line perpendicular to the dust band and passing through the optical centre. This type of radio galaxies therefore has in all four radio sources; two on the optical disc while the other two symmetrically placed on either sides of the optical component as they are in the two-component structures. Figure 19.4 describes the radio structure of NGC 5128.

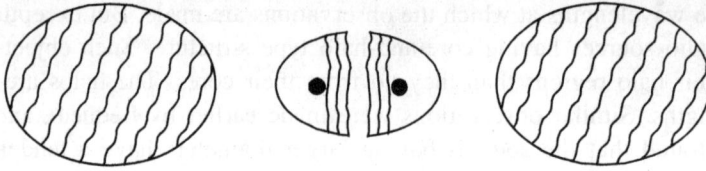

FIGURE 19.4   The structure of NGC 5128 or Centaurus A.

The Third type of radio structure observed is one having a strong but small radio source located at the centre of the optical galaxy. The example of such structure is NGC 1068, shown schematically in Fig, 19.5. This spiral galaxy emits 10^{40} erg s^{-1} of radio energy from a distance of 13 Mpc.

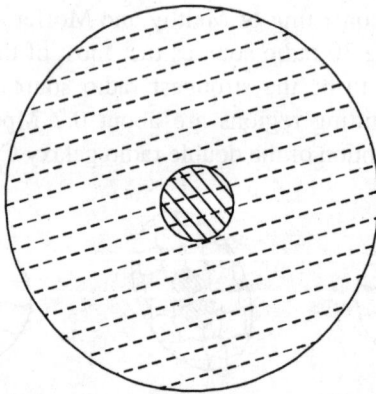

FIGURE 19.5   The radio galaxy NGC 1068 or 3C 71 or M 77.

The Fourth type of radio galaxy is represented by the structure of M 87. This type of radio galaxies has two concentric radio sources, both of which are smaller than the optical image (Fig. 19.6). But the smaller of the sources is very intense. M 87 ejects a jet from the central zone of intense radiation. It emits 10^{41} erg s^{-1} of radio energy from a distance of about 13 Mpc.

The Fifth type of structure of radio galaxies is represented by NGC 5457 or M 101. This type of radio source consists of a single object in which the optical body is smaller in size (Fig. 19.7). This well-developed spiral galaxy NGC 5457 is about 2.6 Mpc away from the Earth, and its radio power is within the limit of 10^{38} erg s^{-1}.

Among the available structures of the radio galaxies, the sixth and last type consists of two radio peaks which are almost symmetrically placed on either side of the optical galaxy and these two components are within the same radio envelope. The example of such structure is 4C 452 as shown in Fig. 19.8. The isophotes of 4C 452 show that it has two intense radio sources on either side of the optical galaxy resembling the radio doubling. These two sources again have a common radio envelope with weaker radio intensity. The structure of 4C 452 as shown in Fig. 19.8 was drawn by Ryle, Elsmore and Neville in 1965 with the help of Cambridge super-synthesis arrangement at 74 cm wavelength.

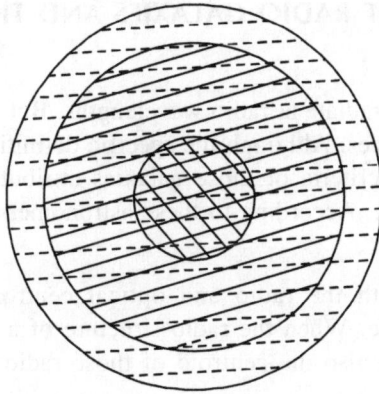

FIGURE 19.6   The structure of the radio galaxy M 87 or 3C 224 or NGC 4486.

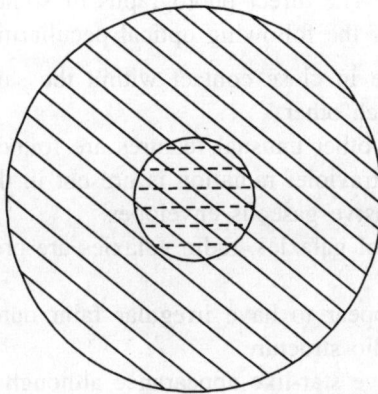

FIGURE 19.7   The structure of NGC 5457 or M 101.

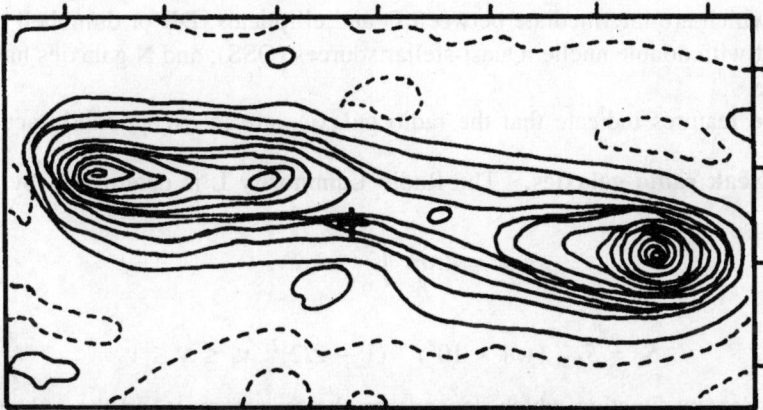

FIGURE 19.8   Isophotes of 4C 452. The peaks are 280 kpc apart as measured from the redshift.

19.4 CLASSIFICATION OF RADIO GALAXIES AND THEIR TYPICAL CHARACTERISTICS

We know that all the galaxies radiate at radio wavelengths. But all of them are not considered as radio galaxies or radio sources. Although the specific definition of a radio galaxy has been given in Section 19.1, a few criteria of those galaxies attributed by Mathews, Morgan and Schmidt will be discussed here. According to these astronomers a radio source should satisfy the following criteria:

(a) *Radio structure.* Both the radio and optical centroid of a radio source should approximately coincide. When the radio structure of a galaxy is double or multiple, the optical centroid is also the centroid of those radio components.

(b) *Optical spectra.* The majority of the sources give strong emission lines in their optical spectra. This phenomenon implies that such sources possess hot gas in turbulent motion.

(c) *Optical peculiarities.* The direct photographs of some of the extragalactic objects exhibit one or other of the following optical peculiarities:

   (i) Two galaxies are in close contact within the same envelope, as if, they are colliding with each other.
   (ii) Jets, plumes and other unusual features are found to be present.
  (iii) The excess of ultraviolet radiation is present in the sources.
   (iv) Some have extensive gaseous envelopes.
   (v) Within a cluster of galaxies, radio galaxies are prominent by their luminosities and sizes.
   (vi) Some sources appear to have irregular faint outer extensions which may be related to the radio structure.
  (vii) Some sources have star-like appearance although they are not actually stars.

On the basis of the observable features of radio galaxies, the following additional classes have been added in the classification of galaxies. Seyfert galaxies which are spirals with extremely bright nuclei, D galaxies, ellipticals with extensive surrounding envelopes; ED or DE galaxies which are intermediate between D and ellipticals; DB or dumbbell galaxies, like D galaxies but with double nuclie, Quasi-stellar sources (QSS); and N galaxies having brilliant starlike nuclie.

All these features indicate that the radio galaxies are in exceptionally active states.

Strong and weak radio galaxies. The Radio Luminosity L is determined by the equation

$$L = 4\pi r^2 \int_{v_1}^{v_2} S_v \, dv,$$

where

$$S_v = S_{400} \, (v/4 \times 10^8)^{-n} \, (1 + Z/2)^2, \quad v_1 \le v \le v_2 \tag{19.3}$$

S_v is the flux density function which depends on the frequency v and the red-shift $Z = \Delta\lambda/\lambda$ and includes two arbitrary constants S_{400} and n for the sources under observation. The values of v_1 and v_2 are taken as 10^7 and 10^{11} Hz respectively.

From the consideration of radio luminosity, the radio galaxies are divided into 'Strong' and 'Weak' sources. The radio, galaxies with luminosity $L > 10^{40}$ erg s^{-1} are called 'Strong' sources while those with $L < 10^{40}$ erg s^{-1} are termed as 'Weak' sources.

Among the radio galaxies those classified as D galaxies (D), dumbbells (DB), N galaxies (N) and Quasi-Stellar Sources (QSS) have $L > 10^{40}$ ergs s^{-1} and as such they are strong sources. Most of these sources have $L > 10^{41}$ ergs s^{-1}. The Quasi-Stellar Sources are the most luminous ones and their L ranges from $\sim 10^{44}$ to $\sim 10^{45}$ erg s^{-1}.

The N galaxies resemble the QSS in respect of their photographic appearances. But difference lies in their radio luminosities. The former are less luminous than the later.

19.5 ENERGY PROCESSES IN RADIO GALAXIES

The most distinctive characteristics of nonthermal radiation are that the intensity of radio emission increases with wavelengths and that the radiation is highly polarized. In thermal radiation caused by free-free transition both these characteristics are absent. In particular, thermal radiation tends to become stronger at shorter wavelengths. The study of the nature of radio radiation received from all the discrete sources exhibits the nonthermal character of the radiation except when the sources are diffuse emission nebulae located within our Galaxy. It was the Swedish Physicists, Alfvén and Herlofson, who first suggested in 1950 that the radio emission received from discrete sources was nonthermal in character, which was subsequently confirmed by the Russian astrophysicist Shklovsky.

Observations have shown that the intensity of the radio radiation from the galactic as well as the extragalactic sources is smoothly proportional to the frequency in the range 1 cm to 10 m. However, an exception is seen around 21-cm wavelengths where the intensity is enhanced due to the 21-cm emission of neutral hydrogen. The *spectral index* of the radiation energy is defined as the slope of the straight line obtained by plotting log (intensity) versus log (wavelength). The spectral indices of all the normal and peculiar galaxies are similar and are found to lie in the range between 0.2 to 1.2 with majority of the galaxies having values between 0.6 to 0.8. For example, the spectral indices of Cygnus A range from 0.5 at 6 m to 1.2 at 3 cm.

In spite of the diverse optical character of the radio galaxies, there are at least two physical criteria common to all of them. First, the similarity in the spectral character of the different sources implies that the radiation is produced by similar radiation mechanism. All the available data at hand at present suggest that the magnetobremsstrahlung is the most obvious candidate. Secondly, there must exist a process to maintain a continuous supply of relativistic electrons. In the absence of such a process, the dissipation of energy by radiation of the electrons would cause a wider range in the spectral indices, since the rate of loss of energy of a relativistic electron caused by synchrotron radiation is proportional to the square of its energy. Again, since the wavelength of radiation is inversely proportional to the square of the electron energy, the loss of energy will be faster at short wavelengths than that at longer wavelengths. So the decay of the energy of the relativistic electrons is accompanied by faster decay of short-wavelength radiation than the long-wavelength counterpart. This would necessarily lead to steeper spectral indices. Since this is not observed, a condition for continuous supply of relativistic electrons must be present in these sources. The condition of continuous supply is essential from the view point of the time-scale of radiation.

An important question that will eventually arise regarding these objects is as to how much energy in radio wavelengths is supplied by such an object in course of its lifetime? The amount will definitely depend on the rate of radiation and the extent of the lifetime. We consider a typical case say 3C 348 or Hercules A. The radio components of this galaxy are separated at a distance of 10^5 to 10^6 light years from its optical centre. If these radio components have travelled at the speed of the relativistic electrons, then they must have taken at least 10^6 years to come at its present position. Since a very strong radio galaxy emits energy at the rate of 10^{45} erg s^{-1}, the total radio energy emitted by a strong source should be at least 3×10^{58} erg. It may be even higher by a factor of 100. We shall see at the end of this chapter that the total energy radiated by a Quasar is higher than this by at least a factor of 10^4. They are the strongest radio emitters so far known.

19.6 RADIO GALAXIES IN EVOLUTIONARY SEQUENCE

There is no well-formulated theory on the origin and subsequent evolution of radio galaxies. But a few evolutionary models of the radio sources suitable to explain their observational properties have been suggested.

The radio galaxies show great diversities in respect of their structures and sizes. Each of these objects are the sits of violent events. So it is quite reasonable to assume that the radio emitting component or components of a galaxy have originated from its nucleus. The radio components then begin to recede upto a distance of 20 to 50 times the galactic diameter, finally leading to a process to be faded away into deep space. The radio emitting components are composed of energetic plasmas which have been termed as 'plasmons' by I.S. Shklovsky. According to his estimation, the plasmons expand with Alfvén velocity

$$v_A = H/(4\pi\rho)^{1/2}$$

where H is the internal magnetic field strength and p is the plasma density. The velocity with which the plasmons are expanding in the radio component must be less than one-fourth of the drift velocity of the component itself separating from the parent body. This is evident from the fact that the plasmon components in the radio double structures are separated by a distance of approximately 3.5 times their diameters.

In the process of double radio source evolution, some time at a later stage, the third and even the fourth component of relativistic plasmas are left behind near the parent galaxy. Or it may be that the later components may have evolved out of a more recent explosion occurred in the nucleus of an older radio galaxy which has already two radio components. From the structure of NGC 5128 or Centaurus A, it appears that this radio galaxy had experienced more recently a second outburst which produced two small intense radio sources on the optical disc. The possibility of a second occurrence is confirmed by indications found in the optical spectrum of the radio galaxies. Almost every radio galaxy exhibits emission line in their optical spectra even though there is no radio emission from the vicinity of the optical galaxy. The presence of emission lines conclusively proves the existence of continued disturbances in the galactic nucleus which may sometimes lead to a titanic event. Several theories have so far been proposed on the evolution of radio galaxies, only three of which are briefly discussed here. These are:

1. Collision hypothesis
2. Supernova activity
3. Early stage of galaxy formation.

Collision Hypothesis

The two radio galaxies NGC 4038–39 which are at a distance of about 22 Mpc away, belong to the class of peculiar galaxies. These galaxies emit 100 times more radio energy than any other normal radio galaxy does. The photographs taken with the 200-inch Hale telescope suggest that the two galaxies are under gravitational interaction. Observations further indicate that the radio energy output of these galaxies is much higher than their energy in optical wavelengths. The photograph of another peculiar radio galaxy 3C 405 or Cygnus A exhibits clearly two galactic nuclei in contact with each other, although their radio emitting regions are separated by a distance of about 200 kpc on either side of the visible galaxy.

The earliest theory for the formation and evolution of radio galaxies was based on such observations. It was believed that radio galaxies originate from two galaxies in collision with each other, as suggested by NGC 4038-39 and Cygnus A. The possible collisions of two galaxies might have generated highly energetic electrons associated with magnetic fields in these galaxies. These galaxies consequently acquired all the properties of a radio source and became radio galaxies. In fact, the optical spectrum of Cygnus A exhibits broad emission lines of highly excited atoms which characterize the existence of turbulent and high-temperature gas generated by the collision of two massive bodies.

The collision hypothesis is however vulnerable to a few questions. Firstly, how the impact takes place between two galaxies which are composed of so thinly spaced stars? Calculations show that the probability of a collision of two stars in the colliding galaxies is negligibly small. As the two galaxies pass through each other, either only a few stars or no star at all, will be subjected to collision with one another. The only possibility remains is that the gas clouds in the galaxies may collide to produce sufficient heat leading to intense radio and optical emissions. Calculations however reveal that the amount of radio energy released by a radio galaxy is far more than that could have been generated by collisions. Again, the radial velocities of the component galaxies that appear to be colliding with each other are found to be the same. It seems hardly possible that two field galaxies moving at random before collision may possess the same radial velocities. Lastly, the study on the structure of many radio galaxies shows that the radio sources existing in these are rather single.

These observed features of some of the radio galaxies weaken the colliding hypothesis and the astrophysicists are inclined to accept some other possible explanations.

Supernova Activity

Burbidge has suggested that radio galaxies may evolve as a sequence of supernova activity of stars located in the densely populated part of the galaxy. According to Burbidge, once a supernova explosion takes place, the intense radiation from the first explosion causes to set nuclear reactions in the atmospheres of other neighbouring stars. The nuclear reactions gradually extends to the interiors of the stars and lead ultimately to other supernova explosions. In this manner a chain of supernova explosions takes place in a galaxy, yielding high energy particles and strong magnetic field. The relativistic electrons that are responsible for the radio emission from a galaxy are produced either by the direct sequence of explosions or by the collisions of interstellar gases with the γ-ray photons produced in the process.

As a consequence of the supernova explosions in the galactic nucleus a single radio emitting component is originated. Then at the second stage the radio sources tend to recede in the form of a dumbbell in as many cases as they may give rise to radio doubling. The ends of the dumbbell are gradually separated and move in the opposite directions upto hundreds of kiloparsecs (Fig. 19.9).

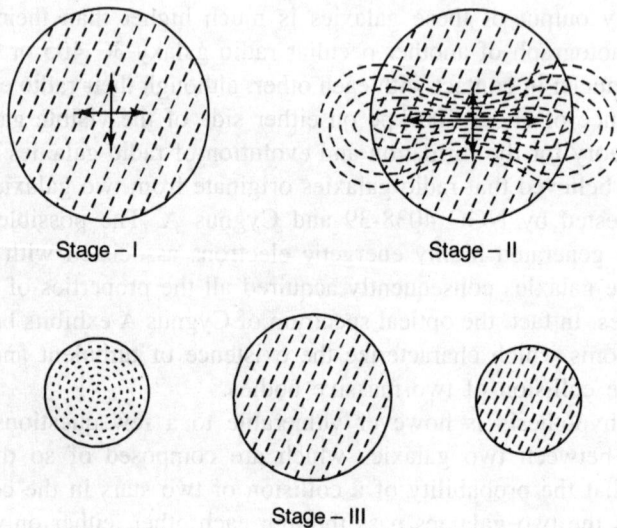

FIGURE 19.9 The different stages of radio doubling formation.

At stage–I, just after the explosion, the magnetic fields are developed and also the high energetic electrons are ejected out of the galaxy. But the electrons are trapped by the newly developed magnetic field and as they spiral along a helical path, they emit radio waves even from invisible regions.

Early Stage of Galaxy Formation

That all the radio galaxies are at the early stage of galactic formation has been propounded by the Soviet physicist V.L. Ginzburg. According to this process of galactic formation, the original gas cloud, said to be a protogalaxy, breaks up as it contracts into gas clouds to form the stars of the galaxy. As the galactic formation proceeds, the gravitational energy is gradually released which in turn produces cosmic rays. At this early stage when the star formation has not proceeded too far, the galaxy still contains enormous quantity of interstellar gas. The collisions between the thickly spaced gas atoms and the newly originated cosmic rays piercing from all directions produce relativistic electrons. The energetic electrons make the source a radio galaxy.

According to this hypothesis the radio emission process is rather of a short duration. The radio emission is supposed to be continued with the early stage of galactic evolution.

Calculations yield that the duration of intense radio emission from a source at 10 cm wavelength is ~ 10^5 to ~ 10^6 years, depending on the field strength. From 10^{-4} gauss field strength of Cygnus A, the age of the radio galaxy has been found to be 4×10^5 years. The age of one of the largest radio galaxies has been estimated from the time lapse of a relativistic electron which moves with Alfvén velocity from the galactic centre to the farthest point of the radio emitting region. It has been found that the minimum age of a large radio galaxy should be ~ 5×10^5 years. Thus it appears that every galaxy, during the process of its evolution, will emerge as a radio galaxy for a relatively short time, possibly of the order of a few times 10^6 years.

19.7 SOME IMPORTANT RADIO GALAXIES

Cygnus A

The first discovered radio galaxy is Cygnus A (3C 405). Cygnus A is also the first radio galaxy in which the radio doubling was found by Jennison and Das Gupta. The two radio emitting regions lie nearly 200 kpc away on either side of the optical nucleus which itself appears like two galactic nuclei in collision. The galaxy emits $10^{44.7}$ erg s^{-1} of energy at radio wavelengths from a distance of nearly 1050 million light years. The optical image of Cygnus A was first obtained by Baade and Minkowski in 1954. They identified the galaxy as the brightest member of a rich cluster and found strong emission lines. The optical form of Cygnus A has been designated by Matthews, Morgan and Schmidt as CD_3. The prefix C implies that Cygnus A is a large D galaxy and suffix 3 signifies the degree of lenticular appearance. The optical galaxy may have single nucleus crossed by dust lane as has been suggested by Shklovsky. But the alternative suggestion of Mathews, Morgan and Schmidt holds that the galaxy is with double nuclei.

Centaurus A

Centaurus A or NGC 5128 is the nearest strong radio galaxy situated at a distance of 5 Mpc from us. The radio envelope of the galaxy is not symmetrical about the axis of rotation and has widely different extensions in different directions measured from rotational axis. As regards the size of the radio galaxy, Centaurus A is extended to 8° by 4°—the largest measure for any such object. The large angular size of Centaurus A enables easy resolution by the pencil beam antennae. This resolution has yielded a more definite radio structure than that of any other radio galaxy. It has been stated already that the galaxy has two pairs of radio sources:

1. A pair of more intense radio sources is located within the optical galaxy.
2. The other pair of less intense but larger radio components are separated by a distance of 600 kpc.

The galaxy emits an energy of 7×10^{41} erg s^{-1} in radio wavelengths and has spectral index of 0.6. The central pair of intense radio sources which is located on the optical disc emits approximately 20 per cent of the total radio emission. An appreciable amount of polarization of light is originating from the north-east component of the central pair of intense radio sources. It is conjectured that the south-west component of the central pair lies behind the galaxy and the radiation from this component is depolarized by its passage through the body of the galaxy. At 10 cm wavelength, the polarization has been found to be 10 to 20 per cent which in places attains 40 per cent.

The dust lane in the optical structure of NGC 5128 is seen to lie from south-east to the north-west diametrically. The Burbidges found that the gases in the dust lane possess recessional velocity of 100 km s^{-1} with respect to the stars of the galaxy. They also found the dust lane rotating at a greater speed than the galaxy itself. The Burbidges suggest that the dust lane is the remnant of the same violent event which created the radio galaxy. The ejected materials in the dust lane are now falling back to the galaxy under the gravitational pull. The radio galaxy NGC 5128 belongs to the optical form of ED-type.

Fornax A

The radio galaxy Fornax A was identified by Shklovsky and Kholopov in 1952 and again by G. de Vaucouleurs in 1953. These authors suggested that the radio source is associated with the optical galaxy NGC 1316. Their identification was shrouded in doubt as the unexpected double structure of unequal radio intensities eluded the actual determination of the location of the radio galaxy. In 1961, Waade was able to depict the basic structure of the radio source. Gardner and Price in 1964 investigated at 74, 21 and 11 cm wavelengths to give a more detailed structure of Fornax A. These latter observers found the structure similar at all the three wavelengths, but the western component was found 50 per cent stronger than its companion. From the isophotes of both the components, variations in the distribution of radio brightness was also detected. The peak of the stronger component is rather fiat and the contours are closely spaced around the peak. The diameters of the radio components are 1.7 times greater than the distance by which they are separated. The radio energy of NGC 1316 is $\sim 10^{42}$ erg s^{-1}.

Virgo A or M87

In 1951, Bolton, Stanley and Slee identified the giant elliptical galaxy M87 with the radio source 3C 274 or Virgo A. The galaxy is at a distance of 16 Mpc and lies as the brightest member of the Virgo cluster at its centre. As it has been already stated, the structure of this radio galaxy is composed of an intense radio core surrounded by an elliptical radio halo covering 12' by 15'— somewhat larger than the optical halo. Recent radio maps indicate that the radio halo possesses a double structure with one large lobe nearly overlapping the other. The radio and optical power radiated by Virgo A amounts to $\sim 10^{43}$ erg s^{-1}, both being caused by synchrotron radiation.

A bright blue jet is seen to be extended from the location of the double core in position angle 290°. Baade measured the polarization of light from the jet to the extent of 30 per cent which suggested that the radiation was due to magnetobremsstrahlung. The length of the jet extends through 20″ or about 1.6 kpc in linear scale in comparison to the diameter of the optical halo of the galaxy of dimension about 30 kpc. The core is well observed at short wavelengths; and as radio spectrum of the halo differs from that of the core, it gives good impression at longer wavelengths. In fact, the spectral index of the core is 0.4 and that of the halo is 1.0. But at 40 cm both the regions emit radiations of equal intensities.

The optical spectrum of the jet shows neither absorption nor emission lines but a continuous spectrum. The galaxy has only weak emission line λ 3727 of [O II] originating from the nucleus. D.E. Osterbrock found that the emission feature is confined to a small region of 1″ only and the feature has been found to be strong and broad with asymmetrical configuration. The main component of the feature moves with the same velocity as the nucleus. The weaker component shows 11 A violet-shift corresponding to a velocity of 900 km s^{-1} relative to the stars of the galaxy. M87 is also the first known galaxy to emit strong X-rays. The X-ray source is extensive—covering almost 1° and radiating an X-ray power of ~ 5 × 10^{43} erg s^{-1}. The mechanism of X-radiation is probably thermal bremsstrahlung from the gas hotter than 10^7 K, forming an extensive halo around the optical galaxy.

19.8 SEYFERT GALAXIES

In 1943, Carl K. Seyfert singled out a group of spiral galaxies such as NGC 1068, NGC 1275 etc. possessing very different characteristics from other normal spiral galaxies. These galaxies possess small and very bright nuclei that give very strong and broad emission lines, although in outer, spiral structure they resemble normal spiral galaxies. These galaxies therefore form a class by themselves and have since been called the Seyfert galaxies after the name of their pioneer investigator. The N galaxies are the ellipticals having Seyfert-like nuclei in appearance, but the two types of nuclei differ in spectral characteristics.

On the basis of differences between the nature of spectral lines of the nuclei, Seyferts have been subdivided into two subclasses: the Seyfert I and the Seyfert II. Seyfert I have broad Balmer lines of hydrogen but narrower high excitation forbidden lines such as [O III] and [Ne III], NGC 5548 is a typical example of this class. Seyfert II are characterized by nearly equal widths of hydrogen and forbidden lines. A typical example of Seyfert II galaxies is NGC 1068. The high excitation forbidden lines indicate an electron density of ~ 10^4 cm^{-3} and electron temperature of ~ 2 × 10^4 K. The widths of the lines are commensurate with Doppler velocities of (1–5) × 10^3 km s^{-1}. Since only one per cent of spiral galaxies are observed to possess Seyfert nature, it is believed that the Seyfert phenomena last for about 10^8 years, that is, one per cent of the Hubble time. Some Seyfert galaxies, such as NGC 4151 is so much fainter than its starlike nucleus that it appears to maintain a link between the normal galaxies on the one hand, and the N type and strong radio galaxies, on the other.

Thus from the point of view of evolutionary sequence, Seyfert galaxies might be the radio galaxies in the making. It is believed that the milder form of explosive events characterize the Seyfert galaxies whereas the radio galaxies are associated with more violent phenomena. The greater the intensity of violent events, the greater the release of energy at radio wavelengths. Such radiation has been found to emanate from a small region of Seyfert galaxies; but in radio galaxies this emission comes from the entire lenticulars or even from a larger area in extreme cases. Thus, it is presumed that the Seyferts are the starting phase in the life history of radio galaxies.

It has been found that the absolute magnitude of the Seyfert galaxies ranges from −18 to −22. The luminosities have been found to be in the range from 10^{43} to 10^{45} erg s^{-1} in the optical range. These galaxies are extremely luminous in infrared, emitting 10^{45-46} erg s^{-1}. This luminosity and the predicted age of 10^8 years imply that a Seyfert galaxy emits energy in the range of 10^{60-61} erg during its lifetime.

In the outer regions beyond the zones of broad emission lines, the line profiles of [N II] λ 6583, [O II] λ 3727 and H reveal the presence of more quiescent gas there. Spiral structures are the principal feature of the outer parts. The nuclei of these galaxies are asymmetrical because of the fact that the gases in the nuclei move in a noncircular motion. Outside the nuclear region nearly circular motion of material prevails. Rotation curves of some of these galaxies have been deduced and masses calculated therefrom yield values a few times 10^{10} $M_\odot$. The turbulent motion of gases in the nuclei attain the velocity much larger than the velocity of escape. That the gases are escaping from the nucleus of NGC 1068 has been found from its photograph taken by H$_\alpha$ filter. In galaxies NGC 1068 and NGC 7469, faint outer rings at a radial distance of 30 kpc are observed lending support to the concept of gas escape.

The galaxy NGC 1275 shows Seyfert criteria. In addition, there exists a mass of gas outside the nucleus. The spectrum of the gas shows discrete set of emission lines. The analysis of these lines shows that these gases are moving with a velocity of 3000 km s^{-1}, relative to the nucleus. It is believed that some violent event at the centre of the galaxy has thrown out the gas with velocities as high as 3000 km s^{-1}. A broad structure of these thrownout materials is now present at a distance of 10 to 20 kpc from the site of the violent event.

Among the Seyfert galaxies only two, NGC 1068 and 1275 satisfy the criteria for a radio galaxy and another four, NGC 3227, 4051, 4151 and 7469 have been found to possess the weak radio sources. The energy emitted at radio wavelengths from NGC 1068 is ~ 10^{40} erg s^{-1} and that from NGC 1275 is ~ 10^{42} erg s^{-1}. The centres of some Seyfert galaxies have been observed to be strong X-ray sources.

We now discuss in some details some of the best observed Seyfert galaxies.

NGC 1068

One of the best studied Seyfert galaxies is their brightest member NGC 1068. The galaxy conforms to the radio source 3C 71 and emits radio energy of ~ 10^{40} erg s^{-1} from a distance of ~ 13 Mpc. The brightness and appearance of the galaxy is nearly like those of a normal spiral. By analyzing the rotation curve Burbidge, Burbidge and Prendergast calculated its mass to be 2.6×10^{10} $M_\odot$ which is centralized within a radius of 2 kpc. From the emission line profiles M.F. Walker concluded that there are discrete clouds of materials moving with velocities ranging from −2500 to +2500 km s^{-1} relative to the centre. Materials are being ejected from the nucleus in discrete clouds of sizes ~ 200 to 350 pc in diameter having masses in the range of 10^6 to 10^7 $M_\odot$.

The spectroscopic investigations reveal that the ultraviolet flux originating from NGC 1068 is more intense man that of any normal spiral galaxy. Viswanathan and Oke were able to resolve two components in continuum of NGC 1068. While one of the components has its origin from stars and gases, the other component is presumed to be of synchrotron radiation. The polarization of the optical spectrum has been studied by several authors. It was noticed that the degree of polarization varies inversely with the wavelengths; as such it is higher in the violet region and decreases with the increase of wavelength.

NGC 1275

The galaxy NGC 1275 has strong radio emission of $\sim 10^{42}$ erg s^{-1}. This radio source is designed as 3C 84 or Perseus A. The galaxy is the brightest member of Perseus cluster having an absolute photographic magnitude of –22.2. The galaxy is of the optical type ED$_2$ and spaced at a distance of 110 Mpc.

The optical structure of NGC 1275 is not a regular spiral—a deviation from all other classical Seyfert galaxies. The central part of the galaxy is bright and extended to 8 kpc in diameter. Outside this bright central part, there are absorbing clouds and also extended features. Studies on the motion of clouds reveal that the gas located at a distance of 10 kpc from the nucleus has a velocity of ~ 3000 km s^{-1} relative to the centre. Most probably, the outer filament has been formed by the gas ejected from the nucleus. It is believed that an explosive event occurred in the nucleus of this galaxy very recently.

The radio structure of the source 3C 84 is complex. The radio studies reveal that the source is extended over a distance of ~ 1 Mpc ranging throughout the Perseus cluster of galaxies. Kellermann and Pauliny-Toth (1968) found as many as three radio sources centred on the galaxy. The largest one of these sources is extended over a size of 80 pc having a typical structure that can be observed in any strong radio galaxy. The size of the second component lies somewhat between 8 to 26 pc, and has a maximum flux density near 800 MHz. M.H. Cohen and co-workers estimated the size of the third component to be larger than 0.25 pc.

NGC 4151

The nucleus of the galaxy NGC 4151 is very bright. A large number of emission lines are observed in the spectrum of this galaxy. In fact, of all the Seyfert galaxies NGC 4151 gives the largest number of lines. These lines have been studied indicating that the widths of the forbidden lines are equivalent to ~ 1000 km s^{-1} while those of Balmer lines correspond to ~ 7500 km s^{-1}.

Studies of the relative intensities of lines by Oke and Sargent reveal that the spectrum of NGC 4151 gives the emission lines in the regions where the average electron density $N_e \approx 5000$ cm^{-3} and the electron temperature $T_e \approx 30000$ K. The total gaseous mass in these regions is about $2 \times 10^5\, M_\odot$. An absorption feature comprising three components corresponding to velocities of –280, –550 and –840 km s^{-1} relative to emission lines was observed. These observations reveal that three discrete shells of gas are being ejected from the nucleus with velocities just mentioned; the rate of ejection of matter has been found to be between 10 to 100 $M_\odot$ per year.

The galaxy is a weak radio source having a radio size of 90 pc located in the nucleus.

EXERCISES

1. Give an account of how the invisible radio universe gradually unfolded itself to the astronomers.

2. Describe the various types of discrete radio sources. In what ways the discrete radio sources of galactic and extragalactic origins differ? Name some prominent galactic radio sources.

3. Why some galaxies are called "radio galaxies"? Discuss the principal differences between a radio galaxy and a normal galaxy. Give five examples of each kind.
   What is the mechanism of energy generation in radio galaxies?

4. The optical identification of extragalactic radio sources is a difficult problem: Narrate briefly how this is done by astronomers.
   (a) Comment on the advantages and disadvantages of each of the techniques used for the purpose.
   (b) Name some of the important centres where these techniques have been highly developed.

5. It has been observed that radio galaxies greatly differ in structure among themselves: Discuss the various kinds of structure of these galaxies indicating prominent features and citing prominent examples in each case.

6. (a) Enumerate the optical peculiarities as are observed in radio galaxies,
   (b) Name different classes of radio galaxies having some prominent optical peculiarity as in (a).

7. Distinguish between *strong* and *weak* radio galaxies. Classify these galaxies into strong and weak sources.

8. How do we know that the energy radiated by radio galaxies is nonthermal? What is the spectral index of this radiation? How can we calculate the total radio energy emitted by a strong radio galaxy?

9. Among the many theories that have been proposed on the evolution of radio galaxies, the following three are considered to be important:
   (a) Collision of two galaxies
   (b) Supernova activity
   (c) Early stage of galaxy formation.
   Briefly describe each of these theories and comment on the strength and weakness of each of these. How can one calculate the age of a radio galaxy?

10. Write what you know of the radio galaxy Virgo A.

11. Discuss briefly in what respects Seyfert galaxies differ from normal spirals. On what basis Seyferts have been subdivided into Seyfert I and Seyfert II?
    How can we estimate the lifetime of Seyfert galaxies?

12. Describe in detail the observable features of the Seyfert galaxies NGC 1068 and NGC 1275. In what respect these two galaxies can be singled out of the class?

SUGGESTED READING

1. Dalgamo, A. and Layzer, D. (Eds.), *Spectroscopy of Astrophysical Plasmas,* Cambridge University Press, Cambridge, 1987.

2. Davis, R.J. and Booth, R.S. (Eds.), *Sub-arcsecond Radio Astronomy,* Cambridge University Press, Cambridge, 1993.

3. Evans, D.E. (Ed.), *External Galaxies and Quasi-Stellar Objects,* IAU Symposium No. 44, D. Reidel Publishing Company, Dordrecht, Holland, 1972.

4. Heeschen, David S. and Wade, Campbell, M., *Extragalactic Radio Sources,* IAU Symposium No. 97, D. Reidel Publishing Company, Dordrecht, Holland, 1982.

5. Hey, J.S., *The Radio Galaxy,* Pergamon Press, New York, 1971.

6. Kurganoff, V., *Introduction to Advanced Astrophysics,* D. Reidel Publishing Company, Dordrecht, Holland, 1980.

7. Moffet, Alan, T., The Structure of Radio Galaxies, *Annual Review of Astronomy and Astrophysics,* 4, p. 145, 1966.

8. Novikov, I.D., *Evolution of the Universe,* Cambridge University Press, Cambridge, 1983.

9. Pearson, TJ. and Zensus, J.A. (Eds.), *Superluminal Radio Sources,* Cambridge University Press, Cambridge, 1987.

10. Roland, Jacques, Sol, Helene and Pelletier, Guy (Eds.), *Extragalactic Radio Sources,* Cambridge University Press, Cambridge, 1992.

11. Shu, Frank, H., *The Physical Universe,* University Science Books, Mill Valley, California, 1982.

12. Smith, Elske V.P. and Jacobs, K.C., *Introductory Astronomy and Astrophysics,* W.B. Saunders Company, Philadelphia, 1973.

13. Steinberg, J.L. and Lequeux, L., *Radio Astronomy,* Bracewell, R.N. (Tr.), McGraw-Hill Book Company, New York, 1963.

14. Verschuur, G.L., *The Invisible Universe,* The English University Press, London, 1974.

15. Verschuur, G.L. and Kellerman, K.I. (Eds.), *Galactic and Extragalactic Radio Astronomy,* Springer-Verlag, 1988.

20 Quasars

20.1 THE DISCOVERY

The most fascinating subject of Astrophysics is attributed to the discovery of a few extra-galactic objects in early 1960s. These objects are starlike in appearance, but contrary to the stellar phenomenon they are found to be strong sources of radio waves. For this, these objects were named *Quasi-Stellar Radio Sources* by Maarten Schmidt and later abbreviated as QUASARS. A more general term Quasi-Stellar Objects (QSO) has been adopted by astrophysicists to include both the Quasi-Stellar Radio Sources (Quasars) and Quasi-Stellar Galaxies (QSG) which are radio-quiet quasi-stellar sources. Although, in optical appearance the two groups of objects resemble each other, while the former is believed to be the strongest sources of radio emission, the latter is not recognized at all as radio sources.

The discovery of quasars was the result of the most successful joint venture of the radio and optical astronomers. With the Radio interferometer of the California Institute of Technology, USA, accurate positions of many radio sources in the Third Cambridge Catalogue (3C Catalogue) were obtained. At the same time, measurement with the *Very Long Baseline Interferometer* (VLBI) at the Jodrell Bank conclusively demonstrated that some of these sources, such as 3C 48, 3C 273 etc. had diameters of the order of 1″ or less. With fairly accurate positions known for the radio sources, photographs taken with 200-inch reflector at Mount Palomar Observatory helped identification of some of these sources with starlike objects. The first source to be so identified was 3C 48 followed quickly by the sources such as 3C 147, 3C 196, 3C 286 and 3C 273. The apparent visual magnitudes of 3C 48 and 3C 273 are 16 and 13 respectively. Two striking characteristics were noticed: *firstly,* these sources have very small angular sizes with appearance like stars and *secondly,* their radio surface brightness was extremely high. These starlike objects drew attention of Sandage who succeeded in obtaining the spectroscopic and photoelectric measures of 3C 48. It was noticed that the continuum light was exceptionally blue and the emission lines were very broad and of extremely peculiar nature—unknown in any other celestial objects.

While the optical identification of 3C 48, 3C 196 and 3C 286 was made by Sandage, the fourth object 3C 273 was identified by Schmidt from one of the 200-inch plates. Schmidt identified 3C 273 with a 13th magnitude stellar object associated with a jet. The radio location used by Schmidt was substantiated by Cyril Hazard and his colleagues. In 1962, within a span of six months, the radio source 3C 273 suffered three lunar occultations. These occultations of 3C 273 enabled Hazard and his co-workers to obtain its accurate radio location in the constellation Virgo. In addition to the high brightness temperature and small angular diameter like 3C 48, these observers found 3C 273 to have two radio components. But the most exciting discovery was performed by Schmidt, In 1963, Schmidt's study revealed that this starlike object showed large redshift in its broad emission lines. These lines were identified with Balmer hydrogen lines and the Mg II λ 2798 line redshifted by an amount $Z = 0.158$. Inspired with this achievement, in the same year, Jesse L. Greenstein and T.A. Mathews re-examined the spectra of 3C 48 previously taken. They found to their expectation that the emission lines of this source were redshifted by an amount $Z = 0.367$. These starlike radio sources were thus known to possess extremely high amount of redshifts hitherto unknown in any other heavenly objects. If the measured redshifts were cosmological to obey Hubble's law, then the corresponding distances are nearly 920 Mpc and 2150 Mpc respectively. The radio and optical powers measured for these objects from such great distances puts their luminosity in the range 10^{44-46} erg s^{-1}, at least 100 times more than any normal galaxy.

20.2 RADIO PROPERTIES

Hazard and his co-workers detected in 1963 for the first time that the source 3C 273 was double with a compact source, less than 1″ in size (3C 273 B), and an extended source (3C 273A) 20″ away. Subsequent investigations by various observers have revealed the double structure of many other quasars. Extensive studies have been carried out in this respect on 3C 147, 3C 9, 3C 204, 3C 280 and some others. In all cases double sources separated by order of 100 kpc distances have been established.

A pair of optical quasars have been discovered by E.J. Wampler and co-workers in 1973. They found that the object was associated with the radio source 4C 11.50. The brighter component 4C 11.50a is a quasar of 17th magnitude having a redshift of $Z = 0.436$. The companion component 4C 11.50b, separated at a distance of 4.8″ has brightness of 19th magnitude and displays much larger redshift of $Z = 1.901$. Hazard and his co-workers found that the source 4C 11.50 is double in radio structure, one of which is superimposed on the optically brighter quasar. Wardle and Miley (1974) studied 30 quasars with the help of NRAO interferometer at 11.1 cm and 3.7 cm wavelengths. They found that out of 22 well-resolved sources 11 have more than two components.

Spectral Index

It has been already stated that spectral index of a radio source is defined from the relation $P(v) \propto v^{-\alpha}$ where $P(v)$ is the flux density related to the source at a frequency v. The median value of the spectral index α which is distributed in the range from 0.2 to 1.3 has been found to be near 0.7 for the quasars. The spectral index is governed by the source dimension which increases with its age. Thus the spectral indices become steeper with the advancement of age of the quasars. The spectral character of the double radio component associated with 3C 273 has been found to be different. The source 3C 273A has a spectral index $\alpha = 0.9$ and in 3C 273B the value of $\alpha = 0.7$.

20.3   OPTICAL PROPERTIES

The quasars are remarkably distinct from the normal stars in respect of having a strong ultraviolet emission. Due to photon absorption by the Balmer lines near 3700, the normal main sequence stars do not show ultraviolet excess like quasars.

The colour-colour diagram of QSOs as compared with that of normal stars is shown in Fig. 20.1. The main sequence stars will yield a curve below the blackbody line AE. The quasars are placed on either sides of the black body line, but all are placed above the main sequence curve. Although more than 90 per cent of all stars occupy the regions close to the curve, the presence of white dwarfs in between the line AE and the curve makes it difficult to isolate quasars from white dwarfs in this region by observation of colour distribution alone. But the quasars placed above the line are unique and this region is not crowded by any other remarkable astronomical objects. Such a unique feature of quasars comes to great help in identifying these among all optical objects.

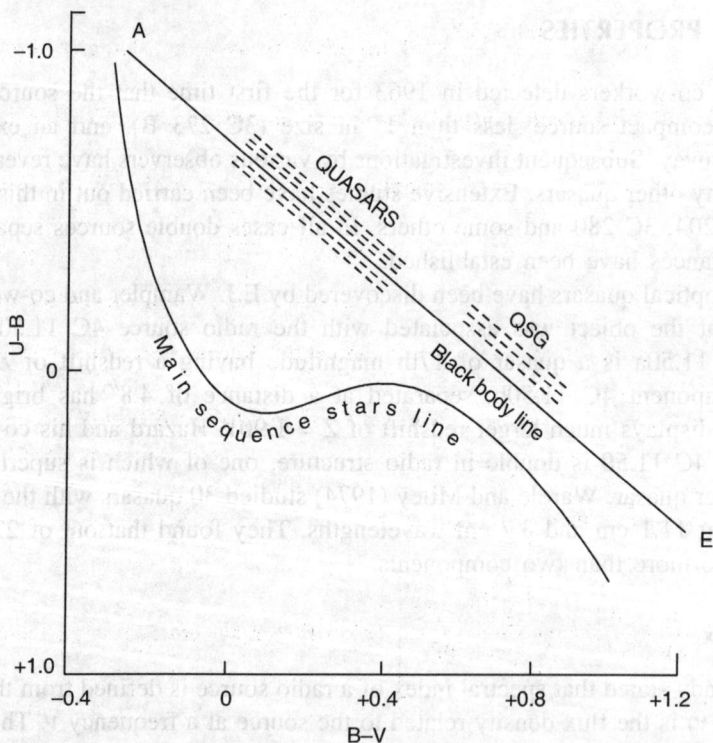

FIGURE 20.1   U-B B-V diagram with the positions of Quasars and Quasi-Stellar Galaxies (QSG).

With the help of this technique, Sandage came to know a new class of radio-quiet objects. In colour-colour diagram these objects lie above the low temperature part of the blackbody line. These radio-quiet Quasi-Stellar Objects are now known as Quasi-Stellar Galaxies (QSGs). These galaxies may be later stage of those objects which have passed through quasar phenomena. According to the present estimation, the Quasi-Stellar Galaxies exceed the quasars in population by a factor of ~100.

The angular diameters of quasars are so small that they are all less than 1″. Nevertheless, these objects have generally been found to vary in optical luminosity. The mode of variation of some of these objects is quite complex, often showing multiplicity in periods in a single object. The fluctuation in optical magnitude by an extent of ~ $0^m.4$ over a period of 13 months was first observed accurately in 3C 48 by Mathews and Sandage. They also found a secondary optical variation of smaller measure, i.e. ~ $0^m.04$ in the same object in a very small time interval of 15 minutes only. Examination of photographic plates taken over 80 years revealed that the brightness of the quasar 3C 273 also suffered variations in brightness by 2 to 3 times over relatively short periods. A general survey by various astronomers, predominantly by Sandage, has established that most of the quasars undergo light variation by different amounts over short periods, such as weeks or months superimposed by longer periods of variation of the scale of years. Such fluctuations clearly imply that the sizes of the sources must be small, of the order of a light year, because no disturbance can be propagated through the entire source with a speed faster than light. Thus, quasars are compact objects surrounded by hot gaseous envelopes where the emission lines originated. The widths and intensities of forbidden lines indicate the measure of density and temperature of the gas, i.e. N_e ~ 10^5 cm^{-3} and T_e ~ 2×10^4 K respectively. Large to relatively small variations in radio fluxes are also common to quasars.

A spectroscopic study of a large number of quasars with large redshifts reveals among other features, strong emission lines of Lyman α, C IV λ 1549 and Mg II λ 2798 in common. In quasars with small redshifts, Balmer lines and forbidden lines of [Ne III], [Ne V], [O II], and [O III] are commonly observed. Although the emission lines are the most prominent feature of quasar spectra, in many of them prominent absorption lines are also observed side by side with emission lines. Large number of absorption lines have been measured in quasars 3C 191, PHL 938, PKS 0119-04, 4C 05.34, PKS 0237-23 etc. in the wavelength range λ 3300 to λ 6000. The most surprising facts revealed from spectroscopic study of quasars are (a) very large redshifts in some of them and (b) widely differing redshifts revealed by different lines in the spectrum of the same object. For example, the spectroscopic investigation of the object PKS 0237-23 by various authors have revealed altogether as many as seven redshift systems by the emission and absorption lines of this object. And the redshifts measured in this object are all, in general, quite large, the largest being Z ~ 2.2015. Still higher redshifts have been measured in some other quasars, and quite a few quasars are now known having $Z > 3$.

20.4 THE REDSHIFT OF QUASARS

One of the best criteria for the identification of quasars is the detection of large redshifts Z, exhibited in their optical spectra.

The redshift can also be deduced from Einstein's theory of general relativity. In this case, the redshift of an object whose mass is M and radius R is given by

$$Z = \frac{GM}{Rc^2} \qquad (20.1)$$

where G is the gravitational constant and c is the velocity of light. Computing the value of G/c^2 in the unit of $M_\odot$, the redshift Z is given by

$$Z = 1.47 \times 10^5 \frac{M}{RM_\odot} \qquad (20.2)$$

The redshift thus estimated from the mass and size of an object is known as the *gravitational redshift*.

Some authors are of the opinion that the large redshifts observed in the spectra of quasars are cosmoiogical, that is, due to the expansion of the Universe. When the receding velocity v has been measured, the distance of the source is given by the Hubble's law (see Eq. 17.8). The observed values of redshift with a range > 3 yield that the quasars are receding with high speeds and lie at very great distances, if the observed redshifts are assumed cosmoiogical.

The observed luminosities combined with distances of the quasars, if assumed cosmological, imply that these objects must be very highly luminous. This fact, combined with their small size as suggested by periods of their variation imply further that they are highly compact objects.

Some authors have argued that the quasars are local objects. The question is how close they may be? They appear to have no proper motions. An object showing no proper motions must be beyond the distance of 30 kpc (~ 10^5 light year). So the quasars, even if local objects, cannot be closer than this distance. Some others have suggested that the seat of the origin of quasars is the centre of our Milky Way Galaxy at a distance of 10 kpc from the Sun. This view has been modified later by shifting the site of explosion from our Galaxy to the nearby galaxies. According to this view, the redshifts of quasars are Doppler shifts, but not due to expansion of the Universe (that is, the redshifts are not cosmological). The quasars, after being ejected from the centres of galaxies, move at relativistic speeds and exhibit Doppler shifts in spectral lines. To explain the fact that the objects show only redshifts and no blueshifts, it has been suggested that due to very large ejection velocities the quasars pass across the observer within a time quite short compared to the time elapsed since their ejection. This will ensure that the quasars which were approaching the observer (producing blueshift) will soon appear to be receding (producing redshift).

Such an explanation of redshifts of quasars is, of course, vulnerable to a few objections. The principal of these are: (a) how such a coherent blob of matter can be ejected at a relativistic speed is not yet well understood, and (b) the event leading to the ejection of such matter should be associated with very large energy release. The energy required to eject a bunch of quasars with observed velocities is ~ 10^{64} erg. It is very difficult to explain events involving such colossal energies.

Some authors have also suggested that the redshifts observed in QSOs are gravitational. Greenstein and Schmidt attributed gravitational redshifts to the QSOs 3C 48 and 3C 273 on the basis of either of the following reasons: (a) these objects are collapsed stars of our Galaxy, and (b) these are collapsed galaxies. But such gravitational redshifts have been suggested to be very unlikely by most authors, Moreover, the emission of forbidden lines remains unexplained under this concept. Hoyle and Fowler in 1967, proposed that the QSOs were clusters of collapsed objects. Such objects can show redshift up to the limit of 3 at the central region.

A combination of cosmological redshifts and gravitational redshifts has been proposed by G.R. Burbidge and F. Hoyle in 1967. They have suggested that QSOs having small redshifts are of the cosmological origin and those having large redshifts are mostly gravitational.

For two decades after their discovery, cosmological redshift for quasars has swayed the opinion of the astronomers. This situation is however changing on the basis of some more recent observations. Some astronomers now believe that the quasars do show anomalous redshifts, that are not accountable on the basis of the Hubble law. Holton Arp has reported in 1987 three quasars having redshifts of 0.34, 0.95 and 2.20 arranged in a triangle around the galaxy NGC 3842 having redshift 0.020. The angular separations of these quasars are of the order of 60–100 arc seconds. Since the quasars are believed to be randomly distributed, the probability of finding three quasars at such small angular separations is $< 10^{-5}$. A more recent (1990) investigation by G.R. Burbidge and his co-workers has shown that quasars having different redshifts have a tendency to cluster around bright galaxies and that the chance for such clustering is extremely low. Earlier in 1980, Arp and Hazard discovered two triplets of quasars aligned in straight lines in one single photographic plate. Again, the chance for such coincidences is $< 10^{-4}$. These and other recent observations have began to challenge the so long cherished concept of the cosmological origin of large redshifts of quasars and galaxies. Some entirely new concepts in physics may have to be requisitioned for satisfactory explanation of the quasar phenomena which have brought forth many puzzles to the physicists and astronomers during the last three decades.

20.5 ACTIVE GALACTIC NUCLEI

In Section 17.8, we had introduced the discussion of the different types of galactic nuclei as classified according to the strength of activities manifested by them. There we had talked of four different types of nuclei, from Type 1 to Type 4, and had found that the activities in the nuclei of Types 1 and 2 are rather mild and those nuclei mostly belong to normal galaxies. Much more violent activities are displayed by nuclei of Types 3 and 4 which we shall discuss in the present section. Because of very strong activities associated with some of these objects displaying spectacular, and sometimes peculiar phenomena, they have been separately called as a class named *Active Galactic Nuclei* (AGNs). These extragalactic objects as a class provide basic observational data for study of the modern observational cosmology.

Seyfert Galaxies and N Galaxies

We have already discussed Seyfert galaxies in fairly good details in Section 19.8. Seyferts are spiral galaxies. The elliptical galaxies possessing the Seyfert-like, small very bright nuclei have been called the N galaxies. It may be emphasized that *all* Seyfert nuclei and N galaxy nuclei are not considered as AGNs. Only the Seyfert 1 nuclei and the nuclei of some N galaxies belong to the class of AGNs. The nuclei of Seyfert 2 galaxies are not generally considered as AGNs. Thus, the nucleus of the Seyfert 1 galaxy NGC 5548 is considered as AGN but that of the Seyfert 2 galaxy NGC 1068 is not. The N galaxies possessing the AGNs have the same type of spectrum as Seyfert 1. The radio source 3C 120 is an example of such galaxies.

Nuclei of Some Radio Galaxies

Most radio galaxies have active nuclei. We have already discussed some of the high energy phenomena in these galaxies in the previous chapter. The jet that emerges from the nucleus of the giant elliptical galaxy M87 (Virgo A), consists of a series of bright knots suggesting a sequence of explosive events in the core of the galaxy. M87 is considered as a radio galaxy although its radio luminosity is not very large. The radio galaxy NGC 5128 (cen A) possesses a very active nucleus emitting radiation over the entire electromagnetic spectrum. Most other radio galaxies also possess strongly active nuclei.

BL Lacertae Objects

A *flat* radio spectrum with other associated peculiarities was found in BL Lacerta, previously thought to be a variable star. The study soon revealed that the object was extragalactic and representative of a class of objects possessing unusual properties. This class of objects have been called *BL Lacertae Objects* or *Blasars*. The principal characteristics of the objects are: (a) a star-like appearance on photographic plates, (b) a featureless optical spectrum, (c) a flat or sometimes an inverted radio spectrum, (d) an extreme and dramatic variability in optical brightness, and (e) high optical polarization. These unusual properties have justified in identifying these objects as a separate class of AGNs.

Quasars

The various peculiar and extreme properties of quasars as already described in the previous sections are all indicative of the fact that these objects are passing through an intensely active state.

Several other classes of objects have also been identified to belong to the category of active galactic nuclei. Radio-Quiet Quasars, LINERs (Low Ionization Nuclear Emission-Line Regions), OVVs (Optically Violent Variables) and Star Burst Galaxies are some of these classes. The inclusion of any such objects in the class of AGNs, however, somewhat depends on the definition and opinion. A discussion of all these lies beyond the format of this book. The reader may consult some of the relevant references given below.

EXERCISES

1. Describe the various steps that led to the discovery of quasars. Why quasars are considered as a very special type of objects in the Universe?

2. What do you know of the radio properties of quasars? Are all quasars radio bright? Discuss the difference between quasars and QSGs.

3. Summarize the principal optical properties of quasars. How would you like to explain the luminosity variation of quasars over short periods?

4. How would you interpret the positions of quasars and QSGs in Fig. 20.1?

5. Explain what is meant by *gravitational* redshift and *cosmological* redshift. Why the redshifts of quasars are thought to be of cosmological origin? What other possible explanations of large redshifts do you know?

6. Describe the various types of active galactic nuclei. What role do the active nuclei may possibly play in the evolution of the parent galaxies?

SUGGESTED READING

1. Arp, H.C., *Quasars, Redshifts and Controversies,* Interstellar Media, Berkeley, 1987.

2. Burbidge, G.R. and Burbidge, E.M., *Quasi-Stellar Objects,* E.H. Freeman Ltd., London, 1967.

3. Duschl, WJ. and Wagner, S.J., *Physics of Active Galactic Nuclei,* Springer-Verlag, Berlin, 1991.

4. Fang, L.Z. and Ruffini, R., *Galaxies, Quasars and Cosmology,* World Scientific, Singapore, 1985.

5. Frank, J., King, A. and Raine, D., *Accretion Power in Astrophysics,* Cambridge University Press, Cambridge, 1992.

6. Hoyle, F., *Galaxies, Nuclei and Quasars,* Harper & Row Publishers, New York, 1965.

7. Kurganoff, V., *Introduction to Advanced Astrophysics,* D. Reidel Publishing Company, Dordrecht, Holland, 1980.

8. Madore, B.F., Sulentic, J. and Bertola, F. (Eds.), *New Ideas in Astronomy,* Cambridge University Press, Cambridge, 1988.

9. Miller, H.R. and Wiita, P.J. (Eds.), *Variability of Active Galactic Nuclei,* Cambridge University Press, Cambridge, 1991.

10. Robinson, A. and Terlevich, R.J. (Eds.), *The Nature of Compact Objects in Active Galactic Nuclei,* Cambridge University Press, Cambridge, 1994.

11. Swarup, G. and Kapahi, V.K. (Eds.), *Quasars,* IAU Symposium No. 119, D. Reidel Publishing Company, Dordrecht, Holland, 1986.

12. Weedman, D.W., *Quasar Astronomy,* Cambridge University Press, Cambridge, 1988.

13. Weeks, T.C., *High Energy Astrophysics,* Chapman and Hall Ltd., London, 1969.

21 Cosmology

21.1 INTRODUCTION

In the earlier chapters we have attempted to introduce ourselves with the physical as well as the dynamical nature of stars and galaxies and also of the interstellar matter, out of which new stars are born. All these objects are the contents of the Universe. Besides these, the Universe also contains the intergalactic matter, radiation, magnetic field and cosmic rays. The question then naturally arises: What possibly is the physical and kinematical nature of the Universe itself, when considered in its entirety? *Cosmology* is the science which attempts to give a satisfactory answer to this fundamental question regarding the understanding of the phenomena behind the *Cosmos*. A definite answer to this question has not, however, yet been found by the scientists. It turns out that the problem is extremely complex and the attack to it has so far become only piecewise and in an indirect manner. The basic requirement in formulating a comprehensive theory of cosmology lies in truly knowing some of the basic facts about the Universe. These are, the shape (geometry) and size, the mass density and the total mass content, the age, the phase of its present dynamical behaviour, and its chemical evolution with time. All these informations are very fundamental for the understanding of the nature of any object. For any individual object or a group of objects occupying any local region of space, these informations are not very difficult to acquire through experiments and theoretical postulates. But for the Universe as a whole, it is precisely these simple facts which pose great difficulties to be known. In fact, none of these basic facts has yet been known with any amount of definiteness when the Universe as a whole is concerned. Even the most powerful optical and radio telescopes available at present are unable to fathom the whole depth of the Universe. And whatever observations at large distances have been obtained, their true interpretations have very often eluded the scientists.

Nevertheless, the observations are very essential; any good theory of cosmology should not only to be based on the latest observations, but also should conform to any new set of observational facts that may be known in future. It is precisely on this line that the attempts have so far been made to build cosmological theories, but only with marginal success. Various models of cosmology have been built by the authors under various assumptions. But none of these models can be singled out as an ideal one. Each appears to be vulnerable to drawbacks when tested with observational results. Not only that, observations themselves have not yet come to the level on the basis of which a final conclusion can be drawn. This is particularly true in the case of objects at very great distances. Evidently, it is these objects that can provide a basis for discrimination between the merits of the different models. But since these distant objects are very faint, not only, that uncertainties creep in the results, but also that the correct interpretation of the observations become increasingly difficult with distance. It appears therefore, that when we attempt to understand the true physical as well as geometrical nature of the Universe, we are inadequately equipped both with the observational as well as with theoretical attainments.

21.2 REDSHIFT AND THE EXPANSION OF THE UNIVERSE

In the early years of this century beginning from 1912, V.M. Slipher of the Lowell Observatory measured large radial velocities in a number of nebulae. A few of these showed velocities of approach, while most others were found to be moving away from us with different velocities ranging as high as 1800 km s^{-1}. These observations could not be correctly interpretted at that time because their extragalactic nature was not known. It was only through the extensive observational works of Edwine P. Hubble in 1920s that a correct explanation of Slipher's observations was revealed. These nebulae were discovered to be external galaxies and those showing approaching velocity were known to be members of the Local Group of galaxies. In the late 1920's Hubble computed the radial velocity of recession of a large number of nearby galaxies on the one hand, and on the other, could ascertain the distances of many of these galaxies by using the various distance indicators (Chapter 17). Hubble's work soon revealed that the velocities of recession of galaxies were, in fact, directly proportional to their distances. The work was further carried out by Hubble and Humason through the subsequent years with fainter and fainter (so more and more distant) galaxies which only confirmed Hubble's original finding. Hubble's law given in Eq. (17.8) was thus established for the galaxies flying away in all directions. This law has often been called the *law of redshifts* since the recessional speeds are revealed by lines in the spectra of galaxies shifted towards longer wavelengths (red end).

Once Hubble's law was established on the observational basis, the main difficulty then rested on the accurate evaluation of the constant of proportionality H. As has already been discussed in Chapter 17, the problem of correctly evaluating H is beset with many difficulties and an absolutely satisfactory value is yet to be determined. Nevertheless, various working values of H have been used from time to time. Hubble himself used a value of 540 km s^{-1} Mpc^{-1} a value too large to be accepted during the subsequent years. Hubble's value was greatly in error as he did not correct the measured distances for interstellar absorption, unknown at the time of Hubble's work. The later values of H usually range from 50 to 150 km s^{-1} Mpc^{-1}. Much work has been done in 1960s and early 1970s and the value of 75 km s^{-1} Mpc^{-1} was obtained by Sandage (1958). Later, he revised this value to nearly 50 km s^{-1} Mpc^{-1} on the basis of latest data (see Chapter 17). Many astronomers are currently in favour of using this last value. It may be useful to mention here that for the evaluation of H one has to observe galaxies lying within a narrow range of distances. The lower limit of the range is determined by the average random velocities of the galaxies in the radial direction. This average is of the order of 200 km s^{-1} and only those galaxies whose recessional velocities supersede this average have to be observed for the purpose. Assuming the value $H = 70$ km s^{-1} Mpc^{-1} which means that galaxies at a distance of 1 Mpc will recede at the rate of 70 km s^{-1}, the minimum distance at which observation will be meaningful for the purpose is ~ 3 Mpc. The upper limit of the distance range is determined by distances to which the distance indicators such as large H II regions, globular clusters, novae, etc. can be used. This distance can be roughly taken as ~ 25 Mpc. Thus, for the evaluation of H one has to compute by some independent method the distances of galaxies lying in the range between 3 to 25 Mpc. The rich cluster of galaxies in Virgo which lies in this range has been widely used for the purpose.

Whatever uncertainties may be there in determining the value of H, the observed redshifts can have only one interpretation, that the galaxies in all directions are receding from us, the velocity of recession being higher the larger the distance to the galaxy. This leads to the conclusion that at least *at the present time the Universe is expanding.* All galaxies and clusters of galaxies are flying away from us with speeds proportional to their distances. This fact is well illustrated by Fig. 21.1 where the redshift $Z = \Delta\lambda/\lambda = v/c$ is plotted against the visual magnitude of brightest cluster galaxies. Here, of course, we have tacitly made two assumptions: First, the brightest galaxies in different clusters have the same intrinsic luminosity, and secondly, the observed redshift is due to expansion. Both these assumptions, particularly, the second one may not be correct. But at present, there is no verified physical law which can replace the expansion hypothesis and therefore, the latter is generally accepted as the correct one. Thus, the redshift being a measure of the recessional speed of a galaxy, Fig. 21.1 indicates a linear distance-velocity relation to exist between the galaxies which are not too far. But whether this

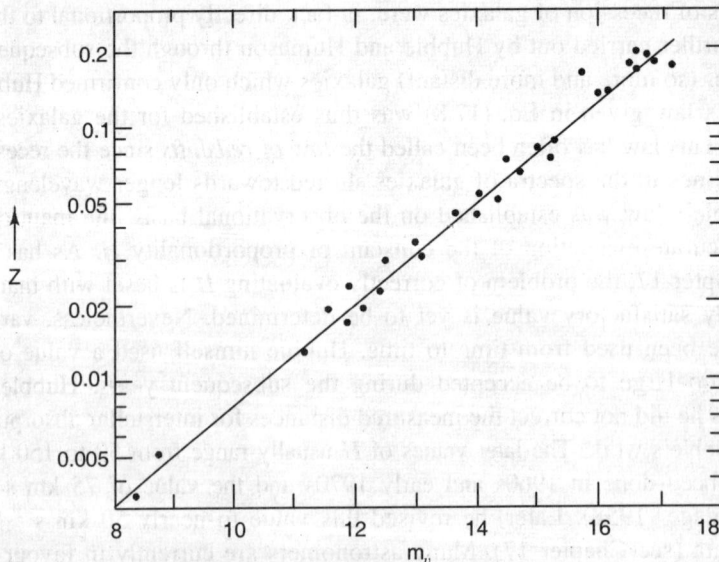

FIGURE 21.1 Redshift-apparent magnitude relation for brightest galaxies in clusters.

linear relation holds also for galaxies at much greater distances cannot be established on the basis of current observations. If the speeds were uniform throughout then it becomes easy to calculate the time-scale of the Universe which is the time elapsed since the expansion began. If t_0 be this time, then $t_0 = D/V = H^{-1}$ which is ~ 1.33×10^{10} years for $H = 75$ km s^{-1} Mpc^{-1} and ~ 2×10^{10} years for $H = 50$ km s^{-1} Mpc^{-1}. Thus, considering the latter value of H, one can conceive that all the matter in the universe was together in a superdense form and at extremely high temperature about 2×10^{10} years ago when, *possibly*, a gigantic explosion took place and as a result material has been flying apart since then. The time t_0 can thus be considered as the *age of the Universe* in its present phase. At this stage the question arises whether the rate of expansion of the Universe remained *constant* throughout or has it undergone a *deceleration* as its age advanced. At present, the cosmologists do not have a definite answer to this question. But according to Newton's law of gravitation, since every mass element attracts every other element, the Universe possibly could not undergo a free expansion. The force of gravitation must have produced a deceleration on the 'flying' galaxies whose speeds of recession must therefore have gradually decreased. If this is true, then the rate of expansion was faster in early days of the Universe than it is now, and its actual age would be somewhat *less* than that calculated now, which may be considered as the extreme upper limit of the age. In the case of deceleration, $t_0 < H^{-1}$. The difference between these two ages is illustrated schematically in Fig. 21.2.

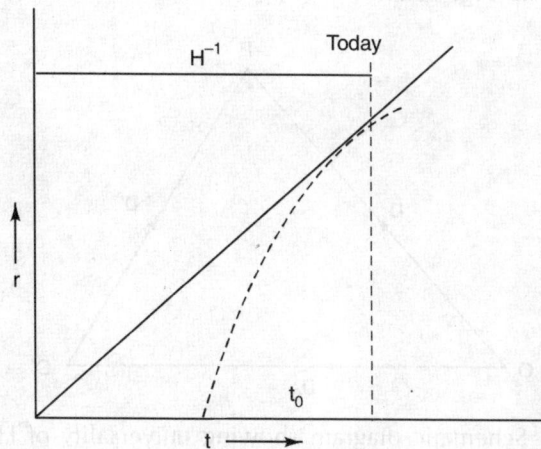

FIGURE 21.2   Age difference of the universe for uniform expansion and expansion with deceleration.

An extreme upper limit of the age of the Universe according to the current observational data combined with theoretical concept of the Universe is thus about 2×10^{10} years, when one assumes that its expansion rate has been uniform since the beginning. This age may be compared with the independently calculated ages of other astronomical objects. For example, all geological studies and radioactive decay analyses have shown that the age of the Earth is < 4×10^9 years. In fact, the greatest age on Earth is predicted by Pb-U dating which yields an upper age limit of 3.5×10^9 years. The upper age limit of lunar rocks borne to the Earth by the Apollo Mission has been calculated to be 4.5×10^9 years. The analyses of meteorite structures have also ascertained the age of 4.5×10^9 years for these interplanetary objects. Curiously enough, this is exactly the order of the age that has been predicted for the Sun on the basis of stellar evolution theory. Thus, all these independent and different studies indicate that the solar system is probably not older than 5×10^9 years—much less than our calculated age of the Universe. But in the Galaxy there are several other objects which are believed to be much older than the solar system. The best examples are globular clusters, old galactic clusters, RR Lyrae variables and subdwarfs. But as is currently understood by astronomers, none of these objects are probably older than 1.5×10^{10} years. From both observational and theoretical viewpoints, therefore, the estimated age of 2×10^{10} years of the Universe with an uncertainty of ± 50 per cent, arising out of the uncertainty in evaluating the Hubble's constant H seems to be quite reasonable.

Lastly, we can attempt to answer in a very simple manner the question: how large is the Universe? The extreme upper limit of the velocity with which a galaxy can move away is c, the speed of light. So the extreme upper limit of distance to which a galaxy can move during the age of the Universe is equal to $ct_0 \simeq 2 \times 10^{28}$ cm or ≈ 6600 Mpc. This is the order of the size (the radius) of the Universe at present.

It must be emphasized that the observed fact that galaxies are receding equally in all directions does not imply that we are occupying the central position of the Universe. In fact, an observer situated at any region of the Universe, will have the similar view. This can be understood by referring to Fig. 21.3.

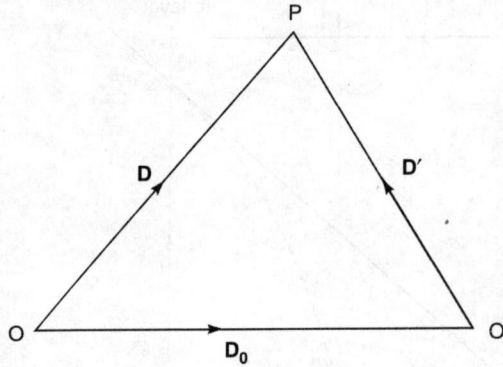

FIGURE 21.3   Schematic diagram showing universality of Hubble's law.

Suppose observers at two different regions O and O' see a galaxy at P where, using vectorial distances,

$$\mathbf{OO'} = \mathbf{D}_0, \ \mathbf{OP} = \mathbf{D} \text{ and } \mathbf{O'P} = \mathbf{D'}$$

Then,

$$\mathbf{D'} = \mathbf{D} - \mathbf{D}_0,$$

Also,

$$\mathbf{V} = d\mathbf{D}/dt = H\mathbf{D} \text{ and}$$

$$\mathbf{V'} = d\mathbf{D'}/dt = H(\mathbf{D} - \mathbf{D}_0) = H\mathbf{D'},$$

by Hubble's law. Thus Hubble's law equally holds for a galaxy at any-point P when it is observed from any two points O and O'. Hence, so far as the recession of galaxies is concerned, the Universe looks the same from all points. We shall further discuss this point in Section 21.4.

21.3 MATTER DENSITY IN THE UNIVERSE AND THE DECELERATION PARAMETER

When distances to galaxies and clusters of galaxies are known accurately, it becomes a simple matter to calculate intergalactic distances and compute the number density of galaxies over a given volume of space. The average mass density will follow from a multiplication of the average number density by the average mass of a galaxy. On the basis of the current observational data, the observed mass density of the universe has been computed to be $\rho_{obs} \sim 1.5 \times 10^{-30}$gm cm^{-3}. The observed density of the luminous matter is ~ 20 per cent of this value.

If the present Universe was created with a gigantic explosion of a supermassive body as the Big Bang theory postulates, the expansion velocity will be *decelerated* by the self-gravitation of the matter. Whether the expansion of the Universe will altogether be halted at some future time is actually determined by computing the magnitude of the *deceleration parameter* defined by

$$q = -\frac{\ddot{R}R}{\dot{R}^2} = -\frac{\ddot{R}}{H^2 R} \tag{21.1}$$

where $\dot{R} = HR = cZ$ which represents the Hubble law. We assume that all the quantities considered are the present-day values, $R(t)$ being the radius of the present-day Universe. The present-day matter density ρ_U is related to q by

$$\rho_U = \frac{3H^2 q}{4\pi G} \tag{21.2}$$

G being the constant of gravitation. For $H = 70$ km s^{-1} Mpc^{-1}

$$\rho_U \approx 2 \times 10^{-29} \, q \text{ gm cm}^{-3}. \tag{21.3}$$

It turns out that q is related to the curvature of space. For the model of the Universe, with positive curvature (one of the Friedmann models) the declaration will have to be sufficient as to bring the expansion to a halt and reverse it. This is then the *closed* or oscillatory Universe with an oscillation time of

$$T = \frac{2\pi q}{H(2q - 1)^{3/2}} \tag{21.4}$$

This time becomes meaningful only when $q > 1/2$, which therefore is the condition for a closed Universe.

The *open* (hyperbolic) model of the Universe is characterized by the values of $q < 1/2$. The transition between *open* and *closed* models of the Universe is characterized by the *flat* (parabolic) model which corresponds to the *critical* value $q = 1/2$. In terms of the mass density it follows from Eq. (21.3) that the *critical mass density* in the present-day Universe is given by Eq. (18.13) which marks the *flat* model of the Universe. The closed and open models are similarly characterized respectively by $\rho_U > 1.0 \times 10^{-29}$ gm cm^{-3} and $\rho_U < 1.0 \times 10^{-29}$ gm cm^{-3}. The currently observed density of $\rho_{obs} \sim 1.5 \times 10^{-30}$ gm cm^{-3} then points in favour of the open model; but as we have already mentioned, the present-day observational status is not sufficient to draw a definite conclusion in this regard. Extensive search, of course, is going on currently in order to establish a more meaningful value of the mass density of the Universe. This is very important as it will determine whether our Universe is open, closed or flat.

The actual discrimination between the open, flat and closed models of the Universe requires the determination of a distinct value of q. This can be achieved only by observations at very great distances where $Z \geq 0.4$. With $H = 70$ km s^{-1} Mpc^{-1}, the above redshift corresponds to a distance of about $D \sim 2000$ Mpc. At these distances, only the brightest galaxies in clusters can be used as distance discriminators. We can try to understand the concepts of the open, flat and closed cosmological models by a quite common analogy derived from Newtonian mechanics. We know that if a body on the Earth is thrown with a kinetic energy corresponding to a velocity less than the escape velocity (11.2 km s^{-1}), it will return back to the Earth. This case is analogous to the closed Universe model in which the sufficiently high density of matter in the Universe ($\rho_U > 1 \times 10^{-29}$ gm cm^{-3} and $q > 1/2$) produces enough gravitational deceleration to halt its expansion. The flat cosmic model is analogous to the case in which the kinetic energy given to the body is just sufficient to provide it the escape velocity which subsequently moves in a parabolic orbit and comes to rest at infinity. The mass density in this model equals the critical value of 1×10^{-29} gm cm^{-3} and $q = 1/2$. If, on the other hand, the kinetic energy given to the body corresponds to a velocity in excess of the escape velocity, it will move to infinity in a hyperbolic orbit with some finite kinetic energy. This base finds anology with the open Universe model which corresponds to $\rho_U < 1 \times 10^{-29}$ gm c^{-3} and $q < 1/2$. The following simple mathematical treatment will make the point more easily comprehensible.

Let us suppose that the Universe is an expanding sphere of constant mass M but whose radius $R(t)$ and density $\rho(t)$ are changing in time as it expands. Then,

$$M = \frac{4}{3}\, \pi\rho(t)\, R^3(t) = \text{constant} \tag{21.5}$$

and its equation of motion is

$$\ddot{R} = -\frac{GM}{R^2} \tag{21.6}$$

Integration of Eq. (21.6) yields the energy integral

$$\frac{1}{2}\dot{R}^2 - \frac{GM}{R} = E \tag{21.7}$$

E being the total energy which is constant. Combining Eqs. (21.1), (21.6) and (21.7) we get

$$q = \frac{1}{2} - \frac{E}{\dot{R}^2} \tag{21.8}$$

where $\dot{R} > 0$ in an expanding Universe. Thus Eq. (21.8) yields the following three cases:

1. $E < 0,\ q > \dfrac{1}{2}$,

2. $E = 0,\ q = \dfrac{1}{2}$,

3. $E > 0,\ q < \dfrac{1}{2}$.

The first case corresponds to the *closed* Universe, the second represents *a flat* Universe while the third represents an *open* Universe. These are shown schematically in Fig. 21.4.

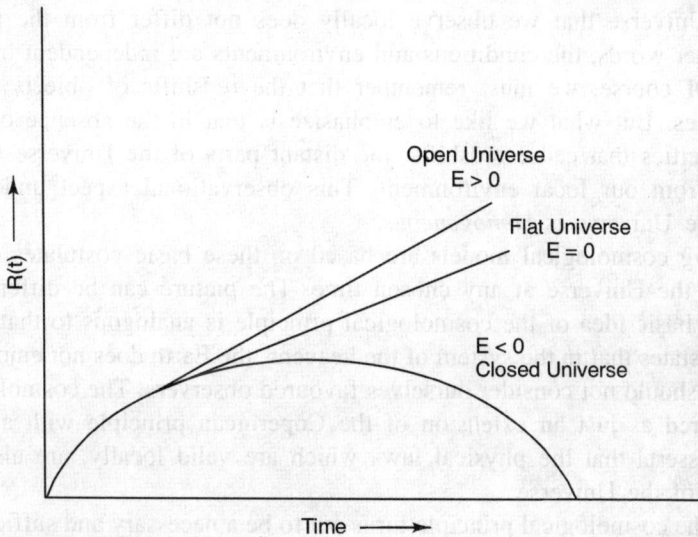

FIGURE 21.4 Models of open, flat and closed Universe.

21.4 THE COSMOLOGICAL PRINCIPLE: THE PERFECT COSMOLOGICAL PRINCIPLE

The primary goal of Cosmology is to construct models of the Universe that will survive the tests of current observations and of those that may be made in future. Many models have so far been constructed which represent various outlooks regarding the overall view of the Universe. All these models are however, based on the validity of a fundamental postulate which has been called the *Cosmological Principle*. The main contention of this principle is that the Universe presents the same picture at any particular epoch in whichever direction we may look from whatever position, except for local irregularities which are of statistical nature. The above statement physically embodies the *isotropy* and *homogeneity* of the Universe. Thus cosmological principle consists in taking for granted the concept that the Universe is isotropic and homogeneous at any particular epoch of observation. The question will then naturally arise as to how far this assumption may be considered valid on the basis of observations?

It turns out that the answer to this basic question is fortunately not beset with much complexity. Observations do really tend to confirm that the large-scale aspect of the Universe is isotropic and homogeneous. We have already seen that the Hubble's law of expansion of the Universe holds irrespective of the position of the observer. Observations further reveal that no matter which direction of the Universe we look to, the type and number density of galaxies are essentially the same. Of course, the colours of galaxies become increasingly redder as we look to greater distances on account of increasing redshifts, but this change in colours is not at all dependent on directions. In whichever direction we may look at, the law of redshifts remains the same. Thus the observable aspects of the Universe are found to maintain complete independence with respect to directions. In other words, the Universe is *isotropic*. No direction is preferred so far as observable aspects of the Universe are concerned.

We next consider the case of the homogeneity of the Universe. This means that the picture of the Universe that we observe locally does not differ from the picture at large distances. In other words, the conditions and environments are independent of the locality of the Universe. Of course, we must remember that the redshifts of objects are different at different distances. But what we like to emphasize is that in the absence of the effects of recessional velocities that cause redshifts, the distant parts of the Universe would not have been different from our local environment. This observational aspect underlies the basic postulate that the Universe is *homogeneous*.

All *evolving* cosmological models are based on these basic postulates of isotropy and homogeneity of the Universe at any chosen time. The picture can be different at different times only. The basic idea of the cosmological principle is analogous to that of *Copernican principle* which states that in the system of the heavens, the Earth does not enjoy any preferred position and we should not consider ourselves favoured observers. The cosmological principle may be considered as just an extension of the Copernican principle with a larger field of application. It asserts that the physical laws which are valid locally, are also valid in any arbitrary region of the Universe.

Although the cosmological principle turns out to be a necessary and sufficient hypothesis for the construction of the evolving models of the Universe, an entirely different type of model, the *Steady-State* model (Section 21.6) makes a more general postulate that the Universe looks the same from any arbitrary position and *at all times*. This implies that the picture of the Universe is not only independent of the distance and direction, *but also of time*. An aspect of the Universe as observed by a local observer at the present epoch will be found unchanged by any other observer situated at an arbitrary location of the Universe at an arbitrary time. This principle, which is vastly more general and is of far-reaching importance, constitutes the foundation on which the *Steady State* model of the Universe has been built. This has been called the *Perfect Cosmological Principle*.

21.5 FUNDAMENTAL EQUATIONS OF COSMOLOGY

In this section, we shall discuss the basic equations which are used to construct cosmological models. Although such models are now-a-days constructed exclusively on the basis of Einstein's theory of General Relativity, attempts were originally made to construct models on the basis of Newton's theory of gravitation. Newtonian Cosmology has, therefore, a place of its own in the history of the development of the subject. The early attempts to tackle cosmological problems with Newtonian theory failed because, over and above in the cosmological principle, the Universe was assumed to be *static*. The last assumption implies that besides the small random motions, the matter in the Universe does not possess any large-scale motions. It was found that under the combined effects of these assumptions, Newtonian theory did not yield any valid solution. The theory itself was therefore discarded for being incapable of correctly tackling the problem. It was later discovered, particularly through the works of E.A. Milne and W.H. Mc Crea in 1934, that not the theory itself but the assumption of a *static Universe* was wrong. By that time, the observational works of astronomers such as V.M. Slipher, E.P. Hubble and M.L. Humason had established the large recessional velocities of distant galaxies which laid the basis of the theory of *Expanding Universe*. The concept of the static Universe had to be abandoned altogether. In this background, the works of Milne and of Mc Crea and Milne revived the interest in Newtonian Cosmology by showing that with suitable interpretation of the terms Newtonian theory could represent many essential aspects of the relativistic cosmology. Not only that, the two theories of cosmology were demonstrated to be at par in many respects. We shall, therefore, first present briefly the essential mathematical steps of Newtonian Cosmology.

Suppose that an observer moving with the origin O of a system of coordinates observes the physical properties at an arbitrary point A, where $\mathbf{OA} = \mathbf{r}$. It can be established that in an expanding Universe in which the cosmological principle strictly holds, the velocity $\mathbf{v}$ of the point P relative to the observer O and the density and pressure at P are given by the relations

$$\mathbf{v} = F(t)\mathbf{r} \tag{21.9}$$

$$\rho = \rho(t) \tag{21.10}$$

$$p = p(t) \tag{21.11}$$

It may be noted that under the assumptions we have made, the velocity $\mathbf{v}$ is a function of the position of the point observed as well as of the time of observation, while the pressure and density depend only on the time. Equation (21.9) which can be looked upon as the equation of motion of the particle occupying the position P can be integrated to yield

$$\mathbf{r} = R(t)\,\mathbf{r}_0, \tag{21.12}$$

where

$$\mathbf{r} = \mathbf{r}_0 \text{ at } t = t_0, \text{ so that } R(t_0) = 1 \tag{21.13}$$

and $R(t)$ is a time-dependent scale factor satisfying the relation

$$\dot{R}(t) = F(t)R(t) \tag{21.14}$$

$R(t)$ thus represents the expansion parameter and Eq. (21.13) shows that, if cosmological principle is assumed to hold strictly in the Universe, its expansion (or contraction) must be uniform. Equations (21.9)–(21.14) describe the character of motion of the particle at P. Such a motion must also satisfy the conservation laws of mass (equation of continuity) and momentum (equation of motion).

The equation of continuity

$$\frac{d\rho}{dt} + \rho \nabla \cdot \mathbf{v} = 0 \qquad (21.15)$$

reduces to

$$\frac{d\rho}{dt} + 3\rho(t)F(t) = 0 \qquad (21.16)$$

in this case by virtue of Eq. (21.9) since div $\mathbf{r} = 3$. Using Eq. (21.14), Eq. (21.16) can be integrated to

$$\rho(t)/\rho(t_0) = 1/R^3(t) \qquad (21.17)$$

which states that as the linear dimensions of the Universe increase by a factor $R(t)$, the density diminishes inversely as the cube of this factor. This is in accordance with the properties of the Eucledian space which only is appropriate for Newtonian Cosmology.

The vector form of the equations of motion is

$$\frac{d\mathbf{v}}{dt} + \frac{1}{\rho} \nabla p - F = 0 \qquad (21.18)$$

where F is the gravitational force per unit mass. With Eqs. (21.9)–(21.11) and using Poisson's equation

$$\nabla \cdot \mathbf{F} = -4\pi G\rho(t) \qquad (21.19)$$

and taking the divergence of Eq. (21.18), we get finally

$$3\left[\frac{d^2F(t)}{dt^2} + F^2(t)\right] = 4\pi G\rho(t) \qquad (21.20)$$

G being the constant of gravitation.

Eliminating $F(t)$ and its derivative and also $\rho(t)$ between Eqs. (21.14), (21.17) and (21.20) we have

$$R^2(t)\,\ddot{R}(t) + \frac{4}{3}\,\pi G\rho(t_0) = 0 \qquad (21.21)$$

For static Universe, $R(t) = R(t_0) = 1$ and the only solution yielded by Eq. (21.21) is the trivial one $\rho(t_0) = 0$ that is, the case of an empty Universe. In order to render the Newtonian Cosmology more general so as to be competitive with the Relativistic Cosmology, an additional term has been arbitrarily added to the right-hand side of Poisson's equation, which can therefore be represented as

$$\nabla \cdot \mathbf{F} = -4\pi G\rho(t) + \Lambda \qquad (21.22)$$

whence,

$$\mathbf{F} = -\frac{4}{3}\,\pi G\rho(t)\mathbf{r} + \frac{1}{3}\Lambda\mathbf{r} \qquad (21.23)$$

Λ used here is the analogue of the *cosmological constant* which is a very important entity in the relativistic theory. The introduction of Λ actually ushers in a change in the Newtonian law of gravitation whose effect is supposed to be felt only at very great distances. With this introduction Eq. (21.21) becomes

$$R^2(t)\,\ddot{R}(t) + \frac{4}{3}\,\pi G\rho(t_0) - \frac{1}{3}\,\Lambda R^3(t) = 0 \tag{21.24}$$

which can be integrated to yield

$$\frac{1}{3}\,\Lambda R^2(t) + \frac{8}{3}\,\pi G\,\frac{\rho(t_0)}{R(t)} - \dot{R}^2(t) = E \tag{21.25}$$

E being the constant of integration, which actually represents the total energy of the moving particle under consideration.

Equation (21.25) which is a differential equation in $R(t)$ and t represents the basic relationship with which the cosmological models can be constructed. The parameters E and Λ are arbitrary each of which may be positive, zero or negative independently of the other. The nature of the models is determined by the actual relationships that hold between $R(t)$ and t which again are dependent on the choice of the values of E and Λ. All the models obtained from Eq. (21.25) have, however, one feature in common: they all just represent the dynamical picture of the Universe in the *classical sense*. This is inherent in Eq. (21.25) which does not contain c, the speed of light, as an entity. But since all our knowledge about the structure and dynamics of the Universe are obtained from light signals, the inclusion of c is relevant in any comprehensive, mathematical formulation of the World models. The limitations of Newtonian cosmology are imposed by the fact that the latter artificially divides the applicability of gravitational and electromagnetic fields into two completely separate domains. So Newtonian dynamical system can predict nothing about the nature of propagation of light which is essential for a World model. The theory has therefore been rightly superseded by Einstein's theory of General Relativity which combines gravitation with the propagation of light. Since both gravitation and electromagnetism play their respective roles unitedly in the theory of General Relativity, the latter has proved vastly more comprehensive in the study of cosmological problems. Almost all of the more recent cosmological models have been built on the basis of this theory.

According to the theory of General Relativity, the space has a curvature and any event is characterized by four general coordinates in space-time labeled by x^i, $i = 0, 1, 2, 3$. In such a four-dimensional space-time, two neighbouring events are separated by an interval ds defined by

$$ds^2 = \sum_{i=0}^{3} \sum_{j=0}^{3} g_{ij}dx^i dx^j \tag{21.26}$$

where g_{ij} are ten functions of the four coordinates, ds^2 is called the *line element*. The concept of *geodesic* is important in the present discussion. Geodesies are orbits such that the distance between two points measured along it is a minimum. The line element ds^2 may be negative, positive or zero. When $ds^2 < 0$, an observer *sees* the two events occurring simultaneously at a mutual separation of $(-ds^2)^{1/2}$. The interval of separation in this case is called *spacelike*. On the other hand, $ds^2 > 0$ means the separation of two events by a *timelike* interval. In such cases, the observer *can travel* from one event to the other. In the absence of electromagnetic fields, in particular, the particles move along geodesies with $ds^2 > 0$. On the other hand, light signals follow geodesies with $ds^2 = 0$.

The term g_{ij} defining the line element in Eq. (21.26) plays an extremely important role in the formulation of the theory of General Relativity. The form of g_{ij} yields two sets of information. First, it gives information about how the gravitational field is distributed. Secondly, the form of g_{ij} tells us what type of coordinate system has been chosen for the problem under investigation. This is vital, because the nature of curvature of space is determined by the particular system of coordinates chosen. The geometry which enables us to separate these two aspects of information from the particular form of g_{ij} is called *Riemannian* whose basic structure is contained in Eq. (21.26). With this brief introduction, we present in the following the basic formulae dealing with Relativistic Cosmology.

Under the assumption of the validity of the cosmological principle, that is, if the Universe is assumed homogeneous and isotropic throughout, then H.P. Robertson and A.G. Walker among others have shown that with a suitable choice of units of length and time the four-dimensional line element ds^2 can be written as

$$ds^2 = dt^2 - R^2(t) \left[\frac{(dx^1)^2 + (dx^2)^2 + (dx^3)^2}{\left\{ 1 + \frac{1}{4} \kappa[(x^1)^2 + (x^2)^2 + (x^3)^2] \right\}^2} \right] \tag{21.27}$$

The above expression is known as the *Robertson-Walker line element* which represents a kinematic model of the Universe. This model is unique but for the arbitrariness of the expansion parameter $R(t)$ and the curvature parameter κ. The curvature of the three space is given by t = constant and the constant κ which can take the values 0, +1 or –1 determines the sign of the space curvature. For any particular value of t, the curvature of space is the same everywhere. The properties of space as determined by particular values of κ are as follows:

1. The space is Eucledian when $\kappa = 0$; in such a space the surface of a sphere of radius r is $4\pi r^2$ and its volume is $4/3\ \pi r^3$.
2. The space is spherical (or elliptical) when $\kappa = +1$; in this case the surface of the sphere of radius r is less than $4\pi r^2$. The space is in this case closed and has a finite volume.
3. The space is hyperbolic and open when $\kappa = -1$. In this case the surface of a sphere of radius r is greater than $4\pi r^2$.

The Einstein field equations are expressed by

$$R_{ij} - \frac{1}{2} R g_{ij} + \Lambda g_{ij} = - K T_{ij} \tag{21.28}$$

which relate the ten functions g_{ij} of the Universe with its material content and the boundary conditions of the problem. $K = 8\pi G/c^2$ is a constant called the *Einstein gravitational constant* and Λ is also a constant called the *cosmological constant.* If the Robertson-Walker line element is assumed to hold, Einstein field equations in relation (21.28) reduce to the following two equations:

$$\frac{2\ddot{R}}{R} + \frac{\dot{R}^2}{R^2} + \frac{8\pi Gp}{c^2} = \Lambda c^2 - \frac{\kappa}{R^2} \qquad (21.29)$$

$$\frac{\dot{R}^2}{R^2} - \frac{8\pi Gp}{3} = \frac{1}{3}\Lambda c^2 - \frac{\kappa}{R^2} \qquad (21.30)$$

It should be noted that under the assumptions made here only two of the ten field equations contained in Eq. (21.28) survive. This is because among the terms in the tensor T_{ij} containing the density, pressure of matter, momentum and energy, only those representing the material density $\rho(t)$ and an isotropic pressure $p(t)$—both of which are functions of time, are relevant in this case. Only those solutions of Eqs. (21.29) and (21.30) are physically meaningful which yield $\rho(t) > 0$ and $p(t) \geq 0$ for a set of values of the parameters κ and Λ and the function $R(t)$.

Solutions to Eqs. (21.29) and (21.30) as such involve much complications. The problem, however, is generally made easier by introducing simplifying assumptions. One obvious simplification is to take pressure $p = 0$ in Eq. (21.29). p, of course, here includes all kinds of pressures such as the pressure of thermal motion of the gas, that of random motions of stars and galaxies and the radiation pressure as well. Observations however reveal that at the present epoch of the Universe the pressure p is very much smaller compared to the magnitude of the density ρ. Thus putting $p = 0$, the above equations can be integrated to yield

$$\frac{1}{3}\Lambda c^2 R^2(t) + \frac{8\pi G\rho R^3(t)}{3c^2 R(t)} - \dot{R}^2(t) = \kappa \qquad (21.31)$$

Since $\dfrac{8\pi G\rho R^3(t)}{3c^2}$ is constant, Eqs. (21.25) and (21.31) are identical in form. Thus, the law of variation of the scale factor $R(t)$ with respect to time t in both Newtonian Cosmology and Relativistic Cosmology (when in the latter the pressure is neglected) is identical. The only difference is that while in the former the constant E on the right-hand side of Eq. (21.25) represents the total energy of the particle, in the latter the constant κ on the right-hand side of Eq. (21.31) represents the spatial curvature. Thus, in fact, the relative importance of the pressure and density in the Universe actually governs the relative difference between the Newtonian and Relativistic Cosmology models. This very important result was established by Milne and Mc Crea as early as in 1934.

A second simplifying assumption which has been extensively utilized is that the cosmological constant Λ has been taken equal to zero. This assumption has been used in constructing most of the relativistic cosmological models. Putting $\Lambda = 0$ and subtracting Eq. (21.30) from (21.29), we get

$$\frac{\ddot{R}}{R} + 4\pi G\left(\frac{p}{c^2} + \frac{\rho}{3}\right) = 0 \tag{21.32}$$

Using the definition of q given in (21.1), Eq. (21.32) reduces to

$$H^2 = \frac{4\pi G}{3q}\left(\frac{3p}{c^2} + \rho\right) \tag{21.33}$$

Since $H = \dot{R}/R$, and $\Lambda = 0$, thus combining Eqs. (21.30) and (21.32) we can write

$$\frac{\kappa}{R^2} = \frac{4\pi G}{3q}\left[\rho(2q - 1) - \frac{3p}{c^2}\right] \tag{21.34}$$

Equation (21.34) shows that at any given epoch the curvature of space is related to the energy content of the Universe as defined by its total density and pressure. Since at the present epoch, $p \approx 0$, relations (21.33) and (21.34) together yield (for the present epoch)

$$\frac{\kappa}{R^2} = H^2(2q - 1) \tag{21.35}$$

and

$$\rho = \frac{3qH^2}{4\pi G} \approx 2 \times 10^{-29}\, q\ \text{gm cm}^{-3} \tag{21.36}$$

with $H = 70$ km s^{-1} Mpc^{-1}. The last two relations show that the Universe will be closed if $q \geq 1/2$ and $p > 10^{-29}$ gm cm^{-3}, a result already obtained in Section 21.3.

21.6 THE CURRENT THEORIES: SOME IMPORTANT MODELS OF THE UNIVERSE

In the preceding section, we have attempted to present the basic mathematical concepts that underlie the Newtonian Cosmology and the Relativistic Cosmology. We have also mentioned that the Newtonian theory is inadequate in constructing reliable world models, since the former fails to consider in its framework the phenomenon of the propagation of light. Since Einstein's theory of General Relativity visualizes a much more comprehensive understanding of the phenomenon of gravitation in relation to the fundamental entity, the *space-time,* of the Universe in its large-scale perspective, most of the world models that have been proposed by physicists and astronomers during the present century are based on General Relativity. The correctness of this theory has been demonstrated in the verification of several of the effects predicted by Einstein on the basis of it. One of the predictions made by Einstein was that because of the warped geometry of space produced by the mass of the Sun, the perihelion of Mercury's orbit will undergo a prograde precession of about 43".16 per century. The actual amount of precession that has since been measured is 43" per century. The agreement is excellent. Secondly, Einstein's prediction that starlight passing close to the limb of the Sun (observable during a total eclipse) should be deflected towards the Sun by 1".75. This has also been observed to agree with fair accuracy. Thirdly, the General Relativity predicts that a photon traversing through a potential difference should suffer a redshift. For example, a photon of energy hv, while traversing through the gravitational potential of the Sun should

suffer a redshift of $\dfrac{GM_\odot}{R_\odot} \dfrac{hv}{c^2}$ which is equivalent to a Doppler effect of 0.64 km s^{-1}. Although

the verification of this gravitational redshift is a very difficult task, such a shift has been formally confirmed by experiments. These and several other observed relativistic effects have proved the correctness of the theory of General Relativity. Two older theories that have been developed for construction of cosmological models are the *Steady-State theory* proposed by H. Bondi and T. Gold and by Fred Hoyle, and the *Scalar-Tensor theory* proposed by C. Brans and R.H. Dicke. Models of the Cosmos based on these last two theories will be briefly discussed at the end of this section. Before that we shall discuss a few World models which have been constructed on the basis of Einstein's theory of General Relativity. On this theory again, two fundamentally different types of models of the Universe have been proposed—the *static* models and the *evolutionary* models.

The Static Model of Einstein

Albert Einstein himself was the first to use his theory of General Relativity in building a World model in 1917 which, of course, due to several of its basic weaknesses has now reduced to one of historical importance only. At that time the expansion of the Universe was not discovered which came actually after a decade from then. So Einstein adopted the cosmological principle and starting with his relativistic equations he calculated the average density of the Universe. The Einstein Universe is spherical with the curvature paramater $\kappa = 1$ and having a constant radius of curvature, R. So space is unbounded but finite in such a Universe, since R is finite. Equations (21.29) and (21.30) therefore yield for Einstein Universe ($\kappa = 1$ and R = constant),

$$\Lambda = \frac{1}{c^2 R^2} + \frac{8\pi Gp}{c^4} \tag{21.37}$$

and

$$\rho = \frac{2}{c^2 R^2 K} - \frac{p}{c^2} \tag{21.38}$$

where we have used Eq. (21.37) to deduce Eq. (21,38) and written $K = \dfrac{8\pi G}{c^2}$ whose numerical value is 1.86×10^{-27} cm gm^{-1}. The cosmological constant Λ actually corresponds to a tension which was introduced by Einstein on an *ad hoc* consideration. The effect of this term is not perceptible over short range of distances as those within the solar system but becomes more and more important with increasing distances so as to be capable of bringing the expansion of the Universe to a halt at a certain stage. This last statement, of course, is not relevant to the Einstein Universe which does not expand at all (Fig. 21.5). Einstein considered Λ as producing an effective repulsion against the gravitational attraction.

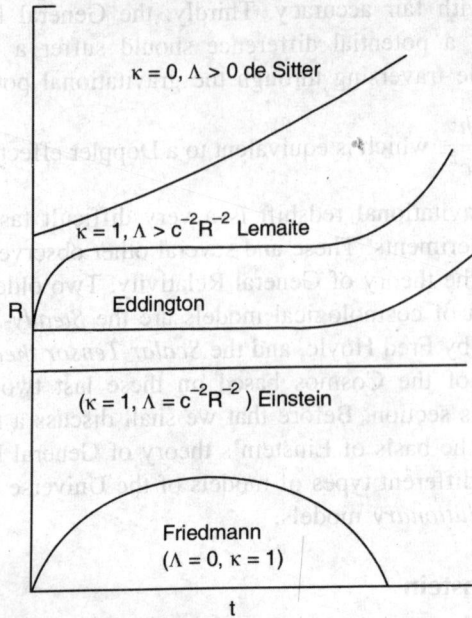

FIGURE 21.5 Schematic diagram for different models of the Universe.

Two basic defects of the Einstein Universe are that he supposed the Universe to be static (not variable with time) and completely devoid of radiation with a temperature of 0 Kelvin. Subsequent observations have proved that both of these assumptions are wrong.

The de Sitter Universe

About the same time in 1917, the Dutch astronomer W. de Sitter proposed another model of the Universe which also was based on the cosmological principle and the relativistic equations. This Universe of de Sitter was expanding in an infinite flat space ($\kappa = 0$) but he visualized an empty Universe ($\rho = 0$). The expanding model of the Universe appeared strange to the physicists at that time, about a decade before Hubble's discovery of receding galaxies. In de Sitter's Universe, $\rho = p = \kappa = 0$, when Eq. (21.30) yields

$$R(t) = R_0 \exp\left[\frac{1}{3} c^2 \Lambda\right]^{1/2} t \qquad (21.39)$$

Since $\dot{R}/R = H$, the Hubble constant, the age of the de Sitter Universe can be expressed by

$$\frac{1}{H} = \left(\frac{3}{c^2 \Lambda}\right)^{1/2} \qquad (21.40)$$

It can be noticed that if in such an empty Universe we assume further that $\Lambda = 0$, then we get a static Universe of undetermined extent. The basic weakness of de Sitter Universe is that it is empty and thus unphysical (Fig. 21.5).

The Lemâitre-Eddington Universe

Around 1930, G.E. Lemâitre and A.S. Eddington discovered independently that the Einstein Universe was unstable against small perturbations. This is because the Einstein Universe is pictured as being held in equilibrium by equal and opposite forces of gravitational attraction and cosmological repulsion (incorporated by Λ). This precarious state of equilibrium, if slightly disturbed, will give rise to instability. The Universe, if disturbed by a slight expansion initially, will subsequently continue to expand with ever-increasing speed. If, on the other hand, the initial disturbance is produced by a slight contraction, the Universe will subsequently, continue to contract with ever-increasing speed. The Lemâitre-Eddington model of the Universe therefore starts with Einstein's static state and undergoes through the first alternative after getting some initial disturbance; that is, it continues to expand subsequently with ever-increasing speed (Fig. 21.5). This Universe, therefore, has an infinite past spent in Einstein state and starting from some finite epoch in the past will continue to expand for an infinite future. The galaxies and stars were formed over a finite period during the initial expanding phase. The works of some authors such as Lemâitre, Mc Crea and Mc Vittie have shown that the static Einstein Universe would expand after the condensation of galaxies started for some reason.

The Lemâitre Universe

The Lemâitre-Eddington model which was well accepted for sometime, however, soon became the subject of various criticisms; particularly, when the formation of elements began to be considered seriously. With a view to get rid of these weaknesses, Lemâitre proposed another model of the Universe also based on the theory of General Relativity. In this model, Lemâitre suggested that the entire matter of the Universe was originally in a single chunk of extremely high density and temperature but occupying a small volume. This body was called by Lemâitre the *Primeval Atom*. Due to some reason this superdense body became unstable, a gigantic explosion took place and material started flowing in all directions at very high speeds. The elements were formed during the early period of rapid expansion. The expansion subsequently slowed down to bring the matter almost to a halt. At this phase, the Universe was like Einstein Universe when the material condensations took place to form galaxies. These condensations triggered the second phase of expansion which has since been continuing with ever-increasing velocity as depicted in Fig. 21.5.

The model proposed by Lemâitre subsequently established the concept of the Big Bang which has since been variously discussed by authors. It envisages the *evolutionary cosmology*. According to this model the Universe started from the explosion of the Primeval Atom at some particular epoch in the past, the galaxies were formed during a particular time period, which therefore have since been aging and evolving together. All the various models constructed on the basis of the various versions of the Big Bang theory are therefore called the evolutionary models of the Universe. The age of the Universe discussed in Section 21.2 is obtained on the basis of this theory.

Lemâitre's model possesses the mathematical characteristics that $\kappa > 0$, $\Lambda > \Lambda_c$, a critical value of Λ defined by

$$\Lambda_c = \frac{\kappa^3}{16\pi^2 G^2 \rho^2(t_0)} \tag{21.41}$$

and $\rho > 0$, but gradually decreasing as the expansion proceeds.

The Friedmann Models

A variety of cosmological models were constructed by the Russian mathematician A. Friedmann in early 1920s. Friedmann used the Einstein relativistic equations with cosmological principle and simplified those equations further by taking $\Lambda = 0$. All the other models which we have enumerated above so far were constructed with non-zero values of the cosmological constant Λ—which is essentially equivalent to a tension against which work has to be done if expansion is to continue. The introduction of Λ by Einstein in his equations in an arbitrary fashion became all along a subject of controversy among cosmologists. Friedmann simplified the matter just by taking $\Lambda = 0$. This was indeed a bold step which greatly simplified the mathematical formalism.

Friedmann has considered models of both closed and open Universe having one characteristic common to all of them, viz., they all start at some time $t = 0$ in an extremely dense state and subsequently evolve in time t in curved space defined by either $\kappa = +1$ (spherical) or $\kappa = -1$ (hyperbolic). In the initial dense state both pressure and density are high which subsequently continue to decrease as the. expansion proceeds. The models with $\kappa = +1$ (and $q > 1/2$) start from dense state, expand to some maximum value of the scale factor $R(t)$, then start contracting again to reach ultimately the initial dense state. Such cycles can be repeated indefinitely giving an *oscillating Universe*. In such a special case the Universe might continue from an indefinite past to an indefinite future, the scale factor $R(t)$ following a sequence of hoops in time t as shown in Fig. 21.6. Both a zero and a non-zero finite value of the scale factor at maximum contraction can be visualized. In the former case, the density becomes infinite, while it is finite in the latter case.

Other Friedmann models with $\Lambda = 0$, $\kappa = -1$ ($0 \leq q < 1/2$) start from dense states and continue to expand indefinitely in the curved hyperbolic infinite space.

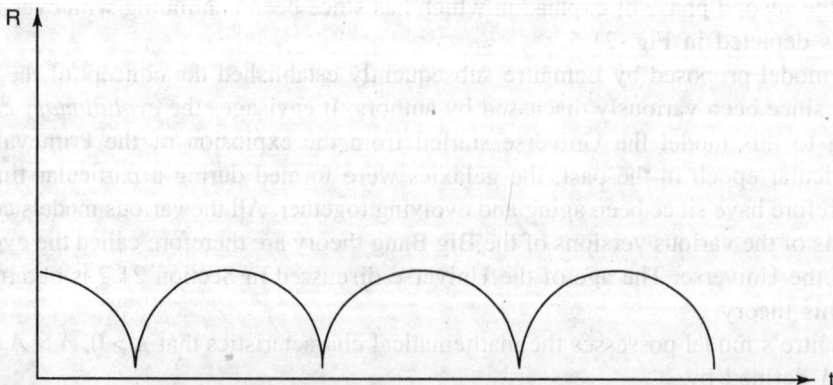

FIGURE 21.6   Oscillating model of the Universe.

The Steady-State Universe

All the models described above are relativistic models of the universe in curved space constructed on the basis of Einstein's equations of General Relativity with various values of Λ and κ. An entirely new model of the expanding Universe was proposed independently by H. Bondi and T. Gold and by Hoyle. This Universe is in flat space ($\kappa = 0$) and is characterized by the decelerating parameter value $q = -1$; that is, the Universe expands with an acceleration. All the relativistic models described so far postulate the *cosmological principle* and the conservation of mass-energy. The steady-state model, on the other hand, goes much beyond in postulating *the, perfect cosmological principle* and discarding the most vital physical law of the conservation of mass-energy in the Universe. The validity of the perfect cosmological principle requires that the Universe is isotropic and homogeneous *at all time*. Since the Universe is expanding, the principle demands that new matter must be *created* to maintain a constant density of the Universe. The newly created matter must replenish that which vanish beyond the limit of the observable Universe. The most remarkable feature of the theory is that the new matter (believed to be hydrogen atom) is supposed to be created out of nothing in a *creation field* called the C-field. Matter, therefore, requires to be *continuously created* in the Universe according to this theory. It turns out, however, that the rate of creation of new matter to replenish the lost amount is very low. The rate of volume expansion of a sphere of radius $R(t)$ is given by

$$\frac{d}{dt}\left[\frac{4}{3}\pi R^3(t)\right] = 4\pi R^2(t)\frac{dR(t)}{dt} = 4\pi R^3(t)H(t) \tag{21.42}$$

by using Hubble's law. If the density is to be maintained at some constant value ρ, then the rate of matter creation within the sphere of radius $R(t)$ becomes $4\pi R^3(t)H(t)\rho$. Therefore, the rate of matter creation per unit volume is $3H(t)\pi$. This is numerically ~ 10^{-47} gm cm^{-3} s^{-1} or about one H atom per cubic kilometre of space every five years. This appears to be such a small rate of creation that it cannot be observationally verified and "it probably can be done by a lesser Being than the Almighty God". But the space is so vast that even this small rate will sum up, when taken over the entire space, to the creation of more than a thousand stars every second. The question therefore naturally arises: where does this enormous amount of energy come from? We do not have any satisfactory answer.

The *continuous creation theory* therefore predicts that in order to satisfy the perfect cosmological principle, new galaxies must be condensed out of the newly created matter where new stars will be formed. These galaxies, taking part in the expansion of the Universe, will separate to greater and greater distances from one another while, at the same time, will age, grow old and eventually proceed toward the end of their lives. Consequently, in any given volume of space there must *always* be found the same proportion of old and young galaxies formed at different ages. The Universe, according to this theory, has neither a beginning nor an end either in space or in time. It is infinitely large and infinitely old, having an infinite future.

Although faced by many difficulties in view of the current observational status of the Universe, the steady-state theory has evoked great interest among the cosmologists since the time of its formulation. Since the relativistic models also are found to be vulnerable against observational tests, the steady-state theory was considered for some time as an alternative, and a very good one at that. The strongest argument against this theory for which many physicists are sceptical about the soundness of it is that its most interesting feature, the continuous creation of matter, violates the law of conservation of energy, which is regarded by the physicists as one of the most sacrosanct laws of physics. But the propounders of the steady-state theory attempt to counteract the objection by arguing that matter in the Universe *has been created* any way at some phase of the Universe. It is no more difficult to conceive the continuous creation of matter than that it was created as the *primodial atom* at some particular time. Moreover, the concept of the continuous creation and of the infinite extension in both space and time of the steady-state Universe has imparted it a great philosophical charm. That is why, in spite of its weaknesses against several observational aspects, the theory has been adhered to by many cosmologists with the introduction of certain modifications from time to time.

The Scalar-Tensor Theory

This theory of the model of the Universe was proposed by C. Brans and R.H. Dicke on the basis of a modification of the relativistic equations of Einstein. As we have already seen, the tensor g_{ij} in Einstein equations of General Relativity given by Eq. (21.28) represents the distribution of the gravitational field. To this tensor field Brans and Dicke have added a small scalar field thereby modifying the relativistic equations. Although the added scalar field is small, its effect becomes far-reaching when considered against the theoretical evolution of the cosmological models. But the observational tests of its validity are quite difficult as very great accuracy of measurements is necessary to observe its small differences from the results obtained on the basis of Einstein's theory. In one case, however, viz., the motion of terrestrial planets, where Einstein's theory has been found to be correct, the Brans-Dicke theory has been found to disagree. Nevertheless, when the cosmologists in their hectic search for a most appropriate theory that will correctly predict all observable aspects of the Universe have been eluded so far, the Brans-Dicke theory must be regarded as a valuable contribution in the race. Whether it will stand the test of time against the background of observational verification, we do not know. But its importance lies in the fact that it has initiated new ideas in cosmology and has encouraged to adopt new lines of observational tests.

21.7   OBSERVATIONAL TESTS OF COSMOLOGICAL MODELS

We have discussed in the last section some of the important World models that have been proposed by cosmologists maintaining different viewpoints. These models are principally of two different types, barring the static model of Einstein and the empty model of de Sitter which now have been reduced to of historical interest only. The two competitive types are the *evolutionary cosmology* and the *steady-state cosmology*. According to the former, the Universe started at a particular epoch some $(1-2) \times 10^{10}$ years ago with a *Big Bang* from an extremely dense and hot state. Whether it started from a singularity or a finite size and whether it will expand indefinitely or fall back again to the original dense and hot state, still remain a controversy. Some authors have tried to show that the Universe might have started from an initial size of the Earth's orbit at a temperature of 10^{12} K. The heavy elements were formed shortly after the time of explosion. The galaxies and stars were all formed at subsequent epochs. They are all aging and evolving since then, i.e. the Universe as a whole is evolving in time. This is in fact, the picture of the Universe depicted, in general, by all evolutionary models of cosmology. These models are all based on the theory of General Relativity. According to the Steady-State Cosmology, on the other hand, the Universe is infinite both in time and extent and the expansion rate of the Universe must increase in time. Any finite volume of the Universe must contain a homogeneous mixture of galaxies of all different ages so that no question of evolution is relevant in this case.

Among such a variety of models, if they are really exhaustive, only one should emerge as true when subjected to observational scrutiny. Unfortunately, very distant galaxies require to be observed for a decisive test, but at such distances several difficulties arise in interpretation of observed properties. One very important test is yielded by the observation of average mass density of the Universe and the corresponding deceleration parameter. This will give a clue to the understanding as to whether the Universe is expanding at a constant rate or the rate of expansion is slowing down. Unfortunately, such a discrimination cannot be made with galaxies for which $Z < 0.2$. For these nearby galaxies the velocity-distance curve is a straight line as shown in Fig. 21.1. The interpretation of such diagrams rests on two basic assumptions as we have observed in Section 21.2. First, the measured redshifts of galaxies are Doppler shifts (also called "Cosmological" redshifts), implying that by measuring shifts of spectral lines we actually measure the velocities of the sources. Secondly, the brightest members in clusters of galaxies have the same absolute luminosity so that by measuring different apparent magnitudes of such members we actually measure their different distances. With these assumptions Fig. 21.1 has been drawn to represent the redshift-magnitude diagram (which is equivalent to the velocity-distance diagram with above two assumptions) for the brightest members of 38 clusters of galaxies. The relation is fairly well represented by a straight line, meaning thereby, that the expansion of the Universe is uniform ($q = 0$). If q is non-zero, the galaxies will be either decelerated ($q > 0$) or accelerated ($q < 0$) and in either case the redshift-magnitude relation will be non-linear. But this nonlinearity is revealed only at very great distances as is illustrated in Fig. 21.7. It is to be noted from the figure that all the four curves corresponding to $q = +2, +1, 0$ and -1 are almost linear and coincident until distances corresponding to about $Z \sim 0.2$ are reached. The curves separate at greater distances and in order to discriminate between the various world models, one has to observe galaxies at these distances. If the distant galaxies are found to lie along the curve marked 0, then an uniformly expanding model of the Universe will be a correct picture. If the galaxies are found to lie along some curve lying intermediate between those marked 0 and +1 then the expansion of the Universe is slowing down. But in this case the Universe will never stop although its rate of expansion will gradually decrease. Both *open* and *flat* models of the Universe are relevant to this case. A peculiar phenomenon will, however, occur if the galaxies are found to lie along the curve marked +1 or above. In this case the Universe will expand to a maximum extension and then retrace back its path in contraction. Calculations reveal that the Universe contracts to a state of high temperature and high density and at the end of contraction it will vanish to a singularity. Some authors however, have suggested that the rotation of the Universe as a whole may spare it from going back to a singularity. The Universe in this case will *bounce back* from a dense state of finite size which subsequently leads it to an unending series of oscillations as depicted in Fig. 21.6.

If, on the other hand, the distant galaxies fall along a curve lying between those represented by 0 and –1, then the Universe is expanding with an increasing rate and will extend to infinity. In particular, if the galaxies are found to lie along the curve $q = -1$, the Universe is in steady-state.

As shown in Fig. 21.7, the present observational status would suggest that the distant galaxies rather lie along the curve for +1. In particular, the correspondence between the galaxies and the curve marked –1 appears poor. A comparatively better fit is suggested between the galaxies and a curve lying anywhere between those marked 0 and +1 which means, that we at present are living in an expanding but decelerating Universe. It appears quite different from a steady-state Universe. But a careful consideration of the entire problem would rather suggest that we will be mistaken to draw any definite conclusion on the basis of the current observational data. Some inherent uncertainty may spoil the entire basis of our inference. The plausible uncertainty may arise from our second basic assumption enumerated above. Galaxies at great distances may not be basically of the same intrinsic luminosity as those close to us. In fact, when we are observing galaxies several billion light years away, we are actually observing them as they were several billion years ago. It may be unlikely that a galaxy will remain essentially unchanged for such a long period. Since the galaxies evolve in time and their stellar content changes with evolution, it seems likely that their intrinsic luminosity also changes with the change of the stellar content. This is perhaps more appropriate for the largest galaxies which are observed for the purpose. Thus the assumption that the largest

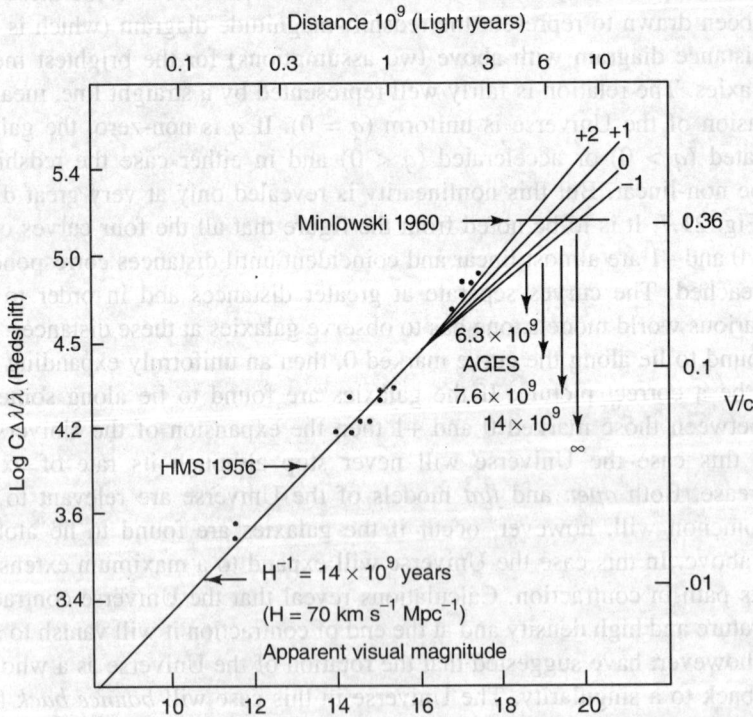

FIGURE 21.7 The theoretical Redshift-Apparent magnitude relation for galaxies.

galaxies in clusters at different distances are essentially of the same intrinsic luminosity, is likely to introduce uncertainty when greater depths of space are concerned. An uncertainty in luminosity even by a factor of two may disturb the entire basis of our conclusion, because galaxies may then fit with a curve lying anywhere between those marked by –1 and +1. In the absence of more accurate observation of distant galaxies, it will be therefore, unwise to draw any firm conclusion regarding the correct model of the Universe.

However, some observational facts have been gathered which most cosmologists believe today, supply evidences against a steady-state Universe. In fact, these observations are believed to favour a Big Bang. But again, in view of the great complexity of things and inherent uncertainties lying in measurements as well as in interpretations, the reader may be warned to be too optimistic in drawing an unambiguous conclusion.

The first observational fact that speaks in favour of a Big Bang (and against a steady-state Universe) is the detection of the so-called 3K (more correctly, 2.7K) isotropic background radiation. It was suggested by Gamow and by Dicke that if the present expansion of the Universe actually started with a Big Bang, some of the radiation released in explosion should be detectable even now. This radiation which was extremely intense and of very high energy during explosion was mostly absorbed when initially flowed through the dense, hot and opaque matter. But as the matter thinned out due to subsequent rapid expansion, some of the radiation escaped encounter with matter and was thus spared. Under these conditions, we should be able to detect a remnant of this radiation that may flow to the Earth. Both the above authors predicted, however, that the original high energy radiation representative of a very hot body, while coming from very far away in space will be very greatly redshifted. As a result, the original radiation in X-rays is likely to be detected in radio waves as if radiated from a cold body at a temperature of a few degrees Kelvin.

This remarkable theoretical prediction was actually verified when A.A. Penzias and R.W. Wilson of Bell Telephone Laboratories detected in 1965 an isotropic flux of background radio radiation at 7.35 cm. The isotropy of this radiation was so perfect that the only plausible explanation required was to assign it to extraterrestrial origin. More detailed and sensitive observations at several more wavelengths not only confirmed the correctness of this assumption but also showed that the measured radiation at these wavelengths correspond to that of a blackbody at about 3K. An additional confirmation of the existence of this isotropic radiation comes from the observation of lines of CN in spectra of the stars ρ Oph and ρ Per. The lines suggest that the level of excitation of CN molecules is the same as it would be if they were *bathed* in radiation corresponding to a wavelength of 2.6 cm which again is characteristic of a blackbody at temperatures around 2.7 K. This remarkable coincidence between the theoretical prediction and observed results has encouraged many cosmologists to believe that in this 3 K radiation, we are actually seeing the Big Bang of the extremely dense and hot *primordial atom* of Lemâitre.

The second observational fact apparently speaks *more against the steady-state Universe than it votes in favour of the Big Bang.* This comes from a study of the distribution of Quasars in space. It is believed that Quasars are objects passing through a particular phase of their evolution. According to the steady-state theory such objects should be uniformly distributed in space, because every volume of space must contain a homogeneous mixture of objects of all ages. But the study of the distribution of a homogeneous sample of Quasars reveals that a large majority of them lie very far away, from us. This implies that the Quasars are objects that mostly existed in early stages of the Universe and are infact reminiscent of the early evolutionary phase of the Universe, the like of which we rarely see to-day. If this interpretation is correct then it certainly violates the concept of the perfect Cosmological Principle which is pivotal to the steady-state cosmology.

We conclude this chapter with a note of warning: The subject of cosmology has been a meeting place of contradictory (or at least alternative) theories and observations. We are not even sure whether the observed redshifts are cosmological. We just *assume it* in the absence of a better alternative. But many physicists have questioned its validity. We do not know yet the true physical nature of the Quasars on the basis of which the soundness of the steady-state cosmology has been questioned. Many authors have suggested that new galaxies do form. The observation of peculiar features in galaxies has confirmed this idea. Groups of galaxies strung along a line and tubular connections between galaxies definitely suggest that they cannot so remain for a very long time, and therefore, must have been of rather recent origin. Such observations therefore lend support to the continuous creation hypothesis. Thus the present observational status does not allow any cosmological model to be either wholly accepted or wholly rejected and we have to keep our choice open for some more time to come.

21.8   THE COSMIC MICROWAVE BACKGROUND RADIATION

The cosmic microwave background radiation was predicted in 1948 by George Gamow, Ralph Alpher and Robert Herman. It was first observed by Arno Penzias and Robert Wilson in 1965 at the Bell Telephone Laboratories in Murray Hill, New Jersy during the calibration of the horn radio antenna devised to track the satellite echo. They found that the noise was independent of the direction of antenna and this indicated that the noise was of cosmic origin. Subsequent studies showed the radiation to have a temperature of 2.7 K and the spectrum was a thermal black body curve. The black body nature of the spectrum, indirectly supports, the big bang theory of the formation of the Universe. One of the profound observations of the 20th century has been that the universe is expanding. This expansion implies a smaller, denser and hotter universe in the distant past. At this high density matter and radiation were in thermal equilibrium. As the universe expanded, both matter and radiation cooled and at 3000 K, electrons joined the atoms breaking the thermal contact and thus matter decoupled from radiation making the universe transparent from an opaque state. Before the 'decoupling era', cosmic microwave background photons easily scattered off electrons. This process of multiple scattering produces what is called a 'thermal' or 'black body' (BB) spectrum of photons. So according to the big bang theory, there should have been a BB spectrum and this was indeed measured with FIRAS (Far Infrared Absolute Spectrophotometer) experiment on NASA's (National Aeronautics and Space Administration) COBE (Cosmic Background Explorer) (Fig. 21.8) satellite. To a first approximation we expect the photos in the universe to have a BB spectrum.

Now there is an another question. If the universe was so uniform, then how were the different structures in it formed that we see today? There must have been some bumps in the early universe that grew to create the structures that we see today. In 1992, COBE detected the bumps (ΔT_b) for the first time.

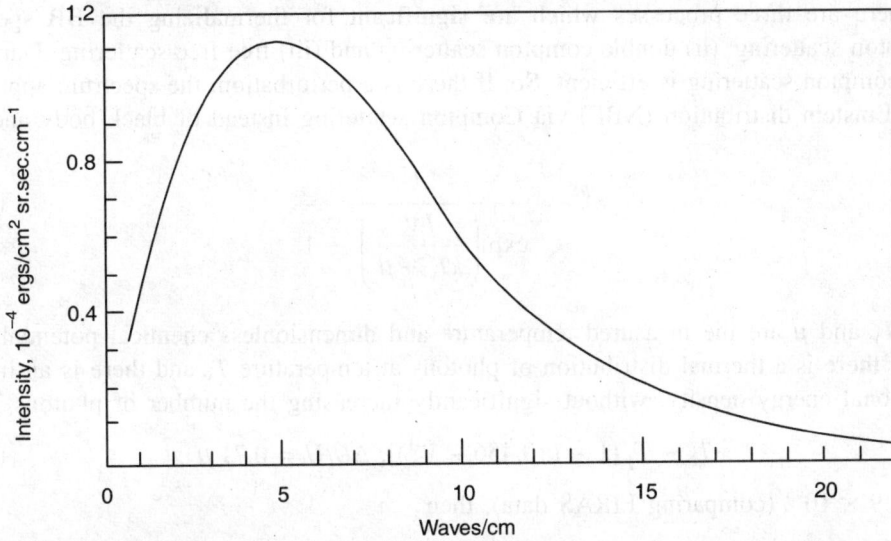

FIGURE 21.8 Cosmic microwave background radiation measured by FIRAS on COBE satellite.

Anisotropy or Bumps (ΔT_b)

Let I_v be the specific intensity of light (incident energy per unit area, per unit solid angle, per unit frequency, per unit time). Then,

$$I_v = \frac{2hv^3}{c^2} n_v \tag{21.43}$$

where v is the frequency, n_v (v) is the quantum mechanical occupation number, i.e. the number of photons (in each polarization state) per unit phase space volume measured in units of h^3, h is Planck's constant. It is assumed that the light is not linearly polarized so that there are an equal number of photons in each polarization state. Therefore, a black body (BB) spectrum

$$n_v = \frac{1}{\exp\left(\dfrac{hv}{kT}\right) - 1} \tag{21.44}$$

where T is the temperature. If there is a small deviation from BB spectrum, and the temperature measured is T_v, which is a good fit to T_b, then, the fluctuation $\Delta T_b = T_b - T_v$ is the differentiation of Eq. (21.43) w.r.t. v and is given by, $\Delta T_b \equiv T_b - T_v, = \dfrac{(e^x - 1)^2}{x^2 e^x} \dfrac{c^2 \Delta T_v}{k_2 v^2}$ where

$$x \equiv \frac{hv}{kT_v} \tag{21.45}$$

There are three processes which are significant for thermalizing the BB spectrum: (i) compton scattering, (ii) double compton scattering and (iii) free-free scattering. During the epoch, compton scattering is efficient. So, if there is a perturbation, the spectrum approaches a Bose Einstein distribution (NBE) via Compton scattering instead of black body and then

$$n^{BE} = \frac{1}{\exp\left(\dfrac{h\nu}{kT_v + \mu}\right) - 1} \tag{21.46}$$

where, T_v and μ are the measured temperature and dimensionless chemical potential. So if initially, there is a thermal distribution of photons at temperature T_v and there is an increase of fractional energy density, without significantly increasing the number of photons, then

$$T_b \sim T_\gamma[1 - \mu(0.456 - x^{-1})], \ \Delta U/U = 0.71 \, \mu \tag{21.47}$$

If, $|\mu| < 9 \times 10^{-5}$ (comparing FIRAS data), then

$$\Delta U/U < 6 \times 10^{-5}, \ 10^5 < Z < 2 \times 10^6$$

Standard Cosmological Model for Anisotropy

Einstein's equation for the evolution of the Universe is,

$$\dot{R}^2(t) = (8/3)\pi G\rho R^{-1}(t) + (\Lambda/3)R^2 - kc^2 \tag{21.48}$$

where R describes the size of the universe, G is the gravitational constant, ρ is the present density of the universe, k is a measure of the curvature of space and Λ is the cosmological constant which can be considered zero energy of a vacuum. If the cosmological term dominates then

$$\dot{R}^2(t) = (\Lambda/3)R^2 \text{ leading to } R(t) \propto \exp\{(\Lambda/3)^{1/2} t \tag{21.49}$$

Therefore, the inflationary theory describes the exponential expansion of space which occurred in the very early Universe. Amplification of initial quantum irregularities then resulted in a spectrum of long wavelength perturbations on scales initially bigger than the horizon size. There is a reasonable agreement that the form of fluctuation spectrum coming out of inflation is

$$|\delta_k|^2 \propto k_n \tag{21.50}$$

where k is the comoving wave number and n is the 'tilt' of the primary spectrum. The latter is predicted to lie close to 1 (Harrison Zeldovich or 'scale invariant' spectrum). The nature of oscillation in the subsequent stage is acoustic in nature. In the photon baryon plasma dominated era, the pressure of photons tends to erase anisotropies, whereas gravitational attraction of the baryons, which are moving at speeds much less than the speed of light makes them tend to collapse to form dense haloes. These two effects complete to create acoustic oscillations which give the CMBR its characteristics peak structure (Fig. 21.9). The peaks (Doppler peaks) correspond to resonances in which the photons decouple when a particular mode is at its peak amplitude. Also, diffusion damping (Silk damping) contribute to the

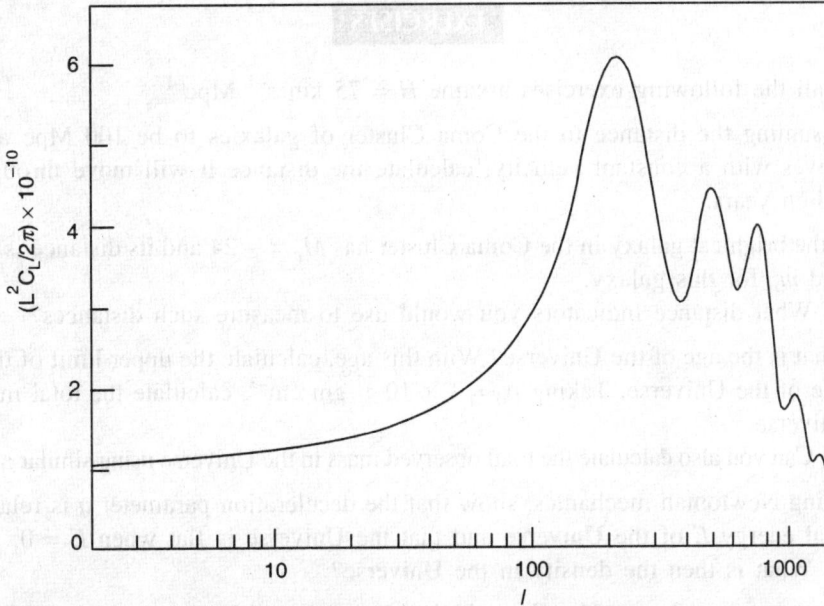

FIGURE 21.9 Power spectrum for standard CDM. Parameters assumed are $\Omega = 1$, $H_0 = 50$ km s^{-1} Mpc^{-1} and a baryon fraction of $\Omega_b = 0.04$.

suppression of anisotropies on small scales when the treatment of the primordial plasma as a fluid begins to break down. So, on large angular scalars (~2°), CMB spectrum reflects inflation; on intermediate angular scales (~1°) there are series of Doppler peaks and on smaller angular scales (~1°) there is a sharp decline in the amplitude. Incorporating all these phenomena a model, called **standard model**, has been constructed involving inflation together with cold dark matters (CDM). The power spectrum for, $\Omega = 1$, $n = 1$, $H_0 = 50$ km s^{-1} is shown in Fig. 21.9.

The quantities plotted are $(l^2 C_l / 2\pi) \times 10^{-10}$ vs l, where

$$C_l = <|a_{lm}|^2, \quad \Delta T(\theta, \varphi)/T = \Sigma a_{lm} Y_{lm} (\theta, \varphi) \tag{21.51}$$

So, $l(l + 1) C_l$ approximately equals the power per unit logarithmic interval in l. $\theta \approx 2/l$ where θ is the angular scale. The peaks contain interesting physical signatures, e.g. the first peak at $l = 200$ and $\theta \sim 1°$ is the 'accoustic peak or Doppler' or 'Sakharov' peak. Now, $l_{peak} \propto \Omega^{-1/2}$. So, the accurate observed position of the peak gives a rough estimate of the total density of the Universe. Also, height of the peak $\sim \Omega_b H_0$. From nucleosynthesis, there is constraint on $\Omega_b H_0^2$. So using $\Omega_b H_0^2$ and $\Omega_b H_0$, value of H_0 can be determined.

EXERCISES

Note: In all the following exercises assume $H = 75$ km s^{-1} Mpc^{-1}

1. Assuming the distance to the Coma Cluster of galaxies to be 100 Mpc and that it moves with a constant velocity, calculate the distance it will move through in one billion years.

2. If the brightest galaxy in the Coma Cluster has $M_v = -24$ and its distance is 100 Mpc, find m_v for this galaxy.

    What distance indicators you would use to measure such distances?

3. What is the age of the Universe? With this age, calculate the upper limit of the present size of the Universe. Taking $\rho_0 = 1 \times 10^{-29}$ gm cm^{-3}, calculate the total mass in the Universe.

    Can you also calculate the total observed mass in the Universe using similar arguments?

4. Using Newtonian mechanics, show that the deceleration parameter q is related to the total energy E of the Universe and that the Universe is flat when $E = 0$.

    What is then the density in the Universe?

5. What do you understand by *Cosmological Principle* and *Perfect Cosmological Principle*? How far these principles appear to hold on the bases of present day observations?

6. Obtain the Fundamental Equation of cosmology based on Newtonian mechanics and discuss the fundamental weakness of this equation in describing the correct World model.

7. "The relative importance of the pressure and density in the Universe governs the relative difference between the classical and relativistic cosmological models"—Establish this statement.

8. What is cosmological constant Λ? Discuss its importance in constructing the relativistic World models. Assuming $\Lambda = 0$, deduce the density in the Universe at the present epoch.

9. Describe the principal features of the Steady-State Universe. How does it differ from other relativistic models of the Universe? Discuss the merits and demerits of the Steady-State model.

10. What do you know of the Primeval Fireball production? Why the observation of the universal microwave background radiation at 3 K destabilized the Steady-State model of the Universe?

SUGGESTED READING

1. Bondi, H., *Cosmology,* Cambridge University Press, Cambridge, 1961.

2. Bonnor, William, *The Mystery of the Expanding Universe,* The Macmillan Company, New York, 1963.

3. Contopoulos, D. and Katsakis, D., *Cosmology* (Translated from Greek by M. Petrou and P.L. Palmer), Springer-Verlag, Berlin, 1987.

4. Harrison, Edward, R., *Cosmology,* Cambridge University Press, Cambridge, 1981.

5. Maeder, A., Martinet, L. and Tammann, G. (Eds), *Observational Cosmology,* Geneva Observatory, Switzerland, 1978.

6. Narlikar, J., *Introduction to Cosmology,* Cambridge University Press, Cambridge, 1993.

7. Partridge, R.B., *3K: The Cosmic Microwave Background Radiation,* Cambridge University Press, Cambridge, 1995.

8. Peebles, P.J.E., *Physical Cosmology,* Princeton University Press, Princeton, New Jersey, 1971.

9. Roy Choudhuri, A.K., *Theoretical Cosmology,* Oxford University Press, Oxford, 1979.

10. Roy Choudhuri, A.K., Banerjee, S. and Banerjee, A., *General Relativity, Astrophysics, and Cosmology,* Springer-Verlag, Berlin, 1992.

11. Sanchez, F., Collados, M. and Rebelo, R. (Eds.), *Observational and Physical Cosmology,* Cambridge University Press, Cambridge, 1992.

12. Sciama, D.W., *Modern Cosmology,* Cambridge University Press, Cambridge, 1971.

Bioastronomy

22.1 INTRODUCTION

A new discipline of science was given birth to in 1982 at the 18th General Assembly of the International Astronomical Union (IAU) held in Patras, Greece. The new discipline of science has been named *Bioastronomy*. The search for Extraterrestrial Intelligence (SETI) and a new Commission, Commission 51, was created by the IAU for study, research and development of this discipline. The creation of this Commission was a major triumph for bioastronomers. The name Bioastronomy is a discovery of Michael Papagiannis of Boston university who was the first President of the IAU Commission 51 (1982–85). Bioastronomy is emerging as a multidisciplinary science where planetary science, planetary systems science, origin of life studies and the search for extraterrestrial intelligence—all converge. Since according to several estimates up to 0.5% of all stars could have a planet similar to our earth, it is a natural question 'Are there intelligent beings elsewhere in our Galaxy'? There are several controversial aspects to consider, to find a satisfactory answer. The most significant one is N, the number of technological civilizations, if any, in an average spiral galaxy like Milky Way. N has been debated at various meetings and extreme values between 10^{10} and 10^{-24} have been suggested. If N is that large then the lifetime of an average technological civilization, L for so long (~ million years) will suddenly cease to exist. So the extraterrestrial have two choices, either they adjust to the rapidly changing environment, e.g. when the 'suns' become red giants and then a white dwarf or they can all emigrate to other stars. The former possibility is very unlikely, hence arises the concept of 'colonization'. There are many reasons against colonization, e.g. self-destruction of technologies, biological degeneration, complete change of cultural interests, cost effectiveness but in spite of all these, there is no reason to reject few exceptions in a billion. What matters is the far out tail of a highly populated distribution. On the other hand, we are actually alone by some reason not yet understood—maybe too impatient to expect results within only 30 years of some occasional SETI searches. As a snapshot of minimum search time, let L be the required search time to get a signal from an object at a distance D. Then $L = 2D/c$, where c is the speed of light. Then if $D = 2500$ light year, then L is 5000 years !!!

22.2 DRAKE EQUATION

In 1960, Franke Drake formulated his equation to calculate the number of technological civilizations (N) in Green Bank meeting held at Green Bank, West Virginia that established SETI as a scientific discipline. The equation states that

$$N = R^* \cdot f_p \cdot n_e \cdot f_1 \cdot f_i \cdot f_c \cdot L$$

where

R^* is the average rate of star formation in our Galaxy

f_p is the fraction of these stars that have planets

n_e is the average number of planets that can potentially support life per star that have planets

f_1 is the fraction of the above that actually go on to develop life at some point

f_i is the fraction of the above that actually go on to develop intelligent life

f_c is the fraction of civilizations that develop a technology that releases detectable signs of their existence into space

L is the length of time such civilizations release detectable signals into space.

The values of the parameters used by Drake and his colleagues in 1961 are:

$R^* = 10, f_p = 0.5, n_e = 2, f_1 = 1, f_i = 0.01, f_c = 0.01, L = 10,000$ years with which $N = 10$.

The current estimates of the parameters are:

$R^* = 7, f_p = 0.5, n_e = 2, f_1 = 0.33, f_i = 0.01, f_c = 0.01, L = 10,000$ years with which $N = 2.3$.

Estimates of Possible Civilization

Let dQ/dt be equal to the rate of star formation in our Galaxy with masses M and $M + dM$.

Then $dQ/dt = 1.3 \, dM/M^{2.3}$ with a standard Salpeter mass function. If the birth rate is assumed to be constant over the last 10^{10} years (Hubble time, considered as the age of the Universe) then the total number of stars with masses between 1 and 1.26 $M_\odot$ is

$$Q_T = \int_0^{10^{10}} dt \int_1^{1.26} 1.3 \, dM/M^{2.3} = 2.6 * 10^9 \text{ stars}$$

The limits are chosen on the basis of the fact that for $M > 1.26 \, M_\odot$ the stars have main sequence life time $5 * 10^9$ years which is shorter for developing technological civilizations in planets around such stars whereas for $M < 1 \, M_\odot$ the main sequence lifetime is $> 10^{10}$ years. The number of stars still on the main sequence is

$$Q_M = \int_1^{1.26} 1.3 \, dM/M^{2.3} \int_0^{10/M^3} 1.9 * 10^9 \text{ stars}$$

So the number of stars died $= Q_T - Q_M = 7 * 10^8$.

Now we are to estimate the number of civilizations (N^*) which might have existed on planet orbiting this $7 * 10^8$ dead stars. If it is assumed that the technological civilizations form with equal probability around all stars between masses 0.1 and 1.26 $M_\odot$, then the minimum value of N^* is

$$N^*_{min} = 7 * 10^8 \text{ N}/10^{11} \sim 10^{-2} \text{ N}.$$

It is assumed that about one-half of the $2 * 10^{11}$ stars in the mass range 0.1 and 1.26 $M_\odot$ were born within the past $5 * 10^9$ years. So if N is as small as 10 ~ 100 at least one civilization had to face termination of the main sequence evolution of its parent star.

22.3 EXTRA SOLAR PLANETARY SYSTEM

The current belief holds that planetary systems are formed as a result of collapse of interstellar molecular gas and dust. The self-gravitational collapse causes the formation of the protostars, each surrounded by a flattened disc forming the stellar nebula. This nebula may, in some cases develop into a planetary system, the individual planets going around the parent star in Keplerian orbits. Evidence for large-scale gaseous discs in Keplerian rotation around low-mass pre-main sequence stars has been obtained by observations with interferometric arrays of millimetre-wave telescopes. The excess emission in the far-infrared observed around numerous young stellar objects suggests the presence of circumstellar dust discs around these objects.

In the search of other planetary systems, the target stars of greatest interest are the nearby isolated main-sequence stars of spectral classes F, G, K, and M with masses in the range 0.2 to 1.8 solar masses. Stars of secondary interest are the F, G, K, and M stars in wide binary systems where existence of planetary systems appears possible. In the first phase of the search spanning over a period of about 10 years, about 800 to 1000 stars within a distance range of nearly 100 light years from the Earth are likely to be chosen as targets. The four general techniques that are employed for this purpose are astrometry, Doppler spectroscopy, photometry and imaging. The first three belong to *indirect* method and the last one to *direct* method. The advances in instrumentation and perfection in the technology to be developed will depend precisely on the objectives to be achieved in the planetary search programme. If the objective is to search for giant massive planets like Jupiter moving in Keplerian orbits within a distance of, say, 10 A.U., from the parent solar type stars, it can be achieved for about 100 such nearby stars with the currently available ground-based instruments. If, on the other hand, the objective extends to the observation of the Uranus-like planets having medium size and mass within, say, 8 to 10 A.U. from the solar-type parent stars, the required instrumentation must be better developed than the one currently available. This will be the first generation of more sophisticated instruments to be designed and developed for the purpose. Lastly, the discovery and study of the Earth-sized planets within a distance of a few A.U. from the parent Sun-like stars throws a challenge to our present knowledge of the technological development and could be realized only with highly developed instruments and their skilful use. This is yet to be achieved, but may not be far from reality. Interferometric telescopes operated on the lunar surface are likely to yield the desired objective.

A little discussion of the *direct* and *indirect* identification of the extra-solar planets may be relevant here. Indirect detection rests on the capability of measuring the effect of the gravitational pull of the planet on the parent star. Just as the gravity of the star pulls the planet in an orbit around it, similarly the planet also pulls the star. But the response of the star is very small due to the small mass ratio between the planet and the star. For Jupiter-like giant planets the effect on the parent star is large enough to be measured relatively easily, but for the Earth-like planets the gravitational effect is so small as to throw challenges for measuring it on the sensitivity of the instruments and on the efficiency of the achievable expertise. Nevertheless, NASA's scientists have accepted the challenge and are working all out to win the success. Another way of indirect discovery of the extra-solar planets involved the detection and measurement of the radiation from the central star and the manner in which it may be affected by the presence and properties of the orbiting planets. Direct detection involves the identification and measurement of photons from the planets themselves. One can identify and measure the photons reflected by the planet's surface as is the case with the lunar phenomena, or measure the energy reprocessed and emitted as thermal planetary photons. Both the efforts however pose great challenges, because the planetary radiation is very much weak compared to the radiation from the central star which lies at an extremely small angular separation from the planet.

Spectroscopy and astrometry are the two complementary approaches for measuring the reflex motion of the parent star in an orbit around a planet. The effect of the motion of the star around the planet will be superimposed on the effect of its steady mean motion relative to the solar system. Both these effects can be separately measured by measuring the Doppler shifts of the spectral lines of the star. The effect of the motion of the star in a small orbit around the planet will be manifested as a small periodic variation superimposed on the mean steady motion of the star in space. If such a periodic variation can be observed in any star, one may be sure of the existence of a planetary system around that star. The astrometric technique, on the other hand, utilizes the accurate measurements of the stellar positions as seen projected on the plane of the sky. In the presence of a planet the star moves in a small orbit. Its position in the sky will therefore suffer small periodic changes around the line of sight. If after subtracting the effects of the changes due to the annual and daily motions of the telescope on the Earth, any additional small periodic changes in the positions of the central star can be measured, that will conclusively indicate the existence of planet-like bodies around such stars. Now, the amplitude of the Dopper shift and the magnitudes of the shifts in stellar positions from the mean line are both proportional to the size of the ratio of the masses of the planet and the star. Larger the relative mass of the planet, the larger are these amplitudes, and so easier to measure them. The Jupiters are therefore easier to detect than the Uranuses and the Earths. On both counts, the detection becomes easier for nearby stars. While the sizes of the position shifts are larger for nearer stars, the Doppler shifts are easier to be measured for brighter stars, and so nearer stars, when the spectral classes of stars are fixed. This is why the extra-solar planetary search-is best performed with F, G, K and M dwarf stars lying, within a distance of 100 light years or so from the Sun.

22.4 NUMBER OF HABITABLE PLANETS

If a big bang model of the Universe is accepted the simplest models are the Friedmann models with zero cosmological constant and these represent solutions to Einstein field equations. The type of the Universe is specified by Ω which is the ratio of current density and the density required to halt the expansion of the Universe, known as critical density. For $\Omega < 1$ the Universe is negatively curved with infinite volume and will expand for ever. If $\Omega > 1$, the

Universe has positive curvature and closed with finite volume and at some point the expansion will halt and contract again to a dense point. The number of habitable planets in our Galaxy is 10^9 (Bracewell, 1975). The total blue luminosity $L = 1.5 * 10^{10} L_\odot$. The mean blue luminosity $\rho_L \sim 4.7 * 10^7 L_\odot \text{ pc}^{-3}$ (Gott and Turner, 1976). Then cosmological density of habitable planets is $\rho_{HP} = 3 * 10^6 \text{ Mpc}^{-3}$. For $\Omega > 1$ let us take $\Omega = 2$. Then the radius of curvature of the Universe is $a_0 = cH_0^{-1} = 6000$ Mpc (with $H_0 = 50$ km s^{-1} Mpc^{-1}). The volume $V = 2\Pi^2 a_0^3 = 4.3 * 10^{12}$ Mpc^3. So the number of habitable planets is $\rho_{HP}.$ $V = 10^{19}$. If the possibility to develop intelligent life in a habitable planet is greater than 10^{-19} then there will be intelligent species in the Universe. Similarly for $\Omega = 1$ and $\Omega < 1$ the numbers are of the order of $2 * 10^{19}$ and $2 * 10^{21}$ ($\Omega = 0.1$) which are comparable.

On ancient earth the simultaneous presence of three states of matter accentuates different chemical reactions and transport of chemicals, otherwise in a cooler planet than ours where reaction rates were one-fourth as great, will have its sun burn away from the main sequence before it witnessed intelligent life. This concept is strengthened by the outcome of a laboratory experiment in geochemistry (Fox and Dose, 1972). They showed that if free energy is supplied to the primitive molecules of our planets' early atmosphere, it could produce sugars, amino acids, purines, pyrimidines and other life related organic substances. Discovery of numerous interstellar organic molecules by radio astronomers (Robinson, 1976) support the above conclusion.

Also life requires long periods of time with generous supplies of heavy elements for its development. Now the necessary heavy elements form gradually in stellar interiors and the Universe is 2–4 times as Earth. So such a combination of heavy elements as well as long time, seems rare to trigger life. However, most of the stars are rich enough in metals to have terrestrial planets and more than about five billion years old are considerably closer to the centre of the Galaxy than we are if the Galaxy had formed from the inside out (J.M. Scalo, 1992, private communication). Also the observations of average K giant stars in the Galactic bulge reveal twice the solar iron abundance (Frogel, 1988) and older bulge stars support the above conjecture.

22.5 SEARCH FOR EXTRATERRESTRIAL CIVILIZATIONS

The idea of the interstellar communications and search for extraterrestrial intelligence developed among astronomers in 1950s when great advances took place in building the radio telescopes across the continents. Highly sensitive radio receivers and strong radio transmitters were put to use by radio astronomers and communication scientists. Around late 1950's, it became clear that communication over galactic distances were possible using radio waves and available radio transmitters at that time. In a pioneering theoretical work Giuseppe Cocconi and Phillip Morrison suggested that interstellar communication was best carried out at radio wavelengths. Their work also inspired the thought of a search for extraterrestrial civilizations by radio waves. The evolution of life on Earth and the development of a society with a sufficiently advanced scientific and technological skill on this planet may not be unique in the Universe. "Other main sequence stars like the Sun with a lifetime of many billions of years may also have planetary systems, and in a small number of these planets such civilizations may have developed as having scientific interests and technical skills much greater than those now available to us. Such a civilization may be knowing our Sun as a likely site for another scientifically advanced civilization and may have considered the establishment of communications with us sending radio wave and possibly, waiting for a *reply* from us. We, on our part, may attempt to receive their signals and reply to them by transmitting our signals. These highly provocative suggestions by Cocconi and Morrison opened an entirely new and highly challenging horizon of scientific investigation where many branches of science and highly advanced technology could meet. A new branch of science called SETI was thus born which has grown over the past thirty years to enormous possibility. In early 1960s Frank Drake initiated his project Ozma in which he unsuccessfully searched for 21-cm signals from the two nearby stars, τ Ceti and ε Eridani. In the subsequent years routine searches for signals from nearby stars have been carried out by several prominent astronomers and groups of astronomers, but with null results. Finally, NASA took up the task in a very big way which, over the last decade, developed a highly efficient infrastructure for performing the job in collaboration with many other centres around the globe.

The radio frequency band that may be effectively used for interstellar communications must satisfy the criterion that the signals are little affected by the interstellar plasma or by the Earth's atmosphere. The criterion is satisfied by the wide radio-band between ~ 1 MHz to ~ 30 GHz, which therefore may be used for the purpose. In 1973 Drake Sagan (1973) and Gott (1982) proposed a SETI frequency standard of v_0 ~ 56 GHz. Let a transmitting civilization be at a redshift z. Then it will observe a microwave background temperature of $T_1 = T_0 (1 + z)$, where T_0 is the current temperature of the cosmic microwave background and will emit signals at a frequency of $hv_e = kT_1 = kT_0(1 + z)$. Because of the cosmological Doppler shift, we will observe these transmitted photons at a frequency $hv_0 = hv_e(1 + z)^{-1}$, so that $hv_0 = kT_0 (1 + z)$ $(1 + z)^{-1} = kT_0$ ~ 56 GHz is independent of the redshift of the civilization. This means that observers in distant Universe can also use a universal frequency standard to communicate. After the COBE satellite, the temperature of cosmic microwave background has been measured more accurately which is

$$T_0 = 2.726 \pm 0.01 \text{ K } (2\sigma) \text{ (Mather et al. 1993) so that } v_0 = 56.8 \text{ GHz} \pm 0.2 \text{ GHZ } (2\sigma)$$

NASA's powerful SETI programme has been inaugurated on October 12, 1992 which marked the 500th anniversary of the discovery of America by Christopher Columbus. Radio antennas at the Arecibo Radio observatory in Puerto Rico and the Goldstone Tracking station in the Mojave Desert in California were simultaneously switched on, on the Columbus Day and the powerful receivers have started analyzing the received radio waves for signals from the extraterrestrials. The NASA programme of High Resolution Microwave Survey (HRMS) has been divided into two parts: a *Targeted Search* and a *Sky Survey*. The Targeted Search which will focus on a limited number of target stars for signals, being supervised by John Billingham and Jill Tarter of NASA Ames Research Centre, will put into service the largest available telescopes across the globe. The Targeted Search of NASA is the most sensitive research ever conducted. This search will however be aimed at a limited number of nearby target stars and will look for artificial signals from some 1000 sun-like stars within a radius of about 100 light years from the Sun. The search will also include some star clusters and nearby galaxies. The frequency range to be covered by this programme is 1 to 3 GHz, taking a special look at the region of the water hole which lies between the frequency of hydrogen (H) emission at around 1.4 GHz. and of the hydroxyl (OH) emission at around 1.7 GHz. This range of frequency is particularly favoured because on the one hand, the naturally occurring radio noise is weak in this region and on the other, water being likely to be essential for sustenance of the alien life also, the alien civilizations may favour these frequencies for interstellar communications.

The second part of NASA's HRMS programme, the Sky Survey, has been put under the supervision of Michael Klein and Samuel Gulkis of the Jet Propulsion Laboratory in California. The aim of this programme is to scan the entire sky over the broad frequency range 1 to 10 GHz using the most sensitive 34-metre antennas of NASA's Deep Space Network spread over the Northern and Southern Hemispheres. The programme is to map small areas of the sky called *sky frames*. After the completion of the survey all the 25,000 or more sky frames will be assembled into mosaicked maps. The entire sky will be mapped 31 times in the process. The possible artificial signals will be identified and separated from the background noise and radio interference from the terrestrial sources by simultaneously processing the received signals in two million channels. Those signals which will be deemed as promising artificials signals will be isolated and saved for further study and verification.

An all-out adventure has thus been started using the currently available most advanced scientific and technological skills. The challenge for the discovery of extraterrestrial civilizations is a very difficult one, but extremely promising also. Astronomers have accepted the challenge with all sincerity and seriousness. The fate is still unknown. What appears to be a probability now may appear as a possibility in the early decades of the 21st century and finally, may be realized as certainty in decades to come. If we fail we have nothing to lose; if we succeed we shall conquer new worlds for man. We want a definite answer to the question: Are we alone in the Universe? This is a question rising from the depth of our soul. If the answer is 'No', it will gradually transform the whole concept of our societal and philosophical process. Every human endeavour will be guided and modified by the results of our contact with the extraterrestrial intelligent beings. To what extent the transformation will take place we cannot predict now. Let us wait and hope. In fact, the progress of human civilization is the result of man's ceaseless quest for knowing the unknown. We do not know yet what will be the result of the SETI. But let us keep our quest alive and ceaseless. Let the intellectual curiosity to know the truth endure. Then the success will be ours by God's grace.

EXERCISES

1. Which stars around us appear to be most suitable to look for planetary systems in them with the currently available instrumentation? What are the difficulties faced in discovering extra-solar planetary systems? What additional difficulties are encountered in detecting an Earth-like extra-solar planet?

2. What do you understand by *direct* and *indirect* identification? What challenges are posed by each such identification? For what kind of system direct identification is easier than the indirect one, and for what kind the reverse is true?

3. Define *Ecosphere* and *Habitable Planet*. What conditions must be present in the Ecosphere in order that it may harbour a Habitable Planet? Discuss the importance of oxygen and ozone layer in the atmosphere of a Habitable Planet.

4. The Ecosphere of a Sun-like star is bounded between 1 and 4 astronomical units from the centre of the star. Find the volume of this Ecosphere in cubic kilometre. Compare this volume with that of the parent star. Also compare this volume with that of the Earth.

5. Discuss the various odds that stand in the way of the astronomers engaged in the search for Extraterrestrial Civilizations. Which ones of these appear to you to be most difficult to overcome?

6. Suppose that the Extraterrestrial Civilizations have been detected after a long and strenuous search. What would be the next set of nearly unsurmountable difficulties to be tackled by the scientists?

   What possible effects of the discovery may greatly influence the human civilization?

SUGGESTED READING

1. *Bioastronomy News:* A News Bulletin published by *The Planetary Society,* 65N, Catalina Avenue, Pasadena, California 91106–2301.

2. Bracewell, R.N., *The Galactic Club: Intelligent Life in Outer Space, San Francisco:* W.H. Freeman, Distributed by Scribner, NY, 1975.

3. Drake, F, Project Ozma, *Physics Today,* **14**, p. 40, 1961.

4. Frogel, J.A. and Elias, J.H., *Astrophysical Journal,* 324, 823, 1988.

5. Fox, S.W. and Dose, K., *Molecular Evolution and the Orogin of Life,* Chapter 4, W.H. Freeman, San Francisco, 1972.

6. Giuseppe, C. and Morrison, P., Search for Interstellar Communications, *Nature,* **184**, p. 844, 1959.

7. Goldsmith, D. and Owen, T., *The Search for Life in the Universe,* The Benjamin/ Commings Pub. Co., London, 1980.

8. Gott, J.R., *Extraterrestrials, What Are They*? In Hart M.H. and Zuckerman B., p. 122, Pergoman Press, NY, 1982.

9. Gott, J.R. and Turner, *E.L.,* ApJ, 209, 1, 1976.

10. Mather, J. et al., Talk at AAS meeting, *Phoenix,* January, 1993.

11. Morrison, P., Billingham, J. and Wolfe, J. (Eds.), *The Search for Extraterrestrial Intelligence, SETI,* NASA SP-419 Washington, 1977.

12. Papagiannis, M.D., *Strategies for the Search for Life in the Universe,* D. Reidel Publishing Company, Dordrecht, Holland, 1980.

13. Papagiannis, M.D., *The Search for Extra-Terrestrial Life: Recent developments*, D. Reidel Publishing Company, Dordrecht, Holland, 1985.

14. Robinson, B.J., *Proc. Astron. Soc., Aust.*, 3, 12, 1976.

15. Scalo, J.M., Private communication, 1992.

16. *SETI News: A* News Bulletin published by *SETI Institute,* 2035 Landings Drive, Mountain View, California 94043.

Index